Heat Transfer and Heat Recovery Systems

Heat Transfer and Heat Recovery Systems

Editors

Jan Danielewicz
Krzysztof Rajski

MDPI • Basel • Beijing • Wuhan • Barcelona • Belgrade • Manchester • Tokyo • Cluj • Tianjin

Editors
Jan Danielewicz
Department of Air-Conditioning, Heating, Gas Engineering and Air Protection
Wroclaw University of Science and Technology
Wroclaw
Poland

Krzysztof Rajski
Department of Air-Conditioning, Heating, Gas Engineering and Air Protection
Wroclaw University of Science and Technology
Wroclaw
Poland

Editorial Office
MDPI
St. Alban-Anlage 66
4052 Basel, Switzerland

This is a reprint of articles from the Special Issue published online in the open access journal *Energies* (ISSN 1996-1073) (available at: www.mdpi.com/journal/energies/special_issues/Heat_Transfer_and_Heat_Recovery_Systems).

For citation purposes, cite each article independently as indicated on the article page online and as indicated below:

LastName, A.A.; LastName, B.B.; LastName, C.C. Article Title. *Journal Name* **Year**, *Volume Number*, Page Range.

ISBN 978-3-0365-7369-4 (Hbk)
ISBN 978-3-0365-7368-7 (PDF)

Contents

Preface to "Heat Transfer and Heat Recovery Systems"

In the domain of modern engineering, heat transfer is universally present, playing a vital role in boosting the efficiency of energy conversion systems. The mechanisms of heat transfer are primarily responsible for determining the efficiency and overall performance of these energy conversion processes. Thus, a profound understanding of these mechanisms is crucial for cultivating sustainable and efficient energy management practices.

Moreover, the retrieval of waste heat is considered the leading approach to achieving sustainable energy management by reducing energy consumption. The capturing and recycling of waste heat from various sources is an increasingly popular strategy for enhancing energy efficiency. Waste heat recovery technologies are progressively adopted in numerous industrial processes, enabling the repurposing of heat that would have otherwise been lost. This approach holds significant potential in lowering energy consumption and greenhouse gas emissions, promoting sustainable energy management.

This reprint aims to highlight the latest advancements in heat transfer technology and heat recovery systems for sustainable development. Although heat recovery systems have received considerable attention, a persistent demand for innovative solutions to address challenges in this field remains. An extensive analysis of the chapters in this reprint reveals four primary categories that represent the majority of the content: heat transfer and heat exchangers, heat recovery, renewables, and domestic hot water preparation systems.

Jan Danielewicz and Krzysztof Rajski

Editors

energies

MDPI

Editorial

Heat Transfer and Heat Recovery Systems

Krzysztof Rajski * **and Jan Danielewicz**

Faculty of Environmental Engineering, Wroclaw University of Science and Technology, 50-377 Wroclaw, Poland
* Correspondence: krzysztof.rajski@pwr.edu.pl

1. Introduction

Heat transfer is present in all modern engineering applications and plays a fundamental role in improving the efficiency of energy conversion systems. Within these systems, the efficiency and overall performance of various energy conversion processes are predominantly determined by the mechanisms of heat transfer. Hence, a comprehensive understanding of these mechanisms is essential for the development of sustainable and efficient energy management practices.

In addition, the mitigation of energy consumption via the retrieval of waste heat is deemed as the predominant course of action towards accomplishing sustainable energy management. One promising strategy to improve energy efficiency is to capture and recycle waste heat from different sources. Waste heat recovery technologies have been increasingly adopted in several industrial processes to retrieve and repurpose heat that would otherwise be lost to the surroundings. This approach holds immense potential in reducing energy consumption and greenhouse gas emissions, thereby promoting sustainable energy management.

The goal of this Special Issue is to showcase the latest developments in the field of heat transfer technology and heat recovery systems that promote sustainable development. While considerable attention has been paid to heat recovery systems, there remains a persistent demand for innovative solutions to address the vast array of challenges in this field. Therefore, this Special Issue endeavors to emphasize and provide novel solutions for these challenges.

2. A Review of the Contributions in This Issue

A word cloud has been created to offer a more comprehensive grasp of the content featured in the Special Issue, as depicted in Figure 1. This figure presents a visual representation of the frequently used words in the titles and abstracts of the published research papers, revealing insights into the key topics explored in this collection. Upon conducting a thorough examination of the articles featured in this Special Issue, four primary categories that encapsulate the majority of the content were identified. These categories consist of heat transfer and heat exchangers, heat recovery, renewables, and domestic hot water preparation systems.

2.1. Heat Transfer and Heat Exchangers

Baig et al. investigated a slotted fin minichannel heat sink (SFMCHS) to manage high heat flux generated in microprocessors [1]. The SFMCHS was developed by modifying a conventional straight integral fin minichannel heat sink (SIFMCHS). The study numerically compared SFMCHSs with fin spacings of 0.5 mm, 1 mm, and 1.5 mm with SIFMCHSs. Two slots per fin minichannel heat sink (SPFMCHS) reduced the base temperature by 9.20%, 8.74%, and 7.39% for 0.5 mm, 1 mm, and 1.5 mm fin spacings, respectively, compared to SIFMCHSs. The 0.5 mm-spaced SPFMCHS had better heat transfer performance compared to 1 mm and 1.5 mm fin spacings, with a uniform temperature distribution at the heat sink base observed in all cases.

Citation: Rajski, K.; Danielewicz, J. Heat Transfer and Heat Recovery Systems. *Energies* **2023**, *16*, 3258. https://doi.org/10.3390/en16073258

Received: 3 April 2023
Accepted: 4 April 2023
Published: 5 April 2023

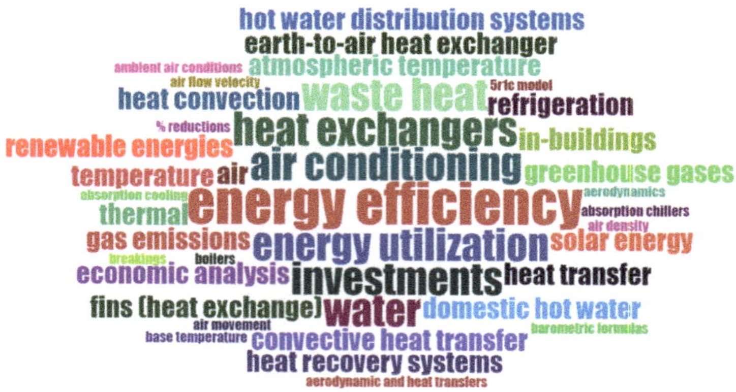

Figure 1. A word-cloud created from the articles published in the Special Issue.

Zhao et al. aimed to optimize the flow and heat transfer performance of channels with frustums of a cone by using Response Surface Methodology (RSM) and multi-objective genetic algorithm [2]. Central composite face-centered design (CCF) was used for the experimental design of channel parameters, and response surface models were constructed. The sensitivity of the channel parameters was analyzed using Sobol's method. Four optimized channels were selected from the Pareto solution set. The response surface models had high accuracy with an RSME of less than 0.25 and R^2 of greater than 0.93.

Basok et al. analyzed the heat transfer and aerodynamics in a high-temperature recuperation system's duct [3]. The research involved a flat duct with a thickness-to-height ratio of 1:10, length of 400a, and one-sided heat input, and mixed air movement. The objective was to determine the average temperatures, air temperature distribution, and heat flow densities along with the heat exchange coefficients and Nusselt numbers. The research method comprised mathematical numerical experimentation and physical modeling. The results showed a difference in the intensity of convective heat transfer between the hot and non-heated surfaces of the duct. The findings suggest that the resistance of the boundary layer in high-temperature heat exchangers and recuperation systems will not exceed 10% of the total pressure losses in the system.

Moreover, the use of plate fin-tube heat exchangers is prevalent in various industries, including air conditioning and refrigeration systems. However, during the manufacturing process, errors can lead to the formation of an air gap between the tube and fin. The effect of the air gap on heat transfer was studied by Łęcki et al. [4] using numerical simulations for a symmetric section of the heat exchanger under periodic flow conditions. The results showed that the fin discontinuity along the circumference reduces the heat transfer rate compared to the perfect fin-tube contact case. The location and size of the gap also affect heat transfer, with the rear gap position being the worst scenario. Reversing the flow direction can lead to up to a 15% increase in heat transfer, especially if the rear gaps are present.

Pavlenko and Koshlak discussed the dynamic interaction between boiling particles in an emulsion, which leads to droplet breakup [5]. They analyzed the forces that determine the breaking of non-boiling and boiling droplets and determine the displacement, deformation, and fragmentation of the dispersed phase. They also studied the dynamics of bubbles in a compressible liquid with consideration for interfacial heat and mass transfer and the effect of standard and system parameters on cavitation processes. The study proposed a new method to assess dynamic effects, which considers all determining factors and accurately represents thermophysical system parameters. The method was validated using superheated emulsion boiling with a sharp decrease in pressure. This study should be considered as the initial stage of rational designs and optimal operating modes of cavitating devices for solving various technological problems.

The process of two-phase expansion, where a fluid undergoes a pressure drop in the liquid-vapor dome, is gaining interest for various processes, especially in low-temperature two-phase heat-to-power cycles. However, modeling and understanding the phenomena in volumetric two-phase expanders are still limited. This topic was reviewed by van Heule et al. [6]. While screw expanders can be modeled using the homogeneous equilibrium model, reciprocating expanders require further investigation. The boiling delay model and homogeneous relaxation model are promising techniques, but more research is needed to apply them. More experimental data on different expander types is also required to understand the impact of design parameters. The review provided a comprehensive overview of available data and modeling techniques, but more research is necessary to achieve efficient volumetric expansion machines.

2.2. Heat Recovery

Liu et al. analyzed the potential of a combined heat and power (CHP) system for heating a building with an indoor swimming pool in Korea, which traditionally uses boilers and heat storage tanks that emit greenhouse gases [7]. The CHP system uses electricity and waste heat from a Phosphoric Acid Fuel Cell (PAFC) system and was analyzed for energy savings, CO_2 reduction, efficiency, and economic feasibility. A dynamic simulation model was developed, and results showed that the CHP system can save up to 15% of energy compared to the conventional heating system, and the FTO model presents the highest efficiency and CO_2 reduction. The FTO model also presents a better economic feasibility than the ELT model.

Manouchehri and Collins validated a model that can predict the performance of drain water heat recovery (DWHR) systems, which are used in residential buildings to recover energy from greywater [8]. The model was implemented into software to perform energy simulations and analyze the impact of different plumbing configurations on energy savings. The study found that the plumbing configuration significantly affects the energy savings expected from DWHR heat exchangers, with the greatest savings achieved in equal-flow configurations. However, the mains temperature could dictate which configuration provides higher energy savings, and designers should consider this when implementing a DWHR heat exchanger. Manufacturers are encouraged to adjust the coil diameter to improve heat transfer rates and increase energy savings when equal flow rates through the heat exchanger are not possible.

The low-grade waste heat generated by data centers is difficult to recover due to its decentralized and low-quality nature, making it a challenge to maximize energy efficiency and utilize heat recovery. However, absorption chiller systems are a promising solution for this issue. In the study by Amiri et al. [9], the feasibility of using an absorption chiller system for waste heat recovery from data centers was analyzed. The proposed system could save 4.3 GWh/year and 13.0 GWh/year of electricity, resulting in a reduction of 3068 and 9208 tons of CO_2 equivalent greenhouse gas emissions annually. The payback period for investors was estimated to be 2.76 and 2.56 years for a 4.5 MW and 13.5 MW data center, respectively. The study highlights the need to consider technical and economic aspects simultaneously. Future research could focus on employing higher-performance absorbers and generators to enhance the capacity of the absorption chiller system.

2.3. Renewables

The depletion of fossil reserves has led to a shift towards renewable energy sources, particularly solar power plants. However, the construction of these plants requires a feasibility study to estimate the solar potential in a particular region. Ayaz et al. [10] compared ground-based measurements of global horizontal irradiance (GHI) and direct normal irradiance (DNI) in Peshawar, Pakistan, with satellite-based model SUNY. The data revealed a significant difference between the two, with maximum differences of 42.90% for GHI and 55.86% for DNI. The study recommends establishing more ground measurement stations across the country to assess solar resources more accurately. It also recommends future studies in different locations with longer time series data. The difference between satellite and ground-based measurements is considerable, and relying solely on satellite-modeled data is not recommended for establishing solar energy programs in Pakistan. Real-time ground measurements are necessary for accurate estimates of solar energy resources.

The feasibility of a desiccant dehumidification-based Maisotsenko cycle evaporative cooling (M-DAC) system for greenhouse air-conditioning in Multan, Pakistan was investigated by Ashraf et al. [11]. Traditional cooling methods were deemed unsuitable for greenhouse production due to their high energy costs and inability to provide an optimum microclimate. The M-DAC system was proposed as an alternative solution. The study aimed to evaluate the system's thermodynamic performance in relation to temperature gradient, relative humidity, vapor pressure deficit (VPD), and dehumidification gradient. Results showed that the M-DAC system achieved a maximum air temperature gradient of 21.9 °C at 39.2 °C ambient air conditions, which was considered optimal for most greenhouse crops. The system also created a dehumidification gradient of 16.8 g/kg at 24 g/kg ambient air conditions, which was within the optimum humidity range for greenhouse growing conditions.

Air conditioning is responsible for consuming around one-fifth of the total power used in buildings worldwide. To reduce energy consumption and its impact on the environment, Earth-to-Air Heat eXchangers (EAHX) are being utilized. Greco et al. reviewed the different applications and peculiarities of EAHX and focused on the hybrid applications where EAHXs are coupled with advanced systems [12]. An IoT-based EAHX control system was proposed to optimize energy efficiency and thermal comfort under different operating conditions. EAHXs utilize geothermal energy, which is renewable and sustainable, and do not emit greenhouse gases. Different parameters such as pipe diameter, length of pipe, and number of pipes affect the performance of EAHX. EAHXs are effective in heating the air during winter and cooling it during summer. Hybrid EAHX systems based on Phase-Change Materials and an Air-Source Heat Pump, finned vertical solar chimney, and placed upstream of air conditioning systems' AHU can achieve improvements in terms of energy performance and reduce greenhouse gas emissions.

Michalak analyzed the impact of air density variation on the operation of earth-to-air heat exchangers (EAHE) coupled with the ventilation system of a residential building [13]. The analysis took into account air density variation with ambient air temperature as Polish hourly typical meteorological years do not contain atmospheric pressure. The energy use for space heating and cooling was computed using the 5R1C thermal network model of EN ISO 13790. Depending on the chosen method, a reduction in annual heating and cooling needs of 7.5% to 8.8% in heating and from 15.3% to 19% in cooling was obtained. The study shows the need to include atmospheric pressure in typical Polish meteorological years and indicates the need for future consideration about an assessment of air humidity impact on the simulated EAHE performance.

2.4. Domestic Hot Water Preparation Systems

A comparison of various methods for calculating the peak power required for domestic hot water (DHW) preparation in buildings was presented by Amanowicz [14]. The study highlighted that DHW has become an important component of energy consumption in buildings, and its peak power can be as high as that required for heating and ventilation. This makes it necessary to select the correct peak power of the heat source, particularly when using renewable energy sources, as it affects the size, investment cost, and economic efficiency of the system. The study showed that accumulative systems with hot water storage tanks are more suitable for modern buildings as they are less sensitive to design errors and result in acceptable peak power for DHW.

Żukowski and Jezierski used a mathematical approach to investigate the impact of various factors on the thermal performance of solar domestic hot water (SDHW) systems in different European climates [15]. They created three deterministic mathematical models using data from computer simulations of SDHW systems in Madrid, Budapest, and Helsinki. The models consider the influence of five factors: volume of heat storage tanks, solar collector total area, maximum efficiency, heat loss coefficient of solar collector, and daily consumption of DHW on the annual useful energy output. The results of the optimization procedure indicate that the maximum annual useful energy output is 1303 kWh/m^2 for Madrid, 918.5 kWh/m^2 for Budapest, and 768 kWh/m^2 for Helsinki.

Żelazna and Gołębiowska proposed a multi-criteria analysis for selecting system parameters such as collector type and solar tank volume [16]. A model of the SHW system was used to calculate possible solutions, ensuring the same comfort of usage for several design options. Three indicators, Simple Payback Time (SPBT), Primary Energy consumption (PE), and IMPACT 2002+, were calculated for the analyzed model. The most favorable solution included a heat-pipe-evacuated tube collector, copper pipes, and a 200 dm^3 water tank. The multi-criteria analysis can be used as a tool for the optimization of selection in the green designing process, significantly improving the environmental balance of analyzed technologies.

3. Conclusions

The heat transfer and heat exchangers category of the Special Issue discusses the development of new heat sink designs and the optimization of existing heat exchangers to improve the efficiency of heat transfer mechanisms in various industrial processes. The heat recovery category highlights the importance of capturing and recycling waste heat from different sources to reduce energy consumption and greenhouse gas emissions. The studies examine combined heat and power systems, acid fuel cells, and drain water heat recovery systems in residential and commercial buildings. The renewables category explores the use of renewable energy sources such as solar thermal energy, biomass, and geothermal energy to provide sustainable energy solutions. The domestic hot water preparation systems category focuses on developing efficient water heating systems for residential buildings. The studies emphasize the importance of reducing energy consumption in water heating processes and developing sustainable solutions.

The articles in this Special Issue provide innovative solutions to address the challenges in heat transfer technology and heat recovery systems, promoting sustainable development and energy efficiency. These solutions can be applied in various settings to improve energy management practices and reduce greenhouse gas emissions.

Author Contributions: Conceptualization, K.R.; writing—original draft preparation, K.R.; writing—review and editing, K.R. and J.D. All authors have read and agreed to the published version of the manuscript.

Data Availability Statement: Data are contained within the article.

Conflicts of Interest: The authors declare no conflict of interest.

References

1. Baig, T.; Rehman, Z.; Tariq, H.A.; Manzoor, S.; Ali, M.; Wadood, A.; Rajski, K.; Park, H. Thermal Performance Investigation of Slotted Fin Minichannel Heat Sink for Microprocessor Cooling. *Energies* **2021**, *14*, 6347. [CrossRef]
2. Zhao, Z.; Xu, L.; Gao, J.; Xi, L.; Ruan, Q.; Li, Y. Multi-Objective Optimization of Parameters of Channels with Staggered Frustum of a Cone Based on Response Surface Methodology. *Energies* **2022**, *15*, 1240. [CrossRef]
3. Basok, B.; Kremnev, V.; Pavlenko, A.; Timoshchenko, A. Aerodynamics and Complicated Heat Transfer with the Mixed Motion of Air in the Flat Duct of a High-Temperature Heat Exchanger. *Energies* **2022**, *15*, 865. [CrossRef]
4. Łęcki, M.; Andrzejewski, D.; Gutkowski, A.N.; Górecki, G. Study of the Influence of the Lack of Contact in Plate and Fin and Tube Heat Exchanger on Heat Transfer Efficiency under Periodic Flow Conditions. *Energies* **2021**, *14*, 3779. [CrossRef]
5. Pavlenko, A.M.; Koshlak, H. Application of Thermal and Cavitation Effects for Heat and Mass Transfer Process Intensification in Multicomponent Liquid Media. *Energies* **2021**, *14*, 7996. [CrossRef]
6. Van Heule, X.; de Paepe, M.; Lecompte, S. Two-Phase Volumetric Expanders: A Review of the State-of-the-Art. *Energies* **2022**, *15*, 4991. [CrossRef]
7. Liu, J.; Kim, S.-C.; Shin, K.-Y. Feasibility Study and Economic Analysis of a Fuel-Cell-Based CHP System for a Comprehensive Sports Center with an Indoor Swimming Pool. *Energies* **2021**, *14*, 6625. [CrossRef]

8. Manouchehri, R.; Collins, M.R. Investigating the Impact of Plumbing Configuration on Energy Savings for Falling-Film Drain Water Heat Recovery Systems. *Energies* **2022**, *15*, 1141. [CrossRef]
9. Amiri, L.; Madadian, E.; Bahrani, N.; Ghoreishi-Madiseh, S.A. Techno-Economic Analysis of Waste Heat Utilization in Data Centers: Application of Absorption Chiller Systems. *Energies* **2021**, *14*, 2433. [CrossRef]
10. Ayaz, A.; Ahmad, F.; Irfan, M.A.A.; Rehman, Z.; Rajski, K.; Danielewicz, J. Comparison of Ground-Based Global Horizontal Irradiance and Direct Normal Irradiance with Satellite-Based SUNY Model. *Energies* **2022**, *15*, 2528. [CrossRef]
11. Ashraf, H.; Sultan, M.; Shamshiri, R.R.; Abbas, F.; Farooq, M.; Sajjad, U.; Md-Tahir, H.; Mahmood, M.H.; Ahmad, F.; Taseer, Y.R.; et al. Dynamic Evaluation of Desiccant Dehumidification Evaporative Cooling Options for Greenhouse Air-Conditioning Application in Multan (Pakistan). *Energies* **2021**, *14*, 1097. [CrossRef]
12. Greco, A.; Gundabattini, E.; Solomon, D.G.; Singh Rassiah, R.; Masselli, C. A Review on Geothermal Renewable Energy Systems for Eco-Friendly Air-Conditioning. *Energies* **2022**, *15*, 5519. [CrossRef]
13. Michalak, P. Impact of Air Density Variation on a Simulated Earth-to-Air Heat Exchanger's Performance. *Energies* **2022**, *15*, 3215. [CrossRef]
14. Amanowicz, Ł. Peak Power of Heat Source for Domestic Hot Water Preparation (DHW) for Residential Estate in Poland as a Representative Case Study for the Climate of Central Europe. *Energies* **2021**, *14*, 8047. [CrossRef]
15. Zukowski, M.; Jezierski, W. New Deterministic Mathematical Model for Estimating the Useful Energy Output of a Medium-Sized Solar Domestic Hot Water System. *Energies* **2021**, *14*, 2753. [CrossRef]
16. Żelazna, A.; Gołębiowska, J. Designing for the Environment: An Example of Multi-Criteria Analysis Used for Solar Hot Water System Selection. *Energies* **2021**, *15*, 65. [CrossRef]

energies

Article

Thermal Performance Investigation of Slotted Fin Minichannel Heat Sink for Microprocessor Cooling

Taha Baig [1,2], Zabdur Rehman [3,*], Hussain Ahmed Tariq [4], Shehryar Manzoor [2], Majid Ali [1], Abdul Wadood [5], Krzysztof Rajski [6] and Herie Park [7,*]

1 Department of Mechanical Engineering, Wah Engineering College, University of Wah, Wah Cantt 47040, Pakistan; taha.baig@wecuw.edu.pk (T.B.); majidmaju995@gmail.com (M.A.)
2 Department of Mechanical Engineering, University of Engineering and Technology Taxila, Taxila 47080, Pakistan; m.shehryar@uettaxila.edu.pk
3 Department of Mechanical Engineering, Air University Islamabad, Aerospace and Aviation Campus, Kamra 43570, Pakistan
4 Department of Mechanical Engineering, Institute of Space Technology, Islamabad 44000, Pakistan; ahmedtariq90@gmail.com
5 Department of Electrical Engineering, Air University Islamabad, Aerospace and Aviation Campus, Kamra 43570, Pakistan; wadood@au.edu.pk
6 Faculty of Environmental Engineering, Wroclaw University of Science and Technology, 50-370 Wrocław, Poland; krzysztof.rajski@pwr.edu.pl
7 Department of Electrical Engineering, Dong-A University, Busan 49315, Korea
* Correspondence: zabd@aack.au.edu.pk (Z.R.); park.herie@gmail.com (H.P.)

Citation: Baig, T.; Rehman, Z.; Tariq, H.A.; Manzoor, S.; Ali, M.; Wadood, A.; Rajski, K.; Park, H. Thermal Performance Investigation of Slotted Fin Minichannel Heat Sink for Microprocessor Cooling. *Energies* **2021**, *14*, 6347. https://doi.org/10.3390/en14196347

Academic Editor: Marco Marengo

Received: 24 August 2021
Accepted: 29 September 2021
Published: 4 October 2021

Abstract: Due to high heat flux generation inside microprocessors, water-cooled heat sinks have gained special attention. For the durability of the microprocessor, this generated flux should be effectively removed. The effective thermal management of high-processing devices is now becoming popular due to high heat flux generation. Heat removal plays a significant role in the longer operation and better performance of heat sinks. In this work, to tackle the heat generation issues, a slotted fin minichannel heat sink (SFMCHS) was investigated by modifying a conventional straight integral fin minichannel heat sink (SIFMCHS). SFMCHSs with fin spacings of 0.5 mm, 1 mm, and 1.5 mm were numerically studied. The numerical results were then compared with SIFMCHSs present in the literature. The base temperatures recorded for two slots per fin minichannel heat sink (SPFMCHS), with 0.5 mm, 1 mm, and 1.5 mm fin spacings, were 42.81 °C, 46.36 °C, and 48.86 °C, respectively, at 1 LPM. The reductions in base temperature achieved with two SPFMCHSs were 9.20%, 8.74%, and 7.39% for 0.5 mm, 1 mm, and 1.5 mm fin spacings, respectively, as compared to SIFMCHSs reported in the literature. The reductions in base temperature noted for three SPFMCHSs were 8.53%, 9.05%, and 5.95% for 0.5 mm, 1 mm, and 1.5 mm fin spacings, respectively, at 1 LPM, as compared to SIFMCHSs reported in the literature. In terms of heat transfer performance, the base temperature and thermal resistance of the 0.5 mm-spaced SPFMCHS is better compared to 1 mm and 1.5 mm fin spacings. The uniform temperature distribution at the base of the heat sink was observed in all cases solved in current work.

Keywords: slotted fin minichannel heat sink; base temperature; thermal management; numerical simulation

1. Introduction

With the rapid development in the information technology sector, the thermal management of electronic devices is becoming essential due to the high generation of heat flux. The world is advancing rapidly towards compact devices; however, this presents the challenging task of the effective and timely removal of unwanted heat. Therefore, this problem has attracted many researchers to create efficient cooling techniques while avoiding any losses in functioning of the devices. The heat transfer can be increased either by enhancing the thermal physical properties of ordinary fluid or by optimizing the heat sink surface to volume ratio. Previously, air was used for the removal of heat from the electronic devices, but it is now unable to remove high heat fluxes. Therefore, attention is now being focused upon liquid cooling techniques, due to their higher efficiency when compared to air.

Saeed and Kim [1] numerically investigated the thermal performance of a water-cooled straight fin minichannel heat sink with varying fin spacings, fin thicknesses, and fin heights. In 1981, Tuckerman and Pease [2], for the first time, circulated water directly through a microchannel. Huang et al. [3] numerically investigated the thermal performance of parallel, staggered, and trapezoidal stagger slotted microchannel heat sinks. Knight et al. [4] proposed a scheme to optimize the geometry of microchannels in both turbulent and laminar regions. Gawali et al. [5] used a straight microchannel heat sink to absorb a large quantity of heat. Wang et al. [6] suggested a numerical model to examine the thermal performance of a microchannel heat sink. They optimized the channel number, channel aspect ratio under a fixed pressure drop, volume flow rate, and pumping power. Hung et al. [7] investigated the thermal performance of a porous microchannel heat sink with rectangular, block, trapezoidal, thin rectangular, and sandwich distributions. Gunnasegaran et al. [8] studied the different geometrical effect on microchannel heat sinks. They found a high heat transfer coefficient using a rectangular shape followed by a trapezoidal and a triangular shape. Qu et al. [9] performed an experimental study on heat transfer enhancement in a trapezoidal silicon microchannel. Kumar and Singh [10] investigated pressure drop and heat transfer in a microchannel heat sink with a trapezoidal shape.

Minichannel heat sinks are popular due to lower pressure drops in comparison to microchannel heat sinks. Xie et al. [11] performed a numerical study for heat sinks with normal, mini, and microchannels with bottom dimensions of 20 mm × 20 mm, by using water as a coolant. Dixit and Gosh [12] performed an experimental study on minichannel heat sinks to investigate the thermal performance of diamond, offset, and straight heat sinks. They observed that the pressure drop for an offset minichannel is lower than in a diamond minichannel. Saini and Webb [13] found that an impinging flow can dissipate 94.4 W as compared to duct flow, which can dissipate 84 W. Naphon and Wiriyasart [14] studied the liquid cooling of a minichannel rectangular heat sink with and without a thermoelectric effect. They found that thermoelectric material has a great impact on CPU temperature, as well as on energy consumption. Saeed and Kim [15] performed both numerical and experimental studies on minichannel heat sinks with fin spacings of 0.5 mm, 1 mm, and 1.5 mm using Al_2O_3-H_2O nanofluids. Al-Taey et al. [16] investigated whether the CPU temperature has a direct relationship with the cooling fluid (water). They also noticed a direct relation of mass flow rate with heat transfer rate and Nusselt number but an inverse relation with thermal resistance. Yu et al. [17] performed both numerical and experimental studies to compare a plate fin heat sink with a plate-pin fin heat sink using air as a cooling fluid. Tariq et al. performed experimental and numerical studies to investigate the thermal performance of cellular structures using air [18], water [19], and nanofluids [20]. Valueva and Purdin [21] numerically investigated the effect of aspect ratio of a rectangular channel on pulsating flow dynamics. Valueva and Purdin [22] developed a numerical model for the heat exchange between stationary and pulsating laminar flow inside a rectangular channel.

Many interrupted fin geometries have reported achieving high heat transfer as compared to integral fin geometries. Khoshvaght-Aliabadi et al. [23] calculated the hydrothermal performance of plate and plate-pins in triangular, trapezoidal, and sinusoidal configuration in a minichannel. Rezaee et al. [24], in their numerical and experimental studies, evaluated the effects of variable pin length and longitudinal pitch in a pin-fin heat sink. Khoshvaght-Aliabadi et al. [25] found that the hydrothermal performance of an interrupted fin is higher as compared to integral fin geometry. Ali et al. [26], in their experimental and numerical studies, investigated the thermal performance of integral-fin and pin-fin heat sinks.

Many studies addressing the thermal performance of minichannel heat sinks can be found in the literature. However, work on minichannel slotted fin heat sinks using a numerical study is rare. In this study, the thermal performance of heat sinks with varying fin spacings of 0.5 mm, 1 mm, and 1.5 mm was numerically investigated. These fin spacings were selected from the literature, and these geometries were optimized by creating slots inside the heat sinks. These were named slotted per fin minichannel heat sink SPFMCHSs. These selected SPFMCHSs were then compared to the straight integral fin minichannel heat sink (SIFMCHS) present in the literature [1] using the same heating power of 325 W. In this study, the conventional SIFMCHS was replaced by a novel SPFMCHS, which had varying fin spacings. To the best of the authors' knowledge, none of the previously mentioned studies have reported the effect of slots and slot thickness in a minichannel heat sink with integral fin spacings of 0.5 mm, 1 mm, and 1.5 mm. Numerical simulations were performed using ANSYS Fluent as a solver. The Reynolds number range for this study was 1900, 2900, and 3800, respectively. Uniform temperature distribution was observed at the base of the heat sink. The selection of both the slot thickness and the number of slots was also discussed in detail.

2. Numerical Model

The problem of the conjugate heat transfer was solved numerically using ANSYS Fluent. Assumptions made in this study were as follows: incompressible flow, turbulent model (k-epsilon), no viscous heating, no inside heat generation, and the same thermal properties during the flow. Absolute velocity formulation was used, along with a pressure-based solver. For this study, transport equation K-ε, with realizable and standard wall functions, was used. Velocity and pressure coupling was controlled by a Semi-Implicit Method for Pressure-Linked Equations (SIMPLE). For pressure, a second-order spatial discretization scheme was used, while for the turbulent dissipation rate, turbulent kinetic energy, and discretization of momentum, a second order upwind scheme was used. The outlet was at zero gauge pressure. For the conjugate problem inside the solid region, velocities were zero everywhere [11,27]. The selected base temperature was 46.424 °C, and the inlet temperature was 27 °C. The following governing equations were used for the conservation of energy, mass, and momentum [28]. For validation, conventional SIFMCHSs with 1 mm spacing were solved numerically by applying the same methodology, and the results were then compared with the literature [1], as shown in Figure 1.

$$\frac{\partial}{\partial x_i}(\varrho u_i) = 0 \tag{1}$$

$$\frac{\partial}{\partial x_i}(\varrho u_i u_j) = -\frac{\partial \varrho}{\partial x_j} + \frac{\partial}{\partial x_i}\left[(\mu + \mu_t)\frac{\partial u_j}{\partial x_i}\right] + \frac{\partial}{\partial x_i}\left[(\mu + \mu_t)\frac{\partial u_i}{\partial x_j}\right], j = 1,\ 2,\ 3 \tag{2}$$

$$\frac{\partial}{\partial x_i}(\varrho u_i T) = \frac{\partial}{\partial x_i}\left[\left(\frac{\lambda}{c_p} + \frac{\mu_t}{\sigma_t}\right)\frac{\partial T}{\partial x_i}\right] \tag{3}$$

$$\frac{\partial}{\partial x_i}(\varrho k u_t) = \frac{\partial}{\partial x_i}\left[\left(\mu + \frac{\mu_t}{\sigma_k}\right)\frac{\partial k}{\partial x_i}\right] + G_k - \rho\epsilon \tag{4}$$

$$\frac{\partial}{\partial x_i}(\varrho \epsilon u_i) = \frac{\partial}{\partial x_i}\left[\left(\mu + \frac{\mu_t}{\sigma_\epsilon}\right)\frac{\partial \epsilon}{\partial x_i}\right] + \frac{\epsilon}{k}(c_1 G_k - c_2 \rho \epsilon) \tag{5}$$

where "k" is kinetic turbulence energy, "$\in$" is turbulence rate of dissipation and "G_k" is the generation of kinetic turbulence energy.

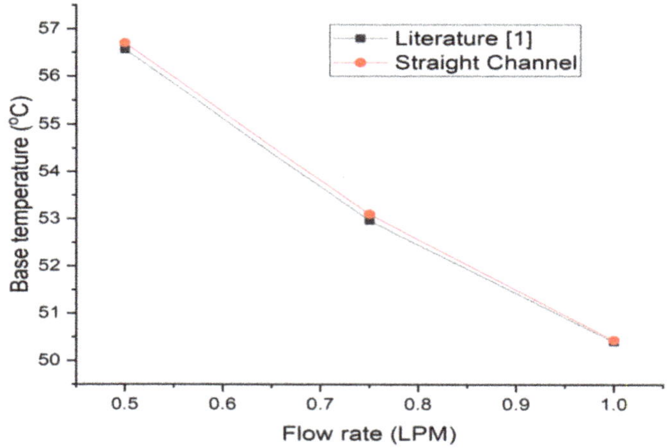

Figure 1. Validation of numerical model.

2.1. Boundary Condition

The following boundary conditions were used to solve the conjugate heat transfer problem:
No-slip conditions at the wall.
Uniform velocity was considered as the inlet of the heat sink

$$\text{At y} = \text{h, u} = 0\text{, v} = -U_{in}\text{, w} = 0 \tag{6}$$

Heat flux was provided at the bottom of the heat sink

$$\text{At y} = 0, -\lambda\frac{\partial T}{\partial y} = q \tag{7}$$

The right and left surface walls were considered as adiabatic.

$$\begin{aligned}\text{At x} = 0,\ \frac{\partial t}{\partial x} = 0\\ \text{At x} = \text{w}_\text{s},\ \frac{\partial t}{\partial x} = 0\end{aligned} \tag{8}$$

The fluid inlet temperature was considered as constant

$$\text{At y} = \text{h},\quad T = T_i \tag{9}$$

2.2. Data Reduction

The following procedure was adopted for data evaluation. The removal of heat from the heat sink was calculated using Equation (6).

$$\dot{Q} = \dot{m}C_p(T_o - T_i) \tag{10}$$

The log of mean temperature difference can be calculated from Equation (7).

$$\text{LMTD} = \frac{(T_b - T_i) - (T_b - T_o)}{\ln\left(\frac{(T_b - T_i)}{(T_b - T_o)}\right)} \tag{11}$$

Thermal resistance of the heat sink was calculated using Equation (8).

$$R_{th} = \frac{\text{LMTD}}{\dot{Q}} \tag{12}$$

2.3. Independent Mesh Study

An intensive study was carried out to ensure that the solution is independent of the mesh. The mesh is considered independent when the temperature difference (between the maximum base temperature and the inlet temperature) shows less than 1% deviation as shown in Figure 2. Six cases with a different number of elements were examined, the number of elements being 2138709, 2358425, 2525411, 2700599, 2912432, and 3102635. The results obtained for case 5 and case 6 were very close to each other, showing a deviation of 0.3% in temperature difference. The number of elements for case 5 was used for the entire study to save computation time and memory.

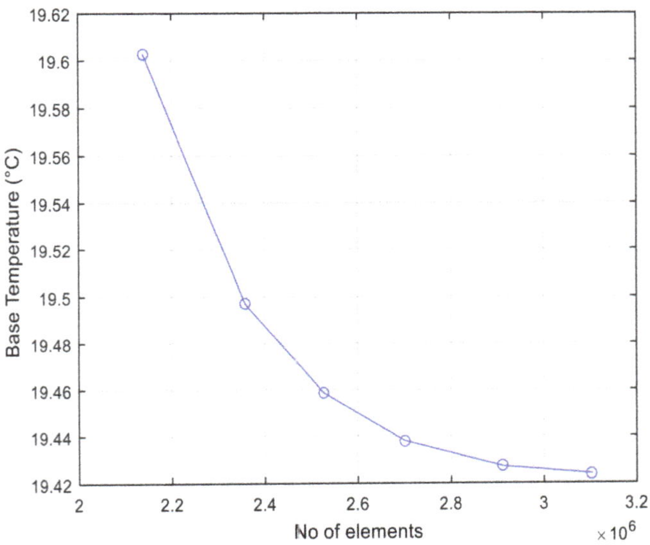

Figure 2. Temperature difference with number of elements.

2.4. Heat Sink

The heat sink was designed on an ANSYS Design Modeler. All of the heat sinks were modeled according to dimensions obtained in the literature [1]. A cubic chip was made at the bottom of the heat sink with dimensions of 28.7 mm × 28.7 mm × 0.5 mm in order to provide 325 W of heat. The dimensions of two SPFMCHSs with 1 mm fin spacing are shown in Figure 3. The details of all the cases solved in this numerical study are provided in Table 1. The isometric view of two SPFMCHSs, three SPFMCHSs, and SIFMCHSs, along with a co-ordinate axis, is shown in Figure 4. SIFMCHSs with varying fin spacings of 0.5 mm, 1 mm, and 1.5 mm are shown in Figure 5. The propagation of flow through two SPFMCHSs from the inlet to outlet is shown in Figure 6.

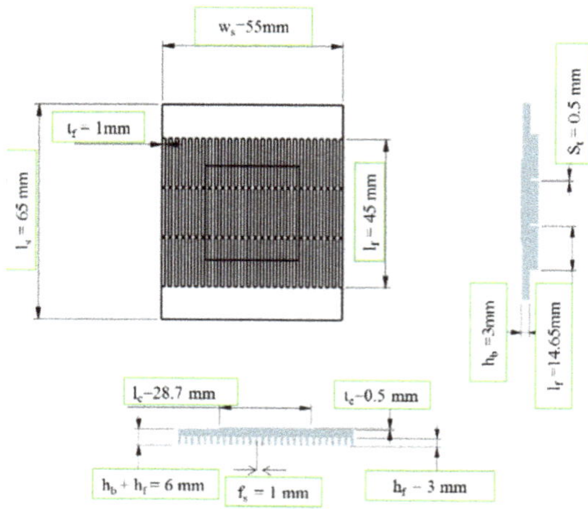

Figure 3. Dimensions of two SPFMCHSs with 1 mm fin spacing.

Table 1. Detail of cases solved with numerical simulations.

No. of Slots Per Fin	f_s (mm)	t_f (mm)	h_f (mm)	S_t (mm)
2	0.5	1	3	0.5
2	1	1	3	0.5
2	1.5	1	3	0.5
3	0.5	1	3	0.5
3	1	1	3	0.5
3	1.5	1	3	0.5

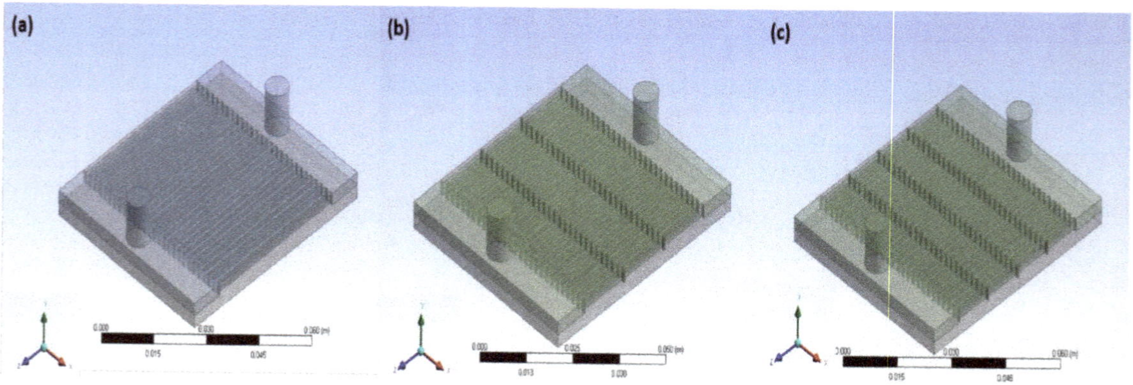

Figure 4. Isometric views for 0.5 mm fin spacing: (**a**) SIFMCHS [1], (**b**) two SPFMCHSs, and (**c**) three SPFMCHSs.

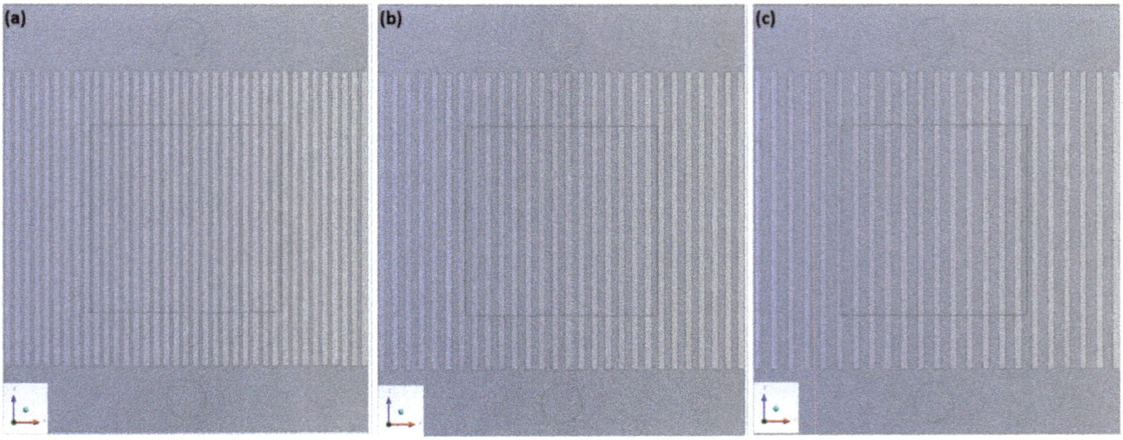

Figure 5. SIFMCHSs: (**a**) 0.5 mm, (**b**) 1 mm, and (**c**) 1.5 mm fin spacing.

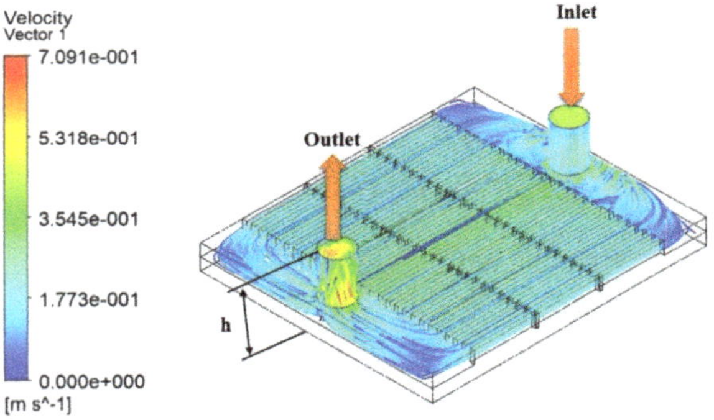

Figure 6. Propagation of flow through two SPFMCHSs.

2.5. Uniformity in Temperature Distribution

For the longer operation and durability of electronic devices, temperature uniformity was essential. All cases that were examined had a uniform temperature distribution inside the heat sink. The contours of the base temperature for two SPFMCHSs and three SPFMCHSs are shown in Figures 7 and 8.

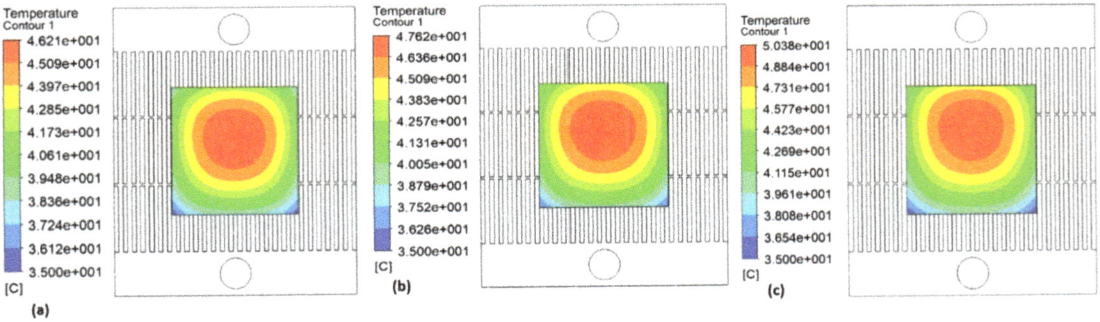

Figure 7. Base temperature distribution of two SPFMCHSs with 1 mm fin spacing: (**a**) 0.5 LPM, (**b**) 0.75 LPM, and (**c**) 1 LPM.

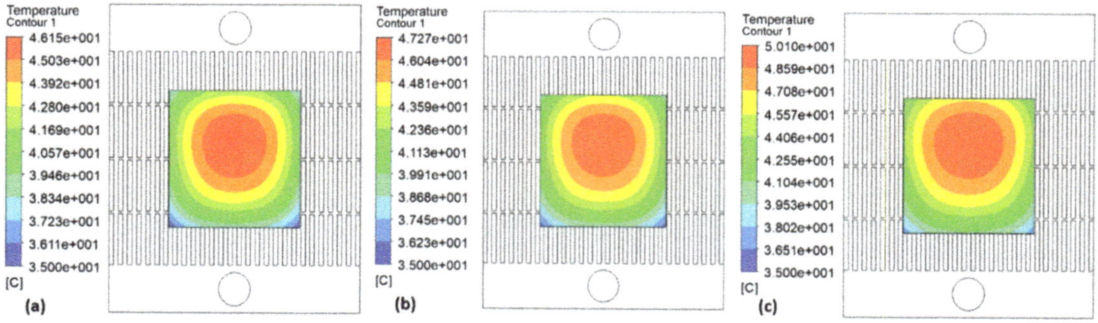

Figure 8. Base temperature distribution of three SPFMCHSs with 1 mm fin spacing: (**a**) 0.5 LPM, (**b**) 0.75 LPM, and (**c**) 1 LPM.

2.6. Selection of Slot Thickness

The flow becomes developed at one quarter of the distance from the entrance between the fins of heat sink. At this stage, the fluid velocity reaches the maximum at the center and the minimum at the boundaries of the flow. To reinitialize the velocity boundary layer, slots were added into the straight integral fins. This made a good contact between the flow and fins. The fully developed flow in between the fins in a 2D plane is shown in Figure 9. The slots were of varying thickness from 0.3 mm up to 1 mm, and the base temperature was recorded against each slot thickness. The thickness of the slot was selected finally as 0.5 mm for both the cases (two and three SPFMCHSs), as the minimum base temperature was recorded using 0.5 mm-thick slots. The formation of vortices took place inside the slots. This led to turbulence and enhanced heat transfer. There was a notable increment in the base temperature recorded for slot thicknesses of 0.5 mm to 0.6 mm. This is the point where the base temperature started to increase instead of decrease. The details are shown in Figure 10. The temperature distribution for two and three SPFMCHSs with 1 mm fin spacing is shown in Figures 7 and 8.

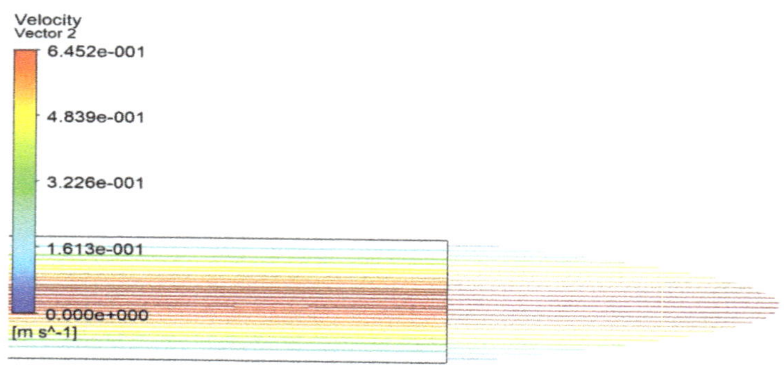

Figure 9. Two-dimensional pictorial view of the fully developed flow in between the fins.

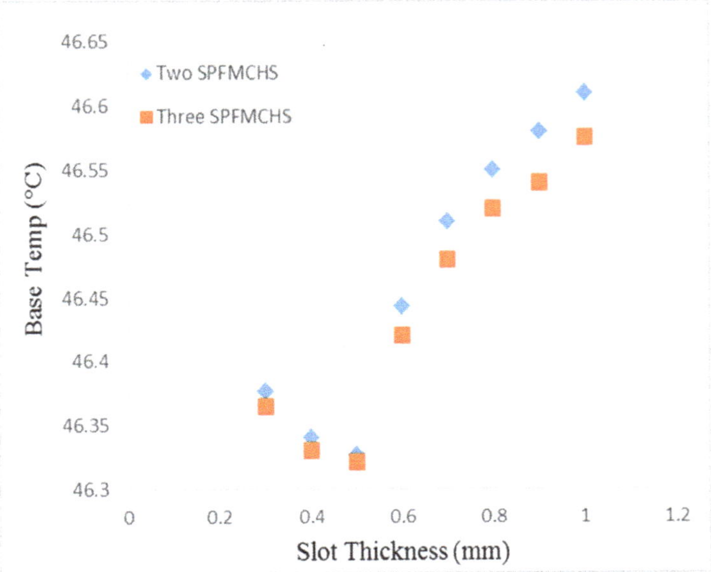

Figure 10. Base temp vs. slot thickness.

3. Results and Discussion

3.1. Heat Transfer

A direct relation was observed between the flow rate and the heat transfer, as shown in Figure 11. The graph shows that heat transfer is dependent on fin spacing and the number of slots. Increasing fin spacing resulted in a reduction in heat transfer. The maximum heat transfer observed in two SPFMCHSs, with 0.5 mm fin spacing, was 316 W at 1 LPM. The minimum heat transfer rate observed in three SPFMCHSs, with 1.5 mm fin spacing, was 280 W at 0.5 LPM. The heat sinks with two SPFMCHSs showed a higher heat transfer as compared to three SPFMCHSs with varying LPM and fin spacings. There was little difference in the heat transfer recorded for two and three SPFMCHSs with 1 mm fin spacing. When comparing two SPFMCHSs to three SPFMCHSs at 1 LPM, the percentage enhancement in heat transfer recorded for 0.5 mm, 1 mm, and 1.5 mm fin spacings was 0.95%, 0.32%, and 0.68%, respectively.

3.2. Base Temperature Drop

The base temperature drop for two and three SPFMCHSs was calculated by subtracting the base temperature of a conventional SIFMCHS [1] from two and three SPFMCHSs with varying fin spacings and LPM, as shown in Figure 12. The maximum temperature drop was recorded in two and three SPFMCHSs with 1 mm fin spacing. The maximum base temperature drop value was 6.31 °C for three SPFMCHSs with 1 mm fin spacing at 0.5 LPM. The minimum base temperature drop value was 3.14 °C for three SPFMCHSs with 1.5 mm fin spacing.

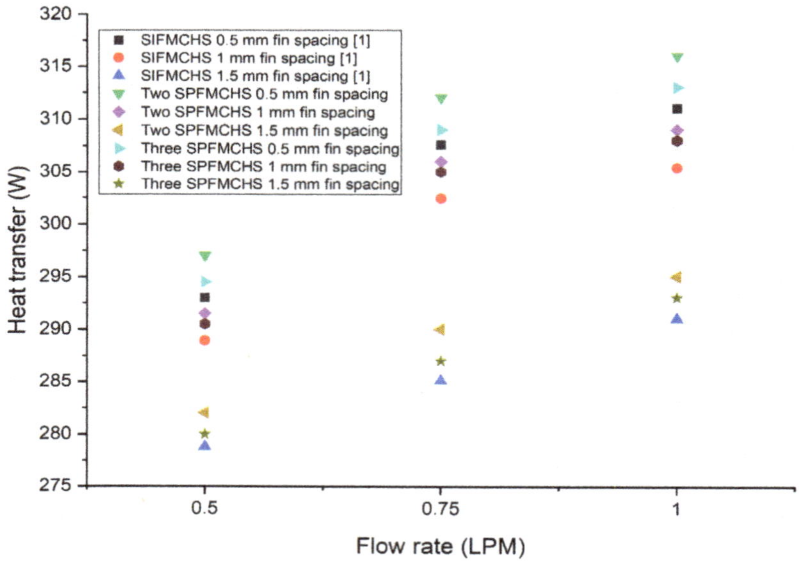

Figure 11. Heat transfer with volumetric flow rate.

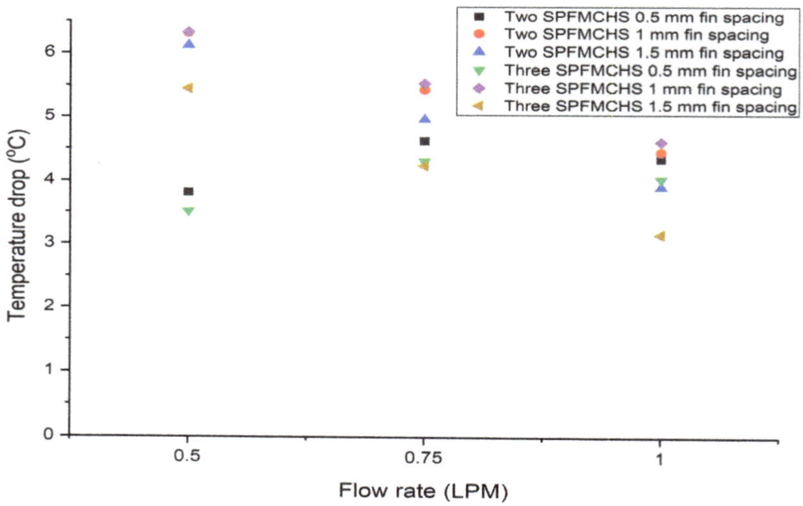

Figure 12. Temperature drop with volumetric flow rate.

3.3. Base Temperature

The base temperatures of SPFMCHSs at different volumetric flow rates are shown in Figure 13. The base temperature increases by increasing the fin spacing as the active area of the heat sink decreases. The minimum base temperature noted for two SPFMCHSs with 0.5 mm fin spacing was 42.81 °C at 1 LPM. The maximum base temperature, observed in two SPFMCHSs with 1.5mm fin spacing, was 53.51 °C at 0.5 LPM. An inverse relation was observed between the flow rate and the base temperature. The reduction in base temperature using SPFMCHSs was greater than that in SIFMCHSs [1]. This clearly highlights the importance of making slots in straight integral fins. The base temperature recorded for two SPFMCHSs with 0.5 mm and 1.5 m fin spacings was lower than for three SPFMCHSs with the same fin spacing. For three SPFMCHSs, the percentage reduction in the base temperature was recorded as 8.53%, 9%, and 5.95% for 0.5 mm, 1 mm, and 1.5 mm fin spacings, respectively, compared to SIFMCHSs [1] at 1 LPM with the same fin spacing. The base temperature recorded in two SPFMCHSs with 0.5 mm, 1 mm, and 1.5 mm fin spacings at 1 LPM was 42.81 °C, 46.36 °C, and 48.87 °C, respectively, which is 9.20%, 8.74%, and 7.39% lower than the reported values in the literature [1] for SIFMCHSs. The base temperature distribution is shown in Figures 7 and 8 for two and three SPFMCHSs, respectively, with 1 mm fin spacing.

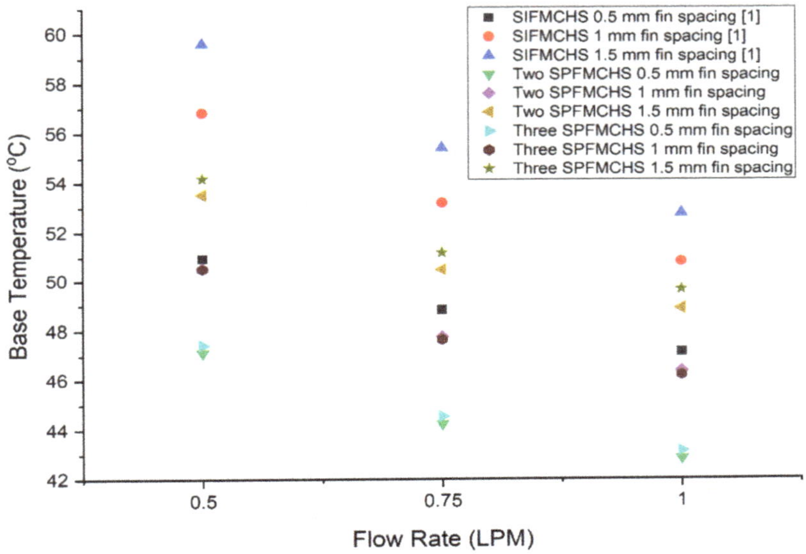

Figure 13. Base temperature with volumetric flow rate.

3.4. Pressure Drop

The pressure drops observed for SPFMCHSs across various volumetric flow rates are shown in Figure 14. By increasing the flow rate, the pressure drop increases. An inverse relationship was observed between the flow rate and the pressure drop. Increasing fin spacing results in a decreased pressure drop at varying LPM. The pressure drops recorded for two and three SPFMCHSs with varying fin spacings were found to be equal. The percentage reduction in the pressure drop observed at 0.5 mm, 1 mm, and 1.5 mm fin spacings for two SPFMCHSs was 0.64%, 20.68%, and 27.12% respectively, when compared to SIFMCHSs at 1 LPM. The percentage difference noted for three SPFMCHSs was 1.50%, 21.26%, and 27.93% at 0.5 mm, 1 mm, and 1.5 mm fin spacings, respectively, when compared to SIFMCHs. The minimum pressure drops recorded for two SPFMCHSs and three SPFMCHSs with 1.5 mm fin spacing were 356.7 Pa and 352.68 Pa, respectively, at 0.5 LPM. The maximum pressure drops recorded for two and three SPFMCHSs with 0.5 mm fin spacing were 1127.65 Pa and 1117.96 Pa, respectively, at 1 LPM. The pumping power increases as the pressure drops to maintain the desired flow rate. The pressure drops recorded for two and three SPFMCHSs with 0.5 mm fin spacing at varying LPM were found to be equal to the reported value in [1] for SIFMCHSs.

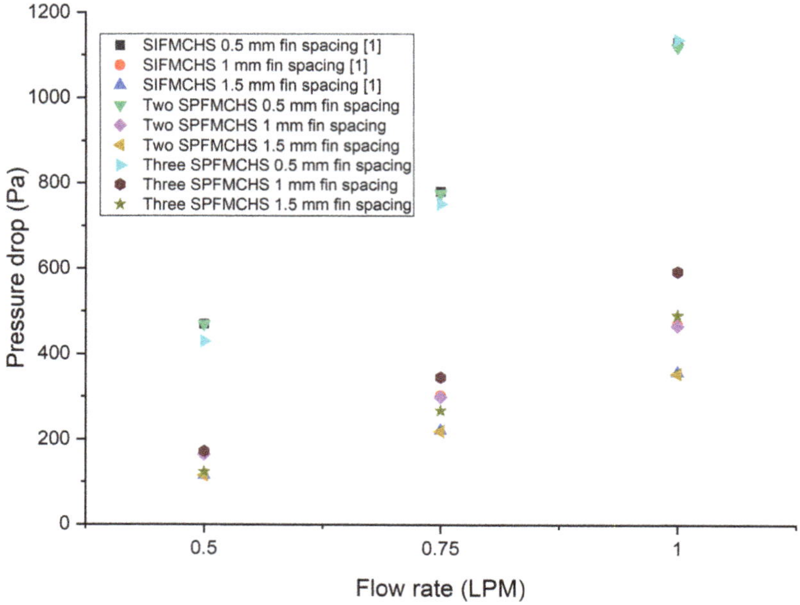

Figure 14. Pressure drop with volumetric flow rate.

3.5. Thermal Resistance

The effect of thermal resistance with volumetric flow rate for SPFMCHSs with two and three SPFMCHSs is shown in Figure 15. The thermal resistance decreases as the flow rate increases. By increasing the fin spacing, the thermal resistance also increases. The minimum thermal resistance recorded in two SPFMCHSs with 0.5 mm fin spacing was 0.036 °C/W at 1 LPM. A similar level of thermal resistance was recorded in two and three SPFMCHSs with various fin spacings. The maximum thermal resistance was recorded in three SPFMCHSs with 1.5 mm fin spacing at 0.5 LPM. The reduction in thermal resistance was recorded as 11.24%, 4.48%, and 8.80% in two SPFMCHSs with 0.5 mm, 1 mm, and 1.5 mm fin spacings, respectively, as compared to the reported values for SIFMCHSs [1] at 1 LPM. The reduction in thermal resistance was observed as 6.31%, 4.83%, and 7.83% in three SPFMCHSs with 0.5 mm, 1 mm, and 1.5 mm fin spacings, respectively, as compared to the reported values for SIFMCHSs [1] at 1 LPM.

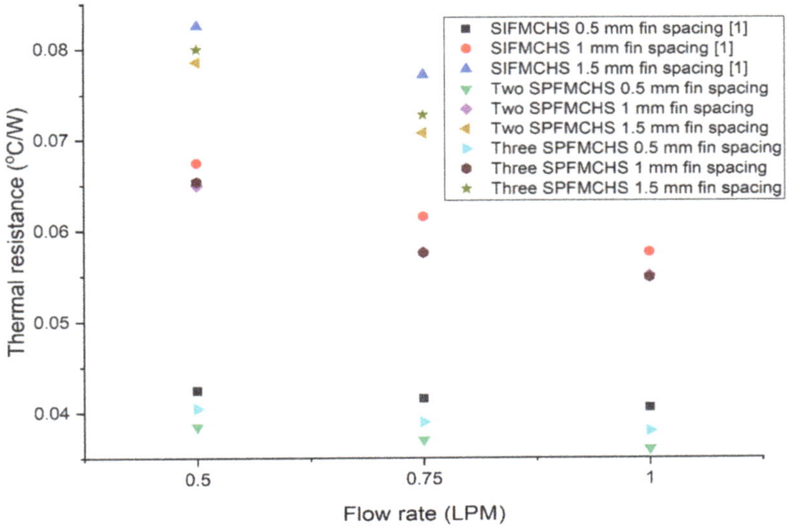

Figure 15. Thermal resistance with volumetric flow rate.

4. Conclusions

A thermal investigation of SPFMCHSs with fin spacings of 0.5 mm, 1 mm, and 1.5 mm was conducted numerically and compared to the SIFMCHS, using water as the cooling fluid, present in the literature. The findings of this detailed numerical study are:

- The base temperature recorded in two SPFMCHSs with 0.5 mm, 1 mm, and 1.5 mm fin spacings at 1 LPM was 42.81 °C, 46.36 °C, and 48.869 °C, respectively. This was 9.20%, 8.74%, and 7.39%, respectively, less than the SIFMCHSs. The minimum base temperature was recorded in two SPFMCHSs with 0.5 mm fin spacing, and the maximum base temperature was recorded in SIFMCHSs with a fin spacing of 1.5 mm at varying LPM. The reduction in the base temperature recorded for 0.5 mm, 1 mm, and 1.5 mm in two SPFMCHSs was 9.2%, 7.61%, and 7.39%, respectively, and in three SPFMCHSs was 8.23%, 7.22%, and 5.95%, respectively, from the reported value in the literature [1].
- The maximum heat transfer was recorded in two SPFMCHSs as compared to three SPFMCHSs with 0.5 mm fin spacing. The minimum heat transfer recorded was for 1.5 mm fin spacing in three SPMCHSs at 0.5 LPM. The percentage increase in the heat transfer recorded for 0.5 mm,1 mm, and 1.5 mm in two SPFMCHSs was 1.6%, 1.16%, and 1.02%, respectively, and for three SPFMCHSs was 0.64%, 0.52%, and 0.44 %, respectively, from the reported value in the literature [1].
- The percentage reduction in the base temperature recorded in three SPFMCHSs with 0.5 mm, 1 mm, and 1.5 mm fin spacings at 1 LPM was 8.53%, 9%, and 5.95%, respectively, when compared to SIFMCHSs.
- The reduction in the thermal resistance was observed as 11.24%, 4.48%, and 8.80% in two SPFMCHSs with 0.5 mm, 1 mm, and 1.5 mm fin spacings, respectively, when compared to SIFMCHSs at 1 LPM. The minimum thermal resistance was recorded in two SPFMCHSs with 0.5 mm fin spacing, and the maximum thermal resistance was recorded in SIFMCHSs with 1.5 mm fin spacing.
- The reduction in the thermal resistance was observed to be 6.31%, 4.83%, and 7.83% in three SPFMCHSs with 0.5 mm, 1 mm, and 1.5 mm fin spacings when compared to SIFMCHSs at 1 LPM. The reduction in the thermal resistance recorded for 0.5 mm, 1 mm, and 1.5 mm fin spacings in two SPFMCHSs was 11.33%, 9.1%, and 8.83%, respectively, from the reported value in the literature [1]. In three SPFMCHSs, the

reduction in the thermal resistance for fin spacings of 0.5 mm, 1 mm, and 1.5 mm was 6.40%, 5.84%, and 4.85%, respectively, from the reported value in the literature.
- The percentage reduction in the pressure drop observed for 0.5 mm, 1 mm, and 1.5 mm fin spacings in two SPFMCHSs was 0.64%, 20.68%, and 27.12%, respectively, as compared to SIFMCHSs at 1 LPM. The percentage difference noted in three SPFMCHSs was 1.50%, 21.26%, and 27.93% at 0.5 mm, 1 mm, and 1.5 mm fin spacings, respectively, when compared to SIFMCHS.

Author Contributions: Conceptualization, Z.R. and H.A.T.; Data curation, T.B., M.A. and H.P.; Formal analysis, M.A.; Funding acquisition, K.R. and H.P.; Investigation, Z.R., H.A.T. and A.W.; Methodology, T.B. and Z.R.; Project administration, H.A.T., S.M. and A.W.; Resources, S.M. and K.R.; Supervision, Z.R.; Validation, Z.R.; Writing—original draft, T.B.; Writing—review and editing, Z.R. and H.P. All authors have read and agreed to the published version of the manuscript.

Funding: This research was funded by the Basic Science Research program through the National Research Foundation of Korea (NRF) funded by the ministry of Education, grant number 2020R1I1A1A01073797.

Conflicts of Interest: The authors declare no conflict of interest.

Nomenclature

C_p	= specific heat, kJ/kg °C
w_s	= width of heat sink, mm
l_s	= length of heat sink, mm
l_f	= length of fin, mm
t_f	= Thickness of fin, mm
S_t	= slot thickness, mm
h_b	= height of heat sink base plate, mm
h_f	= height of fin, mm
l_c	= length of chip, mm
t_c	= thickness of chip, mm
l_t	= total length, mm
f_s	= fin spacing, mm
LMTD	= log of mean temperature difference, °C
$\dot{m}$	= mass flow rate, kg/s
$\dot{Q}$	= heat transfer rate, W
R_{th}	= thermal resistance, °C/W
T_b	= base temperature, °C
T_i	= fluid inlet temperature, °C
T_o	= fluid outlet temperature, °C

Abbreviations

SFMCHS	= slotted fin minichannel heat sink
SIFMCHS	= straight integral fin minichannel heat sink
SPFMCHS	= slots per fin minichannel heat sink
LPM	= liters per minute

References

1. Saeed, M.; Kim, M.H. Numerical study on thermal hydraulic performance of water cooled mini-channel heat sinks. *Int. J. Refrig.* **2016**, *69*, 147–164. [CrossRef]
2. Tuckerman, D.B.; Pease, R.F. High-Performance Heat Sinking for VLSI. *IEEE Electron. Device Lett.* **1981**, *EDL-2*, 126–129. [CrossRef]
3. Huang, S.; Zaho, J.; Gong, L.; Duan, X. Thermal performance and structure optimization for slotted microchannel heat sink. *Appl. Therm. Eng.* **2016**, *115*, 1266–1276. [CrossRef]
4. Knight, R.W.; Hall, D.J.; Goodling, J.S.; Jaeger, R.C. Heat Sink Optimization with Application. *IEEE Trans. Compon. Hybrids Manuf. Technol.* **1992**, *5*, 832–842. [CrossRef]
5. Gawali, B.S.; Swami, V.B.; Thakre, S.D. Theoretical and Experimental Investigation of Heat Transfer Characteristics through a Rectangular Microchannel Heat Sink. *Int. J. Innov. Res. Sci. Eng. Technol.* **2014**, *3*, 8. [CrossRef]
6. Wang, X.D.; An, B.; Xu, J.L. Optimal geometric structure for nanofluid-cooled microchannel heat sink under various constraint conditions. *Energy Convers. Manag.* **2013**, *65*, 528–538. [CrossRef]

7. Hung, T.C.; Huang, Y.X.; Yan, W.M. Thermal performance analysis of porous-microchannel heat sinks with different configuration designs. *Int. J. Heat Mass Transf.* **2013**, *66*, 235–243. [CrossRef]
8. Gunnasegaran, P.; Mohammed, H.A.; Shuaib, N.H.; Saidur, R. The effect of geometrical parameters on heat transfer characteristics of microchannels heat sink with different shapes. *Int. Commun. Heat Mass Transf.* **2010**, *37*, 1078–1086. [CrossRef]
9. Qu, W.; Mala, G.m.; Li, D. Heat tranfer for water flow in trapezoidal silicon microchannel. *Int. J. Heat Mass Transf.* **2000**, *43*, 3925–3936. [CrossRef]
10. Kumar, N.; Singh, N.K. Study and Analysis on Micro Channel Heat Sink in Trapezoidal Shape. *Int. J. Curr. Eng. Technol.* **2017**, *7*, 1115–1118.
11. Xie, X.L.; Tao, W.Q.; He, Y.L. Numerical study of turbulent heat transfer and pressure drop characteristics in water cooled mini channel heat sink. *J. Electron. Packag.* **2007**, *129*, 247–255. [CrossRef]
12. Dixit, T.; Ghosh, I. Low Reynolds number thermo-hydraulic characterization of offset. *Exp. Therm. Fluid Sci.* **2013**, *51*, 227–238. [CrossRef]
13. Saini, M.; Webb, R.L. Heat Rejection Limits of Air Cooled Plane Fin Heat. *IEEE Trans. Compon. Packag. Technol.* **2003**, *26*, 71–79. [CrossRef]
14. Naphon, P.; Wiriyasart, S. Liquid cooling in the mini-rectangular fin heat sink with and without thermoelectric for CPU. *Int. Commun. Heat Mass Transf.* **2009**, *36*, 166–171. [CrossRef]
15. Saeed, M.; Kim, M.H. Heat Tranfer enhancement using Nano fluids (Al_2O_3-H_2O) in mini channel heat sink. *Int. J. Heat Mass Tranafer* **2018**, *120*, 671–682. [CrossRef]
16. Al-Tae'y, K.A.; Ali, E.H.; Jebur, M.N. Experimental Investigation of Water Cooled Minichannel Heat Sink for Computer Processing Unit Cooling. *Int. J. Eng. Res. Appl.* **2017**, *7*, 38–49.
17. Yu, X.; Feng, J.; Feng, Q.; Wang, Q. Development of a plate-pin fin heat sink and its performance comparisons with a plate fin heat sink. *Appl. Therm. Eng.* **2005**, *25*, 173–182. [CrossRef]
18. Tariq, H.A.; Israr, A.; Khan, Y.I.; Anwar, M. Numerical and experimental study of cellular structures as a heat dissipation media. *Heat Mass Transf.* **2019**, *55*, 510–511. [CrossRef]
19. Tariq, H.A.; Shoukat, A.A.; Anwar, M.; Ali, H.M. Water Cooled Micro-hole Cellular Structure as a Heat Dissipation Media: An Experimental and Numerical Study. *Therm. Sci.* **2018**, *2018*, 184. [CrossRef]
20. Tariq, H.A.; Shoukat, A.A.; Hassan, M.; Anwar, M. Thermal management of microelectronic devices using micro-hole cellular structure and nanofluids. *J. Therm. Anal. Calorim.* **2019**, *136*, 2171–2182. [CrossRef]
21. Valueva, E.P.; Purdin, M.S. The pulsating laminar flow in a rectangular channel. *Thermophys. Aeromechanics* **2015**, *22*, 733–744. [CrossRef]
22. Valueva, E.P.; Purdin, M.S. Heat exchange at laminar flow in rectangular channels. *Thermophys. Aeromechanics* **2016**, *23*, 857–867. [CrossRef]
23. Khoshvaght-Aliabadi, M.; Hassani, S.M.; Mazloumi, S.H. Comparison of hydrothermal performance between plate fins and plate-pin fins subject to nanofluid-cooled corrugated miniature heat sinks. *Microelectron. Reliab.* **2017**, *70*, 84–96. [CrossRef]
24. Rezaee, M.; Khoshvaght-Aliabadi, M.; Arani, A.A.A.; Mazloumi, S.H. Heat transfer intensification in pin-fin heat sink by changing pin-length/longitudinal-pitch. *Chem. Eng. Process.-Process. Intensif.* **2019**, *141*, 107544. [CrossRef]
25. Khoshvaght-Aliabadi, M.; Hassani, S.M.; Mazloumi, S.H. Performance enhancement of straight and wavy miniature heat sinks using pin-fin interruptions and nanofluids. *Chem. Eng. Process.* **2017**, *122*, 90–108. [CrossRef]
26. Ali, M.; Shoukat, A.A.; Tariq, H.A.; Anwar, M.; Ali, H. Header Design Optimization of Mini-channel Heat Sinks Using CuO–H_2O andAl_2O_3–H_2O Nanofluids for Thermal Management. *Arab. J. Sci. Eng.* **2019**. [CrossRef]
27. Wang, W.; Li, Y.; Zhang, Y.; Li, B.; Sundén, B. Analysis of laminar flow and heat transfer in an interrupted microchannel heat sink with different shaped ribs. *J. Therm. Anal. Calorim.* **2020**, *140*, 1259–1266. [CrossRef]
28. Tran, N.G.R.; Chang, Y.; Teng, J. A study on five different channel shapes using a novel scheme for meshing and a structure of a multi-nozzle microchannel heat sink. *Int. J. Heat Mass Transf.* **2017**, *105*, 429–442. [CrossRef]

energies

MDPI

Article

Multi-Objective Optimization of Parameters of Channels with Staggered Frustum of a Cone Based on Response Surface Methodology

Zhen Zhao, Liang Xu *, Jianmin Gao, Lei Xi, Qicheng Ruan and Yunlong Li

State Key Laboratory for Manufacturing Systems Engineering, Mechanics Institute, Xi'an Jiaotong University, Xi'an 710049, China; zhaozhen.900803@stu.xjtu.edu.cn (Z.Z.); gjm@mail.xjtu.edu.cn (J.G.); xilei100@mail.xjtu.edu.cn (L.X.); ruanqicheng93@stu.xjtu.edu.cn (Q.R.); ylongli@mail.xjtu.edu.cn (Y.L.)
* Correspondence: xuliang@mail.xjtu.edu.cn; Tel.: +86-(18)-691861949

Abstract: In this study, Response Surface Methodology (*RSM*) and multi-objective genetic algorithm were used to obtain optimum parameters of the channels with frustum of a cone with better flow and heat transfer performance. Central composite face-centered design (*CCF*) was applied to the experimental design of the channel parameters, and on this basis, the response surface models were constructed. The sensitivity of the channel parameters was analyzed by Sobol's method. The multi-objective optimization of the channel parameters was carried out with the goal of achieving maximum Nusselt number ratio (Nu/Nu_0) and minimum friction coefficient ratio (f/f_0). The results show that the root mean square errors (*RSME*) of the fitted response surface models are less than 0.25 and the determination coefficients (R^2) are greater than 0.93; the models have high accuracy. Sobol's method can quantitatively analyze the influence of the channel parameters on flow and heat transfer performance of the channels. When the response is Nu/Nu_0, from high to low, the total sensitivity indexes of the channel parameters are frustum of a cone angle (α), Reynolds number (*Re*), spanwise spacing ratio (Z_2/D), and streamwise spacing ratio (Z_1/D). When the response is f/f_0, the total sensitivity indexes of the channel parameters from high to low are *Re*, Z_1/D, α and Z_2/D. Four optimization channels are selected from the Pareto solution set obtained by multi-objective optimization. Compared with the reference channel, the Nu/Nu_0 of the optimized channels is increased by 21.36% on average, and the f/f_0 is reduced by 9.16% on average.

Keywords: channels with frustum of a cone; multi-objective optimization; Response Surface Methodology; Sobol's method; sensitivity analysis

Citation: Zhao, Z.; Xu, L.; Gao, J.; Xi, L.; Ruan, Q.; Li, Y. Multi-Objective Optimization of Parameters of Channels with Staggered Frustum of a Cone Based on Response Surface Methodology. *Energies* **2022**, *15*, 1240. https://doi.org/10.3390/en15031240

Academic Editors: Jan Danielewicz and Krzysztof Rajski

Received: 11 January 2022
Accepted: 3 February 2022
Published: 8 February 2022

1. Introduction

To cope with severe global climate change and reflect the responsibility assumed by a major country, "carbon peak and carbon neutrality" is an important national strategic goal. Therefore, it is urgent to develop more efficient thermal energy equipment to reduce carbon emissions and improve energy efficiency. Plate air heat exchangers are a compact and efficient piece of heat exchange equipment that is widely used in the chemical industry, electric power, metallurgy, and other industrial sectors.

Selecting appropriate channel parameters can effectively improve the flow and heat transfer performance of heat exchanger channels. Response Surface Methodology (*RSM*) can be used to obtain the combination of design variables under the optimal target response through the explicit functional relationship between design variables and response. Refs. [1,2] used the *CCF* method combined with response surface method to optimize the rib structure of a steam-cooled rectangular channel, and reported that the multi-objective optimization for lower pressure drop and higher heat transfer could be achieved at *Re* of 90,000, α of 41.515, *e/D* of 0.116 and *P/D* of 0.975. Izadi et al. [3] numerically analyzed the natural convection of a porous enclosure under a nonuniform magnetic field using the Local Thermal Non-Equilibrium (*LTNE*)model. The results indicated that the Nusselt numbers of the two phases of porous material converged when increasing the power ratio of the two magnetic sources, and these two thermal indices varied when reducing the power ratio of the two magnetic sources. Jeong et al. [4] numerically analyzed the cooling performance of a cooling channel with a V-shaped rib structure, and then optimized the structural parameters of the V-shaped rib using *RSM*. Seo et al. [5] used the genetic algorithm and *RSM* to optimize a cooling channel with a boot-shaped rib. Mamuria et al. [6] used *RSM* to analyze the sensitivity of four variable parameters of a solar heat exchanger. Bu et al. [7] studied the structural parameters of typical matrix channels by *RSM*, and fitted the polynomial expressions of channel friction coefficient ratio and Nusselt number ratio with respect to rib angle and rib density. Shi et al. [8] used *RSM* to optimize the geometric parameters of microchannels with secondary flow, and selected five optimized structures from the optimized Pareto solution set. Wen Jian et al. [9] combined *RSM* and the multi-objective genetic algorithm to optimize the structure of a serrated fin, studying the effects of fin-height, fin-distance and fin-thickness on flow and heat transfer in the fin channel, finally obtaining three optimized structures. Zheng et al. [10] conducted numerical research on a channel with discrete inclined ribs, studying the effects of Reynolds number, rib-spacing and rib-height ratio on the flow and heat transfer performance of the channel, and analyzed the sensitivity of parameters based on *RSM*.

However, the research on turbulent channels in the above references focuses on the flow and heat transfer performance of the channels, and rarely considers the strength of the turbulent structure. With improvements in the performance of turbulent channels, the problem of fluid excitation can no longer be ignored [11]. High-strength plates can be obtained by the hot stamping forming process [12], and parallel high-strength plates can form a turbulent channel for flow and heat transfer. Considering the fluid excitation in the turbulent channels and the advantages of the hot stamping forming process, our research group proposed a new frustum of a cone structure that can be formed by hot stamping formation [13]. Nevertheless, Ref. [13] only studied a channel with frustum of a cone under a certain structure, and the results showed that the flow and heat transfer performance of the channel was poor. Hence, based on this research, in order to obtain the parameters for a channel with frustum of a cone with better flow and heat transfer performance, *RSM* and the multi-objective genetic algorithm were used to optimize the channel parameters. The experimental design of the channel parameters, including Reynolds number (*Re*), frustum of a cone angle (α), streamwise spacing ratio (Z_1/D), and spanwise spacing ratio (Z_2/D), was carried out by using central composite face-centered design (*CCF*). In addition, on this basis, the second-order polynomial was selected to construct the response surface model. With respect to global sensitivity analysis, Sobol's method was used to analyze the sensitivity of the channel parameters, with the responses being the Nusselt number ratio (Nu/Nu_0) and the friction coefficient ratio (f/f_0) of the channels. To maximize the Nu/Nu_0 and minimize the f/f_0 of the channels, the non-dominated sorting genetic algorithm with elite strategy (*NSGA-II*) was used. These research results can serve as a reference for the parametric multi-objective optimized design of channels with frustum of a cone.

2. Physical Model and Numerical Methods

2.1. Physical Model

The research object of the present study is the channel with staggered frustums of a cone proposed by our research group [13], and the purpose is to carry out multi-objective optimization of the channel parameters. Since the thickness of the channel metal plates is only 1 mm, ignoring the thermal conductivity of the plates, the fluid domain is taken out separately for research. After a certain number of frustum of a cone structures, the fluid will develop into a typical periodic convective heat transfer. For convenience, the minimum periodic model of the channel with frustum of a cone was taken out separately for research. Figure 1 shows the physical model of the periodic channel with staggered frustums of a cone, in which Figure 1a is the diagram of the heat exchanger, Figure 1b is the overall fluid domain model, Figure 1c is the periodic model with central symmetry, and Figure 1d is the side view of the periodic model. As can be seen from Figure 1a, the frustums of a cone have the function of supporting the channel. It can also be seen from Figure 1 that the bottom diameter of each frustum of a cone is D, the height is H, the angle is α, the streamwise spacing and spanwise spacing are Z_1 and Z_2 respectively, and the channel height is $2H$.

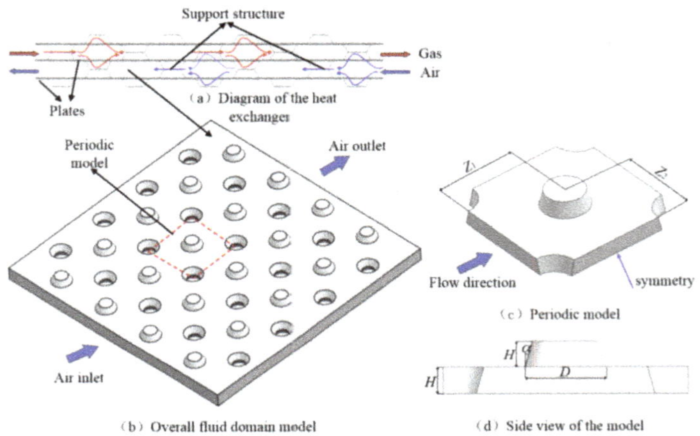

Figure 1. Physical model of the periodic channel with staggered frustums of a cone.

2.2. Numerical Methods

The three-dimensional incompressible Reynolds time-averaged Navier Stokes (*N–S*) equation is solved by FLUENT software. The expressions of continuity, momentum, and energy equations were introduced in detail in Ref. [14]. References [15,16] reported that the Renormalization Group *RNG k-ε* turbulence model can better simulate the flow and heat transfer performance of the turbulent channels. Therefore, the *RNG k-ε* turbulence model was selected to study the performance of the channels with frustum of a cone in this study. In the calculation, the Semi-Implicit Method for Pressure Linked Equation *SIMPLEC* algorithm was used to solve the velocity and pressure coupling equation, and the discrete format of momentum and energy equation was set as the second-order upwind format. It is required that the energy residual converges to 10^{-8}, and the velocity, continuity, k parameter and ε parameter residuals converge to 10^{-6}. The continuity, momentum and energy Equations are as follows:

$$\frac{\partial}{\partial x_i}(\rho u_i) = 0 \tag{1}$$

$$\frac{\partial}{\partial x_i}(\rho u_i u_k) = \frac{\partial}{\partial x_i}(\eta\frac{\partial u_k}{\partial x_i}) - \frac{\partial p}{\partial x_k} \tag{2}$$

$$\frac{\partial}{\partial x_i}(\rho u_i T) = \frac{\partial}{\partial x_i}(\frac{\lambda}{c_p}\frac{\partial T}{\partial x_i}) \tag{3}$$

where ρ is the fluid density; u_i is the vector velocity; η is the viscosity, p is the pressure; T is the temperature; λ is the thermal conductivity of the fluid; c_p is the specific heat capacity at constant pressure; $_k$ is the turbulent kinetic energy equation; ∂ is the diffusion equation.

The *RNG k-ε* turbulence model is as follows:

$$\frac{\partial}{\partial t}(\rho k) + \frac{\partial}{\partial x_i}(\rho k u_i) = \frac{\partial}{\partial x_j}\left(\alpha_k \mu_{eff}\frac{\partial k}{\partial x_j}\right) + G_k - \rho\varepsilon \tag{4}$$

$$\frac{\partial}{\partial t}(\rho\varepsilon) + \frac{\partial}{\partial x_i}(\rho\varepsilon u_i) = \frac{\partial}{\partial x_j}\left(\alpha_\varepsilon \mu_{eff}\frac{\partial \varepsilon}{\partial x_j}\right) + C_{1\varepsilon}\frac{\varepsilon}{k}(G_k + C_{3\varepsilon}G_b) - C_{2\varepsilon}\rho\frac{\varepsilon^2}{k} - R_\varepsilon \quad (5)$$

where α_k and α_ε are the inverse effective Prandtl numbers for k and ε, respectively, G_k is the generation of turbulence kinetic energy, and μ_{eff} is the effective viscosity coefficient.

Figure 2 shows the specific calculation model of the channel with frustums of a cone. Referring to the setting of periodic convective heat transfer boundary conditions in reference [17], the inlet and outlet of the channel were set as periodic boundary conditions, and the mass flow was given. The left and right sides of the channel were also set as periodic boundary conditions. Different from Ref. [17], for convenience, the center of the calculation model was set as a symmetric boundary condition in the present study. The lower wall surface of the channel, including the surface of the frustum of a cone, was the heat transfer wall, which was set to a uniform heat flux of 1000 W × m^{-2}. In addition, the average inlet temperature of the air was set to 300 K.

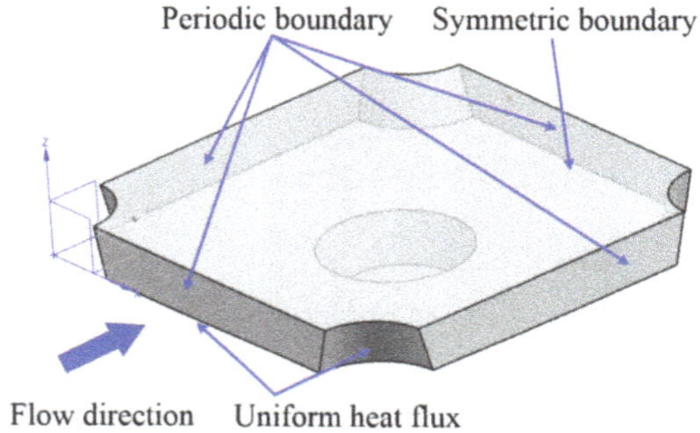

Figure 2. The calculation model of the channel.

Figure 3 shows the grid diagram of the calculation model. The model was divided into hexahedral structured meshes using ICEM software. The mesh near the wall was encrypted. The grid size of the first layer was 0.02 mm, the grid expansion ratio was 1.2, and the maximum grid size was 1 mm. The value of y+(symbol indicating the sparsity of near wall mesh division) near the wall was less than or equal to 1, and the enhanced wall function was used. Such an arrangement can better obtain the flow and heat transfer information near the wall.

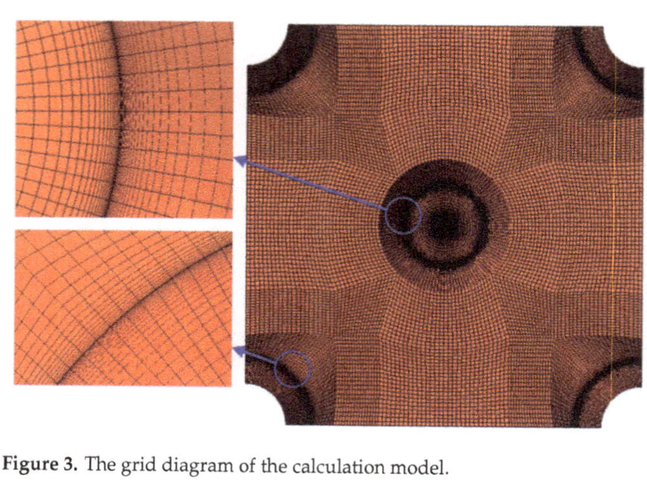

Figure 3. The grid diagram of the calculation model.

The numerical method in the present study was verified using the experimental data for the channel with frustum of a cone presented in Ref. [13]. Figure 4 shows the comparison of the experimental and numerical results of the Nusselt number ratio and friction coefficient ratio of the channel under different Reynolds numbers. It can be seen from Figure 4 that the distribution trends of the experimental and numerical results are basically consistent. Through calculation, the maximum deviation of the Nusselt number ratio is 5.1%, and the maximum deviation of the friction coefficient ratio is 4.3%. These demonstrate the accuracy and feasibility of the numerical method in the present study. Therefore, this numerical method was used in the subsequent numerical studies of the channels with different frustum of a cone structures. In addition, the grid independence of the calculation model was verified. Five sets of grids were divided for the calculation model; when meshing, set the grid size of the first layer to 0.02 mm, and change the number of grids by changing the grid growth ratio (1.5–1.1) and the maximum grid size (0.8 mm–0.4 mm). The total numbers of grids were 75,000, 150,000, 300,000, 600,000 and 1,200,000, respectively. When the number of the grid was 600,000, the deviations of the Nusselt number ratio and friction coefficient ratio of the channel were less than 2%, indicating that the grid met the requirement of independence. The mesh generation strategy mentioned above was adopted in the subsequent calculation models.

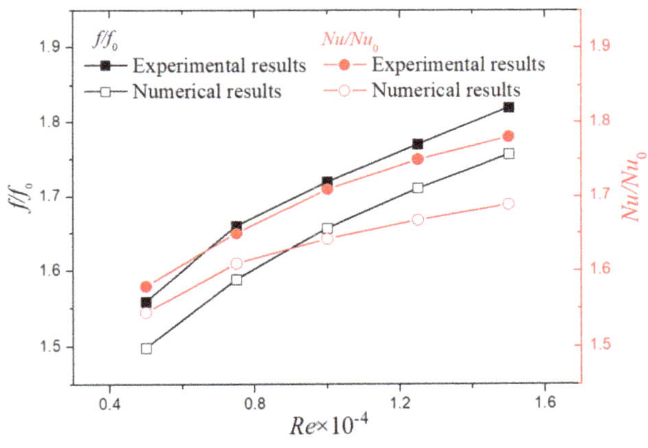

Figure 4. Verification of the numerical method.

2.3. Data Reduction

Reynolds number Re is defined as

$$Re = VP/\nu \tag{6}$$

where V is the average inlet velocity of airflow; P is the characteristic length, and the height of the channel is selected in the present study; ν is the kinematic viscosity of air.

The local Nusselt number Nu_x is defined as

$$Nu_x = q_x P/[\lambda(T_\mathrm{w} - T_\mathrm{f})] \tag{7}$$

where q_x is the local heat flux; λ is the thermal conductivity of air flow; T_w is the local temperature of the wall; T_f is the reference temperature, and its value is the average temperature of inlet and outlet airflow.

The average Nusselt number Nu is defined as

$$Nu = \int Nu_x \mathrm{d}A/A \tag{8}$$

where A is the area of the heat transfer wall.

The friction coefficient f is defined as

$$f = \Delta p P/(2\rho L V^2) \tag{9}$$

where L is the channel length, Δp is the pressure drop at the inlet and outlet of the channel; ρ is the air density.

Nu_0 and f_0 are the average Nusselt number and friction coefficient of smooth parallel plate channel, respectively [18], and the calculation formulas are as follows:

$$Nu_0 = \frac{(f_0/2)(Re - 500)Pr}{1 + 12.7(f_0/2)^{1/2}(Pr^{2/3} - 1)} \tag{10}$$

$$f_0 = (1.58 \ln Re - 2.185)^{-2} \tag{11}$$

3. Optimization Methods

3.1. Response Surface Methodology and Experimental Design

Response Surface Methodology (RSM) can be used to obtain explicit functional relationships between design variables and responses in the global scope by testing local design points. The optimal combination of design variables can be obtained through RSM to achieve the optimal target response. The relationship between system response and design variable is as follows:

$$\boldsymbol{Y} = f(\boldsymbol{X}) + \varepsilon = \boldsymbol{X}\boldsymbol{\beta} + \varepsilon \tag{12}$$

where $\boldsymbol{Y}$ is the response, $\boldsymbol{X}$ is the design variable, $f(\boldsymbol{X})$ is the approximate function of the target, $\boldsymbol{\beta}$ is the fitting coefficient, and ε is the prediction error.

The formula for the sum of squares of system residuals is as follows:

$$SS_E(\boldsymbol{\beta}) = (\boldsymbol{X}\boldsymbol{\beta} - \boldsymbol{Y})^T(\boldsymbol{X}\boldsymbol{\beta} - \boldsymbol{Y}) \tag{13}$$

Through the principle of the least square method, calculate the derivative of function $SS_E(\boldsymbol{\beta})$ with respect to vector $\boldsymbol{\beta}$ and set it to zero. We can draw:

$$\frac{\partial}{\partial \boldsymbol{\beta}} SS_E(\boldsymbol{\beta}) = 2\boldsymbol{X}^T(\boldsymbol{X}\boldsymbol{\beta} - \boldsymbol{Y}) = 0 \tag{14}$$

After finishing Equation (9), the coefficient $\boldsymbol{\beta}$ of the response surface with minimum sum of squares of residuals can be obtained:

$$\boldsymbol{\beta} = (\boldsymbol{X}^T\boldsymbol{X})^{-1}\boldsymbol{X}^T\boldsymbol{Y} \tag{15}$$

The commonly used second-order polynomial response surface model with high accuracy was selected as the approximate function $f(x)$, and the formula is as follows:

$$f(x) = \beta_0 + \sum_{i=1}^{k} \beta_i x_i + \sum_{i=1}^{k} \beta_{ii} x_i{}^2 + \sum_{i=1}^{k-1} \sum_{i<j}^{k} \beta_{ij} x_i x_j \tag{16}$$

where coefficient $\beta = [\beta_0, \ldots, \beta_k, \beta_{11}, \ldots, \beta_{kk}, \beta_{12}, \ldots, \beta \times_{(k-1)k}]^T$; x_i and x_j are the design variables; k is the number of design variables.

To evaluate the fitting accuracy of the obtained response surface model to the test data, the root mean square error ($RMSE$) and determination coefficient (R^2) were selected to evaluate the response surface model. The specific expression of the evaluation index is shown in Ref. [19]. If $RMSE$ is closer to 0, the error of the response surface model is smaller.

If R^2 is closer to 1, the accuracy of the response surface model is higher. $RMSE$ and R^2 are calculated as follows:

$$RMSE = \frac{1}{k\overline{y}}\sqrt{\sum_{i=1}^{k}(y_i - \widetilde{y}_i)^2} \tag{17}$$

$$R^2 = 1 - \frac{\sum_{i=1}^{k}(y_i - \widetilde{y}_i)^2}{\sum_{i=1}^{k}(y_i - \overline{y})^2} \tag{18}$$

where k is the number of sample points, y_i is the test result of the ith sample, $\overline{y}$ is the average of the test results of all samples, and y_i is the calculated value of the response surface model of the ith sample.

The fitting accuracy of the response surface model also depends on the selection of design sample points, so it is necessary to carry out experimental design for design variables. The design variables in the present study are the channel parameters. For convenience of research, frustum of a cone structures with a constant height and bottom diameter were selected, in which the height H was 10 mm and the bottom diameter D was 30 mm. The optimal channel parameter configurations were obtained by changing Reynolds number Re, frustum of a cone angle α, streamwise spacing Z_1 and spanwise spacing Z_2. During the study, the streamwise spacing Z_1 and spanwise spacing Z_2 were dimensionless variables treated as Z_1/D, and Z_2/D. The variation ranges of the design parameters are shown in Table 1.

Table 1. Variation range of channel parameters.

Channel Parameters	Variation Ranges
Reynolds number Re	5000 to 15,000
Frustum of a cone angle $\alpha/°$	0 to 30
Streamwise spacing ratio Z_1/D	1 to 1.5
Spanwise spacing ratio Z_2/D	1 to 1.5

Central composite face-centered design (*CCF*) is one kind of central composite design (*CCD*) that can fit the complete quadratic model and obtain the response surface model with high precision through fewer test points. Therefore, *CCF* was selected for experimental design in the present study. Table 2 shows the central composite face-centered design table, mainly including order, design variables, and responses. Since there are four design variables, 2^4(16)cubic points, 2×4(8)axial points and one center point can be obtained using *CCF* design, so there are 25 sample points in total.

3.2. Parameter Sensitivity Analysis Based on Sobol's method

Many factors affect the flow and heat transfer performance of channels with frustums of a cone, including Reynolds number, frustum of a cone angle, streamwise spacing, and spanwise spacing. Therefore, exploring the impact of the changes of these parameters on the channel performance is of great help to the subsequent parameter optimization. Compared with the local sensitivity analysis method for linear models, the global sensitivity analysis method is suitable for nonlinear complex models. Sobol's method based on variance is a global sensitivity analysis method proposed by the mathematician Sobol [20]. Compared with other global sensitivity analysis methods, Sobol's method can quantitatively analyze the impact of input variables on system output. Its outstanding advantage is that it has no special requirements for analysis function and has a wide range of applications [21].

Table 2. Central composite face-centered design table.

Order	Design Variables				Responses	
	Re	α	Z_1/D	Z_2/D	f/f_0	Nu/Nu_0
1	5000	0	1.50	1.50	1.38	1.39
2	10,000	15	1.25	1.25	1.91	1.65
3	5000	0	1.00	1.50	1.46	1.28
4	15,000	0	1.50	1.50	1.53	1.46
5	10,000	15	1.25	1.00	2.30	1.60
6	10,000	0	1.25	1.25	1.79	1.37
7	5000	0	1.50	1.00	2.10	1.40
8	15,000	15	1.25	1.25	2.09	1.71
9	15,000	30	1.00	1.00	4.96	2.45
10	10,000	15	1.00	1.25	2.65	1.85
11	10,000	15	1.25	1.50	1.64	1.64
12	15,000	0	1.00	1.00	3.26	1.15
13	15,000	0	1.50	1.00	2.38	1.41
14	5000	30	1.00	1.00	2.99	2.04
15	15,000	30	1.50	1.50	2.46	2.35
16	5000	30	1.00	1.50	1.91	1.90
17	15,000	30	1.50	1.00	2.51	1.99
18	15,000	0	1.00	1.50	2.88	1.80
19	5000	15	1.25	1.25	1.69	1.56
20	15,000	30	1.00	1.50	3.29	2.45
21	10,000	15	1.50	1.25	1.74	1.67
22	5000	30	1.50	1.00	2.17	1.93
23	10,000	30	1.25	1.25	3.04	2.37
24	5000	0	1.00	1.00	2.39	1.28
25	5000	30	1.50	1.50	1.49	1.82

The core idea of Sobol's method is to decompose the model into a single parameter and functions composed of parameters. The basic principle of Sobol's method can be seen in Ref. [22]. Assuming that $\boldsymbol{x}= [x_1, \ldots, x_m]$ is an independent input variable defined on I^m, then I^m can be expressed as:

$$I^m = (\boldsymbol{x}|0 \leq x_{i_t} \leq 1; i_t = 1, 2, \cdots, m) \tag{19}$$

Assuming that the model $F(\boldsymbol{x})$ is a square-integrable function, it can be decomposed into the sum of 2_m sub-functions:

$$F(\boldsymbol{x}) = F_0 + \sum_{t=1}^{m} \sum_{i_1 < \cdots < i_t}^{m} F_{i_1 \cdots i_t}(x_{i_1}, \ldots, x_{i_t}) \tag{20}$$

where $1 = i_1 < \ldots < i_t \ldots < i_m = m$, m is the number of variables and F_0 is a constant.

If the conditions are met, it can be obtained:

$$\int_0^1 F_{i_1 \cdots i_t}(x_{i_1}, \ldots, x_{i_t}) \mathrm{d}x_{i_n} = 0, 1 \leq n \leq t \tag{21}$$

The total variance M and partial square deviation $M_{i_1 \ldots i_t}$ of the $F(\boldsymbol{x})$ can be obtained by integrating the squares of Equation (20):

$$M = \sum_{t=1}^{m} \sum_{i_1 < \cdots < i_t}^{m} M_{i_1 \cdots i_t} \tag{22}$$

$$M_{i_1 \ldots i_t} = \int F_{i_1 \cdots i_t}^2 \mathrm{d}x_{i_1} \ldots \mathrm{d}x_{i_t} \tag{23}$$

According to the definition, the sensitivity index of the variable $S_{i_1 \cdots i_t}$ is defined as:

$$S_{i_1 \cdots i_t} = \frac{D_{i_1 \cdots i_t}}{D} \tag{24}$$

Then, the sum of sensitivity indexes of all variables is 1. It can be expressed as follows:

$$\sum_{t=1}^{m} \sum_{i_1<\cdots<i_t}^{m} S_{i_1\cdots i_t} = 1 \tag{25}$$

Equation (25) can be rewritten as follows:

$$\sum_{t=1}^{m} S_{i_t} + \sum_{1\leq q<t\leq m} S_{i_q i_t} + \cdots + S_{i_1\cdots i_m} = 1 \tag{26}$$

According to the definition, the total sensitivity index of an input variable S_{Ti_t} is:

$$S_{Ti_t} = S_{i_t} + \sum_{q\neq t}^{m} S_{i_q i_t} + \cdots + S_{i_1\cdots i_m} \tag{27}$$

where S_{i_t} is the first-order sensitivity index, which represents the impact of input variable x_{i_t} on system output; the second-order sensitivity index $S_{i_q i_t}$ represents the influence of the interaction between input variable S_{i_q} and variable S_{i_t} on the system output. The total sensitivity index S_{Ti_t} indicates the common influence of input variable S_{i_t} and its interaction with other input variables on system output. When the first-order sensitivity index of the input variable S_{i_t} is quite different from the total sensitivity index, it indicates that the interaction between the variable S_{i_t} and other input variables has a great influence on the system output.

3.3. Optimization Process of the Channel with Frustums of a Cone

Figure 5 shows the flow chart for the parameter optimization of the channel with frustums of a cone. As can be seen from Figure 5, the present study optimizes the four design parameters of the channel with frustums of a cone, including Reynolds number *Re*, frustum of a cone angle α, streamwise spacing ratio Z_1/D, and spanwise spacing ratio Z_2/D. *CCF* was used for the experimental design of design parameters, and 25 sample points were obtained. The reliability of the numerical method in the present study was verified by the experimental data. Then, 25 sample points were numerically simulated by *CFD* to obtain the responses (the Nu/Nu_0 and f/f_0 of the channels). On this basis, the response surface models were constructed by the second-order polynomial. Root mean square error (*RMSE*) and determination coefficient (R^2) were selected to evaluate the response surface models. If the models did not meet the requirements, the experimental design of the parameters was optimized, and the simulation was run again according to the above procedure. Then, based on the response surface model, Sobol's method was used to analyze the sensitivity of the parameters. Aiming at achieving minimum f/f_0 and maximum Nu/Nu_0 in the channels, *NSGA-II* was used to carry out the multi-objective optimization of the parameters. Finally, the optimized channels were selected from the Pareto solution set.

Genetic algorithm (*GA*) is a family of global optimization algorithms that simulate natural selection and the genetic mechanism based on biological evolution. Among them, the non-dominated sorting genetic algorithm with elite strategy (*NSGA-II*) was proposed by Deb based on the non-dominated sorting genetic algorithm (*NSGA*). It has the advantages of low computational complexity, fast running speed, and good convergence of solution set [23]. It is worth noting that the multi-objective optimization algorithm does not have a unique global optimal solution, but a set of optimal solutions, namely the Pareto solution set. In practical application, one or more solutions can be selected from the Pareto solution set as the final result.

4. Results Analysis and Discussion

4.1. Construction of Response Surface Model

The second-order polynomial response surface model was obtained based on *CCF* fitting. The specific test design table is shown in Table 2. The design parameters were Re, α, Z_1/D and Z_2/D, and the responses were Nu/Nu_0 and f/f_0. The coefficients of the response surface model are given in Table 3.

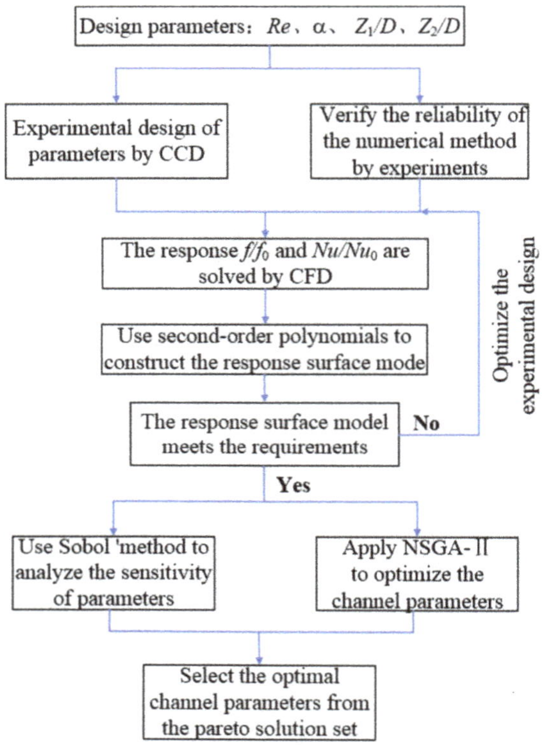

Figure 5. Flow chart of the optimization of the channel parameters.

Table 3. Coefficients of response surface model.

Coefficients	Nu/Nu_0	f/f_0
β_0	0.230	9.66
β_1	0.240×10^{-4}	0.355×10^{-3}
β_2	0.239×10^{-1}	0.800×10^{-3}
β_3	−1.78	−9.75
β_4	3.09	−3.04
β_{11}	0.000	0.000
β_{22}	0.758×10^{-3}	0.188×10^{-2}
β_{33}	0.980	3.28
β_{44}	−1.26	−0.320
β_{12}	0.100×10^{-5}	0.200×10^{-5}
β_{13}	-0.340×10^{-4}	-0.195×10^{-3}
β_{14}	0.650×10^{-4}	0.230×10^{-4}
β_{23}	-0.149×10^{-1}	-0.319×10^{-1}
β_{24}	-0.955×10^{-2}	-0.990×10^{-2}
β_{34}	−0.218	1.76

Figure 6 shows the comparison between the numerically calculated values of the sample points of the experimental design and the corresponding *RSM* predicted values. In the figure, the dotted line represents a deviation of ± 10% from the numerical calculation values, the straight line represents the numerical calculation values, and the scattered points represent the *RSM* prediction values. Figure 6a,b presents a comparison of the values when the responses are Nu/Nu_0 and f/f_0, respectively. It can be seen from Figure 6 that the *RSM* predicted values are distributed near the numerically calculated values, and the errors are basically less than 10%. After calculation, the root mean square error *RMSE* and determination coefficient R^2 of the response surface model can be obtained. The calculation results are shown in Table 4. When the responses are Nu/Nu_0 and f/f_0, the *RMSE* of the models is less than 0.25 and R^2 is greater than 0.93, showing that the fitted response surface model has small error and high accuracy.

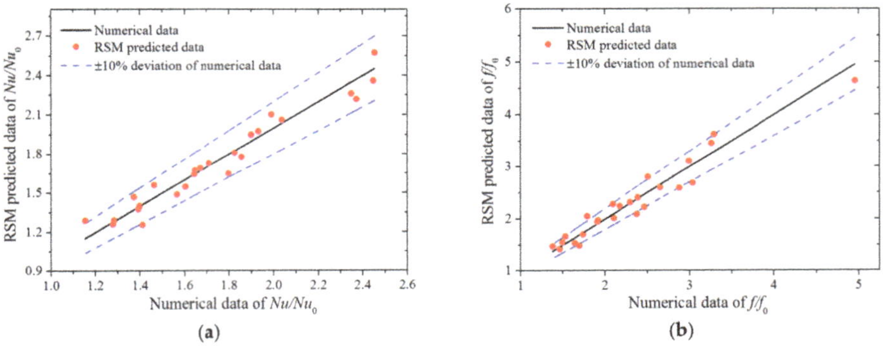

Figure 6. Comparison between numerical calculation and *RSM* prediction: (**a**) Nu/Nu_0; (**b**) f/f_0.

Table 4. Fitting accuracy of response surface model.

Evaluation Index	Nu/Nu_0	f/f_0
RMSE	0.1048	0.2445
R^2	0.9491	0.9399

4.2. Effect of Channel Parameters on Flow and Heat Transfer

The three-dimensional surface and contour map of the Nu/Nu_0 and f/f_0 of the channels are shown in Figures 7 and 8, respectively, to reveal the influence of channel parameters on the heat transfer performance and flow performance of the channels with frustums of a cone. Figure 7a–f show the influence of the combined action of Re–α, Re–Z_1/D, Re–Z_2/D, α–Z_1/D, α–Z_2/D and Z_1/D–Z_2/D on the response in turn.

As can be seen from Figure 7a, when *Re* is constant, increasing α increases the Nu/Nu_0 of the channel, while when α is constant, the increase of the Nu/Nu_0 of the channel along with the increase of *Re* is not very significant. When *Re* is 5000 and α is 0°, the Nu/Nu_0 of the channel reaches its minimum value, while when *Re* is 15,000 and α is 30°, the Nu/Nu_0 of the channel reaches its maximum value. As can be seen from Figure 7b, when *Re* is constant, increasing Z_1/D makes the Nu/Nu_0 of the channel first increase and then decrease. It can be seen from Figure 7c that the increase in Z_2/D under different values of *Re* and the increase in *Re* under different values of Z_2/D can improve the Nu/Nu_0 of the channel. As can be seen from Figure 7d,e, when α is constant, the Nu/Nu_0 of the channel remains basically unchanged with increasing Z_1/D and Z_2/D. When Z_1/D and Z_2/D remain unchanged, increasing α can significantly increase the Nu/Nu_0 of the channel. As can be seen from Figure 7f, when Z_1/D is constant and Z_2/D is increased, Nu/Nu_0 of the channel first decreases and then increases. When Z_2/D is constant, increasing Z_1/D causes the Nu/Nu_0 of the channel to first increase and then decrease.

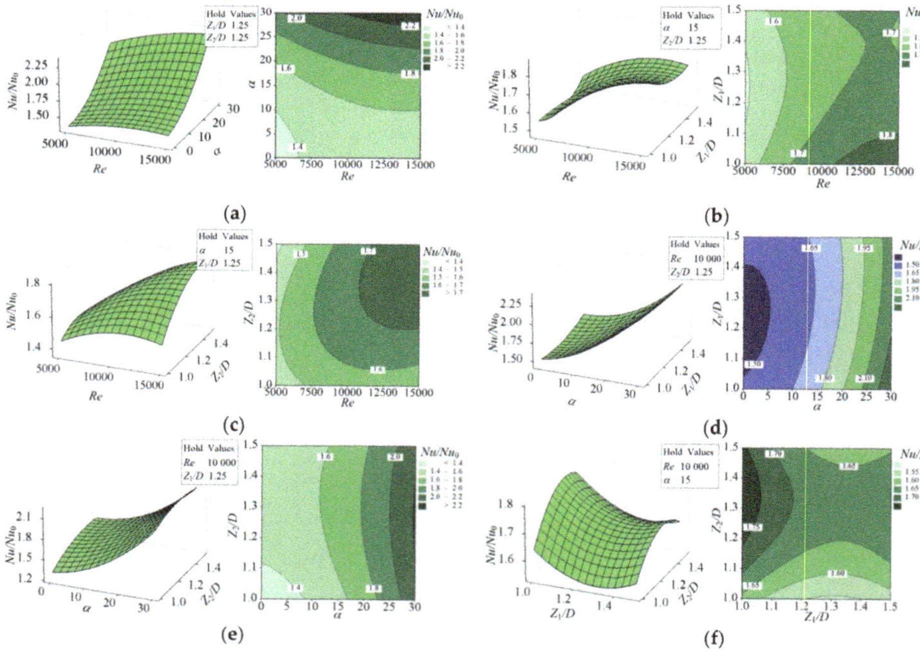

Figure 7. 3D surface and contour map of Nu/Nu_0: (**a**) Re–α; (**b**) Re–Z_1/D; (**c**) Re–Z_2/D; (**d**) α–Z_1/D; (**e**) α–Z_2/D; (**f**) Z_1/D–Z_2/D.

As can be seen from Figure 8a, at low Re, the f/f_0 of the channel first decreases and then increases with increasing α, while at high Re, increasing α leads to an increase in the f/f_0 of the channel. When α is constant, the f/f_0 of the channel increases with increasing Re. It can be seen from Figure 8b,c that increasing Re and decreasing Z_1/D and Z_1/D result in an increase in the f/f_0 of the channel. As can be seen from Figure 8d, when α is constant and Z_1/D is increased, and when Z_1/D is constant and α is increased, the f/f_0 of the channel first decreases and then increases. According to Figure 8e, when α is constant, the f/f_0 of the channel decreases with increasing Z_2/D. When Z_2/D remains unchanged, the f/f_0 of the channel first decreases and then increases with increasing α. As can be seen from Figure 8f, increasing Z_2/D and Z_1/D reduces the f/f_0 of the channel.

The above research shows that when analyzing the flow and heat transfer performance of channels with frustums of a cone, the information obtained limited to a fixed channel parameter is not sufficient to describe the performance of the channels. Building the function of channel performance related to channel parameters based on response surface method is of great significance to studying the influence of channel parameters on channel performance and guiding the parameter optimization and structural design of channels with frustums of a cone.

4.3. Sensitivity Analysis of the Channel Parameters

Figure 9 shows the first-order sensitivity index and total sensitivity index of channel parameters when the response is Nu/Nu_0. The first-order parameter sensitivity index represents the influence of a single parameter on the Nu/Nu_0 of the channels. The total sensitivity index represents the combined influence of a single parameter and its interaction with other parameters on the Nu/Nu_0 of the channels. As can be seen from Figure 9a, when the response is Nu/Nu_0, the first-order sensitivity indexes of the channel parameters from high to low are α, Re, Z_2/D and Z_1/D. Among them, the changes of α and Re have an important influence on the Nu/Nu_0 of the channel. According to Figure 9b, the total parameter sensitivity indexes are α, *Re*, Z_2/D and Z_1/D from high to low, which is the same as the ranking of the first-order sensitivity indexes of the channel parameters. Through calculation, the difference between them is less than 0.02, indicating that the interaction between a single parameter and other parameters of the channels has no significant impact on the Nu/Nu_0 of the channels. In addition, the proportions of α and *Re* in the total sensitivity index are 50.6% and 47.9%, indicating that α and *Re* have a greater impact on the Nu/Nu_0 of the channels compared with Z_1/D and Z_2/D.

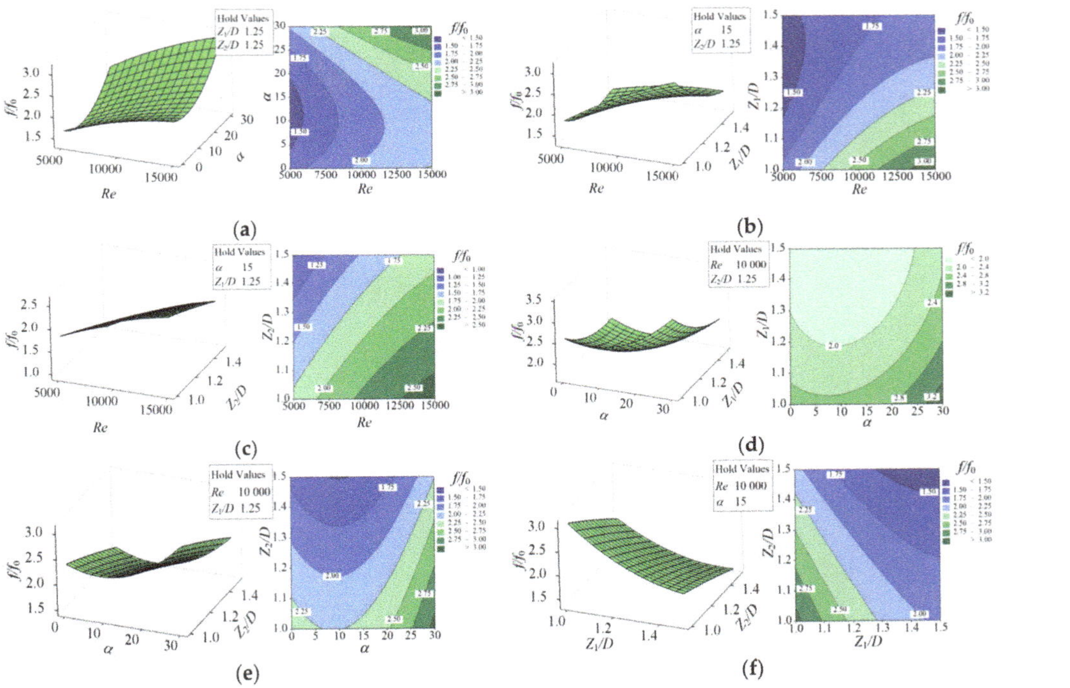

Figure 8. 3D surface and contour map of f/f_0: (**a**) Re–α; (**b**) Re–Z_1/D; (**c**) Re–Z_2/D; (**d**) α–Z_1/D; (**e**) α–Z_2/D; (**f**) Z_1/D–Z_2/D.

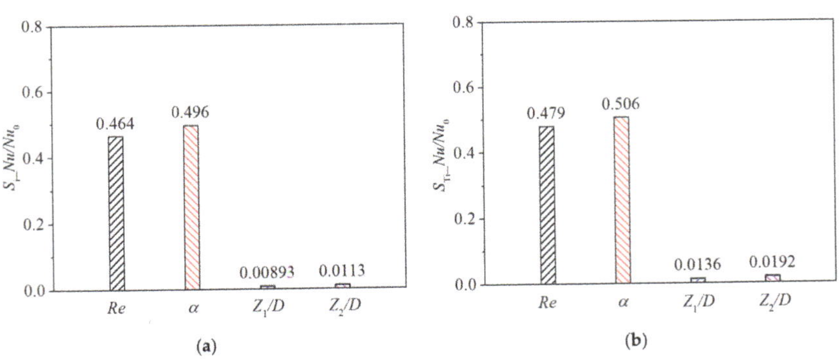

Figure 9. Sensitivity of the parameters when the response is Nu/Nu_0: (**a**) first-order sensitivity index; (**b**) total sensitivity index.

Figure 10 shows the sensitivity index of the channel parameters when the response is f/f_0, where Figures 10a and 10b are the first-order sensitivity index and total sensitivity index, respectively. It can be seen from Figure 10a that when the response is f/f_0, the first-order sensitivity indexes of the channel parameters are *Re*, Z_1/D, α and Z_2/D from high to low. Among them, the first-order sensitivity index of *Re* is significantly higher than other channel parameters, and the first-order sensitivity indexes of α, Z_1/D and Z_2/D are basically the same, all distributed around 0.15. According to Figure 10b, the total sensitivity indexes of the parameters are *Re*, Z_1/D, α and Z_2/D from high to low, which is the same as the first-order sensitivity indexes. Through calculation, it can be seen that the difference between the two is less than 0.021, indicating that the interaction between a single parameter and other parameters of the channels has no significant impact on the f/f_0 of the channels. In addition, the proportion of *Re* in the total sensitivity index is 57.4%, while the proportions of the total sensitivity coefficient of the other three channel parameters are all about 15%, which indicates that *Re* has the greatest impact on the f/f_0 of the channels, while α, Z_1/D and Z_2/D have a fairly small impact.

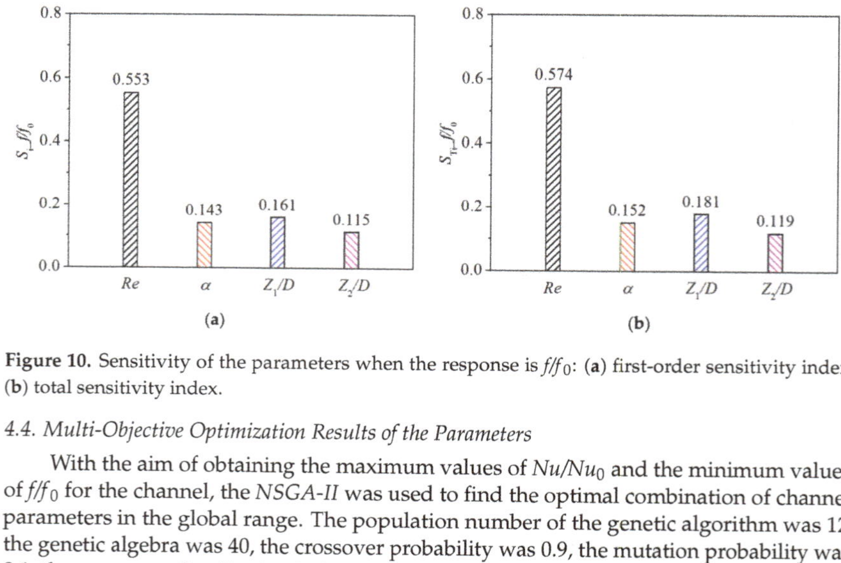

Figure 10. Sensitivity of the parameters when the response is f/f_0: (**a**) first-order sensitivity index; (**b**) total sensitivity index.

4.4. Multi-Objective Optimization Results of the Parameters

With the aim of obtaining the maximum values of Nu/Nu_0 and the minimum values of f/f_0 for the channel, the *NSGA-II* was used to find the optimal combination of channel parameters in the global range. The population number of the genetic algorithm was 12, the genetic algebra was 40, the crossover probability was 0.9, the mutation probability was 0.1, the crossover distribution index was 10, and the mutation distribution index was 20. The specific settings can be found in Ref. [24]. After the operation, a total of 481 solutions were generated, of which the Pareto solution set, the set of optimal solutions, had a total of 130 solutions. Figure 11 shows the solution set of multi-objective optimization. In Figure 11, the blue dots represent all of the solution sets, and the red curve represents the Pareto front connected by the Pareto solution sets. According to Figure 11, when the f/f_0 of the channel is constant, the Nu/Nu_0 of the channel of the point on Pareto front must be at its maximum. Similarly, when the Nu/Nu_0 of the channel is constant, the f/f_0 of the channel of the point on the Pareto front must be at its minimum.

On the basis of the sensitivity analysis of the parameters, the parameter *Re* has the greatest impact in the performance of the channel in terms of the flow and heat transfer performance of the channel. Consequently, the K-means clustering algorithm was used to cluster the Pareto solution sets under different values of *Re*. Figure 12 illustrates the results of K-means clustering of the Pareto solution set. As can be seen from Figure 12, the Pareto solution sets can be divided into four categories—A, B, C and D—under different values of *Re*. Without considering the influence of α, Z_1/D and Z_2/D, when *Re* increases, the Nu/Nu_0 and f/f_0 of the channel under the Pareto solution set increase slightly.

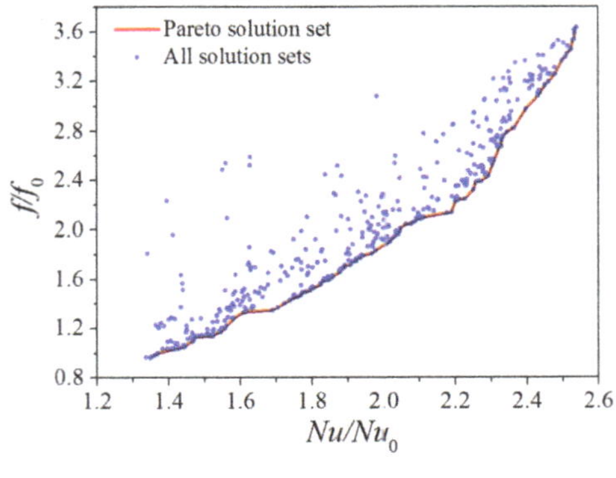

Figure 11. Solution set of multi-objective optimization.

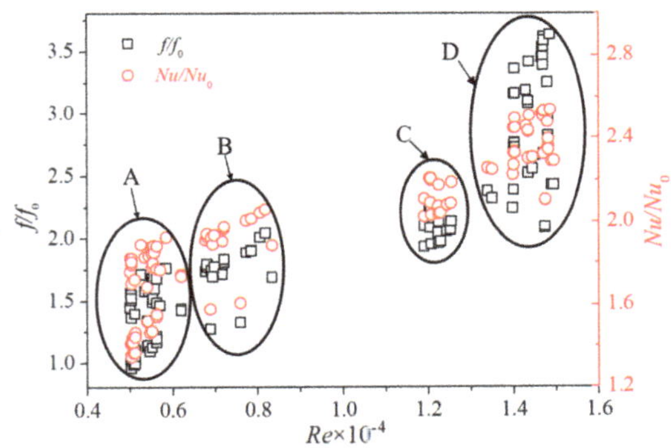

Figure 12. Result of K-means clustering of the Pareto solution set.

An optimal solution was selected from each of the four categories A, B, C and D in Figure 12 for comparative analysis. The specific optimal parameter combinations are provided in Table 5. According to Table 5, the Nu/Nu_0 of optimization points A, B, C and D increased by 9.70%, 21 82%, 26.06% and 27.88%, respectively, compared with the reference channel. In addition, the f/f_0 of the optimization points decreased by 19.89%, 1.05%, –7.85% and –9.42%, respectively, compared with the reference channel. Among them, the Nu/Nu_0 and f/f_0 of optimization points A and B were optimized, while the Nu/Nu_0 of optimization points C and D was considerably improved, but the f/f_0 had increased moderately. This is because the values of *Re* for optimization points C and D are large. When the Nu/Nu_0 increases, f/f_0 will also increase. Overall, compared with the reference channel, the Nu/Nu_0 of the optimized channels increased by 21.36% on average, and the f/f_0 decreased by 9.16% on average. This shows that the optimization results of the channel parameters in the present study are good, and can serve as a reference for the multi-objective optimization of channels with turbulent structures.

Table 5. Optimal channel parameters.

Type	*Re*	α	Z_1/D	Z_2/D	Nu/Nu_0	f/f_0
Reference channel	10,000	15	1.25	1.25	1.65	1.91
Optimization point A	5030	29.94	1.29	1.49	1.81	1.53
Optimization point B	7723	29.88	1.39	1.46	2.01	1.89
Optimization point C	12,433	26.62	1.42	1.44	2.08	2.06
Optimization point D	14,730	26.10	1.40	1.49	2.11	2.09

To further explore the influence of channel parameters on the flow and heat transfer performance of the channels, Figure 13 shows the comparison of surface streamline, temperature distribution, and *Nu* distribution of the heat transfer walls of the reference channel and optimization point C. Figure 13a,c shows the reference channel, and Figure 13b,d gives the optimization points C. It can be seen from Figure 13a,b that the high-temperature area of the heat transfer wall is mainly distributed upstream and downstream of the convex and upstream of the bottom of the concave. In comparison, the temperature of the high-temperature region of the optimized channel is lower and the area with high temperature is smaller. There are large vortexes upstream of the bottom of the concave and upstream and downstream of the convex in the reference channel, while the vortexes in the optimized channel are improved. The improvement of the vortexes will reduce the frictional resistance of the channel and the accumulation of airflow, thus reducing the local temperature of the heat transfer wall and reducing the area with high temperature. According to Figure 13c,d, contrary to the temperature distribution of the heat transfer wall, the high-temperature area had a lower *Nu* and the low-temperature area had a higher *Nu*. In comparison, the *Nu* value of the high-*Nu* region of the optimized channel was higher and the area with high *Nu* was larger. The results show that the optimized channel improves the vortexes at the bottom of the concave and upstream and downstream of the convex, so that the heat transfer wall of the channel has a lower temperature distribution and a higher *Nu* distribution.

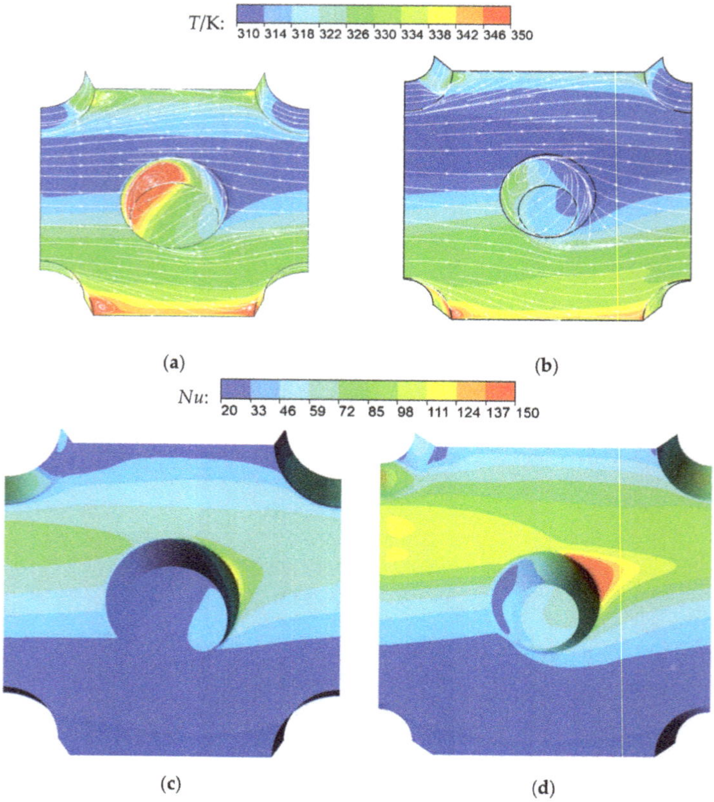

Figure 13. The Comparison of Reference Channel and Optimization Point C: (**a**) Surface Streamline and Temperature Distribution of the Reference Channel; (**b**) Surface Streamline and Temperature Distribution of the Optimized Channel; (**c**) *Nu* Distribution of the Reference Channel; (**d**) *Nu* Distribution of the Optimized Channel.

5. Conclusions

The multi-objective parameter optimization of channels with frustums of a cone was carried out. The design variables were *Re*, α, Z_1/D and Z_2/D. The optimization objective was to maximize the Nu/Nu_0 and minimize the f/f_0. Firstly, the experimental design of the channel parameters was carried out, and the second-order response surface model was constructed. Then, the accuracy of the response surface model was tested. Once the response surface met the requirements, the sensitivity of the channel parameters was analyzed, and the Pareto solution set was obtained. The Pareto solution set was analyzed by K-means clustering, and finally, four optimization channels were selected. The main conclusions can be drawn as follows:

(1) The second-order response surface models obtained by *CCF* have small errors and high accuracy. When the response is Nu/Nu_0, the *RMSE* of the model is 0.1048 and R^2 is 0.9491. When the response is f/f_0, the *RMSE* of the model is 0.2445 and R^2 is 0.9399.

(2) Compared with Z_1/D and Z_2/D, α and *Re* have the greatest impact on the Nu/Nu_0 of the channels. Parameter *Re* has the greatest influence on the f/f_0 of the channels, while α, Z_1/D and Z_2/D have the same, small influence on the f/f_0 of the channels.

(3) By comparing the reference channel with the optimized channel, the results show that the optimized channel improves the vortexes at the bottom of the concave and upstream and downstream of the convex, so that the heat transfer wall of the channel has lower temperature distribution and higher *Nu* distribution.

(4) Compared with the reference channel, the Nu/Nu_0 of the four optimized channels are increased by 9.70%, 21.82%, 26.06% and 27.88%, respectively; the f/f_0 decreased by 19.89%, 1.05%, –7.85% and –9.42%, respectively. The Nu/Nu_0 of the channels increased by 21.36% on average and the f/f_0 declined by 9.16% on average, which demonstrates that the optimization method of the channel parameters in the present study has a good effect.

Author Contributions: Conceptualization, Z.Z. and L.X. (Lei Xi); methodology, Z.Z. and L.X. (Liang Xu); software, Z.Z.; validation, Z.Z., L.X. (Liang Xu) and Q.R.; formal analysis, Z.Z. and J.G.; investigation, Z.Z., Q.R. and Y.L.; resources, Z.Z., L.X. (Lei Xi) and J.G.; data curation, Z.Z. and Y.L.; writing—original draft preparation, Z.Z. and L.X. (Liang Xu); writing—review and editing, L.X. (Liang Xu), Z.Z. and L.X. (Lei Xi); visualization, L.X. (Lei Xi) and Z.Z.; supervision, L.X. (Liang Xu) and J.G.; project administration, J.G.; funding acquisition, L.X. (Liang Xu). All authors have read and agreed to the published version of the manuscript.

Funding: The authors would like to express their sincere gratitude to the National Key R&D Program of China (2018YFB1106400), the National Natural Science Foundation of China (Grant No.51876157), and Natural Science Foundation of Shanxi Province in China (2019JM-096) for providing financial support for this work.

Institutional Review Board Statement: Not applicable.

Informed Consent Statement: Informed consent was obtained from all subjects involved in the study.

Data Availability Statement: Not applicable.

Conflicts of Interest: The authors declare no conflict of interest.

Nomenclature

RSM	Response Surface Methodology
Nu/Nu_0	Nusselt number ratio
f/f_0	Friction coefficient ratic
RSME	Root mean square errors
R^2	Determination coefficients
α	Frustum of a cone angle (°)
Re	Reynolds number
Z_2/D	Spanwise spacing ratio

Z_1/D	Streamwise spacing ratio
NSGA-II	Non-dominated sorting genetic algorithm with elite strategy
CCF	Central composite face-centered design
D	Bottom diameter of the frustum of a cone (mm)
N–S	Navier Stokes
GA	Genetic Algorithm
k	Turbulent kinetic energy equation
ε	Diffusion equation
CCD	central composite design
RNG	Renormalization Group
SIMPLEC	Semi-Implicit Method for Pressure Linked Equation
Z_2	Spanwise spacing (mm)
H	Height of the frustum of a cone (mm)
Z_1	Streamwise spacing (mm)
V	Average inlet velocity of air flow (m/s)
P	Characteristic length (mm)
v	Kinematic viscosity of air (m^2/s)
Nu_x	Local Nusselt number
q_x	Local heat flux (W/m^2)
T_w	Local temperature of the wall (K)
T_f	Reference temperature (K)
Nu	Average Nusselt number
A	Area of heat transfer wall (m^2)
f	Friction coefficient
L	channel length (mm)
Δp	Pressure drop (Pa)
ρ	Air density (kg/m^3)
λ	Thermal conductivity of air (W/m K)
Nu_0	Average Nusselt number of smooth parallel plate channel
f_0	Friction coefficient of smooth parallel plate channel
$\boldsymbol{Y}$	Response
$\boldsymbol{X}$	Design variable
$f(\boldsymbol{X})$	Approximate function of the target
$\boldsymbol{\beta}$	Fitting coefficient
$\boldsymbol{\varepsilon}$	Prediction error
$F(\boldsymbol{x})$	A square integrable function
M	Total variance of the $F(\boldsymbol{x})$
$M_{i_1 \cdots i_t}$	Partial square deviation of the F(**x**)
$S_{i_1 \cdots i_t}$	Sensitivity index of the variable
S_{i_t}	First order sensitivity index
$S_{i_q i_t}$	Second-order sensitivity index
S_{Ti_t}	Total sensitivity index

References

1. Zheng, N.; Liu, P.; Shan, F.; Liu, Z.; Liu, W. Sensitivity analysis and multi-objective optimization of a heat exchanger tube with conical strip vortex generators. *Appl. Therm. Eng.* **2017**, *122*, 642–652. [CrossRef]
2. Xi, L.; Xu, L.; Gao, J.; Zhao, Z.; Li, Y. Numerical analysis and optimization on flow and heat transfer performance of a steam-cooled ribbed channel. *Case Stud. Therm. Eng.* **2021**, *28*, 101442. [CrossRef]
3. Izadi, M.; Mohebbi, R.; Sajjadi, H.; Delouei, A.A. LTNE modeling of Magneto-Ferro natural convection inside a porous enclosure exposed to nonuniform magnetic field. *Phys. A Stat. Mech. Appl.* **2019**, *535*, 122394. [CrossRef]
4. Jeong, H.-S.; Seo, J.-W.; Kim, K.-Y. Multi objective optimization of a slit rib in a rectangular cooling channel. *Heat Transf. Res.* **2018**, *49*, 395–412. [CrossRef]
5. Seo, J.-W.; Afzal, A.; Kim, K.-Y. Efficient multi-objective optimization of a boot-shaped rib in a cooling channel. *Int. J. Therm. Sci.* **2016**, *106*, 122–133. [CrossRef]
6. Mamourian, M.; Milani Shirvan, K.; Mirzakhanlari, S. Two phase simulation and sensitivity analysis of effective parameters on turbulent combined heat transfer and pressure drop in a solar heat exchanger filled with nanofluid by Response Surface Methodology. *Energy* **2016**, *109*, 49–61. [CrossRef]

7. Bu, S.; Yang, Z.; Zhang, W.; Liu, H.; Sun, H. Research on the thermal performance of matrix cooling channel with response surface methodology. *Appl. Therm. Eng.* **2016**, *109*, 75–86. [CrossRef]
8. Shi, X.; Li, S.; Mu, Y.; Yin, B. Geometry parameters optimization for a microchannel heat sink with secondary flow channel. *ICHMT* **2019**, *104*, 89–100. [CrossRef]
9. Wen, J.; Li, K.; Liu, Y.; Wu, M.; Wang, S. Multi-objective optimization of serrated fin in plate-fin heat exchanger by fluid structure interaction. *J. Xi'an Jiaotong Univ.* **2018**, *52*, 130–135. (In Chinese)
10. Zheng, N.; Liu, P.; Liu, Z.; Liu, W. Numerical simulation and sensitivity analysis of heat transfer enhancement in a flat heat exchanger tube with discrete inclined ribs. *Int. J. Heat Mass Transf.* **2017**, *112*, 509–520. [CrossRef]
11. Feng, G. Research progress in the flow induced vibration mechanism of tube bundles in heat exchangers and precautionary measures. *Chem. Ind. Eng. Prog.* **2012**, *31*, 508–512. (In Chinese)
12. Abdulhay, B.; Bourouga, B.; Dessain, C. Experimental and theoretical study of thermal aspects of the hot stamping process. *Appl. Therm. Eng.* **2011**, *31*, 674–685. (In Chinese) [CrossRef]
13. Zhao, Z.; Xu, L.; Gao, J.; Xi, L.; Li, Y. Study on the Flow and Heat Transfer Characteristics of the Heat Exchanger with Cone-Type Vortex Generaters. *J. Xi'an Jiaotong Univ.* **2021**, *10*, 131–143. (In Chinese)
14. Luo, L.; Du, W.; Wang, S.; Wang, L.; Sundén, B.; Zhang, X. Multi-objective optimization of a solar receiver considering both the dimple/protrusion depth and delta-winglet vortex generators. *Energy* **2017**, *137*, 1–19. [CrossRef]
15. Zeng, M.; Zhang, G.; Li, Y.; Niu, Y.; Ma, Y.; Wang, Q. Geometrical Parametric Analysis of Flow and Heat Transfer in the Shell Side of a Spiral-Wound Heat Exchanger. *HTrEn* **2015**, *36*, 790–805. [CrossRef]
16. Liu, J.; Song, Y.; Xie, G.; Sunden, B. Numerical modeling flow and heat transfer in dimpled cooling channels with secondary hemispherical protrusions. *Energy* **2015**, *79*, 1–19. [CrossRef]
17. Kim, H.-M.; Moon, M.-A.; Kim, K.-Y. Multi-objective optimization of a cooling channel with staggered elliptic dimples. *Energy* **2011**, *36*, 3419–3428. [CrossRef]
18. Elyyan, M.A.; Rozati, A.; Tafti, D.K. Investigation of dimpled fins for heat transfer enhancement in compact heat exchangers. *Int. J. Heat Mass Transf.* **2008**, *51*, 2950–2966. [CrossRef]
19. Courand, A.; Metz, M.; Héran, D.; Feilhes, C.; Prezman, F.; Serrano, E.; Bendoula, R.; Ryckewaert, M. Evaluation of a robust regression method (RoBoost-PLSR) to predict biochemical variables for agronomic applications: Case study of grape berry maturity monitoring. *Chemom. Intell. Lab. Syst.* **2022**, *221*, 104485. [CrossRef]
20. Sobol', I.M. Global sensitivity indices for nonlinear mathematical models and their Monte Carlo estimates. *Math. Comput. Simul.* **2001**, *55*, 271–280. [CrossRef]
21. Song, Y.; Wang, Y.; Yang, S.; Wang, S.; Yang, M. Sensitivity analysis and parameter optimization of energy consumption for underwater gliders. *Energy* **2020**, *191*, 116506. [CrossRef]
22. Saltelli, A.; Annoni, P.; Azzini, I.; Campolongo, F.; Ratto, M.; Tarantola, S. Variance based sensitivity analysis of model output. Design and estimator for the total sensitivity index. *Comput. Phys. Commun.* **2010**, *181*, 259–270. [CrossRef]
23. Deb, K.; Pratap, A.; Agarwal, S.; Meyarivan, T. A fast and elitist multiobjective genetic algorithm: NSGA-II. *IEEE Trans. Evol. Comput.* **2002**, *6*, 182–197. [CrossRef]
24. Xu, L.; Ruan, Q.; Shen, Q.; Xi, L.; Gao, J.; Li, Y. Optimization Design of Lattice Structures in Internal Cooling Channel with Variable Aspect Ratio of Gas Turbine Blade. *Energies* **2021**, *14*, 3954. [CrossRef]

energies

MDPI

Article

Aerodynamics and Complicated Heat Transfer with the Mixed Motion of Air in the Flat Duct of a High-Temperature Heat Exchanger

Borys Basok [1], Vyacheslav Kremnev [1], Anatoliy Pavlenko [2,*] and Andriy Timoshchenko [1]

1 Institute of Engineering Thermophysics, National Academy of Sciences of Ukraine, 03057 Kiev, Ukraine; basok@ittf.kiev.ua (B.B.); kremnev@ukr.net (V.K.); a_timoshchenko@ukr.net (A.T.)
2 Department of Building Physics and Renewable Energy, Kielce University of Technology, al. Tysiąclecia Państwa Polskiego 7, 25-314 Kielce, Poland
* Correspondence: apavlenko@tu.kielce.pl; Tel.: +48-883-741-291

Abstract: The purpose of the research is to study the aerodynamics and heat transfer in the duct of a high-temperature recuperation system. The object of the research is a flat duct with a thickness-to-height ratio a:b = 1:10, length c = 400·a with one-sided heat input to the duct surface, complicated heat transfer and mixed air movement in the duct. The objectives of the research are to determine: (a) average temperatures on the duct surfaces; (b) air temperature distribution along the length of the duct; (c) local and average integral temperature values along the length of the duct; (d) local and average integral heat flow densities on the duct surfaces; (e) local and average integral heat exchange coefficients and Nusselt numbers on the duct surfaces; and (f) pressure distribution along the length and total pressure drop in the duct. The research method is based on conducting a mathematical numerical experiment in a stationary three-dimensional Cartesian formulation and physical modeling of processes on a prototype. It is established that the dominant heat transfer between the duct walls and the air is a complicated convective heat transfer, in which forced convection is affected by free convection. There is a 1.5–2.0-fold difference in the values of the heat flow density on the hot and non-heated surfaces of the duct. Generalizing dependences of Nusselt numbers, temperature pressures and friction resistance coefficients as functions of a dimensionless coordinate are obtained.

Keywords: radiation-convective heat transfer; high-temperature recuperation system

Citation: Basok, B.; Kremnev, V.; Pavlenko, A.; Timoshchenko, A. Aerodynamics and Complicated Heat Transfer with the Mixed Motion of Air in the Flat Duct of a High-Temperature Heat Exchanger. *Energies* **2022**, *15*, 865. https://doi.org/10.3390/en15030865

Academic Editor: Jan Danielewicz

Received: 27 December 2021
Accepted: 23 January 2022
Published: 25 January 2022

1. Introduction

Issues of high-temperature recuperation have an attitude to the processes of transferring of substance and energy, which are complicated by the simultaneous action of various factors. These are and different kinds and regimes of substance motion, and different mechanisms of transferring heat in the objects that are investigated. Many works are devoted to studying complicated heat transfers in compact high-temperature heat exchangers of various types [1–4]. The dominant mechanisms of transferring in them are convective and radiative heat transfer [5–10]. The feature of the work consists of the applied nature of the research aimed at studying the conditions of flow and heat transfer in an air-cooled duct, which is the basis of heat recuperation devices for high-temperature industrial plants. The formulation of the research tasks most fully describes the conditions-temperature, magnitude and direction of the heat flow density, the type of coolant and its mass flow, the geometry of the duct and the main physical and mechanical characteristics of its surfaces observed in modern high-temperature processes for the production of glass and stone fibers. The research and processing of the results were carried out using classical methods [11,12]. At the same time, for each of the duct surfaces involved in heat transfer and differing in the magnitude and direction of the supplied heat flow, Nusselt numbers and temperature pressures are determined. The results of the research were compared with the case of convective heat transfer in pipes and ducts with symmetrical heat supply to the duct surfaces, which allowed us to establish the nature and magnitude of the influence of complicated heat transfer on the characteristics of the process, in order to obtain their qualitative and quantitative indicators. The generalized dependences obtained make it possible to calculate the aerodynamic drag, temperature and energy parameters of the process, with complicated heat transfer and mixed motion necessary for the design of high-temperature heat recuperation systems in order to increase the energy efficiency of production processes. Utilization and reduction of heat losses in high-temperature industrial plants in order to increase their energy efficiency is implemented on the basis of the use of innovative heat exchange devices or heat recuperation systems. Heat transfer processes in them are carried out at high temperature heads, one-sided heat input, complicated radiation-conductive-convective heat transfer and mixed motion of air as a heat carrier.

Problem definition.

The simplified scheme of heat flow densities for a fragment of the lining is shown, which is typical for high-temperature heat recuperation systems with an air heat curtain, Figure 1.

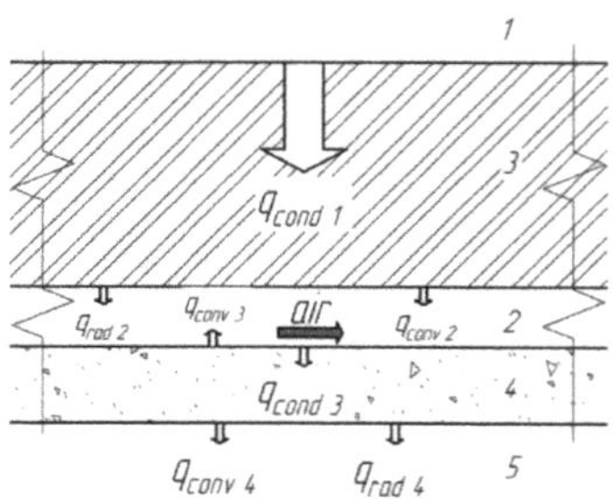

Figure 1. Scheme of heat transfer: 1—internal space of the technological equipment; 2—air duct for recuperation of the heat loss, type the heat curtain; 3—lining; 4—thermal insulation of the lining to reduce dissipative heat loss; 5—environment; $q_{cond\ 1}$, $q_{cond\ 3}$—conductive heat flow; $q_{conv\ 2}$, $q_{conv\ 3}$, $q_{conv\ 4}$—convective heat flow; $q_{rad\ 2}$, $q_{rad\ 4}$—radiation heat flow.

In the summary conductive heat flow through the lining $q_{cond\ 1}$, one consists of losses with convective $q_{conv\ 2}$ and radiation $q_{rad\ 2}$ heat flows from the internal surface of the air duct $q_{cond\ 1} = q_{conv\ 2} + q_{rad\ 2}$; the radiation heat flow $q_{rad\ 2}$ makes up for convective heat transfer from the surface of the duct $q_{conv\ 3}$ and heat loss by conductive $q_{cond\ 3}$ through thermal insulation $q_{rad\ 2} = q_{conv\ 3} + q_{cond\ 3}$. The heat losses through thermal insulation $q_{cond\ 3}$ consists from convective $q_{conv\ 4}$ and radiation $q_{rad\ 4}$ heat flows from external surface of the equipment to environment. As can be seen, the heat losses to the environment comprise complicated radiation-convective heat transfer. For an integral description of all heat flows, the classical expressions presented in [13–16] are used to calculate the value of the convective and radiation heat exchange coefficients. It is assumed that each of the heat transfer mechanisms is independent, and there is no interaction between them.

2. Methodology of the Research

The complicated heat transfer in the air duct is determined by the action radiation and convective mechanisms of the transfer. Cases of such heat transfer do not have universal generalizing correlations, so they should be studied individually for each variant of technological equipment [13,17,18]. In particular, for our problem, the conditions that determine the features of complicated heat transfer in the duct are as follows:

- the Boltzmann criterion is much larger than 1, $Bo \in \left(10^2; 3{\cdot}10^3\right)$, so the
- mechanism of heat transfer between the duct surfaces and the air flow is
- convection;
- the high temperature of the hot surface of duct-up to 700 °C;

- presence of a temperature difference between the duct surfaces;
- radiation transfer of heat from the hot wall of the duct to the cold one;
- turbulent flow regime, $Re \in (4{\cdot}10^3; 10^4)$;
- the combined influence of free and forced convection on heat transfer in the
- duct, $Ra \in (10^5; 6{\cdot}10^6)$.

Research objective. The research objective is to determine: (a) average temperatures on the duct surfaces; (b) distribution of air temperature along the duct length; (c) local and average integral temperature heads along the duct length; (d) local and average integral heat flow density on the duct surfaces; (e) local and average integral heat exchange coefficients and Nusselt numbers on the duct surfaces; and (f) pressure distribution along the length and total pressure drop in the duct.

The object of the research is a flat duct with the ratio of sides of thickness to height a:b = 1:10, length c = 400·a with one-sided heat input (boundary conditions of the 2nd kind) and complicated heat transfer with mixed air motion, Figure 2.

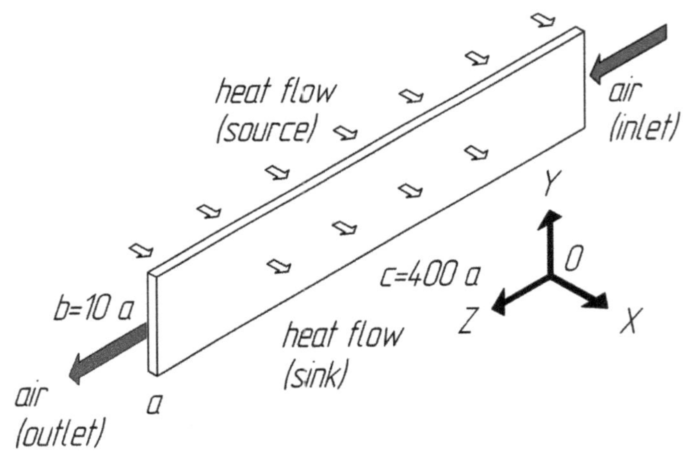

Figure 2. Scheme of the modelling area.

The subject of the research is the aerodynamics of air flow and heat transfer in the duct. The research is based on mathematical and physical modeling of processes. The mathematical numerical experiment is performed on a stationary three-dimensional Cartesian model. Physical modeling of the operation of the air-cooling system is carried out during pilot tests.

3. Physical Model of Complicated Heat Transfer and Mixed Air Motion in the Flat Duct

Features of the problem are: (a) one-sided heat input, as a result of which traditional convective heat transfer in the duct is complicated by radiation heat transfer between the hot surface (source) and other surfaces of the duct that are not heated; (b) air is considered as a diathermal medium; and (c) multidirectional action of vectors of volume forces and pressure forces. The research is conducted using a CFD package for turbulent flow. In particular, to describe radiation heat transfer in the duct, the surface-to-surface (S2S) model of radiation heat transfer in space bounded by gray diffuse surfaces is used. The main assumption in the S2S model-absorption, radiation or scattering by a gas medium-is ignored, and only surface radiation is considered in the analysis. A non-isothermal flow is modeled, turbulent flow regime is considered, the air flow is taken as a Newtonian liquid and the dependence of the thermophysical properties of air on temperature is considered.

Mathematical model of flow and complicated heat transfer in the duct with mixed motion of air. In a general view, the expressions for the S2S model are written as follows [15,16]
energy flow transmitted by other surfaces—$A_k \cdot q_{in,k} = \sum_{j=1}^{N} A_j \cdot q_{out,j} \cdot F_{jk}$;
mathematical formulation of the principle of reciprocity—$A_j \cdot F_{jk} = A_k \cdot F_{kj}$;
heat flow that is gotten on the surface k from surfaces N – $q_{in,k} = \sum_{j=1}^{N} q_{out,j} \cdot F_{jk}$;
factor of mutual influence F_{ij} between the two surfaces with final sizes i and j—

$$F_{ij} = \frac{1}{A_i} \int_{A_i} \int_{A_j} \frac{\cos\theta_i \cdot \cos\theta_j}{\pi \cdot r^2} \delta_{ij} dA_i dA_j;$$

radiation energy flow that transfers from surface k

$$q_{out,k} = \varepsilon_k \cdot \sigma \cdot T_k^4 + \rho_k \cdot \sum_{j=1}^{N} q_{out,j} \cdot F_{jk},$$

where A_k—area of the surface k; F_{jk}—factor which consider geometrical features position of the surfaces k and j and influence on the part of the energy that transmitted from surface j to surface k; $q_{in,k}$—heat flow that get in the surface k from environmental; δ_{ij}—defines the visibility dA_j to dA_i ($\delta_{ij} = 1$ if dA_j is visible from dA_i or 0 otherwise); ε_k—radiant emittance of the surface k; ρ_k—radiant reflectance of the surface k; σ—Stefan-Boltzmann constant. Indexes: in, out—indexes refer to data that get in or get out, correspondingly; i, j, k, N to refer to the surfaces i, j, k, N, correspondingly.

In convective heat transfer, the equations of continuity, momentum transfer, energy, kinetic energy dissipation and kinetic energy dissipation rate were used.

The following boundary conditions were set

$$x = -\frac{a}{2}; \; y \in \left[-\frac{b}{2}; \frac{b}{2}\right]; z \in [0; c], \quad q_0 = -\lambda_{air} \left(\frac{\partial T}{\partial x}\right)_{x=-\frac{a}{2}} + q_{rad,-\frac{a}{2}};$$

$$x = \frac{a}{2}; \; y \in \left[-\frac{b}{2}; \frac{b}{2}\right]; z \in [0; c], \quad 0 = -\lambda_{air} \left(\frac{\partial T}{\partial x}\right)_{x=\frac{a}{2}} - q_{rad,\frac{a}{2}};$$

$$x \in \left[-\frac{a}{2}; \frac{a}{2}\right]; y = \frac{b}{2}; z \in [0; c] \quad 0 = -\lambda_{air} \left(\frac{\partial T}{\partial y}\right)_{y=\frac{b}{2}} - q_{rad,\frac{b}{2}};$$

$$x \in \left[-\frac{a}{2}; \frac{a}{2}\right]; y = -\frac{b}{2}; z \in [0; c] \quad 0 = -\lambda_{air} \left(\frac{\partial T}{\partial y}\right)_{y=-\frac{b}{2}} - q_{rad,-\frac{b}{2}};$$

$$x \in \left[-\frac{a}{2}; \frac{a}{2}\right]; y \in \left[-\frac{b}{2}; \frac{b}{2}\right]; z = 0; \quad G_{inlet} = G_0; T = T_{inlet}; u = v = 0;$$

$$x \in \left[-\frac{a}{2}; \frac{a}{2}\right]; y \in \left[-\frac{b}{2}; \frac{b}{2}\right]; z = c; \; p = 0; G_{outlet} = G_{inlet}; \; u = v = 0; \left(\frac{\partial T}{\partial z}\right)_{z=c} = 0,$$

where $q_0 \in [5000; 6000]$ W/m^2, $G_{inlet} \in [100; 200]$ kg/h, $t_{inlet} = 30$ °C, c = 13.0 m—the parameters of the pilot plant; radiant emittance of duct surfaces $\varepsilon = 0.93$.

4. Data Processing and Displaying Technique

Results of the research are viewed as integral characteristics, namely: average heat exchange coefficient from the duct surface in the i section $\alpha = \frac{q_w}{(t_w - t_a)}$; average heat flow density at the wall of the duct $\overline{q_w} = \frac{1}{b}\int_{-\frac{b}{2}}^{\frac{b}{2}} q_w(y)dy$; average temperature of the wall of the duct $\overline{t_w} = \frac{1}{b}\int_{-\frac{b}{2}}^{\frac{b}{2}} t_w(y)dy$; average mass air temperature $\overline{t_{air}} = \frac{\int_f \rho w t_a df}{\int_f \rho w df}$; average mass air flow rate $\overline{w} = \frac{\int_f \rho w df}{\int_f \rho df}$; average integral heat exchange coefficient over the length of the duct $\overline{\alpha} = \frac{1}{c}\int_0^c \alpha(z)dz$; average integral temperature head $\overline{\Delta t} = \frac{1}{c}\int_0^c (\overline{t_w(z)} - \overline{t_a(z)})dz$;

average integral temperature of air $\overline{t_{air}} = \frac{1}{c}\int_0^c \overline{t_a(z)}dz$; dimensionless average integral temperature head $\overline{\Theta} = \frac{\overline{\Delta t}}{\overline{t_{air}}} = \frac{\int_0^c (\overline{t_w(z)} - \overline{t_a(z)})dz}{\int_0^c \overline{t_a(z)}dz}$; Nusselt average number $\overline{Nu} = \overline{\alpha}\cdot\frac{X}{\lambda}$; pressure losses in the duct caused by friction, Darcy-Weisbach formula, $\Delta p = \overline{\xi}\cdot\frac{c}{X}\cdot\frac{\rho\cdot\overline{w}^2}{2}$; average coefficient of friction resistance $\overline{\xi} = \frac{8\cdot\overline{\tau}_w}{\rho\cdot\overline{w}^2}$ and Nusselt local number $Nu = \alpha\cdot\frac{X}{\lambda}$, where $a,\ b,\ c$—linear dimensions of the duct, respectively: width, height and length; $q_w(y)$—local heat flow density on the wall in the *i* section, is determined by the numerical experiment; t_a—temperature of air, is determined by the numerical experiment; $t_w(y)$—local temperature of the wall in the *i* section, is determined by the numerical experiment; w—air velocity, is determined by the numerical experiment; $x,\ y,\ z$—coordinates along the axes, respectively, OX, OY, OZ; X—characteristic linear duct size; $\alpha(z)$—average heat exchange coefficient from the duct surface; λ, ρ—the coefficient of thermal conductivity and density, respectively, depend on the air temperature; $\overline{\tau}_w$—average tangential stress on the duct wall is determined by the experiment for the hot surface and the unheated surface. Indexes: *air*—value is determined for the air flow; and w—value is determined for the duct wall.

For the analysis of the obtained results of a numerical experiment and the convenience of comparing them with the known results of convective heat exchange in pipes under turbulent flow [19–22], the data are presented in a criterion form. The thermophysical properties that are part of dimensionless complexes are determined for the average mass temperature of the flow in this section. The hydraulic diameter is used as the characteristic linear size of the duct cross-section. The effect of free and forced convection on the heat transfer intensity is determined from the Rayleigh (*Ra*) and Peckle (*Pe*) complexes, respectively. The intensity of heat transfer in the dimensionless view is presented by Nusselt number.

Generalization of the results of the numerical research was carried out by: dimensionless local coordinate—$Z = Re\cdot Pr\cdot\frac{X}{z}$; average dimensionless coordinate $\overline{Z} = \overline{Re}\cdot\overline{Pr}\cdot\frac{X}{c}$; average Reynolds integral number along the length of the calculated region $\overline{Re} = \frac{1}{c}\int_0^c Re(z)dz$; reduced heat transfer intensity from the hot surface $(\overline{Nu_1}/\overline{Nu})$; dimensionless temperature head from the hot surface $\overline{\Theta}_1$; dimensionless temperature head from the unheated surface $\overline{\Theta}_2$; average coefficient of friction resistance, where is: $Re(z)$—local Reynolds number based on the results of a numerical experiment; $\overline{Pr}$—the average Prandtl integral number along the length of the calculated region, determined by analogy with the number $\overline{Re}$. Indexes: 1, 2—the value refers to the hot surface and the non-heated surface, respectively.

Verification of the adequacy of the mathematical model is carried out according to the test research, in which the intensity of convective heat transfer is determined for the described duct geometry under turbulent flow, double-sided heat input and boundary conditions of the 2nd kind. Radiation heat transfer in the duct is not considered. Other characteristics of the test research model correspond to the values taken for the main research. The heat flux density for double-sided heat input is chosen by the condition—$q_1 = 0.5q_0$, where q_0 is the heat flux density for double-sided heat input (main research).

The results of the test research are presented in a generalized form in the $\overline{Z} - \overline{Nu}$ coordinate system and compared with the existing results of the research of convective heat transfer in pipes with turbulent flow at small heat flow densities, when there is still no effect on heat transfer of non-isothermicity through natural convection and there is no change in the thermophysical properties of the heat carrying agent [22]

$$\overline{Nu} = \frac{(\xi/8)\overline{Re}\cdot\overline{Pr}}{K_1 + K_2\sqrt{\frac{\xi}{8}}\left(\overline{Pr}^{\frac{2}{3}} - 1\right)}, \quad (1)$$

where is $K_1 = 1 + \frac{900}{\overline{Re}}$, $K_2 = 12.7$.

The coefficient of hydraulic friction resistance is calculated as follows:

$$\xi = \left(1.821\lg\left(\overline{Re}\right) - 1.64\right)^{-2}.$$

Limits of application of expression (1): $\overline{Re} = 4{\cdot}10^3 \div 5{\cdot}10^6$; $\overline{Pr} = 0.5 \div 5.0$ [22]. The discrepancy between the experimental data and the results of calculations using expression (1) does not exceed 8%.

5. Results of Research Aerodynamics and Heat Transfer for One-Sided Heat Input

The initial data and algorithm for performing a numerical experiment are selected from real experiments of testing a pilot plant, and are shown in Table 1.

Table 1. Plan for performing the numerical experiment with one-sided heat input.

Heat Flow Density, W/m^2	Mass Flow Rate of Air, kg/h			Notes
	100	150	200	
5000	5000_100_1	5000_150_1	5000_200_1	Hot surface
5000	5000_100_2	5000_150_2	5000_200_2	Non-heated surface
5500	5500_100_1	5500_150_1	5500_200_1	Hot surface
5500	5500_100_2	5500_150_2	5500_200_2	Non-heated surface
6000	6000_100_1	6000_150_1	6000_200_1	Hot surface
6000	6000_100_2	6000_150_2	6000_200_2	Non-heated surface

The designations of experiments are given in Table 1, and are used as captions in the figures (Figures 3–18). Figures 3–6 show the velocity and temperature distributions in the air flow, in the vertical ZOY and horizontal ZOX central-section of the duct. The profiles are constructed for cross-sections located at a distance of 10, 100 and 200 calibers from the entry plane. As the caliber, the value of the hydraulic diameter of the duct X is taken.

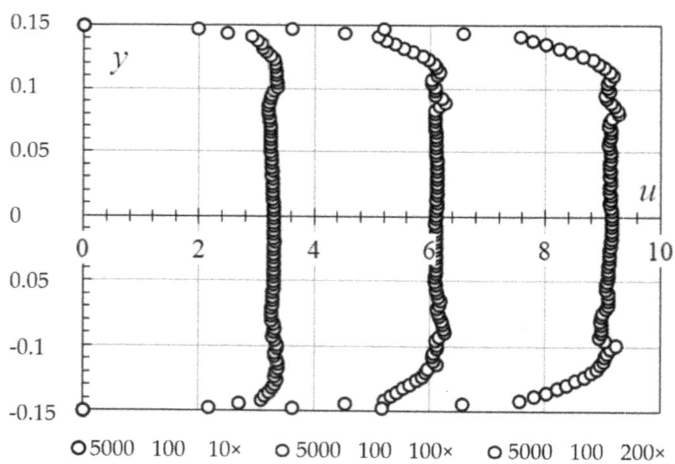

Figure 3. Air velocity profiles in the vertical central-section ZOY at 100 kg/h (y [m], u [m/s]), curves from left to right, respectively: 10×; 100×; 200×.

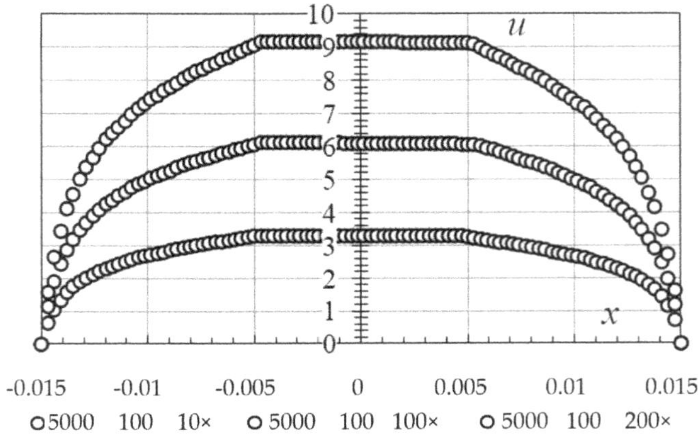

Figure 4. Air velocity profiles in the horizontal central-section of ZOX at 100 kg/h (x [m], u [m/s]), bottom-up curves, respectively: 10×; 100×; 200×.

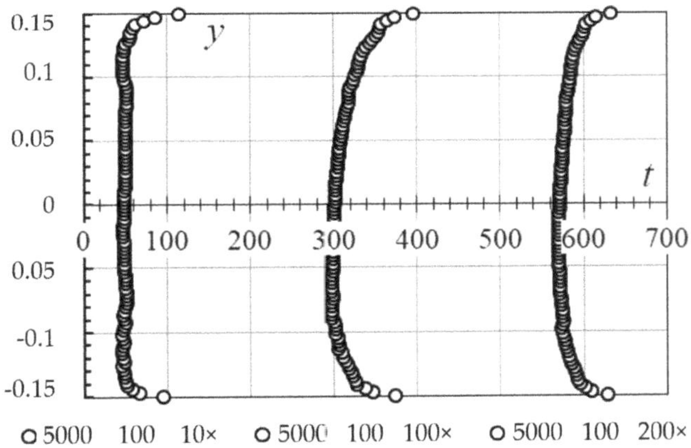

Figure 5. Air temperature profiles in the vertical central-section ZOY, 100 kg/h (t [°C], y [m]), curves from left to right, respectively: 10×; 100×; 200×.

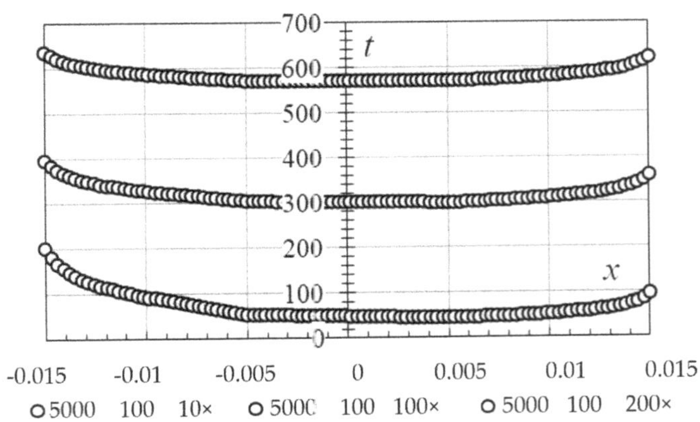

Figure 6. Air temperature profiles in the horizontal central-section ZOX, 100 kg/h (x [m], t [°C]): bottom-up curves, respectively: 10×; 100×; 200×.

For the results obtained for the mass flow rates of air of 100, 150 and 200 kg/h, an asymmetry of the temperature profiles relative to the central axes (cross-sections) of the duct symmetry is observed. The influence of free convection in the central part of the duct on vertical temperature profiles (Figure 5) is observed—the profiles are more filled than in the lower part, below the plane of symmetry. It is observed a higher temperature at the end of duct with a lower air flow rate, than with a higher air flow rate. It is observed decreasing the temperature difference between the duct surfaces as the air temperature increases. The surface temperature of the source (with one-sided heat input) is higher than the temperature of the non-heated surface along the entire length of the duct (Figure 6). Along the perimeter of the duct cross-section, its surface temperature does not have the same value, just as local heat exchange coefficients differ along the duct perimeter. Processing of the results of the research consisted in determining local, average heat exchange coefficients over the duct cross-section. At the same time, considered that the average duct surface temperatures have different values, the heat exchange coefficients are determined separately for the hot surface and the non-heated surface. In Figure 7 shows graphs of changes in the average temperatures (main axis on the left) of the hot surface and the non-heated surface and air along the length of the duct. In addition, this figure shows the temperature heads (auxiliary axis on the right) to the hot surface and the non-heated surface along the length of the duct.

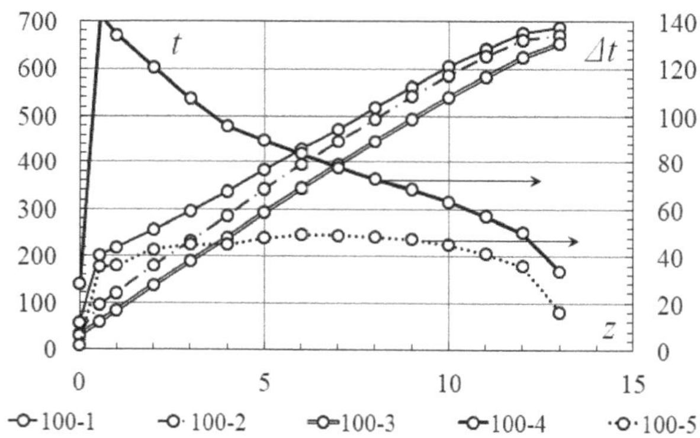

Figure 7. Change of the temperature t [°C] and the temperature head Δt [°C] by the length of the duct z [m] ($G = 100$ kg/h, $q\_0 = 5000$ W/m^2): 100-1—the average temperature of the hot surface; 100-2—the average temperature of the non-heated surface; 100-3—the average temperature of air; 100-4—the temperature head to the hot surface; and 100-5—the temperature head to the non-heated surface.

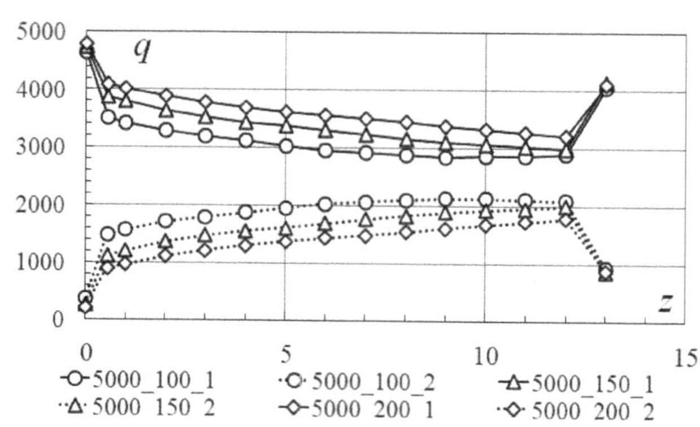

Figure 8. Distribution of the heat flow density q [W/m^2] by the length of the duct z [m]: q_0 – *Const*; G – *Varia*.

The qualitative valuation, which is obtained for one-sided heat input at $q_0 = 5000$ W/m^2, is also observed for cases of heat transfer at $q_0 = 5500$ and 6000 W/m^2.

It is the lower the mass flow rate of air, than is the higher the overall temperature level of the process: (a) higher temperatures of the surfaces of the duct; (b) higher air temperature at the outlet of the duct; (c) smaller temperature difference between the hot surface and the non-heated surface at the outlet of the duct; and (d) smaller difference in temperature heads on the hot surface and on the non-heated surface. It should be expected that at the lower mass flow rate of air, at the end of the duct, the heat transfer rate on the hot surface of the duct and on the non-heated surface will approach a certain equal value. The different pattern of change in the value of the temperature head along the length of the duct on its hot surface and the non-heated surface indicates a qualitative difference in the heat transfer processes on these surfaces. It should also expect different patterns of changes in local heat exchange coefficients on these surfaces along the length of the duct. The distribution of the heat flow density along the length of the duct on the hot surface and the non-heated surface for different conditions of the experiment is shown in Figure 8. The results obtained have well-defined sections with a non-monotonous change in the value of the heat flow density—sections at the inlet/outlet of the duct. The non-monotonicity of the q value at the beginning of the duct-$\frac{z}{X} \leq 10$-is due to the influence of the initial hydrodynamic and thermal regions, as well as the conditions for formulating a numerical experiment. The non-monotonicity of q at the end of the duct-$218 \leq \frac{z}{X} < 236$-is due only to the conditions of the numerical experiment. When processing the results of the research, empirical correlations are approximated by power dependences of the species-$q(z) = Az^B$ (approximation reliability $R^2 > 0.96$). When constructing approximation dependencies, the plots $\frac{z}{X} \leq 10$ and $218 \leq \frac{z}{X} < 236$ are excluded from consideration. Further processing of the research results is carried out using approximation dependencies.

In Figure 9 shows the results of calculating local heat exchange coefficients (α) along the duct length (z), on the hot surface of the duct and on the non-heated surface of the duct. The present results demonstrate that: (a) the behavior of changes in heat exchange coefficients is different, if for the hot surface it qualitatively approaches the process of forced convection, then for the non-heated surface, such an analogy is not observed; (b) the heat exchange coefficients on the non-heated surface are greater than the heat exchange coefficients on the hot surface; and (c) as the air heats up, the difference between the heat exchange coefficients on opposite duct surfaces decreases.

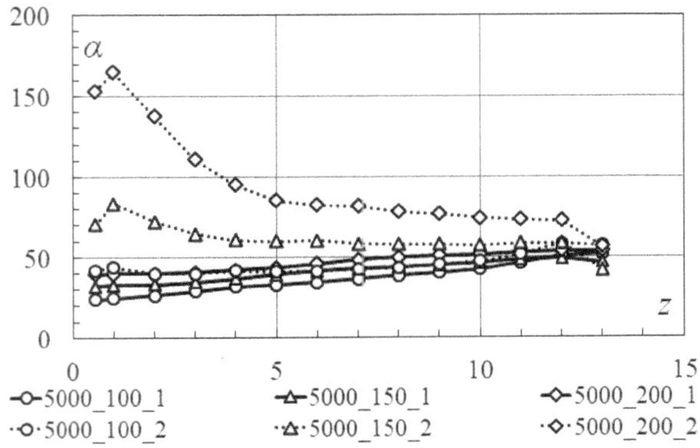

Figure 9. Changing the local heat exchange coefficients α [W/(m^2·°C)] along the length of the duct z [m] $q_0 = 5000$ B /^{2}, *G* – *Varia*.

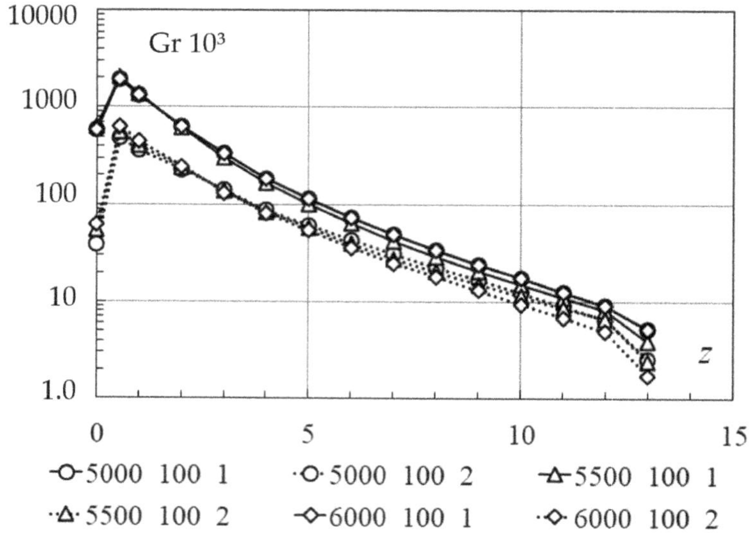

Figure 10. Changing the Grashof numbers along the length of the duct z [m] q_0 – *Varia; G – Const.*

Further generalization of the research results is carried out in the form of dimensionless complexes. Evaluation of the influence of gravitational and inertial forces on the intensity of heat transfer in the duct begins with determining the value of the Grashof *Gr* and Reynolds *Re* numbers along the length of the duct. As noted above, the thermophysical properties that make up dimensionless complexes are determined for the average mass temperature of the flow. The behavior of the Grashof numbers along the length of the duct, on the hot surface of the duct and on the non-heated surface for different regimes of the experiment is shown in Figure 10. The behavior of the Grashof numbers remains the same for all cases provided for in the experiment plan (Table 1). By analogy with the Grashof number, dependencies are also obtained for Reynolds number. The Reynolds numbers dynamics is shown in Figure 11, and is similar for all experiments performed.

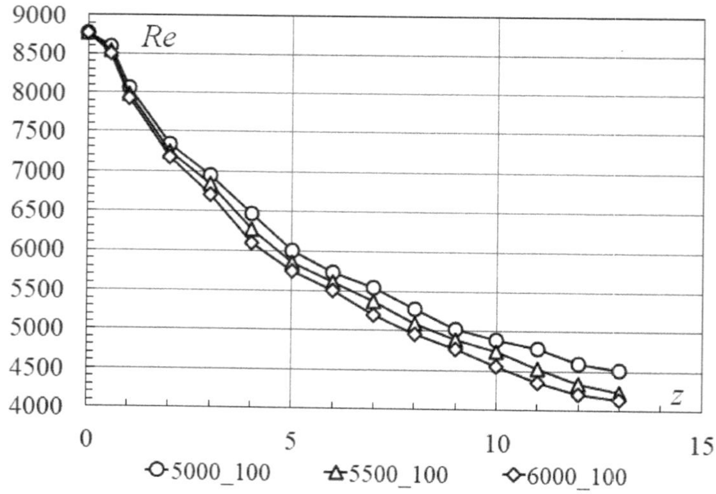

Figure 11. Changing the Reynolds numbers along the length of the duct z [m] q_0 – *Varia; G – Const.*

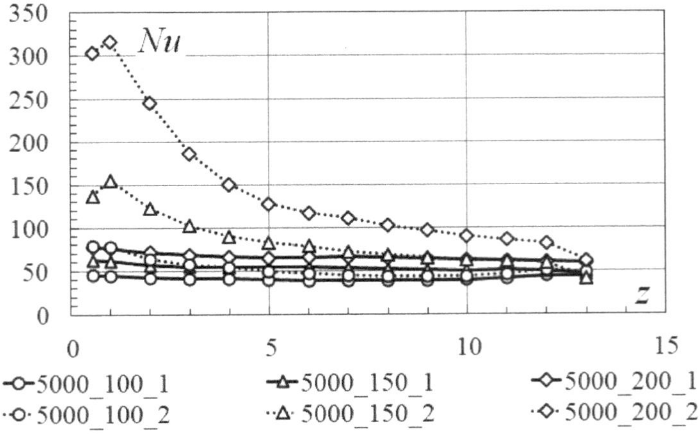

Figure 12. Changing the Nusselt numbers along the length of the duct z [m] $q_0 - Const$; $G - Varia$.

The greatest influence of gravitational forces on the heat transfer intensity is observed in one-third of the duct ($10 \leq \frac{z}{X} < 100$). At the hot surface, the effect of gravitational forces is 1.5–4.0 times higher than at the non-heated surface. The absolute values of the Grashof numbers allow us to assume about the undisputed effect of free convection on the intensity of convective heat transfer in the duct and about the expected differences in the values of heat exchange coefficients for mixed convection from their values for purely forced convection. Absolute values and the behavior of changes in Reynolds numbers indicate a developed turbulent flow regime along the entire length of the duct. The Reynolds number weakly depends on the value of the heat flux density on the hot surface. The difference in local values of the Reynolds numbers increases along the length of the duct, but does not exceed 8% of the initial value.

The dependence of the local Nusselt numbers along the length of the duct on the hot surface and on the non-heated surface is shown in Figure 12. The summary graph of the results of the research of complicated heat transfer in one-sided heat input and mixed motion of air is presented in Figure 13. The results are presented as the dependence of the local Nusselt numbers on the value of the dimensionless coordinate Z, which is determined by the expression $Z = Re \cdot Pr \cdot \frac{X}{z}$.

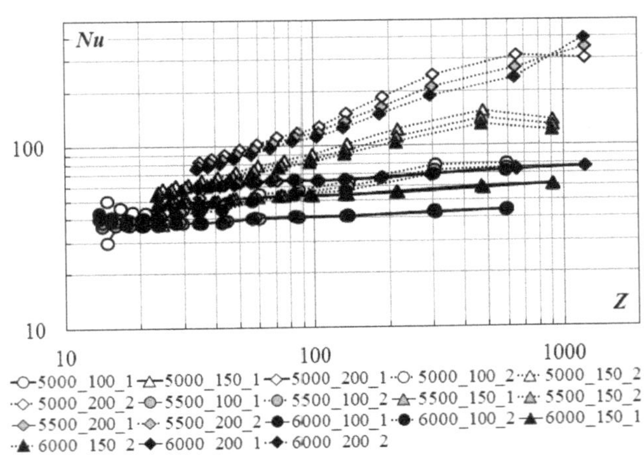

Figure 13. Dependence of the local Nusselt numbers on the dimensionless coordinate.

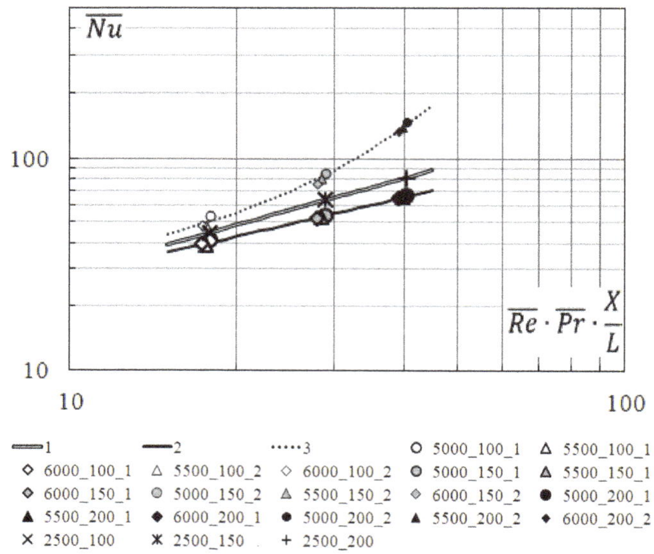

Figure 14. Generalizing dependences of average Nusselt numbers on the dimensionless coordinate. Curve 1—symmetrical two-sided heat input, curve 2—one-sided heat input, the hot surface; curve 3—one-sided heat input, the non-heated surface.

The intensity of convective heat transfer on the hot surface is less than the intensity of ones on the non-heated surface. The intensity of convective heat transfer on the hot surface is more affected by the mass flow rate of air in the duct than by the value of the heat flow density on the surface. On the summary graph Figure 13, it can be seen that the local Nusselt numbers on the hot surface for different values of the q_0 value almost coincide. With sufficient accuracy for technical calculations, it can be argued that in the range of values $q_0 \in [5000;\ 6000]$ W/m^2, the intensities of convective heat transfer are the same. For the non-heated surface, the local Nusselt numbers depend on both the mass flow rate of air in the duct and the value of the heat flow density. In this case, the dependence from the heat flow density q_0 is more significant than for the hot surface. As the q_0 value increases, the local Nusselt number curves on the non-heated surface approach similar curves on a hot surface. As the Z coordinate decreases, the difference in the values of local Nusselt numbers decreases for different heat flow densities q_0 and different air mass flows in duct G.

Before plotting generalizing dependences of the average Nusselt numbers ($\overline{Nu}$) as functions on the dimensionless coordinate ($\overline{Re}{\cdot}\overline{Pr}\cdot\frac{X}{L}$), it is verificated the mathematical model and calculation methods used on the traditional problem of convective heat transfer in the duct with symmetric two-sided heat input.

The research of aerodynamics and heat transfer in symmetric two-sided heat input and turbulent flow regime in the duct (Table 2) has the following goals:

- verification of the mathematical model and computational methods;
- comparison of the results of the research with existing solutions for convective heat transfer in pipes and ducts;
- comparison and generalization of the results of the research of flow and heat transfer in the duct under turbulent flow regime and symmetric two-sided heat input with the results of the research of one-sided heat input, complicated heat transfer and mixed motion.

The following boundary conditions were set

$$x = -\frac{a}{2}; y\epsilon\left[-\frac{b}{2}; \frac{b}{2}\right]; z\,\epsilon\,[0; c] \quad 0 = -\lambda_{air}\left(\frac{\partial T}{\partial x}\right)_{x=-\frac{a}{2}} - q_{x=-\frac{a}{2}}; q_{x=-\frac{a}{2}} = q_0/2;$$

$$x = \frac{a}{2}; y\epsilon\left[-\frac{b}{2}; \frac{b}{2}\right]; z\,\epsilon\,[0; c] \quad 0 = -\lambda_{air}\left(\frac{\partial T}{\partial x}\right)_{x=\frac{a}{2}} - q_{x=\frac{a}{2}}; q_{x=\frac{a}{2}} = q_0/2;$$

$$x\,\epsilon\left[-\frac{a}{2}; \frac{a}{2}\right]; y = \frac{b}{2}; z\,[0; c] \quad 0 = -\lambda_{air}\left(\frac{\partial T}{\partial y}\right)_{y=\frac{b}{2}} - q_{y=\frac{b}{2}}; q_{y=\frac{b}{2}} = 0;$$

$$x\,\epsilon\left[-\frac{a}{2}; \frac{a}{2}\right]; y = -\frac{b}{2}; z\,\epsilon\,[0; c] \quad 0 = -\lambda_{air}\left(\frac{\partial T}{\partial y}\right)_{y=-\frac{b}{2}} - q_{y=-\frac{b}{2}}; q_{y=-\frac{b}{2}} = 0;$$

$$x\,\epsilon\left[-\frac{a}{2}; \frac{a}{2}\right]; y\epsilon\left[-\frac{b}{2}; \frac{b}{2}\right]; z = 0; \quad G_{inlet} = G_0;\ T = T_{inlet};\ u = v = 0;$$

$$x\,\epsilon\left[-\frac{a}{2}; \frac{a}{2}\right]; y\epsilon\left[-\frac{b}{2}; \frac{b}{2}\right]; z = c; \quad p = 0;\ G_{outlet} = G_{inlet} = G_0;\ u = v = 0;\ \left(\frac{\partial T}{\partial z}\right)_{z=c} = 0.$$

where is $q_0 = 5000\ \mathrm{W/m^2}$; $G_{inlet}\epsilon[100;\ 200]$ kg/h; $t_{inlet} = 30$ °C; c = 13.0 m.

Table 2. Plan for performing the numerical experiment with two-sided heat input.

Heat Flow Density, W/m^2	Mass Flow Rate of Air, kg/h			Notes
	100	150	200	
	Two-Sided Heat Input			
2500	2500_100	2500_150	2500_200	Symmetrical heating

The results of the test computational experiment are presented in a dimensionless form. They are compared with the results of the computational experiment on the complicated heat transfer and mixed motion of air (Figures 14–16). Figure 14 shows local Nusselt numbers on the hot surface, the non-heated surface and surfaces with symmetric two-sided heat input, respectively. Figure 15 shows the reduced Nusselt numbers on the hot surface and the non-heated surface, respectively. Figure 16 shows dimensionless average integral temperature heads for the three cases described above. Figures 14 and 16 obtained the qualitatively identical situation—the objective function under symmetric two-sided heat input occupies the intermediate position between the values of the corresponding objective functions on the hot surface and the surface that is non-heated.

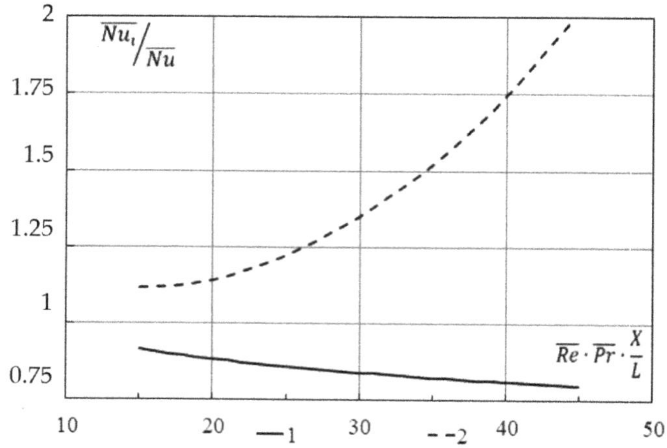

Figure 15. Dependence of relative Nusselt numbers on the dimensionless coordinate. Curve 1—one-sided heat input, hot surface; curve 2—one-sided heat input, non-heated surface.

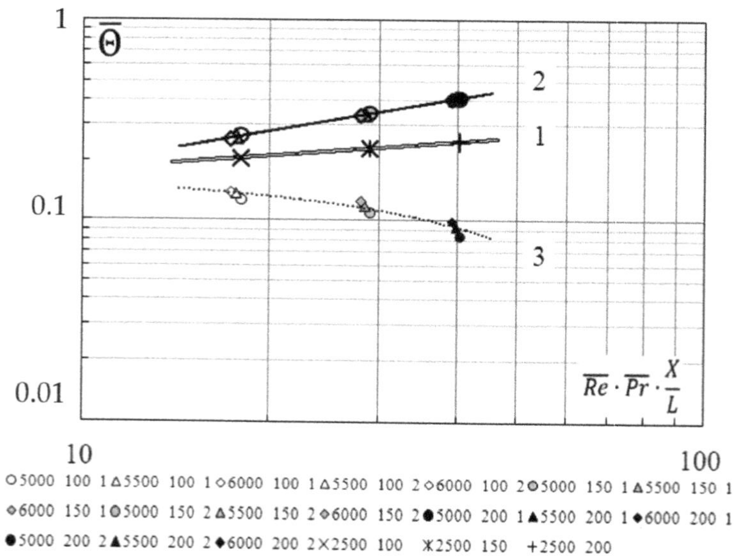

Figure 16. Generalizing dependence of dimensionless average integral temperature heads on the dimensionless coordinate. The designation of curves is similar to Figure 15.

Further final processing of the research results consisted in plotting the generalizing dependence of the average Nusselt numbers ($\overline{Nu}$) on the value of the dimensionless coordinate ($\overline{Re}\cdot\overline{Pr}\cdot\frac{X}{L}$). On the graph, Figure 14 presents the results of researches of complicated heat transfer with the mixed motion of air on the hot surface (curve 2) and the non-heated surface (curve 3); the results of the test research and the results of convective heat transfer in pipes and ducts under turbulent flow regime, published in [22] (curve 1).

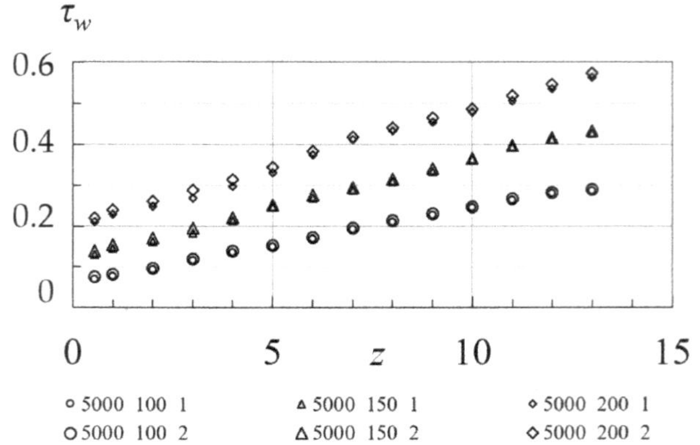

Figure 17. Distribution of tangential stresses τ_w [Pa] on the duct walls along its length z [m] (q_0 – *Const; G – Varia*).

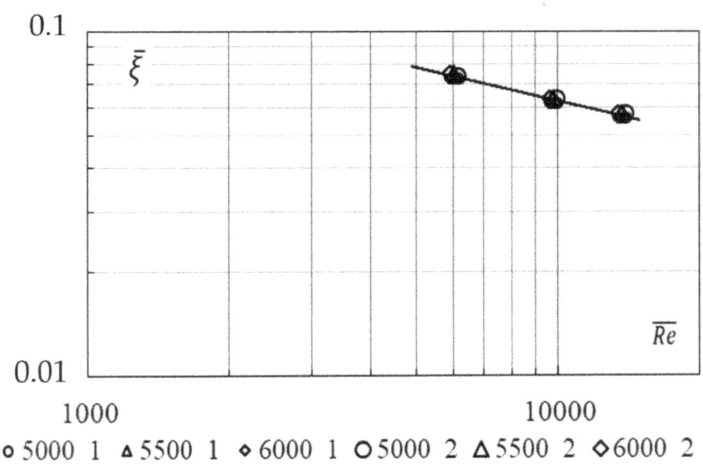

Figure 18. Generalized dependence of average coefficient of resistance of friction $\bar{\xi}$ on the dimensionless coordinate $\overline{Re}$.

The average Nusselt numbers ($\overline{Nu}_1$) and ($\overline{Nu}_2$) are calculated using the known dependence for the average Nusselt number ($\overline{Nu}$), which describes convective heat transfer in pipes and ducts under turbulent flow regime, but using correction factors

$$(\overline{Nu}_1/\overline{Nu}) = \frac{5}{3.9}\cdot\left(\overline{Re}\cdot\overline{Pr}\cdot\frac{X}{L}\right)^{-\frac{1}{8}}, \quad (2)$$

$$(\overline{Nu}_2/\overline{Nu}) = 0.99\cdot e^{\left[4.26\cdot10^{-4}\cdot\left(\overline{Re}\cdot\overline{Pr}\cdot\frac{X}{L}\right)^{1.95}\right]}, \quad (3)$$

where is $\overline{Nu}_1$—average Nusselt number on the hot surface, curve 2; $\overline{Nu}_2$—average Nusselt number on the non-heated surface, curve 3; $\overline{Nu}$—average Nusselt number on the symmetrical two-sided heat input, curve 1, Formula (1). The accuracy approximation is ($R^2 \geq 0.98$).

The change in the dimensionless average integral temperature head along the length of the dimensionless coordinate for the hot surface (curve 2) and the non-heated surface (curve 3) is shown in Figure 16. For comparison, the same graph shows the dependence of the average integral temperature head for the case of symmetric two-sided heated input (curve 1).

$$\Theta_1 = \frac{1}{19} \cdot \left(\overline{Re} \cdot \overline{Pr} \cdot \frac{X}{L} \right)^{\frac{1}{1.8}}, \tag{4}$$

the accuracy approximation is ($R^2 > 0.99$).

$$\Theta_2 = \frac{1}{5.28} \cdot e^{\left[-\frac{1}{55.87} \cdot \left(\overline{Re} \cdot \overline{Pr} \cdot \frac{X}{L} \right) \right]}, \tag{5}$$

the accuracy approximation is ($R^2 = 0.90$).

The distribution of tangential stresses on the walls along length of the duct, for different experimental conditions ($q_0 = 5000$ W/m^2, $G = 100$; 150; 200 kg/h), is shown in Figure 17.

The distribution plotting of the value τ_w obtained for the value of the heat flux density $q_0 = 5000$ W/m^2 remains relevant for another values q_0. It can be seen that there is the minimal difference in the values of τ_w obtained for different duct surfaces. The observed difference may not be considered, and the values of tangential stresses on the walls may be considered the same.

The results of the research are presented in the form of the generalizing dependence of the average coefficient of friction resistance on the value of the dimensionless coordinate (Figure 18). The generalizing graphical dependence is satisfactorily approximated by the power dependence (6), the accuracy approximation is ($R^2 > 0.99$).

$$\overline{\xi} = \left(\frac{3.633}{\overline{Re}^{3.14}} \right)^{0.1}. \tag{6}$$

6. Conclusions

It is established that in the considered range of values of heat flow density and mass flow rate of air, the dominant heat transfer between the walls and air is complicated convective heat transfer, in which forced convection ($Re > 4000$) is affected by free convection ($10^3 < Gr < 10^7$).

There is the difference in the intensity of convective heat transfer on the hot surface and on the non-heated surface. The difference in the density of heat flux transmitted by convection to air on the hot surface, and the non-heated surface is proportional to the expression $((\overline{Nu}_1 \cdot \overline{\Theta}_1)/(\overline{Nu}_2 \cdot \overline{\Theta}_2))$ and is 1.5–2.0.

Generalizing dependences are obtained that allow calculating the average Nusselt numbers and average temperature heads on the duct surfaces. Limits for applying dependencies (2)–(5): $Re = 4 \cdot 10^3 \div 10^4$; $Pr = 0.68 \div 0.70$; $10^3 < Gr < 10^7$.

The generalizing dependence of the value of the coefficient of friction resistance as a function of the value of the dimensionless coordinate (6) is obtained. The obtained values of the resistance value allow us to assume that in the hardware design of high-temperature heat exchangers and heat recuperation systems, the value of pressure losses to overcome the resistance of the boundary layer will not exceed 10% of the total pressure losses in the system.

The research results became the basis for the creation of the engineering methodology for calculating the heat engineering indicators of high-temperature heat exchangers and heat recuperation systems, which are widely used in the production of basalt and glass fibers at the enterprises "Chernivtsi plant of thermal insulation materials" (Chernivtsi, Ukraine) and "Innovation Basalt Technology Sp. z.o.o."(Gdansk, Poland). Their use in heat technologies allows you to return part of the heat with heated air to the technological cycle, provide controlled cooling of the equipment lining and obtain secondary energy resources in a form that is convenient for further use for one's own technological needs or sanitary needs of production.

Author Contributions: Conceptualization, B.B. and A.P.; methodology, V.K.; software, A.T.; validation, B.B., V.K. and A.P.; formal analysis, B.B.; data curation, A.P.; writing—original draft preparation, B.B.; writing—review and editing, A.P.; visualization, A.T.; supervision, A.T.; project administration B.B.; funding acquisition, A.P. All authors have read and agreed to the published version of the manuscript.

Funding: This research received no external funding.

Data Availability Statement: Not applicable.

Conflicts of Interest: The authors declare no conflict of interest. The funders had no role in the design of the study; in the collection, analyses or interpretation of data; in the writing of the manuscript or in the decision to publish the results.

References

1. Govern, J.C.; Herman, C.V.; Nagle, D.C. High Temperature Heat Exchangers (HTHX) for Nuclear Applications. In Proceedings of the ASME 2006 International Mechanical Engineering Congress and Exposition, Chicago, IL, USA, 5–10 November 2006; Volume 1, pp. 713–718.
2. Alkasassbeh, M.; Omar, Z.; Mebarek-Oudina, F.; Raza, J.; Chamkha, A. Heat transfer study of convective fin with temperature-dependent internal heat generation by hybrid block method. *Heat Transfer-Asian Res.* **2019**, *48*, 1225–1244. [CrossRef]
3. Shah, P.K. *Heat Exchangers for Fuel Cell Systems, in Compact Heat Exchangers and Enhancement Technology for the Process Industries*; Begell House: New York, NY, USA, 2003; pp. 205–215.
4. Marzougui, S.; Mebarek-Oudina, F.; Magherbi, M.; Mchirgui, A. Entropy generation and heat transport of Cu–water nanoliquid in porous lid-driven cavity through magnetic field. *Int. J. Numer. Methods Heat Fluid Flow* **2021**. ahead-of-print. [CrossRef]
5. Djebali, R.; Mebarek-Oudina, F.; Rajashekhar, C. Similarity solution analysis of dynamic and thermal boundary layers: Further formulation along a vertical flat plate. *Phys. Scr.* **2021**, *96*, 085206. [CrossRef]
6. Kumada, M. A Study on the High Performance Ceramic Heat Exchanger for Ultra High Temperatures. *Heat Transfer Enhanc. Heat Exch.* **1999**, *IV*, 301–324. [CrossRef]
7. McDonald, C.F. Recuperator considerations for future higher efficiency microturbines. *Appl. Therm. Eng.* **2003**, *23*, 1463–1487. [CrossRef]
8. Rafidi, N.; Blasiak, W. Thermal performance analysis on a two composite material honeycomb heat regenerators used for HiTAC burners. *Appl. Therm. Eng.* **2005**, *25*, 2966–2982. [CrossRef]
9. Tsuji, H.; Gupta, A.; Hasegawa, T.; Katsuki, M.; Kishimoto, K.; Morita, M. *High Temperature Air Combustion; from Energy Conservation to Pollution Reduction*; CRC Press LLC: New York, NY, USA, 2003. [CrossRef]
10. Dadheech, P.K.; Agrawal, P.; Mebarek-Oudina, F.; Abu-Hamdeh, N.H.; Sharma, A. Comparative Heat Transfer Analysis of $MoS_2/C_2H_6O_2$ and SiO_2-$MoS_2/C_2H_6O_2$ Nanofluids with Natural Convection and Inclined Magnetic Field. *J. Nanofluids* **2020**, *9*, 161–167. [CrossRef]
11. C Ranganayakulu, K.N.S. *Compact Heat Exchangers: Analysis, Design and Optimization Using FEM and CFD Approach*; ASME Press: New York, NY, USA, 2018; 546p, ISBN 9781119424185. [CrossRef]
12. Wong, H. *Basic Formulas and Data on Heat Transfer for Engineers*; Addison-Wesley Longman Ltd.: New York, NY, USA, 1977; 236p, ISBN-10:0582460506.
13. Kakaç, S.; Liu, H.; Pramuanjaroenkij, A. *Heat Exchangers: Selection, Rating, and Thermal Design*, 2nd ed.; CRC Press: Boca Raton, FL, USA, 2002. [CrossRef]
14. Minkowycz, W.J.; Sparrow, E.M.; Abraham, J.P.; Gorman, J.M. (Eds.) *Advances in Numerical Heat Transfer: Numerical Simulation of Heat Exchangers*, 1st ed.; CRC Press: Boca Raton, FL, USA, 2016. [CrossRef]
15. Howell, J.R.; Mengüç, M.P.; Daun, K.; Siegel, R. *Thermal Radiation Heat Transfer*, 7th ed.; CRC Press: Boca Raton, FL, USA, 2020. [CrossRef]
16. Basok, B.; Davydenko, B.; Pavlenko, A. Numerical Network Modeling of Heat and Moisture Transfer through Capillary-Porous Building Materials. *Materials* **2021**, *14*, 1819. [CrossRef] [PubMed]
17. Asli, K.H.; Aliyev, S.A.O. *Applied Research in Hydraulics and Heat Flow*, 1st ed.; Apple Academic Press: New York, NY, USA, 2014. [CrossRef]
18. Pavlenko, A.M.; Koshlak, H. Application of Thermal and Cavitation Effects for Heat and Mass Transfer Process Intensification in Multicomponent Liquid Media. *Energies* **2021**, *14*, 7996. [CrossRef]
19. Kremnev, V.; Basok, B.; Timoshchenko, A.; Tymchyshyn, S. Energy Saving in Construction by Wide Application of High-Quality Insulation Based on Basalt Fibers. *J. Mod. Phys.* **2018**, *09*, 1724–1734. [CrossRef]
20. Kremnev, V.; Basok, B.; Davydenko, B.; Timoshchenko, A.; Timoshchenko, A. Flow and Heat Transfer of Basalt Melt in the Feeder of the Smelter Furnace. *J. Appl. Math. Phys.* **2019**, *07*, 2555–2563. [CrossRef]
21. Oleinik, O.A.; Samokhin, V.N. *Mathematical Models in Boundary Layer Theory*, 1st ed.; Chapman & Hall/CRC: London, UK, 1999. [CrossRef]
22. Merkin, J.H. Mixed convection in a Falkner–Skan system. *J. Eng. Math.* **2016**, *100*, 167–185. [CrossRef]

energies

Article

Study of the Influence of the Lack of Contact in Plate and Fin and Tube Heat Exchanger on Heat Transfer Efficiency under Periodic Flow Conditions

Marcin Łęcki *, Dariusz Andrzejewski, Artur N. Gutkowski and Grzegorz Górecki

Faculty of Mechanical Engineering, Institute of Turbomachinery, Lodz University of Technology, 219/223 Wolczanska Street, 90-924 Lodz, Poland; dariusz.andrzejewski@p.lodz.pl (D.A.); artur.gutkowski@p.lodz.pl (A.N.G.); grzegorz.gorecki@p.lodz.pl (G.G.)

* Correspondence: marcin.lecki@p.lodz.pl; Tel.: +48-42631-23-20

Abstract: Plate fin-tube heat exchangers are widely used in air conditioning and refrigeration systems and other industry fields. Various errors made in the manufacturing process can result in the formation of an air gap between the tube and fin. Several numerical simulations were carried out for a symmetric section of plate fin-tube heat exchanger to study the influence of air gap on heat transfer under periodic flow conditions. Different locations and sizes of an air gap spanning 1/2 circumference of the tube were considered for the range of airflow velocities. Velocity and temperature fields for cases with air gap were compared with ideal thermal contact cases. Blocking of heat flow by the gap leads to the reduction of heat transfer rate. Fin discontinuity in the front of the tube causes the smallest reduction of the heat transfer rate in comparison to the ideal tube-fin contact, especially for thin slits. The rear gap position is the worst in the smallest gap range. Therefore, reversing the flow direction can lead to up to a 15% heat transfer increase, if mainly the rear gaps are present. The introduction of a thin slit in the front of the tube leads to convective heat transfer enhancement, which should be further investigated.

Keywords: heat exchanger; plate-fin tube; tube-fin contact; CFD

Citation: Łęcki, M.; Andrzejewski, D.; Gutkowski, A.N.; Górecki, G. Study of the Influence of the Lack of Contact in Plate and Fin and Tube Heat Exchanger on Heat Transfer Efficiency under Periodic Flow Conditions. *Energies* **2021**, *14*, 3779. https://doi.org/10.3390/en14133779

Academic Editor: Angelo Zarrella

Received: 17 May 2021
Accepted: 17 June 2021
Published: 23 June 2021

1. Introduction

Plate-fin and tube heat exchangers are most widely used as evaporators and condensers in refrigeration systems (commercial refrigeration) and in other applications where heat transfer between liquid or evaporating/condensing fluid and gaseous working fluid is required. Proven technology, reliability, and relatively low cost of manufacture make plate-fin and tube heat exchangers very popular, not only as components of refrigeration units but also as gas coolers/heaters, recuperative heaters, dryers or heat pipe, thermosyphon radiators [1–3]. Nowadays, when detailed numerical simulations of fluid flow with heat transfer are within reach, many of the researchers carry out Computational Fluid Dynamics (CFD) studies to optimize the geometry of plate-fin and tube heat exchangers [4,5]. The majority of these works assume an ideal tube fin contact that in practice has to be created during the manufacturing process. The contact is usually achieved by a tube expansion (plastic deformation of the tube by moving the expansion die or by applying pressure at its inner surface). These mechanical methods of expansion cause additional thermal resistance at the tube–fin interface, due to the gaps between the contact surfaces of the fin collars and tubes [4]. Formation of the gap can be also a result of the nonideal joint formation process or assembly requirements. For the egg-crate-type heat exchangers, which are used as evaporators in domestic refrigerators, slits between the tubes are introduced to allow one to pull the whole tube coil through the plate-fin packet [6]. In this case, the lack of contact is always present at approximately 1/2 perimeter of the tube, and it is a feature essential for the assembly of the heat exchanger.

The gaps can be also created because of the wear of the heat exchanger, e.g., as the result of corrosion by exposure to difficult weather conditions or by strong chemical interactions. In Figure 1, one can see a picture (courtesy of Blygold company [7]) of a corroded heat exchanger surface with gaps on the circumference of the tube. In this case, fin discontinuities are just one of the factors affecting heat transfer; there are also white corrosion products around the tubes that impede heat transfer (AlOH thermal conductivity is smaller than for Al).

Figure 1. Gaps in the fin on the circumference of the tube created by a corrosion process, marked areas in the Figure (photo provided by courtesy of Blygold company [7]).

The quality of the joint can be improved by brazing the tube to the fin (the soldering process and its impact on heat transfer are discussed in detail in [8–10]). Because of the additional amount of energy needed for the soldering process and some technical difficulties (ensuring high purity of soldered surfaces), mechanical methods of expansion are most frequently chosen. Non-ideal tube fin connection can cause significant thermal contact resistance which deteriorates heat transfer. Therefore, some of the researchers have undertaken studies taking into account the realistic fin-tube thermal contact and its impact on the thermal efficiency of heat exchangers. Aside from the experimental research aimed at estimating the thermal contact resistance [11,12], its value is also obtained through semiempirical methods, partly based on numerical analysis. Taler and Oclon [13] and Taler and Cebula [14] changed iteratively the value of contact resistance in their CFD model in such a way that the thermal and flow quantities obtained from the numerical solution agreed with the measured ones. Less attention has been paid to cases where there is a lack of contact with the plate-fin along a part of the tube perimeter (no contact between the fin collar and tube). Recently, the topic was addressed by Singh et al. [15] who proposed a CFD model of a finned heat exchanger with a variable contact surface area between the tube and the fin. They considered several cases of gaps of different sizes. They found that the presence of gaps influenced heat transfer and flow (pressure drop) in a significant way. In contrast to [15], where several triangular gaps are distributed evenly along the joint circumference, our work presents a CFD analysis of plate-fin and tube heat exchanger, where there is a lack of contact between the tube and the fin at the half of the tube's circumference. The present study aims to numerically investigate the effect of the lack of contact on the energy efficiency of the heat exchanger for variable parameters, such as different air velocities and various positions of the gap relative to the flow direction. The CFD model is developed with the assumption of periodic flow boundary conditions. The novelty of the present work lies mostly in the consideration of the different locations and sizes of the gaps between the tube and fin relative to the flow direction. Because of the nonuniform heat transfer coefficients on the fin and tube surfaces, the expected degree of heat transfer deterioration is a function of the position of the cavity. Only in [15] was a similar CFD analysis done, but evenly distributed triangular voids excluded the possibility of dependency on the flow direction; additionally, the authors of [15] considered a slightly different geometry of the plate-fin tube heat exchanger than the typical one (with slitted fins). The present work assumes more realistic positions of gaps resulting from wear (corrosion) or erroneous tube expansion in the classic type of plate-fin tube heat exchanger. It contributes to the current area of research by expanding knowledge about the magnitude of heat transfer deterioration (fin efficiency) caused by the lack of contact between fin and tube depending on flow conditions. It can help to understand and predict the thermal effects of wear and manufacture errors on the functioning of compact heat exchangers.

2. Numerical Model of a Heat Exchanger

The heat exchanger under consideration works as a condenser in a refrigeration system. It is a plate-fin and tube heat exchanger, in which tubes are arranged in-line. It was recognized that the location of the gap relative to the flow direction is an important parameter because of the variation in the local heat transfer coefficients on the surface of the fin and the locally variable velocity field [16]. The gap location along tube-fin joint circumference should therefore have a significant impact on heat exchange and flow patterns. Figure 2 shows the analyzed configurations.

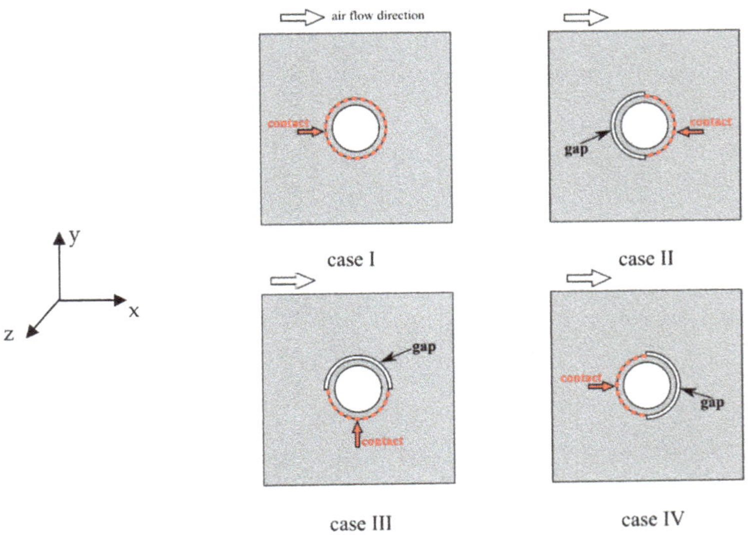

Figure 2. Different positions of the gap with respect to airflow: Case I—without gap; Case II—the gap in the front of the tube; Case III—the gap in the top position; Case IV—gap in the back of the tube.

Case I (without a gap—thickness of discontinuity $\delta = 0$ mm) is treated as a reference. The slit position is nondimensionalized as $\theta/180°$, where θ is an angular position. In the second case, the gap spans the front side of the circumference ($\theta = 0°$). In Case III, it is placed on the upper half perimeter ($\theta = 90°$), in the last IV variant—in the back of the tube ($\theta = 180°$). The angles indicate the centers of the gaps, taking into account the slit circumference span the corresponding angle ranges are front gap ($90° \leq \theta \leq 270°$), top gap ($0° \leq \theta \leq 180°$), and back gap ($-90° \leq \theta \leq 90°$). Four gap thicknesses were taken into account: 0.25, 0.5, 1.0, and 2 mm, corresponding to the following dimensionless values (relative to tube diameter): 0.03125, 0.0625, 0.125, and 0.25. Numerical simulations were carried out for airflow through the heat exchanger equal to 1, 2, 3, 4, and 5 m/s average velocities at the flow area directly upstream of the entrance to the device under consideration.

2.1. Geometry of the Computational Domain

The computational domain is a repeatable slice of the heat exchanger geometry, which is shown in Figure 3. The length and width of the domain are equal to the distance between axes of the tubes L_f. The thickness of the fin in the computational domain is half of the actual fin thickness δ_f, as the temperature field is symmetric at the opposite half of the fin (see the symmetry plane in Figure 3). Tube length in the domain equals $\frac{1}{2}$ of the distance between fins. Taking the half of inter-fin spacing is a consequence of the flow symmetry assumption. The actual domain boundary is marked in Figure 3 by a dashed line and spans up to the symmetry plane. The specific dimensions of the domain are given in Table 1.

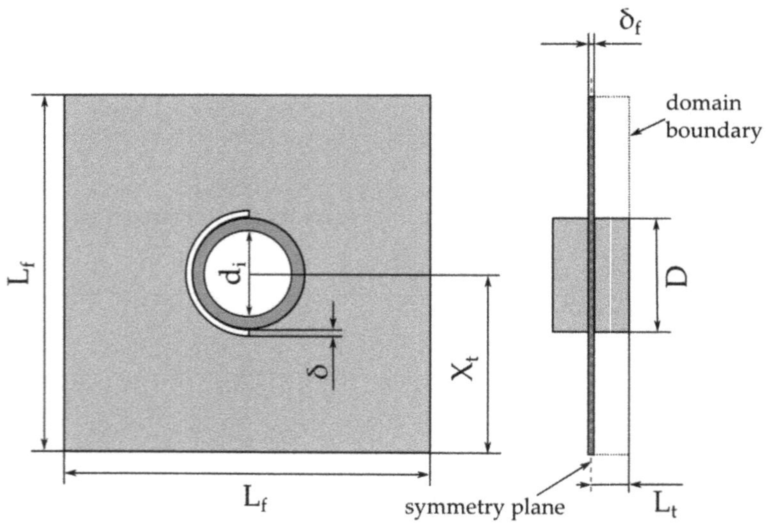

Figure 3. The geometry the simulated slice of the heat exchanger.

Table 1. The dimensions of the domain for the analyzed cases.

Description	Symbol	Value [mm]
Length and width of the fin	*Lf*	25
Tube length	*Lt*	2.5
Outside diameter of the tube	*D*	8
Internal diameter of the tube	*di*	6.2
Fin thickness	δ_f	0.05
Gap thickness	δ	0.25–2.0
Tube spacing	X_t	12.5

2.2. Boundary Conditions and Governing Equations

The shear stress transport (SST) turbulence model was chosen to solve Reynolds (time) averaged Navier-Stokes equations (RANS). SST combines the best features of k-ε

and k-ω models. In the near-wall region, approximately up to half of the boundary layer k-ω formulation is used, which does not require any damping functions, while for the outer zone of the boundary layer k-ε is the governing turbulence model—it eliminates the problem of k-ω sensitivity to values of ω in the freestream (outside the boundary layer). The other advantage of SST is the accurate prediction of the boundary layer separation by application of the turbulence production limiter in stagnation regions [17]. The SST model is also characterized by very good accuracy of the numerical solutions for wall-bounded, complicated geometry flows, obtained with relatively low computational power. Recently, Lindqvist et al. have performed an extensive analysis of RANS turbulence models in the application to numerical simulation of airflow and heat transfer in plate-fin and tube heat exchangers [18]. The k-ω SST was chosen as optimal and used for further validation against experimental data. It was utilized in many works where CFD analysis of the heat transfer for finned and enhanced surfaces was carried out [4,13,19–21].

The following simplifying assumptions were made in the present work:

- Steady-state fluid flow and heat transfer;
- Fluid flow and heat transfer are periodic (fully-developed), meaning that pattern of flow/thermal solution has a periodically repeating nature (this condition is generally fulfilled for tube rows greater than the fourth row);
- Thermophysical properties of air are temperature-dependent (ideal gas);
- Natural convection was not considered as the highest Richardson number calculated for simulation conditions is $Ri = (g \cdot \beta \cdot \Delta T \cdot L)/u^2 = 0.025$ (for $Ri < 0.1$ mechanism of natural convection can be typically considered negligible).

Numerical calculations were carried out using the commercial ANSYS CFX code capable of solving the Navier-Stokes equations [22]:

$$\frac{\partial u_i}{\partial x_i} = 0, \tag{1}$$

$$\rho \frac{\partial}{\partial x_j}(u_i u_j) = -\frac{\partial p}{\partial x_i} + \frac{\partial}{\partial x_j}\left(\tau_{ij} - \rho \overline{u_i u_j}\right) \tag{2}$$

where τ is a stress tensor in the fluid, and $\rho \overline{u_i u_j}$ are Reynolds stresses (turbulent), which depend on average velocity gradients (Boussinesq hypothesis):

$$-\rho \overline{u_i u_j} = \mu_t \left(\frac{\partial u_i}{\partial x_j} + \frac{\partial u_j}{\partial x_i} \right) - \frac{2}{3}\left(\rho k + \mu_t \frac{\partial u_k}{\partial x_k} \right) \delta_{ij} \tag{3}$$

where k—turbulence kinetic energy and μ_t—turbulent viscosity.

The SST turbulence model allows for obtaining turbulent viscosity by using two transport equations of kinetic turbulence energy k and specific dissipation rate ω. Due to the extensive definition of the model, it will not be presented here; its formulation can be found in [17]. The turbulent viscosity is calculated using the equation:

$$\mu_t = \rho \frac{k}{\omega} \tag{4}$$

The energy equation for fluid is given by:

$$\frac{\partial}{\partial x_i}\left[u_i (\rho \cdot e + p) \right] = \frac{\partial}{\partial x_j}\left[\left(\lambda + \frac{c_p \mu_t}{Pr_t} \right) \frac{\partial T}{\partial x_j} \right] \tag{5}$$

where e—total fluid energy, T—temperature, λ—thermal conductivity, c_p—specific heat at constant pressure, and Pr_t—turbulent Prandtl number for energy (assumed value: 0.9). In the solid body (fin and tube domain), the heat is exchanged by means of steady conduction:

$$\frac{\partial}{\partial x_j}\left(\lambda\frac{\partial T}{\partial x_j}\right)=0 \tag{6}$$

The computational domain is divided into three parts: air (fluid), fin, and tube (solids). It was assumed that the thermal conductivity of the solid (aluminum) is constant in the tested temperature range, which is λ = 202.4 W/m/K. A periodic flow was solved with the following boundary conditions and parameters:

- Average air temperature at the inlet: T_{ave} = 22 °C,
- u = 0 m/s on the fin and tube surfaces (no-slip condition),
- Uniform temperature at the inner surface of the tube wall: T_w = 40 °C—which corresponds to the condensation conditions of the working fluid flowing inside the tube.

On the other surfaces of the computational domain, a symmetry boundary condition was assigned, which can be expressed mathematically as:

$$-\overline{n}\cdot\dot{q}=0,\ u\cdot\overline{n}=0,\ \nabla k\cdot\overline{n}=0,\ \nabla\omega\cdot\overline{n}=0 \tag{7}$$

where $\overline{n}$ is a normal vector to the surface on which the boundary condition was set. Additionally, on the solid–fluid contact surfaces, a coupled boundary condition was present, ensuring the energy balance is satisfied between domains.

The periodic condition of the velocity in the x-direction can be written as:

$$u_1(\overline{r})=u_1(\overline{r}+\overline{L})=u_1(\overline{r}+2\overline{L})=\ldots \tag{8}$$

where $\overline{r}$—position vector and $\overline{L}$—the periodic length vector (equations in the rest of the Cartesian directions are analogous).

For viscous flows, the pressure field is considered periodic if the following condition is met:

$$\Delta p=p(\overline{r})-p(\overline{r}+\overline{L})=p(\overline{r}+\overline{L})-p(\overline{r}+2\overline{L})=\ldots \tag{9}$$

The pressure gradient is divided into two parts: a periodic gradient and a linear gradient:

$$\nabla p(\overline{r})=\beta\frac{\overline{L}}{|\overline{L}|}+\nabla\widetilde{p}(\overline{r}) \tag{10}$$

where $\widetilde{p}(\overline{r})$ is so-called periodic pressure and $\beta\frac{\overline{L}}{|\overline{L}|}$—linearly variable pressure component. Periodic pressure is the remainder after subtracting the linear pressure. On the other hand, the linearly variable component results in a force acting on the fluid in the momentum equations. During the solution process, the value of the coefficient β is iteratively changed until the desired convergence between the set and the calculated mass flow is obtained.

The temperature field is analogously divided into two components for periodic simulation:

$$T(\overline{r})=\sigma|\overline{r}|+\widetilde{T}(\overline{r}) \tag{11}$$

where:

$$\sigma=\frac{\dot{Q}}{\dot{m}\cdot c_p\cdot L} \tag{12}$$

and $\dot{Q}$—net heat transfer rate in a periodic domain.

At the inlet and outlet, the periodic boundary condition has the form:

$$T(\overline{r})=\widetilde{T}(\overline{r}+\overline{L}),\ p(\overline{r})=p(\overline{r}+\overline{L}),\ u_1(\overline{r})=u_1(\overline{r}+\overline{L}) \tag{13}$$

The above-mentioned conditions for the periodicity of the airflow are practically implemented in the Ansys-CFX CFD software. The boundaries named in Figure 4 as periodic inlet and periodic outlet are coupled as two flow interfaces; the target mass flow can be specified, which is obtained by iteratively changing the pressure gradient β. Negative heat flux has to be fixed at the periodic outlet boundary to balance the heat transfer rate gained by air flowing through the domain, as the outflowing fluid is "returned" to the inlet (Figure 4). If a heat sink is not introduced, the air temperature rises infinitely during the solution process of the numerical simulation, causing the solver to crash after reaching some maximum temperature level. In the case of numerical simulations carried out within this research, the domain was properly thermally balanced to keep the inlet temperature at an approximately constant level of 22 °C. The symmetry boundary condition is applied at domain boundaries that were not marked in Figure 4 (for both solid and fluid surfaces).

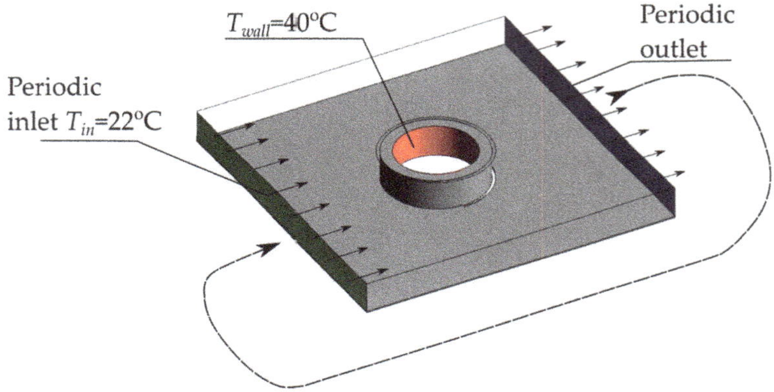

Figure 4. Boundary conditions for the computational domain.

2.3. Computational Grid

Numerical simulations of the fluid flow with coupled heat exchange process were carried out using ANSYS CFX CFD software. The computational grid for the case I is shown in Figure 5. It was created for the repetitive slice of heat exchanger with the use of ANSYS-Meshing application.

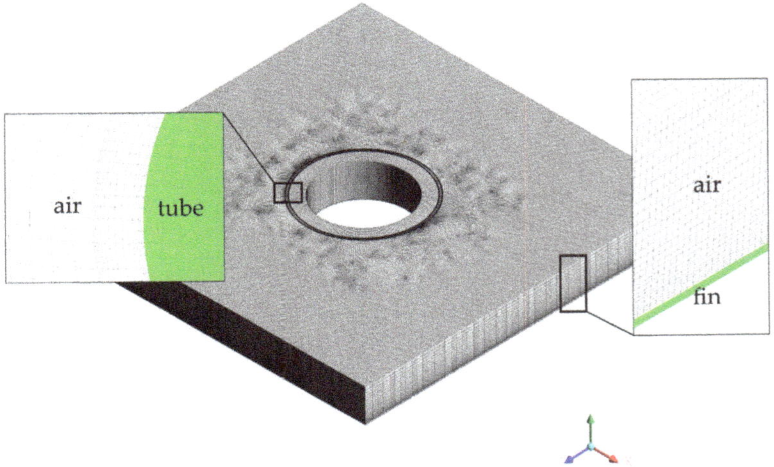

Figure 5. View of the computational mesh.

The grid independence study was performed using a methodology similar to used in works [23,24]. The five grids were tested, namely, coarse, average, fine, very fine, and the finest corresponding to mesh sizes: 0.2×10^6, 1.8×10^6, 2.14×10^6, 2.48×10^6, and 2.80×10^6 nodes. For all of the five meshes, the prism layer elements number at the wall was gradually increased, starting from 10 for the coarse grid by 25, 30, 35, up to 40 for the finest one. The Y^+ parameter and percent change of mean heat transfer coefficient $h_{\%}$ were presented in Figure 6. The $h_{\%}$ is the relative difference between the obtained average heat transfer coefficient for the ith mesh refinement level and $i-1$ refinement (the one-step coarser mesh). The change does not exceed 0.5%; therefore, the average mesh (1.8×10^6 nodes) was chosen as optimal because it also satisfies the maximum $Y^+ < 1$ condition (maximum $Y^+ = 0.44$, average $Y^+ = 0.24$).

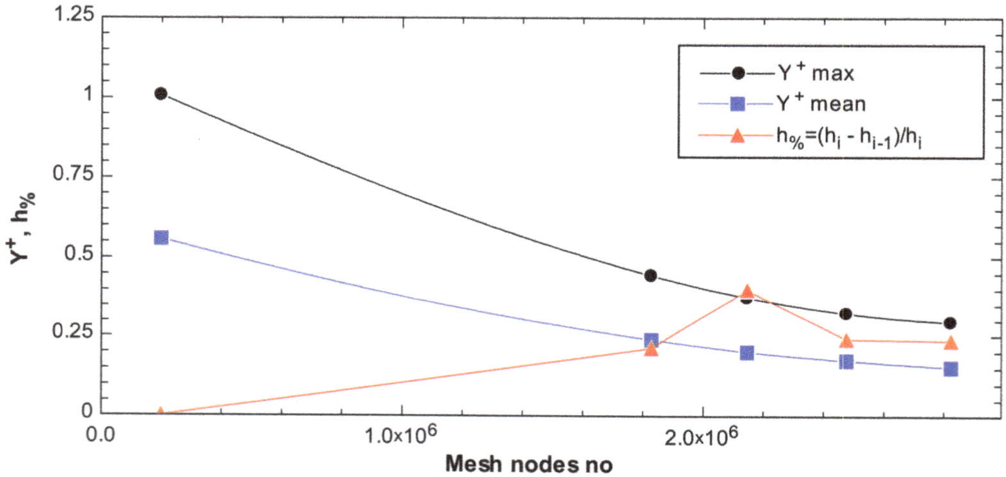

Figure 6. Dependence of an average and maximum Y^+ in the computational domain, and percent change of mean heat transfer coefficient $h_{\%}$ versus the number of mesh nodes.

2.4. Validation of the Numerical Model

The computational model was validated against data published in the work of Saboya and Sparrow [16] as their paper is widely acknowledged as the standard for validation of the studies of heat transfer from finned surfaces. They used the naphthalene sublimation technique [25] to obtain local mass transfer coefficients (Sherwood numbers Sh) on the fin surface of the one-row plate-fin tube heat exchanger. The Sherwood number distribution can be converted to heat transfer coefficients by utilization of the heat-mass transfer analogy:

$$h_{mass} = Sh \cdot \left(\frac{Pr}{Sc}\right)^n \cdot \left(\frac{\lambda}{d_h}\right) \tag{14}$$

where Sc—Schmidt number, d_h—hydraulic diameter according to Kays and London [26], and n—heat-mass transfer analogy exponent. Rosman et al. [27] have chosen the value of the $n = 0.4$ based on his experiment and measurements available in the literature. However, the latest careful measurements have reaffirmed that the most suitable value of the exponent for both laminar and turbulent boundary layers is approximately $n = 0.33$ [23]. Prandtl number and thermal conductivity are assumed as for pure air ($Pr = 0.7$, $\lambda = 0.026$ W/m/K)—the concentration of naphthalene is too low to significantly influence the thermophysical properties of the fluid. Saboya and Sparrow [16] assumed $Sc = 2.5$, yet it is an outdated value, and for room temperature the $Sc = 2.28$ is now recommended [25]. Local Sherwood numbers [16] were converted to heat transfer coefficients according to (14) and prepared to validate the CFD simulation. On the left side of Figure 7, the distribution of heat transfer coefficients on the fin surface is shown obtained from [16] for $Re = 214$. The distribution of local h is only known along the horizontal black lines where the mass transfer coefficients are measured—a 2D contour map is a result of an inverse distance interpolation made in Tecplot 360 EX software [28]. On the right side, there are results from numerical simulation for analogous flow and thermal conditions as in the experimental work. This setup allows a direct comparison between two contour maps. As can be seen, the ranges of heat transfer coefficient values are nearly the same. Near the front edge of the fin, the convective heat transfer seems to be more intensive in the case of the numerical results, but the lower h are a consequence of the interpolation of the Saboya and Sparrow's [16] data from the location significantly upstream of the edge of the fin.

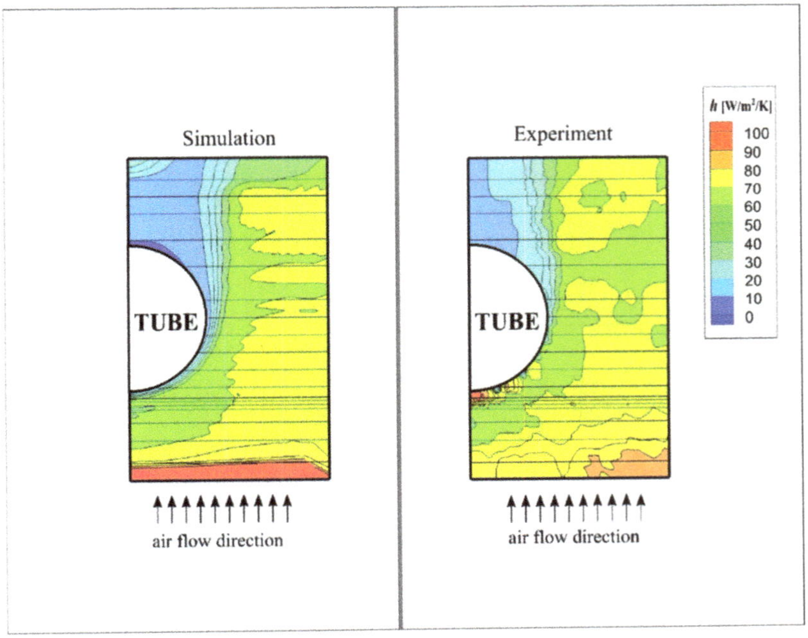

Figure 7. Comparison of local h on the surface of the fin between numerical simulation and experiment [16] for $Re = 214$.

The accordance of the simulation's local h with the experiment is shown in Figure 8. Four lines were chosen, where the measurements were taken exactly—at the front (line 1), back (line 4), and middle segments (lines 2, 3) of the fin. Most of the experimental data is in very good agreement with the CFD solution (maximally ±20% and better). The greatest discrepancies are noted in the near tube region, but they can be caused by the fact that in the experiment [16] the tube surface was adiabatic. The uncertainty of the experimental h was also plotted in the form of error bars. It was based on the estimation of the uncertainty of Sherwood number measurements done in [29] (uncertainty within 7.3%).

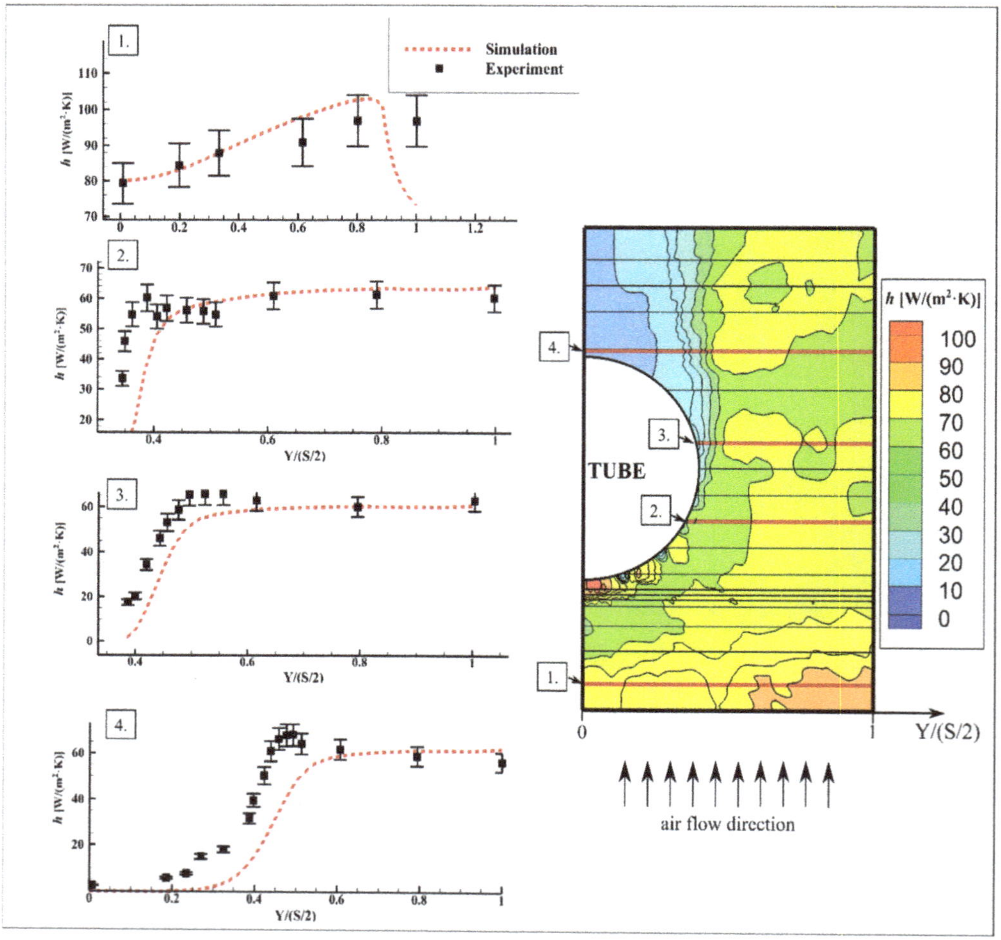

Figure 8. Local heat transfer coefficients from experiment versus CFD simulation at four lines transverse to the airflow at the front (line 1), back (line 4), and middle segments (lines 2, 3) of the fin for *Re* = 214.

The validation presented above was made to test the ability of the model to predict local coupled heat and fluid flow phenomena. Because the experimental flow arrangement (one-row tube-fin) is not adequate to the simulation settings (periodical flow—more than four-row inline tube-fin geometry), additional validation was performed. Mean heat transfer obtained from the simulation was compared against VDI correlation [30], which is true for three rows inline heat exchangers:

$$Nu_D = 0.20 Re_D^{0.6} \left(\frac{A}{A_{t0}} \right)^{-0.15} Pr^{1/3} \tag{15}$$

A comparison between simulation heat transfer coefficients and VDI correlation is presented in Figure 9. The two data sets' values are very close. The relative difference varies from 4.9% for the lowest *Re* to 1.8% for the biggest airflow.

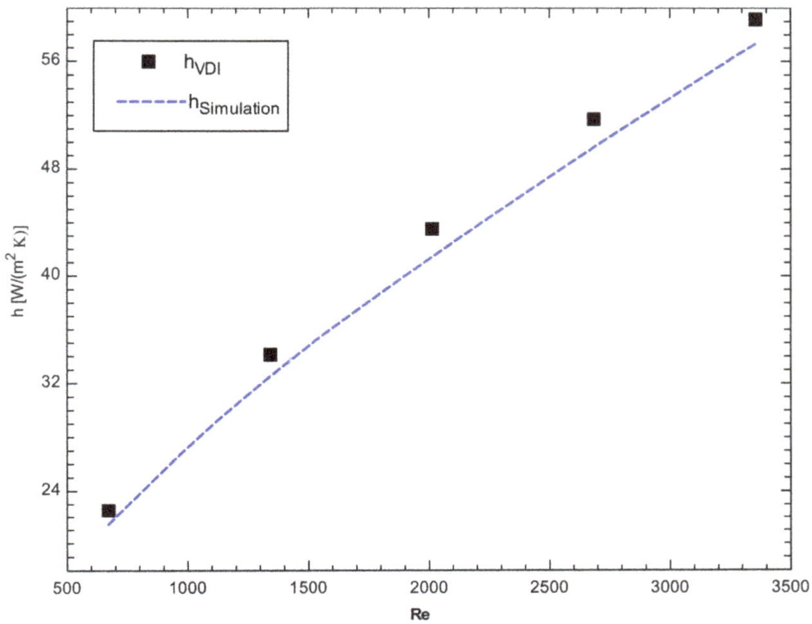

Figure 9. Comparison between heat transfer coefficients obtained from numerical simulation to the ones calculated with VDI correlation valid for three-row inline heat exchangers.

3. Results and Discussion

The results of the numerical simulations were put together in a form of a series of graphs, allowing for the evaluation of the thermal efficiency of the cases with the presence of the gap to perfect fin-tube contact example. In Figure 10, the ratios of heat transfer rates (discontinuity to no-slit heat transfer rate) are presented in the function of the dimensionless gap thickness (gap thickness to diameter). For each Reynolds number, a curve approximating the numerical data is presented. For the gap in the back and front, the minimal ratio of heat transfer rate is approx. 0.75 for the considered airflow span, whereas the side gap attains a slightly lower minimum of 0.72. For the front and side gap localization, there is a bigger spacing between the curves (stronger influence of the airflow velocity) than for the rear position. This effect is the strongest for smaller gaps. For $\delta/D > 0.1$, there is no significant difference between curves spacing for all examined cases. The shapes of the curves are similar for all the arrangements, although for the fin discontinuity in the rear of the tube the heat transfer ratio drop from unity ($\delta/D = 0$) is much steeper than for the other locations. Heat transfer rate ratios are significantly lower for back orientation, especially for small gap thickness and small *Re*. Usually the higher the Reynolds number the lower value of $\dot{Q}/\dot{Q}_0$, although there is an exception for the lowest considered Reynolds number curve for nearly all gaps, excluding the biggest fin discontinuity.

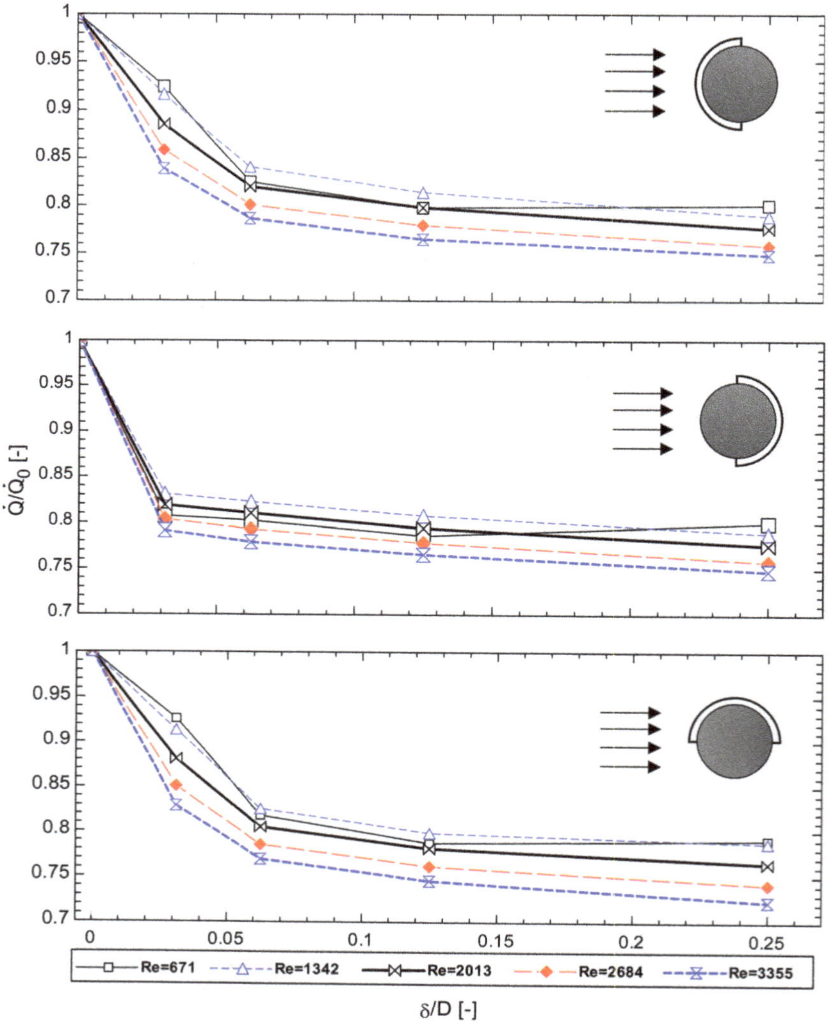

Figure 10. Heat transfer rate ratio in the function of dimensionless gap thickness.

For the rear-positioned slit at Re = 671 $\dot{Q}/\dot{Q}_0$ values are below these for Re = 2013. One can see a nearly monotonic decrease of the heat transfer rate ratio with increasing gap thickness, being the sharpest in the lower δ/D range and reaching practically a plateau for $\delta/D > 0.1$. The data points are connected by a piecewise linear function, instead of polynomial fit, because higher-order curve fitting produced slight inflection in the plateau range, which was a misleading artifact of the method.

A clearer picture of how the flow velocity influences heat transfer emerges from Figure 11. A decreasing trend of $\dot{Q}/\dot{Q}_0$ is seen with increasing Reynolds number, excluding the smallest Re departing from this pattern. In the range of $Re > 1342$, there is a nearly linear negative slope. However, in the span of Re = 671–1342, the decrease of heat transfer rate ratio is smaller or even there is a reverse trend (increase), indicating there can be a change of the flow regime. The curve for the smallest gap is placed remarkably higher than the other ones for the side and front location. Other δ/D cases, for all slit positions, exhibit similar trends, with the δ/D = 0.125 curve approaching or even crossing δ/D = 0.25 with decreasing *Re*.

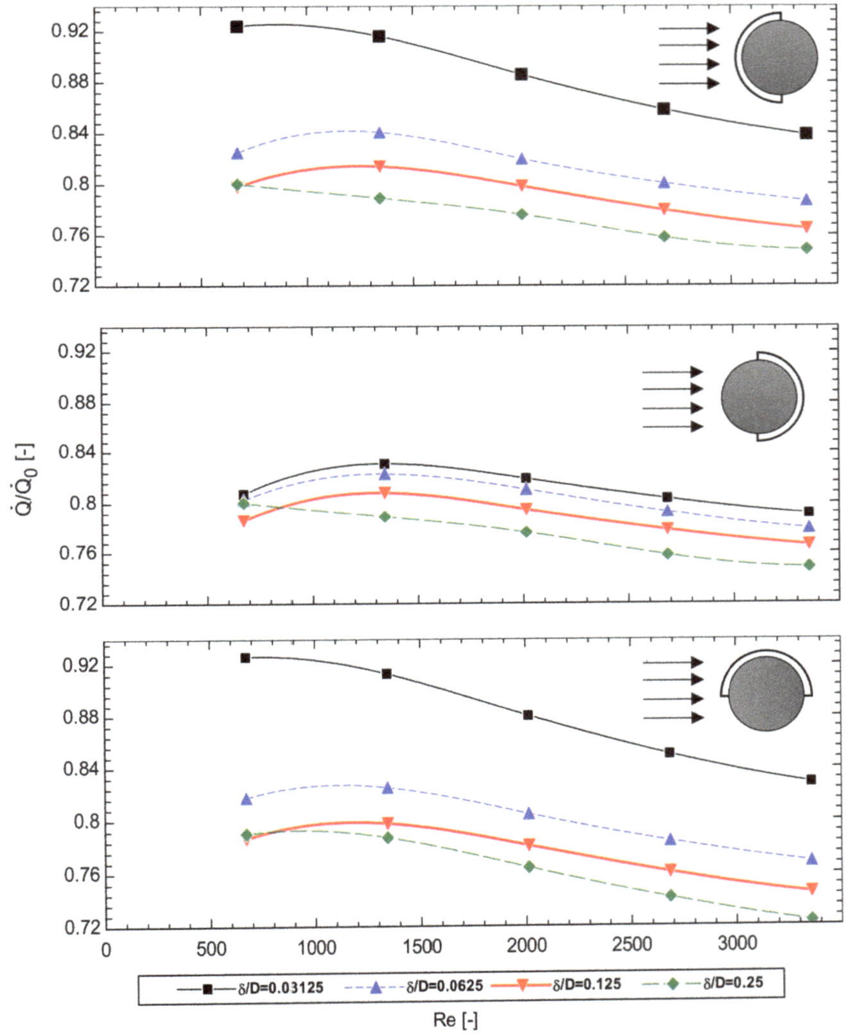

Figure 11. Heat transfer rate ratio in the function of Reynolds number.

The influence of the dimensionless gap position $\theta/180°$ on the $\dot{Q}/\dot{Q}_0$ is addressed in Figure 12. In the case of the smallest gap, the front and side locations exchange nearly equal heat transfer rates, with a significant decrease that can be seen for slit in the back ($\theta/180°$ = 1). For the rest of the considered gap thicknesses, there are very small differences between the tube circumference and gap locations. The weak dependence in the range of δ/D = 0.0625–0.25 is the most evident for the highest *Re* = 3355. The back and front slit cases indicate moderately higher, nearly equal thermal throughputs relatively similar to the side example.

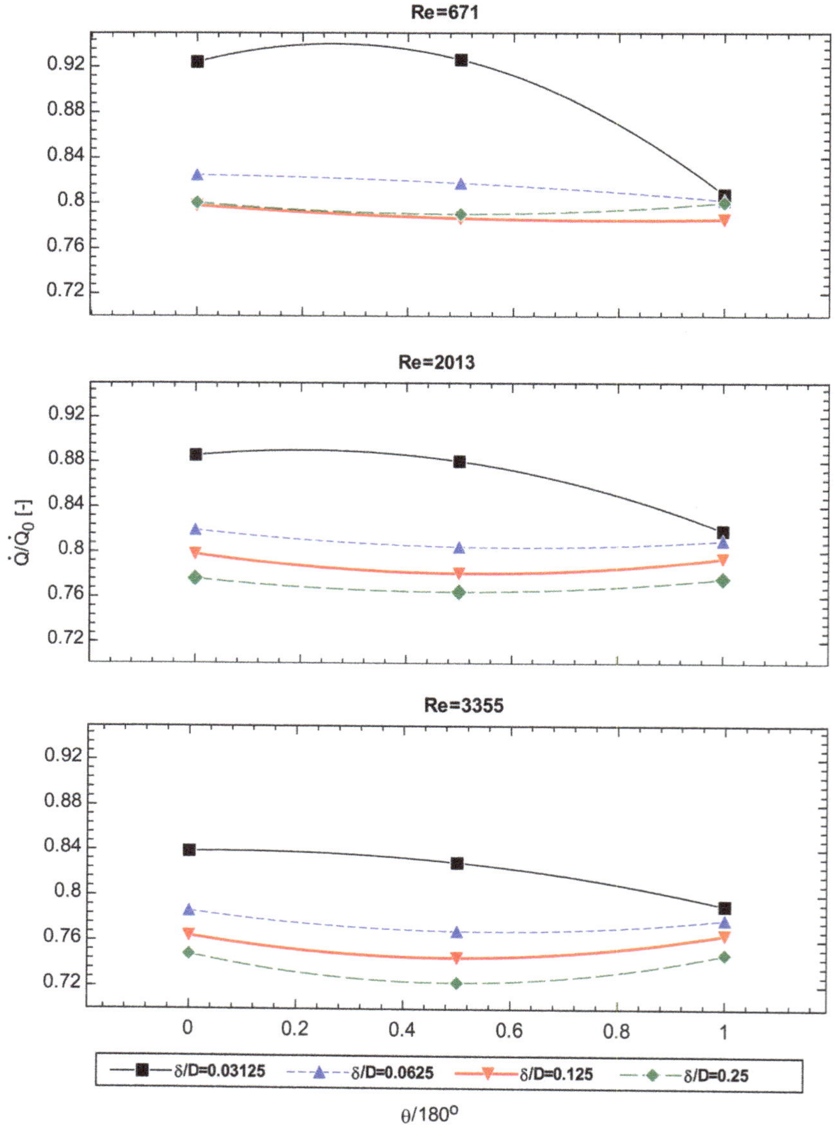

Figure 12. Heat transfer rate ratio in the function of gap position.

In Figure 13, the ratio of mean heat transfer coefficient with the gap presence to heat transfer ratio without any fin discontinuity is presented in the δ/D function. The front and side location results in enhancement of the mean convective heat transfer for the smallest discontinuity. This effect is not present in the rear location. For the remaining δ/D values, h/h_0 is below 1.0 (decrease of h). Besides, the local maximum at $\delta/D = 0.03125$ heat transfer coefficients ratio decreases nonlinearly with the increasing gap thickness, except for $\delta/D = 0.25$ where a slight increase can be observed. For the rear slit location, there is no local maximum and h/h_0 is at an approximately constant level of 0.9.

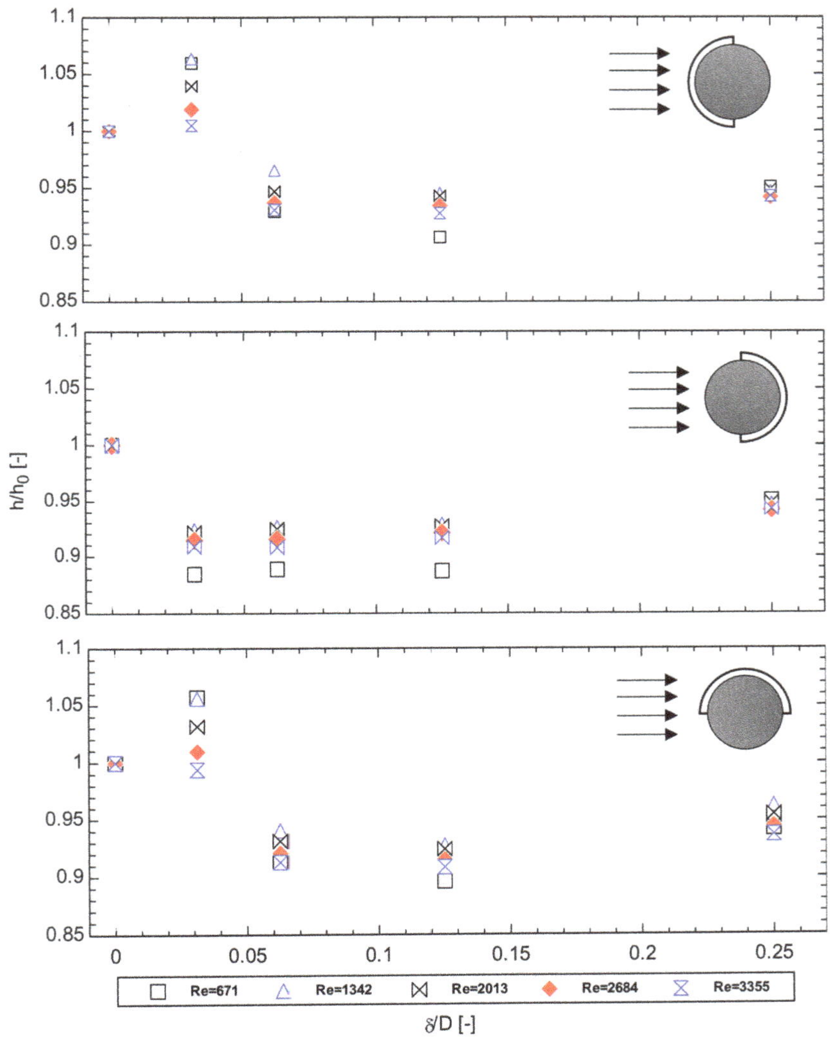

Figure 13. Heat transfer coefficient ratio in the function of dimensionless gap thickness.

In Figure 14, the h/h_0 dependence on the Reynolds number is shown. For the side and front cases at $\delta/D = 0.03125$, the convective heat transfer enhancement is existing at the whole range of Re numbers. For all slit thicknesses, there is a weak decrease of h/h_0 with the increase of the flow velocity. The smallest discontinuity heat transfer coefficients diverge from the wider gap cases, being at least 4% higher, whereas for the rest of the example values are adjacent to each other. There is not any enhancement for the rear position; the highest h/h_0 are noted for $\delta/D = 0.25$ at a nearly constant value of $h/h_0 = 0.95$ for the whole range of Reynolds numbers, diverging from the rest of the considered cases. At the lowest Re, for the remainder examples, the h/h_0 reaches a minimum of approx. 0.88, with a local maximum at $Re = 1342$, and for the rest of the volumetric flow range there is a decrease with the increase of Re.

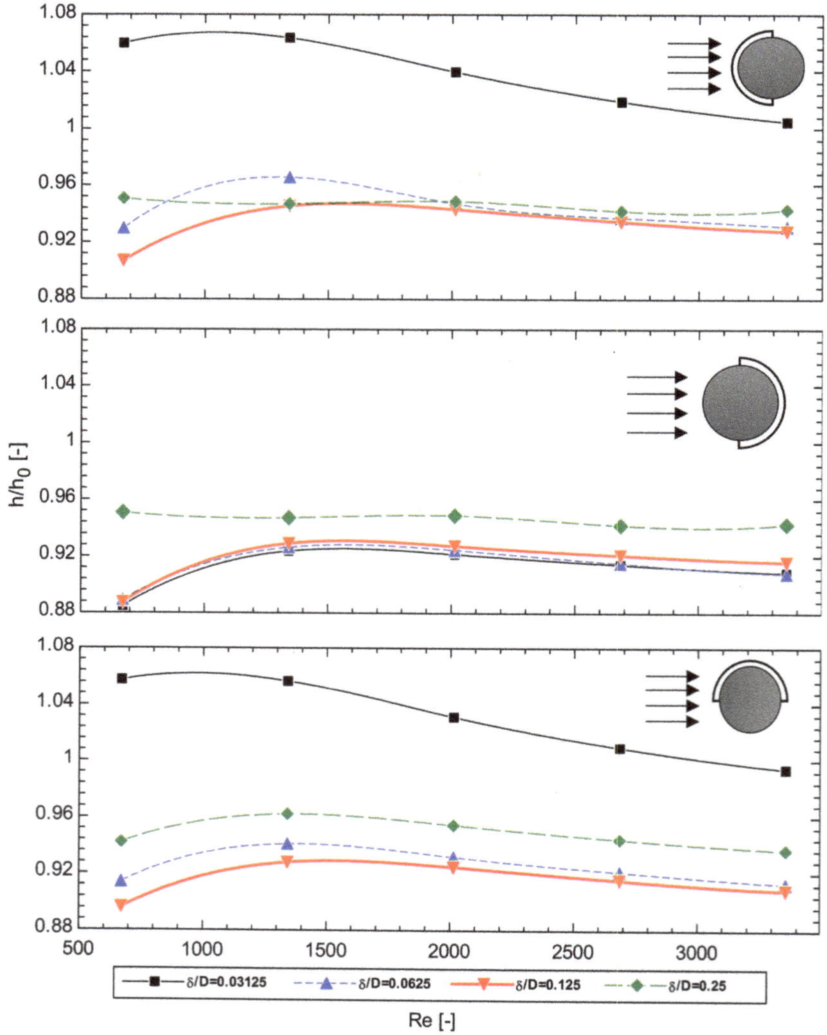

Figure 14. Heat transfer coefficient ratio in the function of Reynolds number.

The dependence on the gap position is shown in Figure 15. At δ/D = 0.03125, there is the highest difference between the front, side, and back locations. The remaining slit thicknesses exhibit a very weak dependence on $\theta/180°$, typically slightly reducing, with a maximum for the front and minimum for the back case.

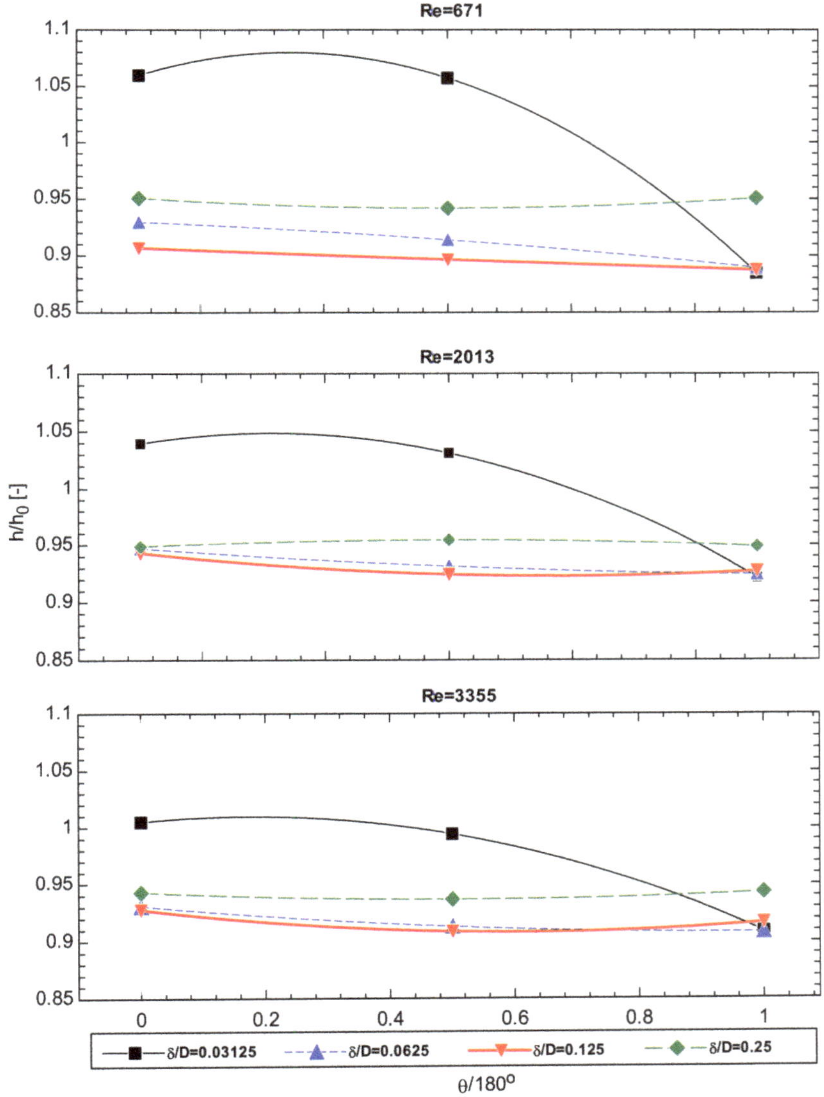

Figure 15. Heat transfer coefficient ratio in the function of the gap position.

In Figure 16, the contour map of a local heat transfer coefficient, which is defined as:

$$h_x = \frac{\dot{q}}{T_f - T_{in}}, \tag{16}$$

where T_{in} is periodic inlet air temperature, is presented for different gap sizes at the lowest Reynolds number (Re = 671) for front fin discontinuity location. Figure 16 was arranged to investigate the convective heat transfer enhancement relatively to no gap case. Intensification is present for the thinnest slit (δ/D = 0.03125)—for bigger gaps, this effect diminishes (Figure 10). The contours are similar for all cases shown; therefore, the analysis starts with a description of common map characteristics. There is a high h_x zone at the inlet edge of the fin—this is the region where the thermal boundary layer is the thinnest. The area of this region is reduced at the center because of the presence of the fluid circulation flow spanning from the leading edge of the fin to the surface of the tube (Figure 17). This circulation is caused by the trace flow downstream of the upstream tube—being a result of boundary layer detachment. In Figure 17, one can see the circulation regions visualized by streamlines. The circulation area is wider in the rear region of the fin (behind the tube) than upstream of the tube (air flows from the bottom to the top in Figure 17). Because the circulating fluid moves with reduced velocity, local heat transfer coefficients are significantly lower. On the sides of the tube, air flows with increased velocity (Figure 17)—this corresponds to the high h_x region in Figure 16 (contour level: 4, h_x = 18–24 W/m^2/K).

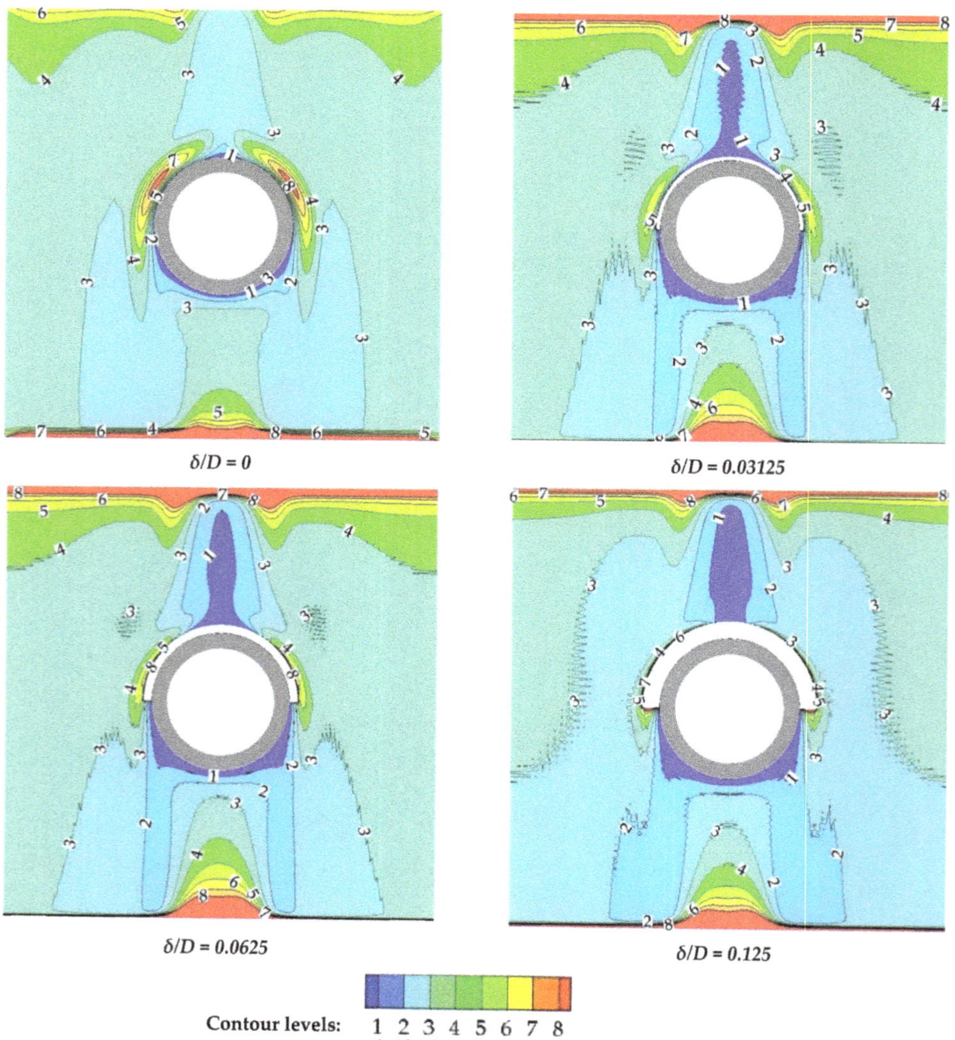

Figure 16. h_x map on the surface of the fin for various gap sizes at *Re* = 671 for the front location of the gap.

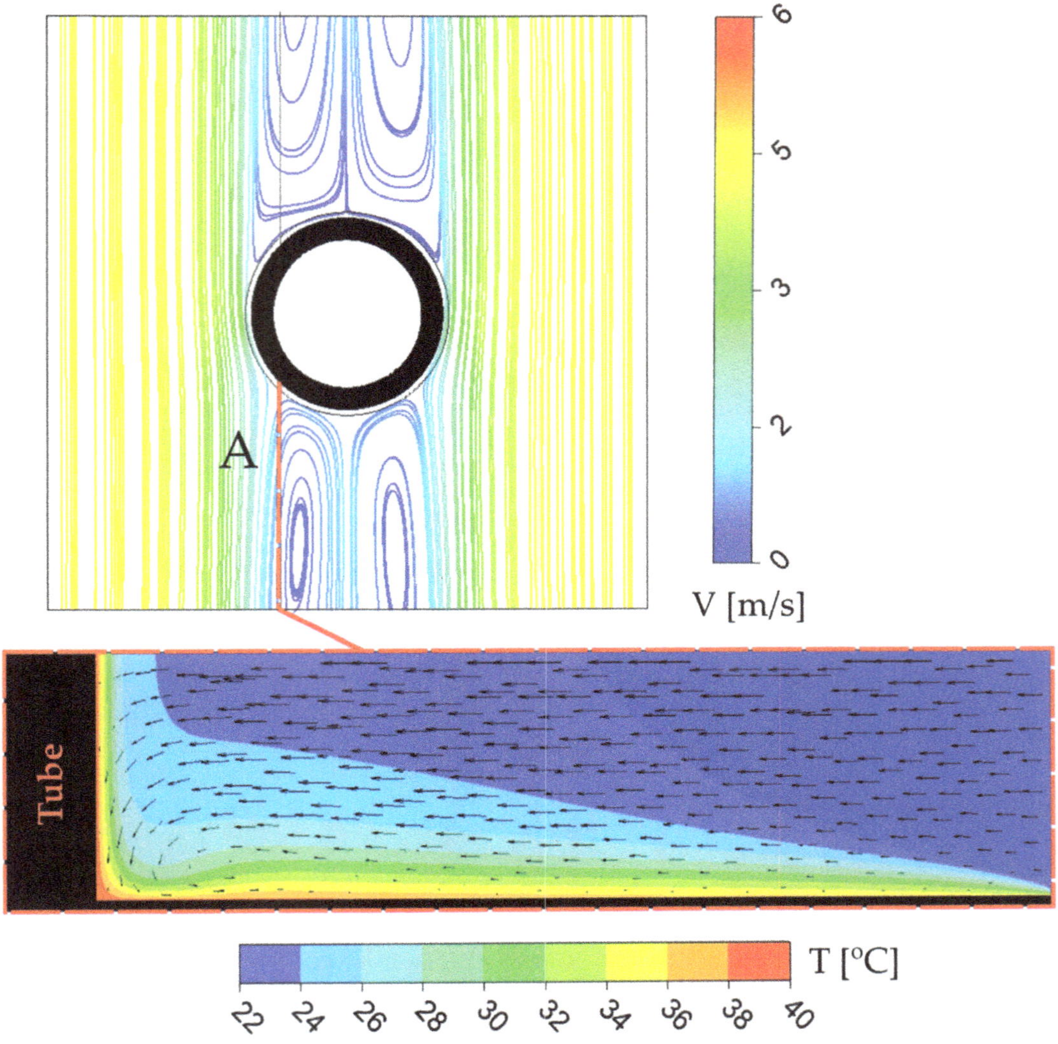

Figure 17. The top view of streamlines with the vector field showed on the face A upstream of the tube, at Re = 2013, for the ideal tube-fin contact case.

At the fin surface zone near the front part of the tube circumference, there are two bands of intensified heat transfer (placed symmetrically on the sides). This local enhancement is caused by a horseshoe vortex, which is visible in Figure 17, where the velocity vector field is visualized at plane A. The swirl of the fluid can be observed at the base of the tube. For the one-row or staggered arrangement plate-fin tube heat exchangers, the horseshoe vortex spans the whole half of the tube's circumference. In the present analysis, periodic flow simulations are equivalent to the inline arrangement. Therefore, the wake behind the preceding tube divides the horseshoe vortex in two (Figure 16—$\delta/D = 0$). There is also a thin region of enhanced heat transfer coefficient at the trailing edge, widening at the middle of the fin. This is the effect of the implementation of periodic heat transfer conditions in the simulations. Because the negative heat flux is set on the outlet, the small backflow of colder fluid causes this "artificial" heat transfer intensification. This effect is practically of the same magnitude for all cases—it cancels in the comparative analysis. Even the h_x contour maps share the same characteristics; for $\delta/D = 0$, the strength of the heat transfer enhancement caused by the horseshoe vortex is the greatest. With the increase of the gap thickness, the horseshoe intensification becomes weaker. The thinnest slit covers partially the area of vortex action, whereas in the $\delta/D = 0.125$ case the vortex is present completely over the gap, which results in only moderately increased local h on the edges of the fin discontinuity. The introduction of the gap improves the heat transfer on the leading edge of the fin and changes somewhat the flow and heat transfer characteristics in the region downstream of the tube.

For $\delta/D = 0.03125$ and $\delta/D = 0.0625$, this alteration is not significant, yet for $\delta/D = 0.125$ there is a dramatic change—low heat transfer coefficient region (contour 3 in Figure 16) doubles its size and merges with the smaller one at the upstream part of the tube. The combined effect of the enhancement on the leading edge with still strong horseshoe vortex intensification can explain the observed augmentation of mean HTC for the thinnest slit relatively to no-gap example (Figure 13). The mean of heat transfer enhancement with thin slits cut in the front of the tube (for low airflow velocities) should be further studied. It could be a promising method of heat transfer intensification in compact heat exchangers.

The enhancement of h_x, by the presence of the slit, relative to ideal fin-tube contact example is not enough to cancel the negative effects of the fin discontinuity at its base (tube-fin junction)—in Figure 11 one can observe at least 6% lower heat transfer rate in the respect to $\delta/D = 0$. In Figure 18, the most obvious consequence of the introduction of the gap can be seen—the blockage of heat flow in the fin area adjacent to the discontinuity. This effect is visualized by dimensionless temperature excess Θ contour map, where:

$$\Theta = \frac{T_f - T_{in}}{T_w - T_{in}}. \tag{17}$$

The fin is the most isothermal for the no-gap example, and the contour map is the most symmetrical. In Figure 18, the heat flux blockage is evident for the cases with the gap—Θ is significantly lower for nearly half of the fin on the side of the slit. For the gap in the front, the reduction of temperature is the smallest, while for the side location it is the highest (lowest fin efficiency). The mean h on the sides of the fin is the highest, and that is the reason for the severe temperature decrease. The Θ is the biggest on the average for the front case because the heat flux is directed backward (low h_x region) and the enhancing effect of the horseshoe vortex in the front of the tube is partially blocked. Analyzing the back gap position, the Θ is significantly lower on average. Heat flux is directed towards the front where the heat transfer intensification regions cause effective cooling of the fin. The result is substantially lower fin temperature in the back. In a conclusion, the gap placement in the front blocks the heat flow in the direction of the enhanced convective heat transfer area, which is not beneficial, although it results in higher fin efficiency than for the gap in the back. This can explain why there is a sudden heat transfer rate ratio drop for the rear slit position. The effect of low fin efficiency seems to combine with poorer convective heat transfer relatively similar to other discontinuity placement cases (Figure 10). The side gap position is the worst of the considered positions in terms of the fin temperature contour. The colder fin area is slightly bigger than the gap in the back. As the heat flux is directed to the side where high-velocity flow increases the intensity of heat transfer, the fin is efficiently cooled there. As the mean heat transfer coefficient is nearly as high as for the gap in the front, $\dot{Q}/\dot{Q}_0$ is slightly lower than the front gap for thin slits. The enhanced convective heat transfer compensates for the low fin efficiency. Figure 19 shows the influence of the flow Reynolds number on the h_x. The h_x contour maps are presented at four Reynolds numbers for a 1 mm thick gap in the back. The heat transfer intensification caused by the presence of the horseshoe vortex is less significant for the lowest Re, and it gets increasingly distinctive as the airflow velocity increases. The thermal boundary layer becomes thinner at the leading edge of the fin with the raising Re, which is visible as the wider high h_x area. As the Re increases, the h_x contours change their values proportionally, but the overall topology (shape of the contours) is analogous for all airflow velocities. In Figure 14, one can see that for Re = 671 h/h_0 is the lowest—this is mainly caused by the low activity of the horseshoe vortex (Figure 19).

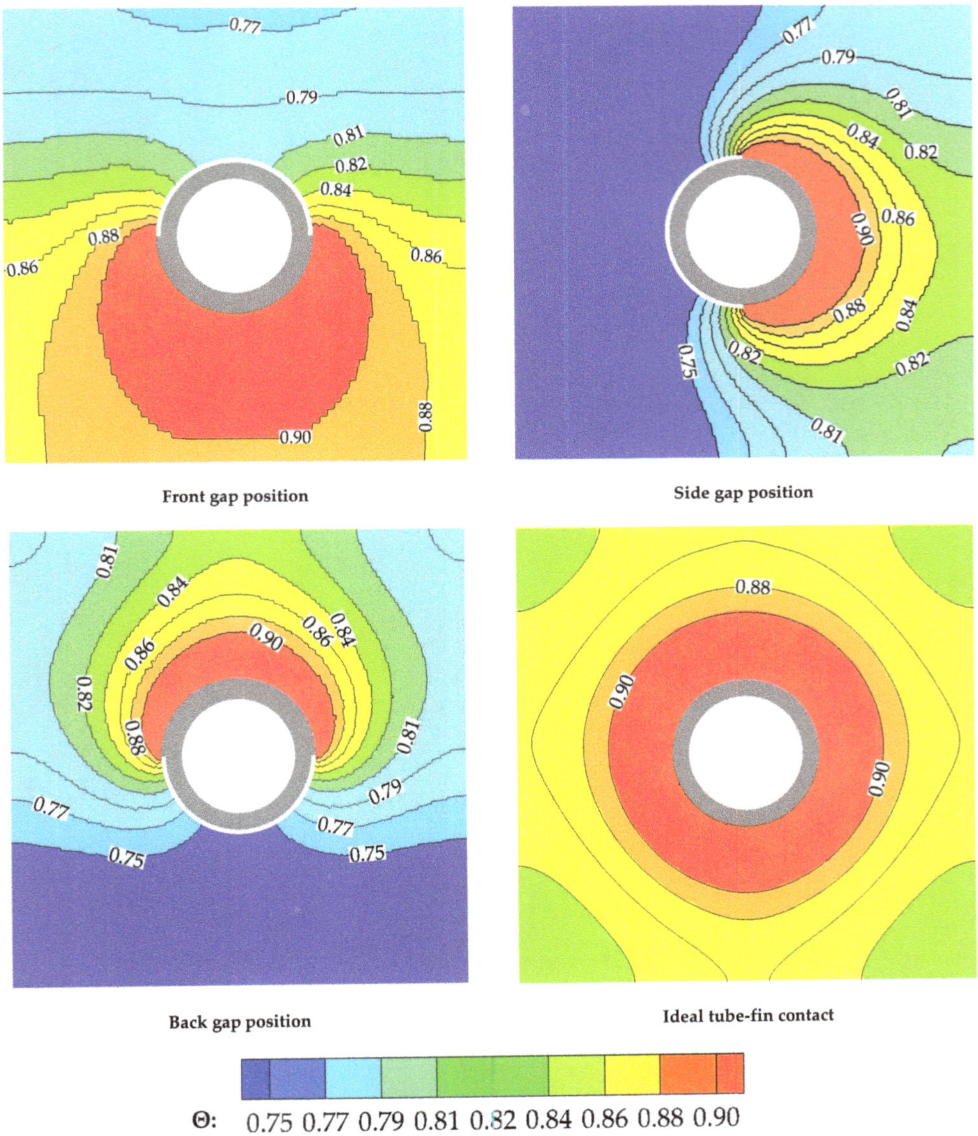

Figure 18. Dimensionless temperature excess Θ contour at Re = 2013 and δ/D = 0.03125 for different gap positions.

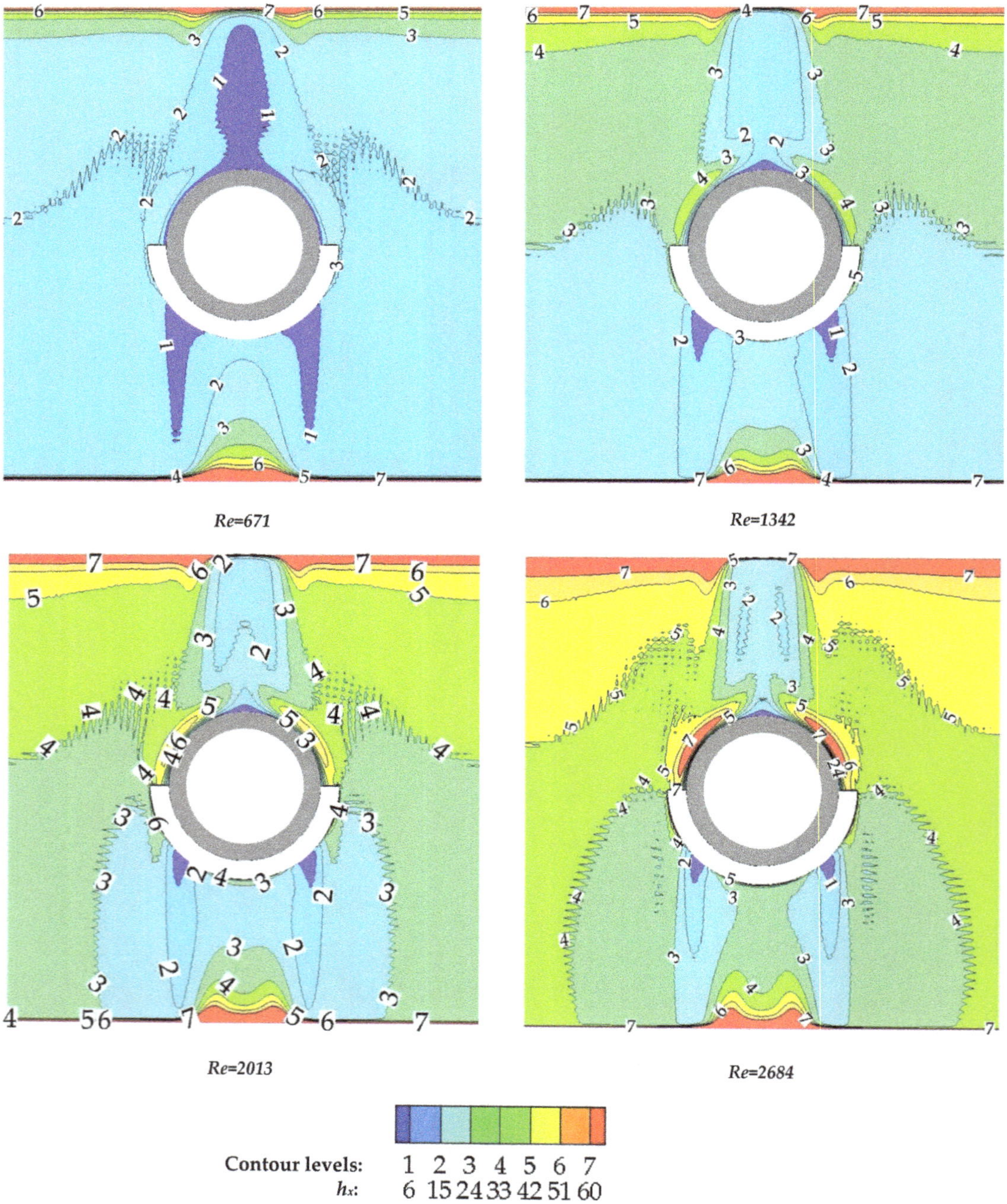

Figure 19. h_x map on the surface of the fin for various Reynolds numbers, with the gap size: δ/D = 0.125 and back gap position.

4. Conclusions

To sum up the key findings from the numerical simulation, the fin discontinuity along the circumference reduces the heat transfer rate relatively to the perfect fin-tube contact case. The placement of the gap reduces the skin friction locally to zero, which can promote beneficial flow structures, which causes enhancement concerning the no-gap case, especially for small slit sizes. However, the convective heat transfer enhancement is overpowered by the heat flux blockage effect which leads to $\dot{Q}/\dot{Q}_0 < 1$ for all the cases. Fin discontinuity in front of the tube causes the smallest reduction of the heat transfer rate in comparison to the ideal tube-fin contact, especially for thin slits. Even the heat transfer rate ratio minimum is obtained for the gap at the side position (approx. 0.72), it performs slightly worse for the small gap range than the front location. The rear gap is characterized by the mean $\dot{Q}/\dot{Q}_0 = 0.8$ for all the gap sizes and therefore is considered the worst scenario, without the advantage of the smallest gap range. For gaps wider than $\delta/D > 0.1$, the heat transfer rate reduction becomes nearly constant for a specified Reynolds number. Generally, the higher the airflow, the $\dot{Q}/\dot{Q}_0$ becomes lower. For multirow plate-fin tube heat exchangers with small fin discontinuities, reversing the flow direction can result in an even 15% heat transfer increase if most of the gaps are in the rear of the tube (low airflow velocity). Realistically, obtaining this maximum is not very probable, although the present study shows that reversing the flow can have a positive effect on the plate-fin tube heat exchanger performance, and its longevity depends on the nature of the origin of the discontinuities (are they, and how fast they enlarge). The convective heat transfer enhancement for thin slits in the front should be further investigated to test the possibility of the introduction of fin cut-outs as heat transfer intensification structures in compact heat exchangers.

Author Contributions: Conceptualization, A.N.G. and M.Ł.; methodology, A.N.G. and M.Ł.; software, G.G.; validation, M.Ł., A.N.G. and D.A.; writing—original draft preparation, M.Ł. and D.A.; writing—review and editing, G.G.; visualization, D.A.; supervision, A.N.G. All authors have read and agreed to the published version of the manuscript.

Funding: This research received no external funding.

Data Availability Statement: Not applicable.

Conflicts of Interest: The authors declare no conflict of interest.

Nomenclature

A	finned side heat transfer surface, m^2
A_{t0}	bare outside tube surface, m^2
CFD	computational fluid dynamics
c_p	specific heat at constant pressure, J/kg/K
d	diameter, m
d_h	$(4\cdot(L_f - D)\cdot L_t\cdot L_f)/(2\,[L_f^2 - (\pi\cdot D^2/4)] + \pi\cdot D\cdot L_t)$—hydraulic diameter, m
D	outer diameter of the tube
e	total fluid energy, J/kg
g	gravitational acceleration, m/s^2
h	heat transfer coefficient, W/m^2/K
h_x	local heat transfer coefficient
$h_{\%}$	$(h_i - h_{i-1})/h_i$, relative change of heat transfer coefficient, %
k	$0.5\cdot u_i'\cdot u_i'$—kinetic energy of turbulence m^2/s^2
L	length, m
$\overline{L}$	periodic length vector, m
$\dot{m}$	mass flow, kg/s
n	heat-mass transfer analogy exponent
$\overline{n}$	vector normal to a surface
Nu	Nusselt number

p	pressure, Pa
$\widetilde{p}$	periodic pressure, Pa
Pr	Prandtl number
Pr_t	turbulent Prandtl number
$\dot{q}$	heat flux (vector), W/m^2
$\dot{Q}$	heat transfer rate, W
$\bar{r}$	position vector, m
Re	$V{\cdot}d_h/\nu$—Reynolds number
Ri	$(g{\cdot}\beta{\cdot}\Delta T{\cdot}L)/V^2$—Richardson number
Sc	Schmidt number
Sh	Sherwood number
SST	shear stress transport
T	temperature, K
$\widetilde{T}$	periodic temperature, K
V	average velocity (scalar), m/s
u	velocity vector, m/s
$\overline{u}$	time-averaged velocity vector, m/s
x	cartesian coordinates vector, m
X_t	tube spacing, m
Y^+	non-dimensional distance between the first mesh node and the wall
Greek symbols	
β	volumetric expansion coefficient, 1/K
δ	gape thickness, thickness, spacing, m
δ_{ij}	Kronecker delta
Δ	difference of a quantity
θ	the gap placement angle
Θ	temperature excess
λ	thermal conductivity, W/m/K
μ	dynamic viscosity, kg·m/s
ν	kinematic viscosity, m^2/s
ρ	density, kg/m^3
$\rho\overline{u_i u_j}$	Reynolds stresses, Pa
σ	linear temperature gradient, K/m
τ_{ij}	stress tensor, Pa/m
ω	specific dissipation rate, 1/s
Subscripts	
0	zero gap thickness
ave	average
d	based on the tube outside diameter
f	fin
g	gap
i	inside, inner, for i-th mesh refinement
i, j, k	indices (1, 2, 3)
in	inlet
o	outlet, outer
t	tube, turbulent
w	wall

References

1. Gorecki, G. Investigation of two-phase thermosyphon performance filled with modern HFC refrigerants. *Heat Mass Transf.* **2018**, *54*, 2131–2143. [CrossRef]
2. Chai, L.; Tassou, S. A Review of Airside Heat Transfer Augmentation with Vortex Generators on Heat Transfer Surface. *Energies* **2018**, *11*, 2737. [CrossRef]
3. Taler, D.; Taler, J.; Trojan, M. Experimental Verification of an Analytical Mathematical Model of a Round or Oval Tube Two-Row Car Radiator. *Energies* **2020**, *13*, 3399. [CrossRef]
4. Lotfi, B.; Sundén, B.; Wang, Q. An investigation of the thermo-hydraulic performance of the smooth wavy fin-and-elliptical tube heat exchangers utilizing new type vortex generators. *Appl. Energy* **2016**, *162*, 1282–1302. [CrossRef]

5. Välikangas, T.; Karvinen, R. Conjugated Heat Transfer Simulation of a Fin-and-Tube Heat Exchanger. *Heat Transf. Eng.* **2018**, *39*, 1192–1200. [CrossRef]
6. Bansal, P.; Wich, T.; Browne, M. Optimisation of egg-crate type evaporators in domestic refrigerators. *Appl. Therm. Eng.* **2001**, *21*, 751–770. [CrossRef]
7. Blygold Blygold—Example of Galvanic Corrosion Damage. Available online: https://www.blygold.com/services/coil-coating/ (accessed on 15 May 2021).
8. Critoph, R.; Holland, M.; Turner, L. Contact resistance in air-cooled plate fin-tube air-conditioning condensers. *Int. J. Refrig.* **1996**, *19*, 400–406. [CrossRef]
9. Jannick, P.; Meurer, C.; Swidersky, H. Potential of Brazed Finned Tube Heat Exchangers in Comparison to Mechanical Produced Finned Tube Heat Exchangers. In Proceedings of the International Refrigeration and Air Conditioning Conference, West Lafayette, IN, USA, 16–19 July 2002.
10. Zhao, H.; Salazar, A.J.; Sekulic, D.P. Analysis of Fin-Tube Joints in a Compact Heat Exchanger. *Heat Transf. Eng.* **2009**, *30*, 931–940. [CrossRef]
11. Jeong, J.; Nyung Kim, C.; Youn, B.; Saeng Kim, Y. A study on the correlation between the thermal contact conductance and effective factors in fin–tube heat exchangers with 9.52 mm tube. *Int. J. Heat Fluid Flow* **2004**, *25*, 1006–1014. [CrossRef]
12. Cheng, W.; Madhusudana, C. Effect of electroplating on the thermal conductance of fin–tube interface. *Appl. Therm. Eng.* **2006**, *26*, 2119–2131. [CrossRef]
13. Taler, D.; Ocłoń, P. Thermal contact resistance in plate fin-and-tube heat exchangers, determined by experimental data and CFD simulations. *Int. J. Therm. Sci.* **2014**, *84*, 309–322. [CrossRef]
14. Taler, D.; Cebula, A. A new method for determination of thermal contact resistance of a fin-to-tube attachment in plate fin-and-tube heat exchangers. *Chem. Process Eng.* **2010**, *31*, 839–855.
15. Singh, S.; Sørensen, K.; Condra, T.J. Influence of the degree of thermal contact in fin and tube heat exchanger: A numerical analysis. *Appl. Therm. Eng.* **2016**, *107*, 612–624. [CrossRef]
16. Saboya, F.E.M.; Sparrow, E.M. Local and Average Transfer Coefficients for One-Row Plate Fin and Tube Heat Exchanger Configurations. *J. Heat Transf.* **1974**, *96*, 265. [CrossRef]
17. Menter, R.F. Zonal Two Equation Kappa-Omega Turbulence Models for Aerodynamic Flows. In Proceedings of the 23rd Fluid Dynamics, Plasmadynamics, and Lasers Conference, Orlando, FL, USA, 6–9 July 1993.
18. Lindqvist, K.; Skaugen, G.; Meyer, O.H.H. Plate fin-and-tube heat exchanger computational fluid dynamics model. *Appl. Therm. Eng.* **2021**, *189*, 116669. [CrossRef]
19. Taler, D.; Ocłoń, P. Determination of heat transfer formulas for gas flow in fin-and-tube heat exchanger with oval tubes using CFD simulations. *Chem. Eng. Process. Process Intensif.* **2014**, *83*, 1–11. [CrossRef]
20. Jasiński, P.B. Numerical study of the thermo-hydraulic characteristics in a circular tube with ball turbulators. Part 2: Heat transfer. *Int. J. Heat Mass Transf.* **2014**, *74*, 473–483. [CrossRef]
21. Ocłoń, P.; Łopata, S. Study of the Effect of Fin-and-Tube Heat Exchanger Fouling on its Structural Performance. *Heat Transf. Eng.* **2018**, *39*, 1139–1155. [CrossRef]
22. Ansys Inc. *ANSYS CFX Theory Guide*; Ansys Inc.: Canonsburg, PA, USA, 2013.
23. Shevchuk, I.V.; Jenkins, S.C.; Weigand, B.; Von Wolfersdorf, J.; Neumann, S.O.; Schnieder, M. Validation and analysis of numerical results for a varying aspect ratio two-pass internal cooling channel. *J. Heat Transf.* **2011**, *133*. [CrossRef]
24. Siddique, W.; El-Gabry, L.; Shevchuk, I.V.; Fransson, T.H. Validation and Analysis of Numerical Results for a Two-Pass Trapezoidal Channel With Different Cooling Configurations of Trailing Edge. *J. Turbomach.* **2012**, *135*. [CrossRef]
25. Goldstein, R.J.; Cho, H.H. A review of mass (heat) transfer measurements using naphthalene sublimation. *Exp. Heat Transf. Fluid Mech. Thermodyn.* **1993**, *10*, 21–40. [CrossRef]
26. Kays, W.M.; William, M.; London, A.L.; Alexander, L. *Compact Heat Exchangers*; Krieger Pub. Co.: Malabar, FL, USA, 1998; ISBN 1575240602.
27. Rosman, E.C.; Carajilescov, P.; Saboya, F.E.M. Performance of One- and Two-Row Tube and Plate Fin Heat Exchangers. *J. Heat Transf.* **1984**, *106*, 627. [CrossRef]
28. Tecplot Inc. *Tecplot.360 EX User's Manual*; Tecplot Inc.: Bellevue, WA, USA, 2018.
29. Saboya, S.M.; Saboya, F.E.M. Transfer coefficients for plate fin and elliptical tube heat exchangers. In Proceedings of the 6th Brazilian Congress of Mechanical Engineering, Rio de Janeiro, Brazil, 15–18 December 1981.
30. Stephan, P. B1 Fundamentals of Heat Transfer. In *VDI Heat Atlas*; Springer: Berlin/Heidelberg, Germany, 2010; pp. 15–30.

energies

MDPI

Article

Application of Thermal and Cavitation Effects for Heat and Mass Transfer Process Intensification in Multicomponent Liquid Media

Anatoliy M. Pavlenko * **and Hanna Koshlak**

Department of Building Physics and Renewable Energy, Kielce University of Technology, 25-314 Kielce, Poland; hkoshlak@tu.kielce.pl

* Correspondence: apavlenko@tu.kielce.pl; Tel.: +48-883-741-291

Abstract: In this paper, the authors consider the processes of dynamic interaction between the boiling particles of the dispersed phase of the emulsion leading to the large droplet breakup. Differences in the consideration of forces that determine the breaking of non-boiling and boiling droplets have been indicated in the study. They have been determined by the possibility of using the model to define the processes of displacement, deformation, or fragmentation of the inclusion of the dispersed phase under the influence of a set of neighboring particles. The dynamics of bubbles in a compressible liquid with consideration for interfacial heat and mass transfer has also been analyzed in the paper. The effect of standard and system parameters on the intensity of cavitation processes is considered. Physical transformations during the cavitation treatment of liquid are caused not only by shock waves and radiated pressure pulses but also by extreme thermal effects. At the stage of ultimate bubble compression, vapor inside the bubble and the liquid in its vicinity transform into the supercritical fluid state. The model analyzes microflow features in the inter-bubble space and quantitatively calculates local values of the velocity and pressure fields, as well as dynamic effects.

Keywords: breaking; desorption; force; speed; acceleration; pressure; dispersed phase

Citation: Pavlenko, A.M.; Koshlak, H. Application of Thermal and Cavitation Effects for Heat and Mass Transfer Process Intensification in Multicomponent Liquid Media. *Energies* **2021**, *14*, 7996. https://doi.org/10.3390/en14237996

Academic Editors: Jan Danielewicz and Krzysztof Rajski

Received: 8 November 2021
Accepted: 24 November 2021
Published: 30 November 2021

1. Introduction

Cavitation phenomena have been studied by many authors [1–6] and can be used to intensify heat and mass transfer processes [7–9]. In order to develop an effective method of a controlled cavitation effect on technological processes, one should have a clear understanding of the physical nature of cavitation mechanisms and their relevant description in spatial and temporal scales.

For years, cavitation has been considered one of the areas of hydrodynamics associated with the elucidation and elimination of negative causes of erosion destruction of the surfaces of propellers, hydraulic turbine blades, and hydraulic devices [10–12]. It was determined that surfaces are eroded under the action of shock waves and cumulative jets formed by the micro-explosions of bubbles, considered as hollow caverns in hydrodynamics [13–15]. Within these mechanisms alone it is difficult to explain the destruction of molecular and colloidal structures and other cavitation effects at the microlevel in the liquid itself. When cavitation is analyzed within hydrodynamics, the impact of thermal factors on the development and specific manifestation of cavitation effects is excluded. To a certain extent, such an approach has been valid to date, although it has been determined that at the maximum compression stage, the vapor–gas medium temperature inside the cavitation bubble exceeds 5000 K, and the pressure increases up to 1000 MPa [16–18]. It is assumed that the thermal energy released in the bubble volume is sufficient for the excitation, ionization, and dissociation of water vapor molecules and gases present in the bubble, the formation of free radicals, acceleration by 5–6 orders of magnitude of the rates of gas-phase chemical reactions [19–21], and even initiation of nuclear fusion reactions in cavitation bubbles [22,23]. That is, under certain conditions, various physical or chemical processes can be initiated in local zones of the liquid while concentrating energy to abnormally high values. The following problem is still to be solved: What conditions are required for the appearance of such effects? At the same time, high-temperature processes in the gas phase (in the cavitation region) do not explain the cavitation effect on disperse inclusions present in the liquid phase (droplets of another liquid in emulsions, particles in suspensions, etc.), assuming the existence of alternative mechanisms.

Reliable theoretical analysis of cavitation mechanisms is hindered, in the authors' opinion, by the lack of generalizing mathematical models with maximum consideration of the main physical factors. Modeling is usually performed with respect to the analysis of a specific experiment without considering the interfacial heat exchange and kinetics of phase transitions; the bubble compression process is assumed to be adiabatic [2,24,25]. The principles of a mathematical model for the dynamics of bubble formation and development in emulsions based on the laws of cavitation described above, with the effect of thermodynamic processes in the cavitation region, are considered in [26,27]. The model validation in the analysis of experimental results defines this approach as having potential.

The purpose of this study is to investigate the mechanisms of the cavitation effect on water emulsions (Figure 1) containing micro-dispersions, molecular and colloidal structures relevant to improving the design of cavitating devices, and substantiating their optimal operating modes.

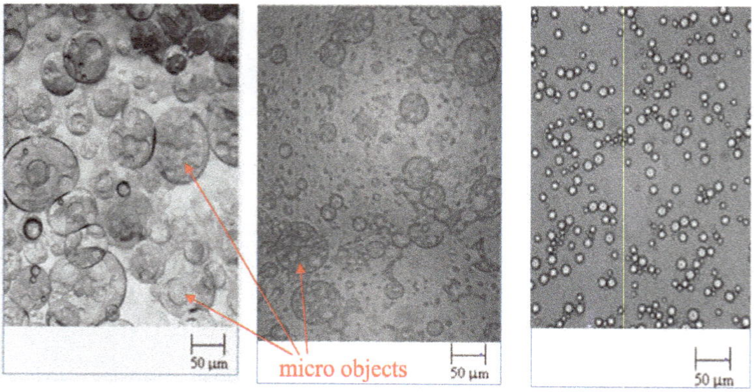

(a) (b) (c)

Figure 1. Photos of emulsion: (**a**) fuel emulsion "water–fuel oil" before thermal cavitation treatment, (**b**) after hydrodynamic cavitation, and (**c**) after thermal cavitation treatment.

Figure 1 shows one example of how thermal cavitation treatment can be applied to homogenization technologies for liquid fuels.

The tasks aimed at the efficiency increase of heat and mass transfer processes are very important for almost all heat technologies. One of the most efficient methods used to increase the intensification of heat and mass transfer processes is the use of emulsions as operating media [28–32], containing various mutually insoluble liquids with certain thermophysical characteristics. An increase in the emulsion temperature is accompanied by their structure change, the breakage of dispersed fluid droplets, or their merging. When thermolabile liquid boils, heat and mass transfer processes are intensified in the emulsion due to the abnormally rapid change in the vapor phase volume and turbulence transition of the medium flow; it undoubtedly increases the efficiency of these processes. The phenomena similar to those described above (cavitation) occur in the vapor layer of the boiling thermolabile part. They have been studied by many researchers [33–35], but these phenomena require further consideration in terms of heat and mass transfer efficiency. Moreover, the processes which induce abnormal values of pressures and temperatures in the case of thermodynamic cavitation and their application represent a separate important problem. One of the possible issues to be solved using such thermal cavitation initiation is homogenization technology.

It is worth mentioning that there is no general approach to the management of cavitation processes to obtain the expected disperse structure. Our proposed thermal cavitation method may be used to evaluate such effects. Its main ideas are described below.

2. Assessment of Dynamic and Thermal Cavitation Effects

Despite different and often contradictory definitions of cavitation, we will adhere to the following general formulation. Cavitation is a set of phenomena in a liquid when the pressure p_e drops below the saturation pressure at temperature p_s (T), and then it rises sharply above this level. If $p_e < p_s$ (T), it is possible to observe the formation and growth of vapor bubbles both during hydrodynamic cavitation and thermal cavitation, when the pressure in a liquid, preliminarily superheated under excess pressure, is sharply reduced. Then, vapor bubbles appear at the interface of the dispersed inclusions of the emulsion. At $p_e > p_s$ (T) they get compressed and subsequently destroyed (explosion). In both cases, a nonequilibrium process of the bubble system proceeds to the state of thermodynamic equilibrium, making it possible to describe bubble boiling and cavitation within the framework of one model, using the same Equations but under different initial conditions.

This concept forms the basis of the mathematical model of bubble dynamics, presented in [26,27].

The entire period of the cavitation effect development is accompanied by the growth and subsequent destruction of bubbles. A pulsating cavitation cluster inevitably appears in hydraulic turbines, pumps, and in various devices where the above conditions of pressure drop are realized [36].

Let us consider the process of vapor bubble compression in water under condition $p_e > p_s$ (T). At the beginning of compression, the vapor pressure inside the bubble is $p_{int} \approx p_s$ (T) and the pressure in a liquid at the bubble wall is $p_i = p_{int} - 2\sigma\ (T)/R - \mu(T){\cdot}v/R$, where R is the bubble radius, v is the liquid radial rate at the bubble interface, and $\sigma(T)$ and $\mu(T)$ are surface tension and viscosity coefficients. As a result of the pressure difference $p_e - p_i$, the liquid moves rapidly towards the bubble center, resulting in a sharp increase in vapor pressure and temperature in the collapsing bubble. The compression rate is regulated by interfacial heat transfer and vapor condensation on the "cold" bubble wall with condensation heat release. When vapor pressure p_{int}, compressed in the bubble, exceeds external pressure p_e, the liquid begins to decelerate, and its kinetic energy is converted into the potential energy of the compressed vapor and into the potential energy of the compressed liquid around the bubble. When kinetic energy is completely transformed, the bubble compression stops while pressure in the liquid at its boundary reaches its maximum value. The hydraulic hammer effect is realized, a compression wave propagates from the bubble into the liquid volume at the velocity of sound and its amplitude decreases with distance. Normally, when cavitation processes are modeled, the liquid is considered incompressible, while the hydraulic hammer phenomenon is not considered. To analyze the degree of the hydraulic hammer effect, the model [27] was modified taking into account liquid compressibility. The liquid compressibility coefficient $k_\beta = f\ (T, p)$ and the temperature coefficient of adiabatic compression $\alpha\ (T, p) = \partial T/\partial p$ are introduced into the model as new parameters. Taking into account the compressibility, the motion Equation, included in the model's basic system of Equations, is presented in the following form:

$$\frac{dv}{d\tau} = \frac{p_{int} - p_{ext} + 1.5(p_{int} - p_{ext})^2 k_\beta - 1.5\rho v^2}{\rho R} \tag{1}$$

where $\rho(T)$ is the liquid density. The derived Equation (1) differs from the classical Rayleigh–Plesset Equation usually used to describe cavitation processes in an incompressible liquid, by the term $1.5(p_{int} - p_e)^2 k_\beta$ in the numerator, considering the potential energy contribution of a compressed liquid, which is then realized in a force impulse form.

After the transition of the vapor parameters in the compressed bubble to the supercritical region, ($T_{steam} > T_{critical}$ and $p_{steam} > p_{critical}$), the liquid–vapor interface disappears (σ ($T_{critical}$) = 0) and a quasi-spherical local zone is formed in place of a bubble with R_{min} radius within which water passes from a supercritical state to a condensed one. Similar effects are observed when the above is exposed to high pressures, even if the water temperature is below critical [37]. We can assert that the thermal cavitation effect is not limited by the volume of a bubble with R_{min} radius where, as it is believed, only the ionization of vapor and gas molecules takes place, though covering a wider area.

When cavitation conditions are realized in a liquid, numerous vapor–gas bubbles forming a cavitation cluster grow and then simultaneously explode. The cavitation action effect is determined by the combined influence of all the cluster bubbles.

The superposition of shock waves from each individual bubble forms a single destructive shock wave. However, this concept fails to explain the destruction of micro objects (micro objects in Figure 1) within the cluster itself. When cavitation cluster features are modeled and analytical calculation methods for cavitation reactors are developed, the temperature is often not even included in the model parameters [14,15]. In [27], regarding the analysis of boiling processes, a model of the bubble ensemble dynamics is considered based on a model of the single bubble dynamics. Considering the liquid compressibility, we propose for it to be included in Equation (1), as this model makes it possible to predict the evolution of a cavitation cluster at all stages of its development and to describe the nature of microflows in the interbubble space of the cluster associated with bubble expansion or contraction. Nevertheless, in this work, a uniform distribution of gas bubbles in a liquid is assumed, which is practically unrealizable in real processes. Moreover, if we consider the problem with the temperature effect, then cluster sizes should be different, as shown in Figure 1 since the clusters have a more complex structure.

Figure 2 shows the distribution of microflow velocity vectors at local points around vapor bubbles, modeled using the technique proposed in [27] (gas bubble radii R = 10–100 µm (in the diagram, the size of each bubble increases by 10 µm)), with the maximum liquid rate at the interface with bubbles being v_i = 220 m/s (superheating temperature is 110 °C).

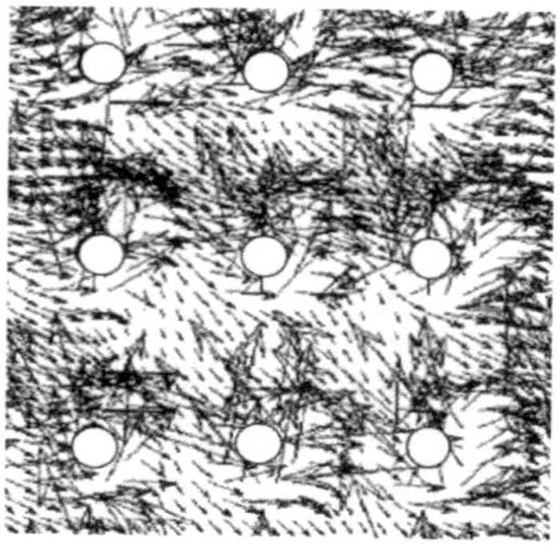

Figure 2. Distribution of velocity vectors inside the cavitation cluster.

Due to the superposition of radial flows of all the bubbles in the cluster, the pattern is quite complex. However, in real emulsions (Figure 1), with distance from the bubble, the rate should change sharply in magnitude and direction, since the dynamics of bubble development depend on its size and distance to the neighboring bubbles, as well as their sizes.

Calculations show that shear rates in the interbubble space at local points at the boundaries of the interaction between flows can reach 10^4 s^{-1}, thus corresponding to a shear stress of about 0.8 kPa if the dispersion phase is water (in these calculations, a water emulsion in fuel oil was used). However, if we use a procedure that focuses on a uniform distribution of low-boiling liquid droplets, as shown in Figure 2, we will not get the resultant flux. The dynamic effects are identical around each droplet and they will be mutually compensated. It does not correspond to the effects in real emulsions, where under similar conditions a circulating flow in the emulsion volume would occur. It is thought that the values of local velocities and accelerations will be different as they should also affect adjacent droplets. Since they have different sizes and are arranged in a chaotic manner, these effects will differ in level and direction. The resulting effect can be much more significant when the dynamic effects are directed towards the opposite droplets. It is also important to bear in mind that depending on the size of the droplets, dynamic effects can occur at different times. For example, if a small droplet evaporates very quickly, resulting in a strong dynamic effect, it may already be realized before the effect of a relatively larger droplet occurs; the resulting effect can be turbulization of the fluid volume, an increase in heat flow, or emulsion homogenization.

It is essential that even at the stage of monotonic cluster expansion when the bubble growth rate does not exceed 2 m/s, the shear rates in liquid local zones are quite high—in the order of 7×10^3 s^{-1}, comparable to the shear rates achieved in homogenizers. Thus, if a high level of thermal action is realized only at the stage of maximum cluster bubble compression, at a time interval of tens of nanoseconds, then the effective influence of shear stresses manifests itself during the entire cluster lifetime. We can consider the additional effect of liquid temperature on the above processes in terms of, for example, homogenization technologies. We assume that the creation of cavitation effects through increasing the superheated pressure with a subsequent sharp decrease will create anomalous dynamic effects at the boundaries of growing vapor bubbles (at the moment of their explosion), which make it possible to crush all the dispersed inclusions in a liquid. We refer to this method as the thermal cavitation homogenization; Figure 1 shows the effect of such an initiation (Figure 1c), as a result of which a homogeneous emulsion is obtained.

However, the method described in [27], used for the assessment of microflow intensity and resulting dynamic effects, does not ensure a result ready for practical use but only reflects some of the physical processes inherent in the model. To assess the real practical effects, it is necessary to modify the model, as proposed in [27], taking into account the real structure of the liquid.

Thus, cavitation effects can also be used for breaking secondary phases in liquids. To do this, it is necessary to develop a quantitative assessment for these phenomena, which would also be suitable for their management.

3. Determination of a Sufficient Level of Supplied Energy to Intensify Heat and Mass Transfer Processes

There is a fundamental difference in the application of methods to affect a dispersed system to crush the dispersed phase and intensify transfer processes. Dispersion breaking is a one-stage action that can either take place instantly, if the intensification level is high enough, or it will not proceed at all, even with an infinitely long energy impact, if the intensification level is below a critical value. Heat and matter transfer in the system, on the contrary, proceed even at the lowest levels of intensification, and any increase in external action only proportionally increases the speed of these processes. However, transfer processes at a certain energy level can also intensify a mass transfer (breaking processes subject to the conversion of thermal energy into kinetic energy).

If, for example, the problem is solved using a certain mixing device to create an emulsion with a given particle size of the dispersed substance in a small container, then you can stir the liquid mixture as long as you like and spend a huge amount of energy, but the required result cannot be achieved if the speed of the mixer (introduced power W_0) is not high enough. At $\Delta\tau \to \infty$ energy, consumption is $E = W \times \Delta\tau \to \infty$. It is enough to slightly increase the number of revolutions to achieve the required power level W_{eff} and under specified conditions, an emulsion with a given particle distribution will be obtained in a very short time $\Delta\tau_{min}$ with low energy consumption $E = W_{eff} \times \Delta\tau_{min}$.

In this example, the result is achieved with the mandatory exceeding of a certain power level W_{eff} in the working volume of the apparatus, whilst W_{eff} and $\Delta\tau_{min}$ are mutually independent parameters.

Thus, with the intensification of mass transfer processes, a maximum power value is also obtained. Therefore, when intensification methods are used to break up dispersions, the specific power level must exceed a certain value $\widetilde{W}_{eff}$ to ensure a positive effect of the operation. To obtain the required effect with minimal energy consumption, it is necessary to provide a specified power level ($W \geq \widetilde{W}_{eff}$) for a short time ($\Delta\tau \rightarrow \Delta\tau_{min}$) or, in other words, convert the input energy into short but powerful impulses.

In this work, within the framework of the model of cavitation bubble growth, the authors studied velocity fields and pressures in the liquid phase (in emulsion) at the final stage of destruction (explosion) of the cavitation cluster, i.e., at the moment of transformation of the liquid thermal energy into the kinetic energy of motion at the vapor cluster boundary. These processes take place in zones of low surface tension—at the interface between emulsion phases (Figure 3). When the vapor cavity expands, the pressure decreases and, accordingly, the vapor temperature. Compression is accompanied by an increase in temperature and pressure as well as the corresponding phase transitions. These processes are considered by the authors to be an "internal" task, well addressed in previously published works (e.g., [26,27]). An abrupt expansion of the vapor region boundary causes a dynamic effect on the neighboring droplets and, as a result, they can split into smaller (thermodynamically stable) clusters if the condition $W \geq \widetilde{W}_{eff}$ is met. The combined effect of these phenomena and the external effects, described in Sections 1 and 2, provide a significant dynamic effect that can be used practically in homogenization technologies.

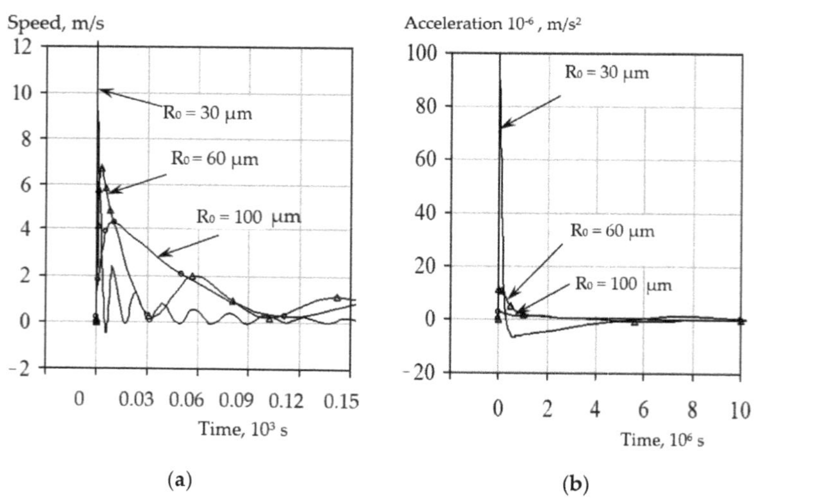

Figure 3. The rate (**a**) and acceleration (**b**) changes of the liquid–vapor interface for different initial droplet radii.

The mathematical models, presented in the works cited above, enable qualitative assessment of the described phenomena; however, they cannot be used to determine the effective power in homogenization technologies and to predict the effect when such a power is introduced into the emulsion.

4. An Estimation Procedure for Dynamic Effects in a Cluster during the Liquid Boiling Estimation of the Energy Input Level

We first demonstrate how to estimate the effects that can be achieved when energy is input into the emulsion volume, e.g., by preheating and rapid pressure release, for a single low-boiling liquid droplet.

Calculations have been shown in Figure 3, using the mathematical model given in [27], taking into account the modified Equation (1) to determine the acceleration of the vapor–liquid interface and the conditions below. The superheating temperature of the primary emulsion (Figure 1) is 180 °C, for droplet sizes of the dispersed phase shown in Figure 2.

The resulting acceleration, taking into account (1), is calculated as follows:

$$g(x,\tau) = \frac{dv}{d\tau}; \tag{2}$$

$$g(x,\tau) = \sum_{i=0}^{n} \left[p_{int} - p_{ext} + 1.5(p_{int} - p_{ext})^2 k_\beta - 1.5\rho v^2 \right] \frac{(x_i - x)R_i}{\rho \times d^3} \tag{3}$$

where x_i are the initial coordinates of droplet centers; $d_i = |x_i - x|$ is the radius vector; x is a coordinate of the acceleration calculation point; v is the oil–vapor interface motion rate for the i-th droplet; and R_i is the radius of the oil–vapor interface of the i-th droplet. The higher the acceleration of the phase interface, the higher the force that occurs when water starts boiling.

The graphs show that the smaller the initial droplet radius is, the more intensively the vapor phase grows, i.e., the interface rate and acceleration are higher. Therefore, in real emulsions where dispersed particles of different diameters are present, remote from each other at different distances, the dynamic effects during boiling of the thermolabile emulsion part (water in our example) will have different dynamics of cavitation effects. In this case, the idealized picture of dynamic effects shown in Figure 2 and, accordingly, their assessment methods, cannot be applied. At different droplet radii, both joint maximums (minimums) of the rate (acceleration) and significant differences in these rates (accelerations) can be observed. Considering that rates (accelerations) for different initial radii have different amplitudes and frequencies of variation in time, it can be assumed that at some point in time Kelvin–Helmholtz instability is possible, and, for acceleration, Rayleigh–Taylor instability.

Let us analyze the conditions that can result in a certain instability. By joint analysis of the Bond and Weber criterion [27], as well as their critical values, with the Rayleigh–Plesset Equation, it follows that critical forces, resulting in Rayleigh–Taylor or Kelvin–Helmholtz instability, respectively, are as follows:

$$F_{Bo}^{cr} = 40\pi\sigma R_i \tag{4}$$

$$F_{We}^{cr} = 30\pi\sigma R_i. \tag{5}$$

Comparison of (4) and (5) shows that $F_{Bo}^{cr} = 1.33 F_{We}^{cr}$, but these forces are caused by various factors, and they cannot be identified.

Deformation and breaking of both the vapor layer and water droplets in water–oil emulsions can be caused by different directions of both the acceleration vector and velocity vector. Let us assume that deformation, breaking, or displacement will occur only if the acceleration vector has a positive direction and, regardless of it, the rate vector is also positive. If we consider the mutual influence of droplets No. 1 and No. 2 (Figure 4) located at a distance l, i.e., $x_1 = 0$, $x_2 = l$, then the acceleration tending to destruct the interface of droplet No. 1 is

$$g(x,\tau)_{p1} = \begin{cases} \sum\limits_{i=1}^{2} g(x,\tau)_i; k_1 \geq 0; k_2 \geq 0; \\ -\sum\limits_{i=1}^{2} g(x,\tau)_i; k_1 \leq 0; k_2 \leq 0; \\ g_{(x,\tau)2}; k_1 < 0; k_2 > 0; \\ -g(x,\tau)_2; k_1 > 0; k_2 < 0. \end{cases} \tag{6}$$

where g_i is determined by Equation (7);

$$k_i = \left[p_{int} - p_{ext} + 1.5(p_{int} - p_{ext})^2 k_\beta - 1.5\rho v^2\right]; \ d_i = |x_i - R_i|. \tag{7}$$

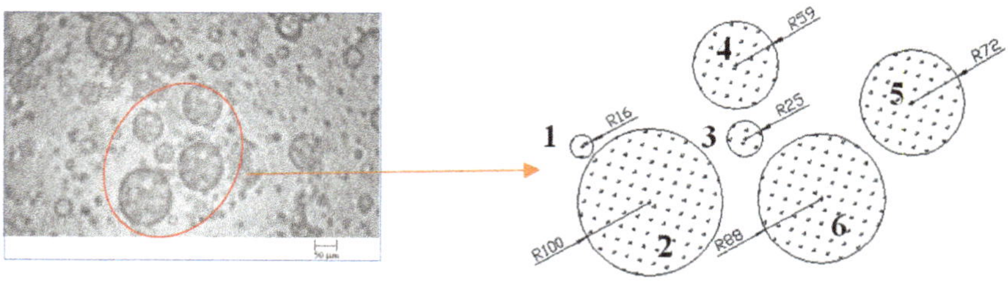

Figure 4. To the calculated model of droplet breakup of the emulsion dispersed phase (characteristic dimensions in microns).

Then, the force caused by the flow acceleration or deceleration is

$$F_{Bo_1} = 4\pi\rho_1 g_1 R_1^3 \tag{8}$$

Similarly, it can be written for the rate

$$v_{p1} = \begin{cases} \sum\limits_{i=1}^{2} v_{ri}; v_1 \geq 0; \ v_2 \geq 0; \\ -\sum\limits_{i=1}^{2} v_{ri}; v_1 \leq 0, \ v_2 \leq 0; \\ v_{r2}; v_1 < 0, v_2 > 0; \\ -v_{r2}; \ v_1 > 0, v_2 < 0. \end{cases} \tag{9}$$

where $v_{ri} = v_i R_i^2 \frac{(x_i - R_i)}{d_i^2}$.

Then the dynamic head force is

$$F_{We1} = 6\pi\rho_1 R_1^2 |v_{p1}| v_{p1} \tag{10}$$

As is known, the capillary force F_σ, regardless of a deformed droplet shape, is always directed so as to restore the spherical shape, i.e., in order for Gibbs free energy to be minimal. Therefore, if a deformed droplet takes the shape of an oblate ellipsoid, the capillary force opposes the external force. If it is deformed into an oblong ellipsoid, then the capillary force coincides in direction with the external force and the droplet will be unstable, while the shape of an oblate ellipsoid can be quasi-stable.

The minimum of a given capillary force is possible. We will assume that this minimum is determined by the Laplace force:

$$F_{\sigma 1} = 8\pi\sigma R_1 \tag{11}$$

Thus, if the dynamic action from the neighboring droplet exceeds the force F_σ, the droplet will be deformed or broken; if the forces are equal, equilibrium will occur; and if $F_\sigma > F_{Bo,We}$, repulsion without deformation will take place.

Differences between $F_{Bo,We}$ and F_σ are equal:

$$\Delta F_{Bo} = F_{Bo} - F_\sigma;\ \Delta F_{We} = F_{We} - F_\sigma \tag{12}$$

Following the example of the force, resulting from flow acceleration or deceleration, the following conditions can be written: $\Delta F_{Bo} \geq F^{cr}_{Bo}$—breaking; $\Delta F_{Bo} \geq 0$—deformation; $\Delta F_{Bo} = 0$—equilibrium; and $\Delta F_{Bo} < 0$—displacement.

The system of Equations (7)–(11) for particle No. 2, located at a distance of l from the first particle, will take the following form:

$$g(x,\tau)_{p2} = \begin{cases} -\sum\limits_{i=1}^{2} g(x,\tau)_i; k_1 \geq 0; k_2 \geq 0; \\ \sum\limits_{i=1}^{2} g(x,\tau)_i; k_1 \leq 0; k_2 \leq 0; \\ g_{(x,\tau)}; k_1 < 0; k_2 > 0; \\ -g(x,\tau)_1; k_1 > 0; k_2 < 0; \end{cases} \quad v_{p2} = \begin{cases} -\sum\limits_{i=1}^{2} v_{ri}; v_1 \geq 0; v_2 \geq 0; \\ \sum\limits_{i=1}^{2} v_{ri}; v_1 \leq 0, v_2 \leq 0; \\ v_{r1}; v_1 < 0, v_2 > 0; \\ -v_{r1}; v_1 > 0, v_2 < 0; \end{cases} \tag{13}$$

where $g(x,\tau)_i = k_i \frac{R_i(x_i - j)}{\rho d_i^3}$; $v_{ri} = v_i R_i^2 \frac{(x_i - j)}{d_i^3}$; and $d_i = |x_i - j|; j = l - R_2$.

$$F_{Bo_2} = 4\pi\rho_2 g_2 R_2^3 \tag{14}$$

$$F_{We2} = 6\pi\rho_2 R_2^2 |v_{p2}| v_{p2} \tag{15}$$

$$F_{\sigma 2} = 8\pi\sigma R_2 \tag{16}$$

Thus, dynamic effects from neighboring droplets of low-boiling liquid can make them move or break if critical values of the ΔF_{Bo} criterion are reached. If opposing forces are approximately equal, abnormal pressure can be observed, as pointed out by the authors of [16–18]. However, each liquid droplet forms its own local dynamic effect during boiling and therefore the effect of all cluster particles shall be considered.

5. Condition for the Formation of a Superposition of Dynamic Effects and the Resulting Effect

The above considerations are valid for the determination of the resulting force sufficient to break the droplet. However, different situations of cluster development are possible. For instance, if equal forces act on the droplet (No. 3, Figure 4) from all sides, or their resultant does not exceed F^{eff}_{Bo}, in this local vapor volume, it is possible to achieve the anomalous pressure and temperature values mentioned at the beginning of this paper. It is also important to evaluate the influence of the dynamic effect angle of action on the droplet surface.

In this part, breaking processes of dispersed phase droplets, surrounded by other droplets during their boiling up, will be analyzed. We take the emulsion shown in Figures 1, 4 and 5 as a study object. This figure shows that there are "small" droplets between "large" droplets: the former serve as sources of increased dynamic forces at boiling up, thus initiating breaking processes of larger droplets due to significant differences in accelerations and growth rates between them.

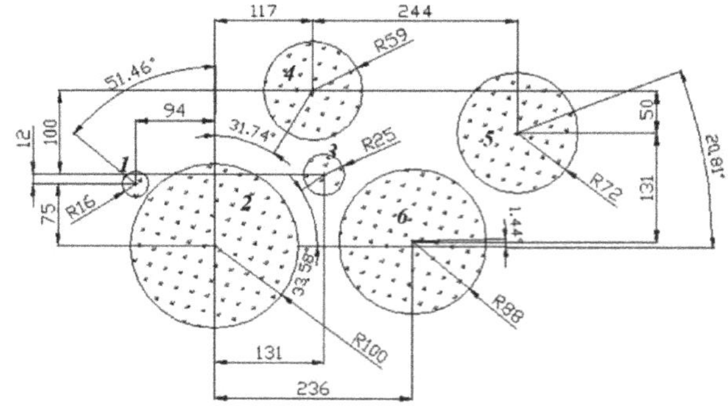

Figure 5. To the calculated model of droplet breakup of the emulsion dispersed phase (characteristic dimensions in microns).

The main types of instability emulsion structure will be considered: Kelvin–Helmholtz instability caused by the difference in rates and Rayleigh–Taylor instability, resulting from the difference in accelerations [27].

When analyzing the possibility of dynamic effects, one can assume that the main role will be played by the forces acting along a normal component to the droplet surface, i.e., forces directed either towards the droplet center or away from it. Thus, the acceleration and rate, acting on the particle surface and directed towards its center (from the center) can be determined from the following Equations:

$$g_{nr}(x_0, y_0) = g(x_0, y_0)k_{nr} \tag{17}$$

$$v_{nr}(x_0, y_0) = v(x_0, y_0)k_{nr} \tag{18}$$

where x_0, y_0 are coordinates of the droplet surface point and k_{nr} is a correction for the normal component.

Correction k_{nr} can be determined by Figure 6:

$$k_{nr} = cos\gamma = \frac{(x_i - x_0)sin\beta + (y_i - y_0)cos\beta}{d_i(x_0, y_0\,)}. \tag{19}$$

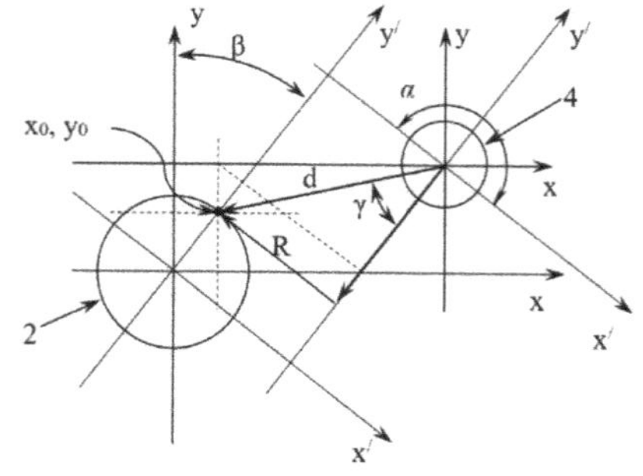

Figure 6. Determination of correction for the normal component and angle α.

The authors assume that acceleration and its resulting force are positive if they are directed towards the center and negative if they are directed away from the center. The rate (acceleration) vector of the liquid–vapor interface of the considered boiling up particle is considered positive if it is directed from the droplet center and negative if it is directed to the center.

The acceleration and velocity of the phase interface movement are determined by the following expressions:

$$g_{nr}(x_0, y_0) = \sum_{i=1}^{N-1} g_i(x_0, y_0) k_{nr_i}; \tag{20}$$

$$v_{nr}(x_0, y_0) = \sum_{i=1}^{N-1} v_i(x_0, y_0) k_{nr_i}, \tag{21}$$

where N is the total number of boiling up dispersed phase droplets.

Taking angle β as a variable (Figure 6), the coordinates of the required opposite points of the particle surface may be represented, where dynamic effects are directed to

$$\begin{gathered} x_{s1} = x + \Delta x;\ y_{s1} = y + \Delta y; \\ x_{s2} = x - \Delta x;\ y_{s2} = y - \Delta y; \\ \Delta x = Rsin\beta;\ \Delta y = Rcos\beta;\ 0 \le \beta \le 180^{\circ} \end{gathered} \tag{22}$$

where x, y are the particle coordinates; x_{s1}, y_{s1} are coordinates of the droplet side surface ($0 \le \alpha \le 180$) from Figure 6; and x_{s2}, y_{s2} are the opposite side coordinates ($180 \le \alpha \le 360$).

Then, correction k_{nr} will be determined by the following expressions:

$$k_{nr_1} = \frac{(x_i - x_{s1})sin\beta + (y_i - y_{s1})cos\beta}{d_i(x_{s1}, y_{s1})}; 0 \le \beta < 180^{\circ}; \tag{23}$$

$$k_{nr_2} = \frac{(x_i - x_{s2})sin\beta + (y_i - y_{s2})cos\beta}{d_i(x_{s2}, y_{s2})};\ 180 \le \beta < 360^{\circ}. \tag{24}$$

In Equation (24), x_{s2}, y_{s2} are calculated, taking into account $0 \le \alpha \le 180$ and k_{nr} at $\beta = \beta + 180$.

Let us consider how the magnitude and direction of the force acting on the surface of the analyzed droplets (Figures 4 and 5) changes. Its value is determined by the acceleration of the vapor cavity boundary when the pressure of the emulsion preheated to $t_0 \approx 105$ °C decreases, to an atmospheric pressure equal to 10^5 Pa. For all droplets, a vapor layer equal to 1 μm at the initial moment of time is assumed. The acceleration change pattern on the surface of the analyzed droplets shows in which direction and how the resulting force will change. Calculation results for water–oil emulsion according to Equations (20)–(24) are shown in Figures 7 and 8.

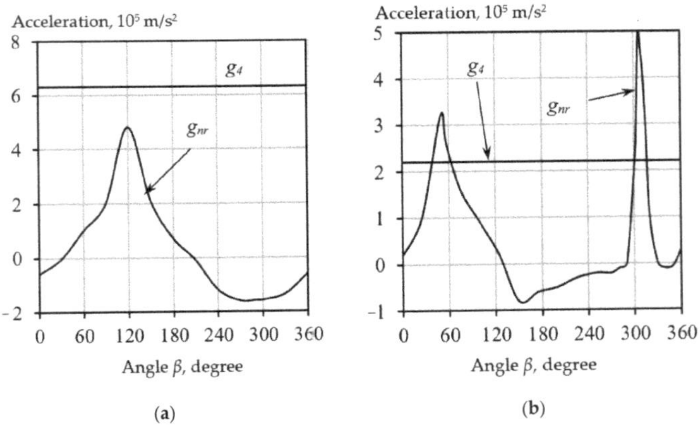

Figure 7. The pattern of acceleration, acting on the surfaces of droplet No. 1 (**a**) and droplet No. 2 (**b**) (Figure 4) and the resulting acceleration flow from neighboring boiling up particles (g_{nr}), acceleration of the oil–vapor interface of the particle itself (g_4) at initial pressure relief from 0.125 MPa down to 0.1 MPa.

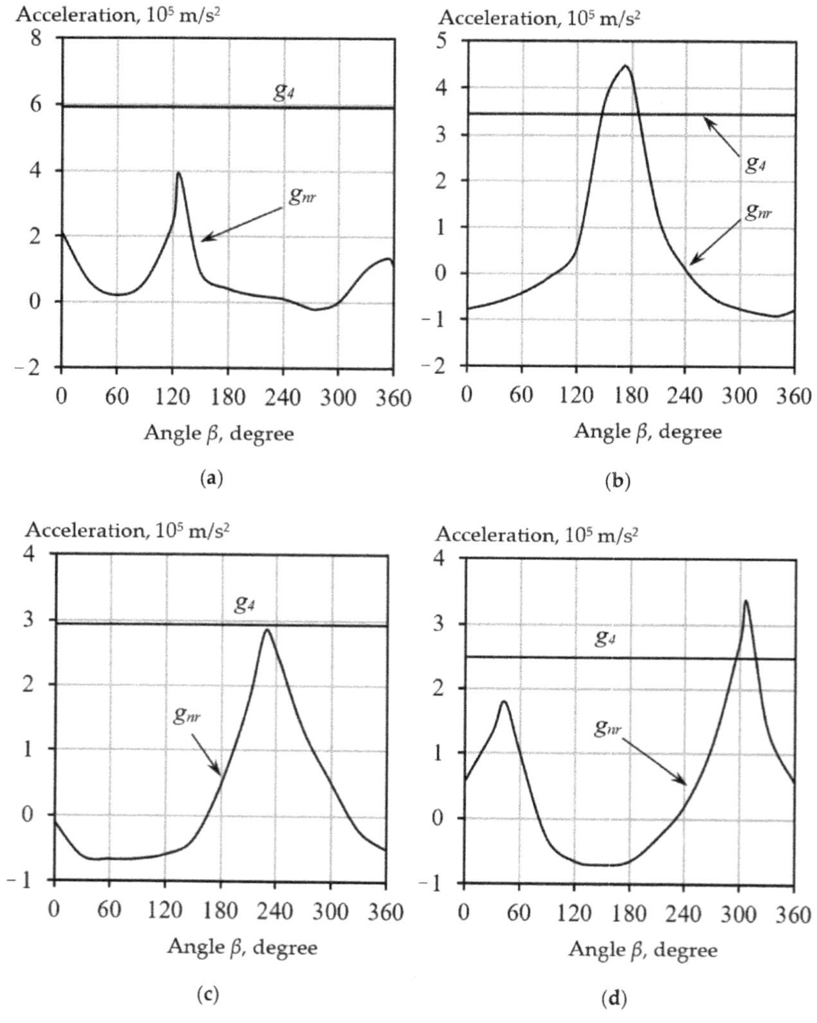

Figure 8. The pattern of acceleration of the interface near the droplet No. 3 (**a**), No. 4 (**b**), No. 5 (**c**), and No. 6 (**d**) with a decrease in pressure.

The graphs clearly show the relationship between the angles of the droplets and the acceleration values of the interface. Thus, for example, for droplet No. 2 (Figure 4) there are two acceleration maximums at once: angle $\beta \approx 55^\circ$ and $\beta \approx 315^\circ$, corresponding to angles of the nearest neighboring droplet No. 1 and No. 3 arrangement, respectively. As the Figures show, there are always several acceleration (force) maximums.

Let us analyze a boiling emulsion droplet (Figure 9) affected by external forces.

If the maximum of two forces acting on opposite sides is $F' = F'_{s1} + F'_{s2}$ and is at an angle of β_1 to axis y, the maximum force acting on one side is F_{s1} and at an angle of β_2 from axis y ($F' > F$, where $F = F_{s1} + F_{s2}$). The force acting from the droplet side is F, with $F > F'_{s1}$, $F > F'_{s2}$, $F > F_{s2}$, $F < F_{s1}$. Then, the opposing force will be able to mitigate all forces, except F_{s1}. Therefore, only one difference will determine the resulting force, which can cause either deformation or breaking. This force is equal to $\Delta F = F_{s1} - F$.

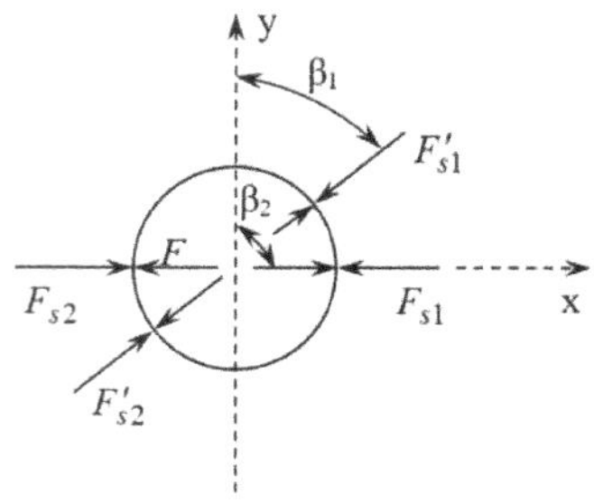

Figure 9. Forces acting on the emulsion droplet (designations are given in the text).

Thus, it is obvious that though the maximum of two opposing forces is at angle β_1 to axis y, the determining maximum is the one caused by force F_{s1} (angle β_2).

This is the distinguishing feature of our method used for the assessment of dynamic effects from those cited in Sections 1 and 2. They basically do not consider the dynamic effects from all cluster droplets and the superposition principle implies the coincidence of the force vector direction, as shown in Figure 2. However, a (secondary) liquid droplet can be deformed or even destructed when tangential stresses arise inside the cluster. Therefore, the issue of what initiation level should be used to destruct the clusters and what sizes of secondary liquid droplets in the emulsion will be formed under such an action is still to be solved.

For a non-boiling droplet, the force acting on its surface, considering the assumed conditions, is determined by the following acceleration:

$$\Delta g = \begin{cases} g_{s1} + g_{s2}; z > 0; \\ g_{s1}; z < 0; \ |g_{s1}| > |g_{s2}|; \\ g_{s2}; z < 0; \ |g_{s2}| > |g_{s1}|, \end{cases} \tag{25}$$

where $z = \frac{g_{s1}}{g_{s2}}$ is a coefficient, considering the direction of the acceleration action; g_{s1} is acceleration, acting on the inclusion surface at $0 \leq \alpha < 180^{\circ}$; and g_{s2} is acceleration, acting on the inclusion surface at $0 \leq \alpha < 360^{\circ}$.

The force that can result in breaking of a non-boiling inclusion of a dispersed phase is determined by the maximum acceleration:

$$g_{\dot{a}} = max\{|\Delta g|\}, \ 0 \leq \alpha < 180^{\circ} \tag{26}$$

For a boiling up particle, two force maximums are determined due to accelerations, acting on one side of the inclusion:

$$g_{d_1} = max\{|g_{si}|\}, \ g_{si} = \begin{cases} g_{s1}; \ |g_{s1}| > |g_{s2}|; \\ g_{s2}; \ |g_{s1}| < |g_{s2}|; \ 0 \leq \beta < 180^{\circ}; \end{cases} \tag{27}$$

and on opposing sides of the inclusion:

$$g_{s1,s2} = \begin{cases} g_{s1} + g_{s2}; \ z > 0; \\ g_{s1}; z < 0; \ |g_{s1}| > |g_{s2}|; \\ g_{s2}; \ z < 0; \ |g_{s2}| > |g_{s1}|; \end{cases} \quad g_{d_2} = max\{|g_{s1,s2}|\}, \ 0 \leq \beta < 180^{\circ} \tag{28}$$

The acceleration maximums g_{d_1} and g_{d_2} correspond to accelerations g_{s1} and g_{s2}, respectively.

Considering that the droplet dynamic effect cannot destruct the droplet itself, it is possible to write Equations to calculate the acceleration and rates of flow in the cluster resulting in its destruction. If the resulting acceleration or rate is positive, then they cause instability, and if they are negative, then there is no destabilizing effect. One may write

$$g_{p1} = \begin{cases} g_{s1} - g_4;\ g_4 \geq 0;\ g_{s1} \geq 0; \\ -g_{s1};\ g_4 > 0;\ g_{s1} < 0; \\ g_{s1};\ g_4 < 0;\ g_{s1} > 0; \\ g_4 - g_{s1};\ g_4 \leq 0;\ g_{s1} \leq 0, \end{cases} \tag{29}$$

where g_{p1} is the resulting acceleration, acting on one side of the inclusion and g_4 is the acceleration of the oil–vapor inclusion interface.

For the opposite side, one can write an Equation similar to (29) but inserting g_{p2} and g_{s2} instead of g_{p1} and g_{s1}, respectively.

Then, the resulting acceleration will be determined by the following Equation:

$$\Delta g_d = g_{p1} + g_{p2} \tag{30}$$

with the following required conditions:

$$\begin{aligned} g_{p1} &= \begin{cases} 0; z < 0;\ |g_{s1}| - |g_{s2}| < 0; \\ 0; z > 0: \begin{cases} g_4 > 0;\ g_{s1} > 0;\ g_{s2} > 0; g_{s1} - g_4 < 0;\ g_{s2} - g_4 > 0; \\ g_4 < 0;\ g_{s1} < 0;\ g_{s2} < 0; g_{s1} - g_4 > 0;\ g_{s2} - g_4 < 0; \end{cases} \end{cases} \\ g_{p2} &= \begin{cases} 0; z < 0;\ |g_{s1}| - |g_{s2}| > 0; \\ 0; z > 0: \begin{cases} g_4 > 0;\ g_{s1} > 0;\ g_{s2} > 0; g_{s1} - g_4 > 0;\ g_{s2} - g_4 < 0; \\ g_4 < 0;\ g_{s1} < 0;\ g_{s2} < 0; g_{s1} - g_4 < 0;\ g_{s2} - g_4 > 0. \end{cases} \end{cases} \end{aligned} \tag{31}$$

If the force maximums (accelerations, rates) do not coincide, acting on the opposite sides of the inclusion and on the one side, we consider that the total force that is of greater importance is determined by

$$g_d = \begin{cases} \Delta g_{d1};\ \Delta g_{d1} > \Delta g_{d2}; \\ \Delta g_{d2};\ \Delta g_{d1} < \Delta g_{d2}, \end{cases} \tag{32}$$

where $g_{d1}, \Delta g_{d2}$ is the resulting acceleration.

Similarly, we can get an Equation for the rate. At the same time,

$$\Delta v = k\sqrt{|v_p|} \tag{33}$$

where $k = \begin{cases} 1,\ v_p \geq 0; \\ -1,\ v_p > 0;\ v_p = v_{p1}|v_{p1}| + v_{p2}|v_{p2}|. \end{cases}$

It is generally accepted that the structural stability of the emulsion is determined by the instability according to the Bond criterion [27]. The force, acting at $0 \leq \alpha < 180^{\circ}$, is equal to

$$F_{g1} = 4\pi R_4^3 \rho g_{s1} \tag{34}$$

For $180^{\circ} \leq \alpha < 360^{\circ}$,

$$F_{g2} = 4\pi R_4^3 \rho g_{s2} \tag{35}$$

The opposing force from the oil–vapor interface of the droplet itself is

$$F_g = 4\pi R_4^3 \rho g_4 \tag{36}$$

The resulting force acting on the droplet is

$$F_{Bo} = 4\pi R_4^3 \rho g_d \tag{37}$$

The critical force, resulting in Bond instability, is

$$F_{Bo}^{eff} = 40\pi\sigma R_4. \tag{38}$$

The force determining a droplet deformation or displacement is

$$\Delta F_{Bo} = F_{Bo} - 8\pi\sigma R_4 \tag{39}$$

For each of the analysed droplets, the calculation is performed either until the condition

$$F_{Bo} \geq F_{Bo}^{eff} \tag{40}$$

is met, or until the mutual meeting of droplets, determined by condition

$$\Delta R_{i,j} \leq 0, \tag{41}$$

where $\Delta R_{i,j} = d_{i,j} - (R_{4i} + R_{4j})$, $d_{i,j} = \sqrt{(x_i - x_j)^2 + (y_i - y_j)^2}$; $i = 1, 2, \ldots, N$; $j = i + 1, i + 2, \ldots, N$; and i, j are droplet numbers with corresponding coordinates $x_i, y_i; x_j, y_j$.

6. Results and Discussion

The calculation results are presented in Figures 10–12. The performed calculations (t_0 = 110 °C) showed that only droplets No. 2 and No. 6 will be destroyed at the initial moment of pressure release. This means that the level of the supplied effective power by emulsion superheating to the specified temperature is sufficient to break all emulsion droplets larger than droplet No. 6. To break smaller droplets, the superheat temperature must be increased.

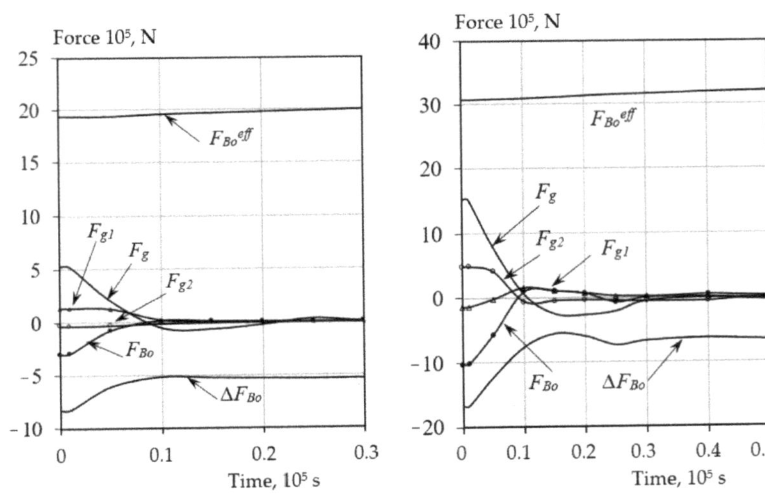

Figure 10. Changes in the forces acting on the surfaces of droplets No. 1 (**a**) and No. 3 (**b**) (Figure 4) in time.

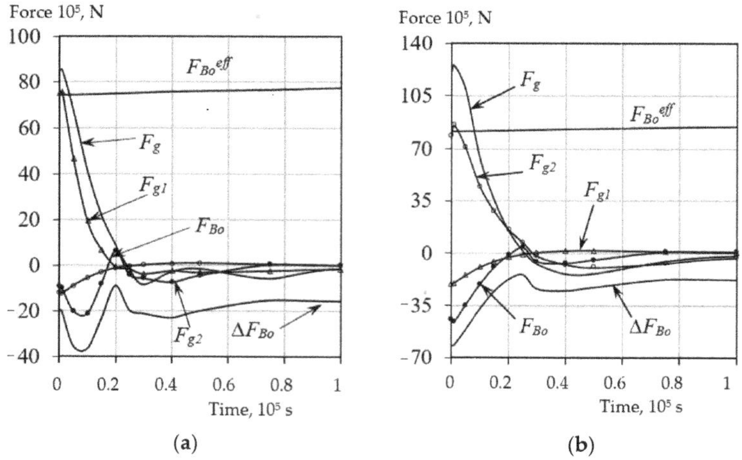

Figure 11. Changes in the forces acting on the surfaces of droplets No. 4 (**a**) and No. 5 (**b**) (Figure 4) in time.

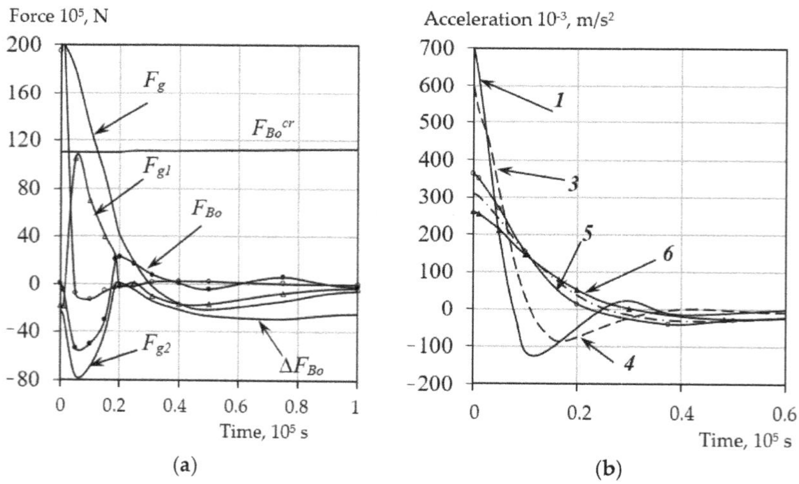

Figure 12. Change in the forces acting on the surface of droplet No. 6 (**a**) and change of oil–vapor interface acceleration (**b**) of analysed droplets (Figure 4) in time.

The joint consideration of all the dynamic effects in a cluster makes it possible to determine the force F_{Bo} that can correspond to the level of effective impact on the cluster, being the cause of the dispersed phase breaking. The figures show that for droplets No. 1, 3, 4, and 5 with a positive value of force F_{Bo}, only their mutual displacement is possible, as indicated by a negative sign (at $\tau \approx 2{\cdot}10^{-6}$ s).

To destroy these droplets, it is necessary to increase the supplied energy to the effective one. It can be achieved by raising superheating temperature and pressure (thermal cavitation initiation). The same effect can be obtained by using mixing devices, where local cavitation zones can appear on the blades at certain flow rates.

The calculations performed for overheating temperatures t_o = 130 °C and t_o = 170 °C demonstrated that, at the moment of instantaneous pressure release, almost all large droplets will be broken, while at t_o = 110 °C only two droplets will be broken due to boiling small neighboring particles.

As for the intensification of heat transfer processes, attention should be paid to the effect of heat accumulation by the high-boiling liquid. With a sudden drop in pressure of the superheated emulsion, the low-boiling liquid, e.g., water, evaporates partially or completely, depending on superheating level. The vapor temperature decreases during expansion. At the same time, a high-boiling fraction of the emulsion transfers the stored energy to vapor and water. Thus, we get an additional impulse of energy from the high-boiling phase. Temperature T can be determined if we know the total liquid volume, the proportion of emulsion phases, and the size and number of the low-boiling fraction droplets. For example, if emulsion volume V is known to contain 30% water and 70% fuel oil, the number of water particles is

$$N = \frac{0.75V}{\pi R^3} \tag{42}$$

where R is the water droplet radius.

Then variation of fuel oil temperature in time is

$$\frac{dE}{d\tau} = 4\pi R_3^2 N \frac{Q}{cm} \tag{43}$$

where m, c are fuel oil mass and heat capacity and Q is the fuel oil heat flow to be determined by the methods proposed in [38].

For example, for an emulsion volume $V = 0.3 \times 10^{-3}$ m^3 with 30% water and an average droplet size of $R = 100$ μm, the number of these droplets is approximately $N = 1.8 \times 10^7$. The calculation results are shown in Figure 13.

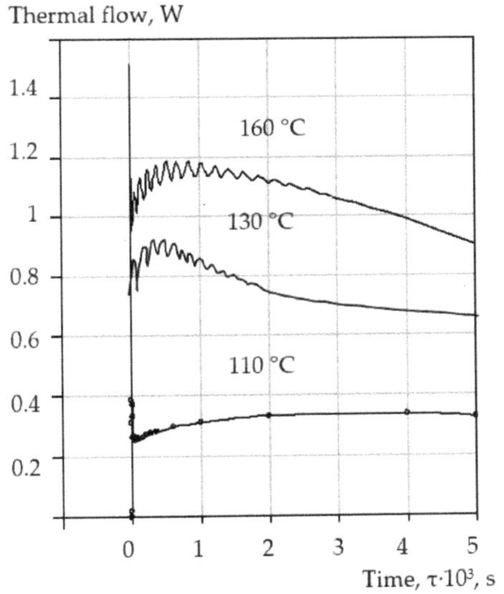

Figure 13. Local effect of the change in heat flow from high-boiling emulsion fraction in time.

Figure 13 shows that heat flow for the considered example is approximately 1 W per particle. As the number of particles is $N = 1.8 \times 10^7$, we can conclude that the heat exchange process is high-intensive, which would allow more heat to be removed or supplied in standard heat exchangers. Clearly, such effects can only be achieved with certain structural parameters of the emulsion, which can easily be determined using this method.

7. Conclusions

The study of hydrodynamic cavitation phenomena in relation to their targeted and effective use for technological process stimulation and intensification requires a unified approach that equally considers the hydrodynamic and heat and mass transfer aspects of these phenomena. Within the framework of this concept, we proposed a new method to assess dynamic effects. This method is based on an advanced model, considering all determining factors and an accurate representation of thermophysical system parameters, which relevantly describes the behavior of bubbles and bubble ensembles in the boiling and cavitation processes. The study results show that the cavitation effect analysis will involve liquid transition around the bubble, as well as the vapor–gas mixture in the bubble itself, into the supercritical fluid state. To assess the spatial and time boundaries of the supercritical region and to analyze the effects of a spherical hydraulic hammer in local zones of the cavitation cloud, the liquid compressibility will be considered in modeling.

The method for assessment of thermal cavitation effects was validated using superheated emulsion boiling with a sharp decrease in pressure. At the interface of the liquid phases (in the region of low surface tension, water in our example), vapor fluids are formed with rapidly increasing volumes. Thus, conditions for dynamic effects on the emulsion structure are provided. This method of local supply of energy W_{eff} makes it possible to implement the task of heat and mass transfer or homogenization process intensification. Moreover, the effect level is easy to regulate using the temperature and pressure of the emulsion preheating. The same phenomena can be used in jet devices.

This study should be considered as the initial stage of substantiation of rational designs and optimal operating modes of cavitating devices as a solution for solving various technological problems.

Author Contributions: Conceptualization, A.M.P. and H.K.; methodology, A.M.P.; validation, A.M.P. and H.K.; data curation, A.M.P.; writing—original draft preparation, A.M.P.; writing—review and editing, A.M.P.; visualization, A.M.P. All authors have read and agreed to the published version of the manuscript.

Funding: This research received no external funding.

Institutional Review Board Statement: Not applicable.

Informed Consent Statement: Not applicable.

Data Availability Statement: Not applicable.

Conflicts of Interest: The authors declare no conflict of interest.

References

1. Fuster, D.; Dopazo, C.; Hauke, G. Liquid compressibility effects during the collapse of a single cavitating bubble. *J. Acoust. Soc. Am.* **2011**, *129*, 122–131. [CrossRef] [PubMed]
2. Albanese, L.; Ciriminna, R.; Meneguzzo, F.; Pagliaro, M. Energy efficient inactivation of Saccharomyces cerevisiae via controlled hydrodynamic cavitation. *Energy Sci. Eng.* **2015**, *3*, 221–238. [CrossRef]
3. Zevnik, J.; Dular, M. Cavitation bubble interaction with a rigid spherical particle on a microscale. *Ultrason. Sonochem.* **2020**, *69*, 105252. [CrossRef]
4. Gogate, P.R.; Pandit, A.B. Engineering design methods for cavitation reactors II: Hydrodynamic cavitation. *AIChE J.* **2000**, *46*, 1641–1649. [CrossRef]
5. Kanthale, P.M.; Gogate, P.R.; Pandit, A.B. Dynamics of gravitational bubbles and design of a hydrodynamic cavitational reactor Ultrason. *Sonochem.* **2005**, *12*, 441–452. [CrossRef] [PubMed]
6. Li, J.; Cheng, P. Bubble cavitation in a microchannel. *Int. J. Heat Mass Transf.* **2004**, *47*, 2689–2698. [CrossRef]
7. Melikhov, V.; Yakush, S.; Le, T. Evaluation of energy and impulse generated by superheated steam bubble collapse in subcooled water. *Nucl. Eng. Des.* **2020**, *366*, 110753. [CrossRef]
8. Roesle, M.; Kulacki, F. An experimental study of boiling in dilute emulsions, part A: Heat transfer. *Int. J. Heat Mass Transf.* **2012**, *55*, 2160–2165. [CrossRef]
9. Podnar, A.; Dular, M.; Sirok, B.; Hocevar, M. Experimental Analysis of Cavitation Phenomena on Kaplan Turbine Blades Using Flow Visualization. *J. Fluids Eng.* **2019**, *141*, 071101. [CrossRef]

10. Chernin, L.; Val, D.V. Probabilistic prediction of cavitation on rotor blades of tidal stream turbines. *Renew. Energy* **2017**, *113*, 688–696. [CrossRef]
11. Pham-Thanh, N.; Van Tho, H.; Yum, Y.J. Evaluation of cavitation erosion of a propeller blade surface made of composite materials. *J. Mech. Sci. Technol.* **2015**, *29*, 1629–1636. [CrossRef]
12. Dietzel, D.; Hitz, T.; Munz, C.-D.; Kronenburg, A. Expansion Rates of Bubble Clusters in Superheated Liquids. In Proceedings of the Germany ILASS–Europe 2017, 28th Conference on Liquid Atomization and Spray Systems, Valencia, Spain, 6–8 September 2017. [CrossRef]
13. Warjito, W.; Budiarso, B.; Syahputra, M.F.; Nasution, S.B. The Effect of Blades Gap on Propeller Openflume Picohydro Turbine Performance. *Int. J. Fluid Mach. Syst.* **2021**, *14*, 122–131. [CrossRef]
14. Tran, B.N.; Jeong, H.; Kim, J.-H.; Park, J.-S.; Yang, C. Effects of Tip Clearance Size on Energy Performance and Pressure Fluctuation of a Tidal Propeller Turbine. *Energies* **2020**, *13*, 4055. [CrossRef]
15. Adhikari, R.C.; Vaz, J.; Wood, D. Cavitation Inception in Crossflow Hydro Turbines. *Energies* **2016**, *9*, 237. [CrossRef]
16. Janssen, D.; Kulacki, F. Flow boiling of dilute emulsions. *Int. J. Heat Mass Transf.* **2017**, *115*, 1000–1007. [CrossRef]
17. Ganesan, B.; Martini, S.; Solorio, J.; Walsh, M.K. Determining the Effects of High Intensity Ultrasound on the Reduction of Microbes in Milk and Orange Juice Using Response Surface Methodology. *Int. J. Food Sci.* **2015**, *2015*, 1–7. [CrossRef]
18. Chandrapala, J.; Oliver, C.; Kentish, S.; Ashokkumar, M. Ultrasonics in food processing – Food quality assurance and food safety. *Trends Food Sci. Technol.* **2012**, *26*, 88–98. [CrossRef]
19. Sun, X.; Chen, S.; Liu, J.; Zhao, S.; Yoon, J.Y. Hydrodynamic Cavitation: A Promising Technology for Industrial-Scale Synthesis of Nanomaterials. *Front. Chem.* **2020**, *8*, 259. [CrossRef]
20. Prajapat, A.L.; Gogate, P.R. Depolymerization of carboxymethyl cellulose using hydrodynamic cavitation combined with ultraviolet irradiation and potassium persulfate. *Ultrason. Sonochem.* **2019**, *51*, 258–263. [CrossRef] [PubMed]
21. Merzkirch, W.; Rockwell, D.; Tropea, C. *Orifice Plates and Venturi Tubes*; Springer International Publishing: Cham, Switzerland; Heidelberg, Germany; New York, NY, USA; Dordrecht, The Netherlands; London, UK, 2015. Available online: https://link.springer.com/content/pdf/bfm%3A978-3-319-16880-7%2F1.pdf (accessed on 17 July 2015).
22. Nigmatulin, R.I.; Taleyarkhan, R.P.; Lahey, R.T. Evidence for nuclear emissions during acoustic cavitation revisited. *Proc. Inst. Mech. Eng. Part A J. Power Energy* **2004**, *218*, 345–364. [CrossRef]
23. Sun, X.; Wang, Z.; Xuan, X.; Ji, L.; Li, X.; Tao, Y.; Boczkaj, G.; Zhao, S.; Yoon, J.Y.; Chen, S. Disinfection characteristics of an advanced rotational hydrodynamic cavitation reactor in pilot scale. *Ultrason. Sonochem.* **2021**, *73*, 105543. [CrossRef] [PubMed]
24. Albanese, L.; Baronti, S.; Liguori, F.; Meneguzzo, F.; Barbaro, P.; Vaccari, F.P. Hydrodynamic cavitation as an energy efficient process to increase biochar surface area and porosity: A case study. *J. Clean. Prod.* **2019**, *210*, 159–169. [CrossRef]
25. Badve, M.; Alpar, T.; Pandit, A.B.; Gogate, P.R.; Csoka, L. Modeling the shear rate and pressure drop in a hydrodynamic cavitation reactor with experimental validation based on KI decomposition studies. *Ultrason. Sonochem.* **2015**, *22*, 272–277. [CrossRef]
26. Pavlenko, A. Dispersed phase breakup in boiling of emulsion. *Heat Transf. Res.* **2018**, *49*, 633–641. [CrossRef]
27. Pavlenko, A. Energy conversion in heat and mass transfer processes in boiling emulsions. *Therm. Sci. Eng. Prog.* **2019**, *15*, 1–8. [CrossRef]
28. Gasanov, B.M.; Bulanov, N.V. Effect of the droplet size of an emulsion dispersion phase in nucleate boiling and emulsion boiling crisis. *Int. J. Heat Mass Transf.* **2015**, *88*, 256–260. [CrossRef]
29. Roesle, M.L.; Lunde, D.L.; Kulacki, F.A. Boiling Heat Transfer to Dilute Emulsions from a Vertical Heated Strip. *J. Heat Transf.* **2015**, *137*, 8. [CrossRef]
30. Rozentsvaig, A.K.; Strashinskii, C.S. The Growth of Vapor Bubbles in the Volume of Superheated Drops, Dispersed in High-Boiling Liquid. *Appl. Math. Sci.* **2014**, *8*, 7519–7528. [CrossRef]
31. Rozentsvaig, A.K.; Strashinskii, C.S. Regimes of Heat Transfer during Boiling Emulsions with Low-Temperature Dispersed Phase. *Appl. Math. Sci.* **2015**, *9*, 5593–5601. [CrossRef]
32. Califano, V.; Calabria, R.; Massoli, P. Experimental evaluation of the effect of emulsion stability on micro-explosion phenomena for water-in-oil emulsions. *Fuel* **2015**, *117*, 87–94. [CrossRef]
33. Mura, E.; Massoli, P.; Josset, C.; Loubar, K.; Bellettre, J. Study of the micro-explosion temperature of water in oil emulsion droplets during the Leidenfrost effect. *Exp. Therm. Fluid Sci.* **2012**, *43*, 63–70. [CrossRef]
34. Aktershev, S.P.; Ovchinnikov, V.V. Modelling of boiling up of a metastable liquid with appearance of the evaporation fronts. *Mod. Sci. Res. Ideas Results Technol.* **2013**, *1*, 77–82.
35. Shinjo, J.; Xia, J.; Megaritis, A.; Ganippa, L.C.; Cracknell, R.F. Modeling temperature distribution inside an emulsion fuel droplet under convective heating: A key to predicting microexplosion and puffing. *At. Sprays* **2016**, *26*, 551–583. [CrossRef]
36. García, M.C.; Gucker, S.N.; Foster, J.E. Understanding the plasma and power characteristics of a self-generated steam bubble discharge. *J. Phys. D Appl. Phys.* **2015**, *48*, 355203. [CrossRef]
37. Feng, J.; Muradoglu, M.; Kim, H.; Ault, J.T.; Stone, H.A. Dynamics of a bubble bouncing at a liquid/liquid/gas interface. *J. Fluid Mech.* **2016**, *807*, 324–352. [CrossRef]
38. Pavlenko, A.M. Change of emulsion structure during heating and boiling. *Int. J. Energy Clean Environ.* **2019**, *20*, 291–302. [CrossRef]

energies

MDPI

Review

Two-Phase Volumetric Expanders: A Review of the State-of-the-Art

Xander van Heule [1], Michel De Paepe [1,2,*] and Steven Lecompte [1,2]

1 Department of Electromechanical, Systems and Metal Engineering, Ghent University, Sint-Pietersnieuwstraat 41, 9000 Ghent, Belgium; xander.vanheule@ugent.be (X.v.H.); steven.lecompte@ugent.be (S.L.)
2 FlandersMake, Gaston Geenslaan 8, 3001 Leuven, Belgium
* Correspondence: michel.depaepe@ugent.be

Abstract: Two-phase expansion is the process where a fluid undergoes a pressure drop through or in the liquid–vapor dome. This operation was historically avoided. However, currently it is studied for a multitude of processes. Due to the volume increase in volumetric expanders, a pressure drop occurs in the fluid resulting in flashing phenomena occurring. These phenomena have been studied before in other processes such as two-phase flows or static flash. However, this has not been extensively studied in volumetric expanders and is mostly neglected. Even if data has shown this is not always neglectable depending on the expander type. The thermal non-equilibrium occurring can be modeled on different principles of flashing flows, such as the mixture model, boiling delay model, and homogeneous relaxation model. The main application area in current literature for volumetric two-phase expansion machines, is in low-temperature two-phase heat-to-power cycles. These cycles have shown benefit over classic options if expanders are available with efficiencies in the range of at least 75%. Experimental investigation of expanders in two-phase operation, though lacking in quantity, has shown that this is an achievable goal. However, the know-how to accomplish this requires more studies, both experimentally and in modeling techniques for the different phenomena occurring within these expanders. The present work provides a brief but comprehensive overview of the available experimental data, applicable flashing modeling techniques, and available models of volumetric two-phase expanders.

Keywords: two-phase; volumetric expander; review

Citation: van Heule, X.; De Paepe, M.; Lecompte, S. Two-Phase Volumetric Expanders: A Review of the State-of-the-Art. *Energies* **2022**, *15*, 4991. https://doi.org/10.3390/en15144991

Academic Editors: Jan Danielewicz and Krzysztof Rajski

Received: 31 May 2022
Accepted: 30 June 2022
Published: 8 July 2022

Publisher's Note: MDPI stays neutral with regard to jurisdictional claims in published maps and institutional affiliations.

1. Introduction

Expansion processes generating work from two-phase liquid–vapor flows are typically avoided. Common issues ranged from condensation in steam engines [1] to liquid impact erosion in turbomachinery [2]. Yet, this review handles exactly the particular topic of utilizing two-phase fluid expansion, its possibilities, and the current status of experimental knowledge.

1.1. Two-Phase Expander Applications

Applications for two-phase expanders are where a single working fluid two-phase liquid–vapor mixture undergoes a pressure drop. Such as in liquid injection cycles [3] or instead of a throttling valve in a heat pump cycle [4–6]. This throttling valve could also be substituted by a two-phase ejector, which will not be included within this review as it was elaborately discussed by [7,8].

Another important application of two-phase expanders is found in the scope of low-grade heat-to-power conversion. These low-grade heat sources range from geothermal or solar heat to the residual heat of industrial processes. The recuperation of low-grade heat-to-power is commercially available technology through the use of the Organic Rankine Cycle (ORC). The ORC can be a basic Rankine cycle or one of the many variations with, for example, multiple pressure levels, the addition of a recuperator or, among others, can include two-phase expansion. Ref. [9] recommends the investigation of two-phase expanders to increase the performance of heat-to-power cycles. However, this technology is not commercially available. Based on the modeling work of [10,11] the authors concluded that optimized two-phase operation of expanders can improve the system power by nearly 50% in specific operating conditions.

An ORC with two-phase expansion occurs when saturated liquid enters the expander of the ORC architecture, the cycle is called a trilateral cycle (TLC) or a trilateral flash cycle (TFC). The inlet state of the expander can also be in the two-phase region. If this is the case, then the cycle is referred to as a partial evaporation organic Rankine cycle (PEORC) or a wet organic Rankine cycle (WORC). A schematic diagram and a temperature-entropy diagram of the TLC are represented in Figure 1. It consists of four parts: a condenser, an evaporator, a pump, and an expander. The difference with a basic ORC is that the evaporator no longer evaporates the working fluid in the TLC. Instead, the fluid entering the expander is in a saturated liquid state and during the expansion process, the fluid is in the two-phase region as can be seen by point 3 on the T-s diagram in Figure 1b. There is thus a need for an efficient expander capable of two-phase expansion. The heat exchanger that adds heat to the working fluid no longer evaporates the fluid, yet it is still referred to as the evaporator in literature.

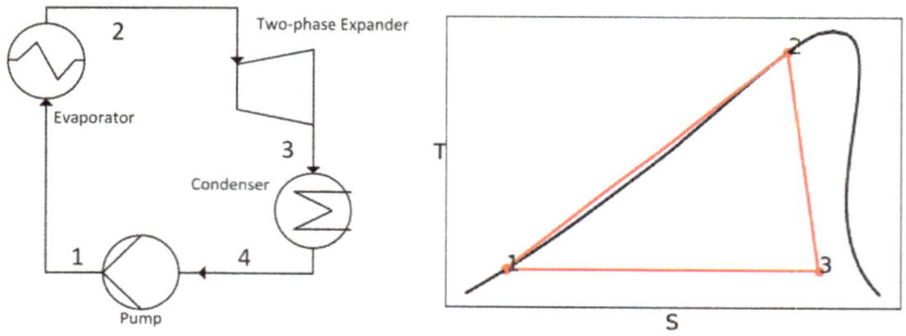

(a) (b)

Figure 1. Trilateral Cycle (TLC) [12]. (**a**) Schematic diagram (**b**) T-s diagram.

It was concluded by [13] that two-phase expansion devices should have an isentropic efficiency of at least 75% to be interesting for power generation applications, which is a realistic design goal. The main difficulty is the lack of practical and efficient two-phase expanders capable of receiving a liquid or a single working fluid two-phase liquid–vapor mixture and the absence of the methods to design these.

Ref. [12] showed that the TLC obtains a larger net power output, thermal efficiency, and exergy efficiency compared to the basic ORC with an increase of 37%. Yet, for evaporation temperatures greater than 135 °C the TLC does perform worse from a financial aspect due to the extra requirements of the heat exchangers, although it does perform better thermodynamically. Ref. [14] also compared the exergy efficiency of power production of the TLC with the ORC. This was defined by the authors as the ratio of the produced power to the incoming exergy flow of the heat carrier which is also referred to as the second law efficiency. They conclude that the exergy efficiency of power production for the TLC is between 14% to 29% higher than the ORC. Yet, because the authors used water as the working fluid, the volume flow rate of the TLC was higher compared to the ORC by a factor of 2.8 or more. This is expected to drop when the working fluid is changed to current refrigerants. The TLC has been practically tested with a small-scale power plant [15] which resulted in 20–40% more power production compared to a conventional ORC. This power plant applied the VPT turbine mentioned before.

As mentioned earlier, another option is to partially evaporate the working fluid up to an optimal vapor quality. This was investigated by [16] where the PEORC was compared to the TLC. The authors also compared the PEORC to the subcritical ORC and the transcritical ORC [17]. The authors show that the PEORC can improve the net power output in comparison to the TLC and that it has a greater second law efficiency than both the subcritical and transcritical ORCs.

1.2. Principles of Flashing

During expansion of a two-phase fluid, flashing will occur. Flashing [18,19] is a physical phenomenon in liquids. It represents a change from liquid to vapor as a result of a sudden decrease in the pressure below the saturation pressure. The heat cannot be contained in the liquid as sensible heat and a metastable superheated liquid occurs [20]. A rapid transfer of heat and mass occurs while the fluid returns to a stable thermodynamic state [21]. This metastable liquid also has a maximum value. This limit is known as the spinodal curve which separates the metastable region from the unstable region in equations of state (EOS) [22,23]. The process of flashing is found and studied in many places such as static flash, boiling liquid expanding vapor explosions (BLEVE) and safety valves, flashing jets, pressurized water reactors, and many more [24–26]. However, the phenomena known through these areas of study have not been directly applied to the case of volumetric two-phase expanders. That is where this review will focus hereafter.

1.3. Expander Types

When considering what type of expander is best suited for an application in single-phase operation, the similarity parameters, specific speed, and specific diameter are often used as defined, respectively, in Equations (1) and (2).

$$N_s = N \cdot \sqrt{\frac{\dot{V}}{(\Delta h)^{3/2}}} \tag{1}$$

$$D_s = \frac{D \cdot (\Delta h)^{1/4}}{\sqrt{\dot{V}}} \tag{2}$$

where N is the rotational speed of the expansion machine, $\dot{V}$ is the volumetric flow rate, D is the characteristic diameter and Δh is the enthalpy difference across the machine. These two numbers can be calculated readily from the design requirements and, by presenting the different types of expansion machines on the same diagram, the most feasible turbine type can be chosen [27]. Balje diagrams can be created for every expander type [28,29] and later employed in design.

Yet, in the event where two-phase fluid is expected, it is not as straightforward to choose turbomachine expanders because the blades could be at risk of impact erosion. However, a plethora of applications where two-phase expansion can be used is suited well to volumetric machines.

1.4. Volumetric Expander Types for Two-Phase Expansion

Volumetric expanders are preferred for the low-temperature heat-to-power application as they have low rotational speeds with low flow rates.This matches well with the current requirements of ORCs. Furthermore, they can cope with relatively high-pressure ratios compared to single-stage turbo expanders and can also inherently handle liquid–vapor mixtures. Both of these traits are beneficial to the TLC and PEORC.

Each type of volumetric expander has some benefits and drawbacks, these are discussed by [30,31]. Currently, only two types are primarily considered for the use of two-phase expanders. These are the screw or Lysholm expander and the piston or reciprocating expander. The most promising type for two-phase expansion is the Lysolm expander. The main benefit of this type is that both phases can be considered to be mixed well. Giving acceptable results when thermodynamic equilibrium is assumed [32–34]. This implies that a simple homogeneous mixture model in thermodynamic equilibrium could be considered. The different modeling approaches for the fluid state are further discussed in Section 2.

Other types of screw expanders, such as the single screw, were deemed too mechanically complex [35]. However, even these are appearing as a possible alternative [36,37].

The reciprocating expander is interesting for applications with higher pressure ratios and lower velocities. The main benefit of this expander type is the built-in volume ratio (BVR). The reciprocating expander is the type that has the largest BVR of all volumetric expander types. This trait is in accordance with the requirements for two-phase expanders which require large volume ratios to accompany the flashing process. A drawback of using this type of expander is the modeling of the expansion process. Assuming thermodynamic equilibrium does not give acceptable results with actual experimental results [38] as it did for Lysholm expanders. It requires a better understanding of the flashing process taking place during expansion.

In volumetric expanders, this sudden depressurization is a result of the increasing working volume. Due to the shape of a reciprocating expander, the flashing process could be compared to static flash experiments [19]. Static flash stands for the case where the fluid remains static in the horizontal direction during the flashing process [39]. These processes have been researched before. For example, as occurring in the desalination process [40] of drinking water production from seawater or as a process taking place in safety valves [41]. The phenomena and data found in literature about static flashing could [20] prove to also be useful in describing the phenomena taking place in volumetric expanders, and most notably reciprocating expanders. To the author's knowledge, other volumetric expander types, such as scroll and vane expanders, have not been found in two-phase expansion research.

1.5. Positioning of This Review

Implementing two-phase expansion technology requires methods to accurately predict the expansion process in various thermodynamic cycles and expander types. Only with this fundamental data, efficient two-phase expanders can be designed. In the following, the review will first shortly describe the possible techniques to model the working fluids' thermodynamic state during the expansion process. Afterward, a summary of the available experimental data is given. Lastly, the specific models applicable to a volumetric expander are also detailed. The gaps in current literature are highlighted and recommendations for future work are given.

2. Methods for Modeling Two-Phase Expansion

There is a multitude of modeling techniques and codes developed for the simulation of flashing. They are often subdivided based on the made assumptions. The assumptions that can be made are that the phases are in thermal equilibrium and/or that they are in mechanical equilibrium. In the context of volumetric expanders, the mechanical equilibrium is assumed as both phases occupy the same expansion chamber or chambers. If thermal equilibrium is assumed as well, then the homogeneous equilibrium model (HEM) can be applied, in the other case, the homogeneous relaxation model (HRM), boiling delay model, or mixture model is used. These techniques will be elaborated upon in the following sections. For other modeling techniques where mechanical equilibrium is not assumed (i.e., in continuous flows), the authors refer to [26] for a review on flashing flow modeling.

2.1. Homogeneous Equilibrium Model

As stated previously, in the homogeneous equilibrium model it is assumed that there is thermal equilibrium between the phases. This implies that both phases share the same saturation temperature:

$$T_g = T_l \tag{3}$$

In which the subscript g stands for the vapor or gas phase and the subscript l for the liquid phase. From this, it follows that the two-phase mixture behaves such as a pseudo single-phase in which the thermodynamic properties can be expressed commonly as in Equation (4). In which x is the mass based vapor quality and y can be any fluid property [26].

$$y = x \cdot y_g + (1 - x) \cdot y_l \tag{4}$$

The HEM consists of three equations: the mass, momentum, and energy conservation equations [42] where the fluid is modeled as a homogeneous mixture of gas and liquid.

Due to its simplicity, it was often used in the past and chosen as the basis for many system codes. However, this approach induces large errors when the equilibrium assumptions do not apply, such as processes in very small time intervals. This is, for example, the case for the reciprocating expander. Even though the expansion process in screw expanders has a similar time constant compared to reciprocating expanders, the HEM can still be used to describe this type of volumetric expander as was shown by [43]. It is speculated that this approach is valid due to the better mixing of the phases within the chambers of a screw expander in comparison to reciprocating ones. Ref. [34] modeled a screw expander with liquid injection as well. The authors also found that assuming thermal equilibrium gives good results compared to the experiments. Ref. [44] used the HEM assumptions within a commercial software environment to model the screw expander within a TLC cycle. This model was studied by [45], in which the authors concluded that the intake phenomena have a large impact on the overall performance of the machine. The constructed model showed that isentropic efficiencies can go up to 83.1%, depending on the operation conditions.

Lastly, comprehensive research on screw expander TLC is found in the works of [33,46]. Due to developments in screw expander technology, they conclusively showed that with the right choice of working fluid a two-phase screw expander can be constructed for use in a TFC. However, water is not found a suitable working fluid within this type of cycle. The net output from the cycle is 10 to 80% greater when compared to a basic ORC. The machines retain a similar size as well.

2.2. Mixture Model

This methodology solves the continuity equations for both the phases and the mixture. The vapor generation rate is postulated to arise from interphase heat transfer. One equation that is often applied to calculate the vapor generation rate is the interfacial exchange model. Given in Equation (5).

$$\mathcal{T} = A_i \cdot \frac{\dot{q}}{h_{gl}} \tag{5}$$

wherein $\mathcal{T}$ is the vapor generation rate, A_i the interfacial area, $\dot{q}$ the total heat flux, from both phases, transferring to the phase interface and h_{gl} is the latent heat of vaporization.

This model is accurate but, depending on the complexity, requires several closure equations. Constitutive equations for the determination of friction force and nucleation rate can be implemented. This model was applied by [38,47], which was evaluated with their previously mentioned experiments [48,49]. The authors considered that the vaporization was mainly driven due to the heat transfer from the liquid to the phase change interface. The model is capable of predicting their data, both the pressure drop and adiabatic efficiency are within an accuracy window of about 5% and this considering both with and without charge and discharge processes. The adiabatic efficiencies are predicted to reach 84%.

2.3. Boiling Delay Model

The boiling delay model states that the boiling only happens when a certain degree of superheat in the liquid is reached. Then the nucleation of bubbles starts to take place. This liquid superheat implies that there is no thermal equilibrium assumption. The bubble nucleation and growth also limits the eventual vapor generation rate. There is a need for two-phase mixture conservation equations beyond the flashing inception, including property calculations for the metastable liquid thermodynamic properties. This modeling technique could be used in the future for very detailed reciprocating expander models. However, due to the need for sufficiently accurate analytical expressions for bubble nucleation and growth rate, there are currently no examples of its use in volumetric expanders. However, Ref. [50] did apply droplet modeling in the stator nozzles of the wet-to-dry cycle, taking into account both thermal and mechanical non-equilibrium. As explained before, this cycle avoids erosion on the blades by only having liquid in the nozzles and superheated vapor in the rotor. This can be easily modeled in equilibrium but [50] showed that this is also feasible when taking non-equilibrium into account.

2.4. Homogeneous Relaxation Model

The homogeneous relaxation model (HRM) is similar to the HEM in the sense that it makes use of the same three continuity equations. However, there is no thermal equilibrium assumption. Instead, the following equation is added [42] which is the vapor mass balance equation.

$$\frac{\partial x}{\partial t} + w\frac{\partial w}{\partial z} = \frac{\mathcal{T}}{\rho} \tag{6}$$

Herein, $\mathcal{T}$ represents the vapor generation rate and x is the actual vapor quality. An implication of not assuming thermal equilibrium is that the liquid and vapor phase temperatures are different, this implies that the total state cannot be represented by Equation (4) as the liquid is in a superheated state. The thermodynamic properties are instead defined as in Equation (7) [42].

$$y = x \cdot y_{s,g} + (1 - x) \cdot y_{m,l} \tag{7}$$

In which the subscripts s and g stand for saturated vapor which is only dependent on the saturation pressure, the subscripts m and l stand for metastable liquid which is dependent on both the pressure and the liquid temperature. From the previous, it thus follows that the fluid is assumed to be a mixture of saturated vapor and superheated liquid, both on the same pressure. Ref. [42] adopted the relaxation equation to the simplest linear approximation, as was successfully performed in the past by other researchers. This results in the following equation.

$$\frac{Dx}{Dt} = \frac{\mathcal{T}}{\rho} = -\frac{x - \bar{x}}{\theta} \tag{8}$$

Herein, x is the actual vapor quality and $\bar{x}$ is the unconstrained equilibrium value of the vapor quality. θ is the local relaxation time and represents the time needed for the fluid to reach equilibrium, for which a relation is constructed. Lastly, Ref. [42] also presented correlations for θ but they were only verified for water with the Moby-Dick experiments. The first correlation gives comparatively good results for small pressures, up to 10 bar.

$$\theta = \theta_0 \cdot \epsilon^{-0.257} \cdot \psi^{-2.24} \tag{9}$$

$$\epsilon = \frac{x \cdot \mu_{s,g}}{\mu} \tag{10}$$

$$\psi = \frac{P_s(T_{in}) - P}{P_s(T_{in})} \tag{11}$$

where θ_0 is equal to 6.51×10^{-4} s. For higher pressures, greater than 10 bar, Equation (9) is substituted for Equation (12).

$$\theta = \theta_0 \cdot \epsilon^{-0.54} \cdot \phi^{-1.76} \tag{12}$$

$$\phi = \frac{P_s(T_{in}) - P}{P_c - P_s(T_{in})} \tag{13}$$

In Equation (12), θ_0 is equal to 3.84×10^{-7} s. P_c is the pressure at the critical point. These equations were tested by [51] and they concluded that this rather simple equation, which only describes one possible mechanism for mass exchange between the phases, can still represent complicated processes. The authors do note that the flashing rate is very sensitive to the value of θ.

Ref. [52] also defined an expression for the relaxation time. Instead of using dimensional analysis and construction of a correlation out of experimental data, the authors evaluated the relaxation time theoretically based on the heat flux around vapor bubbles in the superheated liquid. The authors notice that an increase in the initial void fraction corresponds to a decrease in the relaxation time θ and that the relaxation time decreases for higher values of liquid superheat.

As mentioned before, the HRM model was found to be suited for reciprocating expanders while the HEM can be used for describing the expansion process in screw expanders. This observation is also made in the work of [53]. The authors evaluated the thermodynamic disequilibrium loss, of which the effects increase when the rotation speed increases or if the injection temperature decreases. In these scenarios, with high expander speeds and low heat source temperatures, the authors consider the screw expander to be more suitable. In [54], the authors develop a model for estimating these losses. This model is based on the intake ratio, as the temperature difference between the liquid and vapor phases mainly appears at the initial stage of the two-phase expansion process.

The HRM model has also not yet been fully implemented in an expander model, but [55] has started with the design of a reciprocating expander test-rig. The goal hereby is to gain a better understanding of the flashing principles during variable volume flashing. The model which is being created is based on the HRM model. Initially [56] the model did not yet include an equation for the relaxation variable, instead it was assumed that the evaporation happens instantly. In subsequent work, Ref. [57] the HRM model was included in the reciprocating expander model, but without fitting of the experimental parameters as there is no data available.

A similar method is known as the Delayed Equilibrium Model (DEM) [58] which works on the same principles but uses a three-phase mixture instead of a two-phase mixture. The three phases are saturated vapor, saturated liquid, and superheated liquid states.

2.5. Conclusions and Comparison

Many models have been constructed and used in literature to describe two-phase phenomena. These are mostly applied to flows and other continuous processes and can be subdivided based on taken assumptions. For the specific use of describing the flashing process in a volumetric expander, the possibilities are limited. Because, up until now, the working fluid is assumed to be in mechanical equilibrium, imposing an equilibrium pressure in the two phases, similar to the flashing process where thermal non-equilibrium is used to characterize the system. The most used technique assumes complete equilibrium due to its simplicity and ease of implementation in software. It was shown that this methodology suffices to describe the Lysholm expander, this is speculated to be due to the good mixing within this type of expander. When thermal non-equilibrium is assumed, three methodologies are possible. The mixture model, which solves a given set of continuity equations, was applied by [47] after the authors measured a superheated liquid. The authors were able to predict their experiments by the implementation of the interfacial heat and mass transfer model. Secondly, the boiling delay model would also be applicable, but this has up until now never been tried in literature. This is likely due to the importance of accurate bubble formation and growth equations, which are also impacted by short term mechanical non-equilibrium [26]. Lastly, the homogeneous relaxation model is a rather straightforward extension of the complete equilibrium assumption. It has been successfully used to describe complex systems. Recently, it has been implemented to describe the flashing process in a reciprocating expander [57]. However, experimental validation has to be conducted.

3. Experimental Available Data

The experimental data that have been published on two-phase expansion is hereafter summarized in this section. Firstly, the Lysholm expander will be elaborated on, as this type of expander has received the most attention in literature. Afterward, a brief look will be taken at static flash, as this type of flashing is closely related to what is happening in a reciprocating expander. Lastly, the available experiments on reciprocating expanders will be reviewed last.

3.1. Lysholm Expander

Table 1 shows the available experimental data on two-phase Lysholm expander. The used working fluid, power range, vapor quality range, and measured efficiencies. In the following, they will be compared to one another.

Table 1. Available literature about experimental Lysholm two-phase expanders.

	Working Fluid	Efficieny	Power	Vapor Quality
[59] (1975)	Water	49–55%	16–30 kW	0.12–0.34
[60] (1982)	Water	15–53%	1–60 kW	0.08–0.27
[61] (1982)	Brine	45%	1 MW	0–0.99
[46] (1993)	R113	over 70%	40 kW	-
[32] (1988)	R12	30–60%	10 kW	-
[43] (2013)	R134a	73–92%	50 kW	0.7–1

The screw (or Lysholm) expander is the most prevalent expander that has been studied experimentally in two-phase conditions. Ref. [59] was one of the earliest that published a dataset in 1975, several powers were tested ranging from 16 kW up to 30 kW. The efficiencies that were found for this engine do not come near the theoretical required value of 75% found by [13]. However, Lysholm machine technology has much improved over the years, primarily due to its compressor applications. The authors did not mention whether the vapor quality has an impact on the efficiency within the tested range of 0.12 to 0.34. Ref. [60] tested lower vapor qualities and only mentions an influence of the rotational speed of the screws and the back pressure on the measured efficiencies. Ref. [61] did field tests on a 1 MW screw installation over a range of vapor qualities from 0 to 1. The resulting efficiencies are on average around 45%, which are similar results as [59,60]. All previous experiments mentioned were performed with water as the working fluid. In the following, the reported data comes from the use of refrigerants. Ref. [32] used R12 while [62] executed their experiments with R113. They found efficiencies ranging from 40 to 70%. They concluded that the thermodynamic modeling techniques that are used for single-phase expanders could also be used for two-phase Lysholm expanders with some modifications. They also showed that similar efficiencies can be reached for single-phase and two-phase Lysholm machines. When comparing their efficiencies with single-phase experiments the same conclusions were made by [63]. Lastly, Ref. [43] performed tests with R134a as working fluid. The authors proposed a method to determine the adiabatic efficiency of two-phase operation based solely on single-phase experimental data. This in function of the adiabatic efficiency with saturated vapor and the peak adiabatic efficiency together with the inlet vapor quality. The authors tested their theory on the data of [60–62]. They concluded that for these cases their theory was correct, but that the amount of data is not sufficient to make any firm claims.

3.2. Static Flash

Before going to the specific experiments with reciprocating expanders or piston expanders, static flashing will be introduced. These are similar to reciprocating expander experiments which will be discussed later. This means that the only difference between the reciprocating experiments and the static flash experiments is the load profile which dictates the pressure drop over time. Table 2 shows the experimental static flash data which will be discussed later in this section. Ref. [18] used R12 as the working fluid. The authors installed seven capacitance measurement units over the height of the flashing chamber to measure the void fraction profiles in the cylinder as a function of time when sudden depressurization occurs. It was concluded that the void fraction only linearly increases with respect to the height of the cylinder after around 1.5 s have passed, by which time most thermodynamic disequilibrium has been reduced. Ref. [20] performed experiments on flashing water with a constant initial liquid level of 15 mm and was able to construct a relation of proportionality between the final amount of flashed mass and the initial superheat of the liquid. This parameter is also found as an important quantity in the modeling techniques mentioned earlier. The authors also show that this proportionality can be derived from the heat balance within the flashing chamber. Their results are in good agreement with the results of [64] which performed similar experiments but with a higher initial liquid level. Later [21] also modeled the static flash phenomena and compared it with experiments on a wide variety of conditions in a rectangular enclosure. The authors concluded that the initial pressure has little effect on the flashed mass itself, it only affected the onset of flash evaporation, which is in accordance with previous research. Ref. [65] studied the amount of exergy that is lost due to flash evaporation, the authors concluded that most exergy loss can be avoided if the flashing process takes longer and has a lower superheat temperature of the flashing liquid, this result is beneficial for reciprocating expanders as the pressure drop takes place over a longer time period compared to static flash.

Table 2. Available literature about experimental static flashing.

	Working Fluid	Starting Pressure	Initial Liquid Level
[18] (1984)	R12	10.7 bar	600 mm
[20] (2002)	Water	0.05–0.2 bar	15 mm
[64] (1973)	Water	0.07-0.47 bar	196–225 mm
[21] (2019)	Water	0.05–0.3 bar	0.23–2.3 mm
[65] (2015)	Water	0.08–0.36 bar	100–300 mm

3.3. Reciprocating Expander

Only [38] have performed two-phase expansion experiments with a reciprocating engine. The working fluid used by the authors was water and later ethanol. These experiments are also very similar to static flash experiments with a slower pressure decrease. In the static flash experiments, the piston was suddenly connected to another larger vessel under vacuum conditions, resulting in larger depressurization rates compared to the experiments for reciprocating expanders. Here, the cylinder was loaded with liquid before the piston was set in motion using a linear actuator. The adiabatic efficiencies reached are in the range of 80–95% [38,48]. To compare with [43], the authors theoretically imposed mechanical and electrical losses, achieving a maximum adiabatic efficiency of around 80% which is similar as the Lysholm expander [49]. The authors also conclude that a two-phase reciprocating expander achieves similar efficiencies as a single-phase reciprocating expander.

3.4. Conclusions of Two-Phase Experimental Data

In general, the amount of experimental data on two-phase expanders is quite low considering the theoretically proven potential of some of its applications. Most of this available data was performed on the lysholm (screw) expander. These expanders are based on compressor design methods without a lot of adaptions. This is one area where more research has to be performed, the difference in the design of compressor and expander machines. The available data shows that the equilibrium assumption is sufficient to describe the process in these machines while this is not the case for the reciprocating type. Why this is the case is also only speculated and not yet conclusively shown. To study the flashing expansion phenomena itself, more research is required on other machine types where the liquid superheat is measurable.

4. Modeling of Two-Phase Volumetric Expanders

In the following section, an overview of models of two-phase volumetric expanders in open literature is summarized but first, the inlet and outlet flow valves are considered here shortly. These are often not taken into account or with a straightforward, often equilibrium, process. However, it is possible to also apply the non-equilibrium models here. For example, Ref. [66] used the HRM for this modeling of expander valves, where the authors were able to accurately predict the operation of the inlet and outlet valves independent of the vapor fraction and oil content.

4.1. Lysholm Expander

One of the first analytical models for the two-phase Lysholm expanders is described by [32]. The authors mostly assume thermodynamic equilibrium and thus apply the HEM. The authors compared their analytical model with their performed experiments. Both methods showed an increase in internal efficiency with the rotor speed. Importantly, the authors also looked into a flashing delay. Included by adding a certain time delay between the start of the intake stroke and instantaneous flashing to equilibrium. The authors conclude that this can be omitted as the results for no delay time correspond just as well to their experiments. A similar methodology to [32] was applied by [33]. However, compared to a broader experimental dataset. The authors showed that the working fluid choice and correct rotor profiles are important in the design of TLC with a twin screw expander. Ref. [45] created a chamber model for a twin-screw expander within commercial software with the capability of integration in an entire TLC cycle. This methodology is specific to the machine as it requires geometrical data regarding cell volume evolution, suction and discharge ports as well as the multiple leakage paths in Lysholm expanders [44]. This model also uses thermodynamic equilibrium properties, and thus applies the HEM. The authors found a significant impact of the intake manifold expansion on the overall machine performance. Therefore, the authors designed a Lysholm expander with a variable built-in volume ratio [67]. The simulations show higher total power outputs for smaller BVR due to the higher mass flow rates, while the specific power decreases. Lower BVR also results in higher volumetric efficiencies but lower isentropic efficiencies due to under-expansion. In general, the simulation is capable to optimize the total power output in different operation conditions by varying the BVR. Lastly, Ref. [34] also assumed thermodynamic equilibrium in their initial simulation procedure. Additionally, they also assume that the work is performed by the vapor phase. Later, Ref. [68] modeled the chambers with flash vaporization based on an equation found by experimental spray flash evaporation. This model assumes a minimum superheat of 1 K before evaporation occurs. The amount of evaporation is a fraction of the evaporation needed to achieve equilibrium which was experimentally determined. Thus a boiling delay model was used. The model with thermal disequilibrium predicted slightly lower internal power and isentropic efficiency, but only in the order of 3%. This is in line with the general finding that the equilibrium assumption can be used for Lysholm expanders.

4.2. Reciprocating Expander

After experimental determination of a temperature difference between the liquid and vapor phase, Ref. [47] constructed a five equation flashing model consisting of three energy continuity equations (of the liquid, vapor, and housing) combined with the mass continuity and the interfacial exchange model Equation (5) to predict the experimental data. Ref. [21] compared the model proposed by [47] with the HEM for different expander frequencies and initial superheats. The authors find discrepancies of up to 8% in the isentropic efficiency and expansion work between the two models. These are attributed to the disequilibrium losses by the authors. Ref. [69] proposed a design for a two-phase reciprocating expander consisting of a cyclone separator and the piston itself. The applied modeling for this design was also based on equilibrium assumptions. The authors found isentropic efficiencies in the range of 65 to 85% depending on the working fluid and operational regime. They also noted lower efficiencies for higher engine speeds. Ref. [54] used this design with the model of [47] to estimate the intake losses. For this design, the authors found a linear relation between the intake losses and the intake ratio, defined as the intake time to the expansion time. Ref. [19] also modeled a two-phase expander with a cyclone separator. The authors applied an evaporation model based on static flash pool evaporation and separated the phases in the cyclone and piston parts. Due to the better predicted performance at higher evaporation rates, the authors consider rotary expanders a better match for the two-phase expansion process.

4.3. Conclusions of Two-Phase Expander Modeling

Screw expanders have seen the most modeling of their use with two-phase expansion. Most of these assumed thermal equilibrium throughout expansion as these conform well with the available experimental data. The impact of the inlet port was found to have a non-negligible impact on the operation of the machine. More research about the design of the inlet manifold for two-phase expanders would thus be advised instead of basing the design on the machines compressor operation. When flash evaporation within the chamber was assumed, a discrepancy of only 3% was found with the equilibrium model. Reciprocating expanders on the other hand do require some modeling technique that takes into account the metastable conditions. Only one model of this type was available which takes the metastable condition in the working chamber into account via the mixture model. Other models split up the working volume in the expansion chamber and a phase separator cyclone. This process will have to be studied further, and described with more methods, to better understand the phenomena. These insights could make it possible to better design two-phase volumetric expanders.

5. Conclusions

The current review shows the possibilities and state-of-the-art of two-phase volumetric expansion machines. Research is scarce but promising results are presented. The most notable application for volumetric two-phase expanders is in low-temperature heat-to-power cycles. These cycles have around 20% higher exergy efficiency for power production comapred to the basic organic Rankine cycle if efficient volumetric expansion machines are available. The required efficiencies of 75% are achievable, as was experimentally shown. However, consistent design and simulation models are still unavailable. This requires more research and a better understanding of the process and the occurring phenomena. The modeling of these types of machines is complex. For screw expanders, it was experimentally shown that it can be assumed that the phases are in thermal equilibrium, which implies that the vapor and liquid have the same temperature. Researchers simulate this type of two-phase expander with the use of the homogeneous equilibrium model. The available experimental data is predicted well by these models. On the other hand, it was shown that the thermal equilibrium cannot be assumed for reciprocating expanders. Experimental data and models are very limited for this type of expander. The only available data for reciprocating expanders was described with the use of the mixture model. Another rather simple technique to describe the metastable condition is the homogeneous relaxation model which has been shown to give promising results in other processes, including over inlet and outlet valves. Lastly, the boiling delayed model could also be applied, but presently this was not implemented due to the complexity of the technique and required understanding of the phenomena occurring.

In the near future more research and progress are expected in this area. This increase in interest is likely due to its main predicted application and its need in current times. More work in understanding the principles of flashing in the specific case of volumetric expanders should be performed. Initially, this can be inspired by methods of other flashing phenomenon. Primarily, those where mechanical equilibrium is assumed. For example, the boiling delay model has never been applied to any volumetric expander. Not only does the theoretical understanding need more research, experimental research requires more work as well. Currently, it mostly occurs on lysholm expanders and the experiments are few and far between. More experimental data is required and on a multitude of different expander types. The impact of design parameters on these experiments has to be studied as well.

Funding: This research was funded by FWO-Flanders grant number 1SD9721N.

Acknowledgments: I would like to thank the Research Foundation-Flanders for their financial support through FWO-Flanders grants (1SD9721N).

Conflicts of Interest: The authors declare no conflict of interest.

Abbreviations

The following abbreviations are used in this manuscript:

BLEVE	Boiling Liquid Expanding Vapor Explosion
EOS	Equation of State
LNG	Liquefied Natural Gas
VPT	Variable Phase Turbine
ORC	Organic Rankine Cycle
TLC	Trilateral Cycle
PEORC	Partial Evaporation Organic Rankine Cycle
BVR	Built in Volume Ratio
HEM	Homogeneous Equilibrium Model
HRM	Homogeneous Relaxation Model
WORC	Wet Organic Rankine Cylce

References

1. Cylinder Condensation. *J. Am. Soc. Nav. Eng.* **1909**, *21*, 294–297. Available online: https://onlinelibrary.wiley.com/doi/abs/10.1111/j.1559-3584.1909.tb04491.x (accessed on 30 May 2022). [CrossRef]
2. Di, J.; Wang, S.S.; Yan, X.J.; Cai, L.X.; Xie, Y.H. Experimental investigation on effect of surface strengthening process and roughness on water droplet erosion behavior in turbomachinery. *Tribol. Int.* **2020**, *153*, 106647. [CrossRef]
3. Eyerer, S.; Dawo, F.; Rieger, F.; Schuster, A.; Aumann, R.; Wieland, C.; Spliethoff, H. Experimental and numerical investigation of direct liquid injection into an ORC twin-screw expander. *Energy* **2019**, *178*, 867–878. [CrossRef]
4. Galoppi, G.; Secchi, R.; Ferrari, L.; Ferrara, G.; Karellas, S.; Fiaschi, D. Radial piston expander as a throttling valve in a heat pump: Focus on the 2-phase expansion. *Int. J. Refrig.* **2017**, *82*, 273–282. [CrossRef]
5. Mahmoud, A.M.; Sherif, S.A.; Lear, W.E. Frictional and internal leakage losses in rotary-vane two-phase refrigerating expanders. *J. Energy Resour. Technol. Trans. ASME* **2010**, *132*, 0210071. [CrossRef]
6. Xia, C.; Zhang, W.; Bu, G.; Wang, Z.; Shu, P. Experimental study on a sliding vane expander in the HFC410A refrigeration system for energy recovery. *Appl. Therm. Eng.* **2013**, *59*, 559–567. [CrossRef]
7. Sumeru, K.; Nasution, H.; Ani, F.N. A review on two-phase ejector as an expansion device in vapor compression refrigeration cycle. *Renew. Sustain. Energy Rev.* **2012**, *16*, 4927–4937. [CrossRef]
8. Ringstad, K.E.; Allouche, Y.; Gullo, P.; Ervik, Å.; Banasiak, K.; Hafner, A. A detailed review on CO2 two-phase ejector flow modeling. *Therm. Sci. Eng. Prog.* **2020**, *20*, 100647. [CrossRef]
9. Panesar, A.S.; Bernagozzi, M. Two-phase expander approach for next generation of heat recovery systems. *Int. J. Renew. Energy Dev.* **2019**, *8*, 203–213. [CrossRef]
10. Lemort, V.; Quoilin, S.; Cuevas, C.; Lebrun, J. Testing and modeling a scroll expander integrated into an Organic Rankine Cycle. *Appl. Therm. Eng.* **2009**, *29*, 3094–3102. [CrossRef]
11. Giuffrida, A. Improving the semi-empirical modelling of a single-screw expander for small organic Rankine cycles. *Appl. Energy* **2017**, *193*, 356–368. [CrossRef]
12. Li, Z.; Lu, Y.; Huang, Y.; Qian, G.; Chen, F.; Yu, X.; Roskilly, A. Comparison study of Trilateral Rankine Cycle, Organic Flash Cycle and basic Organic Rankine Cycle for low grade heat recovery. *Energy Procedia* **2017**, *142*, 1441–1447. [CrossRef]
13. DiPippo, R. Ideal thermal efficiency for geothermal binary plants. *Geothermics* **2007**, *36*, 276–285. [CrossRef]
14. Fischer, J. Comparison of trilateral cycles and organic Rankine cycles. *Energy* **2011**, *36*, 6208–6219. [CrossRef]
15. Hays, L. Demonstration of a Variable Phase Turbine Power System for Low Temperature Geothermal Resources. *Energent Corp.* **2010**, 1–57. [CrossRef]
16. Lecompte, S.; van den Broek, M.; De Paepe, M. Thermodynamic analysis of the partially evaporating trilateral cycle. In Proceedings of the 2nd International Seminar on ORC Power Systems, Proceedings, Rotterdam, The Netherlands, 7–8 October 2013; p. 4.
17. Lecompte, S.; Huisseune, H.; van den Broek, M.; De Paepe, M. Methodical thermodynamic analysis and regression models of organic Rankine cycle architectures for waste heat recovery. *Energy* **2015**, *87*, 60–76. [CrossRef]
18. Friedel, L.; Purpst, S. Phase distribution in vessels during depressurisation. *Int. J. Heat Fluid Flow* **1984**, *5*, 229–234. [CrossRef]
19. Wu, W.; Wang, Q.; Zhang, Z.; Wu, Z.; Yang, X.; Xu, L. Influence of evaporating rate on two-phase expansion in the piston expander with cyclone separator. *Therm. Sci.* **2020**, *24*, 2077–2088. [CrossRef]

20. Saury, D.; Harmand, S.; Siroux, M. Experimental study of flash evaporation of a water film. *Int. J. Heat Mass Transf.* **2022**, *45*, 3447–3457. [CrossRef]
21. Wang, C.; Xu, R.; Chen, X.; Jiang, P.; Liu, B. Study on water flash evaporation under reduced pressure. *Int. J. Heat Mass Transf.* **2019**, *131*, 31–40. [CrossRef]
22. Lorenzo, M.D.; Modelling, M.D.L.; Lorenzo, M.D.; Lafon, P. *Modelling and Numerical Simulation of Metastable Two-Phase Flows;* To Cite This Version: HAL Id: Tel-01889103 Modélisation et Simulation Numérique des Ecoulements Diphasiques Métastables; HAL Open Science: Lyon, France, 2018.
23. Debenedetti, P.G. *Metastable Liquids: Concepts and Principles;* Princeton University Press: Princeton, NJ, USA, 1997; Volume 1, pp. 1–400.
24. Polanco, G.; Holdø, A.E.; Munday, G. General review of flashing jet studies. *J. Hazard. Mater.* **2010**, *173*, 2–18. [CrossRef][PubMed]
25. Pinhasi, G.; Ullmann, A.; Dayan, A.; Israel, T.A. Modeling of Flashing Two-Phase Flow Gad A. Pinhasi Department of Chemical Engineering and Biotechnology. *Rev. Chem. Eng.* **2005**, *21*, 133–264. [CrossRef]
26. Liao, Y.; Lucas, D. Computational modelling of flash boiling flows: A literature survey. *Int. J. Heat Mass Transf.* **2017**, *111*, 246–265. [CrossRef]
27. Balje´, O.E. A Study on Design Criteria and Matching of Turbomachines: Part A—Similarity Relations and Design Criteria of Turbines. *J. Eng. Power* **1962**, *84*, 83–102. [CrossRef]
28. Francesconi, M.; Dori, E.; Antonelli, M. Analysis of Balje diagrams for a Wankel expander prototype. *Appl. Energy* **2019**, *238*, 775–785. [CrossRef]
29. Narasimhan, A.K.; Wickramaratne, C.; Kamal, R.; Goswami, D.Y.; Singh, P. Mapping scroll expander performance for organic working fluids using dimensionless parameters in Ns-Ds diagram. *Energy* **2019**, *182*, 739–752. [CrossRef]
30. Dumont, O.; Dickes, R.; Lemort, V. Experimental investigation of four volumetric expanders. *Energy Procedia* **2017**, *129*, 859–866. [CrossRef]
31. Imran, M.; Usman, M.; Park, B.S.; Lee, D.H. Volumetric expanders for low grade heat and waste heat recovery applications. *Renew. Sustain. Energy Rev.* **2016**, *57*, 1090–1109. [CrossRef]
32. Taniguchi, H.; Kudo, K.; Giedt, W.H.; Park, I.; Kumazawa, S. Analytical and Experimental Investigation of Two-Phase Flow Screw Expanders for Power Generation. *J. Eng. Gas Turbines Power* **1988**, *110*, 628–635. [CrossRef]
33. Smith, I.K.; Stošić, N.; Aldis, C.A. Development of the trilateral flash cycle system. Part 3: The design of high-efficiency two-phase screw expanders. *Proc. Inst. Mech. Eng. Part A J. Power Energy* **1996**, *210*, 75–92. [CrossRef]
34. Vasuthevan, H.; Brümmer, A. Thermodynamic Modeling of Screw Expander in a Trilateral Flash Cycle. In Proceedings of the International Compressor Engineering, Refrigeration and Air Conditioning, and High Performance Buildings Conferences, Virtual Conference, 23–28 May 2016; pp. 1–10.
35. Ohman, H.; Lundqvist, P. Srcew expanders in ORC applications, review and a new perspective. In Proceedings of the 3rd International Seminar on ORC Power Systems, Brussels, Belgium, 12–14 October 2015.
36. Xia, G.D.; Zhang, Y.Q.; Wu, Y.T.; Ma, C.F.; Ji, W.N.; Liu, S.W.; Guo, H. Experimental study on the performance of single-screw expander with different inlet vapor dryness. *Appl. Therm. Eng.* **2015**, *87*, 34–40. [CrossRef]
37. Liu, X.; Zhang, H.; Wang, F.; Xia, G.; Li, Z.; Zhu, C.; Zhang, H. Numerical investigation of flow behavior and film thickness in the single screw expander. *Int. J. Mech. Sci.* **2021**, *190*, 106047. [CrossRef]
38. Kanno, H.; Shikazono, N. Experimental and modeling study on adiabatic two-phase expansion in a cylinder. *Int. J. Heat Mass Transf.* **2015**, *86*, 755–763. [CrossRef]
39. Zhang, D.; Chong, D.; Yan, J.; Zhang, Y. Study on steam-carrying effect in static flash evaporation. *Int. J. Heat Mass Transf.* **2012**, *55*, 4487–4497. [CrossRef]
40. Spiegler, K.; El-Sayed, Y. The energetics of desalination processes. *Desalination* **2001**, *134*, 109–128. [CrossRef]
41. Mayinger, F. Two-Phase Flow Phenomena with Depressurization-Consequences for the Design and Layout of Safety and Pressure Relief Valves Phgnomene zweiphasiger Striimung bei Druckentlastung-Folgerungen fiir Konstruktion und Auslegung von Sicherheits-und Druckentlast. *Chem. Eng. Process. Process. Intensif.* **1988**, *23*, 1–11. [CrossRef]
42. Downar-Zapolski, P.; Bilickp, Z.; Bolle, L.; Franco, J. The Non-equilibrium relaxation model for one-dimensional flashing liquid flow. *Int. J. Multiph. Flow* **1996**, *22*, 473–483. [CrossRef]
43. Öhman, H.; Lundqvist, P. Experimental investigation of a Lysholm Turbine operating with superheated, saturated and 2-phase inlet conditions. *Appl. Therm. Eng.* **2013**, *50*, 1211–1218. [CrossRef]
44. Bianchi, G.; Kennedy, S.; Zaher, O.; Tassou, S.A.; Miller, J.; Jouhara, H. Two-phase chamber modeling of a twin-screw expander for Trilateral Flash Cycle applications. *Energy Procedia* **2017**, *129*, 347–354. [CrossRef]
45. Bianchi, G.; Kennedy, S.; Zaher, O.; Tassou, S.A.; Miller, J.; Jouhara, H. Numerical modeling of a two-phase twin-screw expander for Trilateral Flash Cycle applications. *Int. J. Refrig.* **2018**, *88*, 248–259. [CrossRef]
46. Smith, I.K. Development of the trilateral flash cycle system: Part 1: Fundamental considerations. *Proc. Inst. Mech. Eng. Part A J. Power Energy* **1993**, *207*, 179–194. [CrossRef]
47. Kanno, H.; Shikazono, N. Modeling study on two-phase adiabatic expansion in a reciprocating expander. *Int. J. Heat Mass Transf.* **2017**, *104*, 142–148. [CrossRef]
48. Kanno, H.; Shikazono, N. Experimental study on two-phase adiabatic expansion in a reciprocating expander with intake and exhaust processes. *Int. J. Heat Mass Transf.* **2016**, *102*, 1004–1011. [CrossRef]

49. Kanno, H.; Hasegawa, Y.; Hayase, I.; Shikazono, N. Experimental Study on Adiabatic Two-Phase Expansion in A Cylinder for trilateral cycle. In Proceedings of the 3rd International Seminar on ORC Power Systems, Brussels, Belgium, 12–14 October 2015.
50. White, M.T. Investigating the wet-to-dry expansion of organic fluids for power generation. *Int. J. Heat Mass Transf.* **2022**, *192*, 122921. [CrossRef]
51. Banaszkiewicz, M.; Kardaś, D. Numerical calculations of the Moby Dick experiment by means of unsteady relaxation model. *J. Theor. Appl. Mech.* **1997**, *35*, 211–232.
52. Bilicki, Z.; Kwidzilqski, R.; Ali Mohammade, S. Evaluation of the relaxation time of heat and mass exchange in the liquid-vapour bubble flow. *Int. J. Heat Mass Transf.* **1996**, *39*, 753–759. [CrossRef]
53. Wang, Q.; Huang, H.; Xu, L.; Yang, X.; He, Z.; Wu, W. Analysis of the thermodynamic disequilibrium loss of the two-phase expansion process and its influence on the performance of a reciprocating expander. In Proceedings of the IOP Conference Series: Materials Science and Engineering, London, UK, 9–11 September 2019; Institute of Physics Publishing: London, UK, 2019; Volume 604. [CrossRef]
54. Wang, Q.; Wu, W.; He, Z.; Ziviani, D. Analysis of the intake process and its influence on the performance of a two-phase reciprocating expander. *Appl. Therm. Eng.* **2019**, *160*, 113943. [CrossRef]
55. Lecompte, S.; Van Den Broek, M.; De Paepe, M. Design of an Optical-access Expansion Chamber for Two-phase Expansion. *Int. J. Heat Mass Transf.* **2012**, *86*, 755–763.
56. Lecompte, S.; Van Den Broek, M.; De Paepe, M. Initial Design of an Optical-Access Piston Expansion Chamber for Wet-Expansion. *Energy Procedia* **2017**, *129*, 307–314. [CrossRef]
57. van Heule, X.; Skiadopoulos, A.; Manolakos, D.; De Paepe, M.; Lecompte, S. Modelling and design of a two-phase reciprocating expander for integration in a trilateral flash cycle. In Proceedings of the 6th International Seminar on ORC Power Systems, Virtual Conference, 11–13 October 2021; Technical University of Munich: Munich, Germany, 2021.
58. Delayed Equilibrium Model (DEM) of Flashing Choked Flows Relevant to LOCA and Implementation in System Codes, Thermal Hydraulics. *Int. Conf. Nucl. Eng.* **2014**, *2B*, V02BT09A040. Available online: https://www.researchgate.net/publication/263652479_Delayed_Equilibrium_Model_DEM_of_Flashing_Choked_Flows_Relevant_to_LOCA_and_Implementation_in_System_Codes (accessed on 30 May 2022). [CrossRef]
59. Weiss, H.; Steidel, R.; Lundberg, A. Performance Test of a Lysholm Engine. Report. 3 July 1975. Available online: https://digital.library.unt.edu/ark:/67531/metadc1061143/ (accessed on 30 May 2022).
60. Steidel, R.F., Jr.; Weiss, H.; Flower, J.E. Performance Characteristics of the Lysholm Engine as Tested for Geothermal Power Applications in the Imperial Valley. *J. Eng. Power* **1982**, *104*, 231–240. [CrossRef]
61. McKay, R. *Helical Screw Expander Evaluation Project*; Final Report; Jet Propulsion Laboratory, US Department of Energy: La Cañada Flintridge, CA, USA, 1982.
62. Smith, I.K.; Stosic, N.; Aldis, C.A. Lysholm machines as two-phase expanders. In Proceedings of the International Compressor Engineering Conference, West Lafayette, IN, USA, 19–22 July 1994; Purdue University: West Lafayette, IN, USA, 1994.
63. Merigoux, J.M.; Pocard, P. *Solar Power Units with Screw Expanders*; Pergamon Press Inc.: Oxford, UK, 1980; Volume 3, pp. 1293–1317.
64. Miyatake, O.; Murakami, K.; Kawata, Y.; Fujii, T. Fundamental experiments with flash evaporation. *Heat Transf.-Jpn. Res* **1973**, *2*, 89–100.
65. Zhang, D.; Yan, J.; Liu, Y.; Zhao, B. Preliminary exergy analysis of static flash of pure water. *Int. J. Heat Mass Transf.* **2015**, *86*, 377–387. [CrossRef]
66. Schweizer, F.; Fürst, M.; Wachtmeister, G. Modeling and Validation of a Two-Phase Flow Valve for Expanders in Waste Heat Recovery. *J. Therm. Sci. Eng. Appl.* **2021**, *13*, 031026. [CrossRef]
67. Bianchi, G.; Marchionni, M.; Miller, J.; Tassou, S.A. Modelling and off-design performance optimisation of a trilateral flash cycle system using two-phase twin-screw expanders with variable built-in volume ratio. *Appl. Therm. Eng.* **2020**, *179*, 115671. [CrossRef]
68. Vasuthevan, H.; Brümmer, A. Theoretical investigation of flash vaporisation in a screw expander You may also like Theoretical investigation of flash vaporisation in a screw expander. *IOP Conf. Ser. Mater. Sci. Eng.* **2017**, *232*, 012077. [CrossRef]
69. Steffen, M.; Löffler, M.; Schaber, K. Efficiency of a new Triangle Cycle with flash evaporation in a piston engine. *Energy* **2013**, *57*, 295–307. [CrossRef]

energies

MDPI

Article

Feasibility Study and Economic Analysis of a Fuel-Cell-Based CHP System for a Comprehensive Sports Center with an Indoor Swimming Pool

Jie Liu, Sung-Chul Kim and Ki-Yeol Shin *

School of Mechanical Engineering, Yeungnam University, 280 Daehak-ro, Gyeongsan-si 712-749, Korea; lauj@ynu.ac.kr (J.L.); sungkim@ynu.ac.kr (S.-C.K.)
* Correspondence: shinky@ynu.ac.kr; Tel.: +82-53-810-3060

Abstract: Unlike a general commercial building, heating for a building with an indoor swimming pool is highly energy-intensive due to the high energy demand for swimming water heating. In Korea, the conventional heating method for this kind of building is to use boilers and heat storage tanks that have high fuel costs and greenhouse gas emissions. In this study, a combined heat and power (CHP) system for such a building using the electricity and waste heat from a Phosphoric Acid Fuel Cell (PAFC) system was designed and analyzed in terms of its primary energy saving, CO_2 reduction, fuel cell and CHP efficiency, and economic feasibility. The mathematical model of the thermal load evaluation was used with the 3D multi-zone building model in TRNSYS 18 software (Thermal Energy System Specialists, LLC, Madison, WI, USA) to determine the space heating demand and swimming pool heat losses. The energy efficiency of the fuel cell unit was evaluated as a function of the part-load ratio from the operating data. The fundamental components, such as the auxiliary boiler, thermal storage tank, and heat exchanger are also integrated for the simulation of the system's operation. The result shows that the system has a high potential to improve the utilization efficiency of fuel cell energy production. Referring to the local condition of the energy market in Korea, an economic analysis was also carried out by using a specific FC-CHP capacity at 440 kW. The economic benefit is significant in comparison with a conventional heating system, especially for the full-time operating (FTO) mode. The net profit made by comparison with the conventional energy supply system is about 178,352 to 273,879 USD per year, and the payback period is expected to be 6.9 to 10.7 years under different market conditions.

Keywords: fuel cell (FC); phosphoric acid fuel cell (PAFC); combined heat and power (CHP); wasted heat recovery system (WHRS); strategic energy management planning; economic analysis

Citation: Liu, J.; Kim, S.-C.; Shin, K.-Y. Feasibility Study and Economic Analysis of a Fuel-Cell-Based CHP System for a Comprehensive Sports Center with an Indoor Swimming Pool. *Energies* **2021**, *14*, 6625. https://doi.org/10.3390/en14206625

Academic Editors: Jan Danielewicz and Krzysztof Rajski

Received: 31 August 2021
Accepted: 10 October 2021
Published: 14 October 2021

1. Introduction

The main motivations to seek more clean and efficient methods of energy production are the increasing cost of fuel and the need for the reduction of CO_2 and harmful emissions [1]. A combined heat and power (CHP) system has many benefits when it is used to provide electrical and thermal energy for commercial buildings, such as increasing power reliability; the reduction of primary energy consumption, greenhouse gas emissions and costs; and the improvement of power quality [2]. Furthermore, the owner of the CHP system can benefit from the higher efficiency of the fuel energy conversion, the lower cost of fuel per unit of energy, high-quality power, and the rapid variation of the distribution between electrical and thermal loads [3]. For the application of commercial buildings, CHP is decentralized electrical power generation coupled with thermally activated components [4]. In the CHP part, the interest in the improvements of the primary generator over a typical thermal engine to new is growing more, especially in Japan, and more recently in Europe [5]. Comparing with those other CHP technologies, such as Stirling engines or gas turbines, fuel cells—as an electrochemical generator—can achieve substantially higher efficiencies and potentially compete even the most energy-efficient large-scale power plants in terms of electric efficiencies [6,7], due to their wide range of capacities which can be well matched with the energy demand, thereby increasing the energy utilization.

According to the electrolyte and operating temperature, fuel cells can be categorized into the following five major types: alkaline fuel cells (AFC), proton exchange membrane fuel cells (PEMFC), solid oxide fuel cells (SOFC), phosphoric acid fuel cells (PAFC), and molten carbonate fuel cells (MCFC). A comparison of the characteristics of different types of fuel cells is listed in Table 1 [8]. Referring to the features of different types of fuel cells, several studies about fuel-cell-based CHP systems were carried out. In Ivan Verhaert et al.'s study, an AFC based micro-CHP system was compared with other micro-CHP technologies, and it was shown that for buildings with a decreasing heating demand and increasing electricity demand, a fuel-cell-based micro-CHP has a better result, especially, in terms of thermal performance [9]. PEMFC has a major application in transportation due to its potential impact on the environment, such as the control of greenhouse gases emissions [10]. However, due to its high efficiency and low pollution in comparison with conventional combustion-based power generation technologies, PEMFC is also considered to be a prospective alternative power source for distributed energy or CHP applications [11]. A thermodynamic analysis for a combined cooling, heating, and power (CCHP) system based on PEMFC as a prime mover has been performed, and the results indicated that the energy and exergy efficiencies of the CCHP system are 81.55% and 54.5% [12]. SOFC has several advantages, such as high electrical efficiency, high-quality heat supply, small installation footprints, the flexibility of its fuel use, and the use of economical materials [13]; it has successfully been verified in long-term stationary power generation up to the Megawatt-scale, generally for the use of commercial CHP systems. A 175 kW SOFC-CHP system was analyzed, and it successfully decreased annual utility costs by up to 14.5% over a baseline HVAC system [14]. Techno-economic analyses of a PEMFC- and SOFC-based micro-CHP system for a residential application with [15] and without a heat pump [16] were carried out by Marta Gandiglio et al., and the results indicate that a SOFC–CHP system has a better total energy efficiency (up to 81%), and the PEMFC-CHP system presents a total efficiency of 75% and a 3-year payback period. PAFC has a relatively low operating temperature, low-electrolyte cost, and high durability in comparison with the other types of fuel cell technologies. Therefore, the PAFC is believed to be one of the most complete fuel cell technologies, and is easy to commercialize [17,18]. As early as 10 years ago, it was demonstrated that the PAFC has high reliability, efficiency, and flexibility for a variety of applications, especially for distributed power generation [19]. Compared with SOFC, the PAFC cannot be effectively used to drive a heat engine due to its relatively low working temperature. However, the operating temperature of PAFC is higher than PEMFC, and the waste heat recovered from PAFC includes a considerable amount of available energy [20]. The economic and environmental potential of a CHP system based on a 400 kW scale PAFC was introduced by a simulation-based analysis under two different operation strategies, and the results indicated that the electrical load following (ELF) model would be the advisable operation mode for this CHP system in the residential sector [21]. Salvador Acha et al. investigated the feasibility of a 460 kW PAFC-based CHP system in the application of a commercial building. The results indicated that the FC-CHP system is financially competitive against the internal combustion engine [22].

Table 1. Comparison of fuel cell technologies.

	AFC	MCFC	PEMFC	PAFC	SOFC
Operating temperature	60–90 °C	600–700 °C	<120 °C	150–200 °C	500–1000 °C
Electrical efficiency	45–60%	45–60%	45–60%	40%	60%
Typical capacity	<100 kW	300 kW–3 MW	<1–100 kW	5–440 kW	1 kW–2 MW
Applications	Military; Backup power;	Electric utility; Distributed generation;	Distributed generation; Transportation; Specialty vehicles;	Distributed generation; CHP;	Electric utility; Distributed generation;

From the above literature, for the analysis of the operating and economic performance of a fuel-cell-based CHP system, relatively accurate load data plays a vital role, as the CHP performance is highly dependent on the load matching condition. The analysis without complete annual load data [12,13] can just present the performance of the CHP system under specific operating conditions, but is limited to reflect the fluctuation of the load, which has the biggest impact on the performance of the CHP system. Some of the studies used measured data or standard profiles of electricity and thermal loads [5,15,16,21]. It is effective for the analysis of existing building or if the building has a relatively standard load profile according to their functions, such as a school, office, commercial buildings, etc. However, for buildings that are under construction and have a special load profile, such as a commercial building with an indoor swimming pool, the better way is to obtain the load through numerical simulation. The commercial software eQUEST [14] and EnergyPlus [23] were found to be used for the thermal load evaluation for the analysis of a fuel-cell-based CHP system. However, the effective verification of the simulation results of the load was not found. Therefore, in order to overcome the aforementioned limitations, a weather-data-based dynamic computation model for the fuel-cell-based CHP system, which includes the thermal load evaluation of a comprehensive sports center with an indoor swimming pool is developed in this study, and the obtained thermal load is validated by the measured data from a reference building.

The comprehensive sports center with an indoor swimming pool is planned and under construction in Donghae City, Gangwon Province, South Korea. Unlike general commercial buildings, a building with an indoor swimming pool is highly intensive in the use of thermal energy, due to the high heat loss from the pool water. The conventional strategy of energy supply for this kind of building is to provide electrical energy from the main grid and thermal energy from a combustion boiler. Normally, for a general commercial building in Korea, the thermal–electrical load ratio is around 0.5 [24]. However, for a building with an indoor swimming pool, it could be larger than 0.7. This characteristic will improve the utilization rate of waste heat from the fuel cell. The thermal–electrical ratio of the PAFC output is from 0.7 to 1.2 under different part-load conditions which are closer to the aimed-for building than the SOFC (0.5). Moreover, by considering its lower temperature of operating and waste heat, the PAFC is the better choice for the design building than SOFC, although the SOFC has a higher electrical efficiency. In this study, two operating strategies of the FC-CHP system are introduced and analyzed from the following aspects: primary energy consumption, fuel cell and CHP efficiency, and economic feasibility. In the aspect of the energy load, the measured data of the electricity load from a similar building is modified and used. The thermal energy demand, including the heating load, swimming pool heat loss, and domestic hot water demand, is evaluated by the commercial software TRNSYS, which is also used for the dynamic simulation of the system's operation. Figure 1 depicts the main configuration of the designed FC-CHP system for the building. In the PAFC system, the fuel cell stack produces both electrical and thermal energy from the hydrogen produced through the reforming of natural gas. The main electrical and thermal load are provided by fuel cell production. Simultaneously, the electricity from the main power grid and the heat from an auxiliary boiler are also required as a supplement. The performance of a CHP system is highly dependent on the operating strategies due to the unstable characteristics of the load over time. From the perspective of exergy efficiency, there are two representative regimes of operating strategies: electrical load tracking (ELT) and thermal load tracking (TLT) [23,25]. However, from the economic point of view, more electricity production means more revenue, especially in a country with an open electricity market, because the excess electricity production could be sold onto the grid. The TLT model presents a significant shortage in the total amount of power generation in the summer season. Therefore, the TLC model is not considered to be an operating strategy. Additionally, the operating strategy of so-called full time operating (FTO) is carried out in this study. This operating model is always applied in a large scale fuel cell plant [26,27], but is not common in CHP applications. The performance and economic feasibility should be analyzed and compared with the ELT model and traditional energy supply method. In these regards, the main objectives of this study are as follows:

- The evaluation of the thermal loads—which includes the space heating load, swimming pool loss, and hot water demand—by developing a weather-data-based dynamic simulation model, and validation using reference data.
- The evaluation of the performance of the 440 kW PAFC-based CHP system in terms of primary energy saving, system efficiency, and CO_2 reduction with ELT and FTO models.
- The economic feasibility of the 440 kW PAFC-based CHP system in the aspects of net profit (NP) and payback period (PP) under different market conditions.

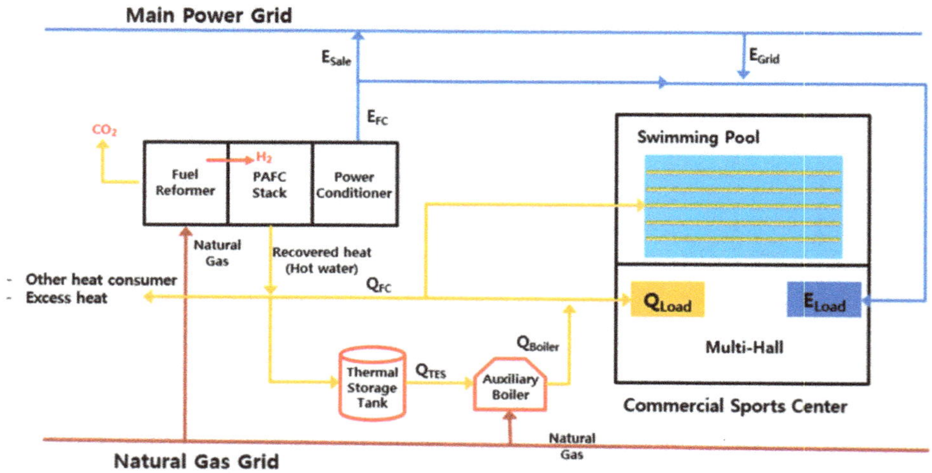

Figure 1. Configuration of the designed FC-CHP system.

2. Model Approach

Figure 2 describes the operating strategies of the designed FC-CHP system. For the ELT model, the fuel cell will operate following the building electricity demand. The power from the main city grid will be a supplement when the building electricity load is less than the minimum capacity (100 kW) or larger than the maximum capacity of the fuel cell (440 kW). In the meantime, the fuel cell's waste heat will be recovered and supply the building thermal energy demand directly, or will be re-stored in a thermal storage tank. When the fuel cell waste heat supply is insufficient, the heat in the heat storage tank will be released to meet the heating demand. When the system is operating under the FLO model, both the power and heat are the maximum generated. Unlike the ELT model, much more excess power is generated and could be sold onto the main power grid. An auxiliary boiler is used for some extreme conditions under both of the two operating models. The dynamic simulation results of the above two operating strategies of the FC-CHP system will be compared with the conventional energy supply strategy.

The dynamic simulation of the FC-CHP system based on the measured electricity load, evaluated thermal load, and fuel cell performance data was also produced in the TRNSYS software. Figure 3 presents the configuration of the model in the TRNSYS simulation studio. The components are modeled by parameters, inputs, and outputs, and are linked with each other by following the operating strategy. The parameters of the main components in the system are introduced in Table 2.

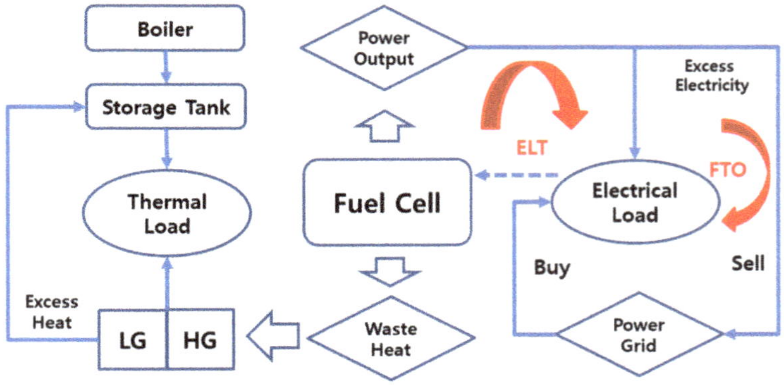

Figure 2. Operating strategy of FC-CHP system.

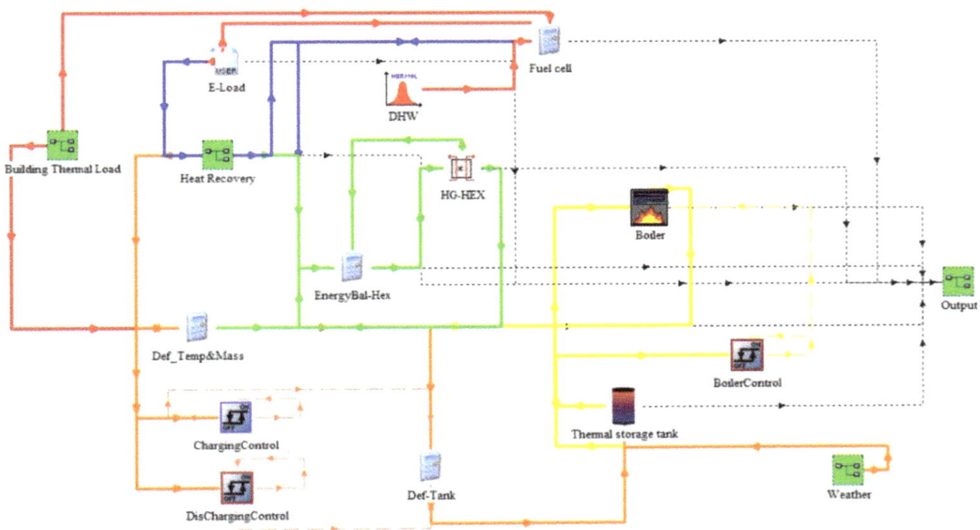

Figure 3. Interface of the fuel-cell-based CHP system simulation in the TRNSYS software.

Table 2. Parameters of the main components used in the system simulation.

Components	Parameters	Values
Fuel cell	Rated electrical capacity, kW	440
	Electrical efficiency, 100%	0.24–0.45
	Thermal efficiency, 100%	0.27–0.49
	Wasted water temperature, °C	60, 121
Heat exchanger	Efficiency, 100%	0.85
Thermal storage tank	Size, m^3	50
Auxiliary boiler	Rated capacity, kW	100

2.1. Performance Data of the PAFC

The performance data of the PAFC—which can present the correlations between the fuel cell's operating part load and the power output, waste heat recovery, and the fuel energy consumption—was obtained from the fuel cell manufacturer [28], and is described in Figure 4. The PAFC can provide the maximum power of 440 kW, and was suggested to operate with a turndown ratio of 0.227, which means the PAFC will not operate at less than 22.7% (about 100 kW) of its rated power. The recovered waste heat can be supplied to the thermal load of the building at two levels of water temperature: high-grade (HG, about 120 °C), and low-grade (LG, about 60 °C). Among them, the LG heat can be recovered over all of the operating time, and the HG heat can be recovered only when the part-load is larger than 51.1%. The electrical power efficiency and thermal efficiency of the fuel cell system are defined as:

$$\eta_{E-FC} = E_{FC}/F_{FC} \tag{1}$$

$$\eta_{Q-FC} = Q_{FC}/F_{FC} \tag{2}$$

where E_{FC} is the electric power produced by the fuel cell unit, Q_{FC} is the total recovered thermal energy, and F_{FC} is the fuel (natural gas) energy consumption of the fuel cell. Thus, the overall efficiency of the fuel cell system is given as:

$$\eta_{FC} = \eta_{E-FC} + \eta_{Q-FC} \tag{3}$$

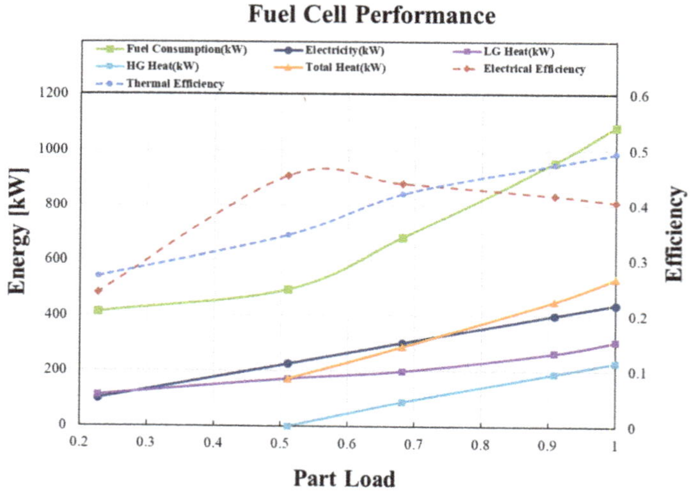

Figure 4. Correlation performance data for a 440 kW PAFC.

For the FTO model, it is suggested that the fuel cell is operated with the maximum power output all the time, even when the electrical power efficiency is not the highest. As shown in Figure 4, the $\eta_{E\text{-}FC}$, $\eta_{Q\text{-}FC}$ and η_{FC} of the suggested PAFC with the maximum electrical power output reach 40.6%, 49.4%, and 90%, respectively. For the approached ELT model, the fuel cell energy production and efficiency will be calculated with the correlations by responding to the building energy load.

2.2. Electricity Load

In this study, a comprehensive sports center with an indoor swimming pool, which is under construction in Donghae City, Korea, is planned to use a 440 kW PAFC-based CHP system for its electrical and thermal energy supply. We first investigated the measured hourly electricity load of a referenced commercial building which has a similar location, construction, floor area, building function, and activity schedule for one year as the input data for the dynamic simulation. As shown in Figure 5a, high electricity demand is present in the summer season due to the large cooling energy which is consumed by the electric air conditioner, and after sorting the hourly load in descending order, it can be seen that there are about 2400 h of electricity demand in a year, which is higher than the maximum capacity of the fuel cell (440 kW), and thus requires auxiliary supply from the main power grid if the fuel cell is operating with the FTO model. If the fuel cell is working with the ELT model, there is still about 3800 h of electricity demand that requires assistance from the main power grid, when the electricity load is lower than the minimum capacity of the fuel cell (100 kW). From a general calculation, about 4.18% and 13.28% electricity demand is required from the main power grid when the fuel cell is working with the FTO and ELT model respectively. The detailed electricity load distribution for the first week in January shown in Figure 5b presents a significant load profile of a commercial building. The measured data reflects the seasonal and daily characteristics of the electrical load well, and will be used as the input data for the dynamic system simulation.

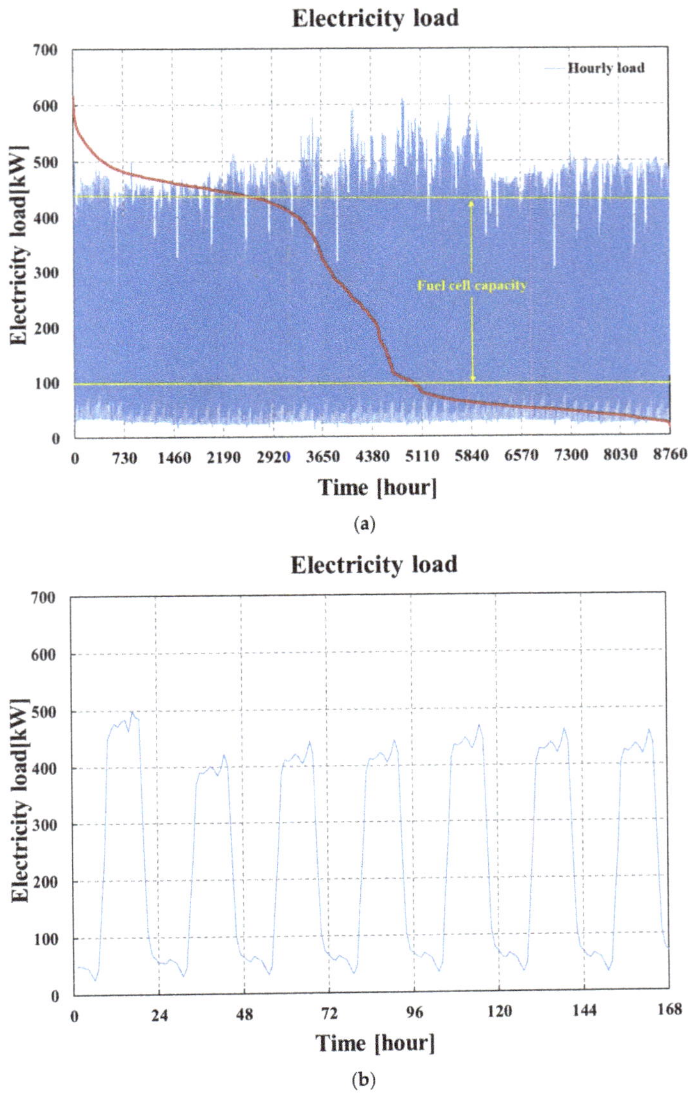

Figure 5. Measured electricity load ((**a**) annual; (**b**) weekly).

2.3. Thermal Load Evaluation

The thermal energy demand of the building is composed of three parts: space heating, swimming pool heat loss, and domestic hot water. A general method to investigate the thermal energy load is to account for the fuel consumption. However, in this study, it is difficult to obtain completed data on fuel consumption, especially in hours. Therefore, in this study, the commercial software TRNSYS—which has been approved by several researchers—is used to evaluate the thermal energy demand.

The 3D building model in TRNSYS is used for the space heating thermal load evaluation. The building geometry was first modelled using Google SketchUp, see Figure 6, then the material properties of the envelope were incorporated in TRNBuild, a 3D building-modelling program integrated into TRNSYS. After that, the weather data (ambient temperature, solar radiation, wind speed, soil temperature, etc.) were input to the TRNBuild component, and the hourly space heating load was calculated in TRNSYS Simulation Studio. The detailed configuration of thermal load macro that was integrated in the CHP system model is shown in Figure 7.

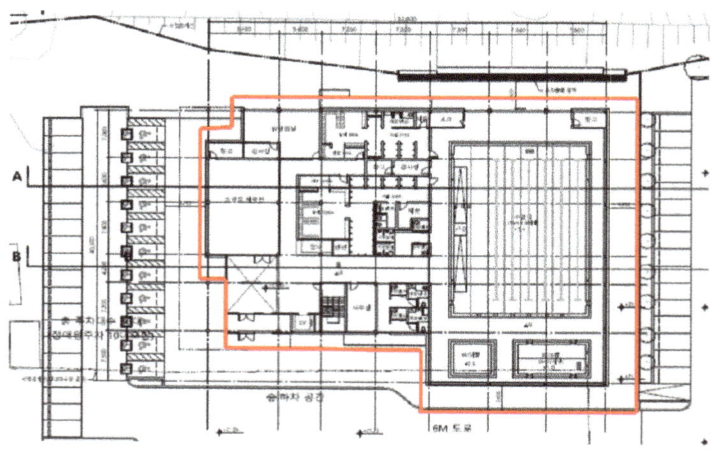

(a)

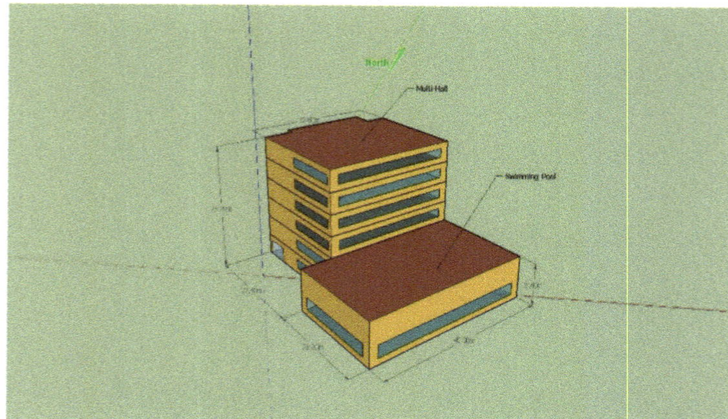

(b)

Figure 6. An architectural plan of the comprehensive sports center with an indoor swimming pool (a) and the 3D modeling (b).

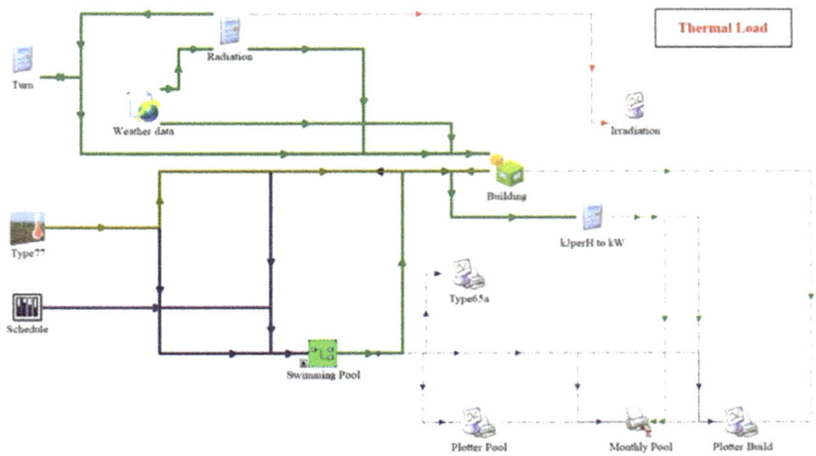

Figure 7. Interface of the building thermal load simulation macro in the TRNSYS software.

The evaluation of the heat loss of the indoor swimming pool is generally based on Thomas Auer's model [29] and is integrated into the TRNSYS thermal load model. The heat loss of the swimming pool Q_{loss} includes four parts:

$$Q_{loss} = Q_{evp} + Q_{conv} + Q_{cond} + Q_{rad} \tag{4}$$

where Q_{evp} is the heat loss by evaporation from the swimming pool surface, which can be calculated by:

$$Q_{evp} = m_{evp} h_{evp} \tag{5}$$

where h_{evp} is the evaporation heat of the water at setting temperature (kJ/kg, 2439.3 kJ/kg for 26 °C water), and m_{evp} is the amount of evaporation water (kg/m^{2}h), which can be calculated by [30]:

$$m_{evp} = A_s(4.08 + 4.28v)(x_{sa,m} - x_a) \tag{6}$$

where $x_{sa,m}$ and x_a are the maximum humidity ratio of saturated air at the same temperature as the water surface (kg/kg, 26 °C is 0.0212858 kg/kg) and the humidity ratio of ambient air (kg/kg, setting in TRNSYS). A_S is the surface area of the swimming pool.

The convection heat loss of swimming pool water Q_{conv} can be calculated on the basis of Newton's formula:

$$Q_{conv} = hA_s(T_{sw} - T_{air}) \tag{7}$$

where T_{sw} and T_{air} are the temperatures of the swimming pool water and the indoor air, respectively. The convective heat transfer coefficient (W/m^2 °C) can be expressed as a linear function of the indoor air speed v, as follows [31]:

$$h = 2.8 + 3.0v \tag{8}$$

Due to the poor heat conductance of soil, the conductive heat loss through the side walls and bottom of the swimming pool is relatively small, and can be calculated by:

$$Q_{cond} = UA_{pw}(T_{sw} - T_{pw}) \tag{9}$$

where the A_{pw} is the wall and bottom area of the swimming pool. U is the conductive heat transfer coefficient (W/m^2 °C).

The heat transfer by long-wave radiation with the wall surfaces of the hall can be calculated on the basis of the Stefan–Boltzmann law (Equation (10)). For an indoor swimming pool, the pool area can be assumed to be completely enclosed.

$$Q_{rad} = \sigma \varepsilon A_s (T_{sw}^4 - T_{cel}^4) \tag{10}$$

where σ is the Stefan–Boltzmann constant (5.67×10^{-8} W/m^2 °C^4), and ε is the emissivity of the swimming pool water surface (0.9 in this study).

The domestic hot water is mainly consumed by the showering of the customers, and is given as normally distributed random values with a mean value of 25 kW in the simulation; the parameters for the TRNSYS simulation are listed in Table 3.

Table 3. Building and swimming pool parameters.

Components	Parameter with Units	Values
Building	Total floor area, m^2	7418.96
	Wall u-value, W/m^{2}K	0.651
	Window u-value, W/m^{2}K	1.1
	Window g-value, 100%	0.62
	Ground u-value, W/m ^{2}K	0.295
	Indoor setting temperature, °C	25
	Indoor humidity ratio, 100%	0.5–0.7
Swimming pool	Surface area, m^2	25 × 15
	Deepth, m	1.5
	Wall heat transfer coefficient, W/m^{2}K	0.25

3. Simulation Results

Thermal load validation is always required to validate a simulation result against measured data. In our case, due to the objective building being under plan, we collected the thermal energy consumption of the building which used to provide the electricity load. It is necessary to mention that unlike the electricity load, it is very difficult to obtain a complete fuel consumption, especially in hours. The daily fuel consumption data for the referenced commercial building is only available for the month of January 2020. The heat supply to the building is from a gas boiler with an average efficiency of 0.85 [32]. The loss from the delivery system is assumed to be 10%. The error indicators used in this study are the MBE (Mean Bias Error) and CV-RMSE (Coefficient of Variance of the Root Mean Square Error), defined as [33]:

$$\text{MBE} = \frac{\sum(\text{Measured}_{\text{day}} - \text{TRNSYS}_{\text{day}})}{\sum(\text{Measured}_{\text{day}})} \times 100\% \tag{11}$$

$$\text{CV} - \text{RMSE} = \frac{\text{RMSE}_{\text{day}}}{\text{Mean}(\text{Measured}_{\text{day}})} \times 100\% \tag{12}$$

$$\text{RMSE} = \sqrt{\frac{\sum(\text{Measured}_{\text{day}} - \text{TRNSYS}_{\text{day}})^2}{(\text{Numberofday}) - 1}} \tag{13}$$

Figure 8 compares the daily heating load predicted by TRNSYS with the measured data. Due to the missing measured data for the 1st, 23th, and 24th, the simulation result for these 3 days was also removed for the validation. After the processing, the MBE was within the acceptable range (less than 5%), but the CV-RMSE was above the criteria for a calibrated model (less than 15%). In consideration that the measured data was only evaluated from the normal record of boiler fuel consumption, not a strict experiment, we can say this is an acceptable error, and the thermal load evaluated by TRNSYS can be used in the dynamic simulation of the FC-CHP system. Figure 9 presents the results of the evaluation of the space heating load and swimming pool heat loss based on the weather data of Donghae city. The peak load values are about 330 kW and 120 kW for space heating and swimming pool heat loss. For the swimming pool, high heat loss is also presented even in summer, due to the evaporation heat loss accounting for the largest proportion of the total heat loss.

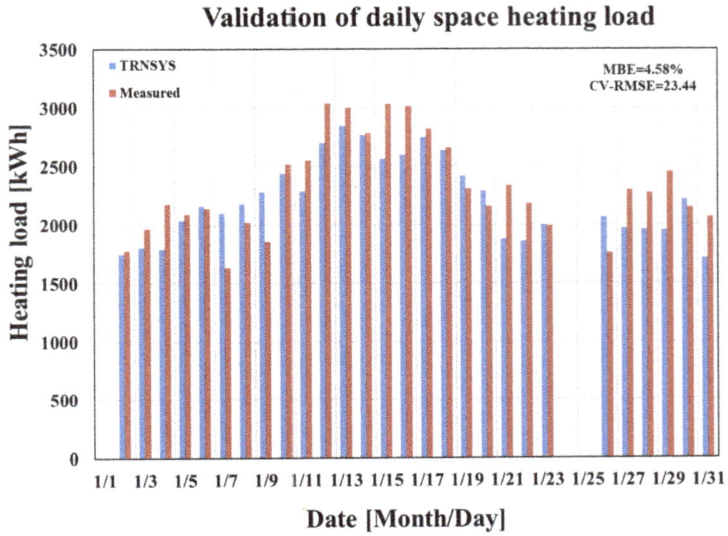

Figure 8. Validation of the daily heating energy consumption between the simulation data and measured data.

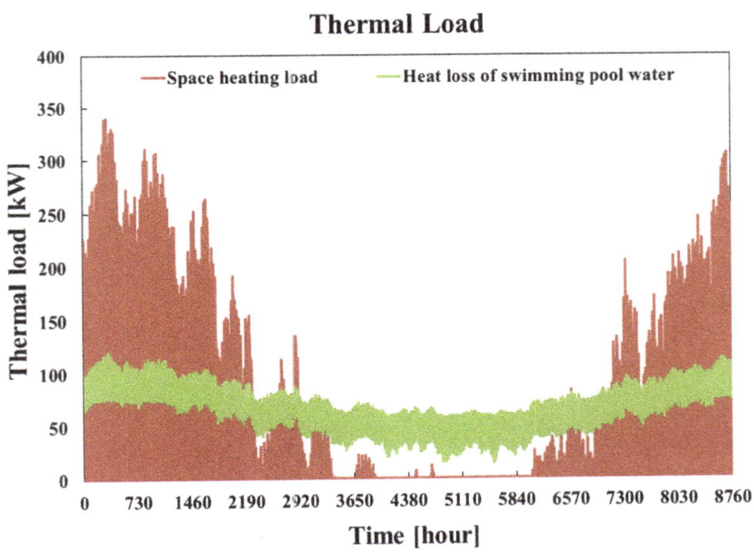

Figure 9. Building space heating load and swimming pool heat loss evaluated by the TRNSYS simulation.

Figure 10 presents the simulation results of the energy demand and supply of the FC-CHP system with the ELT (a) and FTO (b) FC operating strategy, including the building electricity load, thermal load, fuel cell electric power generation, heat generation (high grade and low grade), and effective heat. For the ELT model, the fuel cell will not work during the night time due to the electricity load being less than 100 kW, which is the minimum electric power capacity of the fuel cell. For the FTO model, the fuel cell is working at the maximum capacity all of the time, so that both the electric power and thermal heat generation are presented as the constant maximum value. Here, it has to be mentioned that, for the thermal heat supply for a commercial building, the high-grade waste heat from a fuel cell that has a temperature of about 120 °C cannot be supplied to the building directly. A heat exchanger is recommended to transform the high-grade heat temperature to a low grade. Therefore, considering the heat exchanger efficiency, the effective heat is necessary to evaluate and be used for the analysis of the primary energy saving and economic feasibility.

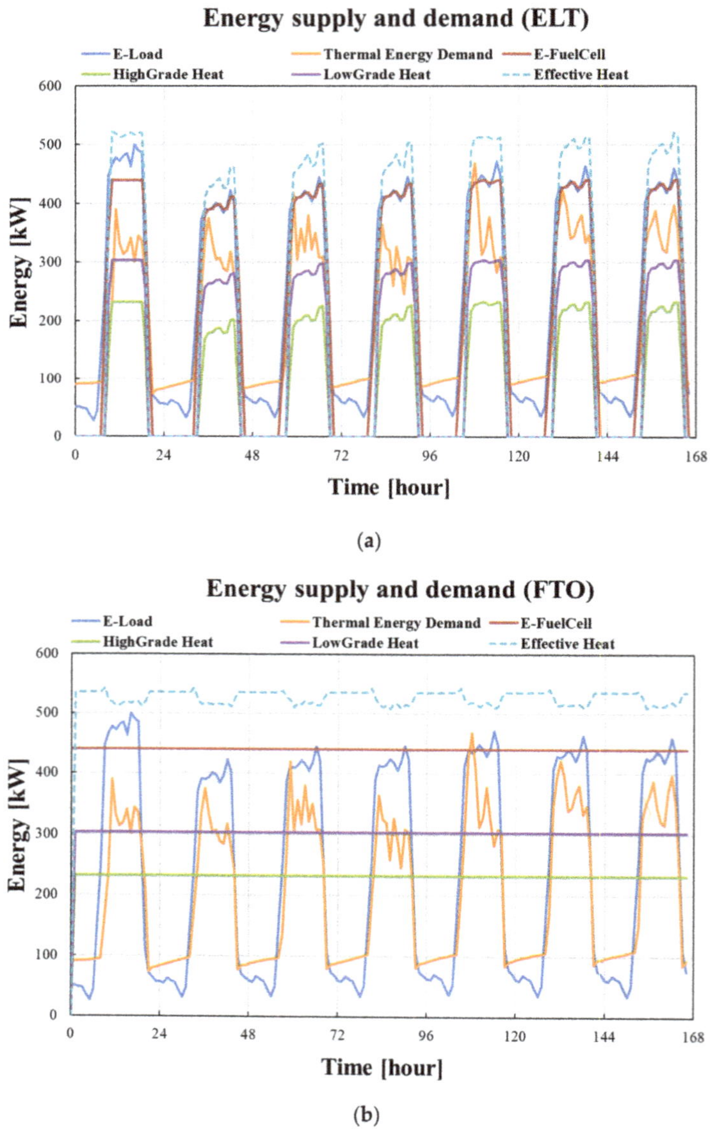

Figure 10. Energy demand and supply of FC-CHP systems for the first week of January ((**a**) ELT model; (**b**) FTO model).

The primary energy consumption for both the ELT and FTO models are evaluated from the annual system simulation. The results are compared with a reference case with a conventional power and heat supply method. The primary energy consumption can be evaluated as:

$$PEC = F_{FC} + F_b + F_{GE} \quad (14)$$

where F_{FC}, F_b, and F_{GE} are the fuel energy consumption of the fuel cell, boiler, and electric power supplied from the main power grid. For the FC-CHP system, the fuel cell and boiler fuel energy consumption can be evaluated from the simulation directly. In the case of the conventional energy supply method, the efficiency of the gas boiler and grid power are assumed to be 0.85 [32] and 0.35 [34,35], respectively.

Figure 11 presents the monthly primary energy consumption in the case of the ELT and FTO models in the FC-CHP system, and the conventional energy supply. The ELT model has lower primary energy consumption than the other two cases, and compared with the reference case, it can save about 10% to 15% fuel energy in the winter season. During the summer, even the large heat loss of the swimming pool can be covered by the fuel cell waste heat; the energy saving is not obvious due to the fact that the fuel cell is always operated with a relatively low efficiency under the ELT model. The FTO model consumes much more fuel energy than the two other cases due to its longer working time and high electrical output capacity. However, from a macro point of view, the FTO model still has its own advantages, such as high electrical efficiency and high waste heat generation.

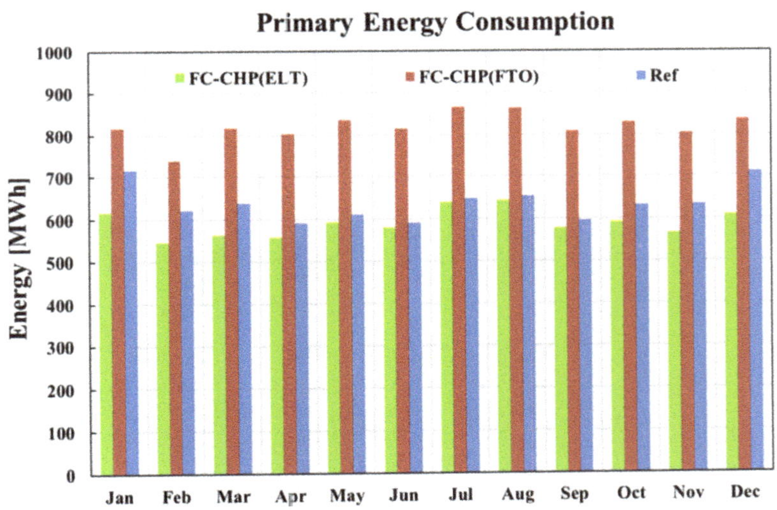

Figure 11. Primary energy consumption for the three case studies (ELT and FTO models in FC-CHP systems and the conventional energy supply model).

The CHP system can play an important role in both the primary energy saving and the reduction of CO_2 emissions. The evaluation of CO_2 emissions is complex due to its being highly dependent on the type of primary energy source. For the fuel cell system used in this study, from the manufacturer's report, the CO_2 emission factor is 0.453 tCO_2/MWh [28]. The CO_2 emission factors of coal and natural gas are assumed to 0.9 and 0.4 tCO_2/MWh, respectively [36]. For the natural gas boiler, the emission factor is set to be 0.2 tCO2/MWh. Therefore, by considering the fraction of electricity generation by the source types, the CO_2 emissions can be evaluated. As can be seen from Figure 12, whether the CHP system is operating in the ELT model or FTO model, the CO_2 emissions of the system are significantly lower than the traditional energy supply mode. Even the FTO model has a large primary energy consumption; the CO_2 emissions are still much lower than the conventional energy system due to the high fuel cell efficiency.

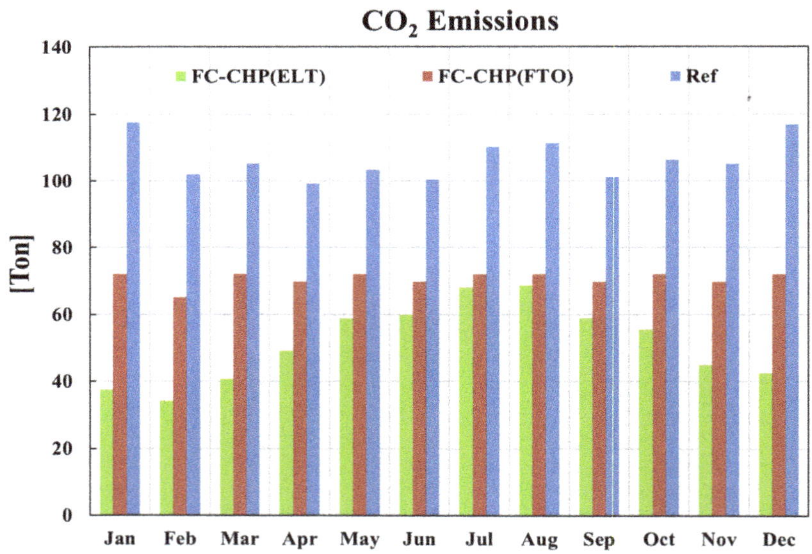

Figure 12. CO_2 emissions for the three case studies (ELT and FTO models in FC-CHP systems and the conventional energy supply model).

The fuel cell efficiency was defined in Equations (1)–(3). The fuel cell is supported to operate with a maximum capacity for all of the time under the FTO operating model so that it can give a significantly high efficiency than the ELT model which is operating by following the electricity load. Here, it has to be mentioned that, for the investigation of the fuel cell efficiency, only the total generation of power and heat are considered. Figure 13 shows the total efficiency of the fuel cell under both operating models. By considering the utilization of energy, the efficiency of the FC-CHP system can be evaluated as follows:

$$\eta_{FC-CHP} = (E_{FC} + Q_{FC-B})/F_{FC} \quad (15)$$

where the E_{FC} is the electrical power generation from the fuel cell, $Q_{FC\text{-}B}$ is the recovered heat from the fuel cell and supply to the building, and F_{FC} is the fuel (natural gas) energy consumption of the fuel cell. Here, the electric power generation for the FTO model includes both amounts which will be supplied to the building and sold to the grid. From Figure 14, we can see that for both the FTO and ELT models, the FC-CHP system presents higher efficiency in winter (about 55% to 60%) than in summer (40% to 50%), because of the high heating demand in winter. From the point of view of the overall efficiency of the fuel cell system, the FTO model has a large advantage over the ELT model. However, at the same time, a large amount of heat generated under the FTO operating model cannot be supplied to the building, but is released to the ambient environment. This is why, when it comes to the system's efficiency, there is not much difference between the two models.

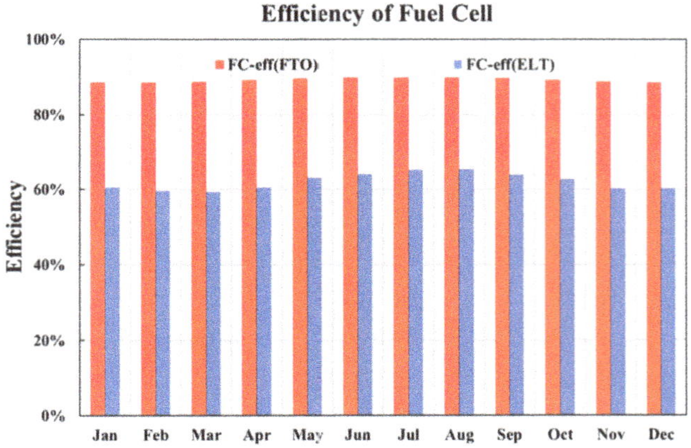

Figure 13. Total efficiency of the fuel cell under the FTO and ELT operating models.

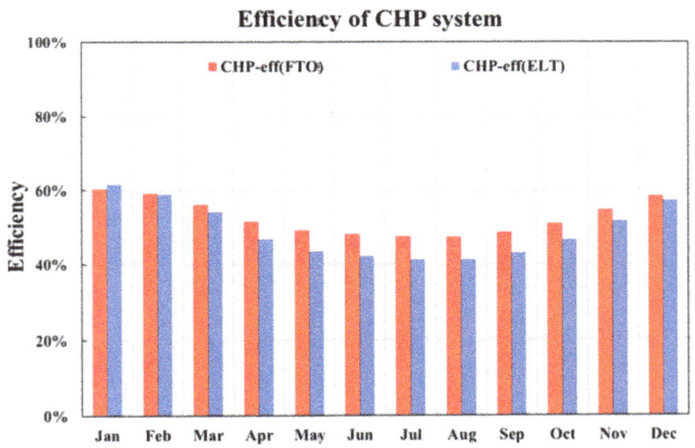

Figure 14. Efficiency of the FC-CHP system under the FTO and ELT operating models.

4. Economic Feasibility

From a general point of view, the fuel cell system is far from being economically feasible due to its high capital and operating cost [37]. However, a great potential has been presented for improving the utilization of energy productions, especially of thermal energy, due to the high heat demand of the commercial building in this study. The operating cost of the power and heat supply for the objective commercial building with an indoor swimming pool can be evaluated as:

$$C = C_f + C_{GE} - S_{EG} \tag{16}$$

where C_f is the cost of the fuel consumption of the fuel cell and boiler for both FC-CHP and conventional energy supply system. C_{GE} is the electric power cost from the main power grid, and S_{EG} is the income created by selling the excess electric power to the main power grid. The price of liquid natural gas (LNG) for industry and fuel cell applications is given as 44.98 and 44.13 USD/MWh (VAT included), respectively [38]. Unlike the price of natural gas, the selling price of fuel cell electricity to the grid is relatively unstable, and should be evaluated as:

$$p_{sell} = \text{SMP} + c_{FC}\text{REC} \tag{17}$$

In Korea, when selling electricity produced by fuel cell, the price is determined by the System Marginal Price (SMP) [39] and the unit price of Renewable Energy Certificates

(REC). REC certifies that power generators produced and supplied electricity by using new and renewable energy facilities, and the weighting factor of fuel cell (c_{FC}) application in Korea is given to be two [40]. Due to the fact that the SMP and the price of REC may fluctuate in real-time, which will have big impacts on the selling price of the electricity, a sensitivity analysis for the effect of the SMP and REC price was carried out by setting the selling price to low, medium and high levels. In Table 4, the monthly average value of the SMP and REC from January 2019 to present are evaluated and set to be the medium level of the price. The low and high level of the price are evaluated by raising and lowering the medium level price by 15%, respectively.

Table 4. SMP and REC prices.

	SMP (USD/MWh)	REC (USD/MWh)
Low (L)	60.24797	36.17903
Medium (M)	70.87996	42.56356
High (H)	81.51196	48.94809

Finlay, the net profit (NP) for the FC-CHP system compared with the conventional energy supply system is evaluated as:

$$\mathrm{NP} = C_{FC-CHP} - C_{REF} \tag{18}$$

The payback period (PP) is another important indicator for the evaluation of economic feasibility. Unlike a general economic model, the main benefit of applying the FC-CHP system is to obtain energy and cost saving by methods other than the traditional energy supply mode. Therefore, the payback period of the FC-CHP system could be simply evaluated by

$$\mathrm{PP} = C_0/(\mathrm{NP} - C_{MD}) \tag{19}$$

Here, C_0 and C_{MD} are the initial cost of the fuel cell system installation, and the maintenance and depreciation cost, respectively. The investment costs, including the fuel cell cost, installation cost, maintenance and depreciation costs are listed in Table 5. Here, it has to be mentioned that for both of the CHP and traditional system, the boiler and storage tank are needed. In this study, the cost saving by the reduction of the storage tank and boiler capacities are ignored.

Table 5. Investigation of the investment and incentive.

Total cost for fuel cell CHP system	$1,800,000
Installation cost	$600,000
Maintenance and depreciation costs	$50,000/year
Incentive from local government	$500,000
Total initial cost (with incentive)	$1,900,000
Total initial cost (without incentive)	$2,400,000

Table 6 summarizes the monthly operating cost and the net profit of the FC-CHP system under the ELT by comparing it with the conventional system. For the ELT model, the economic benefits are not significant. Even in March, April, May, September, and October, there is not only the profit but also the amount of loss presented. In the case of the FTO model, a sensitivity analysis for the economic feasibility is needed due to the large fluctuation of the SMP and REC prices. As shown in Table 4, the SMP and REC price are evaluated into low (L), medium (M) and high (H) levels. Thus, a total of nine market conditions of the electricity selling price are defined by the combination of the SMP and REC prices in different levels. For example, the condition of "LM" indicates that the electricity selling price will be calculated with the low level of the SMP and the medium level of the REC price. Table 7 presents the monthly and annual operating costs of the FC-CHP system under the FTO model and the net profit (NP) by comparing them with the conventional energy supply system in MM condition. A significant net profit is obtained due to that the fuel cell operating for the full time within its rated capacity. A large amount of excess electricity could be sold to the grid. The annual net profit is about 226,097 USD in the MM condition.

Table 6. Operating cost of the FC-CHP model under the ELT operating model and the net profit (NP), in comparison with the conventional energy supply system.

Caparison Data Sets of Operating Cost ($: USD)							
Month	**New System with FC-CHP (ELT)**			**Reference Data from Conventional System**			**NP**
	Fuel Cost	**e-Cost**	**Total**	**Fuel Cost**	**e-Cost**	**Total**	
Jan	23,835	2861	26,696	9446	25,103	34,549	7853
Feb	21,217	2572	23,789	8049	22,489	30,538	6748
Mar	22,363	1930	24,293	7504	17,832	25,337	1043
Apr	21,782	2212	23,994	5174	17,993	23,168	−827
May	22,888	2458	25,347	4156	18,886	23,042	−2305
Jun	22,280	3605	25,885	3544	26,995	30,540	4655
Jul	23,710	3449	4655	3344	28,338	31,682	27,026
Aug	23,695	4830	28,525	3295	28,419	31,714	3188
Sep	22,247	2434	24,680	3780	18,642	22,422	−2259
Oct	22,896	2401	25,297	5033	19,252	24,285	−1012
Nov	21,905	3006	24,911	6604	24,344	30,948	6037
Dec	23,214	3394	26,608	8548	25,641	34,189	7581
Sum	272,032	35,152	284,680	68,478	273,934	342,412	57,728

Table 7. Operating cost of the FC-CHP model under the FTO operating model and net profit (NP), in comparison with the conventional energy supply system.

Caparison Data Sets of Operating Cost ($: USD)								
Month	**New System with FC-CHP (FTO)**				**Reference Data from Conventional System**			**NP**
	Fuel Cost	**e-Cost**	**e-Sale**	**Total**	**Fuel Cost**	**e-Cost**	**Total**	
Jan	35,432	843	26,948	9327	9446	25,103	34,549	25,222
Feb	32,004	849	25,275	7578	8049	22,489	30,538	22,960
Mar	35,432	810	29,030	7212	7504	17,832	25,337	18,125
Apr	34,289	1104	27,468	7925	5174	17,993	23,168	15,242
May	35,432	1268	27,105	9596	4156	18,886	23,042	13,447
Jun	34,289	1857	25,853	10,294	3544	26,995	30,540	20,246
Jul	35,432	2881	25,409	12,904	3344	28,338	31,682	18,777
Aug	35,432	2780	25,074	13,139	3295	28,419	31,714	18,575
Sep	34,289	1250	25,538	10,001	3780	18,642	22,422	12,421
Oct	35,432	1122	26,630	9925	5033	19,252	24,285	14,359
Nov	34,289	1291	26,899	8682	6604	24,344	30,948	22,267
Dec	35,432	1514	27,214	9733	8548	25,641	34,189	24,456
Sum	417,189	17,569	318,442	116,315	68,478	273,934	342,412	226,097

Figure 15 presents the net profits and payback period in nine price conditions when the system is operating with the FTO model. The net profit made in comparison with the conventional energy supply system is about 178,352 to 273,879 USD per year. As we investigated, the initial costs for a 440 kW PAFC base CHP system with and without incentives are around 1,900,000 and 2,400,000 USD, respectively. If the system is operating with the FTO model, the payback periods with and without incentives from the local government are 6.9 to 10.7 years and 8.8 to 13.5 years, respectively.

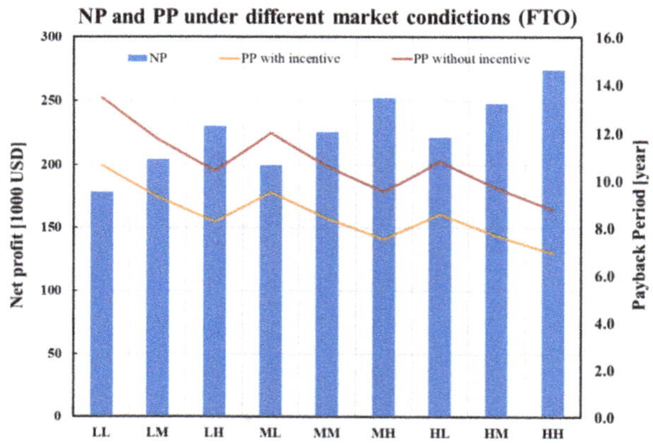

Figure 15. Net profit and payback period under the FTO model with different electricity market conditions.

5. Conclusions

In this study, a comprehensive sports center with an indoor swimming pool which is under construction in Donghae City, Korea is supported by a 440 kW PAFC-based CHP system for its electrical and thermal energy supply. The dynamic simulation of the energy supply system based on the investigated fuel cell performance data, the electrical load of a reference commercial building, and the evaluated thermal load was carried out for the cases with the ELT and FTO operating model. From the results, several key conclusions can be derived as follows:

1. A weather-data-based dynamic simulation model of the FC-CHP system including the components of the 3D-building load, fuel cell system, back-up boiler, heat exchanger, and storage tank was developed. The thermal load was obtained from the dynamic simulation, and was well-validated by the measured data. It can provide a reliable basis for the system simulation, thereby enhancing the credibility of the simulation results.
2. The FTO model was applied as one of the strategies for the fuel cell operating in a CHP system, and was simulated by the dynamic model and compared with the ELT model—which has been widely used in the CHP systems—and the conventional energy supply system in the aspects of primary energy consumption, fuel cell and system efficiency, and CO_2 emissions. From the simulation results, the FTO model presents the highest primary energy consumption and fuel cell efficiency, due to its long operating time and high capacity. The ELT model presents the lowest primary energy consumption, and can save 10% to 15% of the energy compared to the conventional energy system. However, the ELT model has a relatively low fuel cell efficiency compared to the FTO model. For both of FTO and ELT models, the FC-CHP system presents higher efficiency and CO_2 reduction.
3. In the analysis of the economic feasibility, the FTO model presents much better potential than the ELT model. The net profit made in comparison with the conventional energy supply system is about 178,352 to 273,879 USD per year, and the payback period is expected to be 6.9 to10.7 years under different market conditions.
4. Based on the analyses of the performance and economic feasibility of the system, the FTO model is suggested to be the operating strategy for the design of the fuel cell CHP system. However, more case studies with novel design parameters and operating strategies should be tested in future work.
5. As the objective building in this study is in construction now, the results obtained in this study should be validated by the real measured data after the system's completion.

More verification and testing will be helpful to improve the computational model, and thus make it more widely utilized.

Author Contributions: J.L., performed the system modeling and numerical analyses, and drafted the manuscript. S.-C.K., discussed the results. K.-Y.S., collected the measured data, organized the overall evaluation, and reviewed the manuscript. All authors have read and agreed to the published version of the manuscript.

Funding: This research received no external funding.

Institutional Review Board Statement: Not applicable.

Informed Consent Statement: Not applicable.

Data Availability Statement: Not applicable.

Acknowledgments: This work was funded by Donghae City, Gangwon Province, and supported by Donghae Coal-Fired Power Plant in Korea East-West Power Company and A1 Engineering, as a project partner. The authors really appreciate their support during this project.

Conflicts of Interest: The authors declare no conflict of interest.

Nomenclature

A	Area (mm^2)
AFC	Alkaline fuel cell
c	Impact factor of renewable energy
C	Cost ($USD)
CHP	Combined heat and power
CV-RMSE	Coefficient of variance of the root mean square error
E	Electric power, energy (kW, kWh)
F	Fuel energy consumption
FC-CHP	Fuel-cell-based combined heat and power
h	Convective heat transfer coefficient (W/mK);
h_{evp}	Evaporation heat (kJ/kg)
m_{evp}	Mass of evaporation water (kg/m^{2}h)
MBE	Mean bias error
MCFC	Molten carbonate fuel cell
NP	Net profit (USD/year)
p	Price (USD/kWh)
PAFC	Phosphoric acid fuel cell
PEMFC	Proton exchange membrane fuel cell
PP	Payback period (year)
Q	Heat (kW)
REC	Renewable energy certificates
S	Economic income
SMP	System marginal price (USD/MWh)
SOFC	Solid oxide fuel cell
T	Temperature (°C)
U	Conductive heat transfer coefficient (W/mK)
v	Indoor air speed (m/s)
x	Humidity radio (kg/kg)
Greek symbols	
ε	Emissivity
σ	Stefan–Boltzmann constant (W/m^{2}K^4)

Subscripts

0	Initial
a	Air
b	Boiler
cel	Celling
cond	Conduction
conv	Convection
evp	Evaporation
EG	Electric power sold to the main power grid
f	Fuel
FC	Fuel cell
FC-B	Fuel cell to building
FC-CHP	Fuel-cell-based CHP system
GE	Electric power supplied from the main power grid
pw	Wall surfaces of the swimming pool
rad	Radiation
REF	Reference
S	Surface of the swimming pool
sw	Swimming pool water

References

1. Arsalis, A. A comprehensive review of fuel cell-based micro-combined-heat-and-power systems. *Renew. Sustain. Energy Rev.* **2019**, *105*, 391–414. [CrossRef]
2. Mago, P.J.; Smith, A.D. Evaluation of the potential emissions reductions from the use of CHP systems in different commercial buildings. *Build. Environ.* **2012**, *53*, 74–82. [CrossRef]
3. Ahmed, S.; Papadias, D.D.; Ahluwalia, R.K. Configuring a fuel cell based residential combined heat and power system. *J. Power Sources* **2013**, *242*, 884–894.
4. Chamra, L.M.; Mago, P.J. Micro-CHP power generation for residential and small commercial buildings (Ch. 2). In *Electric Power Research Trends*; Schmidt, M.C., Ed.; Nova Science Publisher, Inc.: New York, NY, USA, 2007.
5. Gandiglio, M.; Lanzini, A.; Santarelli, M.; Leone, P. Design and optimization of a proton exchange membrane fuel cell CHP system for residential use. *Energy Build.* **2014**, *69*, 381–393.
6. Löbberding, L.; Madlener, R. Techno-economic analysis of micro fuel cell cogeneration and storage in Germany. *Appl. Energy* **2019**, *235*, 1603–1613. [CrossRef]
7. Staffell, I.; Ingram, A. Life cycle assessment of an alkaline fuel cell CHP system. *Int. J. Hydrogen Energy* **2010**, *35*, 2491–2505. [CrossRef]
8. U.S. Department of Energy. Available online: https://www.energy.gov/sites/prod/files/2016/06/f32/fcto_fuel_cells_comparison_chart_apr2016.pdf (accessed on 15 March 2021).
9. Verhaert, I.; Mulder, G.; De Paepe, M. Evaluation of an alkaline fuel cell system as a micro-CHP. *Energy Conv. Manag.* **2016**, *126*, 424–445.
10. Wang, Y.; Chen, K.S.; Mishler, J.; Cho, S.C.; Adroher, X.C. A review of polymer electrolyte membrane fuel cells: Technology, applications, and needs on fundamental research. *Appl. Energy* **2011**, *88*, 981–1007. [CrossRef]
11. Kang, K.; Yoo, H.; Han, D.; Jo, A.; Lee, J.; Ju, H. Modeling and simulations of fuel cell systems for combined heat and power generation. *Int. J. Hydrogen Energy* **2016**, *41*, 8286–8295.
12. Chahartaghi, M.; Kharkeshi, B.A. Performance analysis of a combined cooling, heating and power system with PEM fuel cell as a prime mover. *Appl. Therm. Eng.* **2018**, *128*, 805–817.
13. Park, Y.J.; Min, G.; Hong, J. Comparative study of solid oxide fuel cell-combined heat and power system designs for optimal thermal integration. *Energy Conv. Manag.* **2019**, *182*, 351–368. [CrossRef]
14. Naimaster IV, E.J.; Sleiti, A.K. Potential of SOFC CHP systems for energy-efficient commercial buildings. *Energy Build.* **2013**, *61*, 153–160.
15. Sorace, M.; Gandiglio, M.; Santarelli, M. Modeling and techno-economic analysis of the integration of a FC based micro-CHP system for residential application with a heat pump. *Energy* **2017**, *120*, 262–275.
16. Napoli, R.; Gandiglio, M.; Lanzini, A.; Santarelli, M. Techno-economic analysis of PEMFC and SOFC micro-CHP fuel cell systems for the residential sector. *Energy Build.* **2015**, *103*, 131–146.
17. Zhang, H.; Lin, G.; Chen, J. Multi-objective optimisation analysis and load matching of a phosphoric acid fuel cell system. *Int. J. Hydrogen Energy* **2012**, *37*, 3438–3446. [CrossRef]
18. Wu, M.; Zhang, H.; Zhao, J.; Wang, F.; Yuan, J. Performance analyzes of an integrated phosphoric acid fuel cell and thermoelectric device system for power and cooling cogeneration. *Int. J. Refrig.* **2018**, *89*, 61–69. [CrossRef]
19. Sammes, N.; Bove, R.; Stahl, K. Phosphoric acid fuel cells: Fundamentals and applications. *Curr. Opin. Solid State Mat. Sci.* **2004**, *8*, 372–378. [CrossRef]

20. Chen, X.; Wang, Y.; Cai, L.; Zhou, Y. Maximum power output and load matching of a phosphoric acid fuel cell-thermoelectric generator hybrid system. *J. Power Sources* **2015**, *294*, 430–436.
21. Ito, H. Economic and environmental assessment of phosphoric acid fuel cell-based combined heat and power system for an apartment complex. *Int. J. Hydrogen Energy* **2017**, *42*, 15449–15463. [CrossRef]
22. Acha, S.; Le Brun, N.; Damaskou, M.; Fubara, T.C.; Mulgundmath, V.; Markides, C.N.; Shah, N. Fuel cells as combined heat and power systems in commercial buildings: A case study in the food-retail sector. *Energy* **2020**, *206*, 118046. [CrossRef]
23. Mago, P.J.; Fumo, N.; Chamra, L.M. Performance analysis of CCHP and CHP systems operating following the thermal and electric load. *Int. J. Energy Res.* **2009**, *33*, 852–864. [CrossRef]
24. Im, Y.H.; Liu, J. Feasibility study on the low temperature district heating and cooling system with bi-lateral heat trades model. *Energy* **2018**, *153*, 988–999. [CrossRef]
25. Hawkes, A.D.; Leach, M.A. Cost-effective operating strategy for residential micro-combined heat and power. *Energy* **2007**, *32*, 711–723. [CrossRef]
26. Yan, H.L.; Wang, G.P.; Lu, Z.W.; Tan, P.; Kwan, T.H.; Xu, H.R.; Chen, B.; Ni, M.; Wu, Z. Techno-economic evaluation and technology roadmap of the MWe-scale SOFC-PEMFC hybrid fuel cell system for clean power generation. *J. Clean Prod.* **2020**, *255*, 120225. [CrossRef]
27. Al-Khori, K.; Bicer, Y.; Aslam, M.I.; Koç, M. Flare emission reduction utilizing solid oxide fuel cells at a natural gas processing plant. *Energy Rep.* **2021**, 5627–5638. [CrossRef]
28. Doosan Fuel Cell. Available online: https://www.doosanfuelcell.com/en (accessed on 21 May 2021).
29. Thomas Auer, Assessment of an Indoor or Outdoor Swimming Pool, TRNSYS-TYPE 144. Available online: https://sel.me.wisc.edu/trnsys/components/type144-manual.pdf (accessed on 20 March 2021).
30. Smith, C.C.; Lof, G.O.G.; Jones, R.W. Rates of evaporation from swimming pools in active use. *ASHRAR Trsns* **1998**, *104*, 14–23.
31. Lam, J.C.; Chan, W.W. Life cycle energy cost analysis of heat pump application for hotel swimming pools. *Energy Conv. Manag.* **2001**, *42*, 1299–1306. [CrossRef]
32. Liu, J.; Chunga, D.H.; Chung, M.; Im, Y.H. Development of load models and operation simulator for a building complex with mixtures of multi-type engines and renewable devices. *Energy Build.* **2018**, *158*, 831–847. [CrossRef]
33. Djunaedy, E.; Van Den Wymelenberg, K. Targeted Calibration of Energy Models for Existing Building. In Proceedings of the ASHRAE 2014 Conference, Seattle, WA, USA, 28 June–2 July 2014.
34. IEA (2021). Available online: https://www.iea.org/countries/korea (accessed on 30 May 2021).
35. Hong, J.H.; Kim, J.; Son, W.; Shin, H.; Kim, N.; Lee, W.K.; Kim, J. Long-term energy strategy scenarios for South Korea: Transition to a sustainable energy system. *Energy Policy* **2019**, *127*, 425–473. [CrossRef]
36. Nam, H.; Nam, H.; Lee, D. Potential of hydrogen replacement in natural-gas-powered fuel cells in Busan, South Korea based on the 2050 clean energy Master Plan of Busan Metropolitan City. *Energy* **2021**, *221*, 119783. [CrossRef]
37. Sevencan, S.; Lindbergh, G.; Lagergren, C.; Alvfors, P. Economic feasibility study of a fuel cell-based combined cooling, heating and power system for a data center. *Energy Build.* **2016**, *111*, 218–223. [CrossRef]
38. Korea City Gas Association. Available online: http://www.citygas.or.kr/index.jsp (accessed on 25 June 2021).
39. Ministry of Trade, Industry and Energy. Available online: http://english.motie.go.kr/www/main.do (accessed on 30 September 2021).
40. Korea Energy Agency. Available online: https://dco.energy.or.kr/renew_eng/main/main.aspx (accessed on 30 September 2021).

energies

Article

Investigating the Impact of Plumbing Configuration on Energy Savings for Falling-Film Drain Water Heat Recovery Systems

Ramin Manouchehri * and Michael R. Collins

Solar Thermal Research Laboratory, Department of Mechanical and Mechatronics Engineering, University of Waterloo, Waterloo, ON N2L 3G1, Canada; mike.collins@uwaterloo.ca
* Correspondence: ramin.manouchehri@uwaterloo.ca

Abstract: Falling-film drain water heat recovery (DWHR) systems are heat exchangers utilized in residential buildings for recovering energy from greywater. A recent publication by the authors contained a validated model that can be used to predict the performance of DWHR heat exchangers under variable flowrates and temperatures, and this work shows the implementation of the model into Transient System Simulation Tool (TRNSYS) software to perform energy simulations. This work aims to show the different plumbing configurations in which DWHR heat exchangers could be installed, and to simulate their performance under various conditions. The results show that plumbing configuration has a significant impact on energy savings expected from DWHR heat exchangers, and maximum savings are achieved in equal-flow configuration. However, other plumbing configurations provide significant savings, and the mains temperature could dictate which configuration provides higher energy savings.

Keywords: falling-film drain water heat recovery; heat exchanger; TRNSYS; energy simulations; variable plumbing configuration

Citation: Manouchehri, R.; Collins, M.R. Investigating the Impact of Plumbing Configuration on Energy Savings for Falling-Film Drain Water Heat Recovery Systems. *Energies* **2022**, *15*, 1141. https://doi.org/10.3390/en15031141

Academic Editor: Jan Danielewicz

Received: 31 December 2021
Accepted: 31 January 2022
Published: 3 February 2022

Publisher's Note: MDPI stays neutral with regard to jurisdictional claims in published maps and institutional affiliations.

1. Introduction

Since 2000, water heating has consistently been the second largest contributor to the total energy consumption in residential buildings in Canada. In the year 2018, 281.3PJ of energy consumed in the Canadian residential sector was attributed to water heating; this represents 17.4% of the total energy consumption and 19.2% of greenhouse gas emissions [1]. Similarly, domestic water heating accounted for 14.8% of the total energy consumption in the residential sector in Europe [2]. Evidently, hot water usage is a significant part of modern housing, and seeking to reduce its impact on the environment is worth investigating.

According to a recent study by Chen et al., an estimated 34.8% of total hot water used in residential buildings within the United States is attributed to showers [3]. Furthermore, on average, drain water holds 80–90% of its thermal energy relative to the mains water supply temperature [4]. Clearly, a significant amount of energy that is consumed towards domestic water heating is not fully utilized during showers, which presents an opportunity to use heat exchangers to recover thermal energy from greywater in residential buildings. This work focuses on the most common type of heat exchanger used for this purpose, namely Falling-Film Drain Water Heat Recovery (DWHR) units.

DWHR heat exchangers consist of a large diameter drainpipe which is wrapped tightly with a coil of smaller tubes. The drainpipe and the tubes are both made of copper. Figure 1 presents a selection of commercially available DWHR heat exchangers, highlighting variations in length, diameter, and coil design for different units. During operation, greywater goes down the drainpipe portion of the heat exchanger, which has a diameter matching the size of the drain stack it replaces. Concurrently, mains water is circulated within the coiled tubes wrapped around the drainpipe which recovers heat from greywater. These heat exchangers are designed to be installed vertically, which implies that greywater forms a falling-film as it traverses down the drainpipe [5]. This falling-film of water ensures that the entire inner surface area of the drainpipe is covered in water, thus providing the heat transfer area for heat recovery to occur. It is worth noting that horizontal designs for such heat exchangers do not rely on a falling-film, and according to a study by Ravichandran et. al., horizontally installed heat exchangers have lower efficiencies than the ones installed vertically [6]. This study only considers heat exchangers that are designed and rated for vertical installation.

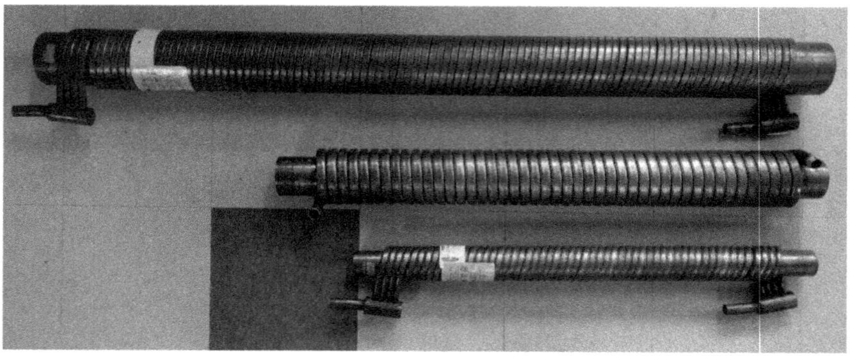

Figure 1. Common designs for falling-film drain water heat recovery (DWHR) heat exchangers.

DWHR systems can be installed in different configurations. Figure 2 shows three variations (A, B and C) which are simplified to only contain a DWHR heat exchanger in addition to a showerhead for water-draw purposes, and each showerhead is equipped with a thermostatic mixing valve. Figure 2A depicts a system where all preheated water from the DWHR system goes to the water heater, Figure 2B depicts a system where all preheated water goes to the mixing valve at the showerhead, and Figure 2C shows a combination of the two previous cases. Lastly, Figure 2D is provided to show the same plumbing system without a DWHR heat exchanger installed.

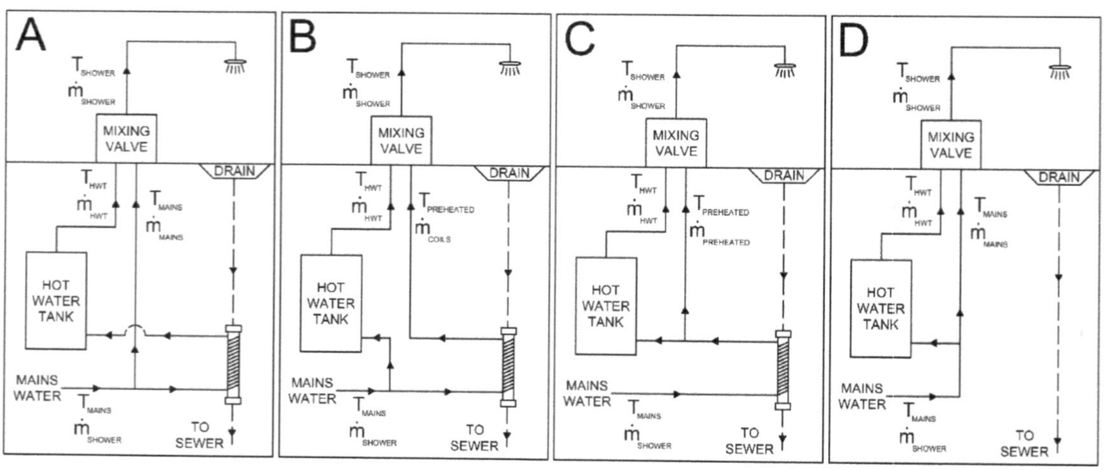

Figure 2. Possible plumbing configurations for DWHR systems in a residential building: (**A**) depicts a system where all preheated water from the DWHR heat exchanger is fed to the water heater; (**B**) depicts a system where all preheated water is fed to the mixing valve at the showerhead; (**C**) is a combination of (**A**) and (**B**); and (**D**) shows the same plumbing system without a DWHR heat exchanger installed.

Configuration C depicts equal flowrates of water through both sides of the heat exchanger, whereas configurations A and B depict unequal-flow conditions. The performance of DWHR systems is a strong function of flowrates [7] and based on the configurations shown, the flowrate of water through the heat exchanger coils is expected to vary based on plumbing configuration to provide the same temperature at the showerhead.

Researchers have been focused on simulating the performance of DWHR heat exchangers under equal-flow arrangements (i.e., config C). This is because the data available on the performance of DWHR heat exchangers is limited to the effectiveness data gathered during the CSA rating process [8]. The rating system is meant to provide a means for comparing different DWHR heat exchangers on an equal basis; it does not and cannot provide any accurate estimates for different configurations in which a heat exchanger could be installed. A study by Slys and Kordana discusses the process for estimating the energy savings associated with different plumbing configurations and highlights the numerous assumptions that need to be made for such calculations [9]. In their work, the authors did not have access to a system model that accounted for varying heat exchanger effectiveness. Without having a model that accounts for the changing effectiveness for the system, there is no way to compare the results numerically or experimentally. This highlights the need for further studies with models that do not rely on assumptions such as a constant effectiveness for the heat exchanger.

It is worth noting that despite being the only configuration covered in standards, configuration C is not always an option, especially in retrofit settings where the plumbing has already been installed and additional modifications are costly. Furthermore, local building codes can impose limitations on which configurations are allowed. For example, configurations B and C are not permitted in Québec, Canada [10]. This restriction is due to the possibility of bacteria growth, such as Legionella, in preheated water that bypasses the water heater. In other words, only plumbing setups in which all preheated water is directed to the water heater are allowed, and that limits the allowable configurations to what was shown in Figure 2A. Clearly, assuming that DWHR heat exchangers are always installed and operated under equal-flow conditions is incorrect, and potentially against local codes. Thus, building simulation models must be created to predict performance of such heat exchangers under unequal-flow conditions.

1.1. Model

A series of studies were undertaken at the University of Waterloo to address the limitations in modelling of DWHR systems. Most notably, the impact of varying inlet temperatures [11] and inlet flow rates [12] were empirically characterized, and the combination of all studies were compiled into a model that can predict the performance of a DWHR heat exchanger under steady-state conditions. A recent publication covers the development and validation of the model in detail [13]. The model is validated experimentally and the mean absolute error between model predictions and experimental results is less than 3% for a total of 135 different validation cases performed in the study. The validation temperatures range from 4 to 50 °C, and the volumetric flow rates range from 4 to 20 L/min.

A detailed description of the model and the principles on which those models are based are not repeated here; instead, this study aims to use the aforementioned model to perform energy simulations for DWHR systems installed in configurations depicted in Figure 2. To this end, the model is implemented into Transient System Simulation Tool (TRNSYS) software as a component and used in simulations described in the following sections.

1.2. TRNSYS

TRNSYS was selected as suitable software for simulations because it allows the performance of simulations while accounting for transient behaviour in components such as water heaters. Transient behaviour of DWHR systems is currently being studied and will be added to model in the future.

A TRNSYS simulation works by allowing interaction between different 'components' on a timestep-by-timestep basis, and this time step is selected by the user at the beginning of each simulation. Each component consists of 'parameters' and 'inputs' which are used to generate 'outputs' for the components for each time step.

Parameters are relevant values that are used by TRNSYS to simulate a component throughout the simulation period. Parameters are independent of time and remain the same throughout the modelling process. In the context of the DWHR component in TRNSYS, the size of the heat exchanger is a parameter in the simulations. On the other hand, inputs to the simulation are variables that can change as a function of time. The inputs to the DWHR component include the inlet water temperatures and flowrates to the heat exchanger. The simulation then uses the inputs and parameters to generate outputs for the component for each time step. The outputs for the DWHR component include the heat recovery rate, heat exchanger effectiveness, and the outlet temperatures for both streams. A summary of the parameters, inputs, and outputs for the DWHR component are shown in Table 1.

Table 1. Parameters, inputs, and outputs associated with the DWHR component created in Transient System Simulation Tool (TRNSYS) software.

Type	Variable Name	Unit	Notes
Parameter	a	min/L	Regression coefficients taken from the characteristic effectiveness curve for the DWHR unit being simulated.
	b	Dimensionless	
Input	$T_{c,i}$	°C	Inlet water temperature to the coils
	$T_{h,i}$	°C	Inlet water temperature to the drainpipe
	$\dot{m}_{coils}$	kg/s	Mass flowrate of water through the coils
	$\dot{m}_{drain}$	kg/s	Mass flowrate of water through the drainpipe
Output	$T_{c,o}$	°C	Outlet water temperature from the coils
	$T_{h,o}$	°C	Outlet water temperature from the drainpipe
	q	kW	Heat recovery rate
	ε	Dimensionless	Heat exchanger effectiveness

Parameters a and b are regression coefficients that are taken from the characteristic effectiveness vs. flowrate curve that is generated during the rating process under the CSA B55.15 standard [8]. Figure 3 shows an example of this characteristic curve, where parameters a and b are determined to be 0.0685 min/L and 1.2796, respectively, and $\dot{V}$ denotes the volumetric flow rate in L/min. The TRNSYS component uses these parameters and the inputs assigned by the user to calculate outputs for each time step.

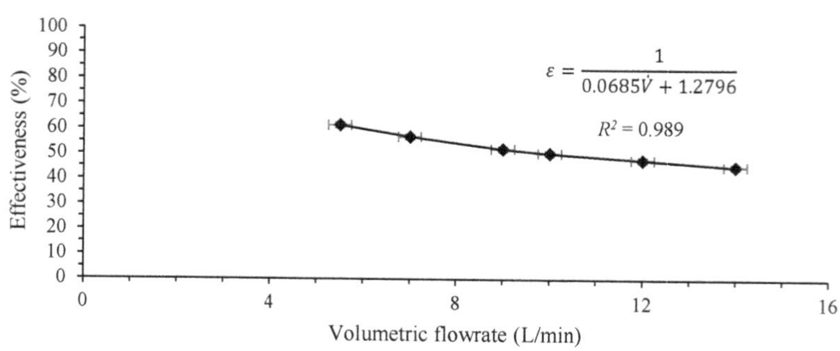

Figure 3. Equal-flow effectiveness vs. volumetric flowrate for a 7.6 cm diameter, 153 cm long DWHR heat exchanger.

2. Materials and Methods

The energy savings associated with a DWHR heat exchanger is highly impacted by the way it is installed in a particular home, and the conditions it is subjected to within the dwelling. The plumbing system within a house is impacted by local codes, the mains temperature depends on the location, and the duration for shower events is decided by the occupants. There are many possibilities for the conditions that a DWHR heat exchanger could be subjected to, and to allow comparison for energy savings under different conditions, a baseline must be established to provide consistency. This baseline is created based on Figure 2 for different plumbing configurations, but prior to running simulations, important simulation constraints, such as the shower condition, the water heater, and the water-draw schedule, must be selected. These key constraints will be discussed.

2.1. Shower Condition

A shower temperature of 38 °C was selected for all shower events. This temperature is based on the temperature used for rating all DWHR systems according to the CSA standard [8].

The flowrate of water through the showerhead is expected to have a significant impact on simulation results, as it would affect both the energy consumption by the water heater, and the energy recovery by the heat exchanger. It is difficult to select one showerhead flowrate to represent all households, as there are many fixtures available on the market, each having its own rated flowrate. In a typical dwelling, this flowrate is also affected by the available water pressure in the plumbing system. Hence, to simplify the matter, three flowrates were selected based on the range of flowrates prescribed by the CSA standard to represent showerheads having low, typical, and high flowrates. These flowrates are 5.5, 9.5 and 14 L/min.

The water draw schedule was chosen for a family of three, with three shower events per day, as noted in Table 2. Shower durations are highly subjective, and the chosen durations were meant to cover both reasonably short and long showers. As these simulations are focused only on savings associated with shower events, no other water draws were added to the daily water-draw schedule. To accommodate the water draws from Table 2, a typical 182 Litre (40-gallon) electric water heater was selected for simulations. The tank volume was found to be sufficient to accommodate all shower durations at all flowrates without depleting the tank. In other words, the shower temperatures do not drop below 38 °C during shower events.

Table 2. Start times and durations for shower events used in the simulations.

Shower Event #	Start Time	Duration
1	6:00 AM	6 min
2	7:00 AM	15 min
3	8:00 PM	9 min

2.2. Electric Water Heater

There are multiple water heaters available for simulation in TRNSYS, and it was crucial to select one that has been validated experimentally. An in-depth study by Allard et al. focused on validating different electric water heaters used in TRNSYS, and the results showed that TRNSYS 'Type 534' accurately depicts the variables of interest, namely the energy consumption and supply temperature during simulations [14]. Type 534 simulates the tank using a nodal approach while accounting for stratification effects. Hence, Type 534 was selected for simulation purposes. This tank was divided into 50 nodes for the simulations, which exceeds the minimum number of nodes recommended by Kleinbach et al. per Equation (1) [15]. Here, N_{FIXED} denotes the minimum number of nodes in the tank, and $Tank_{Turnover}$ is the ratio of daily water draw and total tank volume. $Tank_{Turnover}$ varies based on the showerhead flowrate used in the simulation; hence, N_{FIXED} was calculated for all flowrates. As a result, the tank was divided into 50 nodes, as this number of nodes satisfies the minimum number of nodes required for all flowrates and having a consistent number of nodes in the tank for all simulations allows fair comparison of the results. Table 3 provides a summary of relevant parameters for the water heater used in all simulations in this study.

$$N_{FIXED} = 45.8 \times (Tank_{Turnover})^{-1.218} \tag{1}$$

Table 3. Parameters used to characterize TYPE 534 hot water tank in this study.

Parameter	Value Used in Simulations	Unit
Number of Nodes	50	Dimensionless
Tank Volume	182 (40)	Liters (Gallons)
Tank Height	105 (41.4)	Centimeters (Inches)
Tank Loss Coefficient	3.5	kJ/(hr·m^2·K)
Environment Temperature	20 (68)	°C (°F)
Upper Element	3000	W
Lower Element	3000	W

2.3. Mains Temperature

The energy consumption associated with domestic water heating is directly tied to the mains temperature at the location where the water heater is located. To provide a fair assessment of the impact of mains temperature on the energy savings associated with DWHR heat exchangers, a large range of temperatures were required for simulations. The mains temperatures selected for the simulations were 2, 5, 10, 15, 20, and 25 °C.

The energy savings associated with DWHR systems can be looked at from different perspectives, and it is crucial to clarify how savings are presented in this study. It is insufficient to simply compare the magnitude of energy that was recovered by the DWHR system in each simulation, as there are other components present in domestic water heating. Most importantly, there are heat losses associated with the hot water tank, and the rate of heat recovery is expected to impact the losses, as well as how often the heating elements within the water heater must activate to provide auxiliary heat. Thus, it would be more appropriate to compare the total energy consumed by the water heater instead. For this purpose, the total energy consumed by the water heater in configurations A, B and C are compared to the energy consumption in configuration D (see Figure 2) under identical simulation parameters. This allows presentation of the savings in terms of percentages relative to configuration D.

Equation (2) is used to calculate the energy savings as a percentage of total energy consumed in the reference case (configuration D). E_D denotes the total auxiliary energy consumed by the water heater in configuration D, and E_{Sim} represents the total auxiliary energy consumed by the water heater in configurations A, B or C.

$$Energy\ Savings = \frac{E_D - E_{Sim}}{E_D} \times 100 \tag{2}$$

2.4. Heat Exchangers

Four heat exchangers were selected for this study. This selection spans different heat exchanger lengths and diameters. The corresponding characteristic effectiveness vs. flowrate equation obtained using the CSA standard procedure is also provided for each heat exchanger. See Table 4.

Table 4. Characteristics of the DWHR heat exchangers used in simulations.

	Diameter (cm)	Length (cm)	Curve-Fit Generated Using CSA Flow Rates
Heat Exchanger 1	5.1	91	$\varepsilon = \frac{1}{0.1548\dot{V}+1.7513}$
Heat Exchanger 2	7.6	122	$\varepsilon = \frac{1}{0.0918\dot{V}+1.3109}$
Heat Exchanger 3	7.6	152	$\varepsilon = \frac{1}{0.0681\dot{V}+1.2315}$
Heat Exchanger 4	10.2	152	$\varepsilon = \frac{1}{0.0529\dot{V}+1.2544}$

2.5. Simulation Time Step

Next, appropriate time step was to be determined. Selecting appropriate time steps is crucial for accurate simulations. Smaller time steps often provide higher accuracy at the expense of CPU-time. However, the time domain must be discretized such that all shower events can be appropriately captured by the chosen time step size. To investigate discretization errors, simulations were run for several time step sizes, and for all cases, the energy consumption by the water heater was calculated for a period of a month. A time step of 1 s was used as the reference, and the results from other time steps were compared to the reference case to estimate the discretization error. The results showed that a time step size of 5 min has a discretization error of approximately 17%, which was reduced to below 1% when a 1-min time step was used. Therefore, a time step of 1 min was found to be suitable for simulation purposes.

TRNSYS simulations were set up to perform a month (30 days) of simulation. The results are to be compared using the energy savings metric from Equation (2) for each heat exchanger.

3. Results and Discussion

The results showed that configuration C outperforms the other configurations for all simulations, regardless of flowrate or mains temperature. This is in-line with the expected performance of DWHR heat exchangers, as configuration C has the highest flow rate of water through the heat exchanger's coils during shower events. On the other hand, the mains temperature could dictate which unequal-flow configuration could lead to higher savings. At high mains temperatures, the savings associated with configuration B outpaced those of A, regardless of the heat exchanger being simulated or the showerhead flowrate. Figures 4–6 contain the results for simulations with showerhead flowrates of 5.5, 9.5 and 14 L/min respectively. Each figure contains four plots, which show the energy savings as a function of mains temperature for the three configurations. Note that each point on the plot represents a month of simulation.

It is important to note that significant energy savings can be achieved by installing a DWHR heat exchanger in plumbing systems regardless of the configuration. In situations where the equal-flow configuration is not possible, homeowners should not be discouraged from relying on configurations A or B. Unequal-flow configurations can be thought-of as having a slightly smaller heat exchanger installed relative to the equal-flow configuration, and selecting a larger heat exchanger has a more pronounced impact on energy savings compared to changing the plumbing configuration.

The energy-savings trend for configurations A and B can be explained through closer inspection of the mains side flowrate through the heat exchanger coils for different simulation scenarios. This is best done using the concept of heat capacity ratio, C_r, which is the ratio of heat capacity rates for the heat exchanger, as shown in Equation (3).

$$C_r = \frac{C_{min}}{C_{max}} = \frac{(\dot{m}C_p)_{min}}{(\dot{m}C_p)_{max}} \quad (3)$$

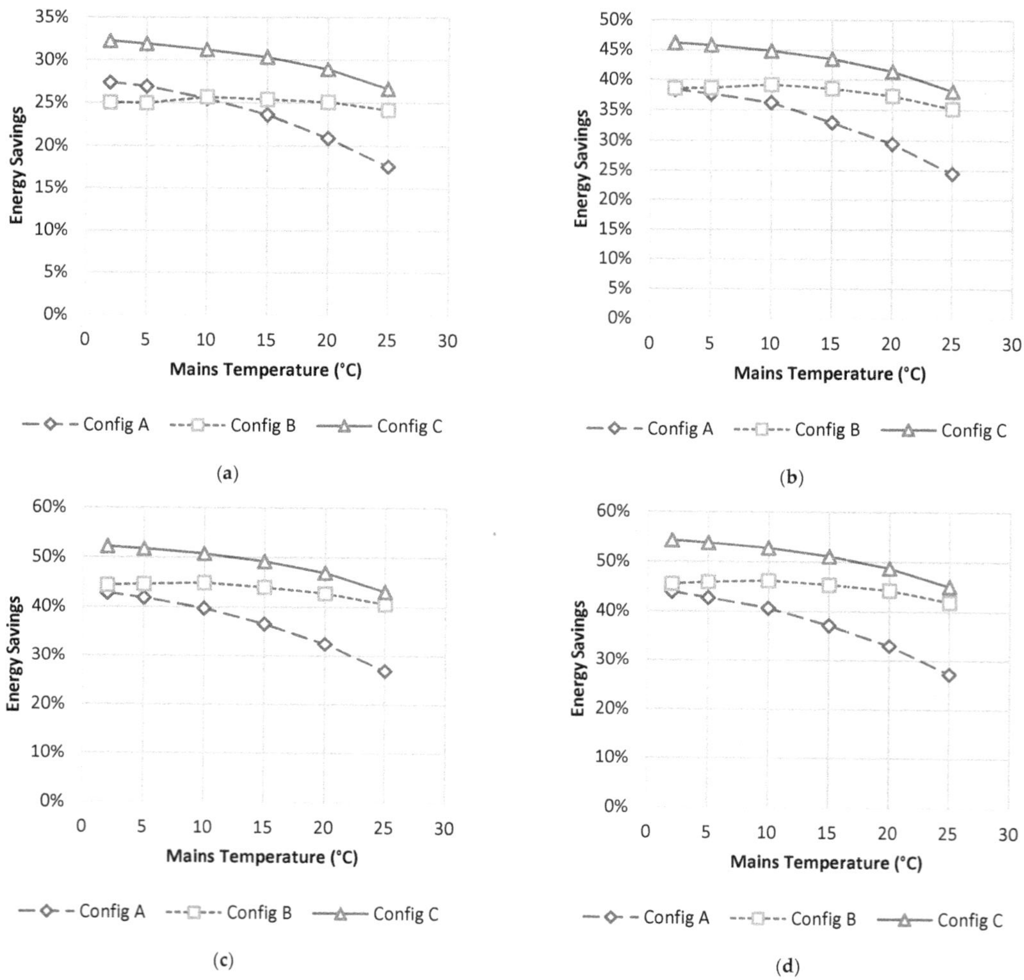

Figure 4. Energy savings as a function of mains temperature for a showerhead flowrate of 5.5 L/min for configurations A, B and C, as labelled. The plots correspond to heat exchanger 1 (**a**), heat exchanger 2 (**b**), heat exchanger 3 (**c**), and heat exchanger 4 (**d**).

In this equation, C represents the heat capacity rate (kW/°C), $\dot{m}$ is the mass flow rate (kg/s), and Cp is the specific heat (kJ/kg°C). Equation (3) is often used in the ε-NTU method for heat exchanger analysis. For this study, constant properties were used, so Equation (3) can be reduced to a ratio of mass flowrates. Note that the maximum flowrate in this study always corresponds to the flowrate of water at the showerhead fixture, which is equal to the flowrate through the drain side of the heat exchanger. The minimum flowrate corresponds to the lower of $\dot{m}_{coils}$ and $\dot{m}_{drain}$. The heat capacity ratio, C_r, is nondimensionalized, and is bound between 0 and 1. When the flowrates through the coils and drain are equal (i.e., Config C), C_r will be equal to 1; however, for the other two configurations, $\dot{m}_{coils}$ will always be lower than $\dot{m}_{drain}$. Thus, C_r can be calculated using Equation (4):

$$C_r = \frac{\dot{m}_{min}}{\dot{m}_{max}} = \frac{\dot{m}_{coils}}{\dot{m}_{shower}} = \frac{\dot{m}_{coils}}{\dot{m}_{drain}} \quad (4)$$

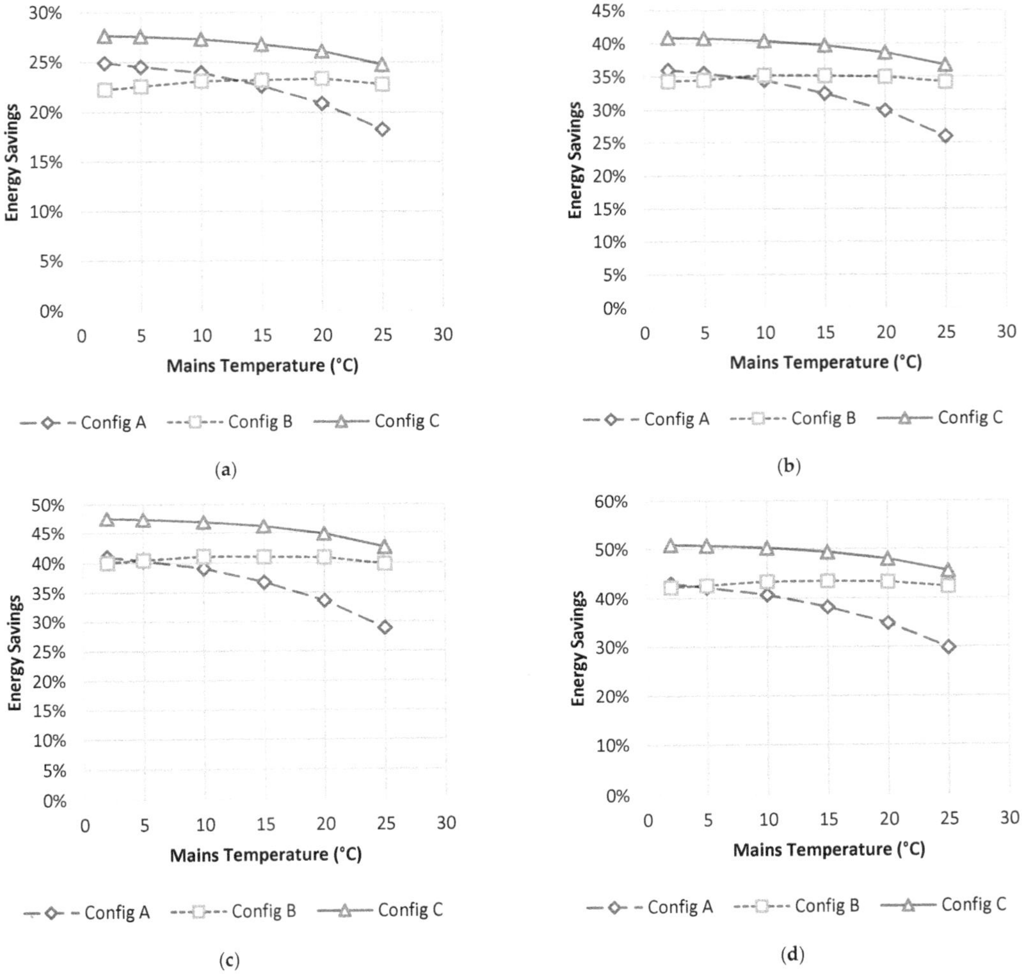

Figure 5. Energy savings as a function of mains temperature for a showerhead flowrate of 9.5 L/min for configurations A, B and C, as labelled. The plots correspond to heat exchanger 1 (**a**), heat exchanger 2 (**b**), heat exchanger 3 (**c**), and heat exchanger 4 (**d**).

For this analysis, the heat capacity ratio was calculated for all time steps in the monthly simulation when shower events occurred, and the results for heat exchanger 1 are shown in Figure 7. In this figure, the x-axis contains six sets labelled as 5.5A, 5.5B, 9.5A, 9.5B, 14A and 14B, where 5.5A implies the set of results are associated with configuration A for a showerhead flowrate of 5.5 L/min, and 5.5B implies the set of results are associated with configuration B for a showerhead flowrate of 5.5 L/min, and so on. Each set of results contains six bars, each of which corresponds to a mains temperature (Tm) between 2 and 25 °C, as labelled in the legend. Note that the heat capacity ratios for configuration C are not plotted here, as they are always equal to 1.

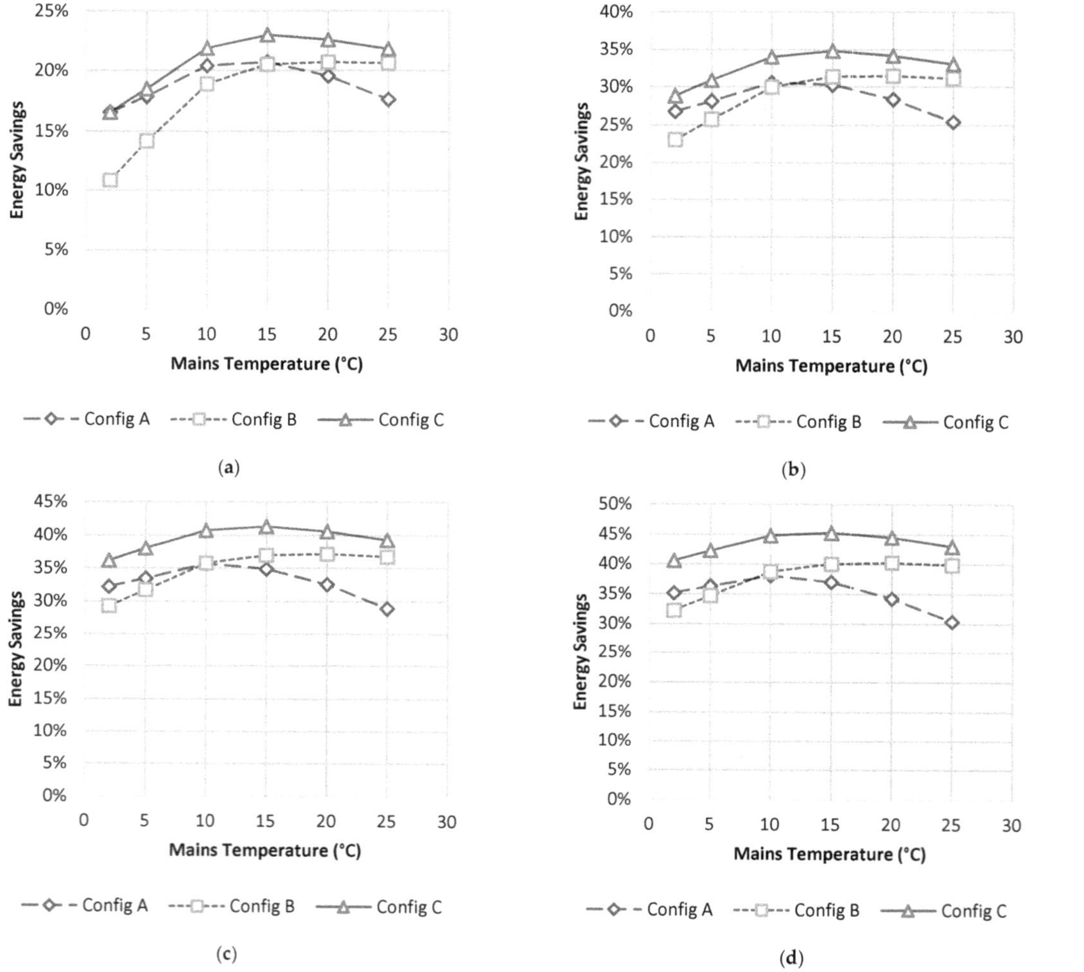

Figure 6. Energy savings as a function of mains temperature for a showerhead flowrate of 14 L/min for configurations A, B and C, as labelled. The plots correspond to heat exchanger 1 (**a**), heat exchanger 2 (**b**), heat exchanger 3 (**c**), and heat exchanger 4 (**d**).

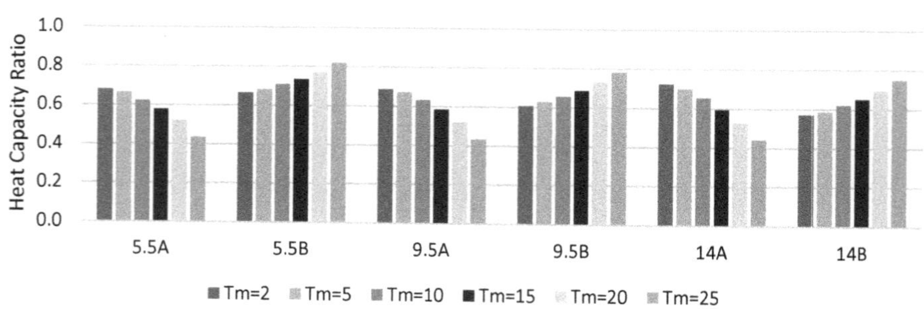

Figure 7. Heat capacity ratios for all simulated cases for heat exchanger 1.

Figure 7 shows that, for configuration A, as the mains temperature was increased in the simulations, the flowrate through the coils was decreased, thereby decreasing C_r. Conversely, configuration B resulted in higher flowrates through the coils as the mains temperature was increased. Figures 8–10 contain the results for heat exchangers 2, 3, and 4, respectively, all of which showed the same trends as Figure 7. Clearly, the mains temperature has a significant impact on $\dot{m}_{coils}$, which is directly tied to how much energy can be recovered by the heat exchangers. Lastly, the results corroborate what was shown previously in Figures 4–6, where the energy savings for the water heater was shown for different configurations.

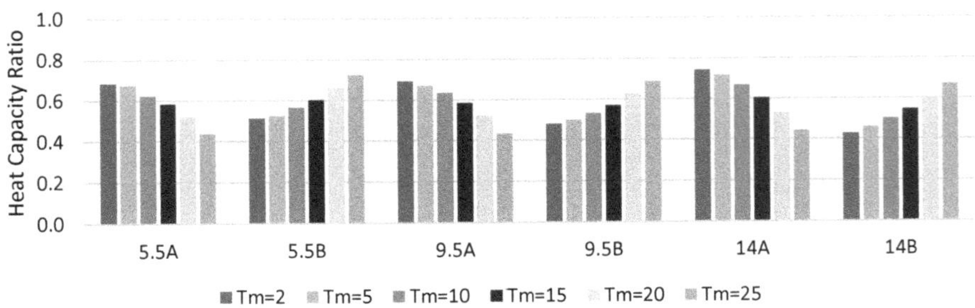

Figure 8. Heat capacity ratios for all simulated cases for heat exchanger 2.

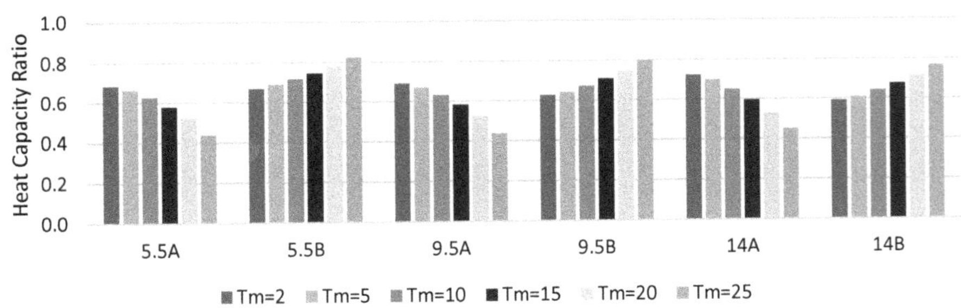

Figure 9. Heat capacity ratios for all simulated cases for heat exchanger 3.

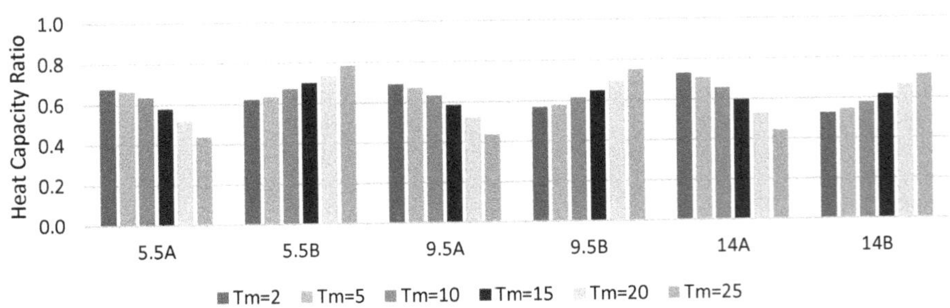

Figure 10. Heat capacity ratios for all simulated cases for heat exchanger 4.

To explain this behaviour, a closer inspection of the plumbing schematic shown in Figure 2 is required. For configuration A, $\dot{m}_{coils}$ is equal to $\dot{m}_{HWT}$, and as the mains temperature is increased, a higher flow rate of water at T_{mains} would flow directly to the mixing valve; thereby decreasing $\dot{m}_{coils}$. In other words, the heat exchangers are less utilized in configuration A as the mains temperature is increased and is closer to the prescribed showerhead temperature. On the other hand, in configuration B, $\dot{m}_{coils}$ is not constrained by $\dot{m}_{HWT}$, and as the mains temperature is increased, higher flowrates of water would flow through the heat exchanger coils to be preheated. As a result, the heat exchanger is better utilized as the mains temperature is increased in configuration B.

The simulation results highlighted that the mains temperature could dictate the flowrate of water that gets preheated by the heat exchanger in unequal-flow configurations, thereby affecting the energy savings. The mains temperature is based on the location where the plumbing system is located and is expected to vary throughout the year. Designers should be mindful of these facts when implementing a DWHR heat exchanger into a residential plumbing system. The results can also be used by manufacturers as guidelines to improve the design of heat exchangers. Currently, DWHR heat exchangers are rated and sold based on data gathered from equal-flow conditions; hence, they are designed to work optimally in such conditions to be competitive on the market. However, the simulation results showed that if the heat exchangers are installed in unequal-flow configurations, the mass flowrate through the coils could be much lower than the rated conditions. Therefore, the manufacturers are encouraged to adjust the coil diameter to improve heat transfer rates, which increases energy savings when having equal flow rates through the heat exchanger is not an option.

The results from this work showed that plumbing configuration has a pronounced impact on energy savings expected from DWHR heat exchangers. This held true regardless of the heat exchanger size or the prescribed temperature and flowrate at the showerhead. It is noted that reporting simulation results without clarifying the plumbing configuration fails to account for variations in flowrate through the heat exchanger coils and is therefore not an advisable approach. It is the authors' hope that future publications in this field will follow suit and include sufficient information regarding the plumbing setup, such that simulations can be replicated.

Author Contributions: Conceptualization, R.M. and M.R.C.; methodology, R.M.; software, R.M.; validation, R.M.; formal analysis, R.M.; investigation, R.M.; resources, R.M. and M.R.C.; data curation, R.M.; writing—original draft preparation, R.M.; writing—review and editing, R.M. and M.R.C.; visualization, R.M.; supervision, M.R.C.; project administration, M.R.C.; funding acquisition, R.M. and M.R.C. All authors have read and agreed to the published version of the manuscript.

Funding: This research was funded by the Natural Sciences and Engineering Research Council of Canada (NSERC), and the Ontario Graduate Scholarship (OGS) program.

Data Availability Statement: Not applicable.

Conflicts of Interest: The authors declare no conflict of interest. The funders had no role in the design of the study; in the collection, analyses, or interpretation of data; in the writing of the manuscript, or in the decision to publish the results.

References

1. Natural Resources Canada, Comprehensive Energy Use Database, Residential Sector, Canada, Table 2: Secondary Energy Use and GHG Emissions by End-Use. 2020. Available online: https://oee.nrcan.gc.ca/corporate/statistics/neud/dpa/showTable.cfm?type=CP&sector=res&juris=ca&rn=2&page=0 (accessed on 1 October 2021).
2. Eurostat, Energy Consumption in Households. June 2021. Available online: https://ec.europa.eu/eurostat/statistics-explained/index.php?title=Energy_consumption_in_households#Energy_consumption_in_households_by_type_of_end-use (accessed on 2 October 2021).
3. Chen, Y.; Fuchs, H.; Schein, J.; Franco, V.; Stratton, H.; Dunham, C. *Calculating Average Hot Water Mixes of Residential Plumbing Fittings*; Energy Analysis And Environmental Impacts Division Lawrence Berkeley National Laboratory: Berkeley, CA, USA, 2020.
4. Cooperman, A.; Brodrick, J. Drain Water Heat Recovery. *ASHRAE J.* **2011**, *53*, 58–62.
5. Beentjes, I.; Manouchehri, R.; Collins, M.R. An investigation of drain-side wetting on the performance of falling film drain water heat recovery systems, Energy and Buildings. *Energy Build.* **2014**, *82*, 660–667. [CrossRef]
6. Ravichandran, A.; Diaz-Elsayed, N.; Thomas, S.; Zhang, Q. An assessment of the influence of local conditions on the economic and environmental sustainability of drain water heat recovery systems. *J. Clean. Prod.* **2021**, *279*, 123589. [CrossRef]
7. Collins, M.R.; van decker, G.W.E.; Murray, J. Characteristic effectiveness curves for falling-film drain water heat recovery systems. *HVACR Res.* **2013**, *19*, 649–662.

8. CSA. *B55.1-15 Test Method for Measuring Efficiency and Pressure Loss of Drain Water Heat Recovery Units*; Canadian Standards Association: Mississauga, ON, Canada, 2015.
9. Słyś, D.; Kordana, S. Financial analysis of the implementation of a Drain Water Heat Recovery unit in residential housing. *Energy Build.* **2014**, *71*, 1–11. [CrossRef]
10. Régie du Bâtiment du Québec, Branchement des Systèmes de Récupération de Chaleur des Eaux de Drainage: Attention aux Légionelles. March 2013. Available online: https://www.rbq.gouv.qc.ca/domaines-dintervention/plomberie/interpretations-et-directives-techniques/branchement-des-systemes-de-recuperation-de-chaleur-des-eaux-de-drainage-attention-aux-legionelles.html (accessed on 6 June 2021).
11. Manouchehri, R.; Collins, M.R. An experimental analysis of the impact of temperature on falling film drain water heat recovery system effectiveness. *Energy Build.* **2016**, *130*, 1–7. [CrossRef]
12. Manouchehri, R.; Collins, M.R. An experimental analysis of the impact of unequal flow on falling film drain water heat recovery system performance. *Energy Build.* **2018**, *165*, 150–159. [CrossRef]
13. Manouchehri, R.; Collins, M.R. Modelling the Steady-State Performance of Coiled Falling-Film Drain Water Heat Recovery Systems Using Rated Data. *Resources* **2020**, *9*, 69. [CrossRef]
14. Allard, Y.; Kummert, M.; Bernier, M.; Moreau, A. Intermodel comparison and experimental validation of electrical water heater models in TRNSYS. In Proceedings of the 12th Conference of International Building Performance Simulation Association, Sydney, Australia, 14–16 November 2011.
15. Kleinbach, E.; Beckman, W.; Klein, S. Performance study of one-dimensional models for stratified thermal storage tanks. *Sol. Energy* **1993**, *50*, 155–166. [CrossRef]

energies

MDPI

Article

Techno-Economic Analysis of Waste Heat Utilization in Data Centers: Application of Absorption Chiller Systems

Leyla Amiri [1,*], Edris Madadian [2], Navid Bahrani [1] and Seyed Ali Ghoreishi-Madiseh [3]

1 Department of Civil and Resource Engineering, Dalhousie University, Sexton Campus, Halifax, NS B3H 4R2, Canada; navid.bahrani@dal.ca
2 Department of Chemical Engineering, University of Waterloo, Waterloo, ON N2L 3G1, Canada; Edris.madadian@uwaterloo.ca
3 Norman B. Keevil Institute of Mining Engineering, University of British Columbia, Vancouver, BC V6T 1Z4, Canada; ali.madiseh@ubc.ca
* Correspondence: leyla.amiri@dal.ca

Abstract: Modern data centers are playing a pivotal role in the global economic situation. Unlike high-quality source of waste heat, it is challenging to recover the decentralized and low-quality waste heat sourced from data centers due to numerous technological and economic hurdles. As such, it is of the utmost importance to explore possible pathways to maximize the energy efficiency of the data centers and to utilize their heat recovery. Absorption chiller systems are a promising technology for the recovery of waste heat at ultra-low temperatures. In fact, the low temperature heat discharged from data centers cannot be retrieved with conventional heat recovery systems. Therefore, the present study investigated feasibility of waste heat recovery from data centers using an absorption chiller system, with the ultimate goal of electrical energy production. To fulfill this objective, a techno-economic assessment of heat recovery using absorption chiller (AC) technique for the data centers with power consumption range of 4.5 to 13.5 MW is performed. The proposed AC system enables saving electricity for the value of 4,340,000 kWh/year and 13,025,000 kWh/year leading to an annual reduction of 3068 and 9208 tons CO_2 equivalent of greenhouse gas (GHG) emissions, respectively. The results of this study suggest an optimum change in the design of the data center while reducing the payback period for the investors.

Keywords: waste heat recovery; absorption chiller; data center; sustainability; thermal pollution

Citation: Amiri, L.; Madadian, E.; Bahrani, N.; Ghoreishi-Madiseh, S.A. Techno-Economic Analysis of Waste Heat Utilization in Data Centers: Application of Absorption Chiller Systems. *Energies* **2021**, *14*, 2433. https://doi.org/10.3390/en14092433

Academic Editors: Jan Danielewicz and Krzysztof Rajski

Received: 8 March 2021
Accepted: 22 April 2021
Published: 24 April 2021

1. Introduction

The incessant growth in the use of conventional fuels and their enormous greenhouse gas emissions necessitate more attention to be brought to environmental awareness. Consequently, many scholars around the world have been encouraged to spend significant amount of work on innovative technologies to supply heating and cooling demands of urban areas via renewable sources of energy. Global technological advancements and development of internet of things (IoT) technologies, big data, and cloud computing have resulted in rapid growth in the number of data centers around the world [1,2]. Owing to the enormous energy consumption resulted from rapidly growing data centers, much attention has been devoted to the possibility of energy recovery from such centers by their authorities as well as energy organizations. The worldwide electricity consumption of data centers is presented in four main categories: (i) Infrastructure uses 44.4% of electricity consumption (mainly ventilation and cooling sectors). (ii) Servers use 38.9% of energy supply, followed by 9.3% and 7.4% of the electricity used by (iii) Communication and (iv) Storage sectors, respectively [3–5].

Over 416 billion kWh of electricity was reportedly consumed by virtue of data processing in around 8 million data centers around the globe in 2019 [6]. A typical data center produces heat values in the range of 3.2 to 6.4 MW comprising 250 sets of cabinets with

42 to 64 server racks [7]. The new generations of data centers follow a growing trend of heat production with recent ones producing 10–250 kW [8]. In the light of the typical design temperature of 25 °C for a data room, this makes data centers as a year-around stable source of heat, and there exists huge potential for waste heat recovery from data centers.

Waste heat recovery and utilization from data centers has been the subject of many research studies within the past decade leading to the introduction of several practices to recover low-grade waste heat. According to the literature, the waste heat can be categorized to four types based on temperature levels: high-grade, medium-grade, low-grade and ultra-low-grade waste heat in the temperature range of >600 °C, 200–600 °C, 80–200 °C, and 45–80 °C, respectively [9,10].

The most common practices include absorption cooling, district heating, and direct and/or indirect power generation [7,11,12]. Although several factors can affect selection of the best method, an overview of available techniques presents district heating as the most viable approach for waste heat recovery in data center. For example, ref. [13] investigated a hot water (60 °C) supercomputer prototype showing energetic and exergetic efficiencies of 80% and 34%, respectively. Huang et al. [6] studied a 1 MW data center showing over 97% of heat recovery, which is equivalent to the heat required of a 30,000 m^2 commercial property. Another study investigated possible strategies for cooling and waste heat recovery of information technology (IT) servers in UK which produced over 2 million tons of CO_2 [14]. They projected that the recovered heat would be sufficient to supply heat demands for some neighborhoods in London. A 4000 ton CO_2 equivalent emission reduction along with $170,000 saving from waste heat recovery of a 3.5 MW data center was estimated. A universal design approach was proposed by Fang et al. [15] for district heating based on the industrial waste heat recovery. They estimated that 122 MW would suffice district heating of a city with a population of 4.6 million in Northern China resulting in reduction of 168,000 tons of CO_2 equivalent. A steady-state model was developed in [16] which utilized the recovered heat from a thermosyphon-assisted air conditioner.

A variety of studies have been published exploring cooling technologies for thermal management in data centers [17–19]. A common practice to form hot and cold aisles in data centers is to place server racks back to back and front to front in an open-aisle (OA) air-cooled configuration [20]. This technique, however, affects the electricity consumption dedicated to air conditioning due to temperature nonuniformity sourced from air circulation between the cold and hot aisle [21]. Similar results was presented in [22], indicating potential weakness of OA systems associated with airflow direction. Alternatively, enclosed aisle (EA) approach was considered by numerous studies to reduce the inhomogeneity of airflow [23,24]. Nevertheless, EA strategy may also raise a new issue with finding the right place to discharge the trapped hot air. Consequently, other promising cooling systems such as free cooling, liquid cooling, two-phase technologies, and building envelope have been subject of many studies up until now. For instance, the authors of [25] modeled the dynamic air conditioning load of a large data center in China to obtain cooling and index. Using simulation of building load characteristics, it appeared that the values of annual cumulative heat load was much higher than its cooling one. A novel heat recovery system composed of data room, refrigeration unit, cooling tower, double conditions heat pump unit, auxiliary boiler, water pump, and pipeline valve refrigeration was proposed.

The low temperature of the waste heat sourced from data centers hinders producing a high-quality energy which requires a reliable technology to make up for this weakness. Among the waste heat recovery technologies reported in the literature, absorption chiller (AC) and Organic Rankine Cycle (ORC) were identified as the most promising technologies for processing waste heat from data centers. Both technologies have shown reliable performance in recovering low-grade source of heats.

The ORC technology has been considered as a valid solution for ultra-low-grade data center waste heat recovery. For instance, the authors of [26] investigated the techno-economic performance of ORC as the means of waste heat recovery from data centers. Using a developed steady-state thermodynamic model, they examined the effect of different types of working fluid (i.e., dry, wet, and isentropic) and the evaporator temperature on the competence of the system. It was concluded that ORC would show the greatest performance in extremely low temperature applications. It was shown that the addition of superheaters in the server cooling cycle demands extra power for the operation which ultimately caused a performance drop, albeit at the higher temperature, the performance of ORC in waste heat recovery improves. A payback period of 4–8 years was estimated for the use of ORC in the data center. In a similar study, the authors of [27] built a 20 kW ORC prototype receiving the waste heat from two common rack servers operating at full capacity and with temperature ranging from 60 °C to 85 °C. The results of the experimental rig demonstrated thermal efficiency of 3.33% which was in a good agreement with the range of efficiency (i.e., 2% to 8%) resulted from the thermodynamic model in the same study. The benefits of the ORC system were not limited to absorbing waste heat from the data center as it sent back the generated electricity to the data center. A recent study [28] investigated heat waste recovery from a simulated condition for a typical data center server rack using a lab-scale ORC. It was demonstrated that for the range of ultra-low waste heat (i.e., 45 °C to 80 °C), thermal efficiencies vary between 1.9% and 4.6%. It was shown that significant fluid temperature differences in the heat exchanger, caused exergy deterioration to the ORC system.

AC systems, on the other hand, are not only able to handle low temperature source of heats, but also adopt waste heats from liquid-cooled and two-phase cooled data centers which further enhance their applicability in the field. The other advantages of AC systems are their low electricity cost, while their requirement for larger cooling tower is considered as their main drawbacks. AC systems mainly address some serious energy and environment issues with the conventional compression chiller (CC) systems. Despite recent achievements in recovering waste heat from data centers, yet underlying concept needs to be further explored broadly. Therefore, the present research investigates feasibility of energy recovery from the heat dissipated by a number of servers to support an AC unit which can further supply cooling requirements for other centers.

2. Methodology

With the purpose of perform the techno-economic study, an inventory of information from a variety of AC manufacturers in the North America were established allowing to analyze utilization and performance of various types of AC systems which are now ready for use for many purposes. Trane and Yazaki [29,30] were selected as AC manufacturing pioneers in the North America. The main reason for the selection of the two manufacturers is their diversity in ultimate application of heat recovered from data centers. Trane develops AC systems working under the low-grade hot water in a range of 70 °C and 110 °C which makes them promising options to recover waste heat in in data centers [29]. Yazaki operates AC units using water–lithium–bromide as the refrigerant which is capable to operate with low quality waste heat (temperature: ~70–95 °C) [30]. The Yazaki AC units will shut down if the inlet hot water temperature exceeds 95 °C. The outlet hot water temperature is ~55 °C. It should be mentioned that the low-grade heat generated through industrial processes can be used by AC systems and generates cooling energy. Otherwise, this low-grade heat will be wasted and discarded through the environment. The other important point about the AC systems is their sensitivity to the temperature of the inlet hot water as an energy source. If the temperature of the inlet hot water is less than the desired temperature, the efficiency and the cooling capacity of the system is reduced as well.

This work provides a substantive methodological guide for district heating using the recovered heat from data centers. Thus, the main idea of this study is to employ an AC unit which is energized by the low-grade heated air in the data centers (Figure 1). When the enough heat load is supplied to AC system, the heat load on the CC unit is reduced. In the other words, a portion of the heated air in the data center will be removed by the AC, which reduces the total workload on the CC. Here, the heated air in the data center has a lower-grade heat compared to the higher-grade heat of the computing elements inside the racks. This low-grade heat air can be effectively cooled down by the AC. Finally, any remaining heat can be removed by the CC.

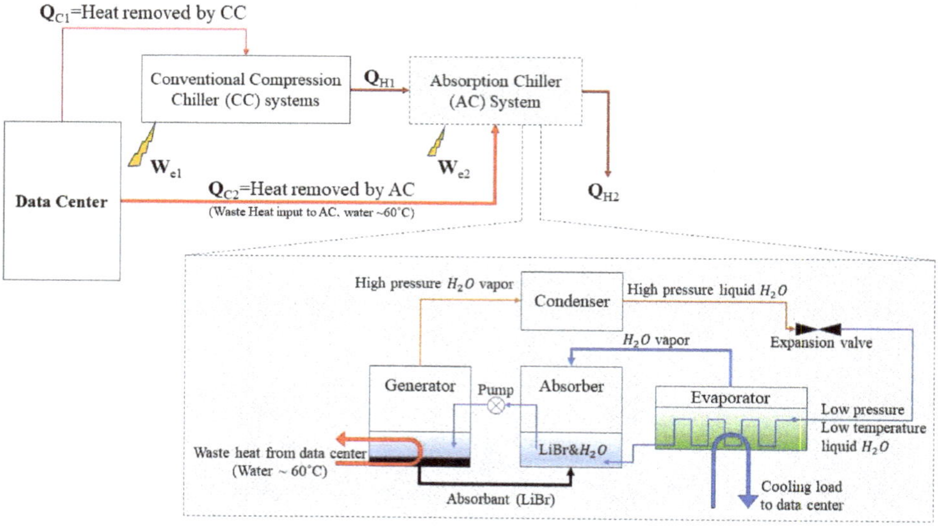

Figure 1. Heat flow of the proposed data center coupled with an absorption chiller system.

Analytical Model

The first step in developing the thermodynamic model for the proposed system is to apply governing equations predominantly conservation of mass and energy. The analytical model is implemented considering CC system as the point of comparison with the proposed AC system. Figure 1 presents the cycle under investigation for the heat flow of the proposed data center. Note that Q_{H1} is the heat rejected from the cycle through the condenser in the CC and can be proportionally considered as a heat source for the AC system. In addition, wet or dry cooling towers should be designed to transfer Q_{H2} (i.e., the heat rejected of the chiller coolant water by absorber and condenser) to the ambient.

According to Figure 1, the total cooling supplied by the hybrid CC system and AC system ($Q_{C-total}$) can be written as (Equation (1))

$$Q_{C-total} = Q_{C1} + Q_{C2} \tag{1}$$

where Q_{C1} and Q_{C2} are the provided cooling by the CC system and AC system, respectively.

Similarly, total electricity consumption by the proposed hybrid CC-AC system (W_{CC-AC}) can be written as follows (Equation (2)):

$$W_{CC-AC} = W_{e1} + \underbrace{W_{e2}}_{\approx zero} \tag{2}$$

where W_{e1} and W_{e2} are the electricity requirement for operating the CC system and AC system, respectively. Note that the energy consumption of the AC system in this integrated proposed system is negligible.

The provided cooling by the CC system (Q_{C1}) is a function of the Coefficient of Performance (COP) of the system as well as the dissipated heat (Q_{H1}) from the system and can be written as follows (Equations (3) and (4)):

$$Q_{C1} = COP_{CC} \times W_{e1} \tag{3}$$

$$Q_{C1} = \frac{COP_{CC}}{1 + COP_{CC}} \times Q_{H1} \tag{4}$$

According to the first and second law of thermodynamics, the dissipated heat and cooling by the CC system can be expressed as (Equations (5) and (6)):

$$Q_{H1} = Q_{C1} + W_{e1} = (1 + COP_{CC}) \times W_{e1} \tag{5}$$

$$Q_{C2} = COP_{AC} \times Q_{H1} = COP_{AC} \times (1 + COP_{CC}) \times W_{e1} \tag{6}$$

Combining Equations (3) and (5), the total cooling provided for the data center and the total electricity consumption by the proposed hybrid system can be rewritten as (Equation (7))

$$Q_{C-total} = Q_{C1} + Q_{C2} = COP_{CC} \times W_{e1} + COP_{AC} \times (1 + COP_{CC}) \times W_{e1} \tag{7}$$

The total electricity consumption by the proposed hybrid CC-AC system can be calculated using Equation (8):

$$W_{CC-AC} = \frac{Q_{C-total}}{COP_{CC} + COP_{AC} \times (1 + COP_{CC})} \tag{8}$$

The total electricity saving of using the proposed system can be calculated using Equations (9) and (10):

$$W_{saving} = W_{CC} - W_{CC-AC} \tag{9}$$

$$\text{Electricity saving } (\%) = \frac{W_{CC} - W_{CC-AC}}{W_{CC}} = 1 - \frac{COP_{CC}}{COP_{CC} + COP_{AC} \times (1 + COP_{CC})} \times 100 \tag{10}$$

A detailed economic analysis including both efficiency and performance for AC systems is carried out for a data center with a power consumption range of 4.5 to 13.5 MW.

In designing cooling system for modern data centers, the waste heat dissipated per rack could be assumed between 10 and 15 kW. Note that if the rack is occupied with supercomputers, the heat generation can be up to 60 kW from each rack [31]. Accordingly, in the calculations, it is assumed that maximum 15 kW of waste heat is dissipated by each server rack (Figure 2). Additionally, water is used as cooling agent with the flowrate of 0.3 to 0.6 l/s. The design temperature of outcoming water is assumed 70 °C which is somewhat lower than the typical temperature of 85 °C. This is ascribed to the heat loss of the intermediate heat exchanger.

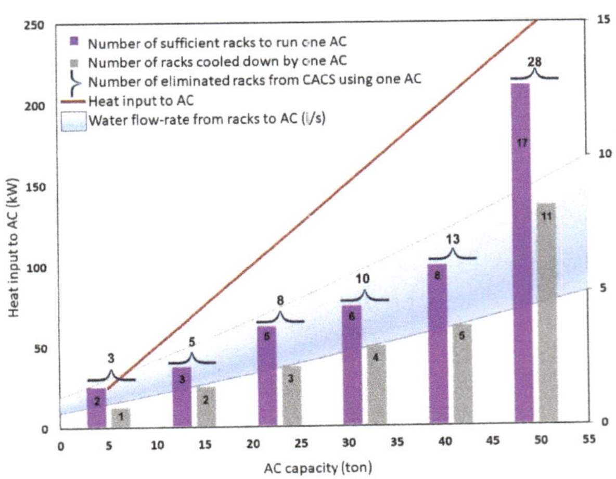

Figure 2. Heat dissipation by server racks and cooling process by AC chillers.

3. Results and Discussion

3.1. Energy Performance

An energy analysis is imperative to make a decision for the selection of the best design. Therefore, Tables 1 and 2 present a detailed energy performance for both 4.5 and 13.5 MW power consumed data centers, respectively. The cooling capacity and the performance of the AC systems depend on many parameters such as the heat source temperature, the flow rate, and the cooling water temperature. For example, for a 10 ton WFC-S10 Yazaki LiBr AC unit, the heat input and the pump power input are 15.2 kW_{th} and 0.18 kW_e, respectively [30,32]. Moreover, the cooling capacity is 9.8 kW_{th} [30,32]. It should be mentioned that kW_e represents the kilowatt electrical while kW_{th} is a representative of kilowatt thermal. Accordingly, COP_{CC} of 4 and COP_{AC} of 0.63 for the CC system and AC system, respectively are assumed when the inlet hot water temperature from data center to AC is 70 °C. We also assume the total of 8760 h data center operates in one year. As a rule of thumb, for a conventional data center energy consumption of cooling system is equivalent to energy consumption of server racks. Electricity consumption for the CC system is equal to the number of racks multiply by power of each rack and total working hours in a year. Accordingly, electricity consumption by the proposed integrated system can be calculated through Equations (9) and (10). Table 1 presents electricity consumption for a 4.5 MW and 13.5 MW data center comprising 300 and 900 server racks, respectively.

Table 1. Energy analysis for cooling a 4.5 MW and 13.5 MW data center comprising 300 and 900 racks with chillers of 575 ton and 1600 ton capacity.

Power Consumption	Conventional Compression Chiller (CC)	Proposed Waste Heat Recovery System	Energy Saving (kWh/Year)
4.5 MW	**Cooling Electricity consumption (kWh/year)**		4,340,000
	$W_{electricity-conventional}$ = 9,855,000	$W_{electricity-hybrid}$ = 5,515,000	
13.5 MW	**Cooling Electricity consumption (kWh/year)**		13,025,000
	$W_{electricity-conventional}$ = 29,565,000	$W_{electricity-hybrid}$ = 16,540,000	

For the 4.5 MW data center, it is estimated that 13 racks can be covered with AC systems of 25 ton capacity. The direct dissipation occurs by means of the AC system generator for eight of the racks, while the remaining five use the evaporator of the AC system. As a consequence, total of 299 server racks can be eliminated from a cold-aisle containment system (CACS) load (13 managed by each chiller) using 23 chillers of 25 ton capacity. The application of CACS is to confine the cold aisle allowing the rest of the data room to become a large hot-air return plenum. As a result, a drastic reduction is observed in the number of server racks cooled by CACS. This change results in a significant reduction in the CACS load. An alternative design for data centers can be considered as number of racks in multiples of 13 without any further need to CACS for cooling purposes. Note that each 25-ton AC unit is capable of saving footprint area of 2 racks.

For the 13.5 MW data center, the results indicate that a 50-ton AC chiller is sufficient to cover 28 racks in the data room. Approximately 60% of the racks (i.e., 17 out of 28) directly disperse the heat into the generator of the AC system. While the remaining 40% use evaporator of the AC system for the heat removal. Thus, it is estimated that 32 chillers of 50 ton capacity enables removal of 896 server racks from the CACS load (i.e., 28 managed by each chiller). Similarly, a significant reduction of 900 to 4 in the number of racks cooled by CACS unit is observed. Similarly, it is possible to design an alternate configuration of racks in multiples of 28 which ultimately drops the any need to cooling by CACS. However, a detailed calculation can be performed based on specific operational needs of the data center. Each 50 ton AC unit is capable of saving footprint area of 3 racks.

Table 2. Economic analysis for a 4.5 and 13.5 MW data center comprising 300 and 900 server racks with chillers of 575 ton and 1600 ton capacity.

Power Consumption	Conventional Compression Chiller (CC)	Proposed Waste Heat Recovery System
4.5 MW	**Electricity cost ($/year)**	
	9,855,000 kWh × 0.12 ($/kWh) = $1,182,500	9,855,000 kWh × 0.12 ($/kWh) = $1,182,600
	Cooling cost ($/year)	
	$1,182,600 (typical value for energy usage of a data center which is proportional to the ones from server racks)	$1,182,600 × (Equation (10)) = $661,800
	Total annual cost ($/year)	
	$2,365,200	$1,844,400
	The prices of the chillers and cooling tower	
		575 * × $2500 = $1,437,500
13.5 MW	**Electricity cost ($/year)**	
	29,565,000 kWh × 0.12 ($/kWh) = $3,547,800	29,565,000 kWh× 0.12($/kWh) = $3,547,800
	Cooling cost ($/year)	
	$3,547,800 (typical value for energy usage of a data center which is proportional to the ones from server racks)	$3,547,800 × (Equation (10)) = $1,984,800
	Total annual cost ($/year)	
	$7,095,600	$5,532,600
	The prices of the chillers and cooling tower	
		1600 * × $2500 = $4,000,000

* Capacity of chillers and cooling towers.

3.2. Economic Performance

An economic analysis is imperative to make a decision for the selection of the best design. Therefore, Table 2 presents a detailed economic analysis for both 4.5 and 13.5 MW power consumed data centers, respectively. The total annual electricity cost for each scenario comprises (1) electricity cost to run the servers and (2) energy cost to operate the cooling infrastructure. Electricity consumption of the data center is calculated based on the average power per racks by the number of servers in each rack. Therefore, the electricity consumption and its associated costs for running the servers in each rack is equal for both scenarios. On the other hand, the electricity associated with the infrastructure equipment (i.e., cooling systems) is calculated for two different scenarios (i.e., the conventional system and the proposed hybrid system).

A rough payback period is calculated based on the given assumptions in Table 2. The rough payback analysis provides how long the obtained savings from the use of the new system will take to pay back for the initial capital cost of the introduced system. As the operating expense and maintenance cost of the AC system is not included in the calculations, it is not possible to have a precise number for the payback period. The simple payback period for the proposed CC-AC system is calculated through Equation (11).

$$Payback\ period\ (year) = \frac{Saving\ Cooling\ Cost\ (\$/year)}{Total\ costs\ of\ the\ AC\ and\ cooling\ tower} \tag{11}$$

Thus, the simple payback estimation of the employment of AC in a 4.5 MW data center is estimated 2.76 years. The payback period for the 13.5 MW data center is, however, calculated between 2.56 years indicating scaling up the data center can enhance the profitability of the low-grade waste heat recovery system.

The results indicate that an AC system with a range of 5 to 50 ton capacity can afford adequate coolth energy to keep one more rack in the safe temperature level by waste heat dissipation from 2 to 17 racks. It is also evident that the AC system provides a faster payback period for a larger scale data center.

3.3. Environmental Performance

As stated earlier, there is growing demand for computing capacity resulting in development of data centers with significant power loads and densities around the globe [33]. Furthermore, the new generation of Central Processing Units (CPUs) can dissipate a remarkable amount of power in excess of 400 watts across the surface area of a credit card [34]. Safe operation of data centers, including numerous pieces of electronic equipment, requires proper temperature management based on the thermal envelopes provided by standardization entities. Full thermo- hygrometric state of the data centers environment must be actively inspected to lessen the potential risk of waste heat discharge to the environment or condensate the moist air when humidity raises. Therefore, management of thermal pollution in data centers is of significant importance.

The principle objective of controlling environment in data centers is to meet the time-varying cooling demand. According to the American Society of Heating, Refrigerating and Air-Conditioning Engineers (ASHRAE) [35], economic considerations have been changing the controlled environmental conditions of the data centers. Cooling data centers can cost up to a quarter of total energy expenses. In its first thermal guidelines for data centers, ASHRAE recommended air temperature envelope for data centers was 20–25 °C based on both reliability and uptime (primary concerns) and energy costs secondary (concerns). As a result, four classes of data center equipment; A1, A2, A3 and A4 (Table 3).

Table 3. Four classes of data centers according to ASHRAE.

Class	Temperature Range (°C)	Humidity Range (%)
A1	15–32	20–80
A2	10–35	20–80
A3	5–40	8–85
A4	5–45	8–90

The majority of today's data centers falls into class A1 and A2, nonetheless, many of modern manufacturers have been commercializing equipment compatible with class A3 and A4 operation.

As shown in Table 1, the amount of energy saving from the proposed systems of 4.5 MW and 13.5 MW power consumed data centers are 4,340,000 kWh/year and 13,025,000 kWh/year, respectively. Considering every kWh emits averagely 7.07×10^{-4} metric tons CO_2 [36], the proposed AC system is estimated to save 3068 t CO_2/year and 9208 t CO_2/year. These amounts of CO_2 are approximately equal to 9066 and 27,190 barrels of crude oil. Furthermore, it is estimate that this GHG emission reduction is equivalent to 358 ha and 1074 ha of additional forest to adsorb this CO_2 from the atmosphere.

4. Conclusions

In the existing energy conversion systems for decentralized source of heat, the majority of primary energy is lost as waste heat. This could be an enormous source of energy particularly in high energy consumption enterprises such as data centers. The highest portion of the energy lost in the data centers are classified as a low-grade waste heat and basically is the most challenging part to be recovered, especially when the temperature of waste heat is below 100 °C. Absorption chiller systems are known as a well-established technology to recover the low-grade waste heat. The advancement of IoT technologies and its interaction with big data and could computing are extensively increasing leading to significant development of data centers. Thus, this research investigated feasibility of heat recovery options and analyzed the techno-economic aspect of an absorption chiller system for the low-grade waste heat recovery from a data center. The results of this study necessitate consideration of both technical and economic aspects at the same time. The proposed hybrid system enables saving electricity for the value of 4.3 GWh/year and 13.0 GWh/year leading to an annual reduction of 3068 and 9208 tons CO_2 equivalent of greenhouse gas (GHG) emissions, respectively. Furthermore, the simple payback of 2.76 and 2.56 years, respectively, for the employment of proposed system in a 4.5 MW and 13.5 MW data center is estimated. Moreover, the environmental analysis also shows that AC system is a promising option for waste heat recovery and save notable equivalent carbon emissions to the environment. It is concluded that enhancing the capacity of the AC system enables waste heat dissipation of higher number of racks while shorten the payback period. Future work could focus on the employment of higher-performance absorbers and generators for designing new absorption cycles.

Author Contributions: L.A.: Conceptualization, investigation, methodology, software, validation, writing—original draft, writing—review and editing. E.M.: conceptualization, validation, writing—review and editing. N.B.: conceptualization, validation, writing—review & editing. S.A.G.-M.: supervision, writing—review and editing. All authors have read and agreed to the published version of the manuscript.

Funding: This research received no external funding.

Institutional Review Board Statement: Not applicable.

Informed Consent Statement: Not applicable.

Data Availability Statement: The data presented in this study are openly available at https://doi.org/10.3390/xxxxx.

Acknowledgments: The first author acknowledge the financial support provided by the Killam Trusts fund scholarships and Dalhousie University.

Conflicts of Interest: The authors declare no conflict of interest.

Abbreviations

CC	Conventional Compression Chiller
AC	Absorption Chiller
CACS	Cold-Aisle Containment System
IoT	Internet of Things
IT	Information Technology
ORC	Organic Rankine Cycle
EA	Enclosed Aisle
OA	Open Aisle
GHG	Greenhouse Gas

References

1. Huang, P.; Copertaro, B.; Zhang, X.; Shen, J.; Löfgren, I.; Rönnelid, M.; Fahlen, J.; Andersson, D.; Svanfeldt, M. A review of data centers as prosumers in district energy systems: Renewable energy integration and waste heat reuse for district heating. *Appl. Energy* **2020**, *258*, 114109. [CrossRef]
2. Koomey, J. *Growth in Data Center Electricity Use 2005 to 2010*; Anal. Press: City, Country, 2011; pp. 1–24. Available online: https://alejandrobarros.com/wp-content/uploads/old/Growth_in_Data_Center_Electricity_use_2005_to_2010.pdf (accessed on 1 January 2020).
3. Dandres, T.; Moghaddam, R.F.; Nguyen, K.K.; Lemieux, Y.; Samson, R.; Cheriet, M. Consideration of marginal electricity in real-time minimization of distributed data centre emissions. *J. Clean. Prod.* **2017**, *143*, 116–124. [CrossRef]
4. Rong, H.; Zhang, H.; Xiao, S.; Li, C.; Hu, C. Optimizing energy consumption for data centers. *Renew. Sustain. Energy Rev.* **2016**, *58*, 674–691. [CrossRef]
5. Van Heddeghem, W.; Lambert, S.; Lannoo, B.; Colle, D.; Pickavet, M.; Demeester, P. Trends in worldwide ICT electricity consumption from 2007 to 2012. *Comput. Commun.* **2014**, *50*, 64–76. [CrossRef]
6. Huang, Q.; Shao, S.; Zhang, H.; Tian, C. Development and composition of a data center heat recovery system and evaluation of annual operation performance. *Energy* **2019**, *189*, 116200. [CrossRef]

7. Ebrahimi, K.; Jones, G.F.; Fleischer, A.S. A review of data center cooling technology, operating conditions and the corresponding low-grade waste heat recovery opportunities. *Renew. Sustain. Energy Rev.* **2014**, *31*, 622–638. [CrossRef]
8. Samadiani, E.; Joshi, Y. Energy efficient thermal management of data centers via open multi-scale design: A review of research questions and approaches. *J. Enhanc. Heat Transf.* **2011**, *18*, 15–30. [CrossRef]
9. van de Bor, D.; Ferreira, C.I.; Kiss, A.A. Low grade waste heat recovery using heat pumps and power cycles. *Energy* **2015**, *89*, 864–873. [CrossRef]
10. Yang, S.; Wang, Y.; Gao, J.; Zhang, Z.; Liu, Z.; Olabi, A.G. Performance Analysis of a Novel Cascade Absorption Refrigeration for Low-Grade Waste Heat Recovery. *ACS Sustain. Chem. Eng.* **2018**, *6*, 8350–8363. [CrossRef]
11. Amiri, L.; Madadian, E.; Hassani, F.P. Energo- and exergo-technical assessment of ground-source heat pump systems for geothermal energy production from underground mines. *Environ. Technol.* **2018**, *40*, 3534–3546. [CrossRef]
12. Amiri, L.; De Brito, M.A.R.; Baidya, D.; Kuyuk, A.F.; Ghoreishi-Madiseh, S.A.; Sasmito, A.P.; Hassani, F.P. Numerical investigation of rock-pile based waste heat storage for remote communities in cold climates. *Appl. Energy* **2019**, *252*, 113475. [CrossRef]
13. Zimmermann, S.; Meijer, I.; Tiwari, M.K.; Paredes, S.; Michel, B.; Poulikakos, D. Aquasar: A hot water cooled data center with direct energy reuse. *Energy* **2012**, *43*, 237–245. [CrossRef]
14. Davies, G.; Maidment, G.; Tozer, R. Using data centres for combined heating and cooling: An investigation for London. *Appl. Therm. Eng.* **2016**, *94*, 296–304. [CrossRef]
15. Fang, H.; Xia, J.; Zhu, K.; Su, Y.; Jiang, Y. Industrial waste heat utilization for low temperature district heating. *Energy Policy* **2013**, *62*, 236–246. [CrossRef]
16. Wang, Z.; Ren, T.; Ma, L.; Zhang, J. Investigations of Ventilation Airflow Characteristics on a Longwall Face—A Computational Approach. *Energies* **2018**, *11*, 1564. [CrossRef]
17. Xie, M.; Wang, J.; Liu, J. Evaluation metrics of thermal management in data centers based on exergy analysis. *Appl. Therm. Eng.* **2019**, *147*, 1083–1095. [CrossRef]
18. Wang, Y.; Sun, X.; Dai, Y.; Wu, G.; Cao, Y.; Huang, D. Numerical investigation of drag reduction by heat-enhanced cavitation. *Appl. Therm. Eng.* **2015**, *75*, 193–202. [CrossRef]
19. Gupta, R.; Moazamigoodarzi, H.; MirhoseiniNejad, S.; Down, D.G.; Puri, I.K. Workload management for air-cooled data centers: An energy and exergy based approach. *Energy* **2020**, *209*, 118485. [CrossRef]
20. Ahmadi, V.E.; Erden, H.S. A parametric CFD study of computer room air handling bypass in air-cooled data centers. *Appl. Therm. Eng.* **2020**, *166*, 114685. [CrossRef]
21. Nadjahi, C.; Louahlia, H.; Lemasson, S. A review of thermal management and innovative cooling strategies for data center. *Sustain. Comput. Inform. Syst.* **2018**, *19*, 14–28. [CrossRef]
22. Schmidt, R.; Cruz, E. Cluster of High-Powered Racks Within a Raised-Floor Computer Data Center: Effect of Perforated Tile Flow Distribution on Rack Inlet Air Temperatures. *J. Electron. Packag.* **2004**, *126*, 510–518. [CrossRef]
23. Erden, H.S.; Koz, M.; Yildirim, M.T.; Khalifa, H.E. Experimental Demonstration and Flow Network Model Verification of Induced CRAH Bypass for Cooling Optimization of Enclosed-Aisle Data Centers. *IEEE Trans. Compon. Packag. Manuf. Technol.* **2017**, *7*, 1795–1803. [CrossRef]
24. Demetriou, D.W.; Khalifa, H.E. Optimization of Enclosed Aisle Data Centers Using Bypass Recirculation. *J. Electron. Packag.* **2012**, *134*, 020904. [CrossRef]
25. Yu, J.; Jiang, Y.; Yan, Y. A simulation study on heat recovery of data center: A case study in Harbin, China. *Renew. Energy* **2019**, *130*, 154–173. [CrossRef]
26. Ebrahimi, K.; Jones, G.F.; Fleischer, A.S. The viability of ultra low temperature waste heat recovery using organic Rankine cycle in dual loop data center applications. *Appl. Therm. Eng.* **2017**, *126*, 393–406. [CrossRef]
27. Araya, S.; Jones, G.F.; Fleischer, A.S. Organic Rankine Cycle as a Waste Heat Recovery System for Data Centers: Design and Construction of a Prototype. In Proceedings of the 2018 17th IEEE Intersociety Conference on Thermal and Thermomechanical Phenomena in Electronic Systems (ITherm); Institute of Electrical and Electronics Engineers (IEEE), San Diego, CA, USA, 29 May–1 June 2018; pp. 850–858.
28. Araya, S.; Wemhoff, A.P.; Jones, G.F.; Fleischer, A.S. Study of a Lab-Scale Organic Rankine Cycle for the Ultra-Low-Temperature Waste Heat Recovery Associated WITH Data Centers. *J. Electron. Packag.* **2020**, *143*. [CrossRef]
29. Thermax. Hot Water Driven Vapor Absorption Machine. Instrumentation. Available online: https://www.trane.com/content/dam/Trane/Commercial/global/products-systems/equipment/chillers/absorption-liquid/hotwater_drivenabsorptionchillers.pdf (accessed on 1 January 2020).
30. VAMTEC | YAZAKI Hot Water Absorption Chiller. Available online: https://www.vamtec.com/absorption-chiller/ (accessed on 1 January 2020).
31. Marcinichen, J.B.; Olivier, J.A.; Thome, J.R. On-chip two-phase cooling of datacenters: Cooling system and energy recovery evaluation. *Appl. Therm. Eng.* **2012**, *41*, 36–51. [CrossRef]
32. Haywood, A.; Sherbeck, J.; Phelan, P.; Varsamopoulos, G.; Gupta, S.K. Thermodynamic feasibility of harvesting data center waste heat to drive an absorption chiller. *Energy Convers. Manag.* **2012**, *58*, 26–34. [CrossRef]
33. Dreibholz, T.; Becke, M.; Adhari, H. *Report to Congress on Server and Data Center Energy Efficiency Public Law 109–431;* Tdr.Wiwi.Uni-Due.De 109; 2007; p. 431. Available online: https://www.osti.gov/biblio/929723 (accessed on 1 January 2020).
34. Corporation, I. *Product Specifications, Processors*; 2019.

35. ASHRAE. *Data Center Power Equipment Thermal Guidelines and Best Practices*; Standard, 2016; pp. 1–60.
36. EPA. *AVERT, U.S. National Weighted Average CO_2 Marginal Emission Rate, Year 2017 Data*; U.S. Environmental Protection Agency, 2018.

energies

MDPI

Article

Comparison of Ground-Based Global Horizontal Irradiance and Direct Normal Irradiance with Satellite-Based SUNY Model

Adnan Ayaz [1], Faraz Ahmad [2,*], Mohammad Abdul Aziz Irfan [1], Zabdur Rehman [2,*], Krzysztof Rajski [3] and Jan Danielewicz [3,*]

1 Mechanical Engineering Department, University of Engineering and Technology Peshawar, Peshawar 25120, Pakistan; adnanayaz@uetpeshawar.edu.pk (A.A.); mairfan@uetpeshawar.edu.pk (M.A.A.I.)
2 Department of Mechanical Engineering, Aerospace and Aviation Campus Kamra, Air University Islamabad, Kamra 43570, Pakistan
3 Faculty of Environmental Engineering, Wrocław University of Science and Technology, PL50377 Wrocław, Poland; krzysztof.rajski@pwr.edu.pl
* Correspondence: faraz.ahmad0460@gmail.com (F.A.); zabd@aack.au.edu.pk (Z.R.); jan.danielewicz@pwr.edu.pl (J.D.)

Abstract: Since the fossil reserves are depleting day by day, the trend of modern energy sector is going towards renewable energy. The demand of solar power plants is therefore at the peak nowadays across the globe. However, the construction of these plants is extremely dependent on feasibility study to estimate the real solar potential before installing it in any region. To evaluate the solar energy potential of Peshawar region in Pakistan, Ground-based global horizontal irradiance (GHI) and direct normal irradiance (DNI) were compared with satellite-based model SUNY. Ground measurements were done at the University of Engineering and Technology Peshawar (UET Peshawar) with the help of pyranometer and shadowband irradiometer. Comparison of the data showed that there was a maximum difference of 42.90% in ground and satellite-based GHI in the month of December. Minimum difference in GHI was found for the month of March that was −3.83%. Moreover, ground-based GHI was overestimated in the month of February, March, and April, while in rest of the months, satellite values of GHI exceeded the ground measurements. Similarly, maximum difference of 55.86% was found in the month of November between ground and satellite-based DNI while minimum difference of −3.34% was seen in DNI in the month of March between the two data. Furthermore, satellite-based DNI was underestimated in the months of February, March, and April while in rest of the months it was overestimated compared to ground measurements. In addition to this, correlation of ground and satellite-based GHI and DNI showed R^2 value of 0.8852 and 0.4139, respectively. The results of this study revealed that the difference between ground measurements and satellite values was considerable and hence real time measurements are necessary to properly estimate solar energy resource in the country.

Keywords: global horizontal irradiance; direct normal irradiance; satellite based SUNY model; combined uncertainty

Citation: Ayaz, A.; Ahmad, F.; Irfan, M.A.A.; Rehman, Z.; Rajski, K.; Danielewicz, J. Comparison of Ground-Based Global Horizontal Irradiance and Direct Normal Irradiance with Satellite-Based SUNY Model. *Energies* **2022**, *15*, 2528. https://doi.org/10.3390/en15072528

Academic Editor: Juan Luis Bosch Saldaña

Received: 24 February 2022
Accepted: 28 March 2022
Published: 30 March 2022

Publisher's Note: MDPI stays neutral with regard to jurisdictional claims in published maps and institutional affiliations.

1. Introduction

Utilization of energy has become a vital part of our life and its demand is increasing day by day [1]. Like most of the countries across the globe, Pakistan is also facing a serious issue of energy crisis for past few decades and the problem is getting severe with the passage of time [2]. At the moment, most of the energy demands are met by utilizing fossil fuels like coal, oil, and gas. However, these fossil fuels are rapidly depleting and they will disappear completely in the coming years. Moreover, consumption of fossil fuels strongly affects natural environment by releasing greenhouse gases into the atmosphere. There is a dire need to shift from conventional energy resources to renewable resources of energy. Solar energy is considered to be one of the vital resources of renewable energy [3].

In the past few decades, numerous studies have been conducted to assess the solar energy resource potential and increasing its efficiency using different techniques [4]. For example, Gueymard et al. [5] conducted a study to assess solar energy resource in United States of America with respect to spatial and temporal variation. Data were taken mostly from SUNY model of National Solar Radiation Database (NSRDB) for a time span between 1998 and 2005. A similar kind of research was conducted by Zell et al. [6] in Saudi Arabia in which a solar energy monitoring program was devised by King Abdullah City for Atomic and Renewable Energy (KACARE). The annual average GHI variated from 5700 Wh/m^2 to 6700 Wh/m^2. Same study mentioned that performance of PV might be degraded because of higher temperature which is almost 30 °C on average at various locations of Saudi Arabia. Similarly, a study was carried out by Al Yahya and Irfan [7] which presented discussion on the new solar atlas of Saudi Arabia. This solar atlas consisted of 41 stations which were capable of delivering solar data across the various regions of the country. This study was part of the Renewable Resource Monitoring and Mapping (RRMM) program. Through RRMM, they found that direct normal irradiance (DNI) of the country ranged from 5000 Wh/m^2/day which was recorded in winter to 9000 Wh/m^2/day in summer.

Moreover, Alnaser et al. [8] presented a study in which they presented the data of Solar Radiation Atlas for the Arab World. For a period of 10 years in Arab world, the highest annual mean global solar irradiance was calculated to be 6.7 kWh/m^2/day Nouakchott, Mauritania and 6.6 kWh/m^2/day in Tamenraset, Algeria. However, the minimum mean global irradiance was calculated in Mosul, Iraq which was 4.1 kWh/m^2/day. Similarly, Bachour and Perez [9] analyzed ground-based GHI at Doha International Airport in the time span between 2008 and 2012 which revealed that average daily GHI was 5.61 kWh/m^2/day which in other words becomes 2048 kWh/m^2/year. In addition to this, a maximum monthly average was calculated to be 6.97 kWh/m^2/day for the month of June.

Satellite-based DNI is affected by several parameters out of which the most critical and important factor is aerosol optical depth (AOD) [10]. Aerosols are very crucial for calculation of solar energy resource and play a critical role to estimate surface irradiance from the available satellite-based solar irradiance. A study in Thailand revealed that in majority of the cases, 10 to 15% of total depletion of solar energy by all atmospheric particles occurred due to aerosols [10]. Moreover, it has been found that 5% of the total depletion of solar energy from aerosols occurs due to scattering of solar energy from continental aerosols. Similarly, a research in India was done to incorporate aerosols dataset to get an accurate satellite model for computing DNI [11]. It was noted that the fine aerosol particles could be removed from atmosphere in monsoon season. It was also noticed that due to wind, aerosols also moved from one location to another, hence creating uncertainty in the estimation of DNI. In addition to this, a study compared eight clear sky broadband models [12]. This study concluded that turbidity plays a very important and crucial role to estimate the irradiance received. This study also stated that the important factors that are considered for clearest atmospheric conditions are aerosol loads, water vapor component, and least turbidity. To account for such issues in analyzing the solar irradiance, Mueller et al. [13] devised a satellite-based model called The SOLIS clear-sky module. While making this model, ozone, water vapors, and aerosols were taken into the account. Results obtained from this model were compared with ground-based measurements.

Satellite-based data are acquired on the basis of forecasting techniques. Some satellite-based forecasting techniques use position of clouds to estimate solar energy over a particular location. A study showed that motion vectors were used to predict the upcoming position of the clouds over particular location of the ground and it was noted that they provided fine forecasting accuracy of almost 6 h. Predictions from cloud pose several problems to properly forecast solar irradiance [14]. Therefore, comparison of satellite-based solar data and ground-based measured solar data has been of much importance to many researchers across the globe. Vignola et al. [15] conducted a study at Kimberly, Idaho in which satellite-based solar data and ground-based solar data were analyzed to validate the satellite-based solar model. It was found that mean bias error for satellite-based global irradiance was 5%

and it was 2% for beam irradiance. Similarly, Blanksby et al. [16] conducted a research in Australia to check the accuracy of satellite-based solar model. It was noted that for lesser values of GHI, satellite-based data were overestimated compared to the ground-based data. However, for greater values of GHI, satellite-based data were either overestimated or underestimated compared to the ground-based data.

A study in Brazil compared satellite-based solar irradiance with ground-based measurements [17]. Irradiance was analyzed on the basis of monthly average daily mean values. Comparison of this satellite-based data and ground measurements showed a global root mean square error of 13% for all data points. To estimate the satellite-based direct normal irradiance in the tropical environment of Thailand, a method was devised by Janjai [18] where the satellite selected for this study was MTSAT-1R. First, data of this satellite were used to estimate the global horizontal irradiance and then satellite-based diffused fraction model was devised to estimate the diffuse solar irradiance. Based on global horizontal irradiance and diffuse irradiance, direct normal irradiance was estimated. A similar kind of research was conducted to verify the values of direct normal irradiance of SUNY which is a satellite-based model with the ground-based values in various locations of California [19]. The study focused on finding variation in both the data and also to find accuracy in SUNY model. Mean bias error was noted in the range of −6.39% to 14.21% and correlation coefficient was 0.90 to 0.95 for direct normal irradiance.

Perez et al. [20] compared the satellite-based data taken from GOES 8 with the data which was interpolated and extrapolated with respect to 12 ground-based measurement stations located in New York and Massachusetts in America. A satellite-based model was devised by Janjai et al. [21] to calculate global solar irradiance in tropical areas of Thailand. This study shows that tropical areas in Thailand have high aerosols load. Ambient temperature and relative humidity were used to predict absorption of solar rays due to water vapors in atmosphere. This model also explained how to relate visibility with depletion due to aerosols load. Monthly average hourly global irradiance calculated from both the satellite-based model and ground-based measurements was quite matching with each other by having a root mean square difference of 10% only. An experiment was conducted in the summer months of the year 1977 in United States to check whether ground-based solar irradiance could be obtained through estimation from geostationary satellites [22]. When satellite-based daily insolation was compared to ground-based pyranometers data, the standard error was calculated to be 10% of the mean data. Furthermore, three pyranometers stations were installed at three different locations of Canada which were Ottawa, Toronto, and Montreal [23]. Measurements done at these stations in the summer and spring duration of the year 1978 concluded that on the average, there measurements had 9% variation compared with satellite-based model for daily insolation data in cloudy conditions.

A lot of studies were conducted on ground-based data and satellite-based data [24–29]. Some of studies done in the past showed that there were various errors associated with satellite-based models and that is why there was a difference between solar data obtained from satellite models and ground measurements [30–32]. One such study in Northeastern US showed that there was relative root mean square error of 23% between satellite model (based on data of GOES-8) and 8 years on site measured hourly irradiance [30]. Pixel-to-irradiance conversion error in satellite model was calculated to be 12–13%. Another analysis was done to find long run variation in broadband solar irradiance at surface [31]. Data analyzed were based on three-hourly duration for a time of 18 years. Satellite data of this research were taken from ISCCP (International Satellite Cloud Climatology Project). The data taken were validated with the data based on two years, obtained from Meteosat. Validation of the data showed that when volcanic aerosols were not and were included, annual average of DNI was reduced by 16% while annual average of GHI was reduced by less than 2.2%. Cebecauer and Suri [32] developed a new model based on the data obtained from Meteosat MSG (Meteosat Second Generation). This improved model had the capability to better predict variation and improve accuracy in GHI and DNI, when there were high vapor contents and aerosols load.

Keeping in view the depletion of fossil fuels, the establishment of solar power plants is need of the hour for any country across the globe. However, their construction must be emphasized in order to overcome energy crises in the country. Feasibility study is one of the important phases in establishment of solar power plants. There is a need to estimate the solar energy potential at a particular site in order to have a clear overview to identify if the site is a good choice for the installation of any solar power plant. To identify the best site for the installation of solar power plants, solar energy resource assessment plays a vital role which can be done using various equipment like pyranometer and phyreliometer. These equipment are used to measure ground-based solar energy resource at any location where establishment of solar power plant is desired. Therefore, ground stations are very important to measure the actual solar energy received with the help of these equipment. Besides ground measurements, there are various satellite models in use these days which show an estimation of solar energy potential and they can be widely used to guess the solar energy resource [33]. Problem associated with satellite models is that they are not accurate and there is significant variation between ground measurements and satellite modeled data. Therefore, satellite-based solar data are often compared to ground-based measurements to find the amount of variation between the two [33]. Comparison of ground measurements and satellite modeled solar data provides a clear insight to decide whether or not solar power plants can be established on the basis of satellite modeled data.

Keeping in view of the previous studies, it was clear that in in some cases ground-based GHI and DNI were more accurate than satellite data. To accurately study the difference in results, an experimental study was performed in the current research, which will enable researchers to accurately use solar data in the applications of solar energy. In the present study, ground-based global horizontal irradiance (GHI), and direct normal irradiance (DNI) were measured and compared with satellite modeled data. Ground-based measurements were done at UET Peshawar by using its weather station. Weather station at UET Peshawar consists of various devices among which pyranometer and shadowband irradiometer are used to measure solar irradiance. Satellite-based solar energy was taken from a model called SUNY.

2. Materials and Methods

2.1. Equipment

Various equipment used in this study are part of the established weather station at University of Engineering and Technology (UET) Peshawar. Weather station at UET Peshawar was installed by World Bank under its program called Energy Sector Management Assistance Program (ESMAP). This project of World Bank aimed to map renewable energy resources in Pakistan. Several other weather stations were also established at different locations of Pakistan. Currently, weather station is maintained by UET Peshawar. Figure 1 shows weather station located at UET Peshawar.

Figure 1. Weather Station established at UET Peshawar.

This weather station consists of various equipment which were used in this study. Some of the equipment used in this study are discussed as follows.

2.1.1. Pyranometer

Pyranometer is used to measure solar irradiance received at a surface and there are various types of pyranometers available in the market with different technologies and specifications. However, pyranometers used in this study came from well-known manufacturer called Kipp and Zonen (Delft, The Netherlands) whose model is Kipp & Zonen CMP10. This pyranometer was used to collect global horizontal irradiance (GHI) received at UET Peshawar. Figure 2 shows pyranometer installed at UET Peshawar and its specifications are given in Table 1.

Figure 2. Kipp & Zonen CMP10 pyranometer installed at UET Peshawar.

Table 1. Specifications of Kipp & Zonen CMP10 pyranometer.

Properties	Values
Spectral range (50% Points)	285 to 2800 nm
Sensitivity	7 to 14 μV/W/m^2
Response Time	<5 s
Zero offset A	<7 W/m^2
Zero offset B	<2 W/m^2
Directional response (up to 80° with 1000 W/m^2 beam)	<10 W/m^2
Temperature dependence of sensitivity (−10 °C to +40 °C)	<1%
Operational temperature range	−40 °C to +80 °C
Maximum solar irradiance	4000 W/m^2

2.1.2. Rotating Shadowband Irradiometer (RSI)

Rotating shadowband irradiometer (RSI) is used to measure global horizontal irradiance (GHI), direct normal irradiance (DNI), and diffuse horizontal irradiance (DHI). RSI installed at UET Peshawar came from Concentrating Solar Power Services (CSPS) which is called CSPS Twin-RSI. It has two radiation detectors made from silicon located in the middle of a spherical shaped shadowband. The two radiations sensors called LI-200 came from LI-COR Inc. (Lincoln, NE, USA). When shadowband is below the sensor in rest position, the sensor measures GHI. Shadowband moves after some time and it comes in the path to block DNI. When DNI is blocked, the sensor measures DHI. When GHI and DHI both are known, DNI can easily be calculated. Figure 3 shows RSI installed at UET Peshawar. Specifications of CSPS Twin-RSI are given in Table 2.

Figure 3. CSPC Twin-RSI installed at UET Peshawar.

Table 2. Specifications of Twin-RSI.

Temperature range	−30 to +65 °C
Humidity	0 to 100% Rh
Dimensions	500 × 100 × 200 mm
Weight	2.1 kg
Power demand	<1 W at average
Output signal	≈90 μA per 1000 W/m^2
Response time	10 μs

2.1.3. Data Logger

Data logger is used for data acquisition which come from various components of weather station like pyranometer and RSI. Data which are collected by various components of weather station are stored in data logger and then extracted by connecting it with a computer. Data can also be transferred remotely. The data obtained from data logger contained missing and redundant values. Therefore, two types of quality controlled (QC) tests were performed to make the data ready to use. First, automatic QC tests were performed to identify the missing values, time shift, consistency of data by comparing the redundant values, and the data were compared with maximum and minimum irradiance values. Then, the visual QC tests were performed for flagging the data. A more detailed description of these tests can be found in [34,35].

Data logger used in UET Peshawar's weather station is a product of Campbell Scientific Inc. and its model is CR1000. This data logger is very good for extreme conditions and remote environments. Figure 4 shows the data logger which is part of the weather station located at UET Peshawar.

2.2. Location of UET Peshawar

University of Engineering and Technology is located at Peshawar region of Pakistan having latitude of 34.0017° N and longitude of 71.4854° E. Data used in this study were taken from Weather station established at UET Peshawar. Then, these data were filtered through QC tests discussed in the data logger section.

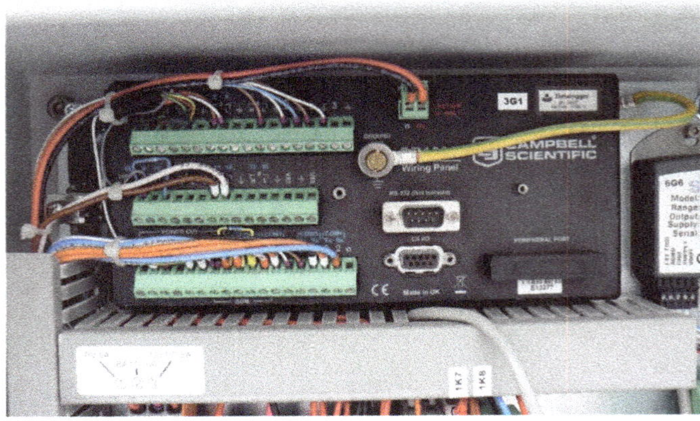

Figure 4. CR1000 data logger installed at UET Peshawar's weather station.

2.3. Methodology Adopted

In this part of the study, satellite-based GHI and DNI are compared with ground measurements for year 2017. As mentioned earlier, ground-based data were taken from weather station established at UET Peshawar.

Ground-based GHI and DNI were measured by pyranometer and Twin-RSI which both are already explained previously. These DNI and GHI were collected from data logger and were further analyzed. Data were available for the whole year 2017 except for the first 16 days of January and the last 19 days of December. Data were received at weather station after each 10 min duration. Available DNI and GHI for each month were processed on the basis of daily total monthly mean. When all the data were processed on the basis of daily total monthly mean, one data point was obtained for each month. In this way, 12 data points were available for the 12 months of 2017.

Satellite model chosen for analysis was SUNY (https://nsrdb.nrel.gov/international-datasets, accessed on 10 February 2021). National Renewable Energy Laboratory (NREL) made National Solar Radiation Database (NSRDB) which developed a model called SUNY. SUNY provides satellite-based data for South Asian countries mostly. SUNY has temporal and spatial resolution of 1 h and 10 × 10 km respectively. DNI and GHI from SUNY model are available after one hour duration. These satellite-based DNI and GHI available from SUNY were analyzed on the basis of daily total monthly mean averaged over 15 years i.e., from 2000 to 2014. Hourly data were summed up to measure the daily total for all 15 years. Then, daily total was further processed to measure daily total averaged over these 15 years and then finally monthly mean of daily total averaged over 15 years was measured. In this way, daily total monthly mean averaged over 15 years for DNI and GHI was calculated on the basis of SUNY satellite model data. For each month, one data point was obtained.

3. Results and Discussions

Satellite-modeled GHI and DNI were compared with ground measurements. Results were plotted and are discussed below.

Table 3 and Figure 5 show the comparison of satellite and ground-based GHI. Ground measurements showed that highest ground-based GHI was 6415 Wh/m^2 and lowest ground-based GHI was 1605 Wh/m^2. Similarly, highest and lowest GHI measured by satellite model were for the months of June and December, respectively. Moreover, GHI showed by satellite model for the month of June was 7177 Wh/m^2 and for the month of December was 2811 Wh/m^2. Comparison of both the data showed a clear trend and the highest difference of 42.90% in GHI was found for the month of December. In other words, satellite-based GHI was overestimated by a value of 1206 Wh/m^2. Similarly, the lowest difference in GHI was noted in the month of March, which was −3.83%. Satellite-based GHI in the month of March was underestimated by a value of −181 Wh/m^2. The average difference in GHI was noted to be 556 Wh/m^2 which means that on average, satellite-based GHI was overestimated by a value of 556 Wh/m^2 for each month. Satellite-based GHI was greater for all the months except February, March, and April where ground measurements exceeded satellite values.

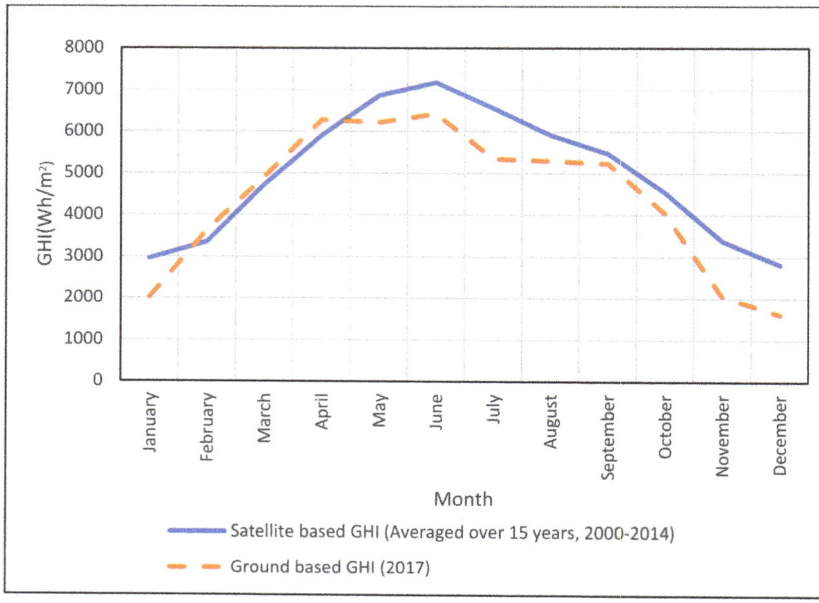

Figure 5. Comparison of ground-based GHI with satellite-based GHI.

Table 3. Comparison of ground-based GHI with satellite-based GHI.

Month	Satellite Value (Wh/m^2)	Ground Value (Wh/m^2)	Difference (Satellite-Ground) (Wh/m^2)	Percent Difference {(Sat-Ground)/Sat}s × 100
January	2955	2013	942	31.87%
February	3350	3627	−277	−8.26%
March	4721	4902	−181	−3.83%
April	5901	6267	−366	−6.20%
May	6861	6210	651	9.48%
June	7177	6415	762	10.61%
July	6553	5341	1212	18.49%
August	5909	5286	623	10.54%
September	5468	5226	242	4.42%
October	4530	4020	510	11.25%
November	3378	2032	1346	39.84%
December	2811	1605	1206	42.90%

The correlation of satellite-based GHI with ground-based GHI has been developed and shown in Figure 6 where each data point represents a month. There is a substantial positive correlation between satellite and ground-based GHI. The value of R^2 for the correlation is 0.8852 which shows that there is high correlation between these two.

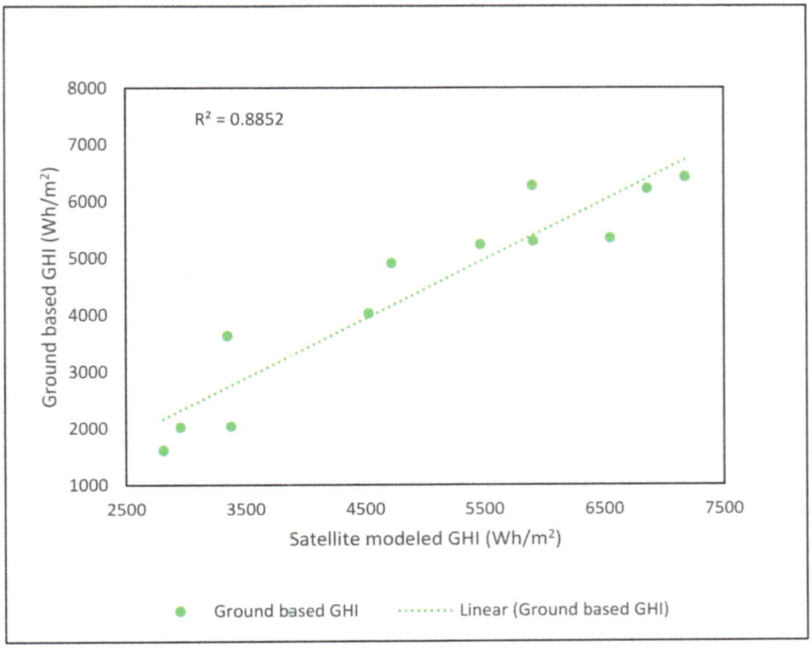

Figure 6. Correlation of satellite modeled GHI with ground-based GHI.

The comparison of ground-based monthly DNI with satellite-based DNI is given in Table 4 and Figure 7 which show that highest and lowest ground DNI were recorded in the months of April and January, respectively. It has been found that ground-based DNI in April was 5884 Wh/m^2 and it was 1718 Wh/m^2 in January. On the other hand, highest and lowest satellite-based DNI were seen in the months of May and February respectively. Satellite-based DNI in May was 6095 Wh/m^2 and it was 3240 Wh/m^2 in February. When satellite modeled and ground measured DNI were compared, satellite-based DNI overestimated ground measurements for all the months except February, March, and April. Maximum difference in DNI was found in the month of November which was 55.86%. Satellite-based DNI in November was overestimated by a value of 2348 Wh/m^2. On the other hand, minimum difference was noted in the month of March which was −3.34%. In March, satellite-based DNI was underestimated by a value of −140 Wh/m^2. Average difference in DNI between both data was 983 Wh/m^2 which means that on the average, for each month, satellite modeled DNI overestimated ground measurements by a value of 983 Wh/m^2. Ground measurements showed lower values of DNI in comparison with satellite modeled value for all months except February, March and April where ground measurements exceeded satellite-based values of DNI.

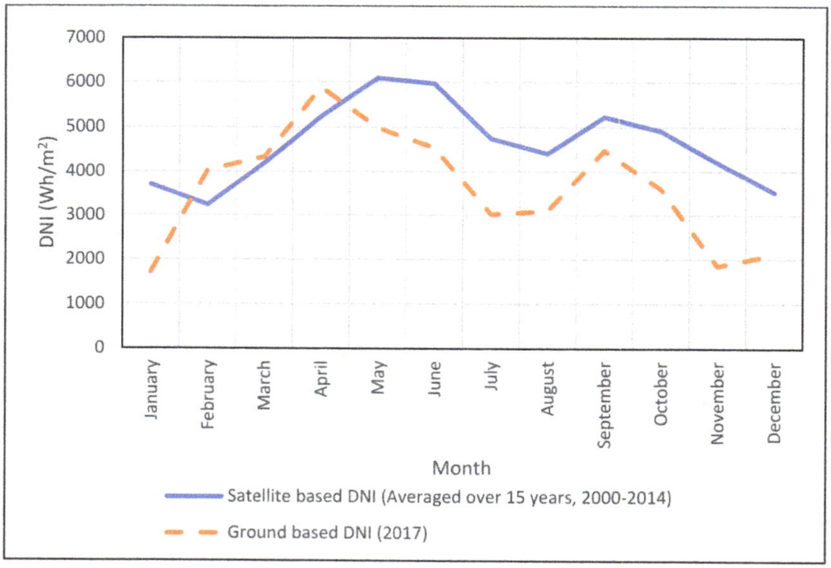

Figure 7. Comparison of ground-based monthly DNI with satellite-based DNI.

Table 4. Comparison of ground-based monthly DNI with satellite-based DNI.

Month	Satellite Value (Wh/m^2)	Ground Value (Wh/m^2)	Difference (Satellite-Ground) (Wh/m^2)	Percent Difference {(Sat-Ground)/Sat}s × 100
January	3706	1718	1988	53.64%
February	3240	4034	−794	−24.50%
March	4196	4336	−140	−3.34%
April	5243	5884	−641	−12.22%
May	6095	4989	1106	18.15%
June	5969	4545	1424	23.86%
July	4747	3027	1720	36.23%
August	4412	3103	1309	29.67%
September	5233	4478	755	14.43%
October	4920	3599	1321	26.85%
November	4203	1855	2348	55.86%
December	3517	2112	1405	39.95%

The correlation between satellite-modeled DNI and ground-based DNI has been developed which is shown in Figure 8 where each data point represents a month. There exists a moderate positive correlation between satellite and ground-based DNI having R^2 of 0.4139. Moreover, Table 5 shows uncertainties in ground measurements as well as satellite modeled GHI and DNI. Uncertainties in equipment were taken from CMP6/CMP10/CMP11/CMP21-Campbell Scientific (Logan, UT, USA) and Rotating Shadowband Irradiometer CSPS Twin-RSI of CSP Services Inc. (Cologne, Germany). Hourly uncertainties of SUNY model were 9.6% for GHI and 15.9% for DNI which were taken from (A Review of Measured/Modeled Solar Resource Uncertainty). Uncertainties in GHI and DNI of Shadowband irradiometer are <2% for annual sum. Daily uncertainties in GHI and DNI of Kipp and Zonen CMP10 pyranometer are <2%. On the basis of these values, in our study, we have assumed that our calculations have 2% uncertainty. Combined uncertainty of both the ground measurements and satellite model is 11.6% for GHI while maximum difference in GHI was found to be 42.90%. Difference in satellite and ground-based GHI exceeded from combined uncertainty which means the difference was significant and it needed to be taken into the account for calculations of solar data in Peshawar. Moreover, combined uncertainty in DNI of satellite model and ground measurements is 17.9% whereas maximum difference in both the data of DNI is 55.86%. Maximum difference in DNI exceeds the combined uncertainty which means that the difference is significant and both the data have considerable variation.

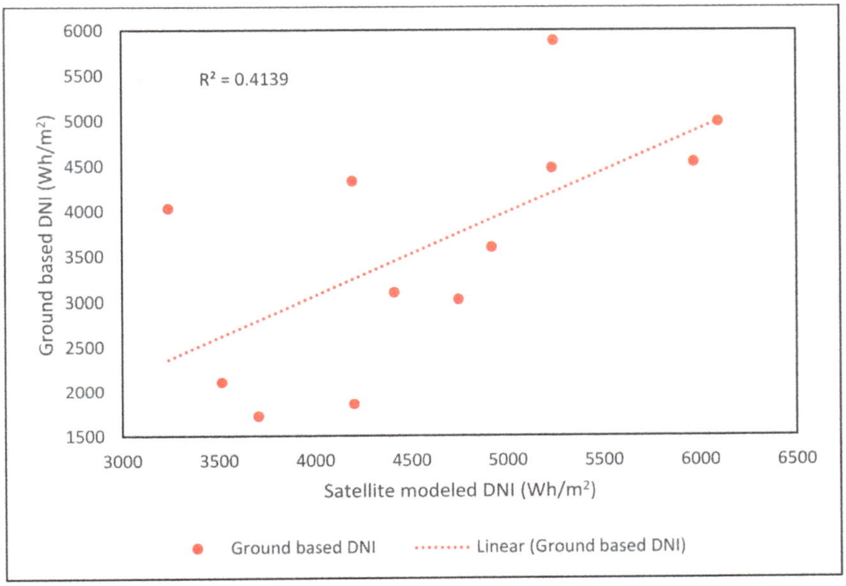

Figure 8. Correlation of satellite-modeled DNI with ground-based DNI.

Table 5. Comparison of uncertainties and maximum differences in GHI and DNI.

Property	Equipment Uncertainty (%)	SUNY Uncertainty (%)	Combined Uncertainty (%)	Maximum Difference (%)
GHI	2	9.6	11.6	42.9
DNI	2	15.9	17.9	55.86

From Table 5, it is evident that maximum difference in GHI and DNI is way greater than combined uncertainty, so there is significant difference which is worth considering in analysis.

Some of the reasons for variation between satellite and ground-based GHI and DNI in literature are [5,8,9]:

- Presence of aerosols in atmosphere;
- Presence of various gases in atmosphere;
- Presence of water vapors in air;
- Poor estimation of satellite model;
- Presence of mountains in vicinity;
- Shading of trees in the surrounding;
- Satellite confuses between clouds and snow.

4. Conclusions

The assessment of solar energy potential in a certain region is a basic step in establishing solar plants in that region. There is solar data available from satellite-based models. However, it is affected by some factors and should be validated using ground-based data. Therefore, there is a need of solar energy measurements based on ground-based data which can be obtained using different equipment like pyranometer and rotating shadowband irradiometer. To evaluate the solar energy potential of Peshawar region in Pakistan, ground-based global horizontal irradiance (GHI) and direct normal irradiance (DNI) were compared with satellite-based model SUNY. Ground measurements were done at (University of Engineering and Technology) UET Peshawar with the help of pyranometer and shadowband irradiometer.

This study concluded that there was significant difference between satellite and ground-based GHI and DNI. In the months of February, March, and April, ground-based GHI and DNI were overestimated compared to satellite-based GHI and DNI. In rest of the months, satellite-based GHI and DNI were greater than ground-based GHI and DNI. Maximum and minimum difference of 42.90% and −3.83% between satellite and ground-based GHI was found in the month of December and March respectively. For DNI, maximum and minimum difference of 55.86% and −3.34% between satellite and ground-based data was found in the month of November and March respectively. The percentage combined uncertainties between satellite and ground-based GHI and DNI were 11.6% and 17.9%. The maximum percent differences were more compared to the combined uncertainties of ground equipment and satellite model. Therefore, the difference was considerable and it is concluded that this difference needs to be taken into account when solar energy resource is assessed in Peshawar. Moreover, correlation of ground and satellite-based GHI and DNI showed squared correlation coefficient R^2 of 0.8852 and 0.4139 respectively.

Moreover, this study recommends that there should be more ground measurement stations across the country to properly assess the solar resource of the country. The difference between ground measurements and satellite values was considerable and hence real-time measurements are necessary to properly estimate solar energy resource in the country. For establishment of any solar energy program in Pakistan and especially in Peshawar, it is not recommended to rely entirely on satellite modeled data, but rather realistic ground measurements are needed. Satellite modeled data can only provide estimation of solar energy resource but not the exact amount of irradiance received at particular location. This study also recommends the researchers and university students to further research in assessment of solar energy resource in various locations of the country.

Furthermore, the data available for ground-based measurements were only for one year (2017) which is a limitation of this study. The study would be far better if the time series was longer enough. Therefore, the authors suggest it as a future recommendation that such studies should be conducted in different locations of the country with a longer time series of data.

Author Contributions: Conceptualization, A.A. and F.A.; software, A.A., Z.R. and F.A.; validation, M.A.A.I. and A.A.; formal analysis, A.A. and F.A.; investigation, A.A., Z.R. and F.A.; resources, A.A., M.A.A.I., Z.R., F.A., K.R. and J.D.; data curation, A.A. and F.A.; writing—original draft preparation, A.A., Z.R. and F.A.; writing—review and editing, A.A., M.A.A.I., Z.R., K.R., J.D. and F.A.; visualization, A.A. and F.A.; supervision, Z.R. and F.A.; project administration, F.A., M.A.A.I. and Z.R.; funding acquisition, Z.R., K.R. and J.D. All authors have read and agreed to the published version of the manuscript.

Funding: This research received no external funding.

Institutional Review Board Statement: Not applicable.

Informed Consent Statement: Not applicable.

Data Availability Statement: Not applicable.

Acknowledgments: The authors acknowledge the support of University of Engineering and Technology (UET) Peshawar to allow us to utilize the weather station installed at UET Peshawar.

Conflicts of Interest: The authors declare no conflict of interest.

Nomenclature

PV	Photovoltaic
UET	University of Engineering and Technology
GHI	Global Horizontal Irradiance
DNI	Direct Normal Irradiance
SUNY	State University of New York
NSRDB	National Solar Radiation Database
RSI	Rotating Shadowband Irradiometer

References

1. Bhattacharya, T.; Chakraborty, A.K.; Pal, K. Effects of Ambient Temperature and Wind Speed on Performance of Monocrystalline Solar Photovoltaic Module in Tripura, India. *J. Sol. Energy* **2014**, *2014*, 817078. [CrossRef]
2. Ayaz, A.; Ahmad, F.; Saher, S.; Noman, M. Solar Thermal Opportunities and Challenges in Pakistan. In *IOP Conference Series: Materials Science and Engineering*; IOP Publishing: Bristol, UK, 2020; Volume 899, p. 012008.
3. Cuce, E.; Cuce, P.M.; Bali, T. An experimental analysis of illumination intensity and temperature dependency of photovoltaic cell parameters. *Appl. Energy* **2013**, *111*, 374–382. [CrossRef]
4. Ayaz, A.; Ahmad, H.; Ahmad, F.; Khan, A.; hasnain Tarmazi, S.M.; Gul, R.M. Self-cleaning of glass surface to maximize the PV cell efficiency. In *IOP Conference Series: Materials Science and Engineering*; IOP Publishing: Bristol, UK, 2020; Volume 899, p. 012006.
5. Gueymard, C.A.; Wilcox, S.M. Assessment of spatial and temporal variability in the US solar resource from radiometric measurements and predictions from models using ground-based or satellite data. *Sol. Energy* **2011**, *85*, 1068–1084. [CrossRef]
6. Zell, E.; Gasim, S.; Wilcox, S.; Katamoura, S.; Stoffel, T.; Shibli, H.; Engel-Cox, J.; Al Subie, M. Assessment of solar radiation resources in Saudi Arabia. *Sol. Energy* **2015**, *119*, 422–438. [CrossRef]
7. AlYahya, S.; Irfan, M.A. Analysis from the new solar radiation Atlas for Saudi Arabia. *Sol. Energy* **2016**, *130*, 116–127. [CrossRef]
8. Alnaser, W.; Eliagoubi, B.; Al-Kalak, A.; Trabelsi, H.; Al-Maalej, M.; El-Sayed, H.; Alloush, M. First solar radiation atlas for the Arab world. *Renew. Energy* **2004**, *29*, 1085–1107. [CrossRef]
9. Bachour, D.; Perez-Astudillo, D. Ground measurements of Global Horizontal Irradiation in Doha, Qatar. *Renew. Energy* **2014**, *71*, 32–36. [CrossRef]
10. Janjai, S.; Laksanaboonsong, J.; Nunez, M.; Thongsathitya, A. Development of a method for generating operational solar radiation maps from satellite data for a tropical environment. *Solar Energy* **2005**, *78*, 739–751. [CrossRef]
11. Gueymard, C.A.; George, R. Gridded aerosol data for improved direct normal irradiance modeling: The case of India. In Proceedings of the American Solar Energy Society Annual Conference, Raleigh, NC, USA, 17–20 May 2011.
12. Ineichen, P. Comparison of eight clear sky broadband models against 16 independent data banks. *Sol. Energy* **2006**, *80*, 468–478. [CrossRef]
13. Mueller, R.; Dagestad, K.; Ineichen, P.; Schroedter-Homscheidt, M.; Cros, S.; Dumortier, D.; Kuhlemann, R.; Olseth, J.; Piernavieja, G.; Reise, C.; et al. Rethinking satellite-based solar irradiance modelling The SOLIS clear-sky module. *Remote Sens. Environ.* **2004**, *91*, 160–174. [CrossRef]
14. Kostylev, V.; Pavlovski, A. Solar power forecasting performance–towards industry standards. In *1st International Workshop on the Integration of Solar Power into Power Systems, Aarhus, Denmark*; Energynautics GmbH Mühlstraße: Langen, Germany, 2011.
15. Vignola, F.; Harlan, P.; Perez, R.; Kmiecik, M. Analysis of satellite derived beam and global solar radiation data. *Sol. Energy* **2007**, *81*, 768–772. [CrossRef]
16. Blanksby, C.; Bennett, D.; Langford, S. Improvement to an existing satellite data set in support of an Australia solar atlas. *Sol. Energy* **2012**, *98*, 111–124. [CrossRef]
17. Pereira, E.B.; Abreu, S.L.; Stuhlmann, R.; Rieland, M.; Colle, S. Survey of the incident solar radiation in Brazil by use of meteosat satellite data. *Sol. Energy* **1996**, *57*, 125–132. [CrossRef]
18. Janjai, S. A method for estimating direct normal solar irradiation from satellite data for a tropical environment. *Sol. Energy* **2010**, *84*, 1685–1695. [CrossRef]
19. Nonnenmacher, L.; Kaur, A.; Coimbra, C.F. Verification of the SUNY direct normal irradiance model with ground measurements. *Sol. Energy* **2014**, *99*, 246–258. [CrossRef]
20. Perez, R.; Seals, R.; Zelenka, A. Comparing satellite remote sensing and ground network measurements for the production of site/time specific irradiance data. *Sol. Energy* **1997**, *60*, 89–96. [CrossRef]
21. Janjai, S.; Pankaew, P.; Laksanaboonsong, J. A model for calculating hourly global solar radiation from satellite data in the tropics. *Appl. Energy* **2009**, *86*, 1450–1457. [CrossRef]
22. Tarpley, J.D. Estimating Incident Solar Radiation at the Surface from Geostationary Satellite Data. *J. Appl. Meteorol.* **1979**, *18*, 1172–1181. [CrossRef]
23. Gautier, C.; Diak, G.; Masse, S. A Simple Physical Model to Estimate Incident Solar Radiation at the Surface from GOES Satellite Data. *J. Appl. Meteorol.* **1980**, *19*, 1005–1012. [CrossRef]
24. Mabasa, B.; Lysko, M.D.; Moloi, S.J. Validating Hourly Satellite Based and Reanalysis Based Global Horizontal Irradiance Datasets over South Africa. *Geomatics* **2021**, *1*, 429–449. [CrossRef]

25. Kheddioui, A.; El Ouiqary, E.M.; Smiej, M. Estimation of the global horizontal solar irradiation GHI for the Moroccan national territory from meteorological satellite images of the Second Generation Meteosat series MSG. *Eur. J. Mol. Clin. Med.* **2021**, *8*, 2814–2826.
26. Jain, S.; Singh, C.; Tripathi, A.K. A Flexible and Effective Method to Integrate the Satellite-Based AOD Data into WRF-Solar Model for GHI Simulation. *J. Indian Soc. Remote Sens.* **2021**, *49*, 2797–2813. [CrossRef]
27. Xing, W.; Zhang, G.; Poslad, S. Estimation of global horizontal irradiance in China using a deep learning method. *Int. J. Remote Sens.* **2021**, *42*, 3899–3917. [CrossRef]
28. Rodríguez-Benítez, F.J.; López-Cuesta, M.; Arbizu-Barrena, C.; Fernández-León, M.M.; Pamos-Ureña, M.Á.; Tovar-Pescador, J.; Santos-Alamillos, F.J.; Pozo-Vázquez, D. Assessment of new solar radiation nowcasting methods based on sky-camera and satellite imagery. *Appl. Energy* **2021**, *292*, 116838. [CrossRef]
29. Alani, O.E.; Ghennioui, H.; Abraim, M.; Ghennioui, A.; Blanc, P.; Saint-Drenan, Y.M.; Naimi, Z. Solar Energy Resource Assessment Using GHI and DNI Satellite Data for Moroccan Climate. In *International Conference on Advanced Technologies for Humanity*; Springer: Cham, Switzerland, 2021; pp. 275–285.
30. Zelenka, A.; Perez, R.; Seals, R.; Renné, D. Effective Accuracy of Satellite-Derived Hourly Irradiances. *Arch. Meteorol. Geophys. Bioclimatol. Ser. B* **1999**, *62*, 199–207. [CrossRef]
31. Lohmann, S.; Schillings, C.; Mayer, B.; Meyer, R. Long-term variability of solar direct and global radiation derived from ISCCP data and comparison with reanalysis data. *Sol. Energy* **2006**, *80*, 1390–1401. [CrossRef]
32. Cebecauer, T.; Suri, M. Accuracy improvements of satellite-derived solar resource based on GEMS re-analysis aerosols. In Proceedings of the Conference SolarPACES, Perpignan, France, 21–24 September 2010.
33. Suri, M.; Cebecauer, T. Satellite-based solar resource data: Model validation statistics versus user's un-certainty. In Proceedings of the ASES SOLAR 2014 Conference, San Francisco, CA, USA, 7 July 2014; pp. 7–9.
34. Maxwell, E.; Wilcox, S.; Rymes, M. *User's Manual for SERI QC Software, Assessing the Quality of Solar Radiation Data*; Solar Energy Research Institute: Golden, CO, USA, 1993.
35. Younes, S.; Claywell, R.; Muneer, T. Quality control of solar radiation data: Present status and proposed new approaches. *Energy* **2005**, *30*, 1533–1549. [CrossRef]

energies

MDPI

Article

Dynamic Evaluation of Desiccant Dehumidification Evaporative Cooling Options for Greenhouse Air-Conditioning Application in Multan (Pakistan)

Hadeed Ashraf [1,†], Muhammad Sultan [1,*,†], Redmond R. Shamshiri [2,*,†], Farrukh Abbas [1,†], Muhammad Farooq [3], Uzair Sajjad [4], Hafiz Md-Tahir [5], Muhammad H. Mahmood [1], Fiaz Ahmad [1], Yousaf R. Taseer [6], Aamir Shahzad [7] and Badar M. K. Niazi [8]

1 Department of Agricultural Engineering, Bahauddin Zakariya University, Bosan Road, Multan 60800, Pakistan; hadeedashraf15@gmail.com (H.A.); farukh_56@hotmail.com (F.A.); hamidmahmood@bzu.edu.pk (M.H.M.); fiazahmad@bzu.edu.pk (F.A.)
2 Department of Engineering for Crop Production, Leibniz Institute for Agricultural Engineering and Bioeconomy, 14469 Potsdam-Bornim, Germany
3 Department of Mechanical Engineering, University of Engineering and Technology, Lahore 39161, Pakistan; engr.farooq@uet.edu.pk
4 Mechanical Engineering Department, National Chiao Tung University, Hsinchu 30010, Taiwan; energyengineer01@gmail.com
5 College of Engineering, Huazhong Agricultural University, Wuhan 430070, China; m.tahir@webmail.hzau.edu.cn
6 Department of Building and Architectural Engineering, Bahauddin Zakariya University, Bosan Road, Multan 60800, Pakistan; engr.yousafraza@bzu.edu.pk
7 Office of the Deputy Director Agriculture (Water Management), Punjab Agriculture Department, Officers Colony, Muzaffargarh Road, Jauharabad, Khushab 41000, Pakistan; ofwm_khb@yahoo.com
8 Agricultural Engineering Institute, National Agricultural Research Centre, Park Road, Islamabad 44000, Pakistan; badar.khan.niazi@gmail.com
* Correspondence: muhammadsultan@bzu.edu.pk (M.S.); rshamshiri@atb-potsdam.de (R.R.S.); Tel.: +92-333-6108888 (M.S.)
† These authors contributed equally to this work.

Citation: Ashraf, H.; Sultan, M.; Shamshiri, R.R.; Abbas, F.; Farooq, M.; Sajjad, U.; Md-Tahir, H.; Mahmood, M.H.; Ahmad, F.; Taseer, Y.R.; et al. Dynamic Evaluation of Desiccant Dehumidification Evaporative Cooling Options for Greenhouse Air-Conditioning Application in Multan (Pakistan). *Energies* **2021**, *14*, 1097. https://doi.org/10.3390/en14041097

Academic Editors: Jan Danielewicz and Krzysztof Rajski

Received: 27 December 2020
Accepted: 14 February 2021
Published: 19 February 2021

Abstract: This study provides insights into the feasibility of a desiccant dehumidification-based Maisotsenko cycle evaporative cooling (M-DAC) system for greenhouse air-conditioning application. Conventional cooling techniques include direct evaporative cooling, refrigeration systems, and passive/active ventilation. which are commonly used in Pakistan; however, they are either not feasible due to their energy cost, or they cannot efficiently provide an optimum microclimate depending on the regions, the growing seasons, and the crop being cultivated. The M-DAC system was therefore proposed and evaluated as an alternative solution for air conditioning to achieve optimum levels of vapor pressure deficit (VPD) for greenhouse crop production. The objective of this study was to investigate the thermodynamic performance of the proposed system from the viewpoints of the temperature gradient, relative humidity level, VPD, and dehumidification gradient. Results showed that the standalone desiccant air-conditioning (DAC) system created maximum dehumidification gradient (i.e., 16.8 g/kg) and maximum temperature gradient (i.e., 8.4 °C) at 24.3 g/kg and 38.6 °C ambient air conditions, respectively. The DAC coupled with a heat exchanger (DAC+HX) created a temperature gradient nearly equal to ambient air conditions, which is not in the optimal range for greenhouse growing conditions. Analysis of the M-DAC system showed that a maximum air temperature gradient, i.e., 21.9 °C at 39.2 °C ambient air condition, can be achieved, and is considered optimal for most greenhouse crops. Results were validated with two microclimate models (OptDeg and Cft) by taking into account the optimality of VPD at different growth stages of tomato plants. This study suggests that the M-DAC system is a feasible method to be considered as an efficient solution for greenhouse air-conditioning under the climate conditions of Multan (Pakistan).

Keywords: desiccant dehumidification; evaporative cooling; Maisotsenko cycle; greenhouse air-conditioning; Pakistan

1. Introduction

Closed-field production of crops and vegetables by means of greenhouses with different covering materials depends on the efficiency of the climate control system to provide optimum growth conditions with a reasonable energy cost. The extensive solar radiation during the daytime causes a greenhouse effect, resulting in high air temperature and relative humidity that can significantly exceed the optimum conditions. This can lead to a significant loss of yield and fruit quality in the absence of a proper ventilation and dehumidification system. Additionally, the plants and vegetables are more prone to pest/fungus/disease attack in higher relative humidity conditions than normal, which may also affect the production level of the greenhouse. To reduce the impact of suboptimal microclimate conditions on production, passive and active ventilation, in addition to direct evaporative cooling systems by means of pad-and-fans or swamp coolers, are conventionally used by greenhouse growers in Pakistan. The problem with these solutions is that they do not allow control over the relative humidity level of the greenhouse environment, hence leaving the plants and vegetables exposed to pests, fungus, and disease attacks. Therefore, a more appropriate air-conditioning (AC) solution that is cost-efficient and can be implemented in large scale commercial production is required. In this regard, desiccant dehumidification-based indirect evaporative cooling air-conditioning systems are gaining research attention that needs to be evaluated for AC application in different closed-field plant production environments.

The focus of this study is the city of Multan, Pakistan, which lies in a warm desert climate where direct evaporative cooling methods are used in combination with passive and active ventilation methods inside agricultural greenhouses. For the purpose of this research, a desiccant dehumidification-based Maisotsenko cycle evaporative cooling (M-DAC) system is proposed and evaluated as an alternative solution for controlling humidity levels to achieve the optimum vapor pressure deficit (VPD). The objective was to provide a systematic thermodynamic analysis of the proposed M-DAC system from the viewpoints of temperature gradient, relative humidity levels, optimum VPD, and dehumidification gradient. A review of the relevant published literature is summarized in Section 2 to identify the most recent achievements and the gaps. Modeling of the M-DAC and dynamic evaluation by means of the optimality degree and comfort ratio models are described in Section 3. The proposed solution is validated and discussed in Section 4 based on the thermodynamic analysis and the microclimate models. Section 5 concludes the study and highlights the potential aspects for further research and development.

2. Background and Literature Review

Evaporative cooling (EC) systems can potentially be classified into direct evaporative (DEC), indirect evaporative (IEC), and Maisotsenko cycle evaporative (MEC) systems. The MEC system is widely studied by the research community for heating, ventilation, and air-conditioning (HVAC) system applications. Pandelidis and Anisimov [1] numerically studied and compared eight different configurations of a standalone Maisotsenko-cycle evaporative heat and mass exchanger. Caliskan et al. [2] studied energy and exergy analyses of the standalone MEC system for building air-conditioning. Zhan et al. [3] studied counter flow standalone MEC system for building air-conditioning application. Caliskan et al. [4] compared traditional vapor compression air-conditioning systems with the MEC system based on energy and exergy analyses. Chua et al. [5] reviewed environment-friendly air-conditioning options for building applications. Pandelidis et al. [6] numerically simulated the MEC heat exchanger for different air-conditioning applications. Rogdakis et al. [7] numerically simulated and experimentally validated the MEC system for building air-conditioning application for the climatic conditions of Greece. Cui et al. [8] studied the MEC system for precooling of ambient air as an energy-saving technique for hot and humid climatic conditions. Riangvilaikul and Kumar [9] numerically investigated the MEC system for different inlet conditions. Sultan [10] studied the MEC system for greenhouse air-conditioning and ventilation applications. Anisimov et al. [11] theoretically and experimentally analyzed the standalone MEC system for air-conditioning and found the MEC system to be an efficient way of indirect evaporative cooling. In addition, results from the study concluded that inlet air velocity impacted the performance of the system. Zhao et al. [12] numerically investigated the performance of the MEC system for climatic conditions of the United Kingdom, which resulted in wet-bulb effectiveness of 1.3. Zube and Gillan [13] experimentally studied a commercial type MEC air-conditioning system and evaluated heat mass transfer parameters inside a heat exchanger for the first time. Weerts [14] reported the power-saving potential of commercially available standalone MEC system compared to the vapor compression air-conditioning (VCAC) system. Maisotsenko and Treyger [15] also reported the energy-saving potential of the standalone MEC system compared to standalone VCAC system. Similarly, [16–18] also studied the standalone MEC system in-depth and concluded that the MEC system can potentially be an environmentally friendly air-conditioning option compared to the VCAC system.

The standalone desiccant air-conditioning (DAC) system has been extensively studied in the literature for various applications. Generally, the DAC system better suits regions with higher relative humidity in the ambient air. The DAC system was studied for the possible application of vehicular air-conditioning [19]. Additionally, the authors used a continuous input and output modulator (i.e., proportional-derivative (PD) controller) to increase the efficiency of the system and concluded that the DAC system coupled with the PD controller was energy conservative compared to conventional AC systems in vehicles [19]. Moreover, the DAC system was successfully installed in a wet market in Hong Kong. The authors numerically simulated the energy loads of a hypothetical wet-market using EnergyPlus for the climatic conditions of Hong Kong and compared results with the actual wet market [20]. The authors concluded that the DAC could potentially be used as a replacement for conventional air-conditioning systems in the wet markets. A solar-assisted open cycle DAC system was experimentally tested for grain storage in Melbourne, Australia [21]. The authors concluded that a 5.85 m^3 solar collector area was sufficient for cooling of up to 200 tonnes of grain. The coefficient of performance (COP) was a function of surrounding conditions and air mass flow rate. A maximum COP of 86.2 was observed under humid conditions for an air mass flow rate of 0.019 kg/s [21]. The standalone DAC system was studied for a greenhouse air temperature and humidity control system [22]. Moreover, the authors studied the greenhouse air-conditioning from a vapor pressure deficit (VPD) point of view. The standalone DAC system with dual desiccant wheels was studied for residual waste heat recovery in marine ships [23]. The authors concluded that the DAC system was 33.4% more power-efficient compared to traditional air-conditioning systems for marine ships. An EC-coupled DAC system was studied in two modes (i.e., ventilation and recirculation modes) [24]. The authors concluded that conventional ventilation mode is more efficient when humidity ratio is below ~10.9 g/kg, whereas the recirculation mode of the air-conditioning system is more efficient for areas where humidity ratio is above ~10.9 g/kg, consuming a higher amount of input energy. Enteria et al. [25] studied the analyses of the first and second law of thermodynamics for a solar thermal electric desiccant air-conditioning system [25]. A simulated solar-operated evaporative cooling-assisted solid desiccant air-conditioning system was developed in a tropical region of Malaysia [26]. The authors used the TRNSYS simulation environment to simulate the desiccant air-conditioning system for climatic conditions of Malaysia. The authors concluded that two-stage ventilation mode desiccant air-conditioning system is better for tropical conditions such as Malaysia, with a temperature gradient of 17.6 °C at 30 °C ambient temperature [26]. Another study investigated a dual-wheel cooling system [27]. Results from the study indicate that higher regeneration temperature (RT) produces higher total COP at both air mass flow rates (i.e., 100 and 200 m^3/h) [27]. Moreover, the performance and energy-saving potential of a two-stage solar-driven rotary wheel DAC system was compared with a conventional vapor compression air-conditioning (VCAC) system for building air-conditioning [28]. Building energy performance simulation (BEPS) was conducted for Berlin and Shanghai. The DAC system was found be to energy-saving and more efficient for both of the cities compared to VCAC system. However, in the case of Shanghai, the RT of the DAC system was 30 °C higher than that of Berlin [28]. Figure 1 shows the schematic diagram and working principle of a conventional vapor compression air-conditioning system, and solar thermal liquid desiccant dehumidification-based air-conditioning system. Moreover, the authors studied desiccant dehumidification systems and evaporative cooling systems in detail for various applications, including greenhouse air-conditioning [29–32], building air-conditioning [33], livestock air-conditioning [34], and agricultural product storage [35–44].

Song and Sobhani [45] studied solar-assisted desiccant air-conditioning system coupled with a Maisotsenko evaporative cooling system and phase change material for building air-conditioning application. The authors concluded that the maximum total COP of 0.404 was achieved using a solar-operated desiccant air-conditioning system coupled with a Maisotsenko evaporative cooler using a phase change material for air-conditioning of an official building in Iran. In another study, a Maisotsenko cycle-assisted desiccant air-conditioning system was studied for the climatic conditions of Japan [46]. Pandelidis et al. [47] numerically studied the Maisotsenko cycle-assisted desiccant air-conditioning system with two modifications, i.e., bypass and bypass with rotary sensible heat exchanger. The authors concluded that all the configurations successfully managed the sensible load of the environment while being highly dependent on the humidity ratio of ambient air. The addition of a rotary sensible heat exchanger in system configuration 3 increased the cooling capacity up to ~2.35 kW at 40 °C ambient temperature generating ~17.25 °C supply air temperature at the same ambient conditions [47]. Table 1 shows the summary of different desiccant-based Maisotsenko cycle evaporative cooling systems for air-conditioning applications. From Table 1 and the reviewed literature, it is evident that there is a gap in the literature regarding OptDeg and Cft plant thermal comfort indices for the proposed M-DAC system. Moreover, this study aims to use a low and easily achievable regeneration temperature (i.e., 50 °C) coupled with different comfort levels of plant growth stages, which is lacking in previous studies.

Table 1. Summary of different desiccant-based Maisotsenko cycle evaporative cooling systems for air-conditioning applications.

System	Findings	Regeneration Temperature	Ref.
Desiccant based indirect evaporative cooling	For 70 °C T_{regen}, supply flow ratio 0.67, indirect EC flow ratio 0.3 results in maximum performance i.e., COP higher than 20	70 °C	[48]
Maisotsenko cycle desiccant evaporative cooling	For outside conditions of 25 °C, lowest temperature gradient was 14.9 °C	70 °C	[1]
Standalone MEC	For Greek climate conditions, MEC system achieved maximum temperature gradient of 20.4 °C in Athens	–	[7]
Maisotsenko cycle desiccant evaporative cooling	Different desiccant materials studied for greenhouse air-conditioning, only limitation was integration of OptDeg and Cft models for different growth stages of crops	Different regeneration temperatures	[10]
Standalone DAC system	Polymer-based sorbents PS-I and PS-II are studied at regeneration temperatures 50 and 80 °C for greenhouse air-conditioning	Different regeneration temperatures	[30]
Standalone DAC system	Activated carbon powder (ACP) and activated carbon fiber (ACF) were studied for greenhouse air-conditioning at regeneration temperatures 41 to 75 °C, only limitation was the detailed analysis of crop growth and plant comfort indices	Different regeneration temperatures	[49]

The study area (i.e., Multan) lies in a warm desert climate of the Köppen–Geiger climatic classification (Figure 2). Therefore, this justifies the need for an air-conditioning system for greenhouses. Usually, pad-type direct evaporative cooling or natural/forced ventilation systems are used for greenhouse air-conditioning. These systems are not feasible throughout the year and fail to achieve the required optimum temperature and relative humidity conditions inside a greenhouse. Moreover, the DEC system fails to handle humidity inside the greenhouse environment, which is crucial to the vapor pressure deficit. The desiccant-based hybrid evaporative cooling systems have not been explored for greenhouse air-conditioning applications in Pakistan.

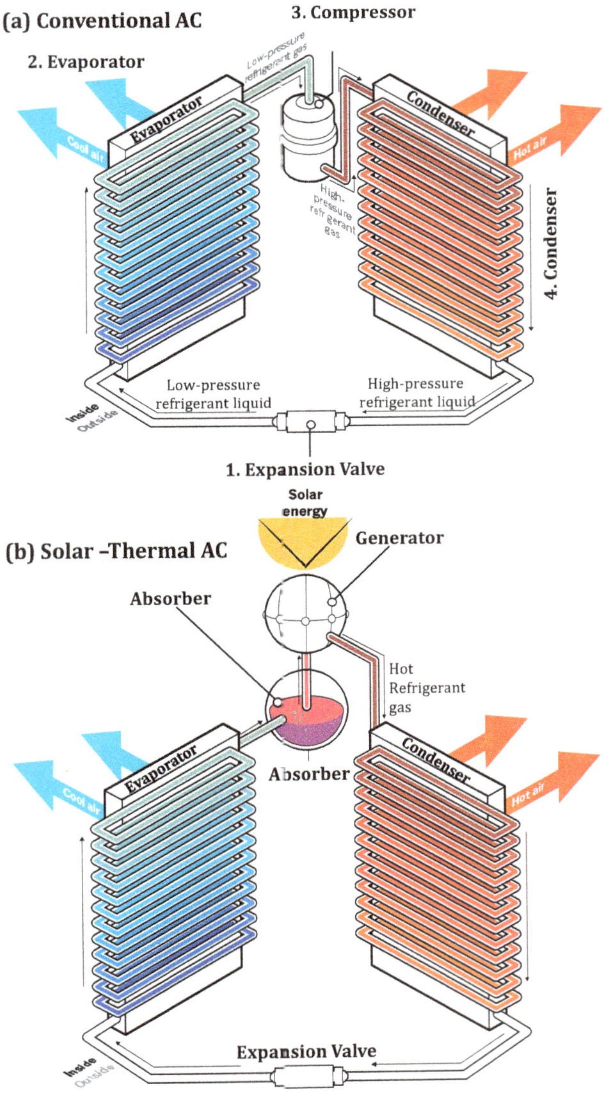

Figure 1. Schematic diagram and working principle of (**a**) conventional vapor compression air-conditioning system, and (**b**) solar thermal liquid desiccant dehumidification-based air-conditioning system, reproduced from [50].

Figure 3 shows the ambient climatic conditions of Multan (Pakistan) and the greenhouse optimum temperature and relative humidity zone. Figure 3 was made using Ladybug Rhino v6 for the climatic conditions of Multan (Pakistan). A typical meteorological EnergyPlus weather file was used in Ladybug Rhino to represent the hourly weather conditions. In Figure 3, the light fill area shows the average hourly variation, whereas the lines show the average daily variation in climatic parameters. From the viewpoint of the literature reviewed above, pad-type direct evaporative cooling systems are only suitable for controlling the temperature, and fail to achieve the optimum humidity level inside the greenhouse. To overcome this issue, an indirect evaporative cooling system could be used, which provides a sensibly cooled product air with no latent transfer from the wet to dry channels of the IEC system. However, due to the low wet bulb effectiveness of the system, it cannot be considered a suitable option for achieving the required temperature level inside the greenhouse. Additionally, higher levels of humidity inside the greenhouse can potentially result in pest, fungus, and disease attacks on the plants and vegetables. Therefore, in this study, a desiccant dehumidification-based Maisotsenko cycle evaporative cooling air-conditioning (M-DAC) system was thermodynamically analyzed for the climatic conditions of Multan (Pakistan). The performance of the system was analyzed for optimum greenhouse air-conditioning conditions from the viewpoints of temperature gradient, relative humidity level, and wet bulb effectiveness of the systems.

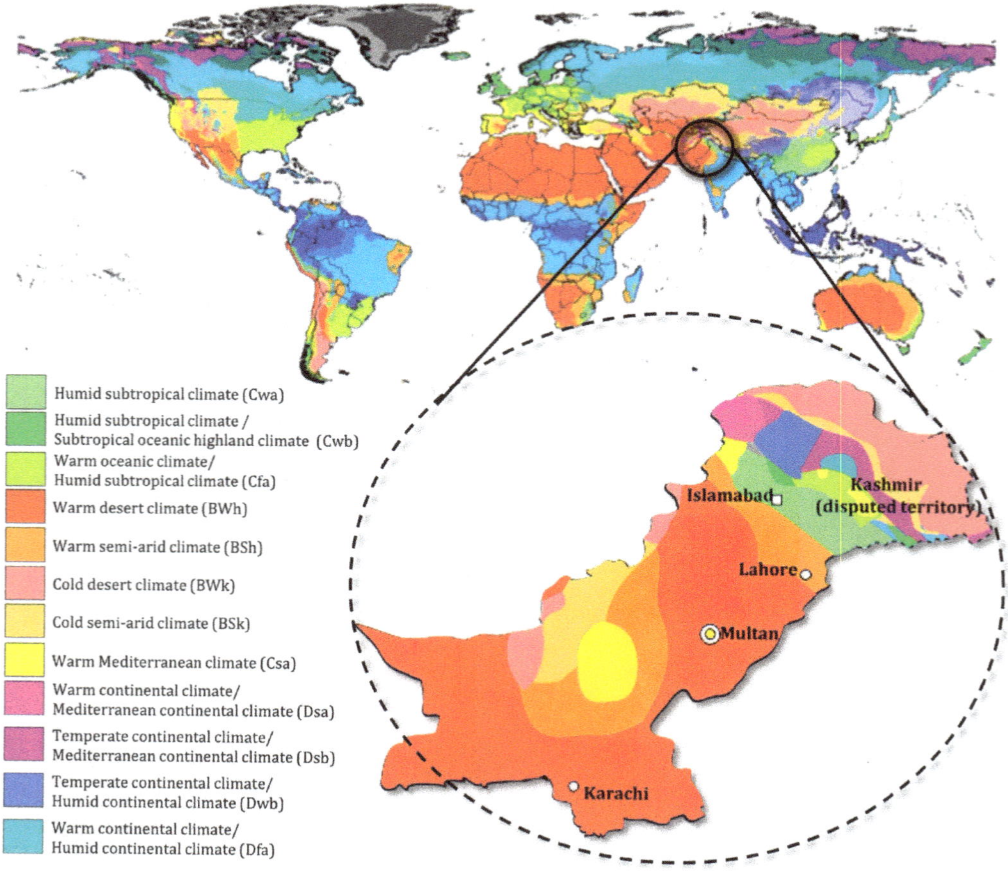

Figure 2. Köppen–Geiger climate classification of Pakistan, reproduced from [51].

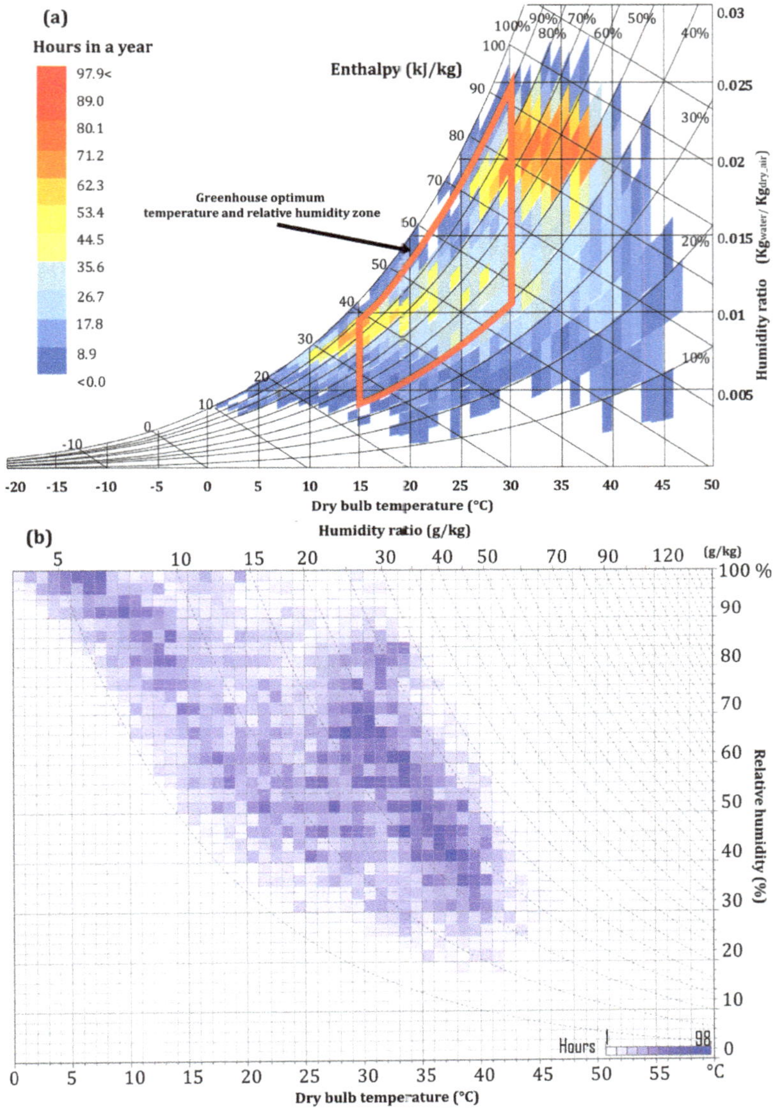

Figure 3. Ambient climatic conditions of Multan (Pakistan) on (a) psychrometric chart, and (b) relative humidity chart.

3. Materials and Methods

3.1. Modelling M-DAC System

A general working principle of the M-DAC system is first shown in Figure 4, followed by a schematic representation in Figure 5 that demonstrates the application of the proposed system for greenhouse air-conditioning for a possible higher yield. Taking into account the climatic conditions of Multan (Pakistan), the performance of a silica gel-based desiccant wheel for M-DAC system was investigated using Equations (1)–(4), as described in the literature [52,53].

$$F_{1,ip} = \frac{A_1}{(T_{ip} + 273.15)^{1.49}} + B_1 \left(\frac{w_{ip}}{1000} \right)^{C_1} \tag{1}$$

$$F_{2,ip} = \frac{(T_{ip} + 273.15)^{1.49}}{A_2} - B_2 \left(\frac{w_{ip}}{1000} \right)^{C_2} \tag{2}$$

$$\eta F_1 = \frac{F_{1,2} - F_{1,1}}{F_{1,8} - F_{1,1}} \tag{3}$$

$$\eta F_2 = \frac{F_{2,2} - F_{2,1}}{F_{2,8} - F_{2,1}} \tag{4}$$

where $F_{1,ip}$ and $F_{2,ip}$ represent the combined potential as a function of the humidity ratio in kg/kg and temperature in °C of the solid desiccant-based air-conditioning system. ηF_1 and ηF_2 represent the efficiencies of the system correlating to the combined potentials. ηF_1 and ηF_2 have typical values of 0.05 and 0.95 for a high-efficiency silica gel-based solid desiccant wheel air-conditioning system. The coefficients A, B, and C are presented in Table 2.

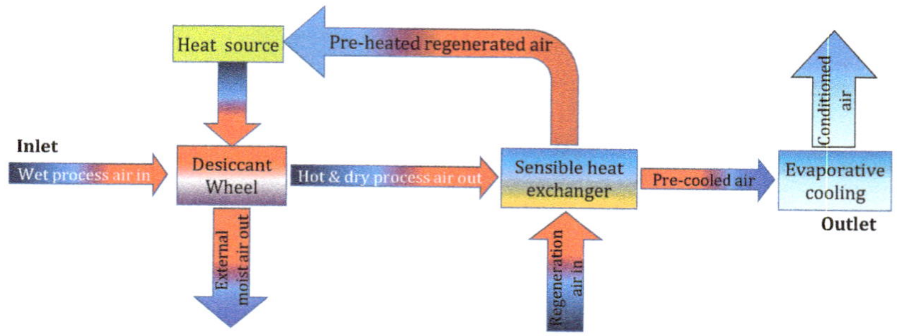

Figure 4. Working principle of a desiccant dehumidification-based evaporative cooling air-conditioning system.

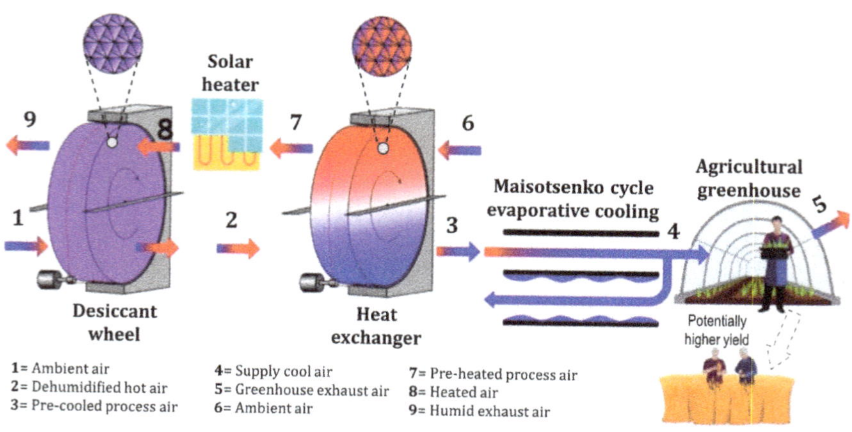

Figure 5. Schematic diagram and working principle of the proposed desiccant dehumidification Maisotsenko cycle evaporative cooling system.

In comparison with the conventional vapor compression air-conditioning systems, the desiccant dehumidification-assisted evaporative cooling air-conditioning system consumes less energy due to the absence of compressor work and is able to operate at lower regeneration temperatures (i.e., 50–90 °C), which are easily attainable using solar thermal, biogas, and waste heat sources. The Maisotsenko cycle evaporative cooling system was modeled using the model of [32]. Equation (5) shows the performance model of the Maisotsenko cycle evaporative cooling system.

Table 2. Coefficients used in the Jurinak model (Equations (1)–(4)).

Coefficient	Value
A_1	−2865
B_1	4.344
C_1	0.8624
A_2	6360
B_2	1.127
C_2	0.07969

$$T_{out} = 6.70 + 0.2630(T_{in}) + 0.5298(w_{in}) \quad (5)$$

where T_{in} represents the inlet temperature of the process air in °C and w_{in} represents the inlet humidity ratio of the process air in kg/kg. Equation (5) models the temperature of product air produced from the Maisotsenko cycle evaporative cooling system. The model was developed by [32] using experimental data of a developed Maisotsenko cycle based evaporative cooling system [11].

The sensible heat exchanger (HX) used in this study was modeled using the heat exchanger standard equation (Equation (6)) provided in the literature by The American Society of Heating, Refrigerating and Air-Conditioning Engineers (ASHRAE) [54].

$$T_3 = T_2 - \varepsilon_{HX}(T_2 - T_1) \quad (6)$$

where T_3 represents the outlet temperature of the sensibly cooled process air resulting from the heat exchanger, T_2 represents the outlet temperature of heated and dehumidified air resulted from the desiccant air-conditioning system, ε_{HX} represents the sensible heat exchange efficiency of the heat exchanger (assumed as 0.9), and T_1 represents the dry bulb temperature of the ambient air inlet into the desiccant air-conditioning system. Figure 5 shows the working principle and schematic of the sensible heat exchanger. Regeneration of the desiccant is done usually through regeneration air at higher than ambient temperature. To heat the regeneration air, the air passes through the heat exchanger which sensibly heats the air to a certain limit. For further heating of the regeneration air before entering the desiccant, a heating unit/source is used. This heating source could be a heating electric coil, solar heater, or biogas-operated thermal heat source. Waste heat from a condenser or any other source in industrial uses can also be used to regenerate the desiccant material and absorb its moisture. For this study, a regeneration temperature of 50 °C was used. This relatively lower temperature is achievable through a solar thermal system, biogas-operated thermal heat source, or waste heat. A flat plate solar collector and heat exchanger were used for this study to provide the required regeneration temperature (i.e., 50 °C). Figure 5 shows the working principle and schematic of a solar thermal heating source.

3.2. Optimality Degree and Comfort Ratio Model

Dynamic assessment of the proposed M-DAC system for greenhouse crop production application was carried out by means of two microclimate models, namely, OptDeg [55] and Cft-ratio [56] that take into account VPD measurements as inputs. The output of the first model, optimality degree of VPD, denoted by $Opt(VPD)_{GS,\,(Light)} = \alpha : VPD \to [0,1]$ is a quantitative value between 0 and 1 that represents how close a VPD reading inside the greenhouse is to the optimum value ($0 \leq \alpha \leq 1$) at a specific growth stage (GS) and light condition. Graphical representation of the membership functions for defining *Opt(VPD)* under different conditions is provided in Figure 6. In this model, a VPD measurement in the greenhouse is first calculated from Equation (7) at the time $t_{m,n}$ (where m and n refer to a specific minute and date of the growing stage), and is then mapped to a value between 0 and 1 that quantifies the optimality degree of the microclimate for greenhouse production. In other words, an optimality-degree equal to 1 refers to a high potential yield with marketable value and high-quality fruit.

$$VPD_{t_{m,n}} = f\left(T_{t_{m,n}}, RH_{t_{m,n}}, \tau_{t_{m,n}}\right) = \frac{C_g 10^{(7.5\tau_{t_{m,n}})/(237.3+\tau_{t_{m,n}})}}{1000} - \frac{C_g 10^{(7.5T_{t_{m,n}})/(237.3+T_{t_{m,n}})}}{1000} \frac{RH_{t_{m,n}}}{100} \quad (7)$$

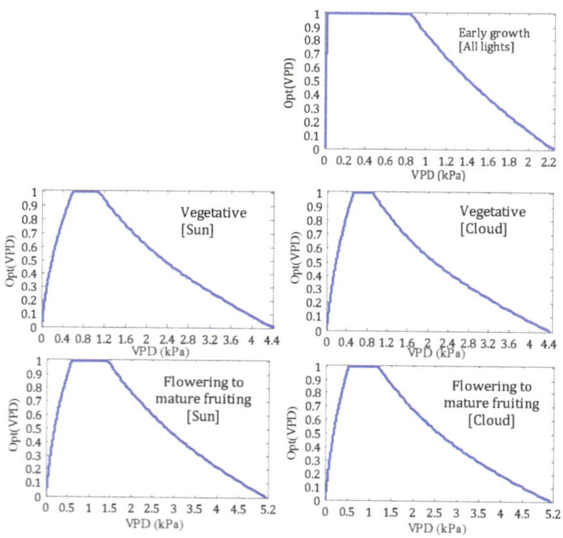

Figure 6. Membership functions for defining optimality degrees of vapor pressure deficit at different light condition and growth stages, reproduced from [40].

It should be noted that there are various functions for calculating VPD that take into accounts different assumptions and constants; however, that presented in Equation (7) is widely accepted and used by meteorologists and commercial greenhouse growers [34]. The first part of Equation (7) represents the vapor pressure of the saturated environment (i.e., the leaf of the plant) in kPa, C_g denotes a coefficient that is equal to 610.7, and τ represents the surface temperature of the leaf, which is usually measured by means of an infrared gun for such experiments. For the sake of simplicity, this parameter was assumed to be equal to the temperature of the air inside the greenhouse at both ambient and evaporative assisted desiccant air-conditioning system supply air temperature. In fact, assuming leaf temperature equal to the surrounding air temperature is normal practice of greenhouse growers. The second part of Equation (7) represents the vapor pressure of the air inside the greenhouse in kPa, T represents the temperature of the air inside the greenhouse in °C, and RH represents the relative humidity of the air inside the greenhouse in (%). Because VPD is the function of temperature of the leaf surface, air temperature, and relative humidity, any slight variation in one of the parameters from the optimal values results in restricted growth of the plants, and can have significant consequences on reducing the overall yield and quality.

The OptDeg model determines the variation of a single VPD measurement from the optimal references in an instant time. Although this model can provide an overview of the performance of the M-DAC system in sample time, a different model called comfort ratio, denoted by Cft-ratio, that incorporates a sample time frame, such as 24 h was used. The comfort ratio of VPD, denoted by $Cft(VPD, t, \alpha_s)_{GS} = \beta$, represents the percent of vapor pressure deficit at a specific time t and growth stage, which lies between the reference conditions of vapor pressure deficit related to α_s. Ideal microclimate conditions inside a greenhouse could potentially be defined as $Cft(VPD, t, 1) = 1$. Self-adjusted optimality degree is denoted by α_s to define the reference conditions required for performance evaluation of the proposed M-DAC system or control for microclimate evaluation. The set of Simulink blocks that were implemented to determine the Cft-ratio of the VPD values for different α_s generated from the M-DAC system are shown in Figure 7. For this research, we selected three levels of α_s as 0, 0.5, and 1, representing failure, marginal, and ideal production. It is worth mentioning that OptDeg and Cft-ratio models were derived with a prime focus on improved fruit quality and yield. The derivations at the back of the models (which can be found in published literature) for greenhouse tomato production were condensed for the sake of simplicity in this study.

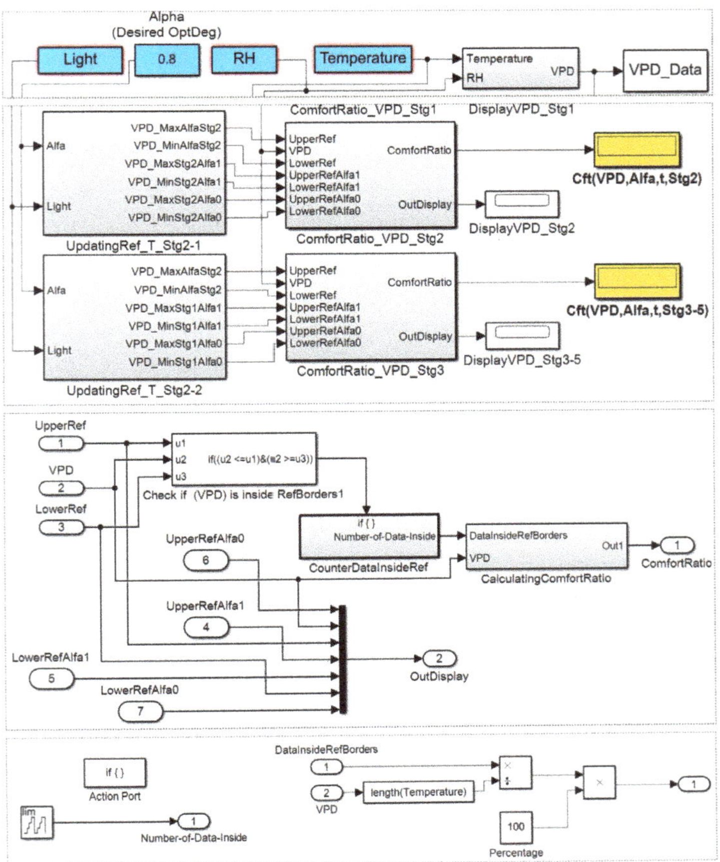

Figure 7. Implementation of the Cft-ratio model using Simulink blocks for validating the performance of the desiccant dehumidification Maisotsenko cycle evaporative cooling (M-DAC) system based on comfort ratios of vapor pressure deficit (VPD) for greenhouse production, reproduced from [56].

4. Results and Discussion

4.1. Validation Based on Thermodynamic Analysis

Figure 8 shows the annual profile of the climatic ambient air conditions of Multan (Pakistan). The daily, hourly, and monthly variation of dry-bulb temperature shows that cooling inside the greenhouse environment is mostly required in summer months (i.e., May to August). However, the applicability of standalone pad-type direct evaporative cooling systems is limited during the monsoon (heavy rainfall season, i.e., July to August, shown in Figure 8, humidity ratio (HR)) which results in excess relative humidity in the air. Consequently, a desiccant dehumidification-based Maisotsenko cycle evaporative cooling system was thermodynamically analyzed for the climatic conditions of Multan. Accordingly, results from the study are presented in Figures 8–13, which show the annual representation of the performance of the proposed system and the profile of the vapor pressure deficit inside the greenhouse.

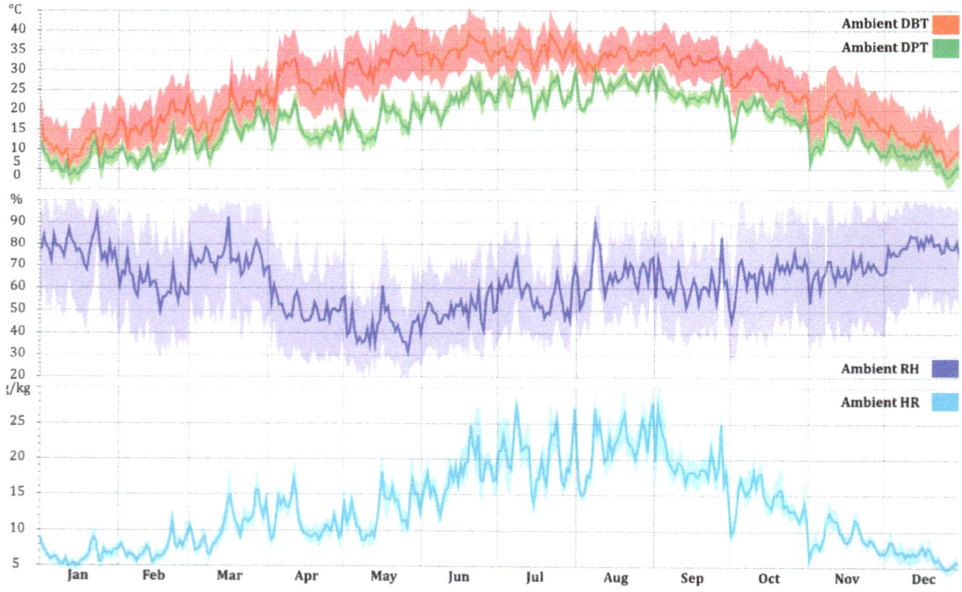

Figure 8. Profile of annual climatic ambient air conditions of Multan (Pakistan).

The standalone DAC system created a maximum temperature gradient, i.e., 8.4 °C more at ambient conditions of 38.6 °C with a humidity ratio of 21.6 g/kg at ambient conditions of 25 g/kg, which lies well outside the required greenhouse optimum temperature and humidity conditions (Figure 9a). It is worth mentioning that the temperature gradient is defined as the difference between the dry bulb temperature of the ambient air and the dry bulb temperature of the product air at the system outlet node. The studied standalone DAC system created unsuitable thermal and humidity conditions inside the greenhouse throughout the year due to a relatively higher temperature and very dehumidified air (Figure 9a). According to Figure 9b, the standalone DAC system coupled with a sensible heat exchanger (DAC+HX) (working efficiency assumed to be 0.9) created a maximum temperature gradient, i.e., 0.77 °C more at ambient conditions of 39.2 °C with a humidity ratio of 16.8 g/kg at ambient conditions of 24.3 g/kg, which lies slightly outside the required greenhouse optimum temperature and humidity conditions (Figure 9b), and is very close to ambient air conditions. The studied DAC+HX system created very close thermal and humidity conditions to ambient conditions inside the greenhouse throughout the year.

Figure 9c shows the psychrometric profile of temperature and relative humidity of the desiccant dehumidification-based Maisotsenko cycle evaporative cooling (M-DAC) system for the climatic conditions of Multan (Pakistan). According to Figure 9c, most of the hourly points of the M-DAC system lie inside the required thermal and humidity conditions of the greenhouse environment. The M-DAC system created a maximum temperature gradient i.e., 21.98 °C. According to Figure 9c, the M-DAC system coupled with a sensible heat exchanger (working efficiency assumed to be 0.9) created a maximum temperature gradient i.e., 21.9 °C more at ambient conditions of 39.2 °C with a humidity ratio of 16.8 g/kg at ambient conditions of 19.6 g/kg, which lies well inside the required thermal and humidity conditions inside the greenhouse environment (Figure 9c).

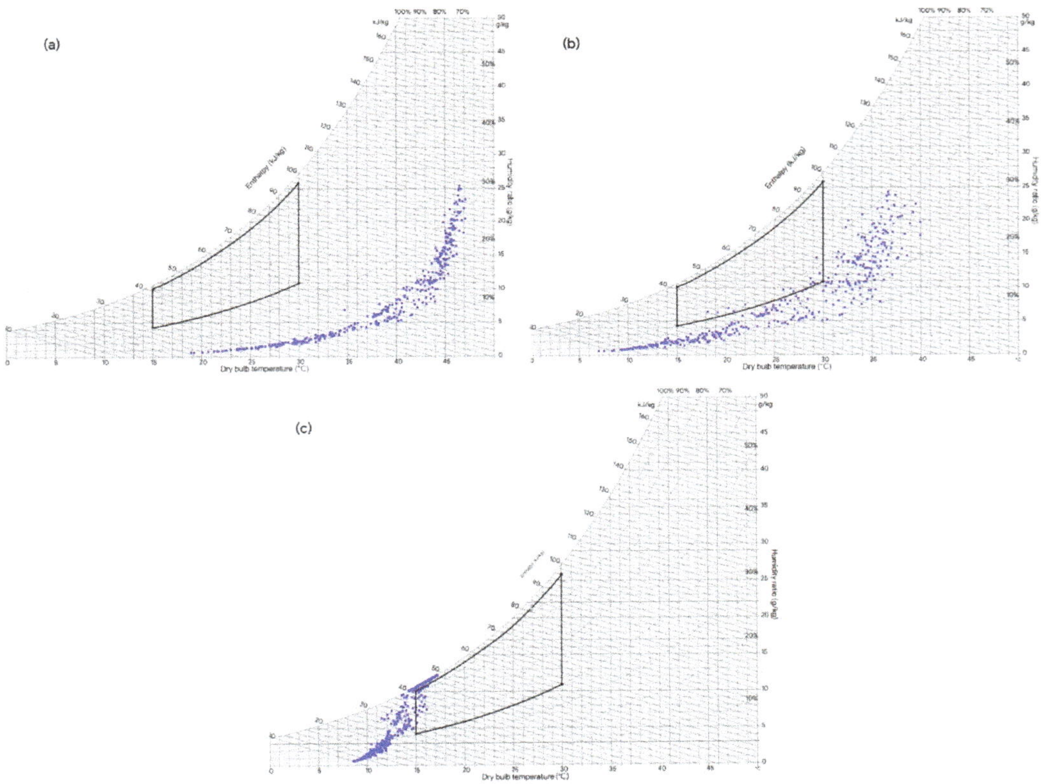

Figure 9. Profile of annual temperature and relative humidity of (**a**) standalone desiccant air-conditioning (DAC) system, (**b**) desiccant air-conditioning coupled with sensible heat exchanger (DAC+HX) system, and (**c**) desiccant dehumidification Maisotsenko cycle evaporative cooling (M-DAC) system, for greenhouse air-conditioning.

Figure 10 shows the vapor pressure deficit (VPD) profile and growth stages of plants inside a greenhouse (for the temperature of the leaf equal to the temperature of the air inside the greenhouse environment). According to Figure 10, the VPD of <0.4>1.6 kPa is considered an extremely dangerous zone for over/under transpiration. The VPD of 0.4–0.8 kPa is considered a slightly low transpiration zone (propagation/early vegetative growth of the plants). The VPD of 0.8–1.2 kPa is considered a healthy transpiration zone (late vegetation/early flowering stage of the plants). The VPD of 1.2–1.6 kPa is considered a high transpiration zone (mid/late flowering stage of the plants).

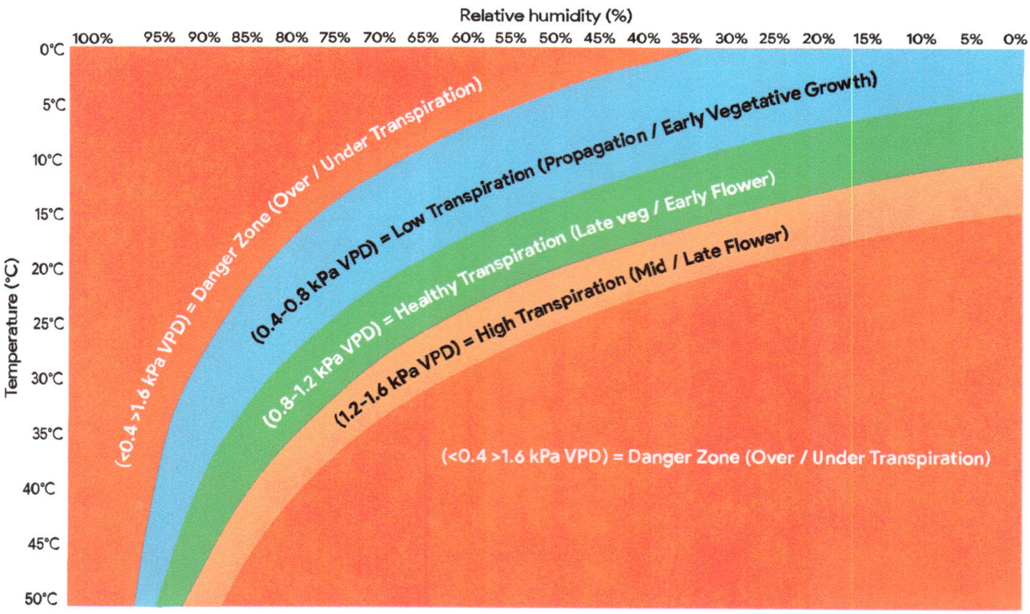

Figure 10. Vapor pressure deficit (VPD) profile and growth stages of plants inside a greenhouse.

Figure 11 shows the temperature profile of the studied systems against the ambient air conditions of Multan (Pakistan) for greenhouse air-conditioning application. The ambient air conditions lie outside the required temperature conditions (marked by the dotted red line, Figure 11) from April to September (i.e., typical summer months) of Multan. Air-conditioning is required inside the greenhouse environment during the summer months. Only M-DAC system created a uniform temperature gradient which lies well inside the required temperature conditions (marked in dotted red line, Figure 11) of the greenhouse environment. Figure 12 shows the temperature gradient of the DAC, DAC+HX, and M-DAC system in the summer months (i.e., May to August) for the greenhouse air-conditioning for the climatic conditions of Multan (Pakistan). The DAC system created higher than ambient temperature conditions throughout the summer months, which is unsuitable for greenhouse air-conditioning requirements (Figure 12). The DAC + HX system created slightly higher/very close to ambient temperature conditions throughout the summer months, which is also not suitable for greenhouse air-conditioning requirements (Figure 12). The M-DAC system created a temperature gradient which is highly suitable for the required temperature conditions throughout the summer months for greenhouse air-conditioning application (Figure 12).

Figure 13 shows the dehumidification profile of the desiccant dehumidification unit of the proposed system for the summer months (i.e., May to August) for the climatic conditions of Multan (Pakistan). The desiccant dehumidification unit created a maximum dehumidification gradient of the ambient air, i.e., 6.61 g/kg at ambient conditions of 14.14 g/kg. It is worth mentioning that the dehumidification gradient is defined as the difference between the ambient air humidity ratio and the humidity ratio of the product air exiting the outlet of the proposed system. Compared to current proposed passive cooling techniques for greenhouse air-conditioning (i.e., pad-and-fan type systems for greenhouse air-conditioning systems), the proposed M-DAC system has the ability to manipulate the moisture inside the greenhouse environment, which is a key factor in defining the vapor pressure deficit inside the greenhouse. In turn, this defines the growth rate and ultimately results in better yield and fruit quality. Moreover, the M-DAC system can easily achieve the required temperature conditions inside a greenhouse compared to passive cooling techniques, i.e., pad-and-fan type AC systems.

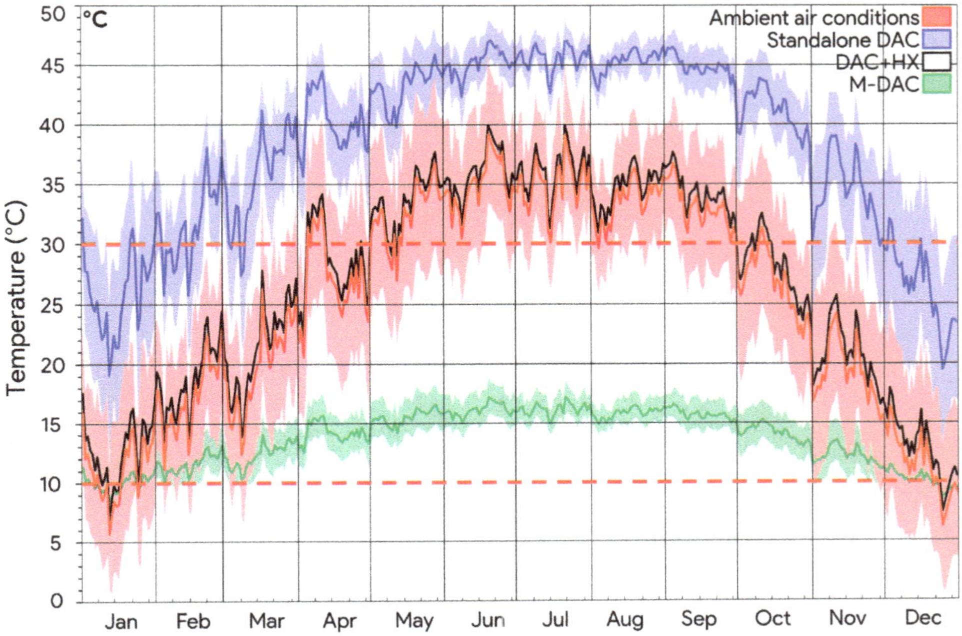

Figure 11. Annual profile of thermodynamic performance of the proposed M-DAC system for greenhouse air-conditioning for the climatic conditions of Multan (Pakistan).

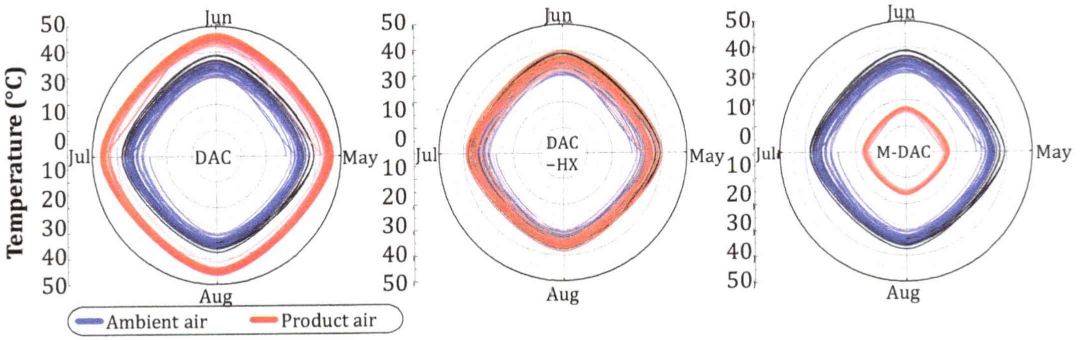

Figure 12. Performance profile of the proposed M-DAC system's temperature gradient in summer months of Multan (Pakistan) for greenhouse air-conditioning application.

4.2. Validation Based on Optimality Degree and Comfort Ratio

Validation of the performance of the proposed M-DC system with respect to the required optimal influential parameters of the greenhouse microclimate (i.e., temperature of the air, inside RH, and vapor pressure deficit) is shown in the plots of Figure 14 for 24 h. Results clearly show that the output of the M-DAC system was very close to the requirement microclimate data generated by the OptDeg model for the vegetative to mature fruiting growth stages. Moreover, the VPD plot implies that during the mid-day hours (i.e., 13:00–17:00) when the air temperature was at the highest peak and relative humidity had the lowest values, the VPD increased to 2 kPa, which is consistent with the output of the M-DAC system. Results of the VPD data generated from the OptDeg model for actual greenhouse tomato cultivation in 130 days versus the simulated VPD data generated from the M-DAC system are plotted in Figure 15. The first derivative of the VPD data sets, plotted in Figure 15a, shows that the optimum VPD values from the OptDeg model and the VPD values from the M-DAC system are not significantly different. This finding was also verified using the one-way analysis of variance (ANOVA) test for each of the five growth stages. Figure 15b shows the deviation of the VPD data of the M-DAC system from 100% optimal condition ($\alpha = 1$), and compares that with actual VPD data of a commercial greenhouse that operated with ventilation and a pad-and-fan evaporative cooling system.

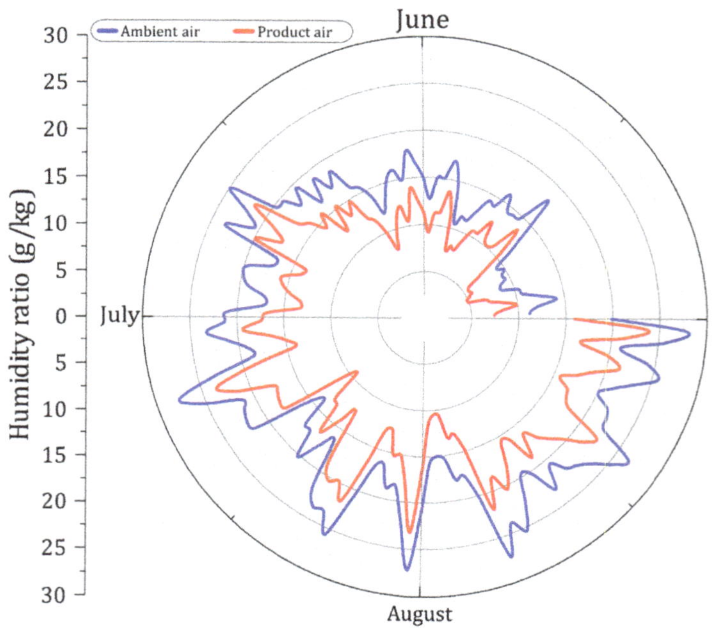

Figure 13. Performance profile of the proposed M-DAC system's humidity ratio in summer months of Multan (Pakistan) for greenhouse air-conditioning application.

A more in-depth result for comparing the performance of the M-DAC system and evaporative pad-and-fan system in a random day of cultivation under fruit formation growth stage is provided by means of the Cft-ratio model in the plots of Figure 16. In Figure 16, the reference conditions related to the optimal, marginal, and failure vapor pressure deficit (i.e., $\alpha_s = 1, 0.5$, and 0, respectively) are presented with a green solid line, blue solid line, and red solid line, respectively. The Cft-ratio Simulink model then determined the percentage of VPD data of each case, shown in Figure 16 that were inside each reference border. From the graphical presentation, it is evident that the VPD data from the M-DAC system resulted in a higher comfort ratio than those from the evaporative pad-and-fan system. In detail, VPD values resulting from the M-DAC system never crossed the marginal reference border, whereas for the conventional pad-and-fan case, during the hours of 13:30–14:30, VPD exceeded this border and increased to 3 kPa, which is considered a failure value for greenhouse production. Sample results of the comfort ratio model for comparing the performance of the evaporative system and M-DAC system in providing optimum vapor pressure deficit for greenhouse production are given in Table 3 for two different growth stages of the tomato plant.

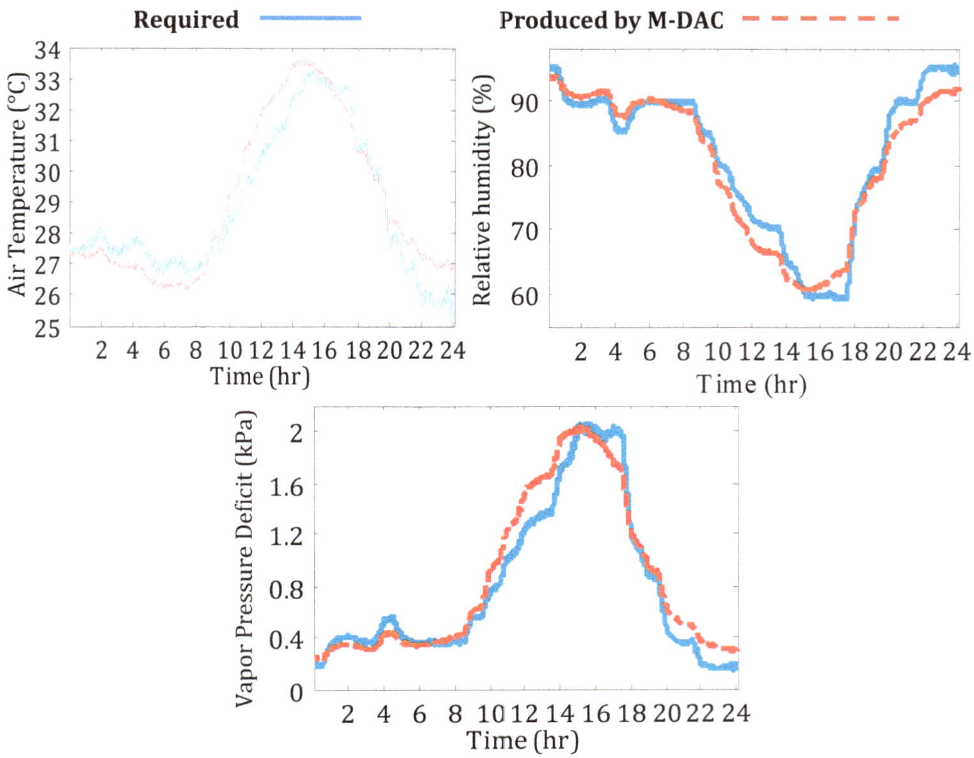

Figure 14. Validation of the performance of the proposed M-DAC with the OptDeg model for 24 h in greenhouse cultivation of tomato under vegetative to mature fruiting growth stages.

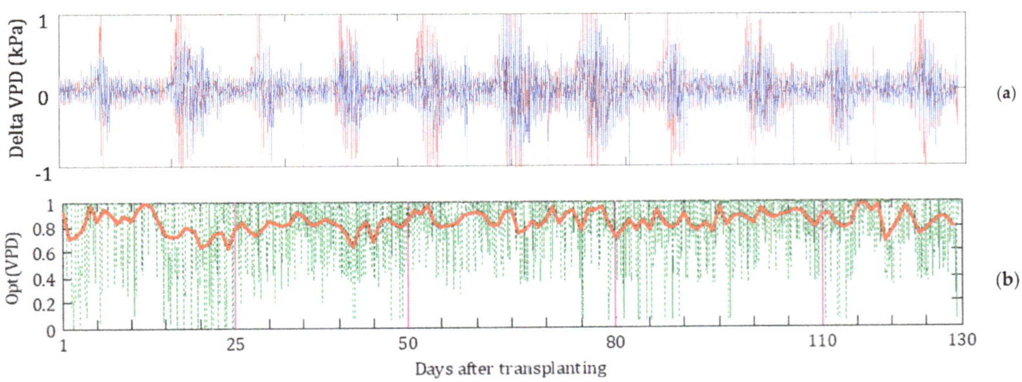

Figure 15. Variation in the VPD data (**a**), and optimality degree of VPD (**b**) resulting from the air temperature and relative humidity data of the proposed M-DAC system simulated for greenhouse cultivation of tomato in 130 days.

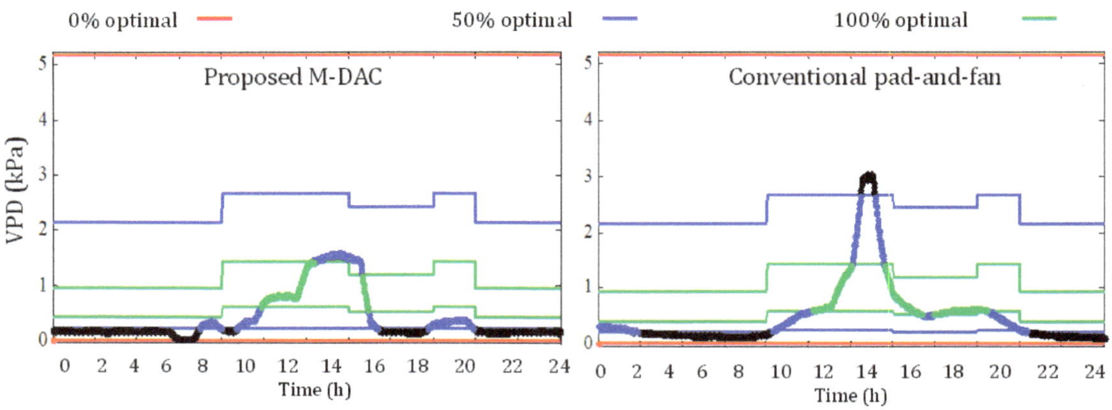

Figure 16. Comparison between comfort ratios of VPD for 24 h resulting from the proposed M-DAC system and VPD data from a conventional evaporative (pad-and-fan) cooled greenhouse.

Table 3. Performance validation of the proposed M-DAC system with the comfort ratio model at α_s = 1 in comparison with experimental data from a tomato greenhouse with evaporative pad-and-fan system [57] for 11 days under different growth stages.

Days	Vegetative Growth Stage		Flowering to Mature Fruiting	
	Pad-and-fan Ref. [57]	M-DAC (This Study)	Pad-and-Fan Ref. [57]	M-DAC (this Study)
1	8.6	21.2	11.1	24.7
2	4.9	17.2	19.6	26.7
3	27	11.9	31.9	18.2
4	13.5	20.4	32.1	24.9
5	21.9	16.4	40.9	31.0
6	25.2	36.0	27.5	43.4
7	4.6	17.5	25.8	39.0
8	32.3	51.4	36.3	65.2
9	5	6.4	8.8	25.5
10	4.3	46.8	8.3	51.0
11	18.8	20.5	32.2	29.4
Mean	15.1	24.1	24.9	34.5
SD	10.4	14.3	11.4	14.0
Min	4.3	6.4	8.3	18.2
Max	32.3	51.4	40.9	65.2

5. Conclusions

The present study aimed to investigate the applicability of desiccant dehumidification-based Maisotsenko cycle evaporative cooling (M-DAC) system for greenhouse air-conditioning application for the climatic conditions of Multan (Pakistan). The study area (i.e., Multan) lies in a warm desert climate of the Köppen–Geiger climatic classification. In this regard, air-conditioning is required in greenhouses to optimize the temperature and humidity conditions, which could potentially increase the production level. Therefore, the present study proposed a desiccant dehumidification-based Maisotsenko cycle evaporative cooling system (M-DAC) for greenhouse air-conditioning application. The standalone DAC system created a (maximum) dehumidification gradient (i.e., 16.8 g/kg) at 24.3 g/kg ambient air conditions, and a (maximum) temperature gradient (i.e., 8.4 °C) at 38.6 °C ambient air conditions. The DAC+HX system created a temperature gradient roughly equal to the ambient air conditions, which are unsuitable for greenhouse air-conditioning. The M-DAC system created a (maximum) temperature gradient (i.e., 21.9 °C) at 39.2 °C ambient air conditions, which lies well within the required optimum temperature conditions of the greenhouse air-conditioning. Additionally, the M-DAC created a dehumidification gradient (i.e., 16.8 g/kg) at 24 g/kg ambient conditions, which also lies inside the range of optimum humidity conditions inside the greenhouse. Moreover, the vapor pressure deficit (VPD) profile of the greenhouse environment related to the growth of plants indicated that a VPD of 0.8–1.2 kPa (for leaf temperature equal to surrounding air temperature) is most suitable for healthy transpiration from the plant leaves at late vegetation and early flowering stages. These results were validated with OptDeg and Cft-ratio microclimate models, which showed that the M-DAC system maintained the VPD of the product air inside the healthy transpiration zone throughout most of the summer months (i.e., May to August). Moreover, the comfort ratio model results of the proposed M-DAC system were compared with experimental results from a tomato greenhouse, indicating that the M-DAC system could potentially achieve the required comfort level compared to a pad-and-fan type air-conditioning system. Thus, the study indicates the M-DAC system is a feasible option for greenhouse air-conditioning for the climatic conditions of Multan (Pakistan).

Author Contributions: Conceptualization, H.A., M.S. and F.A. (Farrukh Abbas); Data curation, H.A. and F.A. (Farrukh Abbas); Formal analysis, H.A., M.S., R.R.S. and F.A. (Farrukh Abbas); Funding acquisition, M.S. and R.R.S.; Investigation, M.S., R.R.S., U.S. and F.A. (Fiaz Ahmad); Methodology, H.A., R.R.S., M.F. and H.M.-T.; Project administration, M.S.; Resources, M.S.; Software, M.H.M.; Supervision, M.S.; Validation, M.S. and R.R.S.; Visual-ization, U.S., H.M.-T., Y.R.T., A.S. and B.M.K.N.; Writing—original draft, H.A., M.S., R.R.S. and F.A. (Farrukh Abbas); Writing—review & editing, M.F., M.H.M., F.A. (Fiaz Ahmad), Y.R.T., A.S. and B.M.K.N. All authors have read and agreed to the published version of the manuscript.

Funding: This research received no external funding.

Data Availability Statement: Data is contained within the article.

Acknowledgments: This research work has been carried out in the Department of Agricultural Engineering, Bahauddin Zakariya University, Multan-Pakistan. Some of this work is part of the M.Sc. (Hons.) research of Farrukh Abbas. This research was funded by Bahauddin Zakariya University, Multan-Pakistan under the Director Research/ORIC grant entitled "Development and performance evaluation of prototypes of direct and indirect evaporative cooling-based air conditioning systems" awarded to Principal Investigator Muhammad Sultan. The authors acknowledge the financial support by the Open Access Publication Fund of the Leibniz Association, Germany, and the partial research funding and editorial supports from Adaptive AgroTech Consultancy International.

Conflicts of Interest: The authors declare no conflict of interest.

References

1. Pandelidis, D.; Anisimov, S. Numerical analysis of the heat and mass transfer processes in selected M-Cycle heat exchangers for the dew point evaporative cooling. *Energy Convers. Manag.* **2015**, *90*, 62–83. [CrossRef]
2. Caliskan, H.; Dincer, I.; Hepbasli, A. A comparative study on energetic, exergetic and environmental performance assessments of novel M-Cycle based air coolers for buildings. *Energy Convers. Manag.* **2012**, *56*, 69–79. [CrossRef]
3. Zhan, C.; Duan, Z.; Zhao, X.; Smith, S.; Jin, H.; Riffat, S. Comparative study of the performance of the M-cycle counter-flow and cross-flow heat exchangers for indirect evaporative cooling—Paving the path toward sustainable cooling of buildings. *Energy* **2011**, *36*, 6790–6805. [CrossRef]
4. Caliskan, H.; Dincer, I.; Hepbasli, A. Exergetic and sustainability performance comparison of novel and conventional air cooling systems for building applications. *Energy Build.* **2011**, *43*, 1461–1472. [CrossRef]
5. Chua, K.J.; Chou, S.K.; Yang, W.M.; Yan, J. Achieving better energy-efficient air conditioning—A review of technologies and strategies. *Appl. Energy* **2013**, *104*, 87–104. [CrossRef]
6. Pandelidis, D.; Anisimov, S.; Worek, W.M. Performance study of the Maisotsenko Cycle heat exchangers in different air-conditioning applications. *Int. J. Heat Mass Transf.* **2015**, *81*, 207–221. [CrossRef]
7. Rogdakis, E.D.; Koronaki, I.P.; Tertipis, D.N. Experimental and computational evaluation of a Maisotsenko evaporative cooler at Greek climate. *Energy Build.* **2014**, *70*, 497–506. [CrossRef]
8. Cui, X.; Chua, K.J.; Yang, W.M. Use of indirect evaporative cooling as pre-cooling unit in humid tropical climate: An energy saving technique. *Energy Procedia* **2014**, *61*, 176–179. [CrossRef]

9. Riangvilaikul, B.; Kumar, S. Numerical study of a novel dew point evaporative cooling system. *Energy Build.* **2010**, *42*, 2241–2250. [CrossRef]
10. Sultan, M. Study on Sorption Characteristics of Water Adsorbents for Agricultural Air-Conditioning Systems. Ph.D. Thesis, Kyushu University, Kyushy, Japan, 2015.
11. Anisimov, S.; Pandelidis, D.; Jedlikowski, A.; Polushkin, V. Performance investigation of a M (Maisotsenko)-cycle cross-flow heat exchanger used for indirect evaporative cooling. *Energy* **2014**, *76*, 593–606. [CrossRef]
12. Zhao, X.; Li, J.M.; Riffat, S.B. Numerical study of a novel counter-flow heat and mass exchanger for dew point evaporative cooling. *Appl. Therm. Eng.* **2008**, *28*, 1942–1951. [CrossRef]
13. Zube, D.; Gillan, L. Evaluating Coolerado Corportion's Heatâ mass exchanger performance through experimental analysis. *Int. J. Energy Clean Environ.* **2011**, *12*, 101–116. [CrossRef]
14. Weerts, B. Coolerado and modeling an application of the Maisotsenko Cycle. *Int. J. Energy Clean Environ.* **2011**, *12*, 287–307. [CrossRef]
15. Maisotsenko, V.; Treyger, I. Way to energy abundance can be found through the Maisotsenko cycle. *Int. J. Energy Clean Environ.* **2011**, *12*, 319–326. [CrossRef]
16. Khalatov, A.; Karp, I.; Isakov, B. Prospects of the Maisotsenko thermodynamic cycle application in Ukraine. *Int. J. Energy Clean Environ.* **2011**, *12*, 141–157. [CrossRef]
17. Anisimov, S.; Pandelidis, D. Heat-and mass-transfer procesess in indirect evaporative air conditioners through the Maisotsenko cycle. *Int. J. Energy Clean Environ.* **2011**, *12*, 273–286. [CrossRef]
18. Rogdakis, E.D.; Tertipis, D.N. Maisotsenko cycle: Technology overview and energy-saving potential in cooling systems. *Energy Emiss. Control Technol.* **2015**, *3*, 15.
19. Nagaya, K.; Senbongi, T.; Li, Y.; Zheng, J.; Murakami, I. High energy efficiency desiccant assisted automobile air-conditioner and its temperature and humidity control system. *Appl. Therm. Eng.* **2006**, *26*, 1545–1551. [CrossRef]
20. Lee, S.H.; Lee, W.L. Site verification and modeling of desiccant-based system as an alternative to conventional air-conditioning systems for wet markets. *Energy* **2013**, *55*, 1076–1083. [CrossRef]
21. Ismail, M.Z.; Angus, D.E.; Thorpe, G.R. The performance of a solar-regenerated open-cycle desiccant bed grain cooling system. *Sol. Energy* **1991**, *46*, 63–70. [CrossRef]
22. Sultan, M.; Miyazaki, T.; Koyama, S.; Saha, B.B. Utilization of Desiccant Air-Conditioning System for Improvement in Greenhouse Productivity: A Neglected Area of Research in Pakistan. *Int. J. Environ.* **2014**, *4*, 1–10.
23. Zheng, G.; Zheng, C.; Yang, G.; Chen, W. Development of a new marine rotary desiccant airconditioning system and its energy consumption analysis. *Energy Procedia* **2011**, *16*, 1095–1101. [CrossRef]
24. Bourdoukan, P.; Wurtz, E.; Joubert, P. Comparison between the conventional and recirculation modes in desiccant cooling cycles and deriving critical efficiencies of components. *Energy* **2010**, *35*, 1057–1067. [CrossRef]
25. Enteria, N.; Yoshino, H.; Takaki, R.; Yonekura, H.; Satake, A.; Mochida, A. First and second law analyses of the developed solar-desiccant air-conditioning system (SDACS) operation during the summer day. *Energy Build.* **2013**, *60*, 239–251. [CrossRef]
26. Dezfouli, M.M.S.; Mat, S.; Pirasteh, G.; Sahari, K.S.M.; Sopian, K.; Ruslan, M.H. Simulation Analysis of the Four Configurations of Solar Desiccant Cooling System Using Evaporative Cooling in Tropical Weather in Malaysia. *Int. J. Photoenergy* **2014**, *2014*, 843617. [CrossRef]
27. Enteria, N.; Yoshino, H.; Satake, A.; Mochida, A.; Takaki, R.; Yoshie, R.; Mitamura, T.; Baba, S. Experimental heat and mass transfer of the separated and coupled rotating desiccant wheel and heat wheel. *Exp. Therm. Fluid Sci.* **2010**, *34*, 603–615. [CrossRef]
28. Ge, T.S.; Ziegler, F.; Wang, R.Z.; Wang, H. Performance comparison between a solar driven rotary desiccant cooling system and conventional vapor compression system (performance study of desiccant cooling). *Appl. Therm. Eng.* **2010**, *30*, 724–731. [CrossRef]
29. Sultan, M.; El-Sharkawy, I.I.; Miyazaki, T.; Saha, B.B.; Koyama, S.; Maruyama, T.; Maeda, S.; Nakamura, T. Water vapor sorption kinetics of polymer based sorbents: Theory and experiments. *Appl. Therm. Eng.* **2016**, *106*, 192–202. [CrossRef]
30. Sultan, M.; Miyazaki, T.; Koyama, S.; Khan, Z.M. Performance evaluation of hydrophilic organic polymer sorbents for desiccant air-conditioning applications. *Adsorpt. Sci. Technol.* **2018**, *36*, 311–326. [CrossRef]
31. Sultan, M.; El-Sharkawy, I.I.; Miyazaki, T.; Saha, B.B.; Koyama, S.; Maruyama, T.; Maeda, S.; Nakamura, T. Insights of water vapor sorption onto polymer based sorbents. *Adsorption* **2015**, *21*, 205–215. [CrossRef]
32. Sultan, M.; Miyazaki, T.; Mahmood, M.H.; Khan, Z.M. Solar assisted evaporative cooling based passive air-conditioning system for agricultural and livestock applications. *J. Eng. Sci. Technol.* **2018**, *13*, 693–703.
33. Noor, S.; Ashraf, H.; Sultan, M.; Khan, Z.M. Evaporative Cooling Options for Building Air-Conditioning: A Comprehensive Study for Climatic Conditions of Multan (Pakistan). *Energies* **2020**, *13*, 3061. [CrossRef]
34. Sultan, M.; Miyazaki, T. Energy-Efficient Air-Conditioning Systems for Nonhuman Applications. In *Refrigeration*; Ekren, O., Ed.; InTech: London, UK, 2017; pp. 97–117.
35. Sultan, M.; El-Sharkawy, I.I.; Miyazaki, T.; Saha, B.B.; Koyama, S. An overview of solid desiccant dehumidification and air conditioning systems. *Renew. Sustain. Energy Rev.* **2015**, *46*, 16–29. [CrossRef]
36. Mahmood, M.H.; Sultan, M.; Miyazaki, T.; Koyama, S.; Maisotsenko, V.S. Overview of the Maisotsenko cycle—A way towards dew point evaporative cooling. *Renew. Sustain. Energy Rev.* **2016**, *66*, 537–555. [CrossRef]

37. Raza, H.M.U.; Ashraf, H.; Shahzad, K.; Sultan, M.; Miyazaki, T.; Usman, M.; Shamshiri, R.R.; Zhou, Y.; Ahmad, R. Investigating Applicability of Evaporative Cooling Systems for Thermal Comfort of Poultry Birds in Pakistan. *Appl. Sci.* **2020**, *10*, 4445. [CrossRef]
38. Kanwal, R.; Ashraf, H.; Sultan, M.; Babu, I.; Yasmin, Z.; Nadeem, M.; Asghar, M.; Shamshiri, R.R.; Ibrahim, S.M.; Ahmad, N.; et al. Effect of 1-Methyl Cyclopropane and Modified Atmosphere Packaging on the Storage of Okra (Abelmoschus esculentus L.): Theory and Experiments. *Sustainability* **2020**, *12*, 7547. [CrossRef]
39. Shamshiri, R.R.; Kalantari, F.; Ting, K.C.; Thorp, K.R.; Hameed, I.A.; Weltzien, C.; Ahmad, D.; Shad, Z.M. Advances in greenhouse automation and controlled environment agriculture: A transition to plant factories and urban agriculture. *Int. J. Agric. Biol. Eng.* **2018**, *11*, 1–22. [CrossRef]
40. Shamshiri, R.R.; Jones, J.W.; Thorp, K.R.; Ahmad, D.; Man, H.C.: Taheri, S. Review of optimum temperature, humidity, and vapour pressure deficit for microclimate evaluation and control in greenhouse cultivation of tomato: A review. *Int. Agrophysics* **2018**, *32*, 287–302. [CrossRef]
41. Shamshiri, R.; Ismail, W.I.W. A review of greenhouse climate control and automation systems in tropical regions. *J. Agric. Sci. Appl.* **2013**, *2*, 176–183. [CrossRef]
42. Shamshiri, R.; Ismail, W.I.W.; bin Ahmad, D. Experimental evaluation of air temperature, relative humidity and vapor pressure deficit in tropical lowland plant production environments. *Adv. Environ. Biol.* **2014**, *8*, 5–14.
43. Sultan, M.; Miyazaki, T.; Koyama, S. Optimization of adsorption isotherm types for desiccant air-conditioning applications. *Renew. Energy* **2018**, *121*, 441–450. [CrossRef]
44. Sultan, M.; Miyazaki, T.; Saha, B.B.; Koyama, S. Steady-state investigation of water vapor adsorption for thermally driven adsorption based greenhouse air-conditioning system. *Renew. Energy* **2016**, *86*, 785–795. [CrossRef]
45. Song, J.; Sobhani, B. Energy and exergy performance of an integrated desiccant cooling system with photovoltaic/thermal using phase change material and maisotsenko cooler. *J. Energy Storage* **2020**, *32*, 101698. [CrossRef]
46. Miyazaki, T.; Nikai, I.; Akisawa, A. Simulation analysis of an open-cycle adsorption air conditioning system—numeral modeling of a fixed bed dehumidification unit and the maisotsenko cycle cooling unit. *Int. J. Energy Clean Environ.* **2011**, *12*, 341–354. [CrossRef]
47. Pandelidis, D.; Anisimov, S.; Worek, W.M.; Drąg, P. Analysis of different applications of Maisotsenko cycle heat exchanger in the desiccant air conditioning systems. *Energy Build.* **2017**, *140*, 154–170. [CrossRef]
48. Goldsworthy, M.; White, S. Optimisation of a desiccant cooling system design with indirect evaporative cooler. *Int. J. Refrig.* **2011**, *34*, 148–158. [CrossRef]
49. Sultan, M.; El-Sharkaw, I.I.; Miyazaki, T.; Saha, B.B.; Koyama, S. Experimental Study on Carbon Based Adsorbents for Greenhouse Dehumidification. *Evergreen* **2014**, *1*, 5–11. [CrossRef]
50. Lim, X. How heat from the sun can keep us all cool. *Nature* **2017**, *542*, 23–24. [CrossRef]
51. Beck, H.E.; Zimmermann, N.E.; McVicar, T.R.; Vergopolan, N.; Berg, A.; Wood, E.F. Present and future Köppen-Geiger climate classification maps at 1-km resolution. *Sci. Data* **2018**, *5*, 180214. [CrossRef]
52. Panaras, G.; Mathioulakis, E.; Belessiotis, V.; Kyriakis, N. Theoretical and experimental investigation of the performance of a desiccant air-conditioning system. *Renew. Energy* **2010**, *35*, 1368–1375. [CrossRef]
53. Panaras, G.; Mathioulakis, E.; Belessiotis, V. Solid desiccant air-conditioning systems–Design parameters. *Energy* **2011**, *36*, 2399–2406. [CrossRef]
54. ASHRAE. *ASHRAE Handbook—Fundamentals*; American Society of Heating, Refrigerating, and Air-Conditioning Engineers: Atlanta, GA, USA, 2017.
55. Shamshiri, R. Measuring optimality degrees of microclimate parameters in protected cultivation of tomato under tropical climate condition. *Measurement* **2017**, *106*, 236–244. [CrossRef]
56. Shamshiri, R.R.; Bojic, I.; van Henten, E.; Balasundram, S.K.; Dworak, V.; Sultan, M.; Weltzien, C. Model-based evaluation of greenhouse microclimate using IoT-Sensor data fusion for energy efficient crop production. *J. Clean. Prod.* **2020**, *263*, 121303. [CrossRef]
57. Rezvani, S.M.; Abyaneh, H.Z.; Shamshiri, R.R.; Balasundram, S.K.; Dworak, V.; Goodarzi, M.; Sultan, M.; Mahns, B. IoT-Based Sensor Data Fusion for Determining Optimality Degrees of Microclimate Parameters in Commercial Greenhouse Production of Tomato. *Sensors* **2020**, *20*, 6474. [CrossRef] [PubMed]

energies

Review

A Review on Geothermal Renewable Energy Systems for Eco-Friendly Air-Conditioning

Adriana Greco [1], Edison Gundabattini [2], Darius Gnanaraj Solomon [3], Raja Singh Rassiah [4] and Claudia Masselli [1,*]

[1] Department of Industrial Engineering, University of Naples Federico II, Piazzale Tecchio 80, 80125 Naples, Italy; adriana.greco@unina.it
[2] Department of Thermal and Energy Engineering, School of Mechanical Engineering, Vellore Institute of Technology (VIT), Vellore 632 014, Tamil Nadu, India; edison.g@vit.ac.in
[3] Department of Design and Automation, School of Mechanical Engineering, Vellore Institute of Technology (VIT), Vellore 632 014, Tamil Nadu, India; dariusgnanaraj.s@vit.ac.in
[4] Advanced Drives Laboratory, Department of Energy and Power Electronics, Vellore Institute of Technology (VIT), Vellore 632 014, Tamil Nadu, India; rrajasingh@vit.ac.in
* Correspondence: claudia.masselli@unina.it

Abstract: Nowadays, air conditioning consumes, on average, around one-fifth of the total power used in buildings globally. The present paper aims to provide the present status on the employment of Earth-to-Air Heat eXchangers (EAHX) to contain the consumption of energy and to reduce the effect on the environment in response to the Montreal and Kyoto protocols in a way to achieve cleaner energy production with a low Global Warming Potential (GWP) and a low ozone depletion potential (ODP). Different peculiarities and applications (direct or hybrid) are critically analyzed and reviewed. Specifically, in this paper, the different hybrid applications presented in the literature, where the Earth-to-Air Heat eXchangers are coupled to advanced systems, are reviewed. Finally, an IoT-based EAHX control system plan is reported and discussed to optimize energy efficiency and thermal comfort to suit operating conditions under different time zones.

Keywords: renewable energy; geothermal; Earth-to-Air Heat eXchangers; ground source heat pumps

Citation: Greco, A.; Gundabattini, E.; Solomon, D.G.; Singh Rassiah, R.; Masselli, C. A Review on Geothermal Renewable Energy Systems for Eco-Friendly Air-Conditioning. *Energies* **2022**, *15*, 5519. https://doi.org/10.3390/en15155519

Academic Editors: Jan Danielewicz and Krzysztof Rajski

Received: 31 May 2022
Accepted: 28 July 2022
Published: 29 July 2022

1. Introduction

Air conditioning is a crucial point to be considered in the building sector (both residential and commercial), both from energy savings and comfort-ensuring points of view. The conventional and most widely spread are HVAC systems (Heating, Ventilation, and Air Conditioning) based on vapor compression [1,2]. Even if they easily ensure the desired thermal comfort in the building environments, they negatively contribute to energy consumptions since HVACs are energy-intensive systems [3,4]. Furthermore, even if the Montreal Protocol [5] and subsequent amendments [6] have been prescribed to progressively phase out ozone depleting refrigerants, the currently employed fluids (HFC) are due to high global warming potential [7–9]. New alternative fluids for vapor compression systems are found on natural fluids or HydroFluoroOlefins (HFO) [10–13], which have problems with flammability and costs.

Approximately 20% of the electricity used in buildings globally today is used to cool spaces using air conditioning units and ceiling fans. In addition to increasing emissions, the growing demand for cooling systems has created a tremendous burden on the energy infrastructure of many nations. There is little question as to whether the requirement for space conditioning on a worldwide scale and the energy necessary to supply it will increase for several upcoming years in the absence of strong legislative initiatives. However, there is a high likelihood of swiftly altering the surge in conditioning energy consumption by implementing measures to increase system performance [14–16]. In developed countries, the building sector impacts 30% of global energy consumption [17], and internally to the building sector, the energy consumption deriving from the HVAC operation is 50% of the overall power consumption of the whole expense.

Moreover, from a more global perspective, considering all fields and possible applications, refrigeration, and air conditioning are accountable, based on projections from the International Institute of Refrigeration, for over 20% of the world's power usage, on average, in 2020. This amount changes significantly among the worldwide countries, based both on the level of change and local climatic circumstances: 14% of the overall electricity demand of the USA is to be attributed to air conditioning, whereas in Mumbai (India), it accounts for 40%. The High-Level Board on Climate Change guesses that, considering the expected climate changes, the residential sector expects to see its energy demand for HVAC systems operating more than 13 times from 2000 to 2050 and more than 30 times up to 2100 [18]. If, on the one hand, climate change causes an increase in energy demand for air conditioning, on the other hand, since most systems are based on vapor compression, this increase contributes to the environmental (global warming) and energy crisis in a vicious circle that is dangerous for our future.

Hence, there is an international concern about containing energy consumption, enhancing the productivity performance of buildings, and minimizing the effect on the environment and human health.

The breaking points of the aforementioned vicious circle are: (i) the development of high-impact strategies for energy savings; (ii) an increase in the energy efficiency of air conditioning and refrigeration systems; and (iii) wide employment of renewable energy sources. In fact, improvements in the performance of HVAC systems have become a top priority for green policies. This is demonstrated, for instance, by the manufacturing industry's increasingly stringent efficiency standards for these systems, the creation of building classification and energy conformity assessments, and the consistency of proactive maintenance procedures for those kinds of systems. In particular, as stated explicitly in the European Directive on Renewable Sources and created by the so-called "green building" criteria, the regulations addressing the use of energy in the building industry aim to promote the use of renewable sources.

The growing interest in discovering substitutes for vapor compression-based HVAC systems has pushed researchers to make efforts to develop solutions based on renewable energies. Renewable energy is energy produced from sources that are available in nature and that may be utilized right away or stored through accumulation technologies. The most prevalent renewable energy sources include geothermal heat-based energy, photovoltaic, wind, biomass, precipitation, tidal energy, and waves [19–22]. Perhaps one of them, solar heating/cooling systems, uses solar energy substantially to heat and cool water for home and industrial purposes [23]. Solar power is used to generate electricity using photovoltaic (pv) systems [24].

As a matter of fact, in the panorama of renewable energy sources, next to solar, geothermal, and wind are also sources due to the big potential for electricity generation. Geothermal energy definitely dominated the renewable energy market in terms of installed electricity power about 30 years ago.

In recent times, the potential of employing geothermal sources for electricity generation has been surpassed and dominated by solar and wind. The data reveal [25] that the surpass on geothermal energy by wind and solar has occurred both in terms of growth rate and installed capacity. This is due to the high cost in charge of geothermal systems as an initial investment that couples with a long payback time. Moreover, the construction time is long, and it is difficult to find available spaces. In some cases, another factor is social acceptance in geothermal power generation that could occur differently from country to country. To deepen these aspects, we suggest referring to the comprehensive and critical analysis provided by Li et al. [25] where the differences among worldwide countries in the use of geothermal have also been treated.

However, in addition to these considerations, geothermal is a largely available renewable energy source that is very suitable for air conditioning applications devoted to buildings [26]. The latter case is referenced to a direct use of geothermal energy: one of the most virtuous worldwide countries is Iceland, where until 2011, 66% of primary energy usage exploited geothermal; currently in Iceland, about 96% of the energy consumption coming from heating and cooling derives from this renewable energy source. Heat pumps are the systems that are best suited for the direct use of geothermal energy; however, on a global balance, the rest of Europe is not so virtuous since the International Energy Agency stated that only 3% of the energy consumption coming from building air conditioning is satisfied by geothermal systems.

A push toward developing efficient geothermal energy-based HVAC systems is more necessary than ever. Two main applications currently exist devoted to this purpose:

- Ground Source Heat Pumps (GSHP);
- Earth-to-Air Heat eXchangers (EAHX).

The present paper aims to provide a state-of-the-art study on the employment of these two applications for air conditioning of buildings. In the next sections, the main aspects related to them are discussed, analyzed, and reviewed.

2. Geothermal-Based Applications

2.1. Generalities

The exploitation of geothermal systems can enhance the energy efficiency of HVAC systems without disturbing indoor comfort settings. Geothermal systems have the potential to realize noteworthy outcomes of eco-sustainability and carbon footprint reduction [27]. A hybrid solar-geothermal conditioning unit is investigated to reduce the Global Warming Potential (GWP) according to the Paris Climate Change Agreement and Kigali Amendment, aiming to contain energy consumption and refrigerant use. The objective was to reduce the cooling energy requirements by 25–40% and reduce the refrigerant requirement by 25–30% [28]. Yang et al. [29] analyzed the cooling and heating energy consumption of a model of a plant factory in China using the simulation tool TRNSYS. The annual energy consumption of the plant factory (both for cooling and heating) was estimated, and the analysis revealed that geothermal-based systems, such as the GSHP designed and modeled in their investigation, are sufficient to supply the energy consumption of the farm and even to guarantee the optimal comfort conditions for cultivating vegetables inside them.

Fernández [30] potentially analyzed the reduction in terms of energy demand, environmental impact, and economic expense if geothermal energy were employed in supermarkets. With respect to the different weather conditions, three case studies were considered: a supermarket placed in Germany, Portugal, and Turkey. They concluded that the employment of renewable energy sources, such as geothermal, led to a reduction in consumptions of 45% in Germany, 30% in Portugal, and 36% in Turkey, respectively.

Cadelano et al. [27] deliberated the energy retrofit of a museum arranged in a historical building through the software TRNSYS. Specifically, the energy consumption and environmental impact were evaluated by investigating the energy performances before and after the introduction of GSHP as an HVAC system. They noticed that the energy consumption in the second case was reduced by 24%.

2.2. Earth to Air-Heat eXchangers (EAHX): Requirements, Design, and Recent Breakthrough Investigations

EAHX is an innovative and efficient passive technology that typically uses low-enthalpy geothermal systems, used in many countries to encourage and strengthen the usage of renewable energy sources and, at the same time, to contribute to the achievement of thermal comfort in buildings. The working fluid in EAHX is air. The EAHX system consists of one or more pipes buried at a depth of at least 1.5–2 m below the surface, i.e., where the soil temperature does not change during the year, known as the Earth Undisturbed Temperature (EUT) [31].

A well-designed EAHX system is utilized independently or in conjunction with a conventional HVAC system to suit the building's heating and cooling needs. The cheapest EAHX could be a pipe of the right size that is buried at the right depth and by which ambient air is blown. The pipe has an outlet that lets air into the inhabited space, and one end that serves as an air input [32]. In summer, hot air is cooled by the soil, which is at a lower temperature; subsequently, it is supplied to the environment to be air-conditioned. Dually, in winter, cold air is heated in the basement and supplied inside to guarantee satisfaction with the thermal comfort standard.

A plant with an EAHX system consists of three main components:

- Suction unit: The air essential for the process of the scheme comes from the suction of outdoor air. The suction unit must be sized according to the connected EAHX tube and the pressure loss that is achieved. Regarding the location of outdoor air intake, it is necessary to comply with the requirements of VDI 6022 [33]. The latter requires that the air intake be of excellent quality. When choosing the location of the air suction, the following points must be considered: the vicinity to the road (road traffic) and other buildings; the proximity to leaf-losing trees/shrubs; proximity to vent openings of any kind; and the main wind direction and location of any systems that can generate annoying odors. The suction towers for EAHX must be made of waterproof material and not pose any risk to health, while the height of the suction entrance must be at a sufficient distance from the earth's surface and from any pollutant emitters. The standard also requires that the suction tower be made of stainless steel.
- Filters: The use of filters in the suction unit can serve several functions. You can use a coarse mesh filter to protect EAHX from pollutant external agents. This is also possible with a medium or fine mesh filter. The use of the coarse mesh filter is suitable for normal operation and is sufficient to ensure compliance with standards and directives. The use of a fine mesh filter is preferable if special health protection measures are to be taken, e.g., in the case of allergies. It should be taken in mind that to increase the useful life of a fine mesh filter, a coarse mesh one is always inserted into it. For the complete system comprising an EAHX and a traditional HVAC system, the directives [33] provide for two levels of filtering, one in the intake unit and the other in the traditional ventilation system. Finally, it should be noted that with the use of a fine mesh filter, the pressure drop for the suction unit increases.
- Pipes: The pipes installed in the EAHX constitute the heart of the system, and it is through them that the transmission of heat amid air and soil takes place. For the material of these pipes, directive VDI 6022 provides for specific requirements: it must be closed cell, watertight, and resistant to corrosion; not harmful to health; the material from which it is made should not store moisture and is due to a system to drain the condensate that forms in the summer.

Hence, the key advantages and disadvantages in the employment of EAHX schemes are listed below.

Advantages:

- The employed fluid used is air (unlimited availability and free);
- Power consumption is lower when compared with the prevailing traditional systems;
- Higher coefficient of performance (COP) when coupled with traditional systems;
- The scheme is simple, so it entails less upkeep and functional costs;
- The eco-friendly EAHX exploits geothermal energy, a renewable energy source, and it does not require the use of refrigerants and compressors.

Disadvantages:

- The installation cost is high;
- The requirement of large available free areas to bury the pipes;
- Condensation may occur in the ducts, which must be removed with the help of small submersible pumps;

- The convective mechanism that is triggered in the tube does not allow it to reach uniform temperatures at the outlet of the exchanger;
- Because air is employed as a refrigerant, the presence of microorganisms could become the main cause of the need to couple the EAHX with a filtered ventilation system, with greater energy consumption and a decrease in air quality.

EAHX's could be used in a huge number of schemes, which vary from one another according to the scheme, the geometry of the exchanger, the quantity of pipes mounting the exchanger and how they are arranged, the material the pipes are made of, etc. The optimal and suitable scheme must consequently consider several vital parameters: technical considerations, such as the availability of material or free land surface; energy considerations, such as the size of the cooling/heating load of the building to be air-conditioned; and economic considerations.

EAHXs can be categorized using a wide range of factors. A suggested categorization of EAHXs is shown schematically in Figure 1 [33]. EAHX systems can be classified into open-loop and closed-loop EAHXs based on their architecture and design. While closed-loop EAHXs circulate air from a building until the appropriate temperature is reached, open-loop EAHXs pull air straight from the surroundings. EAHXs might replace traditional conditioners as a practical option while also providing a straightforward design, easing ecological consequences, and lowering expenses. When contrasted to a closed-loop system, which cycles the same air through the pipes, an open-loop system delivers clean, fresh air that rotates through pipes and fulfils building cooling and heating requirements [34].

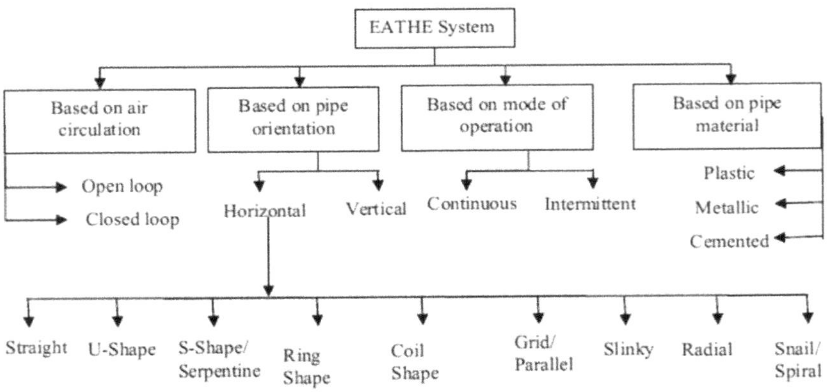

Figure 1. Classification of EAHX systems.

Because of their substantial influence on the lowering of heating/cooling energy loads and the enhancement of interior thermal comfort settings, EAHXs have proven to be appealing technological research. The cooling effectiveness of EAHXs has been researched under diverse geological and environmental circumstances, including relative humidity, air temperature, soil temperature, and air velocity. The EAHX was investigated to see if it might be used to improve energy performance within buildings while also lowering the electrical energy consumption of typical air-conditioning systems. Various upcoming and novel technologies will be useful in augmenting the performance of EAHX with suitable hybrid modes. Hybrid structures comprising hybrid nanofluids and microspray models and by incorporating wavy strips in the structures. These would enhance heat transfer efficiency and provide better thermal comfort as a final outcome [35–39].

2.3. Geometrical Parameters

Rodrigues et al. [40] carried out wide experimental procedures to evaluate the thermal and fluid-dynamic properties of a Y-shaped EAHX. The Energy Performance Indicator (EPI) was used to assess the EAHX's performance. The optimum configuration of the

Y-shaped EAHX boosted the value of EPI while also delivering monthly power savings, according to research [40]. The viability of a novel helical-shaped EAHX over conventional U–shaped EAHX is investigated for evaluating the thermal performances in a restricted ground space. outlet air temperature and effectiveness. The results indicated that the novel design offered an additional 80% of thermally comfortable indoor air conditions, even during peak summer. Similarly, the results also indicated that the novel design is efficient in heating operations to keep interior comfort in the winter spell [41].

Recent studies were also conducted on the impact of functional parameters, together with pipe length and pipe section (circular, rectangular, and elliptical), on the energy performances of EAHX, through sensitivity analysis. The outcomes indicated that the rectangular and elliptical buried pipes with larger surface area are more effective with >88% and >93% in contrast with the conventional (circular) one. The results also indicated that the efficacy of the length of the pipe is better over the pipe cross section [42].

2.4. Physics-Based Parameters

A pure buoyancy-driven EAHX system was recently explored for its practicality and cooling effectiveness by Long et al. [43] in 2022. This system combines solar chimney and thermal mass effects. The testing results revealed that the EAHX system could provide 24 h of varying natural ventilation and passive cooling, as well as significantly improve the indoor thermal environment under hot summer circumstances [43].

Recent studies have indicated that backfilling could augment the thermal potential of an EAHX beyond 400 m in depth to assist emergency ventilation. The backfilling system with low thermal conductivity proved to augment energy efficiency, thermal comfort, and thermal performance up to 100%, in contrast to the original system [44].

2.5. Integrated-Hybrid Systems

In 2022, an integrated passive system comprising a solar chimney with an EAHX (SCEAHX) was investigated by Long et al. [45] for its clean air and cooling capacity.

The SCEAHX system indicated 24 h of natural ventilation in conserving satisfactory indoor thermal comfort and temperature. A SCEAHX system based on Phase-Change Materials (PCM) was also studied to improve the interior thermal environment. The combined PCM-SCEAHX system, according to numerical analysis, not only increased usable cooling capacity but also produced more consistent interior thermal comfort conditions. As a result, the PCM-based SCEAHX system enhanced thermal inertia, regulated airflow rate during the day, and prolonged solar chimney's usage at night [45–47]. A recently developed water-aided EAHX was investigated for its energy efficiency with fins on the internal pipe of the double pipe heat exchanger. The results indicated that the overall heat transfer coefficients were enhanced, changing from 36–68% [48]. Researchers identified that hybrid building integrated photovoltaic/thermal (BIPV/T) and EAHX systems give greater yearly total electrical and thermal energies in contrast to the BIPV/T and EAHX systems. The hybrid system gives around 3% less yearly total thermal and electrical exergy and enviro-economic parameters in contrast with the BIPV/T. The exergo-economic factor of the thermal (BIPV/T) and EAHX systems was >50% of that for the BIPV/T system [49]. Recently, a novel hybrid system was designed comprising a coupled EAHX with a concentrical 2-pipes heat exchanger as a clean energy source. The air is cooled when exchanging heat in the concentrical 2-pipes heat exchanger, having an internal pipe with longitudinal fins filled with water and ice/mixture of ethylene glycol as a PCM. The outcomes indicated that the convective heat transfer coefficient is noteworthy in the hybrid system with an attractive cooling solution, even during peak periods [50]. Another hybrid semi-transparent photovoltaic thermal (SPVT) greenhouse system, together with an EAHX, has been developed to grow plants in a warm season. The impact of air variations, packing factor on photovoltaic (PV) cells, and mass flow rate of air flowing through EAHX, greenhouse gas, and plant temperatures is examined. Outcomes show that the hybrid SPVT-EAHX unit produces >125 kWh of daily inclusive sustainable exergy [51].

2.6. Climatic Conditions

For use in a desert climate, a numerical analysis of the thermal efficacy of an EAHX was built and presented by Maytorena and Hinojosa [52]. The results were confirmed by contrasting them with numerical and experimental data from previous works collected under comparable environmental circumstances. Investigations focused on determining the EAHX's suitable air velocity inlet, causing the optimal chillness in the room with the proper air inlet and output arrangement [52].

Over the years, EAHX technology has proven to be promising for space heating/cooling. In most regions, however, EAHX alone is insufficient to offer interior thermal comfort in all areas worldwide. However, a hybrid system combining an EAHX with an Air-Source Heat Pump (ASHP) has recently been explored to give interior thermal comfort while also conserving energy. The hybrid system reduced average energy consumption and enhanced the value of COP during the summer and winter months when compared to a standalone ASHP for certain ambient air temperatures and humidity ranges [53]. An inventive vertical solar chimney with fins and an EAHX has been examined as a summertime supplementary ventilation and air-conditioning device in a hot, arid environment. The height, diameter, number of fins, depth, length, and diameter of the pipes that make up the EAHX, among other geometric characteristics of the solar chimney, were explored. The research revealed that this system met the ventilation objectives of labs and residential buildings, and that it was able to save 85% usage at its highest potential [54].

2.7. Coupled System of EAHX Upstream of the AHU

D'Agostino et al. [55] estimated the energy efficacy of a peculiar geothermal energy system in which an EAHX is situated upstream of the Air-Handling Unit (AHU) of an office air conditioning unit in Naples (South Italy). Research outcomes indicated that, using the EAHX, the saving of the thermal energy of the coils in the AHU is >40%. Additionally, the efficiency of the EAHX value reached up to 90% with modifications in the tube length and diameter; this value may settle at around 40–50% for the entire year. Li et al. [56] analyzed an Air-to-Air Heat Recovery (AAHR) device and an EAHX (arranged in series) and evaluated the viability of the novel system in severe cold regions. Performance tests indicated that the novel system brings a significant ecological and economic benefit in contrast to the traditional solution. The EAHX significantly reduced greenhouse gas emissions by 17%, which is an indication of energy savings with payback periods of 2.1 and 2.4 years (return rate of 8%).

Kalbasi et al. [57] verified how, by combining an AAHR, EAHX, with an AHU, the cooling energy reduces to an average 33% and 11% reduction in the overall essential power. Research indicated that the coupled system of AHU and EAHX could bring down the total energy falls by an average of 38% with a 100 m long tube, and this value of reductions could even be reduced by 38–49% with the increased length of the tube. Ascione et al. [58] assessed the consequences of using an EAHX in an air-conditioning system on energy efficacy and ecological impact. The results showed that the pre-treating unit connected with the EAHX (pre-cooling in the summer and pre-heating in the winter) resulted in a 30% energy reduction in winter and an average saving rate of 41% in the summer, for a worldwide saving rate of 31% annually.

D'Agostino et al. [59] analyzed a coupled system of EAHX + AHU for the entire year at various locations identified according to the Koppen classification [60]. The results indicated a best outcome for Milan (Italy), with a cooling and heating capacity saving of 55% for a 100 m tube length, indicative of a huge energy savings. The lowest outcome was in Lampedusa (Italy) and Rio de Janeiro (Brazil), where the heating + cooling capacity reduction was possible up to 39% and 24%, respectively. Nevertheless, the best results could be attained for a 100-m tube length in Ottawa (Canada), where the heating + cooling capacity was reduced by 65% when using the EAHX. D'Agostino et al. [61] also investigated an HVAC system composed of EAHX + AHU. They demonstrated the overall energy saving in the city of Naples (Italy) in winter rises from 17% to 47% for the range of tube lengths,

20 m to 140 m. In addition, the power savings in Ottawa (Canada) reached an all-out of 52% for 140 m tube length.

2.8. Ground Source Heat Pump (GSHP) and EAHX

Shimada et al. [62] analyzed the high energy-saving potential GSHP to promote energy and GHG reductions. GSHP designed based experimental work on a pilot facility was conducted. The simulation results indicated an energy savings of 40% for a 3-year GSHP procedure for a small government building in Bangkok. Fernández [30] analysis indicates that the GSHP system of the Portuguese group could exhibit a 30% drop of energy spending and 30% lessening of carbon dioxide emissions in contrast to the conventional space conditioning system. Similarly, GSHP with vertical- and horizontal-type EAHX reduced electricity consumption by 30% and 18%, respectively, in contrast to an ASHP [63,64].

A Gas Injection EAHX (GI-EAHX) with an air conditioning system was analyzed for peak loads by Richter et al. [65]. Experimental results indicated that there was a reduction in the exit temperature of up to 2 °C and cooling power is increased by 26% compared with an equivalent conventional setup. They indicated that the usage of several GI-EAHX within a larger geothermal field could offer reliable peak load handling with a reduced size of a geothermal system with a good thermal capacity [66]. Several researchers optimized the diameter of the EAHX to augment the heat exchange rate up to 3.45%, and a rise of 18.7% was achieved by using ripped inner pipe walls instead of conventional smooth inner walls [67]. Specific details on these aspects were exhaustively summarized by Florides and Kalogirou [68].

Table 1 summarizes the main investigations in which significant energy savings were detected through the employment of geothermal systems. Different geothermal systems have been analyzed, from the simple EAHX configurations toward the GSHP and hybrid solutions.

Table 1. Energy savings through geothermal systems.

Author	Location	Method	Energy Reduction/Savings	CO_2/GHG Reduction
Cadelano et al. [27]	Technical Museum Nikola Tesla in Zagreb	GSHP	48–66%	24%
Fernández [30]	German super markets Portuguese	GSHP	45%/30%	28%/30%
D'Agostino et al. [55]	Office building in Naples (South Italy).	EAHX + AHU	40–50%	–
Li et al. [56]	Cold regions	EAHX + AAHR	8%	17%
Kalbasi et al. [57]		EAHX + AAHR + AHU	38–49%	–
Ascione et al. [58]	Mediterranean climate	EAHX + air conditioning	29–46%	–
D'Agostino et al. [59]	Milan Lampedusa Rio de Janeiro Ottawa	EAHX + AHU	55% 39% 24% 65%	–
D'Agostino et al. [61]	Naples Ottawa	EAHX + AHU	17.5–46% 52%	–
Shimada et al. [62]	Bangkok, Thailand	GSHP	40%	–
Richter et al. [65]	Hamburg University of Technology	GI-BHXs + air conditioning	26%	–

3. Materials Used in Earth-Air Heat Exchangers

The energy converted by the EAHXs depends on design parameters such as the pipe's material, diameter, and depth [69]. The performance of ground-coupled heat exchanger systems is based on air/liquid flow rate, length, material, depth, diameter of pipe, temperature difference, soil temperature, capacity of blower fan, and different layout of pipes [70]. Sakhri et al. [71] reviewed how many researchers have conducted studies on the performance of EAHXs during summer and winter and found the effects of external and internal parameters. The materials employed for building an EAHX are also crucial aspects to consider. In their reviews, Darius et al. [72] and Peretti et al. [73] also considered the influence of the thermal properties of the design materials on the energy performance provided by these geothermal systems. Specifically, pipes made of steel, aluminum, plastics, and copper tubes were reviewed by Peretti et al. [73]. The requirements for pipe materials are resistance coefficients, wall roughness, anti-corrosive, and structurally stable. The functions of the tubes are to transport air and exchange heat with the soil.

In their investigation, Bansal et al. [74] reported that pipe materials do not affect performance much due to their thickness; therefore, cheaper materials can be used in pipes. Lekhal et al. [75] analyzed the performance of EAHX using zinc and PVC. The comparative study was conducted under a warm-temperature climate in northern Algeria. The results show that pipe material affects EAHX performance when EAHX passes from heating to cooling mode. The air outlet temperature provided by zinc pipes is 7.5 °C higher than the outlet temperature provided by PVC. Sakhri et al. [76] used two types of pipe materials: PVC and steel. It was found that PVC is better than steel since PVC is cheap, light in weight, easy to assemble, and modification of shape is possible. Figure 2 shows the variety of materials used in the pipes of the EAHX systems.

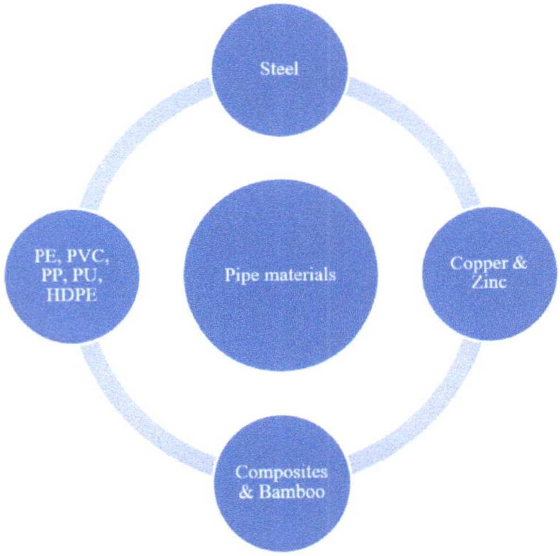

Figure 2. Variety of materials used in Pipes.

Bansal et al. [77] investigated the effects of pipe material and air velocity on the performance of EAHX systems. The investigations on steel and PVC pipes concluded that the velocity of air was found to greatly affect performance. The material having a high coefficient of friction affects the performance of the EAHX system. They concluded that convective heat transfer is dominant compared to conductive heat transfer. The effect of pipe materials on EAHX's performance was investigated. Zinc and PVC materials were used in the study. It was found that zinc EAHX exhibits higher efficiency in a climate with a COP of 9.5 compared to a steppe or arid climate having a COP of 8.2 or 8.1. PVC EAHXs are more suitable in the arid climate with a COP of 9.4 compared to the steppe climate with COPs belonging to the 7.6–8.4 range. The thermal performance of EAHX depends on the geo-climatic conditions and the pipe material [78].

Javadi et al. [79] reported that the most common material used in pipes of EAHXs is polyethylene (PE). Steel is most commonly used after PE. Copper and polyvinyl chloride (PVC) are also common. Polypropylene (PP), polyurethane (PU), composite, high-density polyethylene (HDPE), and polybutylene (PB) are rarely used. 60% of the GHEs use PE, 14% of the GHEs use steel, and 8% of the GHEs use PVC. Composite pipes exhibit better performance compared with the one provided by EAHXs made of HDPE pipes. The addition of aluminum wires to HDPE pipes increases the thermal conductivity. Steel pipes have greater heat transfer coefficients of pipe length than PE ducts.

Svec et al. [80] investigated the heat flow around an EAHX using plastic piping. It was found that the performance of the EAHX was affected when plastic piping was used.

An experimental investigation was performed in the north-eastern region of India [81]. Tunnel pipes were made of bamboo, and a cement–soil plaster mixture was employed to enhance the conductivity of bamboo pipes. Experiments were performed continuously for 7 days, and the delivered air temperature ranged from 25–26 °C with respect to the air temperature at the inlet, which ranged from 35 °C to 42 °C. The underground fresh air system is an easy and economically feasible technique. It reduces the consumption of energy, meets the power demand, and minimizes changes in climatic conditions in buildings.

4. IOT & Control Systems Used in EAHX

4.1. Control Systems Used in EAHX

The concept of EAHX provided by a control system is shown in Figure 3. As discussed in previous sections, EAHX is a green retrofit for cooling and heating buildings [82–84]. The concept of EAHX enables the insufflation of outdoor air directly into the building, providing more favorable conditions in reaching the required thermal comfort (less energy consumption to reach the indoor setting temperature) when coupled to an AHU or another existing air conditioning system [85,86]. Heat exchanges between earth and air increase with decreasing air velocity, altering the air flow rate in the duct and allowing for modification of its output temperature. Such a control would require measuring tools and a centralized control system on an actual structure. As shown in Figure 3, the main parts of the EAHX system IOT controlled are:

- Heat exchanger tubes;
- Sensors and actuators;
- Monitoring systems;
- Data acquisition system;
- Database server and control units;
- Filters, pump, and pump drives;
- Controllers.

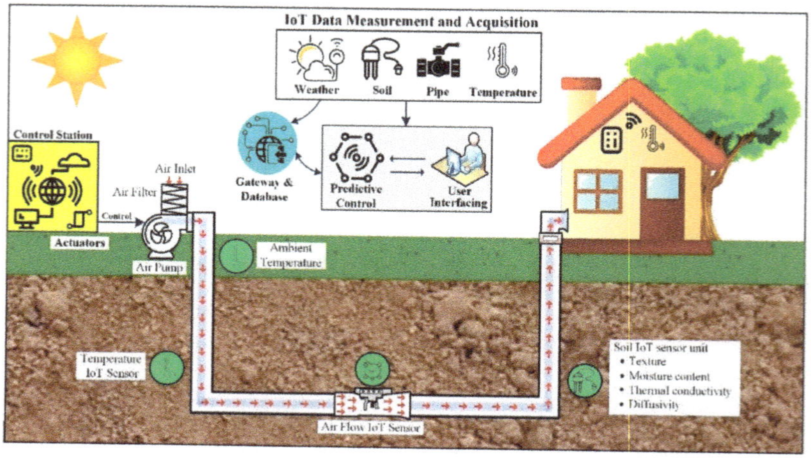

Figure 3. IoT-based control of the EAHX system.

The air pump has on/off control using a temperature sensor for ambient air and extracted room air, which can be set by the user. The aim of this on/off control strategy is to activate the EAHX air-handling unit. The recommended control set temperature for the EAHX air pump is based on user input and predictive control [87].

EAHX systems are designed to ensure hygrothermal effects in interior building environments, and their automated control can significantly improve end-users' comfort. Implementing novel control architectures and enabling the energy enhancement of EAHX systems without compromising building occupants' thermal comfort is of utmost importance. Combining two key modules—an intelligent control process and a smart physical infrastructure—can accomplish this. Energy enhancement and thermal comfort optimization must be the focus of the first module, while the sensing, communication, data storage, and actuation of the EAHX system in response to end-user demands are handled by the second.

The system continues to operate to keep the temperature within the building practically constant until the ideal temperature is reached. As a result, it discovers an air mass flow rate that uses less energy. Several methods have been developed throughout the decades for EAHX systems that may be taken into consideration in the control algorithm module. The majority are intuitive on/off controllers that are unable to account for significant thermal inertia EAHX processes. In addition, PID controllers with complex tuning for thermal inertia EAHX processes, whose performance may degrade during system dynamics. After reaching a steady-state temperature, the PID controller lowers the system's energy usage [88]. A PID controller may be used to decrease the variation in air fan power, which is a cubic function directly connected to the velocity of the air mass flow rate pumped into the building heat exchanger. Appropriate PID gain settings can further reduce this variation, as demonstrated in this study. This is so that the PID controller does not always have to operate at full capacity. When using a PID controller, the system uses less energy after the air inside the building reaches the desired temperature. This is because there is less energy being removed by the heat exchanger at this point because there is less of a difference between the desired temperature and the ground temperature. Furthermore, since fewer energy resources are required for cooling or heating a structure, PID control can improve both the EAHX 's efficiency as well as the sustainability of the globe.

The advantages of two cutting-edge heat exchanger control techniques, H control and μ-synthesis, are described in [89]. The necessity for a design process that explicitly tackles issues with modeling mistakes led to the development of H_∞ optimal control, frequency-domain optimization, and design theory. The H∞ control is developed further in the "μ-synthesis theory," where the uncertainty framework is considered while designing. A more effective predictive control strategy has gained traction over the past 20 years. This control method enables the efficient integration of energy efficacy methods into controller construction, together with problems such as disturbance rejection, constraint fulfillment, and slow-operating dynamic control [90]. Furthermore, the implementation of optimum regulating methods for the improvement of energy efficacy and thermal ease is becoming more abrupt and accessible as a result of the falling prices of smart devices. Advances in information and communication technology (ICT) enable the wide availability of dispersed sensors and data analytics tools. Therefore, it becomes clear that model predictive control is meaningless without being connected to a suitable smart infrastructure that enables the collection and transmission of actual data from the field [91]. By enabling the connecting of sensors, actuators, and other items to the internet and enabling both the awareness of the environment and interaction with it, the Internet of Things (IoT) extends a suitable solution.

4.2. IoT-Based Control of EAHX Systems

Due to the confluence of many technologies, things have changed. Digital intelligence is given to otherwise dumb gadgets by linking all these different things and giving them sensors, enabling them to communicate real-time data without a human being's input. The Internet of Things, which connects the real and digital worlds, is making the environment around us smarter and more responsive. Devices that can sense, analyze, and wirelessly transmit data to an unnamed storage place are used. such as the cloud, which stores, analyses, and presents data in a useful format, the IoT is expected to revolutionize our world by assisting us in monitoring and controlling basic development in our environment [91]. IoT frameworks facilitate communication among "things," with the majority of these frameworks appearing to focus on real-time data solutions. Big data and the Internet of Things are two developing technologies that may be used to generate information and enable applications for energy-efficient buildings [92]. Effective forecasting of heating and cooling demand is crucial to building energy management. One of the key uses of the IoT paradigm is the effective control of heating, ventilation, and air conditioning systems in smart buildings. The inside atmosphere of smart buildings may be improved for energy efficiency and thermal comfort using an IoT-based EAHX control system design [93]. The components are as follows:

It consists of the following elements:

- A gateway connects the nets of sensors and actuators to the Internet;
- An external application programming interface provides weather forecasts;
- An external database server gathers and forwards data from/to the field and from/to the control unit;
- A network of sensors senses environmental conditions and transmits measurements to a gateway;
- A network of actuators controls the pump and connects with the gateway;
- An IP device that serves as the end-user interface and is connected to the database server and host;
- A control unit that connects with the database server and houses the MPC algorithm;
- A dashboard for keeping track of the environment's condition and adjusting the control system mode.

The proposed architecture's mode of operation and information flow are as follows: The network of sensors is used to detect energy usage as well as the internal environmental conditions (temperature, CO_2 level, number of occupants, etc.). The measurements are then transferred in periodic bursts to the control unit, where the Predictive Control algorithm is implemented, and finally to the gateway, which connects to the database server. The control actions provided by this algorithm to the EAHX systems provide the optimal trade-off between energy consumption and comfort for the specified comfort limitations in the foreseeable future. The devices that actuate the EAHX modules get these control commands from the gateway. These control instructions are sent from the gateway to the devices that operate the EAHX modules. The EAHX modules adjust the room's temperature in accordance with the algorithm's predictions. The database server also keeps track of temperature and energy use readings. Through the dashboard on an IP device linked to the database server, these measures are shown to the end user. Users may interact with the control unit through the dashboard and choose the preferred temperature and control mode (i.e., either manual or automatic).

5. Conclusions

The surge in worldwide power consumption and CO_2 emissions is due to air conditioning and refrigeration, which has resulted in a massive increase in global warming and ozone depletion potential. The need for energy increased to achieve the required thermal comfort inside houses and business organizations. Renewable energy has to be utilized instead of conventional energy to reduce pollution in the environment. Alternative methods are being developed to achieve thermal comfort inside buildings. Many passive techniques replace conventional methods to reduce energy consumption. Natural resources supply energy when passive techniques are used.

EAHXs are discussed and critically analyzed as possible solutions for cleaner energy generation and low environmental impact. Specifically, the different hybrid applications presented in the literature, where the EAHXs are coupled to advanced systems, are reviewed.

When it comes to the design of an EAHX, the pipe diameter, length of pipe, and number of pipes are the leading parameters to be decided. As the length of the pipe increases, the pressure reduces, resulting in an increase in thermal performance. A long pipe of a small diameter buried at a large depth with less airflow velocity enhances the performance of EAHX. The temperature at certain depths is the same since the soil's thermal inertial property remains the same throughout the year. EAHX is effectively used to heat the air during the winter and cool the air in summer. The temperature of the Earth remains constant at a depth of 2 m throughout the year. This is called Earth's Undisturbed Temperature (EUT). The EUT is higher than air temperature during winter and lower than air temperature during summer.

EAHX utilizes geothermal energy, which is considered renewable energy and sustainable. Since EAHX does not use a refrigerant, there is no greenhouse gas emission. Soil density, soil properties, conductivity, diffusivity, water, and rock beds are the governing factors for selecting the site for EAHX.

In addition, various integrated hybrid systems are discussed to suit the climate conditions of arid, semi-arid, and congested areas. Hybrid systems indicated higher heat transfer coefficients, higher energy performance, and higher enviro-economic values compared to standalone EAHX systems. Phase change materials is a consolidated and viable resourced for this purpose. It is already employed in vapor compression [94] When compared to a standalone EAHX, a hybrid EAHX based on Phase-Change Materials and an Air-Source Heat Pump lowered average energy consumption and enhanced COP value roundly by 10–15%. At its maximum capacity, EAHX with a finned vertical solar chimney could increase energy savings by up to 60–80%. When EAHX is placed upstream of an air conditioning system's AHU, it may achieve improvements in terms of energy performance of up to 85%. In addition, greenhouse gas emissions were reduced by 17%, and worldwide electricity consumption was lowered by 33–43%. Furthermore, when compared to high-density PE pipes, composite pipes perform better. The heat exchange of HDPE pipes was improved using aluminum wire. Steel pipes, however, have a higher heat exchange/meter of pipe length than PE pipes. Finally, an IoT-based and predictive control algorithm-enabled EAHX system is proposed to optimize energy efficiency and regulate room temperature with thermal comfort. From our point of view, these systems are destined to gain more and more space as a renewable energy-based solution for the next future.

Author Contributions: Conceptualization, E.G., D.G.S., R.S.R. and C.M.; methodology, E.G.; resources, A.G.; data curation, E.G., D.G.S. and R.S.R.; writing—original draft preparation, E.G., D.G.S., R.S.R. and C.M.; writing—review and editing, E.G. and A.G.; visualization, A.G.; supervision, E.G. and A.G.; project administration, E.G. and A.G.; funding acquisition, A.G. All authors have read and agreed to the published version of the manuscript.

Funding: This research received no external funding.

Institutional Review Board Statement: Not applicable.

Informed Consent Statement: Not applicable.

Data Availability Statement: Not applicable.

Conflicts of Interest: The authors have no competing interest.

Abbreviations

AAHR	Air-to-Air Heat Recovery
AHU	Air-Handling Unit
ASHP	Air-Source Heat Pump
BHX	Borehole Heat eXchangers
COP	Coefficient Of Performance
EAHX	Earth-to-Air Heat eXchangers
EPI	Energy Performance Indicator

EUT	Earth Undisturbed Temperature
GHG	Green House Gas
GI-BHX	Gas injection borehole heat exchanger
GSHP	Ground Source Heat Pumps
GWP	Global Warming Potential
HDPE	High-Density Polyethylene
HVAC	Heating, Ventilation and Air Conditioning
IoT	Internet of Things
PB	Polybutylene
PCM	Phase-Change Materials
PE	Polyethylene
PP	Polypropylene
PU	Polyurethane
PVC	polyvinyl chloride
SC	Solar chimney

References

1. World Bank Open Data—Electric Power Consumption (kWh per Capita). Available online: https://data.worldbank.org/indicator/EG.USE.ELEC.KH.PC (accessed on 17 March 2020).
2. Armeanu, D.S.; Gherghina, S.C.; Pasmangiu, G. Exploring the causal nexus between energy consumption, environmental pollution and economic growth: Empirical evidence from central and Eastern Europe. *Energies* **2019**, *12*, 3704. [CrossRef]
3. IEA. *World Energy Outlook*; International Energy Agency: Paris, France, 2009.
4. Aprea, C.; de Rossi, F.; Greco, A.; Renno, C. Refrigeration plant exergetic analysis varying the compressor capacity. *Int. J. Energy Res.* **2003**, *27*, 653–669. [CrossRef]
5. Velders, G.J.; Andersen, S.O.; Daniel, J.S.; Fahey, D.W.; McFarland, M. The importance of the Montreal Protocol in protecting climate. *Proc. Natl. Acad. Sci. USA* **2007**, *104*, 4814–4819. [CrossRef]
6. Birmpili, T. Montreal Protocol at 30: The governance structure, the evolution, and the Kigali Amendment. *Comptes Rendus Geosci.* **2018**, *350*, 425–431. [CrossRef]
7. Aprea, C.; Greco, A.; Rosato, A. Comparison of R407C and R417A heat transfer coefficients and pressure drops during flow boiling in a horizontal smooth tube. *Energy Convers. Manag.* **2008**, *49*, 1629–1636. [CrossRef]
8. Greco, A.; Vanoli, G.P. Evaporation of refrigerants in a smooth horizontal tube: Prediction of R22 and R507 heat transfer coefficients and pressure drop. *Appl. Therm. Eng.* **2004**, *24*, 2189–2206. [CrossRef]
9. Aprea, C.; Greco, A. An exergetic analysis of R22 substitution. *Appl. Therm. Eng.* **2002**, *22*, 1455–1469. [CrossRef]
10. Cabello, R.; Sánchez, D.; Llopis, R.; Catalán, J.; Nebot-Andrés, L.; Torrella, E. Energy evaluation of R152a as drop in replacement for R134a in cascade refrigeration plants. *Appl. Therm. Eng.* **2017**, *110*, 972–984. [CrossRef]
11. Sieres, J.; Santos, J.M. Experimental analysis of R1234yf as a drop-in replacement for R134a in a small power refrigerating system. *Int. J. Refrig.* **2018**, *91*, 230–238. [CrossRef]
12. Aprea, C.; Greco, A.; Maiorino, A.; Masselli, C.; Metallo, A. HFO1234yf as a drop-in replacement for R134a in domestic refrigerators: A life cycle climate performance analysis. *Int. J. Heat Technol.* **2016**, *34*, S212–S218. [CrossRef]
13. Aprea, C.; Greco, A.; Maiorino, A.; Masselli, C.; Metallo, A. HFO1234ze as drop-in replacement for R134a in domestic refrigerators: An environmental impact analysis. *Energy Procedia* **2016**, *101*, 964–971. [CrossRef]
14. Zou, C.; Xiong, B.; Xue, H.; Zheng, D.; Ge, Z.; Wang, Y.; Wu, S. The role of new energy in carbon neutral. *Pet. Explor. Dev.* **2021**, *48*, 480–491. [CrossRef]
15. Franzén, I.; Nedar, L.; Andersson, M. Environmental comparison of energy solutions for heating and cooling. *Sustainability* **2019**, *11*, 7051. [CrossRef]
16. Cabeza, L.F.; Rincón, L.; Vilariño, V.; Pérez, G.; Castell, A. Life cycle assessment (LCA) and life cycle energy analysis (LCEA) of buildings and the building sector: A review. *Renew. Sustain. Energy Rev.* **2014**, *29*, 394–416. [CrossRef]
17. Perez-Lombard, L.; Ortiz, J.; Maestre, I.R. The map of energy flow in HVAC systems. *Appl. Energy* **2011**, *88*, 5020–5031. [CrossRef]
18. Coulomb, D. Air conditioning environmental challenges. *REHVA J.* **2015**, *4*, 30–34.
19. Alahmer, A.; Ajib, S. Solar cooling technologies: State of art and perspectives. *Energy Convers. Manag.* **2020**, *214*, 112896. [CrossRef]
20. Greco, A.; Gundabattini, E.; Gnanaraj, D.S.; Masselli, C. A comparative study on the performances of flat plate and evacuated tube collectors deployable in domestic solar water heating systems in different climate areas. *Climate* **2020**, *8*, 78. [CrossRef]
21. Zhang, Y.; Lin, Y.; Lin, F.; Yang, K. Thermodynamic analysis of a novel combined cooling, heating, and power system consisting of wind energy and transcritical compressed CO2 energy storage. *Energy Convers. Manag.* **2022**, *260*, 115609. [CrossRef]
22. Alsagri, A.S.; Chiasson, A.; Shahzad, M.W. Geothermal energy technologies for cooling and refrigeration systems: An overview. *Arab. J. Sci. Eng.* **2021**, 1–31. [CrossRef]
23. Anand, S.; Gupta, A.; Tyagi, S.K. Solar cooling systems for climate change mitigation: A review. *Renew. Sustain. Energy Rev.* **2015**, *41*, 143–161. [CrossRef]

24. Fouad, M.M.; Shihata, L.A.; Morgan, E.I. An integrated review of factors influencing the perfomance of photovoltaic panels. *Renew. Sustain. Energy Rev.* **2017**, *80*, 1499–1511. [CrossRef]
25. Li, K.; Bian, H.; Liu, C.; Zhang, D.; Yang, Y. Comparison of geothermal with solar and wind power generation systems. *Renew. Sustain. Energy Rev.* **2015**, *42*, 1464–1474. [CrossRef]
26. Giambastiani, B.M.S.; Tinti, F.; Mendrinos, D.; Mastrocicco, M. Energy performance strategies for the large scale introduction of geothermal energy in residential and industrial buildings: The GEO. POWER project. *Energy Policy* **2014**, *65*, 315–322. [CrossRef]
27. Cadelano, G.; Cicolin, F.; Emmi, G.; Mezzasalma, G.; Poletto, D.; Galgaro, A.; Bernardi, A. Improving the energy efficiency, limiting costs and reducing CO_2 emissions of a museum using geothermal energy and energy management policies. *Energies* **2019**, *12*, 3192. [CrossRef]
28. Aggarwal, V.; Meena, C.S.; Kumar, A.; Alam, T.; Kumar, A.; Ghosh, A.; Ghosh, A. Potential and future prospects of geothermal energy in space conditioning of buildings: India and worldwide review. *Sustainability* **2020**, *12*, 8428. [CrossRef]
29. Yang, G.; Shi, H.; Xu, D.; Shen, Z.; Zhang, Z.; Shen, H.; Wang, Z. Research on simulation of energy consumption of ground water-source heat pump air conditioning system in plant factory. *IOP Conf. Ser. Earth Environ. Sci.* **2021**, *770*, 012044. [CrossRef]
30. Fernández, J.C.R. Integration capacity of geothermal energy in supermarkets through case analysis. *Sustain. Energy Technol. Assess.* **2019**, *34*, 49–50. [CrossRef]
31. Bisoniya, T.S. Design of earth–air heat exchanger system. *Geotherm. Energy* **2015**, *3*, 18. [CrossRef]
32. Khandouzi, O.; Pourfallah, M.; Yoosefirad, E.; Shaker, B.; Gholinia, M.; Mouloodi, S. Evaluating and optimizing the geometry of thermal foundation pipes for the utilization of the geothermal energy: Numerical simulation. *J. Energy Storage* **2021**, *37*, 102464. [CrossRef]
33. Agrawal, K.K.; Misra, R.; Agrawal, G.D.; Bhardwaj, M.; Jamuwa, D.K. Effect of different design aspects of pipe for earth air tunnel heat exchanger system: A state of art. *Int. J. Green Energy* **2019**, *16*, 598–614. [CrossRef]
34. Bisoniya, T.S.; Kumar, A.; Baredar, P. Study on Calculation Models of Earth-Air Heat Exchanger Systems. *J. Energy* **2014**, *2014*, 1–15. [CrossRef]
35. Sani, F.H.; Pourfallah, M.; Gholinia, M. The effect of MoS2–Ag/ H2O hybrid nanofluid on improving the performance of a solar collector by placing wavy strips in the absorber tube. *Case Stud. Therm. Eng.* **2022**, *30*, 101760. [CrossRef]
36. Gholinia, M.; Ranjbar, A.A.; Javidan, M.; Hosseinpour, A.A. Employing a new micro-spray model and (MWCNTs–SWCNTs)—H_2O nanofluid on Si-IGBT power module for energy storage: A numerical simulation. *Energy Rep.* **2021**, *7*, 6844–6853. [CrossRef]
37. Shaker, B.; Gholinia, M.; Pourfallah, M.; Ganji, D.D. CFD analysis of Al_2O_3-syltherm oil Nanofluid on parabolic trough solar collector with a new flange-shaped turbulator model. *Theor. Appl. Mech. Lett.* **2022**, *12*, 100323. [CrossRef]
38. Ghobadi, A.H.; Armin, M.; Hassankolaei, S.G.; Hassankolaei, M.G. A new thermal conductivity model of CNTs/$C_2H_6O_2$–H_2O hybrid base nanoliquid between two stretchable rotating discs with Joule heating. *Int. J. Ambient Energy* **2020**, *43*, 3310–3321. [CrossRef]
39. Ghobadi, A.H.; Hassankolaei, M.G. A numerical approach for MHD Al_2O_3–TiO_2/H_2O hybrid nanofluids over a stretching cylinder under the impact of shape factor. *Heat Transf. Res.* **2019**, *48*, 4262–4282. [CrossRef]
40. Rodrigues, M.K.; Vaz, J.; Rocha, L.A.O.; Santos, E.D.d.; Isoldi, L.A. A full approach to Earth-Air Heat Exchanger employing computational modeling, performance analysis and geometric evaluation. *Renew. Energy* **2022**, *191*, 535–556. [CrossRef]
41. Mathur, A.; Kumar, S. Thermal performance and comfort assessment of U-shape and helical shape earth-air heat exchanger in India. *Energy Built Environ.* **2022**, *3*, 171–180. [CrossRef]
42. Benhammou, M.; Sahli, Y.; Moungar, H. Investigation of the impact of pipe geometric form on earth-to-air heat exchanger performance using Complex Finite Fourier Transform analysis. Part I: Operation in cooling mode. *Int. J. Therm. Sci.* **2022**, *177*, 107484. [CrossRef]
43. Long, T.; Li, Y.; Li, W.; Liu, S.; Lu, J.; Zheng, D.; Ye, K.; Qiao, Z.; Huang, S. Investigation on the cooling performance of a buoyancy driven earth-air heat exchanger system and the impact on indoor thermal environment. *Appl. Therm. Eng.* **2022**, *207*, 118148. [CrossRef]
44. Gao, X.; Xiao, Y.; Gao, P. Thermal potential improvement of an earth-air heat exchanger (EAHX) by employing backfilling for deep underground emergency ventilation. *Energy* **2022**, *250*, 123783. [CrossRef]
45. Long, T.; Li, W.; Lv, Y.; Li, Y.; Liu, S.; Lu, J.; Huang, S.; Zhang, Y. Benefits of integrating phase-change material with solar chimney and earth-to-air heat exchanger system for passive ventilation and cooling in summer. *J. Energy Storage* **2022**, *48*, 104037. [CrossRef]
46. Long, T.; Zhao, N.; Li, W.; Wei, S.; Li, Y.; Lu, J.; Huang, S.; Qiao, Z. Natural ventilation performance of solar chimney with and without earth-air heat exchanger during transition seasons. *Energy* **2022**, *250*, 123818. [CrossRef]
47. Bai, Y.; Long, T.; Li, W.; Li, Y.; Liu, S.; Wang, Z.; Lu, J.; Huang, S. Experimental investigation of natural ventilation characteristics of a solar chimney coupled with earth-air heat exchanger (SCEAHX) system in summer and winter. *Renew. Energy* **2022**, *193*, 1001–1018. [CrossRef]
48. Firfiris, V.K.; Kalamaras, S.D.; Martzopoulou, A.G.; Fragos, V.P.; Kotsopoulos, T.A. Improvement of the Performance of an Earth to Air Heat Exchanger for Greenhouse Cooling by the Incorporation of Water Finned Tubes—A Theoretical Approach. *AgriEngineering* **2022**, *4*, 190–206. [CrossRef]
49. Shahsavar, A.; Talebizadehsardari, P.; Arıcı, M. Comparative energy, exergy, environmental, exergoeconomic, and enviroeconomic analysis of building integrated photovoltaic/thermal, earth-air heat exchanger, and hybrid systems. *J. Clean. Prod.* **2022**, *362*, 132510. [CrossRef]
50. Ouzzane, M.; Bady, M. Investigation of an innovative Canadian well system combined with a frozen water/PCM heat exchanger for air-cooling in hot climate. *Appl. Therm. Eng.* **2022**, *213*, 118737. [CrossRef]

51. Yadav, S.; Panda, S.K.; Tiwari, G.N.; Al-Helal, I.M.; Alsadon, A.A.; Shady, M.R.; Tiwari, A. Semi-Transparent Photovoltaic Thermal Greenhouse System Combined with Earth Air Heat Exchanger for Hot Climatic Condition. *J. Therm. Sci. Eng. Appl.* **2022**, *14*, 081007. [CrossRef]
52. Maytorena, V.; Hinojosa, J. Thermal Analysis of a Generic Earth-to-Air Heat Exchanger Coupled with a Room during the Summer Season in a Desert Climate. *J. Energy Eng.* **2022**, *148*, 4022001. [CrossRef]
53. Guo, X.; Wei, H.; He, X.; Du, J.; Yang, D. Experimental evaluation of an earth–to–air heat exchanger and air source heat pump hybrid indoor air conditioning system. *Energy Build.* **2022**, *256*, 111752. [CrossRef]
54. Kalantar, V.; Khayyaminejad, A. Numerical simulation of a combination of a new solar ventilator and geothermal heat exchanger for natural ventilation and space cooling. *Int. J. Energy Environ. Eng.* **2022**, *13*, 785–804. [CrossRef]
55. D'Agostino, D.; Esposito, F.; Greco, A.; Masselli, C.; Minichiello, F. Parametric analysis on an earth-to-air heat exchanger employed in an air conditioning system. *Energies* **2020**, *13*, 2925. [CrossRef]
56. Li, H.; Ni, L.; Yao, Y.; Sun, C. Annual performance experiments of an earth-air heat exchanger fresh air-handling unit in severe cold regions: Operation, economic and greenhouse gas emission analyses. *Renew. Energy* **2020**, *146*, 25–37. [CrossRef]
57. Kalbasi, R.; Ruhani, B.; Rostami, S. Energetic analysis of an air handling unit combined with enthalpy air-to-air heat exchanger. *J. Therm. Anal. Calorim.* **2020**, *139*, 2881–2890. [CrossRef]
58. Ascione, F.; D'Agostino, D.; Marino, C.; Minichiello, F. Earth-to-air heat exchanger for NZEB in Mediterranean climate. *Renew. Energy* **2016**, *99*, 553–563. [CrossRef]
59. D'Agostino, D.; Greco, A.; Masselli, C.; Minichiello, F. The employment of an earth-to-air heat exchanger as pre-treating unit of an air conditioning system for energy saving: A comparison among different worldwide climatic zones. *Energy Build.* **2020**, *229*, 110517. [CrossRef] [PubMed]
60. Lohmann, U.; Sausen, R.; Bengtsson, L.; Cubasch, U.; Perlwitz, J.; Roeckner, E. The Köppen climate classification as a diagnostic tool for general circulation models. *Clim. Res.* **1993**, *3*, 177–193. [CrossRef]
61. D'Agostino, D.; Esposito, F.; Greco, A.; Masselli, C.; Minichiello, F. The energy performances of a ground-to-air heat exchanger: A comparison among köppen climatic areas. *Energies* **2020**, *13*, 2895. [CrossRef]
62. Shimada, Y.; Uchida, Y.; Takashima, I.; Chotpantarat, S.; Widiatmojo, A.; Chokchai, S.; Charusiri, P.; Kurishima, H.; Tokimatsu, K. A study on the operational condition of a ground source heat pump in Bangkok based on a field experiment and simulation. *Energies* **2020**, *13*, 274. [CrossRef]
63. Chokchai, S.; Chotpantarat, S.; Takashima, I.; Uchida, Y.; Widiatmojo, A.; Yasukawa, K.; Charusiri, P. A Pilot Study on Geothermal Heat Pump (GHP) Use for Cooling Operations, and on GHP Site Selection in Tropical Regions Based on a Case Study in Thailand. *Energies* **2018**, *11*, 2356. [CrossRef]
64. Widiatmojo, A.; Chokchai, S.; Takashima, I.; Uchida, Y.; Yasukawa, K.; Chotpantarat, S.; Charusiri, P. Ground-Source Heat Pumps with Horizontal Heat Exchangers for Space Cooling in the Hot Tropical Climate of Thailand. *Energies* **2019**, *12*, 1274. [CrossRef]
65. Richter, F.; Niemann, P.; Schuck, M.; Grabe, J.; Schmitz, G. Comparison of conventional and variable borehole heat exchangers for use in a desiccant assisted air conditioning system. *Energies* **2021**, *14*, 926. [CrossRef]
66. Luo, J.; Rohn, J.; Bayer, M.; Priess, A.; Wilkmann, L.; Xiang, W. Heating and cooling performance analysis of a ground source heat pump system in Southern Germany. *Geothermics* **2015**, *53*, 57–66. [CrossRef]
67. Kurevija, T.; Kalantar, A.; Macenić, M.; Hranić, J. Investigation of Steady-State Heat Extraction Rates for Different Borehole Heat Exchanger Configurations from the Aspect of Implementation of New TurboCollectorTM Pipe System Design. *Energies* **2019**, *12*, 1504. [CrossRef]
68. Florides, G.; Kalogirou, S. Ground heat exchangers—A review of systems, models and applications. *Renew. Energy* **2007**, *32*, 2461–2478. [CrossRef]
69. Bisoniya, T.S.; Kumar, A.; Baredar, P. Experimental and analytical studies of earth–air heat exchanger (EAHX) systems in India: A review. *Renew. Sustain. Energy Rev.* **2013**, *19*, 238–246. [CrossRef]
70. Soni, S.K.; Pandey, M.; Bartaria, V.N. Ground coupled heat exchangers: A review and applications. *Renew. Sustain. Energy Rev.* **2015**, *47*, 83–92. [CrossRef]
71. Sakhri, N.; Menni, Y.; Chamkha, A.; Salmi, M.; Ameur, H. Earth to Air Heat Exchanger and Its Applications in Arid Regions—An Updated Review. *Tec. Ital. J. Eng. Sci.* **2020**, *64*, 83–90. [CrossRef]
72. Darius, D.; Misaran, M.S.; Rahman, M.M.; Ismail, M.A.; Amaludin, A. Working parameters affecting earth-air heat exchanger (EAHX) system performance for passive cooling: A review. In *IOP Conference Series: Materials Science and Engineering*; IOP Publishing: Bristol, UK, 2017; Volume 217, p. 012021.
73. Peretti, C.; Zarrella, A.; De Carli, M.; Zecchin, R. The design and environmental evaluation of earth-to-air heat exchangers (EAHX). A literature review. *Renew. Sustain. Energy Rev.* **2013**, *28*, 107–116. [CrossRef]
74. Bansal, V.; Misra, R.; Agrawal, G.D.; Mathur, J. Performance analysis of earth–pipe–air heat exchanger for winter heating. *Energy Build.* **2009**, *41*, 1151–1154. [CrossRef]

75. Lekhal, M.C.; Mokhtari, A.M.; Belarbi, R. Evaluation of two earth–air heat exchangers efficiency of different pipe materials within a warm temperate climate. In *MATEC Web of Conferences Volume 307 (2020) International Conference on Materials & Energy (ICOME'17 and ICOME'18) Tianjin, China and San Sebastian, Spain, 6–9 July 2017 and 30 April–4 June 2018*; El Ganaoui, M., Liu, B., Bennacer, R., Morsli, S., Darcherif, M., Eds.; EDP Sciences: Les Ulis, France, 2020; p. 01028. [CrossRef]
76. Sakhri, N.; Menni, Y.; Ameur, H. Effect of the pipe material and burying depth on the thermal efficiency of earth-to-air heat exchangers. *Case Stud. Chem. Environ. Eng.* **2020**, *2*, 100013. [CrossRef]
77. Bansal, V.; Misra, R.; Agrawal, G.D.; Mathur, J. Performance analysis of earth–pipe–air heat exchanger for summer cooling. *Energy Build.* **2010**, *42*, 645–648. [CrossRef]
78. Lekhal, M.C.; Benzaama, M.-H.; Kindinis, A.; Mokhtari, A.-M.; Belarbi, R. Effect of geo-climatic conditions and pipe material on the heating performance of earth-air heat exchangers. *Renew. Energy* **2021**, *163*, 22–40. [CrossRef]
79. Javadi, H.; Mousavi Ajarostaghi, S.S.; Rosen, M.A.; Pourfallah, M. Performance of ground heat exchangers: A comprehensive review of recent advances. *Energy* **2019**, *178*, 207–233. [CrossRef]
80. Svec, O.J.; Goodrich, L.E.; Palmer, J.H.L. Heat transfer characteristics of in-ground heat exchangers. *Int. J. Energy Res.* **1983**, *7*, 265–278. [CrossRef]
81. Choudhury, T.; Misra, A.K. Minimizing changing climate impact on buildings using easily and economically feasible earth to air heat exchanger technique. *Mitig. Adapt. Strateg. Glob. Chang.* **2013**, *19*, 947–954. [CrossRef]
82. Bisoniya, T.S.; Kumar, A.; Baredar, P. Energy metrics of earth–air heat exchanger system for hot and dry climatic conditions of India. *Energy Build.* **2015**, *86*, 214–221. [CrossRef]
83. Greco, A.; Masselli, C. The Optimization of the Thermal Performances of an Earth to Air Heat Exchanger for an Air Conditioning System: A Numerical Study. *Energies* **2020**, *13*, 6414. [CrossRef]
84. Maoz, M.; Ali, S.; Muhammad, N.; Amin, A.; Sohaib, M.; Basit, A.; Ahmad, T. Parametric optimization of Earth to Air Heat Exchanger using Response Surface Method. *Sustainability* **2019**, *11*, 3186. [CrossRef]
85. Singh, R.; Sawhney, R.L.; Lazarus, I.J.; Kishore, V.V.N. Recent advancements in earth air tunnel heat exchanger (EATHE) system for indoor thermal comfort application: A review. *Renew. Sustain. Energy Rev.* **2018**, *82*, 2162–2185. [CrossRef]
86. Sobti, J.; Singh, S.K. Earth-air heat exchanger as a green retrofit for Chandīgarh—A critical review. *Geothermal Energy* **2015**, *3*, 14. [CrossRef]
87. Boithias, F.; Zhang, J.; El Mankibi, M.; Haghighat, F.; Michel, P. Simple model and control strategy of earth-to-air heat exchangers. In Proceedings of the 2009 International Conference on Advances in Computational Tools for Engineering Applications ACTEA 2009, Beirut, Lebanon, 15–17 July 2009; pp. 234–239. [CrossRef]
88. Diaz-Mendez, S.E.; Patiño-Carachure, C.; Herrera-Castillo, J.A. Reducing the energy consumption of an earth-air heat exchanger with a PID control system. *Energy Convers. Manag.* **2014**, *77*, 1–6. [CrossRef]
89. Vasičkaninová, A.; Bakošová, M.; Čirka, L'.; Kalúz, M.; Oravec, J. Robust controller design for a laboratory heat exchanger. *Appl. Therm. Eng.* **2018**, *128*, 1297–1309. [CrossRef]
90. Mirakhorli, A.; Dong, B. Market and behavior driven predictive energy management for residential buildings. *Sustain. Cities Soc.* **2018**, *38*, 723–735. [CrossRef]
91. Ruano, A.; Silva, S.; Duarte, H.; Ferreira, P.M. Wireless sensors and IoT platform for intelligent HVAC control. *Appl. Sci.* **2018**, *8*, 370. [CrossRef]
92. Sridharan, M.; Devi, R.; Dharshini, C.S.; Bhavadarani, M. IoT based performance monitoring and control in counter flow double pipe heat exchanger. *Internet Things* **2019**, *5*, 34–40. [CrossRef]
93. Carli, R.; Cavone, G.; Othman, S.B.; Dotoli, M. IoT based architecture for model predictive control of HVAC systems in smart buildings. *Sensors* **2020**, *20*, 781. [CrossRef]
94. Maiorino, A.; Del Duca, M.G.; Mota-Babiloni, A.; Greco, A.; Aprea, C. The thermal performances of a refrigerator incorporating a phase change material. *Int. J. Refrig.* **2019**, *100*, 255–264. [CrossRef]

energies

MDPI

Article

Impact of Air Density Variation on a Simulated Earth-to-Air Heat Exchanger's Performance

Piotr Michalak

Department of Power Systems and Environmental Protection Facilities, Faculty of Mechanical Engineering and Robotics, AGH University of Science and Technology, Mickiewicza 30, 30-059 Krakow, Poland; pmichal@agh.edu.pl; Tel.: +48-126-173-579

Abstract: Due to their simple design and reliable operation, earth-to-air heat exchangers (EAHE) are used in modern buildings to reduce ventilation heat losses. EAHE operation in atmospheric conditions results in variation in ambient air temperature and pressure affecting air density. The paper presents the study on the impact of ambient air density variation on the calculated hourly air temperature at the EAHE outlet and the resulting energy use for space heating and cooling of an exemplary residential building. The ground temperature was computed from the model given in EN 16798-5-1. Then, air density was obtained using five various methods. Energy use for space heating and cooling of the building was computed using the 5R1C thermal network model of EN ISO 13790. Depending on the chosen method and concerning the base case without EAHE, a reduction in annual heating and cooling needs was obtained from 7.5% to 8.8% in heating and from 15.3% to 19% in cooling. Annual heating and cooling gain from EAHE were 600.9 kWh and 628.3 kWh for heating and 616.9 kWh and 603.5 kWh for cooling for the Typical Meteorological Years (TMY) and International Weather for Energy Calculation (IWEC) files, respectively. Unit heating and cooling gains per heat exchanger area were from 34.9 kWh/m^2 to 36.8 kWh/m^2 and from −35.1 kWh/m^2 to −36.3 kWh/m^2. Density variation with temperature from the relevant typical Polish meteorological year at constant pressure, in comparison to the method of EN 16798-5-1, resulted in an hourly difference of that unit gain up to 4.3 W/m^2 and 2.0 W/m^2 for heating and cooling, respectively. The same was true inthe case of IWEC files that resulted in differences of 5.5 W/m^2 and 1.1 W/m^2.

Keywords: earth-to-air heat exchanger; air density; specific heat of air; barometric formula; EAHE; outlet temperature; ground temperature; EN ISO 13790; 5R1C model; hourly simulation

Citation: Michalak, P. Impact of Air Density Variation on a Simulated Earth-to-Air Heat Exchanger's Performance. *Energies* **2022**, *15*, 3215. https://doi.org/10.3390/en15093215

Academic Editor: Marco Marengo

Received: 28 March 2022
Accepted: 26 April 2022
Published: 27 April 2022

1. Introduction

Energy consumption in buildings is of special interest at the national level in many countries due to the economic and environmental reasons [1]. In total, energy demand space heating and cooling have the most significant share [2–4]. Its value depends on several components. One of them, regardless of the building's energy standard, is energy for heating and cooling ventilation air [5–7]. Since ventilation works all year round for hygienic reasons, supplying fresh outdoor air while removing polluted indoor air, its operation significantly affects the thermal demand of a building. Consequently, additional energy is needed for heating or cooling of air supplied to the interior of a building zone, in the cold and warm seasons, respectively.

To reduce the amount of that energy, various techniques of heat recovery from ventilation air are used, usually involving crossflow, counter-current or rotary heat exchangers [8,9]. However, together with the growing popularity of alternative and low-energy solutions, earth-to-air ground heat exchangers (EAHE, EAHX) have been gaining interest recently.

The design of EAHE should be performed carefully to avoid changes after it has been built, since correcting the position of a functioning device requires labour-intensive and costly earthworks. Therefore, numerous studies on the impact of various parameters on the EAHE performance were conducted, including climatic conditions and geographical location, soil type, pipe properties, burial depth and airflow rate [10]. In addition to experiments, theoretical analyses were carried out with the use of numerical simulations. As a result, methods of designing this type of heat exchanger have been developed for use in the heating and cooling of buildings in various climatic zones [11–14].

Since EAHEs are used commonly to reduce losses for heating or cooling the ventilation air, it becomes increasingly important to determine the energy effects of their application. Hence, calculation methods to assess the energy performance of EAHE are of growing significance. For this purpose, various tools have been used in recent studies.

D'Agostino et al. [15] analysed energy savings in an office building equipped with the HVAC system based on fan-coils and primary air, with and without EAHE and an air-to-air heat exchanger (AAHE). The authors used DesignBuilder and EnergyPlus to model the building and an HVAC system. They used the same set of tools in similar studies [16,17] on EAHE connected upstream with an air handling unit (AHU) and fan-coils in an office building. The EAHE model was validated with the two-dimensional finite element method and experimental data provided by other researchers.

Baglivo et al. [18] presented an hourly simulation of an air-cooled heat pump coupled with horizontal EAHE in a residential building. The hourly behaviour of EAHE was simulated in TRNSYS 17. Output results and relevant climatic data were used to compute calculate COP, EER, SCOP and SEER coefficients of the considered heat pump. The same tool was used in [19] to model various options of a ventilation system with EAHE in a residential building. The authors simulated the impact of the pipe numbers, air flow rate and soil thermal conductivity on the building thermal behaviour.

Other tools, as Ansys Fluent [20–22], Ansys CFX [23], Comsol Multiphysics [24], Matlab [25–27] or Scilab [28] were also used. However, although they offer great possibilities, their use requires specialised knowledge and experience. Hence, there is also a need for methods simpler but accurate enough to perform annual simulations with hourly time steps required by current standards and provide reasonable time resolution for thermal phenomena, occupation schedules, weather data and calculation accuracy [29–31], especially in terms of energy certification of buildings.

The EN 16798-5-1 standard [32], among others, gives the calculation method for preheating and precooling with necessary equations for ground temperature, temperature change in EAHE during airflow in the pipe and several auxiliary variables. The method is intended for use with hourly time steps. That standard has been used in several studies recently.

Skotnicka-Siepsiak [33] presented measurements of ground temperature and outlet air temperature of EAHE in a residential building in Olsztyn (northern Poland) in the warm months (May, June, July and August) of 2016, 2017 and 2018. Then, they were compared in the relevant charts with results based on EN 16798-5-1. The model provided comparable results for stable outdoor conditions in July and August. Larger differences were noticed in May and June.

Brata et al. [34] compared outlet temperature and monthly heat and cooling gains measured and calculated from EN 16798-5-1 for EAHE in an existing passive house near Timisoara in Romania. To assess the accuracy of the model authors used a histogram of residuals calculated as the difference between measured and calculated hourly values. They were between −0.9 °C and 2.8 °C. In general, monthly heat gains were slightly underestimated but monthly cooling energy was overestimated.

Michalak [35] coupled the model of EN 16798-5-1 with the thermal network model of a building zone from EN ISO 13790 [36] in an MS Excel spreadsheet to simulate in hourly time step an annual performance of EAHE connected with a ventilation system of a low-energy residential building in south Poland. In this way, hourly, as well as monthly and annual energy use for space heating and cooling and peak power with and without EAHE was estimated simply and efficiently.

It seems that due to its simplicity, this method does not offer as extensive possibilities as professional simulation programs. As mentioned above, they consider many physical properties. The most important of these include soil and pipe material parameters, flow rate and outside air temperature. They determine heat transfer conditions between soil and air flowing through a heat exchanger pipe from its inlet to its outlet. However, the above publications based on the standard model do not consider the influence of air temperature changes on air density [37], which in turn affects outlet air temperature from a heat exchanger. In [34] the air density was assumed constant. In [35,38] no information on the air density was given. Variation of thermal properties of air was considered in Comsol Multiphysics [24] and TRNSYS [38]. However, the review carried out shows that in most cases the authors used a fixed value for the air density and the specific heat of air. It was so in all types of tools, such as TRNSYS [19,39], Matlab [26,27] and Ansys [40,41]. There were also no studies comparing the assessment of these simplifications on outlet air temperature and resulting energy gain.

Therefore, this paper aims to present a detailed simulation study on the influence of changes in the air density and the specific heat depending on atmospheric conditions on a value of the temperature of outlet air from a heat exchanger. The method given in EN 16798-5-1 was used. To better illustrate the scale of the problem, based on the results obtained, the effect in the form of monthly and annual energy demand for heating and cooling of an example residential building was also determined. For this purpose, a well-known thermal resistance-capacitance model of a building zone given in EN ISO 13790 was used. As it is restricted only to sensible heating and cooling, the humidification and dehumidification processes of air were not considered here. Hence, the condensation in the buried pipe was also not analysed.

The following section presents the case building and climatic conditions in the considered location. Then the problem of thermal properties of air, in particular its density and specific heat capacity as a function of air temperature and atmospheric pressure are briefly presented. Moreover, the simulation models to obtain hourly outlet air temperature from EAHE and heating and cooling energy were presented. After them, the results are presented and compared with other studies, and concluding remarks are given.

2. Materials and Methods

2.1. Case Building

For further consideration, a single-story residential building was selected, located in Kraków (south Poland) in the third Polish climatic zone, following the subdivision from zone I to V given in the PN-EN 12831 standard [42] presented in Figure 1.

(a)

(b)

Figure 1. (**a**) Model of the test building; (**b**) location of the building.

The building was built with traditional technology. It has external walls made of 25 cm ceramic blocks, insulated with 10 cm of Styrofoam. A 30° gable roof is isolated with a 15 cm layer of mineral wool. The ground floor was insulated with a 10 cm layer of Styrofoam. Following Polish requirements [43,44] ventilation airflow was set at 90 m^3/h. The building has a total heated floor area of 65 m^2, a total volume of 163 m^3 and is inhabited by four people.

The mean annual temperature for the period 1971–2000 measured in the meteorological station Kraków-Balice and used to build a typical meteorological year (TMY) amounted to 8.20 °C and varied from 2.6 °C in February to 17.5 °C in July and August [45]. The hourly air temperature was from −20 °C on 16 February at 4:00 to 32.9 °C on 7 June at 12:00. Global horizontal solar irradiance was up to 998 W/m^2 on 17 May at 11:00 (Figure 2).

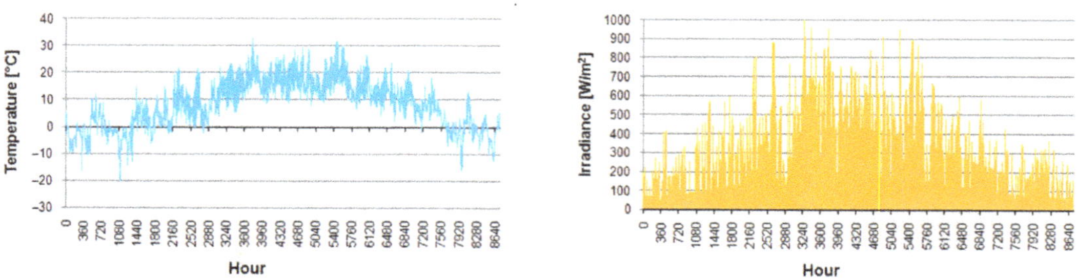

(a) (b)

Figure 2. TMY for Kraków: (**a**) outdoor air temperature; (**b**) global horizontal irradiance.

2.2. Earth-to-Air Heat Exchanger

As the design issues of EAHEs have been discussed by many researchers recently [23,24,46–49], these results were used to omit time-consuming preliminary considerations.

In the study [24] for Stockholm (Sweden) a 10 m long polyethylene pipe of 20 cm diameter, buried at 2 m was used for a volumetric airflow rate of 60 dm^3/s (216 m^3/h). For a residential building located in south Italy [50] authors considered 70 m long polypropylene pipes with a200 mm diameter buried at 1.50 m deep. In another case in a similar location [16] for an office building with a total floor area of 260 m^2 and ventilation airflow rate of 1300 m^3/h authors recommended EAHE from a pipe with 0.2 m diameter, between 80 m and 100 m long and buried between 2 m and 2.5 m. EAHE connected to a ventilation system of a single-family building in Poland presented in [51] consisted of two 25 m long parallel PVC 160/3.6 mm diameter pipes buried with increasing depth from 1.1 m up to 1.6 m at inlet and outlet, respectively. The ventilation airflow rate was 300 m^3/h. The next one [33,52] was located in northern Poland. For a residential building with a usable floor area of 115 m^2 and ventilation airflow rate of 150 m^3/h EAHE was built from a 41 m long pipe with 0.2 m diameter buried at an average depth of 2.12 m.

Based on these experiences and the assumed ventilation rate of 90 m^3/h, a 30 m-long pipe was chosen, with a diameter of0.2 m and the burial depth set at 2 m. The main parameters of EAHE are given in Table 1. Assuming pipe wall thickness of 8.8 mm the inner surface area of the duct is A_s = 17.20 m^2. Ground properties were taken from EN 16798-5-1.

Table 1. Parameters of the earth-to-air heat exchanger.

Parameter	Symbol	Value	Unit
Duct length	L	30	m
Air velocity in the duct	v	0.957	m/s
Outer pipe diameter	d_o	0.200	m
Inner pipe diameter	d_i	0.182	m
Average burial depth	z	2.0	m
Thermal conductivity of the duct	λ_{du}	0.27	W/m·K
Density of the ground	ρ_{gnd}	1800	kg/m^3
Specific heat of the ground	c_{gnd}	1300	J/kg·K
Thermal conductivity of the ground	λ_{gnd}	1.50	W/m·K

The base for the EAHE performance simulation is the calculation of the hourly ground temperature at the burial depth. For this purpose, the method given in EN 16798-5-1 was used. Simplification assumptions of that model were given and discussed in [35,53–55]. The most important are as follows:

- The ground is considered as a semi-infinite, homogenous and anisotropic medium,
- The sinusoidal variation of ground surface temperature is assumed,
- The model does not also take into account various environmental factors, as solar irradiance incident on a ground surface and periodic presence of snow cover on the ground.
- The impact of EAHE operation on the ground temperature was omitted.

The hourly ground temperature is given by the expression:

$$T_{gnd} = T_{e;mn;an} + (T_{e;max;m} - T_{e;mn;an}) \cdot e^{-\xi} \cdot \cos\left(2\pi \frac{t_{an}}{8760} - \xi - f_t\right). \quad (1)$$

The ξ factor is given by:

$$\xi = z \cdot \sqrt{\frac{\pi \cdot \varrho_{gnd} \cdot c_{gnd}}{\lambda_{gnd} \cdot 8760 \cdot 3600}}. \quad (2)$$

The time shift factor f_t is expressed by the formula:

$$f_t = \pi\left(\frac{2 \cdot t_{an;min}}{8760} + 1\right). \quad (3)$$

Finally, the difference between EAHE inlet and outlet air temperature (change in air temperature) can be computed:

$$\Delta T_{sup} = \left(T_{gnd} - T_e\right) \cdot \left[1 - e^{-\left(\frac{U_{du} \cdot A_S}{q_{v;sup} \cdot \varrho_a \cdot c_a}\right)}\right]. \quad (4)$$

Then, ventilation air at temperature:

$$T_{sup} = T_e + \Delta T_{sup} \quad (5)$$

is supplied to an interior of a considered building.

The overall heat transfer coefficient of the EAHE is given in the EN 16798-5-1 standard by the relationship:

$$U_{du} = \left(\frac{1}{2\pi} \frac{1}{\lambda_{du}} \cdot \ln \frac{\frac{d_o}{2}}{\frac{d_i}{2}} + \frac{1}{h_i}\right)^{-1}. \quad (6)$$

However, it should be pointed out here that this equation is incorrect. According to the heat transfer theory [56] the overall heat transfer coefficient for a cylindrical heat exchanger, assuming convection only on an internal surface is given per unit inner surface area of a pipe by the expression:

$$U_{du} = \left(\frac{1}{h_i} + \frac{r_i}{\lambda_{du}} \cdot \ln\frac{r_o}{r_i}\right)^{-1} = \left(\frac{1}{h_i} + \frac{d_i}{2\lambda_{du}} \cdot \ln\frac{\frac{r_o}{2}}{\frac{r_o}{2}}\right)^{-1}. \quad (7)$$

Finally:

$$U_{du} = \left(\frac{1}{h_i} + \frac{d_i}{2\lambda_{du}} \cdot \ln\frac{d_o}{d_i}\right)^{-1}. \quad (8)$$

Then, if all quantities in Equation (4) are expressed as given in "Symbols" section, i.e., U_{du} [W/m^2K], A_s [m^2], $q_{v;sup}$ [m^3/s], ρ_a [kg/m^3] and c_a [J/(kg·K)] the fraction in the power of e in Equation (4) is dimensionless. Physically both denominator and numerator are then heat fluxes.

The inside surface heat transfer coefficient is given by:

$$h_i = \left[4.13 + 0.23\frac{T_{md}}{100} - 0.0077\left(\frac{T_{md}}{100}\right)^2\right]\frac{v^{0.75}}{d_i^{0.25}}. \quad (9)$$

That standard indicates the possibility to set $T_{md} = T_e$ to avoid iterative computations when applying Equation (9).

2.3. The 5R1C Model

The EN ISO 13790 standard introduced the thermal network model of a building zone built from five resistors and one capacitor (Figure 3). It is intended for calculations of sensible energy use for space heating and cooling. Therefore, humidification and dehumidification of air are not considered here. Consequently, condensation processes in the buried pipe of EAHE were not analysed.

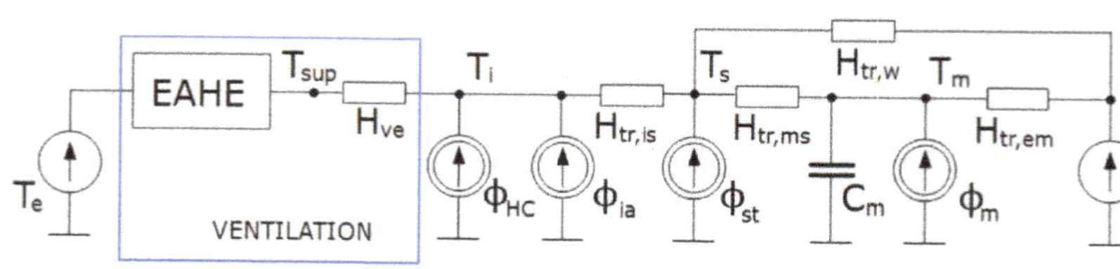

Figure 3. The circuit diagram of the 5R1C thermal network model of a building zone with an earth-to-air heat exchanger to supply ventilation air.

That model was chosen for several important reasons as ease modification for various applications [57–60], low computational requirements [61–63] and its simplicity and possibility to apply in a spreadsheet [64–66] not requiring specialised commercial simulation tools while producing reliable results [67].

External partitions of a building are divided into two categories. The thermally light elements (doors, windows, curtain walls, glazed walls, etc.) are described by the $H_{tr,w}$ thermal transmission coefficient. The thermally heavy elements (walls, ceilings) are included in $H_{tr,em}$ and $H_{tr,ms}$ thermal transmission coefficients, both connected to the thermal capacity, C_m, representing the thermal mass of the building [68]. It is the weak point of this model since this way all the thermal inertia of various elements of a considered zone is considered as a single capacitance.

The ambient environment is represented by the external air temperature (T_e). $H_{tr,is}$ is the coupling conductance between the central node (T_s) and an indoor environment (T_i).

Ventilation air, at temperature T_{sup}, is supplied to a building zone through the heat transfer by ventilation, H_{ve}. To include the operation of EAHE its inlet and outlet were connected to T_e and to T_{sup}, respectively. T_{sup} is calculated from Equation (5).

Heat fluxes from internal sources and solar radiation are split into three parts: φ_{ia}, φ_{st}, and φ_m, connected to the indoor air, the central, and the thermal mass nodes, respectively. Heating or cooling power (φ_{HC}) is supplied to or extracted from the indoor air node. The calculation procedure to obtain heating and cooling power to maintain the required set point temperatures is given in EN ISO 13790 and has been presented in detail recently [69,70].

That procedure requires T_{sup} to be calculated at each time step. In the case when EAHE is not used and external air is directly supplied to a building then:

$$T_{sup} = T_e \tag{10}$$

and ventilation heat flux is given by:

$$\phi_{ve} = H_{ve} \cdot \left(T_{sup} - T_i\right) = H_{ve} \cdot \left(T_e - T_i\right) \tag{11}$$

with the heat transfer by ventilation:

$$H_{ve} = \varrho_a \cdot c_a \cdot q_{v;sup}. \tag{12}$$

If EAHE is connected to the ventilation system, as shown in Figure 3, then an additional heat gain (or loss) is added to the heat exchanger during the flow of ambient air in the duct:

$$\phi_{EAHE} = \varrho_a \cdot c_a \cdot q_{v;sup} \cdot \Delta T_{sup}. \tag{13}$$

The supplying air temperature is calculated from Equation (5) and ventilation flux:

$$\phi_{ve} = \varrho_a \cdot c_a \cdot q_{v;sup} \cdot \left(T_e + \Delta T_{sup} - T_i\right). \tag{14}$$

As presented in Equations (12)–(14), apart from the temperature difference, the product of $\rho_a{\cdot}c_a$, called volumetric heat capacity of air, affects the calculated energy effect of ventilation and EAHE operation. Therefore, the dependence of these parameters on external conditions has to be carefully analysed and is presented in the next section.

Depending on the increase or decrease of the temperature of air passing the heat exchanger ($\Delta T_{sup} > 0$ or $\Delta T_{sup} < 0$), it is possible to talk about the heat or cold gain and to determine their monthly values. From this, including Equation (13), the monthly heat gain in EAHE in the m-th month is given by the expression:

$$\text{if } \Delta T_{sup} > 0 : Q_{gH,m} = \sum_{n=1}^{h} \phi_{EAHE,n} \cdot \Delta\tau_m. \tag{15}$$

Consequently, the monthly cooling energy gain in EAHE in the m-th month is given by the expression:

$$\text{if } \Delta T_{sup} < 0 : Q_{gC,m} = \sum_{n=1}^{h} \phi_{EAHE,n} \cdot \Delta\tau_m. \tag{16}$$

Thermal conductance and single capacitance of the RC network model (Figure 3) are given in Table 2. The thermal and physical properties of materials used were taken from manufacturers. Thermal resistances were calculated following ISO 6946 [71] Thermal bridges were neglected. Thermal capacities were calculated according to the detailed method of ISO 13786 [72] for the calculation period of 24 h. Solar absorptance of the roof and wall surfaces was: α = 0.8 and α = 0.6, respectively. Constant internal gains of 150 W throughout the year and continuous operation of the ventilation system were also assumed.

Table 2. Thermal network model elements of the building.

Element	Value	Unit
$H_{tr,w}$	10.00	W/K
$H_{tr,is}$	983.32	W/K
$H_{tr,ms}$	1490.13	W/K
$H_{tr,em}$	85.22	W/K
H_{ve}	30.00	W/K
C_m	23.56	MJ/K

EN ISO 13790 assumes constant volumetric heat capacity of air: $\rho_a c_a = 1200\ \mathrm{J/(m^3K)}$ as in other building simulations [52,68,73]. Hence, applying Equation (12) we get H_{ve} = 30.0 W/K. This parameter was the same in all simulations to highlight the influence of the variation of air parameters on the operation of the analysed exchanger and not the whole exchanger-building system.

2.4. Thermal Properties of Dry Air in Building Simulations

Since, as stated in the introductory section, humidification and dehumidification of air is omitted in the present study, air shall be considered a dry gas. Within the air temperature and atmospheric pressure variation met in climatic conditions of inhabited areas around the globe it is sufficient to describe the atmosphere by the perfect gas law [74] linking together the absolute temperature (T), absolute pressure (p) and volume (V) for the given mass (M) of gas with a given individual gas constant (R_i):

$$pV = MR_iT. \tag{17}$$

Rearranging the above the unknown density can be calculated:

$$\rho_a = \frac{M}{V} = \frac{p}{R_iT}. \tag{18}$$

According to the International Union of Pure and Applied Chemistry (IUPAC) [75], we assumed the universal gas constant R = 8.3144598 J/(mol·K) and molar mass of dry air μ = 28.96546 g/mol. Then, individual gas constant of dry air:

$$R_i = \frac{R}{\mu} = 287.047394\frac{J}{kgK}. \tag{19}$$

From this, at standard conditions defined by air temperature o T_0 = 273.15 K (0 °C) and barometric pressure of p_0 = 100 kPa at the sea level (altitude h = 0), we get:

$$\rho_{a,0} = \frac{p_0}{R_i \cdot T_0} = 1.2754\frac{kg}{m^3}. \tag{20}$$

In the same way, dry air density can be computed at a given hourly external (ambient) air temperature and pressure.

If the influence of variable atmospheric conditions on selected physical properties of air is to be considered, the most relevant independent variables and the ranges of their variability encountered in practice must be identified first. Taking into account WMO data [76] it seems reasonable to assume atmospheric pressure variation from 950 hPa to 1050 hPa and air temperature from −50 °C to +50 °C. The resulting air density variation for presented assumptions, calculated from Equation (18), is presented in Figure 4.

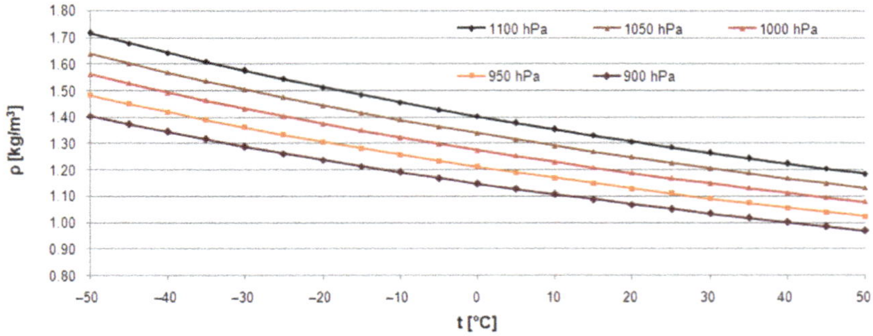

Figure 4. Dry air density variation as a function of its temperature and atmospheric pressure.

Taking a pressure and temperature variation ranges from 950 hPa to 1050 hPa and from −20 °C to +20 °C, respectively, the air density varies from 1.12896 kg/m^3 (950 hPa and +20 °C) to 1.44497 kg/m^3 at 1050 hPa and −20 °C. In relation to 1.27540 kg/m^3 at 0 °C and 1000 hPa it means the corresponding change of −13.0% and 13.3%.

If atmospheric pressure in a considered location is unknown then it can be calculated to the value at the known altitude from the barometric formula [77,78]. For ideal gas two forms of this equation exist [79]. The first one assumes the isothermal atmosphere, i.e., with a constant temperature independent on a height [80]:

$$T(h) = T_0 \tag{21}$$

and:

$$p(h) = p_0 \cdot \exp\left(-\frac{g \cdot h}{R_i \cdot T_0}\right). \tag{22}$$

Hence, the air density, $\rho_{a,h}$, at a given altitude, h, is given by:

$$\rho_{a,h} = \rho_{a,0} \cdot \exp\left(-\frac{g \cdot h}{R_i \cdot T_0}\right). \tag{23}$$

Standard conditions are used usually as reference. From this, assuming gravity acceleration g = 9.80665 m/s^2, we get from Equation (23):

$$\rho_{a,h} = 1.2754 \cdot \exp\left(\frac{-h}{7995.29}\right). \tag{24}$$

According to this equation the air density decreases from 1.27540 kg/m^3 to 1.19808 kg/m^3 (−6.1%), 1.12545 kg/m^3 (−11.8%), and 0.99133 kg/m^3 (−22.1%) at 500 m, 1000 m and 2000 m above sea level, respectively. However, since inhabited areas cover lands from 79 m below to 5516 m above sea level [81], these differences can be more significant when performing calculations for higher altitudes.

The second form of the barometric formula includes the lapse rate [82], i.e., the rate of decrease of temperature with a height. In the troposphere it amounts L = −6.51 K/km [83]. Then, in relation to sea level with h_0 = 0 we get [84]:

$$T(h) = T_0 + Lh, \tag{25}$$

and:

$$p(h) = p_0\left(\frac{T(h)}{T_0}\right)^{-\frac{\mu g}{RL}} = p_0\left(\frac{T(h)}{T_0}\right)^{5.248}. \tag{26}$$

Density at altitude h is given by:

$$\rho_{a,h} = \rho_{a,0}\left(1 + \frac{L \cdot h}{T_0}\right)^{\frac{\mu g}{RL} - 1}. \quad (27)$$

Inserting the aforementioned constants we get:

$$\rho_{a,h} = 1.2754\left(1 - \frac{0.00651 \cdot h}{273.15}\right)^{4.248}. \quad (28)$$

The EN 16798-5-1 standard recommends use of Equation (27) for $T_{a,ref}$ = 20 °C and air density at sea level $\rho_{a,ref}$ = 1.204 kg/m^3 at p = 101,325 Pa. Hence, the above relationship is given in that standard in the following form:

$$\rho_{a,h} = 1.204\left(1 - \frac{0.00651 \cdot h}{293}\right)^{4.255}. \quad (29)$$

Different values of the molar mass of air and universal gas constant were used in that standard which resulted in differences in the power (4.248 and 4.255) of expressions given by Equations (28) and (29).

According to Equation (29), for the given assumptions, air density decreases from 1.204 kg/m^3 to 1.14811 kg/m^3 (−4.6%), 1.09422 kg/m^3 (−9.1%), and 0.99227 kg/m^3 (−17.6%) at 500 m, 1000 m and 2000 m above sea level, respectively. Differences are lower than in the case of Equation (24) but still significant.

Application of Equation (23) at the temperature of 20 °C results in changes of density from 1.20413 kg/m^3 to 1.13113 kg/m^3, 1.06256 kg/m^3 and 0.96734 kg/m^3 in the same order. As the air temperature lapse rate is not included in this relationship, it produces results with the same percentage differences in relation to the sea level as for 0 °C for the same altitudes. A comparison of results from Equation (23) at 0 °C, Equation (23) at 20 °C and Equation (29) is given in Figure 5.

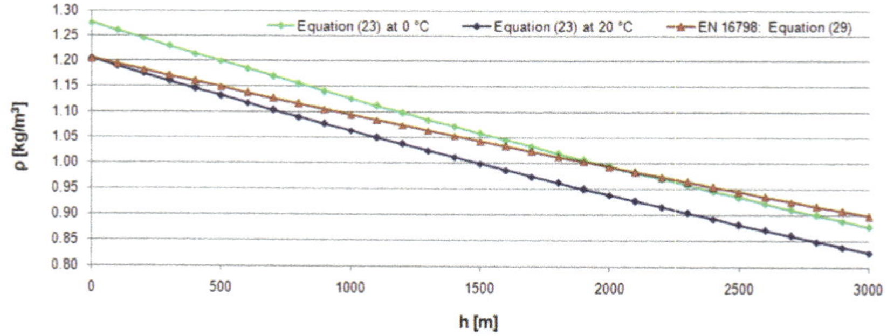

Figure 5. Dry air density variation as a function of altitude.

The impact of air temperature and pressure on the specific heat capacity of air in atmospheric conditions is less significant [85] (Figure 6). Following correlations reported by Lemmon et al. [86] for dry air as an ideal gas, its specific heat at p = 101,325 Pa varies from 1006.03 J/(kg K) at 220 K (−53.15 °C) to 1007.41 J/(kg K) at 320 K (46.85 °C) with the minimum of 1005.68 J/(kg K) from 230 K to 280 K (from −43.15 °C to 6.85 °C). According to the next reference [87] at p = 100 kPa it varies from 1006.1 J/(kg K) at −50 °C to 1007.7 J/(kg K) at 50 °C with the minimum of 1005.7 J/(kg K) at −20 °C. Hence, in the considered range, the maximum value of c_a is only 0.3% higher than the minimum in both cases, and for building energy simulation purposes its variation can be omitted using the constant c_a = 1006.0 K/(kg K) [37].

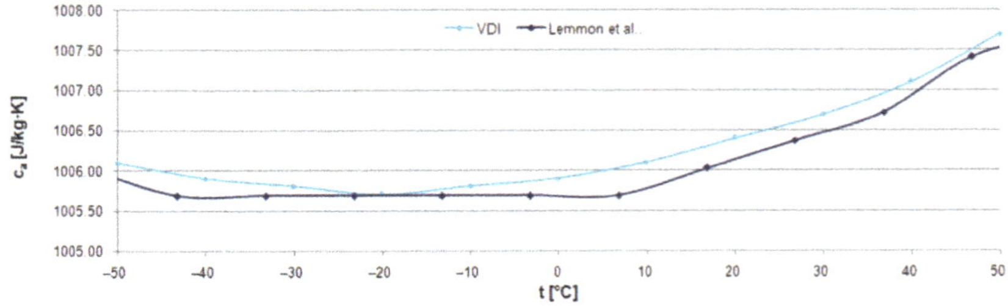

Figure 6. Specific heat of dry air as a function of temperature.

For comparison, it should be noted that the EN 16798-5-1 standard imposes for the specific heat of air use of the constant value c_a = 0.000279 kWh/(kg K) = 1004.4 J/(kg K), i.e., approximately 0.12% lower than the minimum values given.

3. Results and Discussion

3.1. Introduction

To present the problem of the air density variation with atmospheric conditions in the chosen location, several simulations were performed. They were based on the hourly data from the hourly typical meteorological year (TMY) for Kraków [45].

As described in [88] Polish TMYs do not include atmospheric pressure. Hence, it is necessary to determine the conditions for the meteorological station altitude using appropriate relationships. For comparison, the second source of meteorological data was also used. It was the International Weather for Energy Calculations (IWEC) weather data file for Kraków provided on the EnergyPlus website [89]. The IWEC weather files for over 200 locations around the world are freely provided by the U.S. Department of Energy [90,91].

It should be noted that air temperature in these two sources varies within a very similar range: from −20 °C to 32.9 °C and from −20.1 °C to 32.0 °C for TMY and IWEC files, respectively. Moreover, monthly values are very close (Figure 7a). However, hourly data differ more significantly (Figure 7b) as they are prepared from different source data.

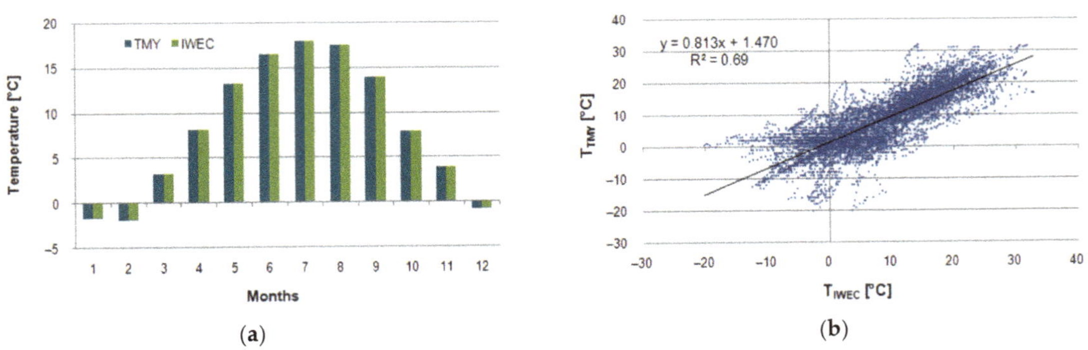

Figure 7. Air temperature in Kraków according to weather data from TMY and IWEC: (**a**) monthly; (**b**) hourly.

Then, the hourly operation of EAHE at the same location was simulated assuming five methods:

(1) Air density and specific heat of air from EN 16798-5-1, TMY;
(2) Constant volumetric heat capacity of air from EN ISO 13790, TMY;
(3) Air density from Equations (18) and (29) using the TMY data;
(4) Air density from Equation (18) using the IWEC data;

(5) Air density and specific heat of air from EN 16798-5-1, IWEC.

As the reference, the first case, i.e., the EN 16798-5-1 standard, was chosen as the recommended one for energy calculations of buildings in Poland. In the second case $\rho_a c_a$ = 1200 J/(m^{3}K) was used. In the third method, constant barometric pressure was obtained from Equation (29) and then varying air density was computed from Equation (18) using ambient air temperature from TMY. In the next method, as pressure and temperature are given in the IWEC file, air density was obtained from Equation (18). In the last method, air density was calculated following EN 16798-5-1 and the ambient air temperature was taken from the IWEC weather file. Hence, in relation to the previous case, there could be fully assessed the impact of the simplification introduced by that standard.

As the presented resistance-capacitance model of a building zone and the EAHE model are designed for calculations in hourly time steps they both were easily linked and simulated in a spreadsheet using hourly weather data for Kraków. This way no commercial program was necessary and an easily available tool was used.

3.2. Thermal Properties of Dry Air

At first, calculations following EN 16798-5-1 are presented. Meteorological station Kraków-Balice is located at the altitude of h = 237 m above sea level. Then from Equation (28) we get:

$$\rho_{a,h} = 1.204\left(1 - \frac{0.00651 \cdot 237}{293}\right)^{4.255} = 1.177254\frac{\text{kg}}{\text{m}^3}, \quad (30)$$

and resulting volumetric specific heat of air $\rho_a c_a$ = 1182.43 J/(m^{3}K). It is lower by about 1.5% from $\rho_a c_a$ = 1200 J/(m^{3}K) recommended by EN ISO 13790.

In the next step, assuming standard conditions given in Equation (18), from Equation (29) air pressure was calculated: p(h) = 97,071.05 Pa. Then, from Equation (18), hourly air density at the considered location, following air temperature variations, was obtained (Figure 8).

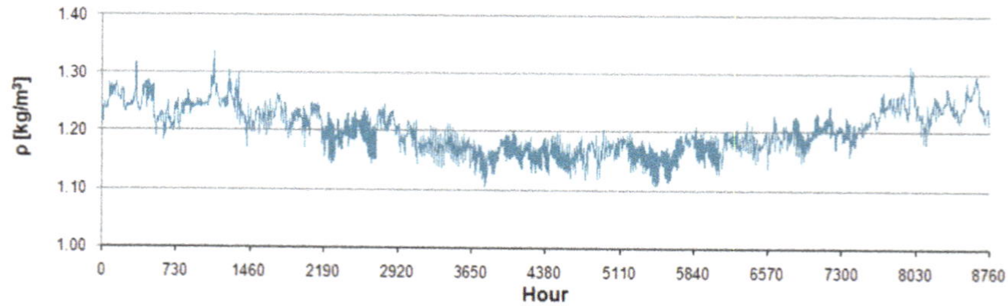

Figure 8. Calculated air density.

It varied from 1.10495 kg/m^3 on 7 June at 10:00 to 1.33691 kg/m^3 on 16 February at 4:00, i.e., from −8.1% to 11.1% in relation to the annual average of 1.20286 kg/m^3, which was only 2.1% higher than 1.177254 kg/m^3 calculated following EN 16798-5-1. The monthly average density was from 1.1610 kg/m^3 in June to 1.2504 kg/m^3 in February.

The air density calculated from the IWEC weather file varied from 1.12822 kg/m^3 on 8 July from 12:00 to 13.00 to 1.38083 kg/m^{3}on 2 February at 6:00, i.e., from −7.9% to 12.7% in relation to the annual average of 1.22518 kg/m^3 (Figure 9a). The monthly average density was from 1.1847 kg/m^3 in August to 1.2704 kg/m^3 in February. Hourly values differed from those of EN 16798-5-1 (Figure 9b).

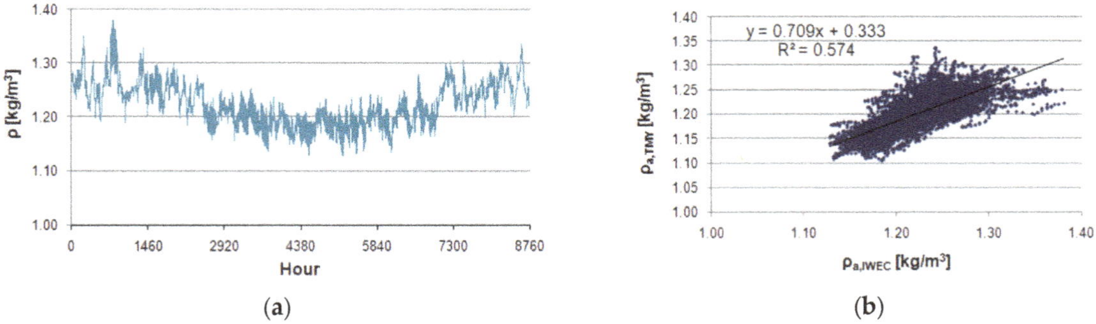

Figure 9. Hourly air density for Kraków calculated from the IWEC data: (**a**) during the year; (**b**) correlation of hourly values in relation to the results from EN 16798-5-1.

Volumetric heat capacity for EN 16798 (method 1) amounted to 1182.43 J/(m^{3}K). In the third method, the hourly value (dark green colour in Figure 10) was from 1111.58 J/(m^{3}K) to 1344.93 J/(m^{3}K) in comparison to the annual average of 1210.08 J/(m^{3}K). Monthly averaged values were from 1167.95 J/(m^{3}K) in June to 1257.87 J/(m^{3}K) in February.

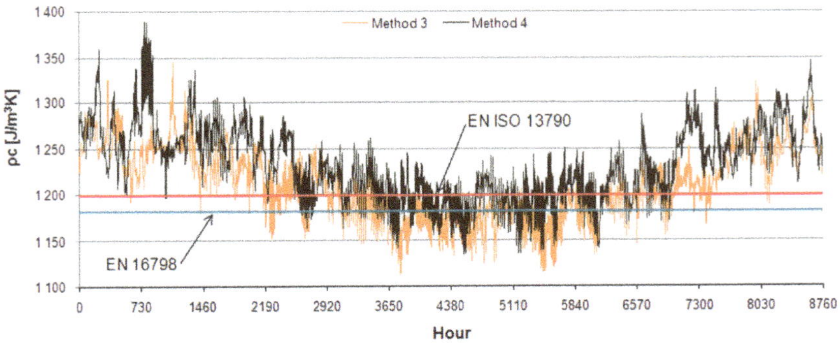

Figure 10. Volumetric specific heat.

The last considered method produced values from 1134.99 J/(m^{3}K) to 1389.12 J/(m^{3}K) with 1232.53 J/(m^{3}K) on average. Monthly averaged values were from 1191.85 J/(m^{3}K) in July to 1277.98 J/(m^{3}K) in February.

3.3. Performance of EAHE

To assess the effect of the considered methods on EAHE performance, we performed hourly simulations of the presented heat exchanger supplying ventilation air to the residential building.

As it could have been predicted from Equations (4) and (5), the influence of the density and specific heat variability on the temperature change in EAHE (ΔT_{sup}) and outlet air temperature from EAHE (T_{sup}) are negligible. These predictions were confirmed by the results of the calculations. In the following paragraphs, they are given in relation to the first method, following Section 3.1, as used in energy calculations in Poland.

Outlet air temperature (Figure 11), calculated according to method (1), in the order presented in Section 3.1, varied from 0.93 °C on 16 February at 6:00 to 15.35 °C on 16 August at 15:00 (Figure 9). For method (2), it was from 0.90 °C on 16 February at 6:00 to 15.37 °C on 16 August at 15:00. In the next case, it was from 0.59 °C on 16 February at 6:00 to 15.25 °C on 16 August at 15:00. According to the fourth method, it varied from 0.65 °C on 3 February at 6:00 to 15.05 °C on 15 August at 15:00. In the last case, similar values were obtained, i.e., from 1.09 °C (3 February at 6:00) to 15.12 °C (15 August at 15:00). This shows that the baseline temperature varied within the same limits taking extreme values on the same days.

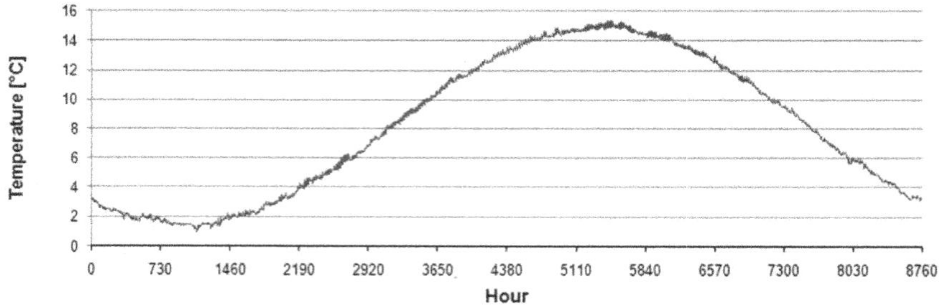

Figure 11. Outlet air temperature (T_{sup}) from EAHE.

Hourly differences between calculated outlet values were also negligible. In relation to method (1), they were from −0.03 °C to 0.03 °C, from −0.02 °C to 0.34 °C, from −0.72 °C to 1.04 °C, and from −0.72 °C to 0.73 °C, respectively. Therefore, strong correlations between them were obtained (Figures 12 and 13).

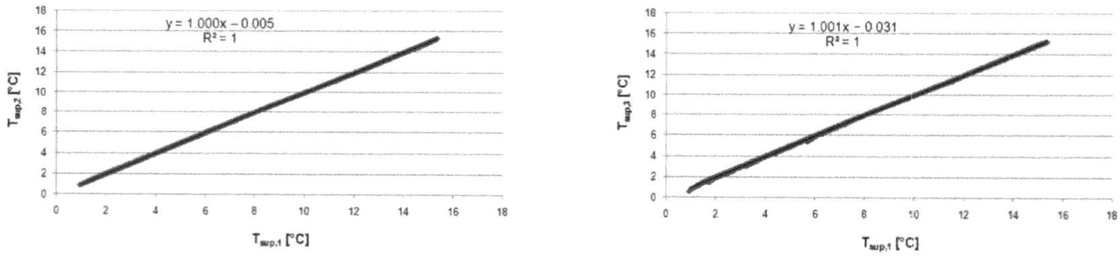

(a) (b)

Figure 12. Hourly outlet air temperature from EAHE from method (1) and: (**a**) method (2); (**b**) method (3).

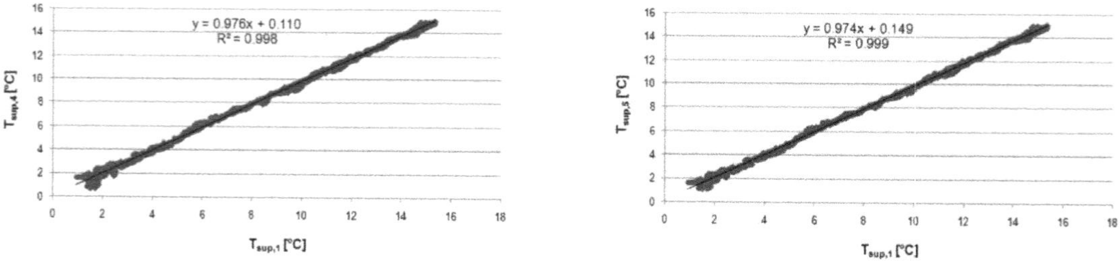

(a) (b)

Figure 13. Hourly outlet air temperature from EAHE from method (1) and: (**a**) method (4); (**b**) method (5).

For the subsequent methods hourly heat flux gain in EAHE varied following the temperature rise in EAHE (Equation (14)) and was from −631.4 W on 7 June at 12:00 to 638.4 W on 29 November at 6:00, from −639.8 W on 7 June at 12:00 to 646.9 W on 29 November at 6:00, from −597.2 W on 7 June at 12:00 to 704.2 W on 29 November at 6:00, from −546.4 W on 22 April at 15:00 to 721.2 W on 2 February at 6:00 and from −554.4 W on 22 April at 15:00 to 627.0 W on 2 February at 6:00.

Hourly differences in heating gain in relation to the first method were up to −8.4 W (20 December at 24:00), −74.4 W (16 February at 6:00), from −721.2 W (2 February at 6:00) to 595.6 W (16 February at 4:00) and from −627.0 W (2 February at 6:00) to 596.6 W (16 February at 5:00). In case of cooling these differences were up to 8.3 W, from −34.2 W (7 June at 12:00) to 3.8 W (1 February at 22:00), from −459.7 W (7 June at 12:00) to 483.1 W (4 August at 15:00), and from −461.0 W (7 June at 12:00) to 500.8 W (4 August at 15:00).

As it can be noticed the largest discrepancies were in the last case because of different meteorological datasets. However, from the point of air density variation the most interesting are two comparisons. The first is between methods 1 and 3, i.e., with constant air density and constant specific heat of air from EN 16798-5-1 and air density varying with ambient air temperature from TMY at constant pressure computed from the same standard (Figure 14). The second is a comparison between methods 5 and 4, both based on the IWEC datasets. These two comparisons should be considered the most important for the aim of this study. Thus, the hourly difference ina heating and cooling flux between the two last methods was up to −94.2 W (2 February at 6:00), and from 11.4 W (11 March at 12:00) to −18.2 W (8 July at 15:00), respectively (Figure 15).

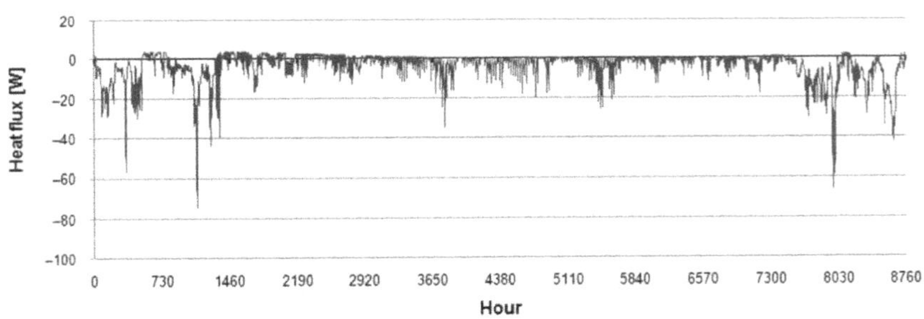

Figure 14. Hourly differences of calculated heat flux from EAHE between methods 1 and 3.

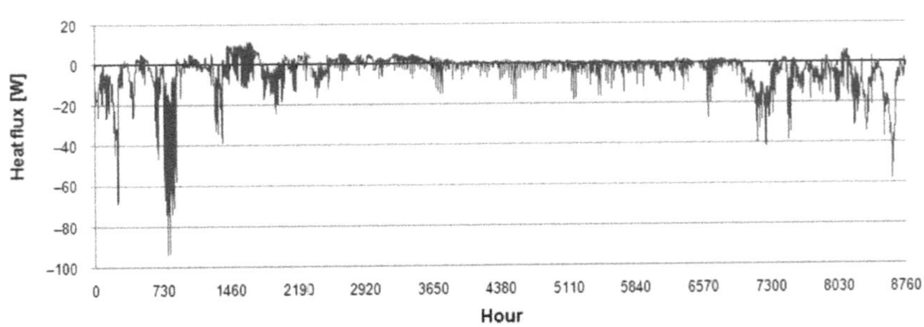

Figure 15. Hourly differences of calculated heat flux from EAHE between methods 5 and 4.

The most significant differences are observed during periods with low outdoor air temperature, between −10 °C and −20 °C, i.e., when ventilation heat loss is significant and heating needs are high. Hence, the design of EAHE in such conditions should include the change in air density with ambient temperature.

Negative differences between simplified and more detailed methods mean that the use of constant air parameters from EN 16798-5-1 results in an underestimation of the resulting heat flux. In the considered cases this effect is especially visible at low outdoor temperatures during the heating period.

In the study [33] for Polish conditions, it was also concluded that the model of EN 16798-5-1 underestimated heat gain and overestimated cooling load. In [34] authors compared the same model and measurements and they stated that annual heating and cooling energy from the model was lower by 20–30% and greater by 8–20% than that from measurements, respectively.

As far as monthly (see Equations (15) and (16)) and annual aggregated values are considered, for the first three methods close monthly gains both for heating and cooling were obtained (Figure 16). However, the application of the last method resulted in more significant discrepancies. Monthly heat gain varied from 1 kWh in June to 128.7 kWh in November, from 1 kWh in June to 130.4 kW in November, from 1 kWh in June to 136.0 kWh in November, from 3.9 kWh in May to 128.5 kWh in December and from 3.7 kWh in May to 119.1 kWh in December.

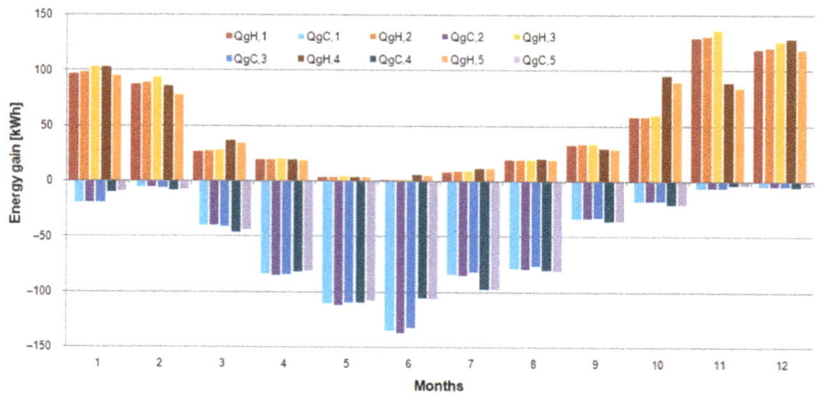

Figure 16. Monthly heating and cooling energy gain in EAHE from methods 1–5.

Nevertheless, annual heat gains were quite similar and amounted to 600.9 kWh, 608.8 kWh, 632.4 kWh, 628.3 kWh and 586.1 kWh. In relation to the internal surface area of EAHE, the annual unit heat was 34.9 kWh/m^2, 35.4 kWh/m^2, 36.8 kWh/m^2, 36.5 kWh/m^2 and 34.1 kWh/m^2.

In the case of cooling the situation was quite similar. Monthly cooling energy gain varied from −4.4 kWh in December to −135.0 kWh in June, from −4.5 kWh in December to −136.7 kW in June, from −4.5 kWh in December to 132.3 kWh in June, from −3.6 kWh in November to −109.5 kWh in May and from −3.5 kWh in November to −108.2 kWh in May. Annual gains amounted to −616.9 kWh, −625.0 kWh, −610.6 kWh, −603.5 kWh and −600.1 kWh. In relation to the internal surface area of EAHE, the annual unit heat was −35.9 kWh/m^2, −36.3 kWh/m^2, −35.5 kWh/m^2, −35.1 kWh/m^2 and −34.9 kWh/m^2.

Simulation study for Swedish conditions [24] with a 10 m long polyethylene pipe of 20 cm diameter, buried at 2 m resulted in the estimated annual energy saving at 525 kWh and 300 kWh for heating and cooling season, respectively. It means the average gain per unit surface area of EAHE at 83.6 kWh/m^2 and 47.8 kWh/m^2.

Łuczak et al. [92] analysed various configurations of EAHE of EAHE connected with an air conditioning system in Polish conditions using the Rehau company program for the selection of ground heat exchangers. Unit heat gain per internal surface area of EAHE during the heating period (from 1 September to 31 March) was 40.7 kWh/m^2, 35.2, kWh/m^2, 34.2 kWh/m^2, 34.1 kWh/m^2 for the Tichelmanns' system (200 mm diameter, 1330 m and 182 m long), for the meandering system (315 mm diameter and 120 m long) and for the ring system ((315 mm diameter and 120 m long), respectively.

Simulations of EAHE (5 m long pipe with a diameter of 200 mm and airflow rate of 150 m^3/h) located in southern Italy with ANSYS Fluent, presented by Congedo et al. [93], resulted in a total monthly heat gain was from 8.95 kWh in February to 174.70 kWh in November and from −26.26 kWh in August to −167.67 kWh in July.

The experimental study of Skotnicka et al. [94] contains valuable comparative results from a laboratory experiment with EAHE conducted from 1 July to 30 September 2016 in Olsztyn (north Poland). To build EAHE a 41 m long polypropylene pipe with 0.2 m diameter was used. The maximum measured and calculated hourly heating and cooling gain was 0.29 kWh, 0.73 kWh, −0.16 kWh and −0.65 kWh, respectively.

Monthly measured heat gain was 37.24 kWh, 62.42 kWh and 82.41 kWh in July, August and September, respectively. Consequently, it means monthly unit heat gain per internal surface area of EAHE at 1.4 kWh/m^2, 2.4 kWh/m^2 and 3.7 kWh/m^2. Total and unit gain for this period was 182.07 kWh and 7.07 kWh/m^2, respectively. Calculated monthly heat gains were similar to measured values and amounted for the same months to 44.00 kWh, 58.27 kWh and 95.22 kWh. Hence, unit gains were 1.7 kWh/m^2, 2.3 kWh/m^2 and 3.7 kWh/m^2.

Measured monthly cooling energy in the same months were−6.42 kWh, −4.45 kWh and −1.35 kWh with unit values of 0.25 kWh/m^2, 0.17 kWh/m^2 and 0.05 kWh/m^2. Simulated values were several times greater: −57.58 kWh, −79.92 kWh and −88.32 kWh with unit energy of 2.2 kWh/m^2, 3.1 kWh/m^2 and 3.42 kWh/m^2.

In the next study [34] authors experimentally analysed the hourly performance of EAHE built from the 35 m long pipe with the external and internal diameters of 200 and 185 mm, respectively. Measured annual heating and cooling energy was 1761 kWh and 1152 kWh in 2013 and 1410 kWh and 1063 kWh in 2014, respectively. It leads to energy gains per unit area of 86.6 kWh/m^2, 56.6 kWh/m^2, 69.3 kWh/m^2 and 52.3 kWh/m^2, respectively. Differences between measured and values calculated according to EN 16798-5-1 of heating, cooling and seasonal energy gain from EAHE were up to 49%, 59% and 79%, respectively.

When analysing the presented outcomes, there should be remembered that humidification and dehumidification processes inside the EAHE were ignored. This simplification is commonly included by other researchers for convenience [23,26,47–49].

However, Niu et al. [95] developed in Matlab the polynomial regression model for predicting the cooling capacity of EAHE including total, sensible and latent cooling capacity as a function of the air temperature, the air relative humidity, the air velocity at the inlet of EAHE, the tube surface temperature and the tube length and diameter.

The presented results showed that the air temperature along EAHE was independent of inlet air humidity. They also showed that in the case of low inlet air relative humidity, below 40%, no condensation along with the length of the tube occurred. Their study was devoted, however, only to cooling demand. In Polish conditions, where heating prevails, the situation can be different.

3.4. Energy Performance of the Building

To estimate the impact of EAHE operation on energy use for space heating and cooling of the considered exemplary residential building the necessary simulations were performed. To estimate the obtained energy effect, the base case was simulated without EAHE at first. Then, the following four cases with EAHE were considered.

In comparison to the base case ($Q_{H,0}$ and $Q_{C,0}$ in Figure 17), the introduction of EAHE resulted in visible energy savings in the considered case regardless of the chosen calculation method. It is especially clear in the months with dominant heating or cooling conditions. However, in varying ambient conditions (March, April, September, and October) and the thermal demand of the building changing during the day from heating to cooling and vice versa, the effect is not so obvious. Then it seems to be more beneficial to introduce a bypass to switch the intake of ventilation air between EAHE and outdoor air [35,50,51,96].

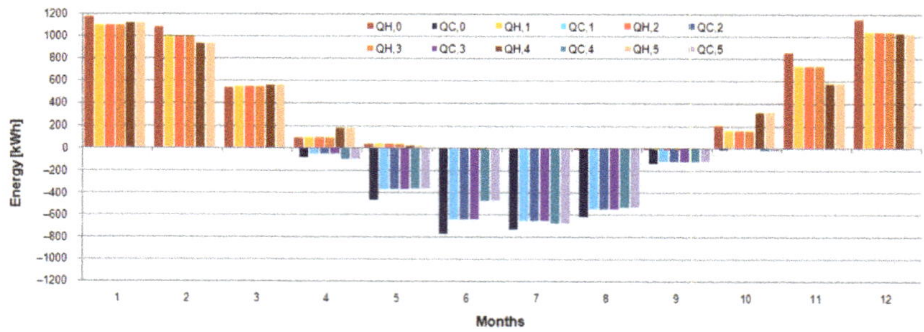

Figure 17. Monthly energy use for space heating and cooling of the building.

Annual heating and cooling energy decreased from 5109 kWh and 2792 kWh in the base case to 4667 kWh and 2364 kWh, 4667 kWh and 2365 kWh, 4670 kWh and 2363, 4724 kWh and 2261 kWh, and 4719 kWh and 2261 kWh according to the successive methods, respectively. Hence, EAHE was responsible for the average annual reduction of 8.8% in heating and 15.3% in cooling energy in the first three methods. For the last one, reductions of 7.5% and 19.0% were obtained.

Figure 17 shows differences in calculated annual energy for space heating and cooling when using different weather datasets. For this reason, the weather files for energy simulations of buildings, especially in the case of energy certification, should be properly prepared. Moreover, atmospheric pressure should be included in Polish typical meteorological years.

Simulations in the Swedish climate [24] resulted in a 5% reduction for heating and a 50% reduction for cooling. In the case of a low-energy building located in Polish climatic conditions [35] EAHE reduced annual simulated heating and cooling needs by 13% and 43%, respectively. Results of long-term measurements of EAHE supplying ventilation air in a residential building located in north Poland presented in [33] showed annual savings of around 20% and 3% for heating and cooling energy, respectively.

The use of EAHE also improved the performance of the thermal energy source. Maximum heating and cooling power decreased significantly (Table 3). Only in the last case, it increase compared to the base case. It was caused by lower outdoor temperatures, especially on 2 February, when it varied from −20.1 °C at 7:00 to −4 °C at 14:00. The peak values of heating power in the remaining cases were noticed on 26 December when the outdoor air temperature was from −9.5 °C at 12:00 to −12.5 °C at 9:00. Moreover, unnecessary ventilation loss was reduced resulting in lower thermal demand for the building.

Table 3. Maximum hourly heating and cooling power.

Parameter	Base Case	1	2	3	4	5	Unit
$\Phi_{H,max}$	3104	2471	2471	2480	3672	3658	W
$\Phi_{C,max}$	−3657	−3021	−3022	−3017	−3026	−3026	W
$\Phi_{ve,max}$	207	0	0	0	0	0	W
$\Phi_{ve,min}$	−1206	−644	−644	−644	−619	−620	W
τ_H	3952	3997	3996	3996	3936	3936	h
τ_C	2415	2365	2365	2365	2342	2342	h

The next important aspect of this solution is the annual operation time of heating (τ_H) and cooling (τ_C). It was reduced in the cooling mode in all cases. In heating, it slightly increased, but the share of hourly power above 2000 W dropped rapidly (Figures 18 and 19).

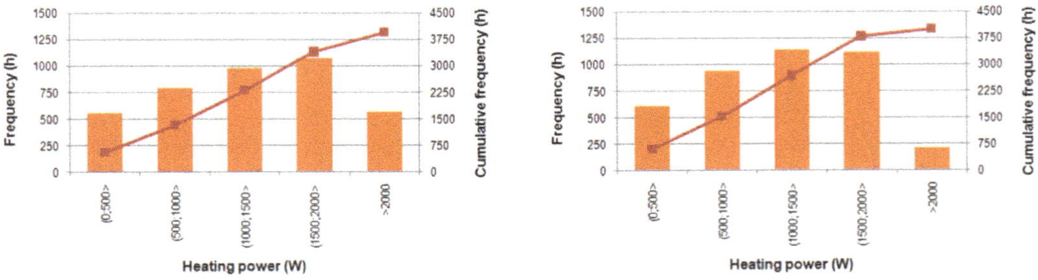

(a) (b)

Figure 18. Histograms of heating power: (**a**) the base case; (**b**) method 1.

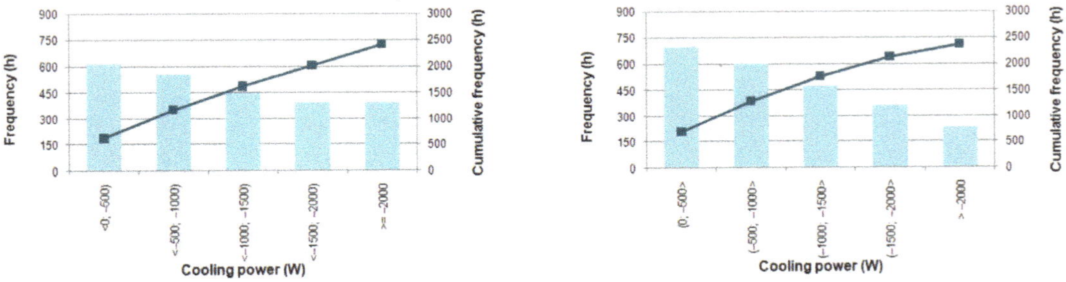

(a) (b)

Figure 19. Histograms of cooling power: (**a**) the base case; (**b**) method 1.

Presented histograms confirm a more stable operation of the heating and cooling source. In relation to the base case operation with heating and cooling power above 2000 W changed from 567 h to 210 h and from 396 h to 232 h. At the same time, power usage below 500 W increased: from 554 h to 611 h and from 618 h to 695 h.

4. Conclusions

In this study, the impact of air density variation with atmospheric conditions on the operation of the earth-to-air heat exchanger coupled with the ventilation system of the residential building in hourly time steps was studied.

As the Polish hourly typical meteorological years, available for 61 locations and intended for energy simulations of buildings, do not contain atmospheric pressure, the analysis took into account air density variation with ambient air temperature. The barometric pressure was computed as constant in relation to the reference conditions at sea level using the barometric formula at the altitude of the meteorological station. Therefore, air density variation analysis is limited only to one factor influencing its value—air temperature. In such a case, the density variation with temperature from the relevant weather files at constant pressure in comparison to the method of EN 16798-5-1 resulted in hourly differences in outlet air temperature up to 0.34 °C which can be treated as insignificant. Despite this, hourly differences in energy gain per unit area of the heat exchanger were up to 4.3 W/m^2 and 2.0 W/m^2 for heating and cooling, respectively.

For comparative purposes also the IWEC file was also used. However, as IWEC files are available only for four locations in Poland they are not used in practice. Instead, Polish TMYs are in common use as prepared for energy calculations of buildings both in monthly and hourly resolution. In this case, where both hourly ambient air temperature and atmospheric pressure were given, hourly differences in relation to the method using the same weather data but constant air properties from EN 16798-5-1 were up to 5.5 W/m^2 and 1.1 W/m^2 for heating and cooling, respectively.

These values seem not to be large, but analysed EAHE was rather small. So, in the case of larger objects hourly heating and cooling loads can differ more significantly. It was confirmed by comparisons of simplified methods with more detailed. In general, the model from EN 16798-5-1 underestimated heat gain and overestimated cooling load.

The presented considerations also confirmed the impact of the weather dataset used in calculations on monthly and annual energy gain from EAHE. The second one amounted to 600.9 kWh and 628.3 kWh for heating and 616.9 kWh and 603.5 kWh in cooling for the first and fourth methods (TMY and IWEC files), respectively. However, within the same weather data files it occurred that aggregated annual values of energy gain from EAHE were not significantly affected by simplifications regarding thermal properties of air, introduced both by EN 16798-5-1 or EN ISO 13790. As the influence of varying air density on ventilation loss and energy demand for heating and cooling of the building was not considered here, it may be analysed in a separate study.

The application of EAHE significantly decreased the energy demand for the heating and cooling of the building, by 8.2% and 16.8% on average, respectively.

As IWEC files are not used in energy audits or certification of buildings, it seems reasonable to introduce atmospheric pressure in typical Polish meteorological years. It would allow assessing the impact of air density variation on the EAHE performance.

This study shows the need to include atmospheric pressure in typical Polish meteorological years. Similarly, as in the case of heat load calculation following PN EN 12831, the design conditions for EAHE dimensioning in Polish climatic conditions could also be developed. It also indicates the need for future consideration about an assessment of air humidity impact on the simulated EAHE performance.

Funding: This research received no external funding.

Institutional Review Board Statement: Not applicable.

Informed Consent Statement: Not applicable.

Data Availability Statement: Not applicable.

Conflicts of Interest: The author declares no conflict of interest.

Symbols and Abbreviations

A_s	inner surface of the ground heat exchanger, m^2
C_m	thermal capacity of the building, J/K
E_A	unit heating and cooling energy use per unit area of a building, kWh/m^2
$H_{tr,em}$	external part of the $H_{tr,op}$ thermal transmission coefficient, W/K
$H_{tr,is}$	coupling conductance, W/K
$H_{tr,ms}$	internal part of the $H_{tr,op}$ thermal transmission coefficient, W/K
$H_{tr,op}$	thermal transmission coefficient for thermally heavy envelope elements, W/K
$H_{tr,w}$	thermal transmission coefficient for thermally light envelope elements, W/K
H_{ve}	thermal transmission coefficient by ventilation air, W/K
L	length of the duct, m
Q_C	annual energy use for space cooling, kWh
$Q_{gC,m}$	cooling energy gain in EAHE in m-th month, Wh
$Q_{gH,m}$	heating energy gain in EAHE in m-th month, Wh
Q_H	annual energy use for space heating, kWh
Q_{HC}	annual energy use for space heating and cooling, kWh
ΔQ_{HC}	percentage savings in annual energy use for space heating and cooling, %
T_e	external (outdoor) air temperature, °C
$T_{e;mn;an}$	mean annual temperature of outdoor air, °C
$T_{e;max;m}$	maximum mean monthly temperature of outdoor air, °C
T_{gnd}	hourly ground temperature, °C

T_i	internal (indoor) air temperature, °C
$T_{i,C,set}$	set-point indoor air temperature for cooling, °C
$T_{i,H,set}$	set-point indoor air temperature for heating, °C
T_m	thermal mass node temperature, °C
T_{md}	average air temperature in the duct, °C
T_s	central node temperature, °C
T_{sup}	supply air temperature, °C
c_a	specific heat of air at constant pressure, J/(kg·K)
c_{gnd}	specific heat of the ground, J/(kg·K)
d_i	inner diameter of the duct, m
d_o	outer diameter of the duct, m
f_t	time shift factor,
h_i	inside surface heat transfer coefficient, W/m^{2}K
h_m	number of hours in m-th month,
$q_{v;sup}$	volumetric airflow rate, m^3/s
t_{an}	annual hour, with t_{an} = 0 h at the beginning of the year; h
$t_{an;min}$	the time of the year when the monthly mean outdoor temperature is minimal, h
v	air velocity in the duct, m/s
z	burial depth of the duct, m
$\Delta\tau_m$	length of the month m, h
λ_{gnd}	thermal conductivity of the ground, W/(m·K)
λ_{du}	thermal conductivity of the duct, W/(m·K)
ξ	damping factor,
ρ_a	air density, kg/m^3
ρ_{gnd}	ground density, kg/m^3
φ_{int}	heat flow rate due to internal heat sources, W
φ_{sol}	heat flow rate due to solar heat sources, W
φ_{ia}	heat flow rate to internal air node, W
φ_{st}	heat flow rate to central node, W
φ_m	heat flow rate to mass node, W
φ_{ve}	heat flow rate by ventilation, W
φ_{EAHE}	heat flow rate from EAHE, W
φ_C	cooling power supplied to or extracted from the indoor air node, W
φ_H	heating power supplied to or extracted from the indoor air node, W
φ_{HC}	heating or cooling power supplied to or extracted from the indoor air node, W

References

1. Buildings Performance Institute Europe. Europe's Buildings under the Microscope. 2011. Available online: https://www.bpie.eu/wp-content/uploads/2015/10/HR_EU_B_under_microscope_study.pdf (accessed on 9 March 2022).
2. Kesten, D.; Tereci, A.; Strzalka, A.M.; Eicker, U. A method to quantify the energy performance in urban quarters. *HVACR Res.* **2012**, *18*, 100–111. [CrossRef]
3. Aksamija, A. Design methods for sustainable, high performance building facades. *Adv. Build. Energy Res.* **2016**, *10*, 240–262. [CrossRef]
4. Kneifel, J.; O'Rear, E. Reducing the impacts of weather variability on long-term building energy performance by adopting energy-efficient measures and systems: A case study. *J. Build. Perform. Simul.* **2017**, *10*, 58–71. [CrossRef]
5. Witkowska, A.; Krawczyk, D.A.; Rodero, A. Investment Costs of Heating in Poland and Spain—A Case Study. *Proceedings* **2019**, *16*, 40. [CrossRef]
6. Chwieduk, D.; Chwieduk, M. Determination of the Energy Performance of a Solar Low Energy House with Regard to Aspects of Energy Efficiency and Smartness of the House. *Energies* **2020**, *13*, 3232. [CrossRef]
7. Ballarini, I.; Costantino, A.; Fabrizio, E.; Corrado, V. A Methodology to Investigate the Deviations between Simple and Detailed Dynamic Methods for the Building Energy Performance Assessment. *Energies* **2020**, *13*, 6217. [CrossRef]
8. Mardiana-Idayu, A.; Riffat, S.B. Review on heat recovery technologies for building applications. *Renew. Sustain. Energy Rev.* **2012**, *16*, 1241–1255. [CrossRef]
9. Xu, Q.; Riffat, S.; Zhang, S. Review of Heat Recovery Technologies for Building Applications. *Energies* **2019**, *12*, 1285. [CrossRef]
10. Abbaspour-Fard, M.H.; Gholami, A.; Khojastehpour, M. Evaluation of an earth-to-air heat exchanger for the north-east of Iran with semi-arid climate. *Int. J. Green Energy* **2011**, *8*, 499–510. [CrossRef]
11. Tan, L.; Love, J.A. A Literature Review on Heating of Ventilation Air with Large Diameter Earth Tubes in Cold Climates. *Energies* **2013**, *6*, 3734–3743. [CrossRef]

12. Greco, A.; Masselli, C. The Optimization of the Thermal Performances of an Earth to Air Heat Exchanger for an Air Conditioning System: A Numerical Study. *Energies* **2020**, *13*, 6414. [CrossRef]
13. Pakari, A.; Ghani, S. Energy Savings Resulting from Using a Near-Surface Earth-to-Air Heat Exchanger for Precooling in Hot Desert Climates. *Energies* **2021**, *14*, 8044. [CrossRef]
14. Peña, S.A.P.; Ibarra, J.E.J. Potential Applicability of Earth to Air Heat Exchanger for Cooling in a Colombian Tropical Weather. *Buildings* **2021**, *11*, 219. [CrossRef]
15. D'Agostino, D.; Marino, C.; Minichiello, F. Earth-to-Air Versus Air-to-Air Heat Exchangers: A Numerical Study on the Ener-getic, Economic, and Environmental Performances for Italian Office Buildings. *Heat Transf. Eng.* **2020**, *41*, 1040–1051. [CrossRef]
16. D'Agostino, D.; Esposito, F.; Greco, A.; Masselli, C.; Minichiello, F. Parametric Analysis on an Earth-to-Air Heat Exchanger Employed in an Air Conditioning System. *Energies* **2020**, *13*, 2925. [CrossRef]
17. D'Agostino, D.; Esposito, F.; Greco, A.; Masselli, C.; Minichiello, F. The Energy Performances of a Ground-to-Air Heat Exchanger: A Comparison among Köppen Climatic Areas. *Energies* **2020**, *13*, 2895. [CrossRef]
18. Baglivo, C.; Bonuso, S.; Congedo, P.M. Performance Analysis of Air Cooled Heat Pump Coupled with Horizontal Air Ground Heat Exchanger in the Mediterranean Climate. *Energies* **2018**, *11*, 2704. [CrossRef]
19. Baglivo, C.; D'Agostino, D.; Congedo, P.M. Design of a Ventilation System Coupled with a Horizontal Air-Ground Heat Exchanger (HAGHE) for a Residential Building in a Warm Climate. *Energies* **2018**, *11*, 2122. [CrossRef]
20. Ahmed, S.F.; Khan, M.M.K.; Amanullah, M.T.O.; Rasul, M.G.; Hassan, N.M.S. Performance assessment of earth pipe cooling system for low energy buildings in a subtropical climate. *Energy Convers. Manag.* **2015**, *106*, 815–825. [CrossRef]
21. Ahmed, S.F.; Amanullah, M.T.O.; Khan, M.M.K.; Rasul, M.G.; Hassan, N.M.S. Parametric study on thermal performance of horizontal earth pipe cooling system in summer. *Energy Convers. Manag.* **2016**, *114*, 324–337. [CrossRef]
22. Cao, S.; Li, F.; Li, X.; Yang, B. Feasibility analysys of earth-air heat exchanger (EAHE) in a sports and culture centre in Tianjin, China. *Case Stud. Therm. Eng.* **2021**, *26*, 101654. [CrossRef]
23. Bisoniya, T.S.; Kumar, A.; Baredar, P. Energy metrics of earth–air heat exchanger system for hot and dry climatic conditions of India. *Energy Build.* **2015**, *86*, 214–221. [CrossRef]
24. Havtun, H.; Törnqvist, C. Reducing Ventilation Energy Demand by Using Air-to-Earth Heat Exchangers. Part 1–Parametric Study. In *Sustainability in Energy and Buildings*; Hakansson, A., Höjer, M., Howlett, R., Jain, L., Eds.; Smart Innovation, Systems and Technologies; Springer: Berlin/Heidelberg, Germany, 2013; Volume 22. [CrossRef]
25. Ahmadi, S.; Shahrestani, M.I.; Sayadian, S.; Poshitri, A.H. Performance analysis of an integrated cooling system consisted of earth-to-air heat exchanger (EAHE) and water spray channel. *J. Therm. Anal. Calorim.* **2021**, *143*, 473–483. [CrossRef]
26. Akbarpoor, A.M.; Poshtiri, A.H.; Biglari, F. Performance analysis of domed roof integrated with earth-to-air heat exchanger system to meet thermal comfort conditions in buildings. *Renew. Energy* **2021**, *168*, 1265–1293. [CrossRef]
27. Zhang, C.; Xiao, F.; Wang, J. Design optimization of multi-functional building envelope for thermal insulation and exhaust air heat recovery in different climates. *J. Build. Eng.* **2021**, *43*, 103151. [CrossRef]
28. Molcrette, V.F.A.; Autier, V.R.B. New expression to calculate quantity of recovered heat in the earth-pipe-air heat-exchanger operating in winter heating mode. *Arch. Thermodyn.* **2020**, *41*, 103–117. [CrossRef]
29. Hong, T.; Buhl, F.; Haves, P.; Selkowitz, S.; Wetter, M. Comparing computer run time of building simulation programs. *Build. Simul.* **2008**, *1*, 210–213. [CrossRef]
30. Piana, E.A.; Grassi, B.; Socal, L. A Standard-Based Method to Simulate the Behavior of Thermal Solar Systems with a Stratified Storage Tank. *Energies* **2020**, *13*, 266. [CrossRef]
31. Patrčević, F.; Dović, D.; Horvat, I.; Filipović, P. A Novel Dynamic Approach to Cost-Optimal Energy Performance Calculations of a Solar Hot Water System in an nZEB Multi-Apartment Building. *Energies* **2022**, *15*, 509. [CrossRef]
32. *EN ISO 16798-5-1*; Energy Performance of Buildings. Ventilation for Buildings Calculation Methods for Energy Requirements of Ventilation and Air Conditioning Systems (Modules M5-6, M5-8, M6-5, M6-8, M7-5, M7-8). Method 1: Distribution and Generation. International Organization for Standardization: Geneva, Switzerland, 2017.
33. Skotnicka-Siepsiak, A. Operation of a Tube GAHE in Northeastern Poland in Spring and Summer—A Comparison of Real-World Data with Mathematically Modeled Data. *Energies* **2020**, *13*, 1778. [CrossRef]
34. Brata, S.; Tănasă, C.; Dan, D.; Stoian, V.; Doboși, I.S.; Brata, S. Energy potential of a ground-air heat exchanger–measurements and computational models. *Sci. Technol. Built Environ.* **2021**, *28*, 84–93. [CrossRef]
35. Michalak, P. Hourly Simulation of an Earth-to-Air Heat Exchanger in a Low-Energy Residential Building. *Energies* **2022**, *15*, 1898. [CrossRef]
36. *EN ISO 13790*; Energy Performance of Buildings—Calculation of Energy Use for Space Heating and Cooling. International Organization for Standardization: Geneva, Switzerland, 2009.
37. Michalak, P. Ventilation heat loss in a multifamily building under varying air density. *J. Mech. Energy Eng.* **2020**, *4*, 97–102. [CrossRef]
38. Bouhess, H.; Hamdi, H.; Benhamou, B.; Bennouna, A.; Hollmuller, P.; Limam, K. Dynamic simulation of an earth-to-air heat exchanger connected to a villa type house in Marrakech. In Proceedings of the 13th Conference of International Building Performance Simulation Association, Chambéry, France, 25–28 August 2013.

39. Ghaith, F.A.; Razzaq, H.U. Thermal performance of earth-air heat exchanger systems for cooling applications in residential buildings. In Proceedings of the ASME International Mechanical Engineering Congress and Exposition, Pittsburgh, PA, USA, 9–15 November 2018; pp. 1–12. [CrossRef]
40. Badescu, V.; Isvoranu, D. Pneumatic and thermal design procedure and analysis of earth-to-air heat exchangers of registry type. *Appl. Energy* **2011**, *88*, 1266–1280. [CrossRef]
41. Serageldin, A.A.; Abdelrahman, A.K.; Ookawara, S. Parametric study and optimization of a solar chimney passive ventilation system coupled with an earth-to-air heat exchanger. *Sustain. Energy Technol. Assess.* **2018**, *30*, 263–278. [CrossRef]
42. *PN-EN 12831-1:2017*; Energy Performance of Buildings–Method for Calculation of the Design Heat Load-Part1: Space Heating Load. Polish Committee for Standardization: Warsaw, Poland, 2017.
43. Ferdyn-Grygierek, J.; Baranowski, A.; Blaszczok, M.; Kaczmarczyk, J. Thermal Diagnostics of Natural Ventilation in Buildings: An Integrated Approach. *Energies* **2019**, *12*, 4556. [CrossRef]
44. Sowa, J.; Mijakowski, M. Humidity-Sensitive, Demand-Controlled Ventilation Applied to Multiunit Residential Building—Performance and Energy Consumption in Dfb Continental Climate. *Energies* **2020**, *13*, 6669. [CrossRef]
45. Ministry of Investment and Development. Data for Energy Calculations of Buildings. Typical Meteorological Years and Statistical Climatic Data for Energy Calculations of Buildings. Available online: https://www.gov.pl/web/archiwum-inwestycje-rozwoj/dane-do-obliczen-energetycznych-budynkow (accessed on 9 March 2022).
46. Pfafferott, J. Evaluation of earth-to-air heat exchangers with a standardised method to calculate energy efficiency. *Energy Build.* **2003**, *35*, 971–983. [CrossRef]
47. Chlela, F.; Husaunndee, A.; Riederer, P.; Inard, C. Numerical Evaluation of Earth to Air Heat Exchangers and Heat Recovery Ventilation Systems. *Int. J. Vent.* **2007**, *6*, 31–42. [CrossRef]
48. Warwick, D.J.; Cripps, A.J.; Kolokotroni, M. Integrating Active Thermal Mass Strategies with HVAC Systems: Dynamic Thermal Modelling. *Int. J. Vent.* **2009**, *7*, 345–367. [CrossRef]
49. Ralegaonkar, R.; Kamath, M.V.; Dakwale, V.A. Design and Development of Geothermal Cooling System for Composite Climatic Zone in India. *J. Inst. Eng. Ser. A* **2014**, *95*, 179–183. [CrossRef]
50. Stasi, R.; Liuzzi, S.; Paterno, S.; Ruggiero, F.; Stefanizzi, P.; Stragapede, A. Combining bioclimatic strategies with efficient HVAC plants to reach nearly-zero energy building goals in Mediterranean climate. *Sustain. Cities Soc.* **2020**, *63*, 102479. [CrossRef]
51. Trząski, A.; Zawada, B. The influence of environmental and geometrical factors on air-ground tube heat exchanger energy efficiency. *Build. Environ.* **2011**, *46*, 1436–1444. [CrossRef]
52. Skotnicka-Siepsiak, A. An Evaluation of the Performance of a Ground-to-Air Heat Exchanger in Different Ventilation Scenarios in a Single-Family Home in a Climate Characterized by Cold Winters and Hot Summers. *Energies* **2022**, *15*, 105. [CrossRef]
53. Gwadera, M.; Larwa, B.; Kupiec, K. Undisturbed Ground Temperature—Different Methods of Determination. *Sustainability* **2017**, *9*, 2055. [CrossRef]
54. Larwa, B.; Gwadera, M.; Kicińska, I.; Kupiec, K. Parameters of the Carslaw-Jaeger equation describing the temperature distribution in the ground. *Tech. Trans.* **2018**, *115*, 67–78. [CrossRef]
55. Larwa, B. Heat Transfer Model to Predict Temperature Distribution in the Ground. *Energies* **2019**, *12*, 25. [CrossRef]
56. Bergman, T.L.; Lavine, A.A.; Incropera, F.P.; DeWitt, D.P. *Fundamentals of Heat and Mass Transfer*, 8th ed.; Wiley: Hoboken, NI, USA, 2018.
57. Panão, M.J.N.O.; Santos, C.A.P.; Mateus, N.M.; da Graça, G.C. Validation of a lumped RC model for thermal simulation of a double skin natural and mechanical ventilated test cell. *Energy Build.* **2016**, *121*, 92–103. [CrossRef]
58. Jayathissa, P.; Luzzatto, M.; Schmidli, J.; Hofer, J.; Nagy, Z.; Schlueter, A. Optimising building net energy demand with dynamic BIPV shading. *Appl. Energy* **2017**, *202*, 726–735. [CrossRef]
59. Shen, P.; Braham, W.; Yi, Y. Development of a lightweight building simulation tool using simplified zone thermal coupling for fast parametric study. *Appl. Energy* **2018**, *223*, 188–214. [CrossRef]
60. Costantino, A.; Fabrizio, E.; Ghiggini, A.; Bariani, M. Climate control in broiler houses: A thermal model for the calculation of the energy use and indoor environmental conditions. *Energy Build* **2018**, *169*, 110–126. [CrossRef]
61. Lauster, M.; Teichmann, J.; Fuchs, M.; Streblow, R.; Mueller, D. Low order thermal network models for dynamic simulations of buildings on city district scale. *Build. Environ.* **2014**, *73*, 223–231. [CrossRef]
62. Horvat, I.; Dović, D. Dynamic modeling approach for determining buildings technical system energy performance. *Energy Convers. Manag.* **2016**, *125*, 154–165. [CrossRef]
63. Elci, M.; Delgado, B.M.; Henning, H.M.; Henze, G.P.; Herkel, S. Aggregation of residential buildings for thermal building simulations on an urban district scale. *Sustain. Cities Soc.* **2018**, *39*, 537–547. [CrossRef]
64. Tagliabue, L.C.; Buzzetti, M.; Marenzi, G. Energy performance of greenhouse for energy saving in buildings. *Energy Procedia* **2012**, *30*, 1233–1242. [CrossRef]
65. Fabrizio, E.; Ghiggini, A.; Bariani, M. Energy performance and indoor environmental control of animal houses: A modelling tool. *Energy Procedia* **2015**, *82*, 439–444. [CrossRef]
66. Jędrzejuk, H.; Rucińska, J. Verifying a need of artificial cooling-a simplified method dedicated to single-family houses in Poland. *Energy Procedia* **2015**, *78*, 1093–1098. [CrossRef]
67. Fischer, D.; Wolf, T.; Scherer, J.; Wille-Haussmann, B. A stochastic bottom-up model for space heating and domestic hot water load profiles for German households. *Energy Build.* **2016**, *124*, 120–128. [CrossRef]

68. Mora, T.D.; Teso, L.; Carnieletto, L.; Zarrella, A.; Romagnoni, P. Comparative Analysis between Dynamic and Quasi-Steady-State Methods at an Urban Scale on a Social-Housing District in Venice. *Energies* **2021**, *14*, 5164. [CrossRef]
69. Bruno, R.; Pizzuti, G.; Arcuri, N. The Prediction of Thermal Loads in Building by Means of the EN ISO 13790 Dynamic Model: A Comparison with TRNSYS. *Energy Procedia* **2016**, *101*, 192–199. [CrossRef]
70. Costantino, A.; Comba, L.; Sicardi, G.; Bariani, M.; Fabrizio, E. Energy performance and climate control in mechanically ventilated greenhouses: A dynamic modelling-based assessment and investigation. *Appl. Energy* **2021**, *288*, 116583. [CrossRef]
71. *ISO 6946:2017*; Building Components and Building Elements—Thermal Resistance and Thermal Transmittance—Calculation Methods. International Organization for Standardization: Geneva, Switzerland, 2017.
72. *ISO 13786:2017*; Thermal Performance of Building Components—Dynamic Thermal Characteristics—Calculation Methods. International Organization for Standardization: Geneva, Switzerland, 2017.
73. Lomas, K.J. Architectural design of an advanced naturally ventilated building form. *Energy Build.* **2007**, *39*, 166–181. [CrossRef]
74. Vestfálová, M.; Šafařík, P. Determination of the applicability limits of the ideal gas model for the calculation of moist air properties. *EPJ Web Conf.* **2018**, *180*, 02115. [CrossRef]
75. Mohr, P.J.; Newell, D.B.; Taylor, B.N. CODATA Recommended Values of the Fundamental Physical Constants: 2014. *J. Phys. Chem. Ref. Data* **2016**, *45*, 043102. [CrossRef]
76. Arizona State University. World Meteorological Organization's World Weather & Climate Extremes Archive. Available online: https://wmo.asu.edu/content/world-meteorological-organization-global-weather-climate-extremes-archive (accessed on 4 March 2022).
77. Berberan-Santos, M.N.; Bodunov, E.N.; Pogliani, L. On the barometric formula. *Am. J. Phys.* **1997**, *65*, 404–412. [CrossRef]
78. Andrews, D.G. *An Introduction to Atmospheric Physics*; Cambridge University Press: Cambridge, UK, 2000.
79. National Ocean and Atmosphere Administration. *U.S. Standard Atmosphere 1976*; Nation Ocean and Atmosphere Administration: Washington, DC, USA, 1976. Available online: https://www.ngdc.noaa.gov/stp/space-weather/online-publications/miscellaneous/us-standard-atmosphere-1976/ (accessed on 14 March 2022).
80. Dubinova, A.A. Exact explicit barometric formula for a warm isothermal Fermi gas. *Tech. Phys.* **2009**, *54*, 210–213. [CrossRef]
81. Cohen, J.E.; Small, C. Hypsographic demography: The distribution of human population by altitude. *Proc. Natl. Acad. Sci. USA* **1998**, *95*, 14009–14014. [CrossRef]
82. Benenson, W.; Harris, J.W.; Stöcker, H.; Lutz, H. *Handbook of Physics*; Springer: New York, NY, USA, 2002. [CrossRef]
83. Jacobson, M.Z. *Fundamentals of Atmospheric Modeling*, 2nd ed.; Cambridge University Press: Cambridge, UK, 2012. [CrossRef]
84. Corstanje, A.; Bonardi, A.; Buitink, S.; Falcke, H.; Hörandel, J.R.; Mitra, P.; Mulrey, K.; Nelles, A.; Rachen, J.P.; Rossetto, L.; et al. The effect of the atmospheric refractive index on the radio signal of extensive air showers. *Astropart. Phys.* **2017**, *89*, 23–29. [CrossRef]
85. Wagner, W.; Kretzschmar, H.J.; Span, R.; Krauss, R. D2 Properties of Selected Important Pure Substances. In *VDI Heat Atlas*; Springer: New York, NY, USA, 2010. [CrossRef]
86. Lemmon, E.W.; Jacobsen, R.T.; Penoncello, S.G.; Friend, D.G. Thermodynamic properties of air and mixtures of nitrogen, argon, and oxygen from 60 to 2000 K at pressures to 2000 MPa. *J. Phys. Chem. Ref. Data* **2000**, *29*, 331–385. [CrossRef]
87. Struchtrup, H. Properties and Property Relations. In *Thermodynamics and Energy Conversion*; Springer: Berlin/Heidelberg, Germany, 2014. [CrossRef]
88. Michalak, P. Modelling of Solar Irradiance Incident on Building Envelopes in Polish Climatic Conditions: The Impact on Energy Performance Indicators of Residential Buildings. *Energies* **2021**, *14*, 4371. [CrossRef]
89. World Meteorological Organization. Weather Data. Available online: https://energyplus.net/weather (accessed on 17 March 2022).
90. Bellia, L.; Pedace, A.; Fragliasso, F. The role of weather data files in Climate-based Daylight Modeling. *Sol. Energy* **2015**, *112*, 169–182. [CrossRef]
91. Costanzo, V.; Evola, G.; Infantone, M.; Marletta, L. Updated Typical Weather Years for the Energy Simulation of Buildings in Mediterranean Climate. A Case Study for Sicily. *Energies* **2020**, *13*, 4115. [CrossRef]
92. Łuczak, R.; Ptaszyński, B.; Kuczera, Z.; Życzkowski, P. Energy efficiency of ground-air heat exchanger in the ventilation and air conditioning systems. *E3S Web Conf.* **2018**, *46*, 00015. [CrossRef]
93. Congedo, P.M.; Lorusso, C.; De Giorgi, M.G.; Marti, R.; D'Agostino, D. Horizontal Air-Ground Heat Exchanger Performance and Humidity Simulation by Computational Fluid Dynamic Analysis. *Energies* **2016**, *9*, 930. [CrossRef]
94. Skotnicka-Siepsiak, A.; Wesołowski, M.; Piechocki, J. Experimental and Numerical Study of an Earth-to-Air Heat Exchanger in Northeastern Poland. *Pol. J. Environ. Stud.* **2018**, *27*, 1255–1260. [CrossRef]
95. Niu, F.; Yu, Y.; Yu, D.; Li, H. Heat and mass transfer performance analysis and cooling capacity prediction of earth to air heat exchanger. *Appl. Energy* **2015**, *137*, 211–221. [CrossRef]
96. Romanska-Zapala, A.; Bomberg, M.; Dechnik, M.; Fedorczak-Cisak, M.; Furtak, M. On Preheating of the Outdoor Ventilation Air. *Energies* **2020**, *13*, 15. [CrossRef]

energies

MDPI

Article

Peak Power of Heat Source for Domestic Hot Water Preparation (DHW) for Residential Estate in Poland as a Representative Case Study for the Climate of Central Europe

Łukasz Amanowicz

Institute of Environmental Engineering and Building Installations, Poznan University of Technology, Pl. M. Skłodowskiej-Curie 5, 60-965 Poznan, Poland; lukasz.amanowicz@put.poznan.pl

Abstract: Due to the energy transformation in buildings, the proportions of energy consumption for heating, ventilation and domestic hot water preparation (DHW) have changed. The latter component can now play a significant role, not only in the context of the annual heat demand, but also in the context of selecting the peak power of the heat source. In this paper, the comparison of chosen methods for its calculation is presented. The results show that for contemporary residential buildings, the peak power for DHW preparation can achieve the same or higher value as the peak power for heating and ventilation. For this reason, nowadays the correct selection of the peak power of a heat source for DHW purposes becomes more important, especially if it uses renewable energy sources, because it affects its size and so the investment cost and economic efficiency. It is also indicated that in modern buildings, mainly accumulative systems with hot water storage tanks should be taken into account because they are less sensitive to design errors (wrongly selected peak value in the context of the uncertainty of hot water consumption) and because they result in acceptable value of peak power for DHW in comparison to heating and ventilation.

Keywords: domestic hot water; peak power; energy performance of buildings; DHW; energy transformation

Citation: Amanowicz, Ł. Peak Power of Heat Source for Domestic Hot Water Preparation (DHW) for Residential Estate in Poland as a Representative Case Study for the Climate of Central Europe. *Energies* **2021**, *14*, 8047. https://doi.org/10.3390/en14238047

Academic Editors: Jan Danielewicz, Chi-Ming Lai and Krzysztof Rajski

Received: 20 September 2021
Accepted: 25 November 2021
Published: 1 December 2021

1. Introduction

The energy consumption in buildings is constantly changing, mainly due to the amendments of legal requirements. Examples of such activities are, for example, energy certification, CO_2 emission limits, local programs supporting the modernization of existing facilities, as well as subsidies for new, energy-efficient buildings and systems. Currently, a lot of emphasis is placed not only on reducing the energy demand itself, but also on increasing the degree of use of renewable energy sources (RES), which favors the reduction of primary energy consumption and reduces the depletion of fossil fuel resources. It can be seen that it is not possible to clearly state how much energy consumption should be reduced, because it depends on many parameters. For example, in the paper [1], the optimum energy usage for residential buildings in developing countries is discussed as vulnerable to the current economic situation in the given country. In the last few decades, much attention has been devoted to the analysis of thermal modernization of buildings and research on the energy efficiency of heating, ventilation and air conditioning systems (HVAC). For example, the article [2] presents the results of long-term research showing energy savings resulting from the use of thermostatic valves, ranging from 7% to 23% for the analyzed cases. The results of the research presented in the article [3] showed that the use of mechanical ventilation systems with heat recovery from the air removed in residential buildings allows for significant savings in the annual amount of energy, and at the same time confirmed that recuperator systems with an integrated wall intake-exhaust devices are safe to use from a hygienic point of view. In the paper [4], the review of the heat recovery methods in ventilation was presented as a way to diminish energy demand for heating the buildings. Another of the possibilities of reducing primary energy consumption by mechanical ventilation systems is the use of earth-to-air heat exchangers (EAHEs), which has been shown in many works, e.g., in a review article [5] or in more detailed works about supporting ventilation systems by means of ground heat exchangers, e.g., [6–8] or mechanical analysis about improving the efficiency of EAHEs, e.g., [9,10]. Building heating systems that use renewable energy sources, such as heat pumps, achieve the highest efficiency in cooperation with low-temperature heaters. In practice, surface heating systems are mainly used. The most common system is floor heating, but in works [11,12], the innovative ceiling heating and cooling panels have been presented, showing the increase in their efficiency thanks to the use of an appropriately corrugated surface. An important role for the energy efficiency of HVAC systems is also played by controlling their efficiency, which was described in the example of wall heating systems with heat pipes in the article [13]. Renewable energy sources that are most often used to meet the energy needs of buildings are the energy of the sun, wind, ground and the use of the so-called waste heat (heat recovery, i.e., from exhaust air in ventilation). The energy of the sun can be harvested by using solar collectors to prepare domestic hot water, such as in 227 houses in California (USA), the experiences of which are described in the paper [14], or those compared in terms of effectiveness in the article [15]. The collectors may be of the tubular type, such as those mentioned just now, or of the flat type, such as those reviewed in [16]. Another way to obtain energy from the sun is to use photovoltaic panels (PV panels) which are used to generate electricity and were reviewed, for example, in the paper [17]. Wind energy is used much less frequently in buildings, as electricity generation is usually carried out by large wind farms, such as offshore wind farms described in [18] or classic ones, the optimization techniques of which are described in [19]. Soil energy can be used, for example, to heat/cool the ventilation air with the aforementioned earth-to-air heat exchangers or by using ground-type heat pumps, such as those described in [20]. There are many more examples of modern systems that are aimed at decreasing the energy demand in building and increasing the efficiency of its usage. Some of them were mentioned to introduce the Reader to the background of the changes that buildings are currently undergoing in terms of energy efficiency and the structure of HVAC systems.

Due to the changes in building envelope and HVAC systems, described above, the relationship between various components of thermal balance of the buildings is constantly changing. In the article [21] one can find that in the last decades the share of energy demand for ventilation in the energy demand for ventilation and heat transfer from the building increased from about 20% to about 60%. Thanks to the better insulation of the building and using the mechanical ventilation systems with heat recovery from exhausted air and earth-to-air heat exchangers [22,23], the specific heat power for heating the building decreased from about 150 W/m^2 to about 20 W/m^2. The heat demand for heating the hot water has also changed in the last decades, mainly because of the decreasing water consumption related to individual accounting of consumption and increasing water and energy prices. The decreased hot water usage caused the higher water consumption irregularity which influences the heating peak power of the heat source. In the literature, one can find articles presenting the results of research on the possibility of increasing the efficiency of domestic hot water preparation systems, such as [24], as well as analyzing the possibility of their integration with renewable energy sources [25], usage of phase change materials in solar domestic hot water systems [26,27] or verifying the influence of technical parameters of the hot water preparation system on its effectiveness [28]. The sheer volume of hot water consumption by building residents is also a subject of analysis. In the article [29], based on the example of several single-family houses, it was shown that the values of hot water consumption taken in the calculation of the building's energy performance may differ from the real consumption by up to 39%. The differentiation of user preferences, as well as the duration and time of hot water consumption, make it impossible to accurately determine its instantaneous, maximum flowrate. In practice, only simplified methods are used, based on empirical experience from the operation of existing systems. For this reason, in paper [30], the authors presented a new approach of quantifying the flexibility potential of residential heat pumps in the context of stochastic character of domestic hot water usage, highlighting the probabilistic meaning of hot water consumption. In the literature on hot water systems, one can find many works that focus on increasing their energy efficiency or comfort of use. Article [31] describes a method to improve control of hot water temperature that can result in significant energy and water savings thanks to using advanced mixing valves. Investigations presented in [32] were focused on research on plate shower heat exchanger to reduce the domestic hot water energy demand. The results of the prototype tests showed the possibility of heat energy savings for a family house up to 300 kWh/year. However, the analysis did not refer to the multitude of calculation methods and the selection of the peak power of devices for DHW preparation. The impact of the use of nanoparticles in the solar fluid on the efficiency of the DHW preparation system was analyzed in [33]. The improvement in the efficiency of the system was not analyzed against the current changes in peak power for heating and ventilation of buildings. In [34], instantaneous heat exchanger with chemical-based disinfection was analyzed, which enables the reduction of hot water temperature in the DHW system, which resulted in a reduction in the circulation heat loss up to 66% for the analyzed multi-family building. The selection of the system power in the context of the amount of hot water demand was not the subject of research. Another way to reduce the energy consumption of a hot water system is to use shower heat exchangers for heat recovery in residential buildings. The possibility of using such a system was presented, for example, in articles [35,36]. Energy savings in DHW systems can also be achieved by appropriate control of their operation, which was demonstrated in the work [37], which focuses on the analysis of the heat storage tank. The improvement of the energy efficiency of the DHW solar system due to the mechanical modification of the heat storage tank structure was the subject of the research presented in [38]. On the other hand, Chandra & Matuska [39] analyzed the quality of temperature stratification in heat storage tanks in the context of the energy efficiency of the DHW system. Heat storage tanks in DHW systems can also take advantage of the phase change phenomenon, if they are filled with PCM materials, as shown, for example, in the papers [40–42]. In turn, the paper [43] presents the importance of the storage tank (cold water inlets and obstacles) structure itself for the energy efficiency of the DHW preparation system, which, if appropriate, may result in up to 15% higher efficiency compared to the non-optimized structure. Optimization of the storage tank structure may also contribute to shortening, by up to 10%, the time of preparing hot water, which improves the comfort of its use and has been analyzed in [44]. In all of these studies on heat storage tanks, the impact of their use on the peak power of the heat source in the context of the available computational methods was not analyzed. The subject of the stochastic nature of the variability of hot water consumption by users was discussed in many articles. In the work [45], neural networks were used to predict the demand for hot water both in single houses and in multi-family buildings. In the analysis presented in the article [46], the DHWcalc and TRNSYS software was used to determine district heating load profiles for domestic hot water preparation. Probabilistic model for predicting occupancy and domestic hot water use was also presented in [47]. In the paper [48], the smart-meter was used to measure the instantaneous consumption of hot water, and the results of the analysis provided in the paper showed that "statistical and machine learning analysis can forecast the seasonal DHW demand". Nevertheless, the tools used in the above-mentioned works go beyond the workshop of typical engineers, as well as scientists who deal with the subject of improving the energy efficiency of DHW systems by improving control systems or changing the design of system components. For this reason, this paper focuses on the comparison of generally available and popular methods of calculating peak power for DHW preparation purposes. The work [49], as well as [50], deal with the topic of life cycle costs in the context of hot water preparation systems. Important issues of the use of materials from their acquisition through the production and utilization process were emphasized; however, it was not analyzed in the context of various methods of selecting the peak power of heat sources and/or DHW system components. An interesting idea for increasing energy efficiency is the combination of an air-conditioning system and hot water preparation, which was presented, for example, in the works [51,52]. Thanks to this solution, waste heat from the cooling process of rooms, especially in warm climates, can be used to prepare hot water for household needs.

Due to the changes in the HVAC systems mentioned above, the current view on the selection of heat source peak power in the context of hot water preparation should be revised. In the review paper [53] on sustainable and energy-efficient domestic hot water systems, many publications and studies have been analyzed. Among them, the analyses that concern the determination of the peak power of a heat source for the purposes of preparing domestic hot water in the context of mentioned above changes in the consumption ratio were not found. In this paper, it is considered as a knowledge gap, and at the same time, an important issue to study from a practical point of view. In this article, this problem is discussed in the context of various methods of determining the peak power of a heat source for the purposes of preparing domestic hot water. Contrary to the work presented in the article [29], the focus is given on the power of the heat source, not on the energy consumed in the annual cycle of its operation. The changes that have taken place in buildings in terms of improving the thermal insulation of partitions and improving the efficiency of HVAC systems, in particular the use of heat recovery from the exhaust air in mechanical ventilation systems, have changed the ratio between the peak thermal power needed to prepare domestic hot water and the power for heating and ventilation. This means the need to revise the calculation methods used so far. The above literature review on DHW systems shows that this issue has not been analyzed. DHW systems are usually analyzed in terms of the possibility of using renewable energy sources or in terms of their efficiency, but little attention is paid to the peak power selection which affects the size, cost, heat loss, efficiency and the very possibility of cooperation with RES. Improper assumption of peak power (usually oversizing) will not allow to achieve the required savings and reduce the effectiveness of even the most theoretically effective solution. On the other hand, undersized systems will not provide users with a sufficient level of comfort.

2. Materials and Methods

2.1. Calculating the Heating Power for Preparing Hot Water

The universal formula for calculating the heating power for preparing domestic hot water can be given as:

$$\dot{Q} = \rho_w \cdot \dot{V}_{hw} \cdot c_w \cdot (t_{hw} - t_w) \text{ [kW]} \quad (1)$$

where:

ρ_w—density of water, [kg/m^3];

$\dot{V}_{hw}$—volumetric flowrate of hot water, [m^3/s];

c_w—specific heat of water, [kJ/(kg·K)];

t_{hw}—temperature of hot water, usually assumed as 55 °C to avoid diseases connected with bacteria Legionella;

t_w—temperature of cold water, usually assumed as 5 °C for surface water sources and 10 °C for deep ground water sources.

In Equation (1), the assumption of the stream of hot water $\dot{V}_{hw}$ (hot water consumption) is crucial.

2.2. Hot Water Consumption—Standard Method

The instantaneous amount of hot water consumption depends on the behavior of users, which cannot be predicted. Older people and people with small children, people who work at home and people who work shifts, etc., use water in different amounts and at different times. In existing buildings, you can measure the current water consumption with a water meter. This information is not sufficient to predict water consumption in newly built facilities whose users are unknown. Moreover, depending on the users (whether the same people still occupy the buildings) and on time (whether the users age, change jobs, habits, etc.), the measured value will change. Therefore, it cannot be defined once or unequivocally. For this reason, some simplifications and assumptions as well as various calculation methods are used in practice. Interestingly, they can result in values that differ significantly from each other, as demonstrated later in this work. The literature review shows that the hot water usage is difficult to predict due to unpredictable user behavior. Hot water consumption can be predicted on the basis of many statistical data from already existing buildings, inhabited by people with a similar consumption profile. For this purpose, probabilistic methods or neural networks are used, as indicated by the authors of the studies presented in the works, respectively, [46,47]. Due to the important role of the method of use for temporary water abstraction, the authors of the studies [48] write clearly not about its calculation, but "prediction", emphasizing the stochastic nature of this consumption, depending on the way the system is used by users. The authors of the review [54] also emphasize a significant impact on energy consumption in buildings, which, although it concerns heating, ventilation and lighting systems, seems to indicate a similar unpredictability of their use, as in the case of DHW systems. The authors of the model for optimizing energy consumption in a building [55], based on user behavior survey, also refer to "prediction" rather than "calculating" the actual values of energy consumption in a building.

Figure 1 shows the example of variability of hot water consumption in the case of a multi-family housing estate for three selected days of the week (Sunday, Thursday, Saturday) at different hours of the day. The chart is taken from a handbook on district heating [56].

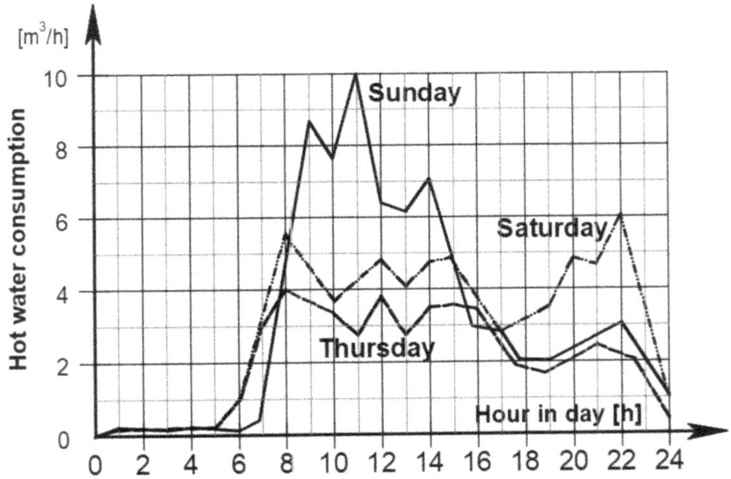

Figure 1. An example of hot water consumption for exemplary residential estate [56].

Due to the high variability of hot water consumption during the day, as well as week, month or year, the method described in the standard PN-B/92-01706 can be used. It allows us to determine the temporary hot water consumption assuming the average daily water consumption per capita and the number of inhabitants. The average daily consumption of hot water is a value that is also subject to fluctuations, not only depending on the users, but also in the long term, e.g., with a change in water and energy prices, a change in the wealth of a given community, or a change in the level of hygiene awareness. Nevertheless, it is a value that is easier to verify, e.g., on the basis of data from water companies. In this method, the daily consumption expressed in dm^3/person/day is then divided by the time of consumption during the day and multiplied by the coefficient of hourly consumption irregularity k_{h1}, which is an increasing factor aimed at calculating the peak (maximum) value of hourly water consumption:

$$\dot{V}_{hw,max} = \frac{\dot{V}_{hw,av} \cdot M}{\tau} \cdot k_{h1} \quad (2)$$

$$k_{h1} = 9.32 \cdot M^{-0.244} \quad (3)$$

where:
$\dot{V}_{hw,max}$—peak (maximum) hourly hot water consumption, [dm^3/s];
$\dot{V}_{hw,av}$—average daily hot water consumption per person, [dm^3/person/day];
M—the number of residents, [−];
τ—time of hot water consumption during the day, assumed usually as 18–20 h = 64,800–72,000 s, taking into account the time for sleeping;
k_{h1}—coefficient of hourly consumption irregularity, [−].

The coefficient k_{h1} decreases with the increase in the number of inhabitants, which reflects the assumption that the more people live in the housing estate, the greater the uniformity of heat consumption, i.e., the less probable a situation is that many people will take a shower or turn on the faucet in the sink at the same time. This assumption means that in the case of a small group of people, the risk of simultaneous consumption, significantly increased compared to the average value, is greater. This approach seems to be justified also due to the greater accumulation and hence the inertia of large domestic hot water preparation systems designed to serve a larger number of inhabitants.

In the standard PN-B/92-01706 the hot water consumption at the level of 110–130 dm^3/person/day was recommended. Due to the water consumption changes (decreased hot water consumption to a level of about 60 dm^3/person/day) in 2011, in articles [57,58], it was recommended to assume adjusted value of the lower water stream and the higher irregularity of the consumption:

$$k_{h2} = 8.8 \cdot M^{-0.167} \quad (4)$$

It should be noted that further changes in domestic hot water consumption will affect the value of peak consumption and the irregularity coefficient will increase. A survey presented in the paper [59] showed that "*the perspective of environmental protection is not a sufficient motivator to save energy for heating domestic hot water*" and therefore only the water and energy price model can drive further changes.

2.3. Hot Water Consumption—Experimental Measurements

The second method, taken into account in this study, is calculating the hot water consumption with approximated formulas based on the experimental data published in a book about hot water preparation systems [60] in 2008. Approximated results enable calculation of average and maximum hot water consumption as a function of the number of residents with the assumption of probability that this value would be exceeded (Equations (5)–(8)). For example, the probability of 1% means that the water stream can be exceeded once in one day for 100 days.

$$\dot{V}_{hw,av,1\%} = 0.006386 \cdot M^{0.5} + 0.00102M \ [\mathrm{dm^3/s}] \quad (5)$$

$$\dot{V}_{hw,av,10\%} = 0.003048 \cdot M^{0.5} + 0.00102M \ [\mathrm{dm^3/s}] \quad (6)$$

$$\dot{V}_{hw,max,1\%} = 0.03017 \cdot M^{0.5} + 0.00102M \ [\mathrm{dm^3/s}] \quad (7)$$

$$\dot{V}_{hw,max,10\%} = 0.01429 \cdot M^{0.5} + 0.00102M \ [\mathrm{dm^3/s}] \quad (8)$$

Due to the fact that hot water consumption is random, the use of probability for a mathematical description of this process seems to be the most reliable method. Moreover, this method allows us to select the comfort level understood as the probability of failure to maintain the continuity of hot water supply within the set parameters (stream and temperature). In this way, in the process of designing the system, it is possible to decide on its size, taking into account the compromise between the comfort (higher consumption values taken for calculations) and the cost of obtaining it (higher cost of purchasing and operating an oversized system), adjusting the solution to the expectations and financial possibilities of users. In this way, one can consciously design a larger system for a five-star hotel and a smaller one for a lower-standard hotel. However, it should be borne in mind that the presented formulas are the result of the approximation of measurement data of housing estates located in a one single city—Szczecin, Poland, Central Europe—for a limited set of buildings and only in the time period covered by the research. This means that the data used for calculations may not be inadequate for other settlements located in other cities or at a different time. It is a specificity related to the estimation of hot water consumption. Consequently, continuous monitoring of hot water consumption should be the subject of research. Recently, there have been published papers that show the changes that have occurred in this area during the COVID-19 pandemic, e.g., [61].

2.4. Hot Water Consumption—Sander's Method

Another method for calculation of hot water consumption, taken into account in this study, is the Sander's method described for example in the book [56]. In this method, the water stream for heating source peak power calculation is assumed on the basis of the hot water volume flowing from the water device, temperature of water and time of use:

$$\dot{V}_{\mathrm{hw}} = \frac{\sum N_{\mathrm{i}} \cdot V_{\mathrm{i}}}{\tau \cdot 60} \cdot \varphi \ [\mathrm{dm}^3/\mathrm{s}] \quad (9)$$

where:

N_{i}—number of each type of devices, [−];
V_{i}—the volume of hot water used during single use, [dm^3];
τ—time of single use, [min];
φ—coefficient of non-simultaneous consumption of hot water, [−].

Input values for calculating different water devices are presented in Table 1. The value of the coefficient of non-simultaneous consumption of hot water φ can be read from Figure 2. The coefficient is used to take into account that not all water devices will be used at the same time. This is especially important in the case of housing estates, where the greater value of peak power should be taken for the selection of heat exchangers (e.g., housing heat exchangers), and smaller for the selection of the central heat source (taking into account the non-simultaneous consumption, reading the coefficient for the total number of water devices supplied by the heat source at the whole estate).

Table 1. Typical values of volume flowing from the water device, temperature of water and time of use for calculating the peak power for hot water preparation using Sander's method.

Device	V [dm^3]	t_{hw} [°C]	τ [min]
Sink	10–15	35–40	2–3
Kitchen sink	30–50	55	5
Bath	150–250	40	15–20
Shower	50	40	6

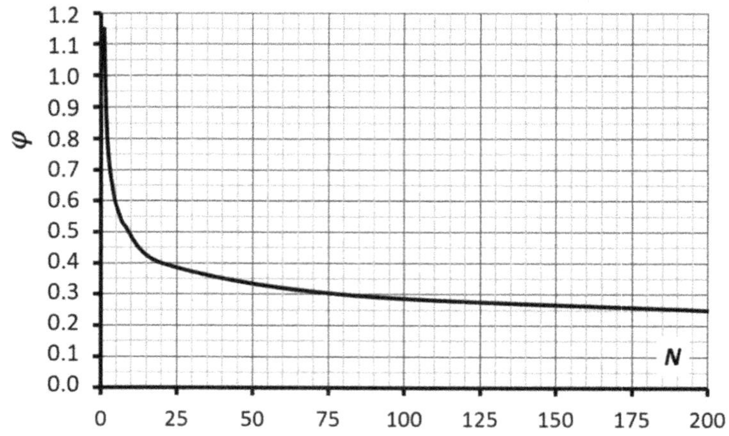

Figure 2. Value of coefficient of non-simultaneous consumption of hot water [56].

In practice for multi-family buildings, it is assumed that only the biggest water device should be taken into account as a representative for each flat. This approach assumes that a temporary and small consumption of hot water from other devices, such as a sink, will not result in a lack of power in the local hot water preparation system, which has a large thermal inertia. Then Equation (9) takes a simplified form, where only the number of baths or showers is taken into account:

$$\dot{V}_{hw} = \frac{N_{bath/shower} \cdot V_{bath/shower}}{\tau \cdot 60} \cdot \varphi \ [dm^3/s] \quad (10)$$

For the bath, it can be assumed that the V_{bath} = 150–200 dm^3 of water at the temperature of 40 °C (100–133 of water at the temperature of 55 °C) is used in a time τ = 15 min. For showers V = 50 dm^3 at the temperature of 40 °C (35 dm^3 at the temperature of 55 °C) in a time of 6 min. In this method, one can easily take into account the presence of hot water storage tanks by setting a longer water preparation time τ, i.e., 60 min. It is possible to discuss the value of the hot water temperature that should be used in the calculations. The results of studies of two single-family houses in Denmark [62] showed that the temperature of the water used in the shower was 35.5 °C to 40.4 °C, and for hand washing 20.5 °C to 26.5 °C. Taking the bath as a representative device and assuming the hot water temperature of 40 °C seems to be justified in the context of the quoted research results.

To calculate the peak power, universal formula (Equation (1)) can be used, but it has to be noted that for different water devices, different values of hot water temperature should be taken into account (see Table 1).

3. Results

3.1. Comparison of Hot Water Consumption

All above-mentioned formulas result in different values of hot water stream. The differences in results are very high, as is shown in Figures 3–5. The assumptions for calculations with Equation (1) are as follows:

- The number of residents per one flat: M/N = 3.5;
- PN-B/92, av.: $V_{hw,av}$ = 120 dm^3/person/day;
- PN-B/92, max.: $V_{hw,av}$ = 120 dm^3/person/day, k_h calculated from Equation (2);
- PN modified 2011, av.: $V_{hw,av}$ = 60 dm^3/person/day;
- PN modified 2011, max.: $V_{hw,av}$ = 60 dm^3/person/day, k_h calculated from Equation (3);
- PN-B/92 and PN modified: time of hot water consumption during the day: 18 h;
- Sander, bath, flow-type: V = 100 dm^3, τ = 15 min;

- Sander, bath, accumulative: $V = 100\ dm^3$, $\tau = 60$ min;
- Sander, shower, flow-type: $V = 35\ dm^3$, $\tau = 15$ min;
- Sander, shower, accumulative: $V = 35\ dm^3$, $\tau = 60$ min.

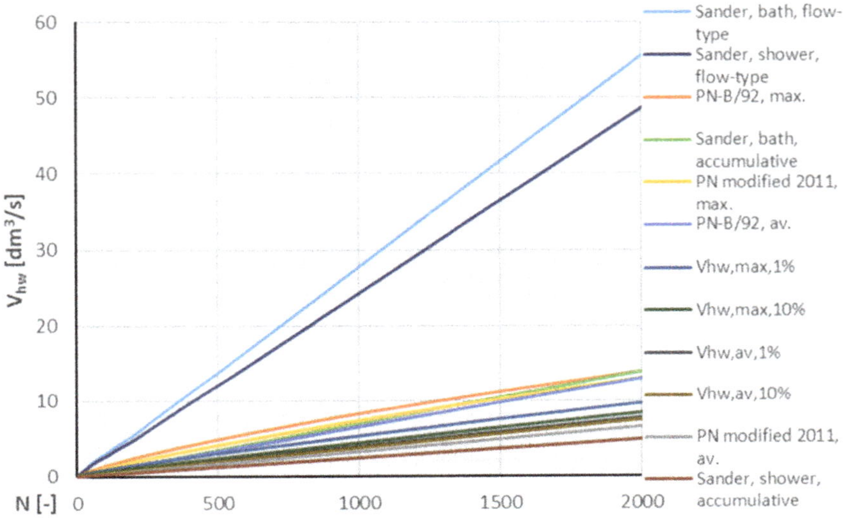

Figure 3. Comparison of hot water consumption calculated with different formulas, the number of flats: $N = 0$–2000.

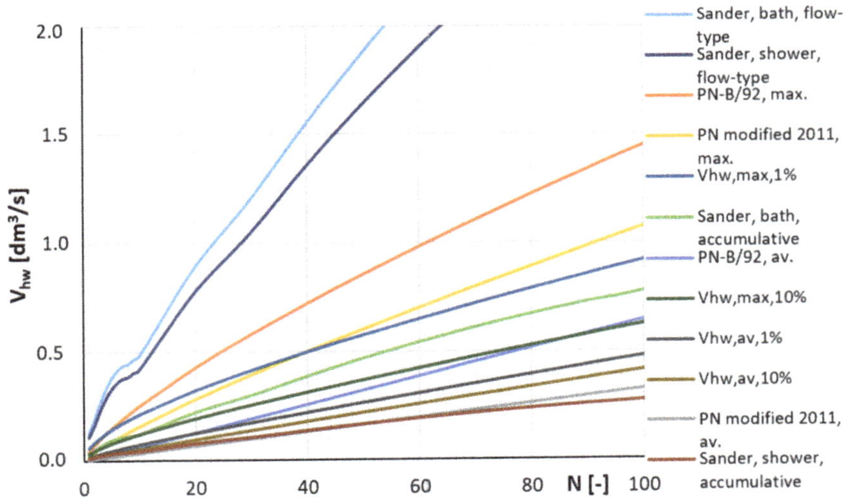

Figure 4. Comparison of hot water consumption calculated with different formulas the number of flats: $N = 1$–100.

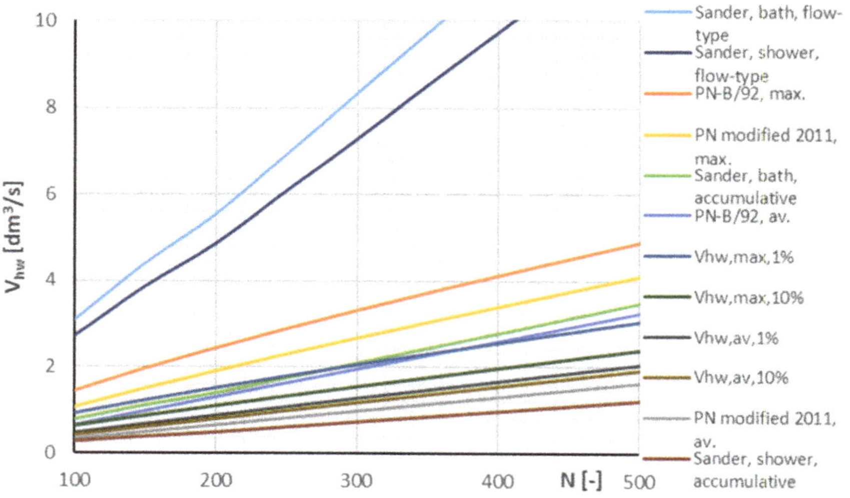

Figure 5. Comparison of hot water consumption calculated with different formulas, the number of flats: N = 100–500.

Results presented in Figures 3–5 show that:

- The highest value of hot water stream is obtained with Sander's method for flow-type system (without accumulation tanks), assuming baths as representative water devices;
- The lowest value of hot water stream is obtained with Sander's method for accumulative-type system (with accumulation tanks), assuming showers as representative water devices;
- Similar value as for Sander's method with accumulation tanks assuming showers was obtained using standard PN-B/92-01706 method and average (not maximum) stream of water (without taking into account coefficient k_h); for the number of residents 0 to 50 the Sander's method results in higher values and for $N > 50$ lower values (about 33%).

3.2. Peak Power of the Heat Source

For calculation of the heat source power for housing estate the power demand for heating the buildings and for hot water preparation should be taken into account. For buildings with high thermal inertia, the priority of hot water preparation can be assumed to reduce the peak power of the heat source. For traditional buildings with a high demand for thermal power for heating, the peak power could be calculated as a sum of heating needs and average hot water needs, but in the contemporary low energy buildings, this approach fails because of low heating needs. In Figure 4, the comparison of heat demand for heating the buildings in different insulation standards is presented. The results were prepared with assumptions:

- $Q_{hw,av}$ calculated from PN-B/92-01706, V_{hw} = 60 dm^3/person/day;
- $Q_{hw,max}$ calculated from PN-B/92-01706, V_{hw} = 60 dm^3/person/day, k_h calculated from Equation (3);
- Q_{co} calculated with assumption: q_{co1} = 100 W/m^2, q_{co2} = 50 W/m^2, q_{co3} = 20 W/m^2;
- The number of residents per one flat M/N = 3.5;
- The average area of a single flat: A = 55 m^2.

In Figure 6, the heating power for preparing hot water calculated from the Sander's method (for shower, flow-type and accumulative) and $V_{hw,max,1\%}$ are also presented.

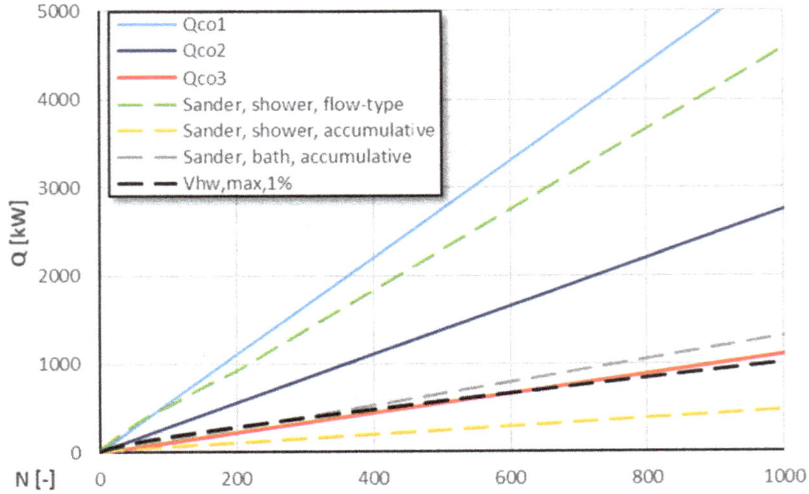

Figure 6. Heat source peak power for heating the buildings in different insulation standards and for preparing the hot water—comparison for *N* residents.

4. Discussion

The introduction to the work presents information on the energy transformation of buildings towards low-energy buildings. In the future, it will lead to the construction of nearly zero energy buildings (nZEB) or even plus energy buildings—as indicated by the case study showing the methodology of designing such facilities [63]. The results of this analysis showed that with the reduction of the thermal power for heating and ventilation purposes, the proportions between the peak power for these purposes and the power necessary for the preparation of domestic hot water changed significantly, as shown in Figure 6. It can therefore be seen that for low-energy multi-family buildings, the priority of hot water can be not sufficient for heating up hot water, especially when the baths are used instead of showers. For such buildings, the storage tanks are recommended to diminish the heat source peak power. In the context of diminished energy demand for heating and ventilation of the building, the priority of domestic hot water cannot be assumed without checking the peak power for hot water preparation. For the energy efficient buildings, the peak power for heating water can be the same or even higher than the peak power for heating and ventilation purposes. This indicates the need for changes in the approach to designing the peak power of a heat source in countries with a climate such as in Central Europe, where so far in traditional construction, the demand for heating and ventilation purposes was dominant.

In this study, the following methods of calculating the peak power for domestic hot water preparation were analyzed:

(1) PN-B/92-01706;
(2) Experimental results;
(3) Sander's method (flow-type and accumulative).

Each of the methods can be calibrated by changing the input data. In the case of the method based on experimental results, there is a possibility to choose a formula with a different probability of exceeding the peak consumption of hot water.

Method (1) PN-B/92-01706 is outdated and is based on non-updated values of daily hot water demand and uses an unverified coefficient of unevenness of hot water usage. Correcting—making real—the value of daily consumption does not guarantee obtaining reliable results. The method should be updated and validated against experimental data, taking into account the large number of objects analyzed. An interesting concept would be to make the results dependent on the chosen usage profile, which would be in line with the contemporary occupant-oriented designing trend.

Method (2) is based on the experimental data from the past that was averaged from measurements for a few selected buildings in a single city. For this reason, it is difficult to consider them universal. The advantage of this approach is to obtain the result of the peak hot water stream with information about the probability of the frequency of its exceeding (1% or 10%). Thus, this method enables the user to choose the degree of comfort of receiving hot water, and thus adjust it to the standard of the facility or the wealth of the users' wallet.

Method (3) Sander's method is based on the basic equations from the theory of heat flow. Therefore, it is the easiest to interpret physically, and it is also the easiest to verify the correctness of the entered data, because it is based on data on the amount of water flowing out at a given time from a given equipment. This method makes it possible to assume the hot water temperature depending on the regarded water device and to perform calculations for flow-type or accumulative systems. The only uncertain element of this method is the assumption of the simultaneous consumption factor, which depends on the number of water devices in a given building/estate. It is a data that can be read from a table or graph and cannot be confirmed personally. Nevertheless, according to the author, this method seems to inspire the greatest confidence, and in the case of the use of accumulation systems, results in peak heat power that is acceptable for modern buildings with low heat demand for other purposes (heating and ventilation).

5. Conclusions

Results of the work presented above show that:

- Different methods of calculating hot water consumption result in very divergent results;
- For the number of flats $n = 1$ to 100, the lowest values of hot water consumption are obtained with sander's method (shower, accumulative) and PN-B/92 (average stream, modified in 2011);
- The hot water consumption in the real buildings with the probability of exceed 1% $v_{hw,max,1\%}$ (recommended in book [58]) is almost two times higher than the lowest ones;
- The average value calculated from PN-B/92 for 1–80 flats is lower than $v_{hw,max,10\%}$ and for $n > 80$ is higher;
- For $n = 1$–80, $v_{hw,max,1\%}$ is the most similar to the maximum water stream calculated from pn-b/92 modified in 2011;
- For $n = 100$–500, $v_{hw,max,1\%}$ is the most similar to sander's (bath, accumulative);
- The maximum water stream calculated from PN-B/92 modified in 2011 for $n = 100$–500 is about 16–34% higher than $v_{hw,max,1\%}$;
- The peak power for heating the hot water in the flow-type system can be a few times higher than the peak power needed for heating and ventilation purposes of the contemporary low-energy building.

The present study shows that a thorough analysis of the domestic hot water system is now necessary, as this element may be decisive in the selection of the peak power of the heat source. This is particularly important when using renewable energy sources (RES), when the cost of purchasing oversized units can frustrate investment and discourage RES use. The results of this study show that the reduction of source peak power should then be achieved by using accumulation systems with hot water tanks, which results in an acceptable peak power for DHW preparation in comparison with the power needed for heating and ventilation.

Funding: This research was funded by the Polish Ministry of Science and Higher Education, grant number 504 101/0713/SBAD/0935.

Institutional Review Board Statement: Not applicable.

Informed Consent Statement: Not applicable.

Data Availability Statement: Not applicable.

Conflicts of Interest: The author declares no conflict of interest.

References

1. Basińska, M.; Koczyk, H.; Szczechowiak, E. Sensitivity analysis in determining the optimum energy for residential buildings in Polish conditions. *Energy Build.* **2015**, *17*, 307–318. [CrossRef]
2. Cholewa, T.; Siuta-Olcha, A.; Balaras, C.A. Actual energy savings from the use of thermostatic radiator valves in residential buildings—Long term field evaluation. *Energy Build.* **2017**, *151*, 487–493. [CrossRef]
3. Ratajczak, K.; Amanowicz, Ł.; Szczechowiak, E. Assessment of the air streams mixing in wall-type heat recovery units for ventilation of existing and refurbishing buildings toward low energy buildings. *Energy Build.* **2020**, *227*, 110427. [CrossRef]
4. Zender-Świercz, E. A Review of Heat Recovery in Ventilation. *Energies* **2021**, *14*, 1759. [CrossRef]
5. Agrawal, K.K.; Misra, R.; Agrawal, G.D.; Bhardwaj, M.; Jamuwa, D.K. The state of art on the applications, technology integration, and latest research trends of earth-air-heat exchanger system. *Geothermics* **2019**, *82*, 34–50. [CrossRef]
6. Wei, H.; Yang, D. Performance evaluation of flat rectangular earth-to-air heat exchangers in harmonically fluctuating thermal environments. *Appl. Therm. Eng.* **2019**, *162*, 114–262. [CrossRef]
7. Żukowski, M.; Topolańska, J. Comparison of thermal performance between tube and plate ground-air heat exchangers. *Renew. Energy* **2018**, *115*, 697–710. [CrossRef]
8. Wei, H.; Yang, D.; Du, J.; Guo, X. Field experiments on the effects of an earth-to-air heat exchanger on the indoor thermal environment in summer and winter for a typical hot-summer and cold-winter region. *Renew. Energy* **2021**, *167*, 530–541. [CrossRef]
9. Liu, Z.; Sun, P.; Xie, M.; Zhou, Y.; He, Y.; Zhang, G.; Chen, D.; Li, S.; Yan, Z.; Qin, D. Multivariant optimization and sensitivity analysis of an experimental vertical earth-to-air heat exchanger system integrating phase change material with Taguchi method. *Renew. Energy* **2021**, *173*, 401–414. [CrossRef]
10. Yang, L.H.; Hu, J.W.; Chiang, Y.C.; Chen, S.L. Performance analysis of building-integrated earth-air heat exchanger retrofitted with a supplementary water system for cooling-dominated climate in Taiwan. *Energy Build.* **2021**, *242*, 110949. [CrossRef]
11. Wojtkowiak, J.; Amanowicz, Ł. Effect of surface corrugation on cooling capacity of ceiling panel. *Therm. Sci. Eng. Prog.* **2020**, *19*, 100572. [CrossRef]
12. Wojtkowiak, J.; Amanowicz, Ł.; Mróz, T. A new type of cooling ceiling panel with corrugated surface—Experimental investigation. *Int. J. Energy Res.* **2019**, *43*, 7275–7286. [CrossRef]
13. Amanowicz, Ł. Controlling the Thermal Power of a Wall Heating Panel with Heat Pipes by Changing the Mass Flowrate and Temperature of Supplying Water—Experimental Investigations. *Energies* **2020**, *13*, 6547. [CrossRef]
14. Sanguinetti, A.; Outcault, S.; Alston-Stepnitz, E.; Moezzi, M.; Ingle, A. Residential solar water heating: California adopters and their experiences. *Renew. Energy* **2021**, *170*, 1081–1095. [CrossRef]
15. Zielinko, P.; Krawczyk, D.A. Comparison of the Efficiency of Solar Collectors in Terms of the Working Medium—Review of Selected Technical Solutions. *Environ. Sci. Proc.* **2021**, *9*, 5. [CrossRef]
16. Alam, T.; Balam, N.B.; Kulkarni, K.S.; Siddiqui, M.I.H.; Kapoor, N.R.; Meena, C.S.; Kumar, A.; Cozzolino, R. Performance Augmentation of the Flat Plate Solar Thermal Collector: A Review. *Energies* **2021**, *14*, 6203. [CrossRef]
17. Bhakre, S.S.; Sawarkar, P.D.; Kalamkar, V.R. Performance evaluation of PV panel surfaces exposed to hydraulic cooling—A review. *Sol. Energy* **2021**, *224*, 1193–1209. [CrossRef]
18. Díaz, H.; Soares, C.G. Review of the current status, technology and future trends of offshore wind farms. *Ocean. Eng.* **2020**, *209*, 107381. [CrossRef]
19. Balasubramanian, K.; Thanikanti, S.B.; Subramaniam, U.; Sudhakar, N.; Sichilalu, S. A novel review on optimization techniques used in wind farm modelling. *Renew. Energy Focus* **2020**, *35*, 84–96. [CrossRef]
20. Wang, Z.; Luther, M.B.; Amirkhani, M.; Liu, C.; Horan, P. State of the Art on Heat Pumps for Residential Buildings. *Buildings* **2021**, *11*, 350. [CrossRef]
21. Amanowicz, Ł.; Szczechowiak, E. Zasady projektowania systemów wentylacji budynków energooszczędnych. *Ciepłownictwo Ogrzewnictwo Wentylacja* **2017**, *48*, 72–78. [CrossRef]
22. Amanowicz, Ł.; Wojtkowiak, J. Validation of CFD model for simulation of multi-pipe earth-to-air heat exchangers (EAHEs) flow performance. *Therm. Sci. Eng. Prog.* **2018**, *5*, 44–49. [CrossRef]
23. Amanowicz, Ł. Influence of geometrical parameters on the flow characteristics of multi-pipe earth-to-air heat exchangers—Experimental and CFD investigations. *Appl. Energy* **2018**, *226*, 849–861. [CrossRef]
24. Cholewa, T.; Siuta-Olcha, A.; Anasiewicz, R. On the possibilities to increase energy efficiency of domestic hot water preparation systems in existing buildings—Long term field research. *J. Clean. Prod.* **2019**, *217*, 194–203. [CrossRef]
25. Dudkiewicz, E.; Fidorów-Kaprawy, N. The energy analysis of a hybrid hot tap water preparation system based on renewable and waste sources. *Energy* **2017**, *127*, 198–208. [CrossRef]
26. Douvi, E.; Pagkalos, C.; Dogkas, G.; Koukou, M.G.; Stathopoulos, V.N.; Caouris, Y.; Vrachopoulos, M.G. Phase Change Materials in Solar Domestic Hot Water Systems: A review. *Int. J. Thermofluids* **2021**, *10*, 100075. [CrossRef]

27. Jachura, A.; Sekret, R. Life Cycle Assessment of the Use of Phase Change Material in an Evacuated Solar Tube Collector. *Energies* **2021**, *14*, 4146. [CrossRef]
28. Zukowski, M.; Jezierski, W. New Deterministic Mathematical Model for Estimating the Useful Energy Output of a Medium-Sized Solar Domestic Hot Water System. *Energies* **2021**, *14*, 2753. [CrossRef]
29. Ratajczak, K.; Michalak, K.; Narojczyk, M.; Amanowicz, Ł. Real Domestic Hot Water Consumption in Residential Buildings and Its Impact on Buildings' Energy Performance—Case Study in Poland. *Energies* **2021**, *14*, 5010. [CrossRef]
30. You, Z.; Zade, M.; Kumaran Nalini, B.; Tzscheutschler, P. Flexibility Estimation of Residential Heat Pumps under Heat Demand Uncertainty. *Energies* **2021**, *14*, 5709. [CrossRef]
31. Rahmatmand, A.; Vratonjic, M.; Sullivan, P.E. Energy and thermal comfort performance evaluation of thermostatic and electronic mixing valves used to provide domestic hot water of buildings. *Energy Build.* **2020**, *212*, 109830. [CrossRef]
32. Vavřička, R.; Boháč, J.; Matuška, T. Experimental development of the plate shower heat exchanger to reduce the domestic hot water energy demand. *Energy Build.* **2022**, *254*, 111536. [CrossRef]
33. Saeed, T.; Ibrahim, M. Improving solar domestic hot water effectiveness using Al_2O_3 and ZnO nanoparticles: Focusing on annual energy-saving. *J. Taiwan Inst. Chem. Eng.* **2021**, *124*, 205–215. [CrossRef]
34. Benakopoulos, T.; Tunzi, M.; Salenbien, R.; Vanhoudt, D.; Svendsen, S. Low return temperature from domestic hot-water system based on instantaneous heat exchanger with chemical-based disinfection solution. *Energy* **2021**, *215*, 119211. [CrossRef]
35. Nagpal, H.; Spriet, J.; Murali, M.K.; McNabola, A. Heat Recovery from Wastewater—A Review of Available Resource. *Water* **2021**, *13*, 1274. [CrossRef]
36. Kordana, S.; Pochwat, K.; Słyś, D.; Starzec, M. Opportunities and Threats of Implementing Drain Water Heat Recovery Units in Poland. *Resources* **2019**, *8*, 88. [CrossRef]
37. Huang, T.; Yang, X.; Svendsen, S. Multi-mode control method for the existing domestic hot water storage tanks with district heating supply. *Energy* **2020**, *191*, 116517. [CrossRef]
38. Erdemir, D.; Atesoglu, H.; Altuntop, N. Experimental investigation on enhancement of thermal performance with obstacle placing in the horizontal hot water tank used in solar domestic hot water system. *Renew. Energy* **2019**, *138*, 187–197. [CrossRef]
39. Pal Chandra, Y.P.; Matuska, T. Numerical prediction of the stratification performance in domestic hot water storage tanks. *Renew. Energy* **2020**, *154*, 1165–1179. [CrossRef]
40. Dogkas, G.; Konstantaras, J.; Koukou, M.K.; Vrachopoulos, M.G.r.; Pagkalos, C.h.; Stathopoulos, V.N.; Pandis, P.K.; Lymperis, K.; Coelho, L.; Rebola, A. Development and experimental testing of a compact thermal energy storage tank using paraffin targeting domestic hot water production needs. *Therm. Sci. Eng. Prog.* **2020**, *19*, 100573. [CrossRef]
41. Abdelsalam, M.Y.; Teamah, H.M.; Lightstone, M.F.; Cotton, J.S. Hybrid thermal energy storage with phase change materials for solar domestic hot water applications: Direct versus indirect heat exchange systems. *Renew. Energy* **2020**, *147*, 77–88. [CrossRef]
42. Thonon, M.; Zalewski, L.; Gibout, S.; Franquet, E.; Fraisse, G.; Pailha, M. Experimental Comparison of Three Characterization Methods for Two Phase Change Materials Suitable for Domestic Hot Water Storage. *Appl. Sci.* **2021**, *11*, 10229. [CrossRef]
43. Malec, A.; Cholewa, T.; Siuta-Olcha, A. Influence of Cold Water Inlets and Obstacles on the Energy Efficiency of the Hot Water Production Process in a Hot Water Storage Tank. *Energies* **2021**, *14*, 6509. [CrossRef]
44. Obstawski, P.; Janaszek-Mańkowska, M.; Ratajski, A. Diagnostics of a Domestic Hot Water Storage Tank under Operating Conditions. *Processes* **2021**, *9*, 1771. [CrossRef]
45. Maltais, L.-G.; Gosselin, L. Predictability analysis of domestic hot water consumption with neural networks: From single units to large residential buildings. *Energy* **2021**, *229*, 120658. [CrossRef]
46. Braas, H.; Jordan, U.; Best, I.; Orozaliev, J.; Vajen, K. District heating load profiles for domestic hot water preparation with realistic simultaneity using DHWcalc and TRNSYS. *Energy* **2020**, *201*, 117552. [CrossRef]
47. Rouleau, j.; Ramallo-González, A.P.; Gosselin, L.; Blanchet, P.; Natarajan, S. A unified probabilistic model for predicting occupancy, domestic hot water use and electricity use in residential buildings. *Energy Build.* **2019**, *202*, 109375. [CrossRef]
48. Lee, J.Y.; Yim, T. Energy and flow demand analysis of domestic hot water in an apartment complex using a smart meter. *Energy* **2021**, *229*, 120678. [CrossRef]
49. Raluy, R.G.; Dias, A.C. Domestic hot water systems: Environmental performance from a life cycle assessment perspective. *Sustain. Prod. Consum.* **2021**, *26*, 1011–1020. [CrossRef]
50. Liu, W.; Chen, C.; Wu, H.; Guo, C.; Chen, Y.; Liu, W.; Cui, Z. Environmental life cycle assessment and techno-economic analysis of domestic hot water systems in China. *Energy Convers. Manag.* **2019**, *199*, 111943. [CrossRef]
51. Congedo, P.M.; Baglivo, C.; Carrieri, L. Application of an unconventional thermal and mechanical energy storage coupled with the air conditioning and domestic hot water systems of a residential building. *Energy Build.* **2020**, *224*, 110234. [CrossRef]
52. Jia, J.; Lee, W.L.; Cheng, Y.; Tian, Q. Can reversible room air-conditioner be used for combined space and domestic hot water heating in subtropical dwellings? Techno-economic evidence from Hong Kong. *Energy* **2021**, *223*, 119911. [CrossRef]
53. Pomianowski, M.Z.; Johra, H.; Marszal-Pomianowska, A.; Zhang, C. Sustainable and energy-efficient domestic hot water systems: A review. *Renew. Sustain. Energy Rev.* **2020**, *128*, 109900. [CrossRef]
54. Chen, S.; Zhang, G.; Xia, X.; Chen, Y.; Setunge, S.; Shi, L. The impacts of occupant behavior on building energy consumption: A review. *Sustain. Energy Technol. Assess.* **2021**, *45*, 101212. [CrossRef]
55. Amasyali, K.; El-Gohary, N.M. Real data-driven occupant-behavior optimization for reduced energy consumption and improved comfort. *Appl. Energy* **2021**, *302*, 117276. [CrossRef]

56. Bagieński, Z.; Amanowicz, Ł. *Ciepłownictwo. Projektowanie Kotłowni i Ciepłowni*; Wydawnictwo Politechniki Poznańskiej: Poznań, Poland, 2018; ISBN 978-83-7775-519-8.
57. Nowakowski, E. Współczynniki godzinowej nierównomierności zużycia ciepłej wody. *Rynek Instalacyjny* **2011**, *7–8*, 76–78.
58. Nowakowski, E. Współczynniki chwilowych nierównomierności zużycia ciepłej wody. *Rynek Instalacyjny* **2011**, *9*, 60–61.
59. Kordana-Obuch, S.; Starzec, M.; Słyś, D. Assessment of the Feasibility of Implementing Shower Heat Exchangers in Residential Buildings Based on Users' Energy Saving Preferences. *Energies* **2021**, *14*, 5547. [CrossRef]
60. Szaflik, W. *Projektowanie Instalacji Ciepłej Wody w Budynkach Mieszkalnych*; Ośrodek Informacji Technika Instalacyjna w Budownictwie: Warsaw, Poland, 2011; ISBN 978-83-7457-049-7.
61. Kim, D.; Yim, T.; Lee, J.Y. Analytical study on changes in domestic hot water use caused by COVID-19 pandemic. *Energy* **2021**, *231*, 120915. [CrossRef]
62. Marszal-Pomianowska, A.; Jensen, R.L.; Pomianowski, M.; Larsen, O.K.; Jørgensen, J.S.; Knudsen, S.S. Comfort of Domestic Water in Residential Buildings: Flow, Temperature and Energy in Draw-Off Points: Field Study in Two Danish Detached Houses. *Energies* **2021**, *14*, 3314. [CrossRef]
63. Radomski, B.; Mróz, T. The Methodology for Designing Residential Buildings with a Positive Energy Balance—Case Study. *Energies* **2021**, *14*, 5162. [CrossRef]

energies

MDPI

Article

New Deterministic Mathematical Model for Estimating the Useful Energy Output of a Medium-Sized Solar Domestic Hot Water System

Miroslaw Zukowski [1,*] and Walery Jezierski [2]

1 Department of HVAC Engineering, Faculty of Civil Engineering and Environmental Sciences, Bialystok University of Technology, Wiejska 45E Street, 15-351 Bialystok, Poland

2 Department of Energy Efficient Construction and Geodesy, Faculty of Civil Engineering and Environmental Sciences, Bialystok University of Technology, Wiejska 45E Street, 15-351 Bialystok, Poland; w.jezierski@pb.edu.pl

* Correspondence: m.zukowski@pb.edu.pl

Abstract: According to the authors of this paper, the mathematical point of view allows us to see what sometimes cannot be seen from the designer's point of view. The aim of this study was to estimate the influence of the most important parameters (volume of heat storage tanks, daily consumption of domestic hot water, optical efficiency, heat loss coefficient, and total area of a solar collector) on the thermal power output of solar domestic hot water (SDHW) system in European climatic conditions. Three deterministic mathematical models of these relationships for Madrid, Budapest, and Helsinki were created. The database for the development of these models was carried out using computer simulations made in the TRNSYS software environment. The SDHW system located at the Bialystok University of Technology (Poland) was the source of the measurement results used to validate the simulation model. The mathematical optimization procedure showed that the maximum annual useful energy output that can be obtained from 1 m^2 of gross collector area is 1303 kWh in the case of Madrid, 918.5 kWh for Budapest, and 768 kWh for Helsinki weather conditions.

Keywords: thermal solar collectors; useful energy output; energy simulations; solar domestic hot water system; deterministic mathematical model

Citation: Zukowski, M.; Jezierski, W. New Deterministic Mathematical Model for Estimating the Useful Energy Output of a Medium-Sized Solar Domestic Hot Water System. *Energies* **2021**, *14*, 2753. https://doi.org/10.3390/en14102753

Academic Editor: Carlo Renno

Received: 7 April 2021
Accepted: 10 May 2021
Published: 11 May 2021

1. Introduction

The effects of climate change are becoming more and more visible around the world. It manifests itself mainly in extreme weather events, loss of biodiversity and acidification of waters. The continuous increase in the global mean temperature directly affects the global rise in sea level. The global action plan to combat far-reaching climate change was set out in the Paris Agreement and signed by 195 countries in December 2015 [1]. The next step towards achieving climate neutrality was the decision of the European Commission (EC) taken in December 2020 [2]. EC proposed a 55% reduction in CO_2 emissions in 2030 compared to 1990 levels.

One way to achieve this difficult goal is the widespread use of the conversion of solar radiation into heat and electricity. The most effective devices implementing this process are thermal solar collectors, which support the heating of domestic hot water (DHW). It is difficult to achieve optimal performance of these systems at the design stage and later in the operating conditions. This is due to the large number of components in this system combined with the time-varying effects of weather conditions. Therefore, there are many methods of designing solar domestic hot water (SDHW) systems. Generally, we can divide them into two groups [3]. The first type of methods are based on analytical formulas describing the heat exchange phenomena occurring in these systems. An example might be one of the most frequently used method, i.e., the f-Chart [4,5]. Based on the value of solar radiation intensity, ambient temperature, and DHW demand, the amount of energy that can be provided by the designed system is calculated. The calculation algorithm of this method is based on the function of two variables whose values are influenced by the design and operational parameters. The error of the f-Chart method largely depends on the accuracy of the data describing a given system and ranges from 5% to 10% [6].

The second way to determine the performance of SDHW systems are computer energy simulations. They allow for very accurate modelling of the equipment characteristics, the system of pipes connecting them, and weather conditions. A large number of parameters describing the entire system requires the use of specialized software. TRNSYS, Design-Builder (based on a computing engine EnergyPlus), GetSolar, T*SOL, RETScreenl, and Polysun are the most popular computer programs often used in scientific analysis.

The authors of this paper focused the most attention on the application of the TRNSYS (Transient System Simulation Program) software package to simulate the operation of the SDHW system during the literature review. This computer program was used in this study.

Essabbani et al. [7] simulated the operation of a SDHW system consisting of 20 m^2 of flat plate solar collectors (FPC), a 300-L storage tank, and an electric heater. The testing facility was located in Oujda (Morocco) and the period of analysis was from the December 2011 to March 2012. The influence of total solar radiation, ambient temperature, and daily DHW consumption on the system performance was analyzed experimentally. Additionally, a numerical simulation of this installation was applied using the TRNSYS software. As a result of this research, it was found that the average share of solar energy in the total energy demand was 45% in the winter season. In March, the solar fraction was higher than in other months. However, the maximum required auxiliary energy was 740 MJ in December.

Ayompe et al. [8] developed and validated a model of a solar water heating system with forced circulation used in typical European residential homes. The tested rig consisted of two flat plate collectors (a total area of 4 m^2), an evacuated tube collector (ETC) with 30 heat pipe tubes, two 300 L hot water storage tanks, a control unit, and a pump station. A quasi-steady state simulation of this SDHW system were made using the TRNSYS software. This model was validated based on the results of measurements performed on a bench located in Dublin (Ireland). As it turned out, the mean absolute error in calculations of the heat collected by the collectors was 14.1% in the case of FPC circuit and 16.8% in the case of ETC loop.

The optimization of SDHW systems intended for single-family housing was performed by Hobbi and Siddiqui [9]. In this analysis, a forced indirect circulation system with a heat exchanger separating the solar circuit from the DHW circuit was considered. A south-facing flat collector with a slope equal to Montreal's latitude (45.5 degrees) was assumed. The share of energy obtained from solar collectors in the total energy demand for water heating was used as the optimization parameter. Computer simulations using TRNSYS software were used in this study. It was found that the designed system can satisfy 83% to 97% (in summer) and 30% to 62% (in winter) of the demand for DHW. It should be noted that the authors of this article did not use any special optimization methods.

The model of the SDHW system consisting of two FPC (5 m^2 of the absorber area), a heat storage tank (volume of 0.4 m^3), and a coil heat exchanger was developed by Mongibello et al. [10]. The operation of this system was simulated using the code written in MatLab environment and using TRNSYS software. The input data was the results of the measurements made at the ENEA Portici Research Center. The results of the calculations from both solvers were compared and it turned out that they are characterized by the good accordance.

Lima et al. [11] simulated a thermosiphon water heating system equipped with FPC, a storage tank, and an auxiliary heater in the climate conditions of Sao Paulo (Brazil). The hourly meteorological data collected by the Institute of Astronomy and Geosciences of the University of de Sao Paulo were used in the calculations. The main conclusion of this analysis was the following: the optimal configuration of this kind of thermosiphon system depends on the investor's concerns and local electricity prices.

The model of the SDHW system equipped with a heat pipe ETC and cooperating with an additional gas heater was made by Mehmood et al. [12] in the TRNSYS software environment. An experimental setup installed at National University of Sciences and Technology, Islamabad (Pakistan) was developed to validate the simulation model. A comparison of the results of calculations and experimental tests showed that the discrepancy between them ranges from 5 to 9%. Based on the computational analysis, it was found that the modelled system can replace conventional gas water heaters, providing 23% to 56% fossil fuel savings annually.

The performance of the ETC and FPC collectors were compared by Mohasseb and Kasaeian [13] in two different climatic conditions: cold weather-Tabriz (Iran) and hot-Bandar-Abbas (Iran). The operation of two SDHW systems with different collectors was simulated using the TRNSYS software. Based on the results of the calculations, it was found that the climatic conditions and inlet temperature had a greater influence on the operation of FPC than ETC. Besides, the annual useful energy from ETC is 30% higher than from FPC in cold climate, and 15% in hot weather conditions.

Mathematical equations that give relationships between the characteristic parameters of the SDHW system were not developed in the analyses described above. Additionally, optimization algorithms were not used to solve the issues related to the selection of the optimal variant.

The main disadvantages of computer simulations is the relatively long time needed for creating a model of the system and the large number of input data. They are dedicated to scientific analysis rather than being a tool used by designers. Therefore, it was decided to apply a deterministic mathematical modelling in order to simplify the method of the energy yield estimation from thermal solar collectors. The article presents a new approach for determining the annual useful energy output based on a multivariable function of volume of heat storage tanks, daily consumption of DHW, and the basic parameters of a solar collector. Climatic conditions are not included as one of the variables because they consist of many factors and cannot be described by one coherent indicator. In optimization analysis, this type of parameter is called uncontrollable and heterogeneous. That is why the numerical experiment was performed separately for three European weather zones. This is a drawback of the method presented here, but on the other hand, the development of a model for a specific climate zone allows to achieve high accuracy of the modelling results.

2. Materials and Methods

2.1. Model of SDHW System Developed in the TRNSYS Software Environment

The development of a simplified method of determining the energy yield from thermal collectors consisted of three main stages. First, the SDHW model of the system was created using the TRNSYS software. The second step was to validate this model based on a measurement database. Deterministic mathematical modelling was the third stage of this study. The selected TRNSYS simulation results created a database for this modelling. The selection of appropriate variants was based on the symmetrical three-level plan.

The modelled system consisted of a circuit with a battery of solar collectors (glycol solution as a heat transfer medium (HTM)) and a DHW circuit (water as a HTM). Both loops were separated by a plate compact heat exchanger. The other elements were heat storage tanks, separate pumps for each loop, a control and regulation system, and an auxiliary water heater. The schematic diagram of the model is shown in Figure 1. The list of the main modules creating this model is presented in Table 1. The model was built on the basis of an example solution presented in [14].

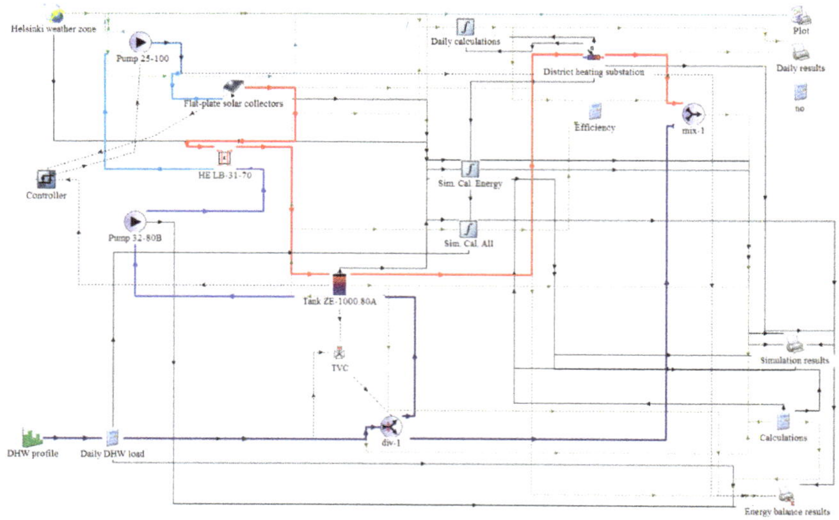

Figure 1. Scheme of the model of SDHW system developed in the TRNSYS software environment.

Table 1. List of the main modules of the SDHW system model.

Type Number	Module Name
1	Solar Collector; Quadratic Efficiency, 2nd Order Incidence Angle Modifiers
11	Controlled flow diverter
14	Time Dependent Forcing Function
15	Weather Data Processor; Combines data reading, radiation processing and sky temperature calculations
24	Quantity Integrator
91	Heat Exchanger with Constant Effectiveness
114	Single Speed Pump
115	Controller for Tempering Valves-Heating Mode
138	Auxiliary Fluid Heater
158	Cylindrical Storage Tank
165	ON/OFF Differential Controller

An important parameter that must be considered when creating this type of model is the consumption profile of DHW for occupants in the housing sector. An example of hourly prediction of DHW consumption was presented in [15]. This research work was based on data from the historical time series of DHW heat use in hotels in Norway. In the review of existing studies on hot water demand [16], we can find structured information on the DHW consumption patterns specified in technical standards for different types of buildings. The authors also noted the important impact of the mains water temperature on the energy consumption for DHW heating. A Time Dependent Forcing Function (TDFF) that allowed to employ the DHW consumption profile was used in this simulation. The repeated pattern of this function is established by a set of time-dependent data points shown for this case in Table 2. In order to reduce the time step during the calculations, linear interpolation is used.

Table 2. DHW profile assumed in the model.

Hour	0–7	8	9	10	11	12	13–17	18	19	20	21	22	23–24
Fraction	0	0.2	0.2	0	0.1	0.1	0	0	0.1	0.1	0.1	0.1	0

Three different locations are considered in in the current analysis. These are Madrid (Spain), Budapest (Hungary), and Helsinki (Finland) for which different mean annual values of the main water temperatures were assumed: 17 °C, 15 °C, and 12 °C, respectively. On the other hand, the temperature of the hot water collected in the storage tank was constant for each variant and amounted to 55 °C.

Presented above the model of SDHW system, due to its characteristics, is the most dedicated to typical multi-family apartment buildings. That is why the title of the article uses the term "medium size".

2.2. Validation of the SDHW System Model

As is well known, even a precisely developed numerical model may be inaccurate. Therefore, it was decided to compare the results of computer simulations with the measurement results. The source of the measurement database was the SDHW system located at the Bialystok University of Technology (Poland). The screenshot of the automatic operating parameters recording system is shown in Figure 2. A more detailed description of this measuring stand can be found in [17].

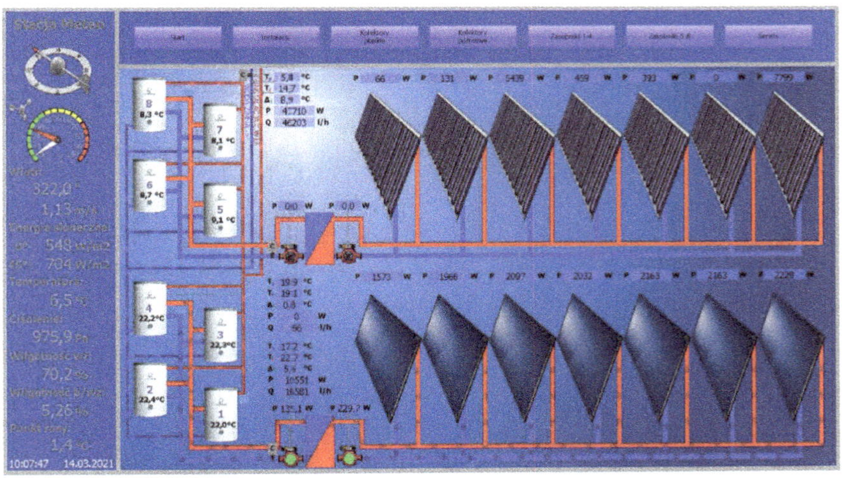

Figure 2. The screenshot of the scheme of the of SDHW system and working online data acquisition system.

This active solar heating system consists of:

- Seven rows of flat plate collectors, five connected in series in each row, with a total gross area of 71.9 m^2;
- Seven rows of evacuated tube collectors, three connected in series in each row, with an active total area of 74.45 m^2;
- Eight heat storage tanks with a volume of 1 m^3 each (four for each type of collectors).

The set of FPC was selected for comparative analysis. These solar panels have the following characteristics:

- Coefficient of collector efficiency—0.784 (-);
- Heat loss coefficient—3.64 $W/(m^2K)$;
- Temperature dependence heat loss coefficient—0.00185 $W/(m^2K^2)$;
- Maximum flow rate—0.000033 m^3/s;
- Gross area of single panel—2.054324 m^2.

In order to estimate the accuracy of the model, the measurement results of an annual useful energy output Q_{SC} (in kWh/m^2) from 5 years (2016–2020), were selected. The measurement of this value was made using a heat meter placed on the solar loop next to the heat exchanger. Figure 3 shows the amount of energy supplied by the solar collectors during the year as a function of the intensity of solar radiation Q_{SR} (in kWh/m^2) incident perpendicular to the plane of the collectors (blue rhombus).

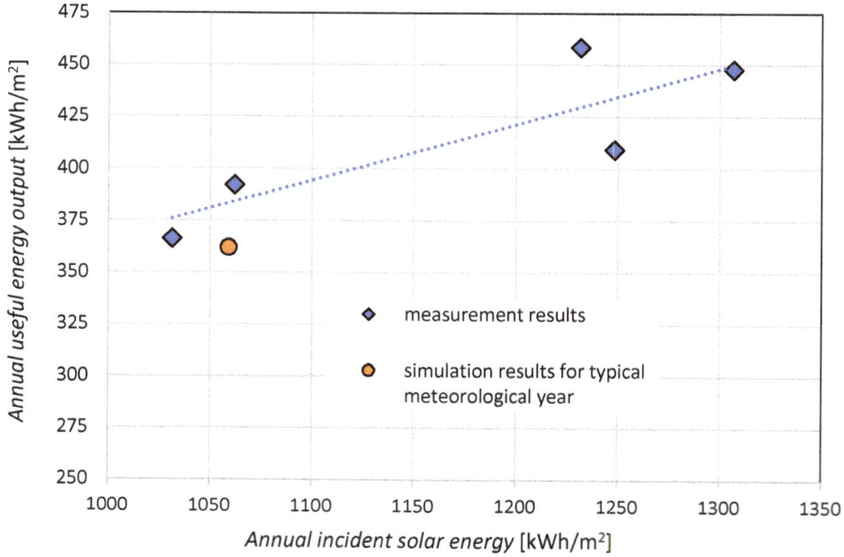

Figure 3. Dependence of annual useful energy output on the intensity of solar radiation for the FPC collector system.

The results of these measurements can be approximated (dashed line in Figure 3) by the following equation: $Q_{SC} = 0.2685 \cdot Q_{SC} + 98.946$.

The error resulting from comparing the result of the computer simulation (circle in Figure 3) for a typical meteorological year (TMY) with the value calculated from the above equation for TMY conditions is 5.87%. Therefore, it should be considered that the developed model of SDHW system reflects the trend of the real complex physical process with relatively good accuracy.

2.3. Mathematical Model for Determining the Solar Collector Useful Energy Output

Mathematical modelling was used as the research method. It allows to describe the operation of the tested object, determine the necessary output data, and determine the optimal values of the input parameters using mathematical relationships. The use of mathematical modelling allows you to resign from physical modelling, reduce the number of samples, and the labor intensity compared to the physical experiment [18]. The main component in such procedure is the mathematical model. Utilitarianism, efficiency, and practical usefulness of this type of model can be ensured by developing short algorithms. In order to obtain the desired information, the most important factors/parameters describing the object under study or phenomenon should be used.

The main goal of this research was to determine the influence of the most important parameters on thermal performance of SDHW systems in European climatic conditions. Thus, it was assumed that the annual useful energy output Q_{SC} of the solar collectors is an objective function Y, and the value of Y depends on the following parameters:

- Volume of heat storage tanks V_{ST} [m^3] (variable X_1),
- Solar collector total area F_{SC} [m^2] (variable X_2),
- Intercept (maximum) efficiency of solar collector η_o [-] (variable X_3),

- Heat loss coefficient a_1 [W/(m^{2}K)] (variable X_4),
- Daily consumption of DHW m [kg/day] (variable X_5).

The useful energy output makes a clear physical sense and is strongly explicit. The selected factors (X_1–X_5) are measurable, controllable, independent, and consistent, i.e., they meet the basic requirements of mathematical modelling [19]. It should be noted that climatic conditions cannot be considered as a parameter in such mathematical models. This factor is presented by a set of various meteorological indicators, which are difficult to combine with one coherent parameter. In optimization, this is the so-called heterogeneous and uncontrollable parameter. As mentioned before, it was decided to choose three different locations: (group I) Madrid (Spain)—($Q_{I,SC} = Y_{I,i}$), (group II) Budapest (Hungary)—($Q_{II,SC} = Y_{II,i}$), and (group III) Helsinki (Finland)—($Q_{III,SC} = Y_{III,i}$), the characteristics of which are shown in Table 3. The parameters listed in this table were calculated using the TRNSYS software using the data for a typical meteorological year. Additionally, three charts that show the variability of the basic meteorological parameters during the year for three selected locations are created. Figure 4 shows the average daily air temperature, Figure 5 shows wind velocity, and the distribution of total horizontal solar radiation is presented in Figure 6.

Table 3. List of the main modules of the SDHW system model.

Group Number	Town	Average Annual Outside Temperature [°C]	Average Annual Wind Velocity [m/s]	Solar Beam Radiation on Horizontal Surface [kWh/m^2]	Solar Diffuse Radiation on Horizontal Surface [kWh/m^2]
I	Madrid	13.9	2.5	1006	654
II	Budapest	10.4	3.0	529	668
III	Helsinki	4.5	3.5	463	483

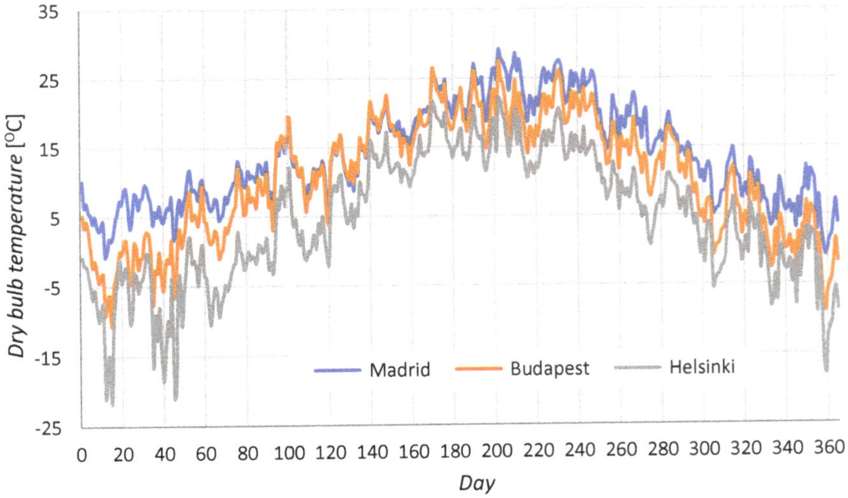

Figure 4. Average daily dry bulb temperature.

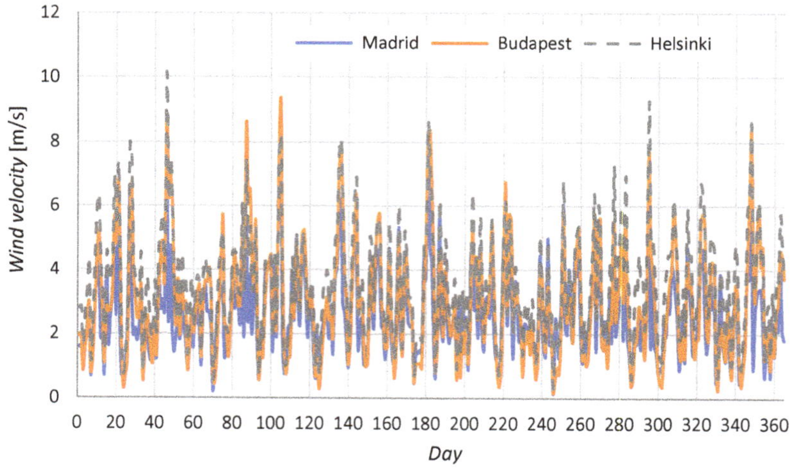

Figure 5. Average daily wind velocity.

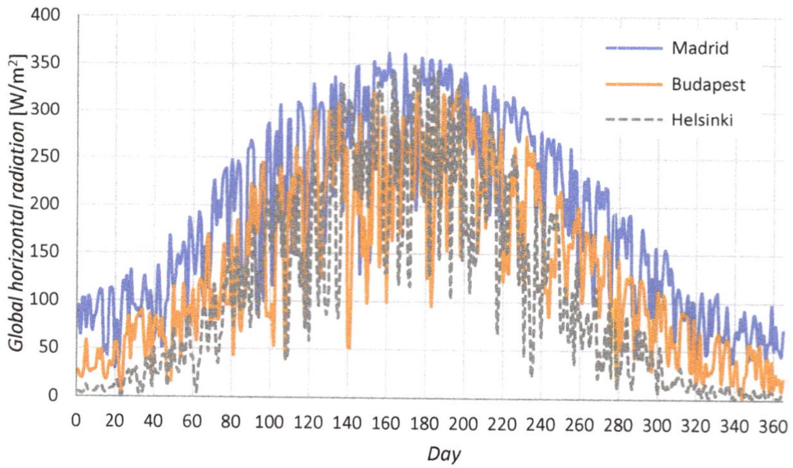

Figure 6. Average daily global solar radiation incident on a horizontal plane.

As can be seen from the characteristics presented in Table 3, the best conditions for converting solar energy into the heat exist in Madrid. This location is characterized by a high average annual temperature of the outside air, relatively low windiness, and a high value of solar radiation with a higher share of beam fraction (61%). Budapest is a city with moderate weather conditions. In this case, the total intensity of solar radiation is relatively high, but a predominance of the diffusion fraction (56%) is observed. The worst conditions in terms of installation of solar collectors are in Helsinki. There is a low outside air temperature, the lowest insolation value, and a relatively high average wind velocity.

Each of these locations differs in latitude and the share of direct and diffused solar radiation. Therefore, the optimal solar panel inclination angle β has been determined for each city separately. The results of the multivariate calculations are presented in the form of three graphs in Figure 7 for Madrid, Figure 8 for Budapest, and Figure 9 for Helsinki. The maximum annual useful energy output allowed to determine the optimal value of β, which is 36 degrees for Madrid, 39.5 degrees for Budapest, and 47.5 degrees for Helsinki. In the case of Madrid, we can observe an interesting phenomenon. It consists in the fact that the optimal inclination of the collector is in a wide range from 35 to 44 degrees with only small deviations. The opposite is Budapest, where we have a very clearly marked maximum.

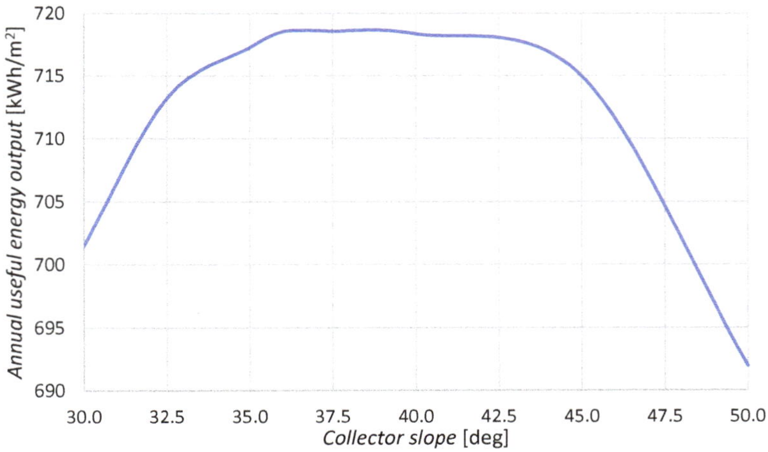

Figure 7. Dependence of the inclination angle of the solar collector on annual useful energy output for Madrid.

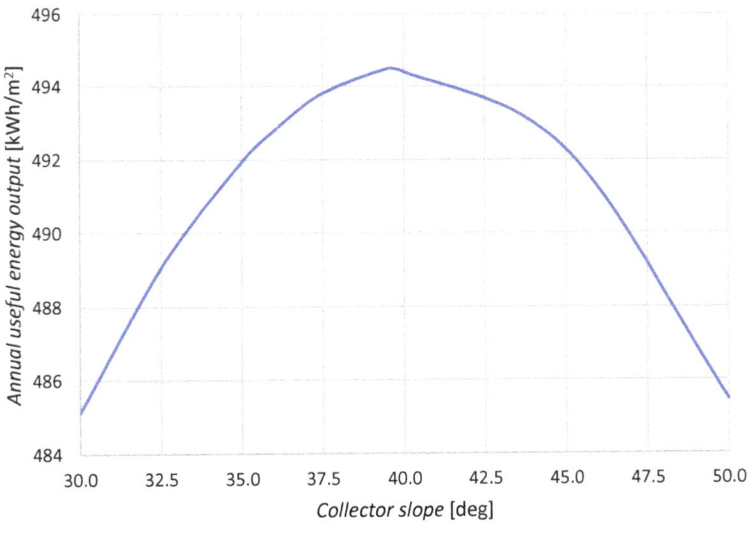

Figure 8. Dependence of the inclination angle of the solar collector on an annual useful energy output for Budapest.

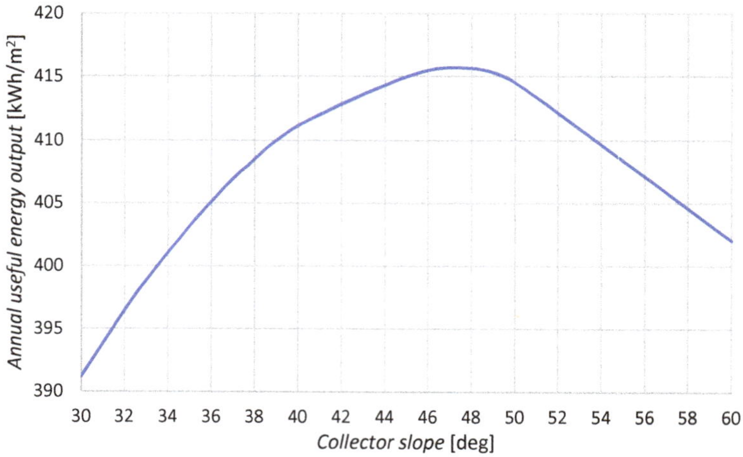

Figure 9. Dependence of the inclination angle of the solar collector on an annual useful energy output for Helsinki.

Thus, due to the heterogeneous nature of the weather factors, three new mathematical models for each of the groups of climatic conditions were developed. It was assumed that the relationship $Y = f\cdot(X_1, X_2, X_3, X_4, X_5)$ can be described as a second-degree polynomial equation in the following form:

$$Y = a_0 + a_1X_1 + a_2X_2 + a_3X_3 + a_4X_4 + a_5X_5 + a_{12}X_1X_2 + a_{13}X_1X_3 + a_{14}X_1X_4 + a_{15}X_1X_5 + a_{23}X_2X_3 + a_{24}X_2X_4 + a_{25}X_2X_5 + a_{34}X_3X_4 + a_{35}X_3X_5 + a_{45}X_4X_5 + a_{11}X_1X_1 + a_{22}X_2X_2 + a_{33}X_3X_3 + a_{44}X_4X_4 + a_{55}X_5X_5 \quad (1)$$

The ranges of variability for each of the factors were assumed at three levels:

- Low-marked as (−1),
- Medium-marked as (0),
- High-marked as (1).

Table 4 presents the natural values of the factors $\dot{X}_1 \div \dot{X}_5$ and the corresponding normalized values $(X_1) \div (X_5)$, which are defined by the following formula [19]:

$$X_i = \frac{\dot{X}_i - 0.5\cdot\left(\dot{X}_{i,max} + \dot{X}_{i,min}\right)}{0.5\cdot\left(\dot{X}_{i,max} - \dot{X}_{i,min}\right)} \quad (2)$$

where: $\dot{X}_i$, $\dot{X}_{i,max}$, $\dot{X}_{i,min}$—current, maximum and minimum natural values of the i-th factor, respectively.

Table 4. Natural and standardized values of the factors.

Factor Level $\dot{X}_i$	V_{ST} [m³] (X_1)	F_{SC} [m²] (X_2)	η_o [-] (X_3)	a_1 [W/(m²K)] (X_4)	m [kg/d] (X_5)
Low (−1)	2.0	30.0	0.58	2.0	1200
Medium (0)	7.0	100.0	0.78	6.0	4800
High (+1)	12.0	170.0	0.98	10.0	8400
Range of variability ΔX_i	5.0	70.0	0.20	4.0	3600

The range of collector surfaces, volumes of accumulation tanks, and consumption of DHW have been selected to represent the SDHW medium-sized system.

In order to create a database for the description of Y relationship (Equation (1)), a series of computer simulations were performed in TRNSYS environment software. Based on a symmetrical three-level plan [18], only 26 series of calculations were made. The $Q_{SC,i}$ (Y_i) estimation results for three different locations are shown in Table 5.

Table 5. Planning matrix and calculation results of annual useful energy output [kWh/m^2].

No	V_{ST}	F_{SC}	η_o	a_1	m	$Q_{I,SC}$	$Q_{II,SC}$	$Q_{III,SC}$
	X_1	X_2	X_3	X_4	X_5	$Y_{I,i}$	$Y_{II,i}$	$Y_{III,i}$
1	2 −1	30 −1	0.58 −1	2 −1	8400 +1	721	504	425
2	12 +1	30 −1	0.58 −1	2 −1	1200 −1	570	407	339
3	2 −1	170 +1	0.58 −1	2 −1	1200 −1	118	108	98
4	12 +1	170 +1	0.58 −1	2 −1	8400 +1	500	349	293
5	2 −1	30 −1	0.98 +1	2 −1	1200 −1	641	545	474
6	12 +1	30 −1	0.98 +1	2 −1	8400 +1	1289	901	758
7	2 −1	170 +1	0.98 +1	2 −1	8400 +1	505	438	373
8	12 +1	170 +1	0.98 +1	2 −1	1200 −1	145	141	133
9	2 −1	30 −1	0.58 −1	10 +1	1200 −1	286	192	161
10	12 +1	30 −1	0.58 −1	10 +1	8400 +1	570	391	331
11	2 −1	170 +1	0.58 −1	10 +1	8400 +1	133	99	85
12	12 +1	170 +1	0.58 −1	10 +1	1200 −1	75	52	45
13	2 −1	30 −1	0.98 +1	10 +1	8400 +1	799	575	490
14	12 +1	30 −1	0.98 +1	10 +1	1200 −1	508	355	300
15	2 −1	170 +1	0.98 +1	10 +1	1200 −1	97	75	64
16	12 +1	170 +1	0.98 +1	10 +1	8400 +1	294	203	171
17	2 −1	100 0	0.78 0	6 0	4800 0	376	272	229
18	12 +1	100 0	0.78 0	6 0	4800 0	450	310	260
19	7 0	30 −1	0.78 0	6 0	4800 0	791	544	455
20	7 0	170 +1	0.78 0	6 0	4800 0	296	207	173
21	7 0	100 0	0.58 −1	6 0	4800 0	330	227	191

Table 5. *Cont.*

No	V_{ST}	F_{SC}	η_o	a_1	m	$Q_{I,SC}$	$Q_{II,SC}$	$Q_{III,SC}$
	X_1	X_2	X_3	X_4	X_5	$Y_{I,i}$	$Y_{II,i}$	$Y_{III,i}$
22	7 0	100 0	0.98 +1	6 0	4800 0	544	384	322
23	7 0	100 0	0.78 0	2 −1	4800 0	661	505	425
24	7 0	100 0	0.78 0	10 +1	4800 0	309	213	178
25	7 0	100 0	0.78 0	6 0	1200 −1	193	159	139
26	7 0	100 0	0.78 0	6 0	8400 +1	509	354	297

The results of these calculations $Q_{I,SC}$, $Q_{II,SC}$, $Q_{III,SC}$ (Table 5) were used to approximate the functions of the examined objects. The least squares method was chosen for this approximation. This procedure was performed using matrix calculus $\mathbf{A} = (\mathbf{X}^T \mathbf{X})^{-1} \mathbf{X}^T \mathbf{Y}$. In this vector notation: **A** denotes the column vector of the coefficients a_i, **X**-experiment plan matrix, **Y**-column vector of the calculation results, and superscript **T** means matrix transpose. In this way, the three mathematical models were developed:

Madrid

$$\begin{aligned} \hat{Y}_I = \; & 444.07 + 40.28X_1 - 222.80X_2 + 84.39X_3 - 115.50X_4 + 149.27X_5 - 20.56X_1X_2 - 16.43X_1X_3 \\ & -24.18X_1X_4 + 21.19X_1X_5 - 54.68X_2X_3 + 24.31X_2X_4 - 23.56X_2X_5 - 2.31X_3X_4 + 38.81X_3X_5 - 44.43X_4X_5 \\ & -31.07X_1X_1 + 99.42X_2X_2 - 7.08X_3X_3 + 40.92X_4X_4 - 93.07X_5X_5, \end{aligned} \quad (3)$$

Budapest

$$\begin{aligned} \hat{Y}_{II} = \; & 313.51 + 16.72X_1 - 152.33X_2 + 71.56X_3 - 96.83X_4 + 98.89X_5 - 13.3X_1X_2 - 20.56X_1X_3 \\ & -8.94X_1X_4 + 12.06X_1X_5 - 39.56X_2X_3 + 14.81X_2X_4 - 9.94X_2X_5 - 11.43X_3X_4 + 26.06X_3X_5 - 24.81X_4X_5 \\ & -22.51X_1X_1 + 61.98X_2X_2 - 8.02X_3X_3 + 45.48X_4X_4 - 51.01X_5X_5, \end{aligned} \quad (4)$$

Helsinki

$$\begin{aligned} \hat{Y}_{III} = \; & 262.68 + 12.83X_1 - 127.66X_2 + 62.02X_3 - 82.94X_4 + 81.67X_5 - 9.75X_1X_2 - 17.37X_1X_3 \\ & -6.63X_1X_4 + 10.10X_1X_5 - 34.12X_2X_3 + 11.38X_2X_4 - 9.25X_2X_5 - 11.25X_3X_4 + 20.63X_3X_5 - 18.62X_4X_5 \\ & -18.18X_1X_1 + 51.31X_2X_2 - 6.19X_3X_3 + 38.81X_4X_4 - 44.68X_5X_5. \end{aligned} \quad (5)$$

A characteristic feature of deterministic models is their unambiguous agreement between the external influence and the reaction to that influence. This principle was applied when the adequacy of the model was checked. The relevance of the Equations (3)–(5) were assessed by comparing the variance of the mean value S_y^2 and the residual variance S_r^2:

$$S_y^2 = \frac{\Sigma\left(Y_i - \overline{Y}\right)^2}{N-1}, \quad (6)$$

$$S_r^2 = \frac{\Sigma\left(\hat{Y}_i - \overline{Y}\right)^2}{N-N_b}, \quad (7)$$

where N is the number of calculations (N = 26) and N_b is the number of coefficients in the regression equation (N_b = 21).

The Fisher criterion [18,20], commonly known as the *F*-test, was used to exam the above models:

$$F = \frac{S_y^2(N-1)}{S_r^2(N-N_b)}. \tag{8}$$

The regression equation approximates the calculation results correctly if the Fisher criterion F is much higher than the tabular value F_t. For Equations (3)–(5), the value of F is 152.47, 30.98, and 32.38 for Madrid, Budapest, and Helsinki, respectively. However, the tabular value of F_t [19] is smaller and equal to 4.52. Therefore, it can be concluded that the developed models are adequate because F_t is many times lower than F. A particularly good fit of the regression predictions to the calculation results is confirmed by the coefficient of determinations R^2, which are in the range from 0.9935 to 0.9942. Additionally, the significance of the coefficients of the regression equations was also checked using the *t*-criterion [19], with a positive result. The description of this procedure was omitted due to its complexity. Therefore, it was found that the models (Equations (3)–(5) should be considered useful for further analysis.

In conclusion, it should be mentioned that the data set for the development of the mathematical model and its subsequent testing was created using the TRNSYS software. The model was validated on the basis of the measurement data collected by the corresponding author of this paper.

3. Results and Discussion

The analysis of the influence of the examined factors $X_1 \div X_5$ on the annual useful energy output Q_{SC} of the collector was performed on mathematical models described by equations Equations (3)–(5). The center of the multivariate-space is formed by the average level of coefficients, i.e., in this case:

- Volume of heat storage tanks V_{ST} = 7 m^3 (X_1 = 0),
- Solar collector total area F_{SC} = 100 m^2 (X_2 = 0),
- Intercept (maximum) efficiency of solar collector η_o = 0.78 (X_3 = 0),
- Heat loss coefficient a_1 = 6 W/(m^{2}K) (X_4 = 0),
- Daily consumption of DHW m = 4800 kg/day (X_5 = 0).

The annual useful energy output of the collector for selected groups of weather conditions are as follows: for Madrid (group I) $Q_{I.SC}$= 444 kWh/m^2, for Budapest (group II) $Q_{II.SC}$= 313 kWh/m^2, and for Helsinki (III group) $Q_{III.SC}$= 263 kWh/m^2. As can be seen from the above results, the Q_{SC} value for the same solar collector fluctuates significantly, i.e., compared to Madrid it decreases by 29.5% for Budapest and as much as 41.0% for Helsinki. It is caused by changes in various climatic parameters, especially by total tilted surface solar radiation, which decrease by 27.6% for Budapest and 38.1% for Helsinki. However, when comparing the decrease in solar radiation on a horizontal plane, these values are slightly different: 27.9% for Budapest and 43.0% for Helsinki. This proves that the use of simple mathematical equations where Q_{SC} is only a function of the radiation intensity can cause some errors in designing SDHW systems.

The influence of each factor on the Q_{SC} was then estimated from a mathematical point of view. In each of the considered models, the level of parameters impact varies. When changing the selected factors from the lower to the higher level, as could be expected, the energy output increases with the increase of the $V_{ST}(X_1)$, $\eta_o(X_3)$, $m(X_5)$ and decreases with increasing factors F_{SC} (X_2) and $a_1(X_4)$. The effects of changes in $\Delta Q_{i.SC}$ (natural and percentage) for each model are given in Table 5. As can be seen, weather conditions do not change the nature of the influence of the factors under consideration, but strongly change the degree of this influence. The most interesting and useful information for designers is the data on the interaction of factors in the obtained models. When analyzing the signs and values of the coefficients with double interactions certain regularities were noticed. They apply to all locations because the nature of the influence of the factors does not change in them.

Analysis of the results from Table 6 shows that increasing the capacity of the storage tank from 2 m^3 to 12 m^3 causes a change in Q_{SC}: the highest increase by 21.6% is achieved in hot weather conditions (Madrid) and the smallest by 11.1% in more severe climatic conditions (Helsinki).

Table 6. Effects of changing $X_1 \div X_5$ factors from lower to higher level on annual useful energy output [kWh/m^2/%].

Group Number	V_{ST}	F_{SC}	η_o	a_1	m
		X_2	X_3	X_4	X_5
I	+80.56 (21.6%)	−445.60 (−58.2%)	+168.78 (+47.9%)	−231.00 (−38.5%)	+298.54 (+148%)
II	+33.44 (+12.2%)	−304.66 (−57.7%)	+143.12 (+61.2%)	−193.66 (−42.5%)	+197.78 (+120.9%)
III	+25.66 (+11.1%)	−255.32 (−57.8%)	+124.12 (+63.8%)	−165.88 (−43.1%)	+163.34 (+119.8%)

Furthermore, it can be stated that increasing the collector area from 30 to 170 m^2 strongly reduces the collector's energy output and shows the highest negative effect among the factors under consideration. Despite the fact that the natural values of this decrease strongly fluctuate in various weather conditions, the percentage decrease in all locations is around −58%.

While change the optical efficiency of the collector η_o from 0.58 to 0.98 significantly increases the collector's energy production, which ranges from 48% for Madrid to 64% for Helsinki. The increase in heat loss of the collector, characterized by the a_1 coefficient, significantly reduces the Q_{SC} value. The average percentage change in this value is about 41% and slightly differs for all meteorological conditions.

The daily DHW consumption has the highest (positive) impact on the collector's energy yield. Q_{SC} increases from 148% (Madrid) to 120% (Helsinki) with a change of m from 1200 kg/day to 8400 kg/day.

The interactions between the five parameters tested in this study are discussed below. Regarding the $V_{ST}(X_1)$, it can be observed that the positive effect of this factor decreases with the increase of the factors $F_{SC}(X_2)$, $\eta_o(X_3)$, $a_1(X_4)$, but increases with the increase of $m(X_5)$. For the two factors $V_{ST}(X_1)$ and $m(X_5)$ there is a synergism phenomenon, because when increasing them together, it affects more strongly than it would be when increasing them separately. The large negative effect of $F_{SC}(X_2)$ increases with the increase of the factors $V_{ST}(X_1)$, $\eta_o(X_3)$, $m(X_5)$, but it decreases with the increase of $a_1(X_4)$.

For the factor $\eta_o(X_3)$, its positive effect decreases with the increase of the factors $V_{ST}(X_1)$, $F_{SC}(X_2)$, $a_1(X_4)$, but increases with the increase of $m(X_5)$. For the two factors $\eta_o(X_3)$ and $m(X_5)$ there is also a synergy, as when each of them increase together the influence is stronger than it would be with a separate increase. For the factor $m(X_5)$, the high positive impact will be the stronger, with the greater factors $V_{ST}(X_1)$ and $\eta_o(X_3)$. For these two pairs of factors $V_{ST}(X_1)-m(X_5)$ and $\eta_o(X_3)-m(X_5)$ there is synergism, as each of them influences more strongly when increased together than with their separate increased. The positive influence of factor $m(X_5)$ will be smaller if the factors $F_{SC}(X_2)$ and $a_1(X_4)$ increase.

Next, the procedure of optimizing the value of energy produced by the collectors was carried out. As could be expected, the optimal values of the SDHW system parameters occur on the boundaries of the factor space. The maximum of the examined functions was achieved with the extreme values of the factors $X_2 = -1$, $X_3 = +1$, $X_4 = -1$, $X_5 = +1$. Only factor X_1 changed the optimal value in different locations.

Thus, the maximum annual useful energy output (Y_I) is 1303 kWh/m^2 for Madrid. $Y_{I,\,max}$ is achieved with the following parameters: the storage tank capacity V_{ST} = 12 m^3, the collector area F_{SC} = 30 m^2, the collector's optical efficiency η_o = 0.98, the linear heat loss coefficient a_1 = 2 W/m^2·K, and the daily DHW consumption m = 8400 kg/day. In the case of Budapest weather zone, the function Y_{II} has a maximum of 918.5 kWh/m^2 for storage tank capacity V_{ST} = 10.5 m^3, and for the function Y_{III} (Helsinki weather conditions) the maximum of 768 kWh/m^2 is obtained for V_{ST} = 10 m^3.

It should be noted that the above analysis makes perfect sense from the point of view of mathematical relations. However, in design practice, sometimes a slightly different approach to such issues is needed. It often happens that in practical conditions we cannot obtain the maximum or optimal value of an objective function $Y_{i,\,max}$ estimated as a result of solving the optimization problem. In the case of SDHW systems, this may be due to the inability to install the appropriate number of collectors due to the insufficient roof area. Sometimes it also happens that it is not possible to design an appropriate volume of hot water storage tanks due to the lack of space in the district heating substations.

Using mathematical models, Equations (3)–(5), we can check whether the designed system is far from the optimal parameters in a given situation. To visualize this issue, three charts have been drawn up. The first one shows the effect of changing the amount of accumulated hot water on the useful energy output of the collectors (Figure 10). The second one shows the change of DHW consumption effects on the same parameter characterizing the collectors (Figure 11). The impact of the collector surface on their performance is presented in Figure 12. The calculations were made for the parameters located in the center of the multivariate-space that is formed by the average level of coefficients, i.e., equal to 0.

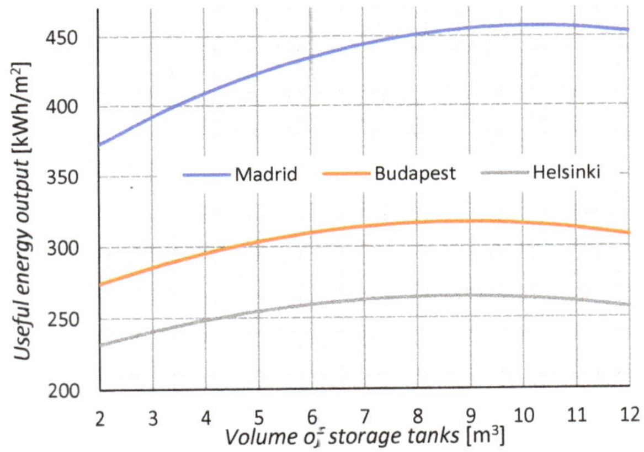

Figure 10. Dependence of the volume of heat storage tanks on an annual useful energy output.

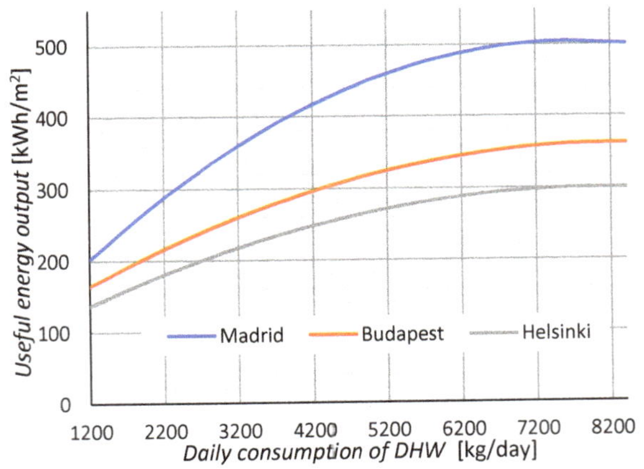

Figure 11. Dependence of the daily consumption of DHW on an annual useful energy output.

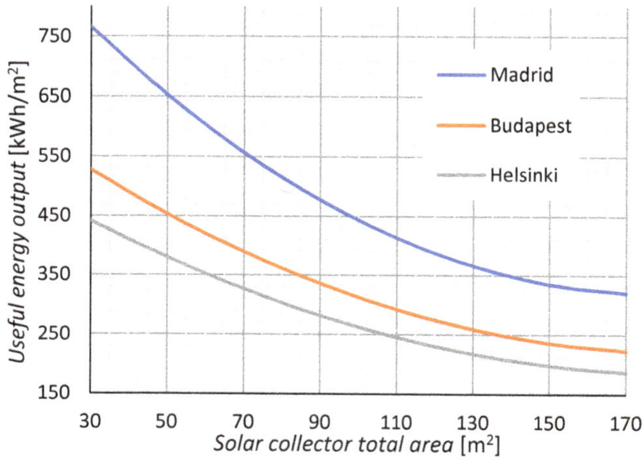

Figure 12. The influence of the collector surface area on an annual useful energy output.

The trend of the influence of the storage tanks capacity on the collector performance (Figure 10) clearly shows that an increase in tank volume above 8 m^3 does not make any practical sense in the case of Madrid, although the optimum is 10 m^3. For the other two locations (Budapest, Helsinki), the maximum volume is between 8 m^3 and 9 m^3. However, as in the previous case, reducing the storage tank capacity by 2 m^3 causes a decrease in useful energy output only by about 2.5%.

As can be seen in Figure 11, the change in the DHW consumption has a significant impact on the useful energy output of the collectors. An almost linear increase in the Q_{SC} of solar collectors in 70% of the range of DHW consumption change can be observed. In operating conditions, it often happens that the energy efficiency of the SDHW system is lower in summer, when the demand for DHW decreases due to holiday trips.

Sometimes it happens that an excessive number of collectors is designed when, for example, there is a large roof area to be used. As can be seen from the charts presented in Figure 12, the amount of energy converted into the heat by 1 m^2 of solar panels significantly decreases with increasing their surface area. So, as we can see, the oversizing of the SDHW system significantly reduces its efficiency.

Figure 13 shows the effect of DHW consumption on the outlet temperature T_{OUT} of the solar collectors. The calculations were made for Madrid, 30 m^3 of collector area and 4 m^3 of storage tank capacity in the period from 1 to 15 September. As could be expected, the highest temperature is achieved with the lowest DHW demand of 1200 kg/day. A 3.5-fold increase in daily consumption lowers T_{OUT} value by an average of 21.2%, while its 7-fold increase reduces this temperature by slightly less, i.e., by 25%. Lowering the temperature of the water flowing out of the solar collectors will be associated of course with the need to provide more energy from a supplementary heat source.

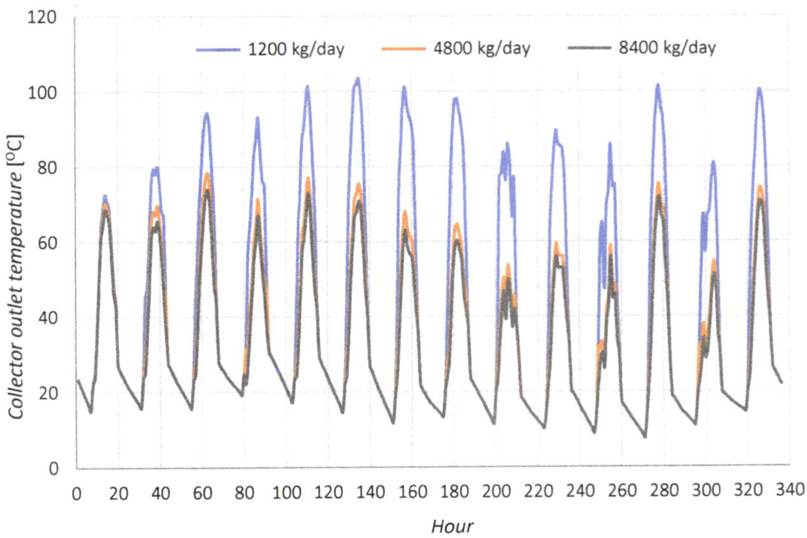

Figure 13. Outlet temperature from the solar collectors depending on the DHW consumption.

4. Conclusions

This article presents a mathematical approach to the issue of the influence of selected parameters on the thermal performance of a solar collector. Based on the results of the computational experiment for hot (Madrid), moderate (Budapest) and cold (Helsinki) climatic zones, three new deterministic mathematical models (Equations (3)–(5) were developed. These equations can be used to determine the influence of five factors: volume of heat storage tanks, solar collector total area, maximum efficiency, and heat loss coefficient of solar collector, and daily consumption of DHW on the annual useful energy output. Based on these models, the degree and nature of the influence of the examined factors were estimated. It should be noted that the analysis presented in this article is the most appropriate for medium-sized SDHW systems.

The change of the examined factors from the lower to the upper level showed significant fluctuations in their impact both for individual factors and for the weather zones (Table 5). The useful energy output Q_{SC} of thermal solar collectors to a large extent depends on the weather conditions and, above all, the intensity of solar radiation. The Q_{SC} value is 29.5% lower for Budapest and 41% lower for Helsinki compared to the most suitable solar conversion zone in which Madrid is located.

The results of the optimization procedure according to the energy criterion showed that the maximum annual useful energy output, which is possible to supply from solar collectors, is 1303 kWh/m^2 for Madrid weather conditions, 918.5 kWh/m^2 for Budapest, and 768 kWh/m^2 for Helsinki conditions.

The above results can be obtained at the optimum collector inclination β for a particular location, i.e., 36 degrees for Madrid, 39.5 degrees for Budapest, and 47.5 degrees for Helsinki. These values were not calculated based on the latitude that designers usually estimate it at. The novelty is the determination of the β value using the method described in this article.

It is planned to develop similar models for selected main cities in Europe. The obtained dependencies will allow to write a simple application that can be used for a relatively precise determination of the annual useful energy output of thermal collectors. It is also planned to carry out a similar mathematical experiment using Solar Fraction as the objective function.

Author Contributions: Conceptualization, M.Z. and W.J.; methodology, M.Z. and W.J.; energy simulations, M.Z.; mathematical modelling, W.J.; validation, M.Z.; formal analysis, W.J. and M.Z.; investigation, M.Z. and W.J.; writing—original draft preparation, M.Z. and W.J.; writing—review and editing, M.Z. and W.J.; project administration, M.Z.; funding acquisition, M.Z. and W.J. All authors have read and agreed to the published version of the manuscript.

Funding: This work was supported by the Bialystok University of Technology and financed by the Ministry of Science and Higher Education of the Republic of Poland [grant numbers WZ/WB-IIS/9/2019, WZ/WB-IIL/5/2019].

Institutional Review Board Statement: Not applicable.

Informed Consent Statement: Not applicable.

Data Availability Statement: Not applicable.

Conflicts of Interest: The authors declare no conflict of interest.

References

1. Paris Agreement. Available online: https://unfccc.int/process-and-meetings/the-paris-agreement/the-paris-agreement (accessed on 15 March 2021).
2. Decision of the European Commission. Available online: https://ec.europa.eu/clima/policies/eu-climate-action/2030_ctp_en (accessed on 15 March 2021).
3. Kulkarni, G.N.; Kedare, S.B.; Bandyopadhyay, S. Determination of design space and optimization of solar water heating systems. *Sol. Energy* **2007**, *81*, 958–968. [CrossRef]
4. Klein, S.A.; Beckman, W.A.; Duffie, J.A. Design Procedure for Solar Heating Systems. *Sol. Energy* **1976**, *18*, 113–126. [CrossRef]
5. Beckman, W.A.; Klein, S.A.; Duffie, J.A. *Solar Heating Design by the F-Chart Method*; Wiley & Sons, Inc.: New York, NY, USA, 1977.
6. Haberl, J.S.; Cho, S. *Literature Review of Uncertainty of Analysis Methods (F-Chart Program)*; Report to the Texas Commission on Environmental Quality; Energy Systems Laboratory, Texas A&M University: College Station, TX, USA, 2004.
7. Essabbani, T.; Moufekkir, F.; Mezrhab, A.; Naji, H. Numerical Computation of Thermal Performance of a Simulation of a Solar Domestic Hot Water System. *Appl. Sol. Energy* **2015**, *51*, 22–33. [CrossRef]
8. Ayompe, L.M.; Duffy, A.; McCormack, S.J.; Conlon, M. Validated TRNSYS model for forced circulation solar water heating systems with flat plate and heat pipe evacuated tube collectors. *Appl. Therm. Eng.* **2011**, *31*, 1536–1542. [CrossRef]
9. Hobbi, A.; Siddiqui, K. Optimal design of a forced circulation solar water heating system for a residential unit in cold climate using TRNSYS. *Sol. Energy* **2009**, *83*, 700–714. [CrossRef]
10. Mongibello, L.; Bianco, N.; Caliano, M.; de Lucad, A.; Graditi, G. Transient analysis of a solar domestic hot water system using two different solvers. *Energy Procedia* **2015**, *81*, 89–99. [CrossRef]
11. Lima, J.B.A.; Prado, R.T.A.; Taborianski, V.M. Optimization of tank and flat-plate collector of solar water heating system for single-family households to assure economic efficiency through the TRNSYS program. *Renew. Energy* **2006**, *31*, 1581–1595. [CrossRef]
12. Mehmood, A.; Waqas, A.; Said, Z.; Rahman, S.M.A.; Akram, M. Performance evaluation of solar water heating system with heat pipe evacuated tubes provided with natural gas backup. *Energy Rep.* **2019**, *5*, 1432–1444. [CrossRef]
13. Mohasseb, S.; Kasaeian, A. Comparing the Performance of Flat Plate Collector and Evacuated Tube Collector for Building and Industrial Usage in Hot and Cold Climate in Iran with TRNSYS Software. In Proceedings of the Ninth International Conference on Engineering Computational Technology, Napoli, Italy, 2–5 September 2014.
14. TRNSYS 18. *TRaNsient SYstem Simulation Program*; Tutorials; Solar Energy Laboratory, University of Wisconsin-Madison: Madison, WI, USA, 2020; Volume 9.
15. Ivanko, D.; Sorensen, A.L.; Nord, N. Selecting the model and influencing variables for DHW heat use prediction in a hotel in Norway. *Energy Build.* **2020**, *228*, 110441. [CrossRef]
16. Fuentes, E.; Arce, L.; Salom, J. A review of domestic hot water consumption profiles for application in systems and buildings energy performance analysis. *Renew. Sustain. Energy Rev.* **2018**, *81*, 1530–1547. [CrossRef]
17. Zukowski, M.; Radzajewska, P. A New Method to Determine the Annual Energy Output of Liquid-Based Solar Collectors. *Energies* **2019**, *12*, 4586. [CrossRef]
18. Durakovic, B. Design of Experiments Application, Concepts, Examples: State of the Art. *Period. Eng. Nat. Sci.* **2017**, *5*, 421–439. [CrossRef]
19. Korzynski, M. Metodyka eksperymentu. In *Planowanie, Realizacja i Statystyczne Opracowanie Wynikow Eksperymentow Technologicznych*; WNT: Warszawa, Poland, 2006.
20. Hartmann, K.; Lezki, E.; Schär, W. *Statistische Versuchsplanung und—Auswertung in der Stoffwirtschaft*; VEB: Leipzig, Germany, 1977.

energies

MDPI

Article

Designing for the Environment: An Example of Multi-Criteria Analysis Used for Solar Hot Water System Selection

Agnieszka Żelazna and Justyna Gołębiowska *

Faculty of Environmental Engineering, Lublin University of Technology, 40B Nadbystrzycka Str., 20-618 Lublin, Poland; a.zelazna@pollub.pl
* Correspondence: j.golebiowska@pollub.pl; Tel.: +48-8153-844-06

Abstract: In the European Union, the building sector accounts for more than 40% of final energy consumption, contributing to the deterioration of the quality of the environment. Among the various solutions that aim to reduce the negative environmental impact caused by the operation of buildings, solar hot water systems (SHW) are popular. The choice of a SHW system is associated with the comfort of use and the access to low-cost energy. The design guidelines include the technical parameters for system operation such as materials, dimensions, sizing and operation temperatures. However, the legitimacy of choosing a particular solution and the available technical parameters are key issues. In the presented study, a multi-criteria analysis was proposed as a basis for the proper selection of system parameters, e.g., collector type, solar tank volume. A model of the SHW system was used to calculate the possible solutions, ensuring the same comfort of usage for several design options. The analyzed model was then used for the calculation of three various indicators: Simple Payback Time (SPBT), Primary Energy consumption (PE) and IMPACT 2002+. The application of a multi-criteria analysis based on a Life Cycle Assessment allowed for beneficial solutions to be found from the point of view of economics, non-renewable resources and environmental protection.

Keywords: solar thermal energy; SHW system; multi-criteria analysis; SPBT; EPBT; primary energy; IMPACT 2002+; LCA; sustainable design

Citation: Żelazna, A.; Gołębiowska, J. Designing for the Environment: An Example of Multi-Criteria Analysis Used for Solar Hot Water System Selection. *Energies* **2022**, *15*, 65. https://doi.org/10.3390/en15010065

Academic Editor: Jan Danielewicz

Received: 24 November 2021
Accepted: 20 December 2021
Published: 22 December 2021

1. Introduction

At present, energy seems to be an essential element for living in a modern society. The change in our lifestyle observed in recent decades resulted in a significant increase in energy consumption, especially in the case of the electricity, but also in other forms such as heat, required to meet the rising housing needs [1–3]. The growing interest in new renewable technologies in recent years is caused by several issues, such as the depletion of fossil resources, necessity to reduce air pollution, response to the problem of global climate change and the social expectations in the field of raising living standards and the comfortable use of buildings [4,5].

Among the renewable energy sources, solar energy is one of the most popular options because of several advantages. The energy supplied from the Sun to our planet exceeds the human demand for energy per year by about 8000 times. In theory, this means that the electromagnetic radiation from the Sun is able to cover the global demand for energy without the additional use of other energy sources. This energy is almost unlimited and does not cause environmental pollution, unlike conventional energy sources. The current level of radiation energy conversion technology does not allow the Sun's energy potential to be utilized in its entirety, which is why there is an aim to improve the devices that convert solar radiation into heat or electricity. An important disadvantage of using solar energy potential is that its absorption requires a large amount of space and favorable conditions [6]. The amount of solar radiation reaching the Earth's surface decreases with many factors, such as increasing cloudiness, as well as air contamination [7,8].

In the scientific literature of the last several years, an increasing interest in the subject of Life Cycle Assessment of solar systems can be observed; this applies to both technological processes and specific devices or materials. Various Life Cycle Impact Assessment (LCIA) methods are used. These include: Global Warming Potential (GWP), Cumulative Energy Demand (CED) and Eco-indicator 99. Some studies deal with the Life Cycle Assessment of solar energy installations and devices, including domestic hot water preparation systems and solar collectors themselves.

Over time, there have been significant advancements achieved by various research groups in the field of solar–thermal systems LCA [9–13]. Some of these include the LCA analysis of flat solar collectors [14] from raw material extraction, through production to obtaining a finished commercial product. The environmental load resulting from the transport of materials to the production plant was also taken into account. This research was based on the energy and ecological characteristics of the input products. The highest value of environmental load was connected with the following impact categories: primary energy consumption and acidification. The energy payback period, understood as the time needed to recover the energy expenditure incurred for the production of the collector, excluding the other elements and energy required for operation, was shorter than 1 year [14].

Another study [15] presented the Life Cycle Impact Assessment carried out for a solar heating installation located in Greece. The production stages of individual elements of the installation and their impact on the environment were considered. The system boundaries include the production of raw materials (steel, glass, copper, and aluminum), operation of the installation and final disposal. The installation consisted of a flat solar collector with a copper absorber (4 m^2) and a domestic hot water tank with a capacity of 200 dm^3. An electric heater was an additional heat source. Studies showed that the system provided 1702 kWh per year, with solar energy accounting for 58.8% of this value. An LCIA analysis was carried out using the Eco-indicator 99 method, which showed that the largest environmental impact came from the solar hot water tank. Moreover, the greatest environmental burden was connected with the following impact categories: soil acidification, smog production and impact on global warming.

An environmental analysis was also used for the assessment of the selective surface for flat-plate solar collectors, which was carried out using the following methods: CED, GHG (Greenhouse Gas Protocol method based on the set of characterization factors by Intergovernmental Panel for Climate Change IPCC GWP 100a) and Eco-indicator 99. The authors compared the conventional method of selective surface production with the latest production method—roll to roll (R2R).

Research showed that the annual energy demand for the production of 200,000 m^2 of selective surface is equal 7.91 TJ. Additional research based on CED allowed the energy payback time for R2R production to be determined, which was only 3–4 days. Among the analyzed factors, copper has the greatest impact on the environment, as it also requires the highest amount of energy in the production of selective surfaces. The results of the analysis clearly showed that the production of the selective surface for flat-plate solar collectors with the conventional method was much more harmful to the environment than the production by means of the roll-to-roll method [16].

A study in Munich (temperate climate conditions) presented an analysis of the life cycle of two hot water preparation systems with glazed solar collectors and unglazed polypropylene collectors. The Eco-indicator 99 method was used to assess the entire life cycle. The main components of both systems were solar panels and a 300 dm^3 buffer tank. The glazed panel consisted of an outer aluminum frame, two layers of mineral wool, a copper absorber and a copper alloy absorber piping system. The whole construction was hermetically closed with 4 mm thick transparent glass. In contrast, the plastic collector considered in this study was a polypropylene strip 7 mm thick, 313 mm wide and of variable length, which allowed the panel to be modular. The duration of the operation of both installations was 10 years. For glazed solar collectors, the calculated index was 49.7 points, while for plastic collectors it was 18.7 points. The largest share here was the use of fossil fuels: 52% and 68%, respectively. On the other hand, the GWP calculated over a period of 100 years was 1670.92 kg CO_{2eq} for a glazed solar collector and 105.06 kg CO_{2eq} for a plastic collector. The economic payback time, in comparison to the natural gas boiler, is in the range 9–11 years/8–13 years for the system with unglazed/glazed panels, respectively [17].

To conclude, it should be emphasized that the analyses carried out so far concern the production of materials, the entire life cycle of the installation or case studies [18–21]; however, attention is rarely paid to proper design parameters. In the literature, few research works address the topic of the life cycle of installations that operate under the temperate climatic conditions, such as the Polish climate, and relate to a long-term perspective. The present study significantly supplements the current state of knowledge due to the implementation of the selected design criteria into the analyzed model of life cycle.

2. Materials and Methods

In the current study, an attempt to conduct a holistic, multi-criteria analysis of a solar thermal system was presented in order to set the designer's guidelines and frames for correct system selection. A typical installation of solar hot water system (SHW), supported by an electric heater in the winter period was chosen as the subject of the analysis. The scheme of the typical solar installation is presented in Figure 1. The analysis included 6 variants of installations located in Central Europe (Poland, Łódź: 51.759247° N, 19.455982° E), with orientation towards the south (azimuth 0°) and inclination of 40°. The daily domestic hot water demand was assumed at constant level of 180 L. The characteristics of the technical parameters for different variants of installations are shown in Table 1. The aim of the modeling was to estimate the influence of selected design parameters (type of collector, piping and water tank capacity) on the economical, energetic and environmental aspects of SHW operation.

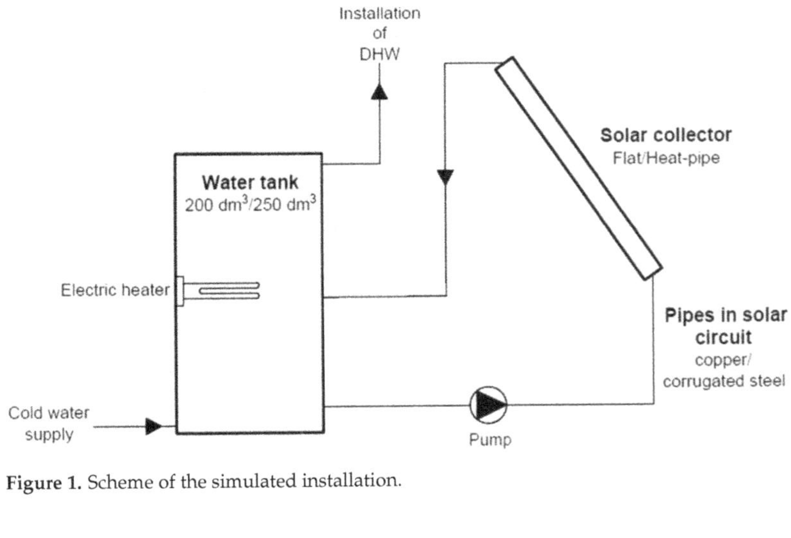

Figure 1. Scheme of the simulated installation.

Table 1. Technical parameters of the analyzed system variants.

Variant of Installation	Solar Collector	Absorber Area	Pipes	Water Tank	Solar Energy	Annual Efficiency
-	-	m^2	-	dm^3	kWh/yr	%
Variant I	Flat-plate, CU absorber	4.66	Copper	200	1602	30
Variant II	Flat-plate, CU absorber	4.66	Copper	250	1627	30.5
Variant III	Flat-plate, CU absorber	4.66	Steel (corrugated)	200	1576	29.5
Variant IV	Flat-plate, CU absorber	4.66	Steel (corrugated)	250	1600	30
Variant V	Heat pipe	4.29	Copper	200	1842	37.5
Variant VI	Heat pipe	4.29	Copper	250	1865	37.9
Variant VII	Heat pipe	4.29	Steel (corrugated)	200	1810	36.8
Variant VIII	Heat pipe	4.29	Steel (corrugated)	250	1834	37.3

The base scenario for comparison with the design options included an electric heater, while the annual electricity consumption equaled 2727 kWh. Modeling of solar energy gains was performed by using GetSolar software (Hottgenroth Software GmbH & Co. KG, Germany), a popular tool using f-chart-based estimation of energy gains from SHW systems in a chosen location (Łódź, Poland), characterized by the total annual solar radiation of 3710.8 MJ/m^2 and the average annual total insolation of 1560.4 h.

The main inputs to the analysis include: solar irradiance on collector plane and ambient temperature data (from WetSyn weather data synthesis program integrated with GetSolar software), type of solar system (single, combined, etc.), collector parameters (including optical efficiency and collector area), heat losses, water tank volume, inclination and azimuth, mass flow of solar fluid, pipes length and pipes diameter. The more detailed description of the f-chart model can be found in [22].

A detailed inventory, based on cost of materials and energy consumption estimated for every variant of system, was then used for the detailed Life Cycle Assessment. Modeling of the life cycle was based on the use of SimaPro v. 8.0.5.13 software (PRE Consultans, Amersfoort, The Netherlands) and Ecoinvent v. 3 database (Ecoinvent Assoiation, Zurich, Switzerland), which was used for the description of unit processes. Life Cycle Impact Assessment was performed by means of the the IMPACT 2002+ and CED v. 3 methods.

The IMPACT 2002+ method allows for the assessment of environmental stress connected with the life cycle of the solar system. The indicator (expressed in points: Pt) links all types of inventory results throughout the fourteen midpoint categories into four main damage categories: Human Health, Ecosystem Quality, Climate Change and Resources. The higher the indicator calculated by using this method, the greater the environmental impact of the system [23]. The functional unit was 1 kWh of energy generated by the system.

Cumulative Energy Demand (CED) method was used to estimate the total primary energy consumption (expressed in MJ) of the solar system throughout its entire lifetime, including the direct and indirect energy streams associated with the particular phases of life cycle [24].

For the validation of final results in CED, Monte Carlo statistical simulations (lognormal distribution) were implemented in order to compare the possible error between the achieved results.

In the multi-criteria analysis, the economic and environmental issues were discussed according to the following scheme. Firstly, Simple Payback Time (SPBT) was calculated for every variant of installation according to the definition, expressing the total cost of investment and operation divided by yearly savings [11].

In the next step, Energy Payback Time (EPBT) was calculated as a basis for technological foundations of the functioning of the system. EPBT was based on the CED results: in this case, statistical processing of data was a basis for the determination of the confidence interval. In the third part of evaluation, IMPACT 2002+ was calculated as an environmental criterion. The results obtained during the mentioned stages were then evaluated to indicate the most advantageous option with the use of the Weighted Sum Method.

3. Results

In the following subsection, the results of the economic, energetic and environmental evaluation of the analyzed SHW systems are presented with special attention paid to the financial and energetic payback times, as well as to the environmental damage categories affected by the production and operation of the systems. All of the detailed calculations are finally used to evaluate the selected installation options based on the ratings awarded in the multi-criteria analysis.

3.1. Solar Hot Water Systems: Economic Evaluation

The economic evaluation of the SHW system in variants I–VIII was based on the on-line price catalogues and average market cost of installation, with the use of direct contact with the owners and investors. The data on the operation of the installation and service costs were obtained from the users of the installation in the form of an interview confirmed by billing documentation.

The pump operation cost was based on the measured energy consumption in a 3-year operation period of heat-pipe collector installation [11] and the price of energy in Poland in 2018 (0.5055 PLN/kWh recalculated to 0.1113 Euro/kWh) according to the Energy Regulatory Office [25].

According to the data presented in Table 2, the Simple Payback Time in the basic option (investment cost covered by the owner only) exceeds 11 years in all of the variants, assuming a constant price of electricity. However, while considering the popular support programs such as Prosument 2, the payback time may be shorter. In this case, financing the investment by taking up credit is a clear approach, which may worsen the presented results. In the case of the support programs financed by the local authorities, where the discount may reach 70%, SPBT decreases by almost threefold to 4.1 years in the most effective options of installation.

Table 2. Results of economic evaluation for the analyzed system variants.

Variant of Installation	Investment Cost	Pump Operation Cost	Servicing/Utilization Cost	Savings	SPBT *	SPBT **	SPBT ***
-	Euro	Euro/yr	Euro	Euro/yr	Years	Years	Years
Variant I	1795.43	13.56	220.26	164.8	12.2	9.8	4.6
Variant II	1998.07	13.77	220.26	167.4	13.3	10.5	4.9
Variant III	1759.91	13.34	220.26	162.1	12.2	9.8	4.6
Variant IV	1962.56	13.54	220.26	164.6	13.3	10.6	4.9
Variant V	1871.53	15.59	220.26	189.5	11.0	8.8	4.1
Variant VI	2074.17	15.78	220.26	191.9	12.0	9.5	4.4
Variant VII	1836.01	15.32	220.26	186.2	11.0	8.8	4.1
Variant VIII	2038.66	15.52	220.26	188.7	12.0	9.5	4.4

*—own investment, **—Prosument 2 program, 30% discount, 10 years credit, ***—local programs with 70% discount.

3.2. Solar Hot Water Systems: Energy Balance

Energy balance of the SHW system was considered with the use of the CED method, as well as the calculation of the energy generated by the solar system. It is worth underlining that the predicted energy consumption in the 20-year operation period was included in the calculation, according to Figure 2.

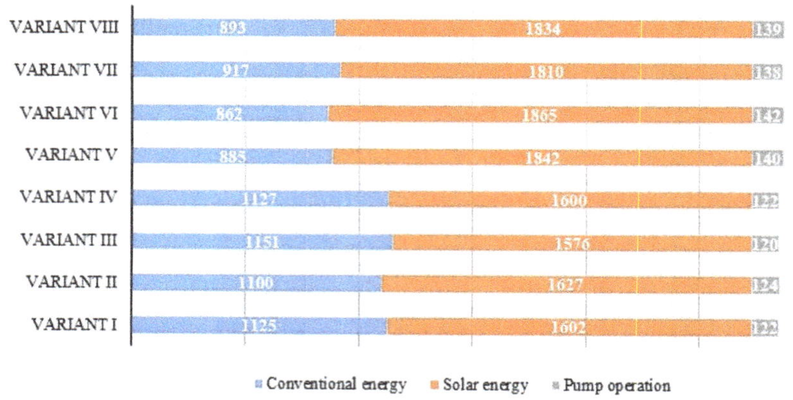

Figure 2. Energy balance of one-year operation of system according to the applied model.

The obtained EPBT results (Table 3) are higher than in some literature studies [17]. This may be caused by both the differences in solar irradiation, as well as the type of system and source of energy used for pumping. In all the considered cases, the EPBT indicator was significantly shorter than the predicted lifespan, assumed as 20 years.

Table 3. Results of energetic and economic evaluation.

Variant of Installation	CED	Energy Generated	EPBT
	MJ	MJ/yr	Years
Variant I	50,513.62	5767.20	8.76
Variant II	51,545.74	5857.20	8.80
Variant III	50,849.69	5673.60	8.96
Variant IV	51,364.05	5760.00	9.00
Variant V	53,830.63	6631.20	8.12
Variant VI	54,819.90	6714.00	8.17
Variant VII	54,038.12	6516.00	8.29
Variant VIII	55,048.82	6602.40	8.34

The results of LCIA conducted using CED indicator were subjected to a sensitivity analysis by means of the Monte Carlo statistical method (Figure 3). All system inputs and outputs were successively changed in the frames of standard deviation in 1000 trials, which allowed the scale of possible error to be shown in the calculations. The confidence level was assumed as 95%. In all the variants, the scale of possible error was similar, and the statistical distribution was positively skewed with a higher mean than the median in all analyzed cases.

3.3. Solar Hot Water Systems: Environmental Impact Indicators

The impact on the environment, considered as ecosystems, resources, human health and climate change, was assessed with the IMPACT2002+ method. The results of the impact assessment stage, recalculated to the functional unit (1 kWh of energy generated by the system) are presented in Figure 4. The share analysis on the level of processes shows that the electricity used for pumping is the most influential process in the SHW system life cycle. Other important elements include solar collectors and water tanks due to the high consumption of materials.

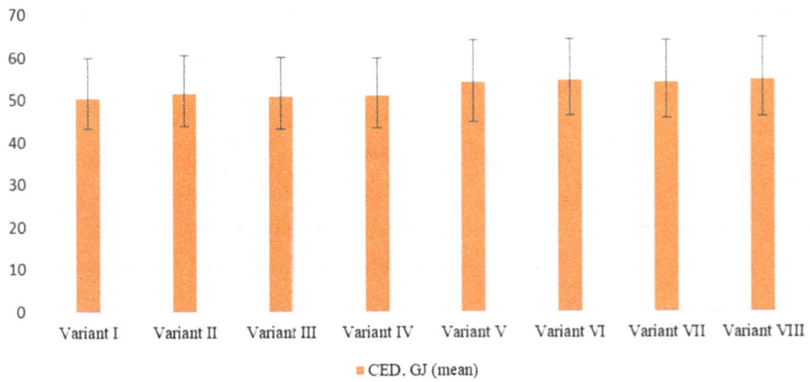

Figure 3. Results of CED calculation with the statistical error based on the Monte Carlo analysis.

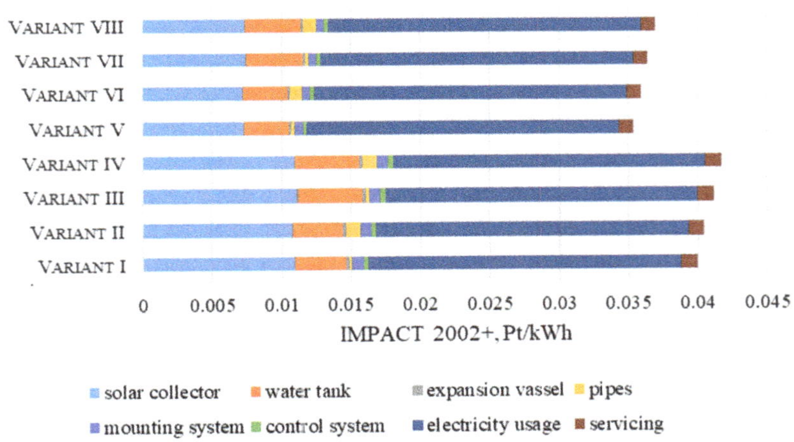

Figure 4. Results of IMPACT 2002+ calculation per functional unit, Pt/kWh.

According to the share analysis on the damage categories level, the results obtained for the analyzed systems are similar for Variants I-VIII, with the highest contribution from the Human health, Climate change and Resources categories. This is caused mostly by the similarities in construction and the supporting conventional energy used (Figure 5).

3.4. Solar Hot Water Systems: Multi-Criteria Analysis

The analysis performed in this study aimed to compare several alternatives of solar hot water system construction on the basis of SPBT, EPBT and IMPACT 2002+ indicators. Assuming equal weights for all the analyzed criteria, the MCA results were presented by means of the Weighted Sum Method indicator as a percentage of the difference between the actual and minimal parameter related to the difference between the maximal and minimal parameter (Table 4).

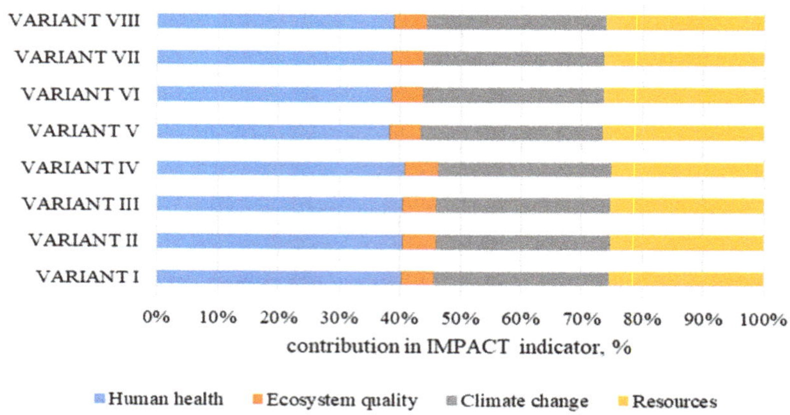

Figure 5. Share analysis of IMPACT2002+ indicator on damage category level.

Table 4. Results of multi-criteria evaluation.

Variant of Installation	SPBT	EPBT	IMPACT 2002+	WSM Result
-	Years	Years	Pt/kWh	%
Variant I	4.6	8.8	0.040	68%
Variant II	4.9	8.8	0.040	85%
Variant III	4.6	9.0	0.041	83%
Variant IV	4.9	9.0	0.042	100%
Variant V	4.1	8.1	0.035	0%
Variant VI	4.4	8.2	0.036	16%
Variant VII	4.1	8.3	0.036	13%
Variant VIII	4.4	8.3	0.037	29%

According to the data in Table 4, the best option is Variant V of installation, where all of the analyzed parameters were minimal due to the high efficiency of the system. Conversely, Variant IV can be assessed as the worst due to the longest payback periods and the highest environmental impact indicator. Following the specification of the analyzed systems, one can state that the most meaningful parameter is the type of solar collector. This contributed to approximately a 60% difference between systems (Variant I–IV vs. Variant V–VIII). The other analyzed parameters, such as water tank capacity (16–17% difference), and pipe material (13–15%) were also important. In turn, compared to the literature values, the analyzed solutions are characterized by higher payback times. This is connected with the method of analysis, e.g., covering the additional conventional energy spent on system operation, as well as various climate conditions.

4. Conclusions

According to the presented results, the most favorable solution included a heat-pipe-evacuated tube collector, copper pipes and 200 dm^3 water tank. Although the selection of a higher-capacity tank led to the optimization of solar energy conversion efficiency, the increase in this parameter was not crucial for the conducted research. However, it should be underlined that the values of the calculated indicators are comparable for all cases. For example, the payback times for both energy and financial inputs differed by no more than 11 months. Therefore, the analyzed systems can be assessed as green solutions, significantly minimizing the environmental impact connected with conventional energy usage.

The multi-criteria analysis, covering several aspects of systems' life cycle, can be used as a tool for the optimization of their selection in the green designing process. Moreover, the additional criteria used for sustainable designing can significantly improve the environmental balance of the analyzed technologies.

Author Contributions: Conceptualization, A.Ż.; methodology, A.Ż.; software, A.Ż. and J.G.; validation, A.Ż.; formal analysis, A.Ż. and J.G.; investigation, A.Ż.; resources, A.Ż.; data curation, A.Ż.; writing—original draft preparation, A.Ż.; writing—review and editing, A.Ż. and J.G.; supervision, A.Ż. and J.G.; project administration, A.Ż. and J.G. All authors have read and agreed to the published version of the manuscript.

Funding: The publication is supported from founds FD-20/IS-6/012 and FD-20/IS-6/042 granted by Faculty of Environmental Engineering, Lublin University of Technology.

Data Availability Statement: Not applicable.

Conflicts of Interest: The authors declare no conflict of interest.

References

1. Van Ruijven, B.J.; De Cian, E.; Wing, I.S. Amplification of future energy demand growth due to climate change. *Nat. Commun.* **2019**, *10*, 1–12. [CrossRef] [PubMed]
2. Jackson, R.B.; Le Quéré, C.; Andrew, R.M.; Canadell, J.G.; Korsbakken, J.I.; Liu, Z.; Peters, G.P.; Zheng, B. Global energy growth is outpacing decarbonization. *Environ. Res. Lett.* **2018**, *13*, 120401. [CrossRef]
3. Żelazna, A.; Gołębiowska, J.; Zdyb, A.; Pawłowski, A. A Hybrid vs. On-Grid Photovoltaic System: Multicriteria Analysis of Environmental, Economic, and Technical Aspects in Life Cycle Perspective. *Energies* **2020**, *13*, 3978. [CrossRef]
4. Asdrubali, F.; Baldinelli, G.; D'Alessandro, F.; Scrucca, F. Life cycle assessment of electricity production from renewable energies: Review and results harmonization. *Renew. Sustain. Energy Rev.* **2015**, *42*, 1113–1122. [CrossRef]
5. Żelazna, A.; Zdyb, A.; Pawłowski, A. The influence of selected factors on PV systems environmental indicators. *Annu. Set Environ. Prot.* **2016**, *18*, 722–732.
6. Olczak, P.; Matuszewska, D.; Zabagło, J. The Comparison of Solar Energy Gaining Effectiveness between Flat Plate Collectors and Evacuated Tube Collectors with Heat Pipe: Case Study. *Energies* **2020**, *13*, 1829. [CrossRef]
7. Mularczyk, A.; Husa, B. Rozwój i perspektywy energii solarnej w Polsce i Województwie Śląskim. *Zesz. Naukowe. Organ. Zarządzanie Politech. Śląska* **2015**, *86*, 361–377. (In Polish)
8. Olczak, P. The comparison of solar installation heat gains and SHW simulation results–case study. *Polityka Energetyczna* **2020**, *23*, 41–54. (In Polish) [CrossRef]
9. Hang, Y.; Qu, M.; Zhao, F. Economic and environmental life cycle analysis of solar hot water systems in the United States. *Energy Build.* **2012**, *45*, 181–188. [CrossRef]
10. Souliotis, M.; Arnaoutakis, N.; Panaras, G.; Kavga, A.; Papaefthimiou, S. Experimental study and Life Cycle Assessment (LCA) of Hybrid Photovoltaic/Thermal (PV/T) solar systems for domestic applications. *Renew. Energy* **2018**, *126*, 708–723. [CrossRef]
11. Saint, R.M.; Pomponi, F.; Currie, J.I. A method for a cradle-to-cradle life cycle assessment of integrated collector-storage solar water heaters. *IOP Conf. Ser. Mater. Sci. Eng.* **2019**, *556*, 012061. [CrossRef]
12. Kalogirou, S. Thermal performance, economic and environmental life cycle analysis of thermosiphon solar water heaters. *Sol. Energy* **2009**, *83*, 39–48. [CrossRef]
13. Żelazna, A. The influence of collector type on emission indicators in solar systems life cycle assessment. *Annu. Set Environ. Prot.* **2013**, *15*, 258–271.
14. Głuszek, A.; Magiera, J.; Neupauer, K. Analiza cyklu życia (LCA) cieczowego kolektora słonecznego. *Inżynieria Apar. Chem.* **2010**, *49*, 35–36. (In Polish)
15. Koroneos, C.J.; Nanaki, E.A. Life cycle environmental impact assessment of solar water heater. *J. Clean. Prod.* **2012**, *37*, 154–161. [CrossRef]
16. Altun-Ciftcioglu, G.A.; Gokulu, O.; Kardigan, F.; Kardigan, M.A.N. Life cycle (LCA) of solar selective surface produced by continuous process and solar flat collectors. *Sol. Energy* **2016**, *135*, 284–290. [CrossRef]
17. Comodi, G.; Bevilacqua, M.; Caresana, F.; Paciarotti, C.; Pelagalli, L.; Venella, P. Life cycle assessment and energy-CO2-economic payback analyses of renewable domestic hot water systems with unglazed and glazed solar thermal panels. *Appl. Energy* **2016**, *164*, 944–955. [CrossRef]
18. Milousi, M.; Souliotis, M.; Arampatzis, G.; Papaefthimiou, S. Evaluating the Environmental Performance of Solar Energy Systems Through a Combined Life Cycle Assessment and Cost Analysis. *Sustainability* **2019**, *11*, 2539. [CrossRef]
19. Mahmud, M.A.P.; Huda, N.; Farjana, S.H.; Lang, C. Environmental Impacts of Solar-Photovoltaic and Solar-Thermal Systems with Life-Cycle Assessment. *Energies* **2018**, *11*, 2346. [CrossRef]
20. De Laborderie, A.; Puech, C.; Adra, N.; Blanc, I.; Beloin-Saint-Pierre, D.; Padey, P.; Payet, J.; Sie, M.; Jacquin, P. Environmental Impacts of Solar Thermal Systems with Life Cycle Assessment. In *The World Renewable Energy Congress*; Linköping University Electronic Press: Linkoping, Sweden, 2011; pp. 8–13. Available online: https://hal-mines-paristech.archives-ouvertes.fr/hal-00668172/document (accessed on 10 May 2021).

21. Barbir, F. Fuel Cell Applications. In *PEM Fuel Cells, Theory and Practice*, 2nd ed.; Barbir, F., Ed.; Academic Press: Cambridge, MA, USA, 2013; pp. 373–434.
22. Astudillo-Flores, M.; Zalamea-Leon, E.; Pelaez-Samaniego, A.B.E.M.; Calle-Siguencia, J. Modelling of solar thermal energy for household use in equatorial latitude by using the F-Chart model. Renew. *Energy Power Qual. J.* **2021**, *19*, 269–275. [CrossRef]
23. Jolliet, O.; Margni, M.; Charles, R.; Humbert, S.; Payet, J.; Rebitzer, G.; Rosenbaum, R. IMPACT 2002+: A New Life Cycle Impact Assessment Methodology. *Int. J. LCA* **2003**, *8*, 324–330. [CrossRef]
24. Frischknecht, R.; Wyss, F.; Knöpfel, S.B.; Lützkendorf, T.; Balouktsi, M. Cumulative energy demand in LCA: The energy harvested approach. *Int. J. Life Cycle Assess.* **2015**, *20*, 957–969. [CrossRef]
25. Energy Regulatory Office. Available online: https://www.ure.gov.pl/en (accessed on 4 May 2021).

MDPI
St. Alban-Anlage 66
4052 Basel
Switzerland
Tel. +41 61 683 77 34
Fax +41 61 302 89 18
www.mdpi.com

Energies Editorial Office
E-mail: energies@mdpi.com
www.mdpi.com/journal/energies

www.ingramcontent.com/pod-product-compliance
Lightning Source LLC
Chambersburg PA
CBHW041943260926
51850CB00018B/679
9783036573694

CONTEXTUAL REINTERPRETATION OF QUANTUM NONLOCALITY

The quantum information revolution has had a huge impact not only on quantum technologies, including quantum computing and cryptography, but also on the foundations of quantum mechanics. This book presents the information viewpoint on the foundations of quantum physics by highlighting the role of complementarity and contextuality and coupling the ideas of the fathers of quantum mechanics, Bohr and Einstein, with the modern quantum information framework. The classical–quantum dilemma is resolved through an appeal to the Bild conception of scientific theories established in the nineteenth century by Hertz and Boltzmann. Bell inequalities are treated from the complementarity–contextuality viewpoint, supporting the attempts to discard nonlocality from quantum physics. Philosophical aspects of the topic are explored from a physicist's perspective, balancing accessibility with scientific rigor. This unique approach to quantum foundations will be of interest to graduates, PhD students, and researchers in fields ranging from quantum information to philosophy.

ANDREI KHRENNIKOV is Professor of Mathematics at Linnaeus University, Växjö, Sweden. Khrennikov is also Director of the International Center for Mathematical Modeling (ICMM), applying mathematical modeling and computer simulations to multidisciplinary studies in physics, cognitive science, economics, social science, and engineering. Andrei has been organizing the "Växjö Conferences" on quantum foundations, information, and probability since 2000, making it the longest continuous series of conferences on the foundations of physics.

CONTEXTUAL REINTERPRETATION OF QUANTUM NONLOCALITY

ANDREI KHRENNIKOV
Linnaeus University

CAMBRIDGE
UNIVERSITY PRESS

CAMBRIDGE
UNIVERSITY PRESS

Shaftesbury Road, Cambridge CB2 8EA, United Kingdom

One Liberty Plaza, 20th Floor, New York, NY 10006, USA

477 Williamstown Road, Port Melbourne, VIC 3207, Australia

314–321, 3rd Floor, Plot 3, Splendor Forum, Jasola District Centre, New Delhi – 110025, India

103 Penang Road, #05–06/07, Visioncrest Commercial, Singapore 238467

Cambridge University Press is part of Cambridge University Press & Assessment, a department of the University of Cambridge.

We share the University's mission to contribute to society through the pursuit of education, learning and research at the highest international levels of excellence.

www.cambridge.org
Information on this title: www.cambridge.org/9781009313476

DOI: 10.1017/9781009313469

When citing this work, please include a reference to the DOI 10.1017/9781009313469

First published 2025

A catalogue record for this publication is available from the British Library.

A Cataloging-in-Publication data record for this book is available from the Library of Congress.

ISBN 978-1-009-31347-6 Hardback

To the memory of Oleg Georgievich Smolyanov, my supervisor at Moscow State University. From him I learned to take easily both mathematics and physics. He ignited my interest in quantum physics and its foundations.

Contents

Preface

The quantum information revolution, also known as the second quantum revolution, has influenced not only the development of quantum technology, but it also has been accelerating studies on foundations of quantum theory. Regardless, the future perspectives of wide implementation, quantum computers, networks, and cryptography are new powerful systems for testing quantum foundations and elaborating novel insights on theoretical description of micro- and macrophenomena and their interrelation. One of the main consequences of the quantum information revolution is that quantum systems are treated as information processors nowadays. Roughly speaking, the main theoretical interest is in quantum software. Of course, the experimental implementation is based on advanced hardware, including highly efficient photodetectors. The Nobel Prize (2022) given to Aspect, Clauser, and Zeilinger "for experiments with entangled photons, establishing the violation of Bell inequalities and pioneering quantum information science" was generated by the quantum foundational studies. And I am proud that the series of the conferences on quantum foundations, information, and probability arranged annually in Växjö (Sweden) since the year 2000 contributed essentially to critical analysis of the "Bell project" [53, 54]. The hot debates on the possible interpretations of quantum mechanics (QM) and especially on the violation of the Bell-type inequalities were supplemented by the talks of leading experimentalists involved in testing of these inequalities. Such annual assembles energized studies on this topic, both theoretical and experimental. One of my organizational contributions to this project was arranging the special session *Big event – Final Bell test* at the Quantum and Beyond conference in 2016. The session included the talks of the representatives of all experimental groups who performed the loophole-free experiments. The session was concluded by the talk of Zeilinger, who presented the future of experiments in quantum foundations.

- A. Aspect (the talk was given by P. Grangier): Closing the door on Einstein and Bohr's quantum debate.
- G. Weihs: Violation of Bell's inequality under strict Einstein locality conditions.
- R. Hanson: From the first loophole-free Bell test to a quantum Internet.
- M. Giustina and M. Versteegh: Significant loophole-free test of local realism with entangled photons.
- K. Shalm: A strong loophole-free test of Bell's inequalities.
- H. Weinfurter: Event-ready loophole-free Bell test using heralded atom-atom entanglement.
- P. Grangier: Violation of Bell's Inequalities in a quantum realistic framework.
- A. Zeilinger: The future of Bell experiments.

Videos of these talks can be found at lnuplay; see [32, 139, 148, 354, 371, 372]. It is important to note that not all participants of the conference agreed with Aspect and Grangier that the Einstein–Bohr debate is closed. During the following conferences the debates on the possible interpretations of the outputs of these loophole-free experiments were continued. This is a good place to recall that Bell himself predicted this situation [53]. I have personally been very engaged in the foundational research and I enjoyed the Växjö conferences, the talks, and discussions with talented people.[1] Only one big dark cloud has always been at the quantum horizon, the cloud of quantum nonlocality. Spooky action at a distance disturbs, but at the same time stimulates my mind.

This metaphor "dark cloud at the quantum horizon" I borrowed from a lunch-talk with Anton Zeilinger, where he told the story about how he became interested in foundations. In the 1950s he participated in a conference which assembled the world-leading experts in quantum physics. The general attitude of participants was that development of quantum theory is more or less completed. There was just one dark cloud at the quantum horizon, the cloud of quantum gravity. As I understood, he was curious about whether there are other clouds at the quantum foundational horizon and this curiosity led him to the Nobel Prize in 2022.

In this nonlocal atmosphere I tried to seek salvation in the circle of QBists led by Fuchs with the active support of Mermin and Schack. Although QBism was widely represented at the conferences in Växjö since the year 2001 and I have always supported plenary talks and special sessions of QBists, I am a strong antagonist of QBism for the subjective probability interpretation of QM and highlighting the personal experiences of "quantum agents." Educated within the Soviet probability school and being lucky to have conversations with Kolmogorov on foundations of

[1] Who are generally talented, and not just in science. I think that many participants remember the great evening when Daniel Greenberger played piano in Teleborg Castle.

probability, I treat the probability objectively. The first time I heard about subjective probability was from Fuchs, and hearing such nonsense annoys me a lot. However, QBism declares that in QM there is no spooky action at a distance; QM is local. So, maybe even QBism is better than nonlocality! This was precisely the message of Hänsch in his talk in Växjö. I remind the reader that he received the Nobel Prize (2005) for "contributions to the development of laser-based precision spectroscopy, including the optical frequency comb technique." Hence, he is an excellent experimentalist and his appeal to QBism is not due to attraction by the subjective viewpoint on QM. He said that he needs a consistent local interpretation of QM and QBism provides it. Hänsch's talk induced hot debates and he was attacked by "quantum realists." I remember the critical comments of Vaidman and especially Elitzur. At one of the conferences, we had the social opinion poll on acceptance of QBism as the basic interpretation of QM. Surprisingly, many of the world's leading experimentalists in quantum foundations (e.g., Weihs) supported QBism. QBism was less popular among theoreticians and philosophers; it was rejected by the majority of the participants. I suspect that Zeilinger is sympathetic to QBism, although he did not declare this openly; but his talk in Vienna at the conference "Quantum [Un]Speakables II: 50 Years of Bell's Theorem" was very much in the spirit of QBism.

I spent a lot of time trying to proceed in quantum foundations and handle the violation of the Bell inequalities without appealing to nonlocality. I have understood that one of the barriers on the way toward liberation of QM from nonlocality is operating with the term "local realism." This is one of the most misleading notions in the history of physics, but it is firmly in the heads of the majority of physicists who are interested in fundamental questions. Two totally different notions, namely, realism and locality, are jointly mixed and one of them shadows the other.

The first signal of this notion ambiguity I received was from the answer of Gisin to my comment on his talk that the notion of quantum nonlocality is misleading. Gisin answered that the most misleading is the notion of local realism and first one should eliminate it from discussions on violation of the Bell inequalities: one cannot keep one term and dismiss the other. Local nonrealism doesn't mean anything, as Gisin wrote in article [136] (see also my note [209]). I remark that Bell did not operate with the notion of local realism; at least in his book [54] summarizing his main works he did not even mention it.

In principle, one can reject both realism and locality. In particular, one famous experimentalist who is well known for his rejection of realism (like many representatives of Vienna's circle) told me that QM would finally stimulate people to reject realism, but then he suddenly added that today it is clear that one should also reject locality. The latter statement was really sudden for me in the light of all our previous conversations. I ignore the double rejection possibility simply by using Occam's razor. But I remember Cromwell's rule and I may be mistaken. By himself Gisin is definitely in favor of rejection of locality.

I recall a similar conversation with a world-famous supporter of the many worlds interpretation. He always positioned himself as a realist (being in one of his numerous worlds). But once he said that QM is nonlocal. I was really shocked; I didn't expect from him such a hit below the belt. When I became a bit calmer, I asked him: "Why?!" And he pointed to nonlocality of the projection postulate.

Realism was so deeply in my mind that, for me, it was difficult to dismiss it and imagine the acausal universe driven by irreducible quantum randomness which was advertised by von Neumann [365]. I spent a few years of my life working on the mathematical models combining realism and locality and constrained by the violation of the Bell inequalities. I interacted with practically all leading experts in foundations who work on similar models, in particular, with Accardi, Kupczynski, Hess and Philipp, De Raedt and Michielsen, Nieuwenhuizen, Volovich, de la Pena, and Cetto. In spite of controversy in the evaluation of such studies which culminated in the famous bet of Accardi with Gill, I have the impression that local realistic treatment of quantum phenomena is possible. I think that it can be done by discarding hidden assumptions not only of the Bell inequality project, but of quantum theory as whole, for example, *the ergodicity assumption.* But in this book I have no possibility of proceeding in this direction. From my viewpoint, the main problem of the local realistic approach to quantum phenomena is clustering into special models working for concrete phenomena (at least as claimed by the authors) without a general theory. For example, the probability interference can be modeled under the assumption that there exists an indivisible quantum of time.

My leavetaking from realism, at least in its naive form, started with studying Bohr's works and discussions with Plotnitsky on the complementarity principle. After a while I understood that the complementarity principle is rooted in contexuality of measurements. Contextuality is understood very widely here as dependence on the complex of experimental conditions. According to Bohr, measurements' outcomes can't be treated as objective properties of a system; they are generated in the complex process of interaction between a system and a measurement device. Bohr operated with the term "conditions," but not contexts. In modern terminology, these are experimental contexts. Such a notion of contextuality is wider than so-called joint measurement contextuality which was invented by Bell in the discussion on possible interpretations of the violation of his inequality and nowadays it is intensively studied in quantum information theory. We recall that Bell's contextuality is about measurement of an observable A in the contexts of joint measurements with other observables, say $B_1, \ldots, B_n$ which are compatible with A. Like Bohr, Bell did not use the term contextuality.

One of the main chapters of this book (Chapter 3) is devoted to Bohr's contextuality and its place in the formulation of the principle of complementarity. This is the basic principle of QM. It is typically reduced to wave–particle duality. However, careful reading of Bohr shows that this principle has a more complex structure. In the light of this book's presentation, it might be better to call it *the principle of contextuality-complementarity.* This is an epistemic principle on structuring knowledge which can be obtained about a quantum system. The ontological basis of the complementarity principle is the quantum postulate declaring the existence of an indivisible quantum of action given by the Planck constant h. In Chapter 5, the direct epistemic counterpart of the quantum postulate is formulated in the form of *the quantum action invariance.* This principle and the complementarity principle are treated as two basic assumptions of QM, similar to Einstein's principles of special relativity (compare this with Zeilinger's search for fundamental principles of QM [353]).

The Bohr complementarity and Heisenberg uncertainty principles are deeply interconnected. Heisenberg also related his principle to the quantum postulate. In this book we don't analyze the Heisenberg uncertainty principle. Such an analysis should include the principle's recent reconsideration within the theory of quantum instruments which was done by Ozawa, but I am not ready to discuss Ozawa's modification of the uncertainty principle.

Now we come to the issue of *quantum nonlocality.* As well as local realism, this is an ambiguous notion. Two different sorts of nonlocality are often mixed. One of them can be called intrinsic quantum nonlocality. This is the apparent nonlocality of a quantum state update based on the projection postulate; another nonlocality is nonlocality of the hidden variables models violating the Bell inequalities. These two sorts of nonlocality should be sharply distinguished, as was done in Aspect's talk at one of the Växjö conferences. He started his lecture with the statement that QM is intrinsically nonlocal, since, for a compound system, the operation of the state projection resulting from a local measurement is nonlocal. Then he said that this nonlocality is so unpleasant for any physicist that one should search for another theoretical formalism which would be free of projection nonlocality. And in this way one introduces hidden variables. Finally, he said that, as is shown by the violations of the Bell inequalities, the subquantum locality (Bell locality) also should be rejected. This is a very consistent presentation of the Bell inequality project (see also [31] and Chapter 2), much better than Bell's original motivation – to explain the perfect quantum correlations [54]. From my viewpoint, correlations are not mystical at all, irrespective of their quantumness or classicality (see Chapter 13 on probabilistic entanglement of observables and also Chapter 14 on classical Brownian entanglement). I prefer Aspect's motivation for going beyond QM as attempting to find a deeper theory which would not suffer from projection nonlocality.

In Chapter 2 projection nonlocality is analyzed in connection to a selection of interpretations of QM. As is well known, QM is characterized by a diversity of interpretations. Some experts consider it as a sign of the crisis in quantum foundations (e.g., as Ballentine put it at the first Växjö conference in year 2000); others rigidly keep to one concrete interpretation without even paying attention to the existence of other interpretations. But the majority of researchers working in quantum physics are not interested in the interpretation problem. I have asked the question: *What is your interpretation of the wave function?* to hundreds of participants of the Växjö conference series. Some of them were not able to say anything consistent and others automatically replied: *I use the Copenhagen interpretation!* However, the meaning of the latter is another delicate problem. This is a good place to recall Plotnitsky's claim that the term "Copenhagen interpretation" unifies a bunch of interpretations; he suggested speaking about the interpretations in the spirit of Copenhagen. My experience of discussions in Växjö supports this suggestion. Following Ballentine, we cluster the interpretations into two classes:

- **SI** the statistical (ensemble) interpretation,
- **II** the individual interpretation.

By **SI** a quantum state describes the statistical features of an ensemble of identically prepared systems; by **II** a quantum state is a state of an individual system, say an electron. Ballentine identifies **II** with the orthodox Copenhagen interpretation. The projection nonlocality issue depends on an interpretation. By following **II** one really confronts (as, e.g., Aspect) with the intrinsic quantum nonlocality. However, by following **SI** one treats the projection postulate as the quantum analog of the Bayes rule for classical probability update.

In fact, the EPR paper was directed against **II**, the orthodox Copenhagen interpretation. Unfortunately, it was not formulated precisely. Later in 1949 Einstein pointed out that there are two choices: (a) the orthodox Copenhagen interpretation and nonlocality, and (b) the statistical (ensemble) interpretation and locality. We note that Einstein considered the intrinsic quantum nonlocality.

By struggling against the orthodox Copenhagen interpretation Einstein did not question the validity of quantum theory. Specifically the reasoning in the EPR paper [115] is crucially based on the validity of the Heisenberg uncertainty principle. I claim that Bell's model with hidden variables contradicts this principle as well as the complementarity principle. To follow Bell, one should from the very beginning reject these principles and also the quantum postulate, the existence of the indivisible quantum of action. This would be the difficult decision, similar to the decision to reconstruct special relativity by rejecting the Einstein postulate on the constancy of the light velocity.

What is the main problem of the Bell-like models with hidden variables? This is the identification of the quantities of a subquantum model with quantum observables, the identification of their outcomes and probability distributions (Chapter 15). This casualty of the Bell project was highlighted by de Broglie, immediately after publication of the Bell works. Unfortunately, his message was completely ignored. De Broglie interpreted his double solution theory as a hidden variables theory. (We remark that this common identification of de Broglie's and Bohm's theories is wrong. In particular, de Broglie did not consider his theory as physically nonlocal; he just pointed to apparent nonlocality of mathematical equations.) For de Broglie, the values and probability distributions of subquantum quantities and quantum observables should not be identified. Hence, in subquantum theories not only variables, but even quantities (functions of hidden variables $\xi = \xi(\lambda)$) and their probability distributions are hidden.

De Broglie's approach to subquantum theories matches perfectly with the *Bild conception* for structuring of scientific theories which was established by Hertz and Boltzmann and later applied by Schrödinger to QM (Chapter 15). By this conception one should distinguish an observational model and a causal theoretical model. As was pointed out by Schrödinger, QM is an observational model and it is characterized by acausality (as was emphasized by von Neumann). One can search for causal theoretical models beyond QM. Due to Hertz and Bolzmann, coupling between causal theoretical and observational models can be tricky and varying. From this viewpoint, Bell suggested just one special coupling. Hence, employing the Bild conception opens the door to creation of a variety of subquantum causal theoretical models, in spite of the negative output of the Bell project. Unfortunately, in quantum foundations the Bild conception is practically unknown. Its partial resemblance can be found in ontic-epistemic structuring of physical theories due to Primas and Atmanspacher.

A promising causal theoretical model for QM is *prequantum classical statistical field theory*; see Chapter 16. This is a model of classical random fields reproducing quantum averages and correlations. Its coupling with QM is trickier than for the Bell model with hidden variables.

Bohr's contextuality is formalized within the contextual measurement model (CMM) developed in Chapter 10 (see also Chapter 11 for CMM presentation of classical and quantum physics). A contextual measurement model is based on the calculus of contextual probabilities which are closely related to generalized probability models. CMM serves as the basis for the *Växjö interpretation* of QM, a contextual statistical interpretation (Chapter 12). By this interpretation QM is the machinery for generally non-Bayesian probability update. One of the fundamental exhibitions of non-Bayesianity is the violation of the classical *formula of total probability* (FTP). Contextual generalization of FTP contains an additional term, the interference term. In QM this is the formula for interference of probabilities, for example, in the two-slit experiment (compare Feynman [119, 120]).

Chapter 6 presents a purely quantum approach to the Bell inequalities which are exemplified as the tests of incompatibility of local variables, or in the mathematical terms, as the tests of noncommutativity: $[\hat{A}_1, \hat{A}_2] \neq 0$ and $[\hat{B}_1, \hat{B}_2] \neq 0$ for Alice's and Bob's observables, respectively. In this book we highlight the role of the original Bell (OB) inequality (Chapters 8 and 9). Interpretationally and experimentally it is more complex than the CHSH-inequality. Classical optics entanglement is analyzed in Chapter 7.

Chapter 17 is devoted to critical analysis of QBism's private experience perspective in light of Ozawa's intersubjectivity theorem based on theory of quantum instruments and the indirect measurement scheme (see Chapter 1 for a brief introduction to this scheme). This chapter is a continuation of my challenging QBism during the last 20 years.

I was one of the pioneers in applications of the methodology and formalism of quantum theory outside of physics [178, 196, 197]. This area is known as *quantum-like modeling* (see Chapter 18). I put a lot of effort into establishing and advertising it and I am thankful to all my friends who contributed to the volume on *quantum-like revolution* [318]. It is important to note that, in spite of my opposition to QBism in quantum physics, I appeal to it [221] as a possible interpretational basis for modeling cognition and decision making with quantum formalism. I can't say that the idea of applying the quantum methods outside of physics was immediately welcomed by the participants of the first Växjö conferences. I remember some hateful comments on my talks devoted to quantum-like modeling at the beginning of this century: "Please don't touch our beautiful quantum theory with your dirty hands!" But by opening the Overton window wider and wider I was able to smooth the reactions of "rigid quantum physicists" on the use of the quantum methods outside of physics, at least in decision making. Later a few special sessions on quantum-like modeling in decision making, economics, and finance were organized in cooperation with Emmanuel Haven who contributed so much in advertising this area of research in Växjö. Nowadays quantum-like modeling is a well-established area of research. One of its problems is attraction of numerous newcomers from various domains of science who are not so well educated in quantum theory. Often "quantum" is considered as a magic word giving the possibility of resolving all problems, like "Open Sesame!" in the tale "Aladdin and the Magic Lamp." But this is merely the symptom of its rapid growth.

Finally, Chapter 19 points to the possibility of exploring noncommutative probability to model classical complex disordered systems. Here experimental contextuality plays a crucial role. This chapter, as well as Chapter 14, contains suggestions for some experiments to check contextuality for classical and semiclassical physical systems.

This book may be of interest even to those who are not excited by quantum mysteries and paradoxes (or scandals [290]) and consider QM as an observational model describing quantum measurements. Quantum mechanics without nonlocality and spooky action at a distance need not seem insipid. One can find plenty of interesting problems: theoretical, experimental, mathematical, and philosophical. Quantum mechanics treated as OM, as a measurement model, works very well and there is no reason to question it. Its apparent acausality is typical for OMs as was highlighted already by Hertz, Boltzmann, and Schödinger. If one wants to go beyond QM, it is natural to appeal to the Bild conception. In particular, the violation of the Bell inequalities should be analyzed within this conception.

Acknowledgments and Recollections

This book summarizes my foundational studies. I was lucky to speak as well with Kolmogorov, concerning foundations of probability and possible generalizations of his axiomatics, as with Mackey regarding the conditional probability structure of QM. Then this viewpoint on quantum probability was discussed with Accardi, Ballentine, and Gudder and these discussions were imprinted in my mind.

My views on the causes of the Bell inequalities violations were influenced by conversations with Lochak, De Baere, and De Muynck. From the bloody confrontation between Accardi and Gill during a few Växjö conferences and being one of the members of the jury for their famous bet, I learned to play the role of an external observer in scientific discussions, often at the cost of my own perspective of the problem, and sometimes even at the cost of my friends. I also learned that some quantum foundational questions are so complex that they would never be been resolved, at least during the lifetime of my generation.

During 24 Växjö conferences I have debated quantum foundations with many bright scientists and often we disagreed, but this was in the spirit of these conferences. I am especially thankful to the members of the "Växjö club," those who participated in the majority of the conferences, especially to Bengtsson, D'Ariano, Elitzur, Fuchs, Genovese, Jaeger, Larsson, Nieuwenhuizen, Plotnitsky, Rauch, Vaidman, and Weihs.

At the beginning of this century Bohmians, especially Goldstein and Hiley, strongly influenced my realist attitude – my first conference had the title "Bohmian Mechanics – 2000." I was impressed by the spiritual strength of Basil Hiley who often struggled alone against a hundred of anti-Bohmians. This was a good lesson on scientific struggle. I and other participants enjoyed the hot battle between Basil Hiley and Marlan Scully on whether Bohmian mechanics can be experimentally distinguished from conventional QM. Only strong pressure from the chairman, Theo Nieuwenhuizen, led to termination of this battle.

The presence in Växjö of Gerard 't Hooft was a breath of fresh air for all kinds of realists. His bravery in his struggle against the orthodox Copenhagen views, multiplied by the weight of the Nobel Prize, induced a wave of enthusiasm. The impression was that his journey beyond quantum would soon lead to establishing of the realist interpretation of quantum theory.

Unfortunately – or fortunately? – it didn't happen. The first signal that ′t Hooft proceeds too quickly without taking care of many basic foundational problems came during one of our conversations; it happens that he "didn't pay attention" that his model conflicts with the Bell theorem. But his reaction to this bit of news was unruffled and it seems that this theorem didn't disturb him so much. (I was really surprised!) In the postconference email exchange his attitude was Bohm-like and, as I understood, nonlocality was not a substantial problem. Later he came with an argument based on superdeterminism and, in particular, rejection of free will. From my present perspective, in principle 't Hooft need not appeal to superdeterminism. The latter, although logically possible, is not so attractive for the majority of the physicists. The model of 't Hooft matches perfectly the Bild conception of Herz, Botzmann, and Schrödinger on the interrelation between observational and causal theoretic models (Chapter 15). Coupling between two levels of the physical description can have a variety of colors. As has already been repeated a few times, on the way beyond quantum one need not follow Bell with his specific rules for such coupling.

Contacts with the Viennese quantum community were vital for the formation of my later views. I will always remember my conversations with Helmut Rauch during many conferences in Växjö and my visits to the Atom Institute. He had his own picture of quantum phenomena, a kind of reality of both the wave function and particle. Helmut Rauch strongly supported the Växjö series of conferences and he attracted to Växjö the top experimentalists from Vienna. I spent a lot of time with Johann Summhammer, both in Vienna and during his visits to Växjö and I was really astonished by his idealist perspective of the physical world. Spending time with him generated first doubts in my "naive realism." I still remember that once I found Johann staying near Växjö lake and looking at it. He told me something like "Look at this beautiful picture which is created in our brain by integrating the clicks of photodetectors."

I was lucky to visit the University of Vienna during realization of the final Bell tests and enjoyed the atmosphere of the coming great scientific event. I spent a few months with Zeilinger's group during both experiments closing the last loopholes: in 2013 when the detection efficiency loophole was closed [137], and in 2015 when the previous experiment was completed to the loophole-free experiment [138]. This is a good place to highlight the contribution of Marissa Giustina in these breakthrough experiments. We spent a lot of time drinking tea and coffee and speaking about the experiments. I was curious about "experimental technicalities." (In fact, I had my own vested interest: I was searching for experimental data that might show some kind of deviations from quantum theory.) I was really surprised by Marissa's experience in the design and methodology of experimenting in quantum foundations and her remarkable understanding of the basic problems of quantum theory. She worked hard on the realization of both experiments and spent a few hundred hours in the basement of Vienna Hofburg Palace (for the last experiment), collecting data and then analyzing these huge data-sets and modifying the experimental setup. Definitely her contribution to closing all loopholes in the Bell test was crucial. As a university professor, I was also interested in the pedagogical practice of Anton Zeilinger: he put a young person at the heart of this fundamental experiment. And Marissa justified his hopes.

Relaxing conversations with Anton Zeilinger and the members of his experimental group, especially with Sven Ramelow, Rupert Ursin, Johannes Kofler, and Bernhard Wittmann, were vital for my leavetaking from "naive realism." In particular, I was excited by Anton's stories regarding his debates with the Dalai Lama. I suspect that a recording of these debates has never been translated and published. And I think that it should be done, sooner or later. In Vienna I learned that a photon is simply a click of a photodetector and nothing more, that the correlations in Bell experiments are natural and can be influenced by a variety of technicalities, and that some parameters of experimental tests cannot be defined on the basis of quantum theory alone but depend a lot on the experience of the experimentalists who select them by hand. Joint article [214] is a nice scientific memory about my days spent with Zeilinger's group.

Consequently, experimentalists are less orthodox than theoreticians and more open to beyond-quantum theories, especially if such theories can be helpful for designing experiments. After my talk at one of the Växjö conferences, Alain Aspect asked me a question about negative probability, its possible interpretations and description of the Bell argument. I was surprised, since I didn't expect such a question from an experimentalist. Moreover, by getting to know that I have a series of papers and books on negative probability and its interpretation (e.g., [177, 179, 181]), he continued the discussion during lunch and told about his own model with negative probabilities violating the Bell inequality; later he sent me this paper, old and practically unknown (cf. with Zeilinger's paper [353]).

Finally, I would like to thank my coauthors Elena Loubenets, Arkady Plotnitsky, Armen Allahverdyan, and Theo Nieuwenhuizen; some results of our joint research are included in this book.

It is interesting that the intensity of the debates in Växjö is going down. The last hot confrontation was between Ozawa and Lahti on Ozawa's rethinking of Heisenberg's uncertainty relation and the possibility of its violation. The cause of the debating decline is not clear. It might be that the basic quantum foundational problems have been resolved or at least the community came to a consensus. But it might be that one part of the quantum community with its specific perspective on foundations became so powerful that this led to disappearance of outsiders from the scene, at least in Växjö. It might be also a signal of decreasing interest in quantum foundations. This is a good place to recall that in the 1970s the interest in QM foundations wasn't so high. During Alain Aspect's PhD defense one of the jury's members made the following remark: "But, where would this bright student find a job?" So, Bell's test strongly stimulated foundational research.

Successful loophole-free experiments have created the widespread illusion that, roughly speaking, there is no other way beyond QM and it is frozen forever. I do not think so. One of the messages of this book is that before trying to go "beyond quantum," one should analyze in more detail a variety of possible ways (see Chapter 15). And one should remember that the main impact of Bell's model was the possibility of its experimental verification. So, a new quantum foundational revolution would be connected to new experimental tests. As an example of such tests, I point to Sorkin's equalities. Weihs' group in Innsbruck spent a few years trying to find violations of these inequalities, but they encountered very difficult technical problems related to detectors' functioning. Up to now, Sorkin's inequalities have not been violated.

Experimentalists working on quantum foundations need new theories and their experimental tests as soon as possible. Otherwise, people may move to the rapidly growing field of quantum technologies.

1

Introduction to Quantum Measurement Theory

1.1 Pure and Mixed States

Let $\mathcal{H}$ be a complex Hilbert space; for simplicity we restrict consideration to finite-dimensional spaces.[1] We recall that a *pure quantum state* can be represented by a normalized vector of $\mathcal{H}$, namely, $||\psi|| = 1$. Two vectors that differ only by a phase, that is, $\psi = e^{i\theta}\phi$, represent the same quantum state. Under consideration of a single state, its phase does not play any role, but by manipulating a few states, the relative phases play a crucial role, for example, in the interference effect.

In physical literature the Dirac notation $|\psi\rangle$ is used for vectors of Hilbert space. We shall use both notations, ψ and $|\psi\rangle$.

The space of linear operators acting in $\mathcal{H}$ is denoted as $\mathcal{L}(\mathcal{H})$. (In the infinite-dimensional case this symbol denotes the space of bounded linear operators.) We remark that $\mathcal{L}(\mathcal{H})$ is a linear space by itself. It can be endowed with the Hilbert space structure

$$\langle \hat{A}|\hat{B}\rangle = \mathrm{Tr}\,\hat{A}^\star\hat{B}. \tag{1.1}$$

(In the infinite-dimensional case the scalar product is defined on the space of Hilbert–Schmidt operators.)

A linear operator $\hat{A}$ is called Hermitian if $\hat{A} = \hat{A}^\star$. The space of Hermitian operators $\mathcal{L}_H$ is the real Hilbert space with scalar product (1.1). A linear operator $\hat{A}$ is called positive semidefinite if $\langle\psi|\hat{A}|\psi\rangle \geq 0$ for any $\psi \in \mathcal{H}$ or symbolically $\hat{A} \geq 0$.

A density operator $\hat{\rho}$ is determined by the following conditions:

- $\hat{\rho} = \hat{\rho}^\star$ (Hermitian operator);
- $\hat{\rho} \geq 0$ (positive semidefinite operator);
- $\mathrm{Tr}\,\hat{\rho} = 1$ (trace one operator).

The space of density operators is denoted by the symbol $D \equiv D(\mathcal{H})$.

[1] Sometimes infinite-dimensional spaces will be considered. We make remarks about this case in the corresponding places.

We note that each pure state $|\psi\rangle$ can be represented by a density operator, projection on the state vector. In the Dirac notation, for a pure state $|\psi\rangle$, the corresponding density operator (projector) is symbolically written as $|\psi\rangle\langle\psi|$.

Typically one says that a density operator represents a mixed quantum state – a statistical mixture of pure states. However, such an interpretation is ambiguous, since the same density operator can be represented by a variety of probabilistically weighted sums of projectors corresponding to pure states.

1.2 Observables, Born's Rule, Projection Postulate

By axiomatics of QM due to von Neumann [365] an observable A (with a discrete range of values) is represented by a Hermitian operator,

$$\hat{A} = \sum_{\alpha} \alpha \hat{E}_A(\alpha), \tag{1.2}$$

where $\hat{E}_A(\alpha)$ is a projection onto the space $\mathcal{H}_A(\alpha)$ of eigenvectors for the eigenvalue α.

Observables are denoted by the symbols $A, B, \ldots$ and the corresponding operators by the symbols $\hat{A}, \hat{B}, \ldots$. To simplify notation, the unit operator is denoted as I. If operators are used as indexes, the hat symbol is omitted, for example, $\hat{\rho}$, but P_ρ.

For a pure state $|\psi\rangle$, the probability of getting the outcome $A = \alpha$ is given by *Born's rule*:

$$P_\psi(A = \alpha) = \|\hat{E}_A(\alpha)\psi\|^2 = \langle \hat{E}_A(\alpha)\psi|\psi\rangle. \tag{1.3}$$

(The latter equality is due to idempotence of projectors, $E_A(\alpha)E_A(\alpha) = E_A(\alpha)$.)

For each state ψ and quantum observable $\hat{A}$, formula (1.3) determines the probability distribution and classical probability theory (Section 1.5) can be applied. So, the mean value (average) of a quantum observable is given by the integral formula (1.22). In quantum terms it can be expressed as

$$\langle A\rangle_\psi = \langle \hat{A}\psi|\psi\rangle. \tag{1.4}$$

A measurement with the outcome $A = \alpha$ generates back-action onto the system's state:

$$|\psi\rangle \to |\psi\rangle_{A=\alpha} = \hat{E}_A(\alpha)\psi/\|\hat{E}_A(\alpha)\psi\|. \tag{1.5}$$

This is the projection postulate, one of the axioms of QM. In this general form it was formulated by Lüders [274]. Often one refers to (1.5) as the von Neumann projection postulate. However, von Neumann explored (1.5) only for observables represented by Hermitian operators with nondegenerate spectra [365]. In the case of degenerate spectra he considered more general state updates, which in the future would lead to quantum instrument theory.

Measurements of another observable, say B, conditioned on the outcome $A = \alpha$ lead to probability:

$$P_\psi(B = \beta | A = \alpha) = \|\hat{E}_B(\beta)\psi_{A=\alpha}\|^2 = \frac{\|\hat{E}_B(\beta)\hat{E}_A(\alpha)\psi\|^2}{\|\hat{E}_A(\alpha)\psi\|^2}. \tag{1.6}$$

This is quantum *conditional probability* (see Chapters 10 and 13 for the mathematical details, interpretation, and applications).

The preceding formulas can be easily generalized to states given by density operators, but we do this directly for quantum instruments; see the following discussion. Observables represented mathematically by Hermitian operators and generating the projection state update are often called von Neumann observables and referred to as performing accurate measurements. Generally quantum instruments describe noisy instruments.

1.3 Quantum Instruments

We consider also linear operators acting in $\mathcal{L}(\mathcal{H})$, *superoperators*. A superoperator is called positive if it maps the set of positive semidefinite operators into itself: for $\hat{\rho} \geq 0,\ \hat{T}(\hat{\rho}) \geq 0$.

The detailed presentation of quantum instruments theory can be found in Section 11.3 (for technical details see [291]-[300]). Now we say just a few words about them. In this book they will be used only in Chapter 10 as an example of a measurement procedure of the nonprojection type and in Chapter 18 devoted to applications of the quantum formalism and methodology outside of physics (quantum-like modeling). The rest of the book is based solely on the von Neumann measurement theory with observables given by Hermitian operators.

Consider an observable A with the following discrete range of values:

$$X = \{x_1, \ldots, x_m, \ldots\}.$$

Any map $x \rightarrow \mathcal{I}_A(x)$, where, for each $x \in X$, the map $\mathcal{I}_A(x)$ is a positive superoperator and

$$\sum_x \mathcal{I}_A(x) \colon D \rightarrow D, \tag{1.7}$$

is called a *quantum instrument*. It represents one of the measurement procedures of an observable A (see Section 11.3).

The probability for the output $A = x$ is given by the generalized Born rule in the form

$$P_\rho(A = x) = \text{Tr}\ [\mathcal{I}_A(x)\hat{\rho}]. \tag{1.8}$$

We note that a measurement with the output $A = x$ generates the state-update by the following transformation:

$$\hat{\rho} \to \hat{\rho}_x = \frac{\mathcal{I}_A(x)\hat{\rho}}{\mathrm{Tr}\,\mathcal{I}_A(x)\hat{\rho}}. \tag{1.9}$$

Thus, the projection postulate is no longer a requirement for the state update. An observable A can be measured by a variety of instruments generating the same probability distribution, but different state updates.

A special class of quantum instruments was introduced by Davis and Levis (see, e.g., [98]) and we call them *the Davis–Levis instruments.* This class is too general for applications in quantum physics (although it might be useful in quantum-like modeling; see Chapter 18). In quantum physics the special class of quantum instruments is used (see, e.g., Ozawa [298]); such instruments are given by *completely positive maps* $\mathcal{I}(x)$.

A superoperator $\hat{\mathcal{T}}\colon \mathcal{L}(\mathcal{H}) \to \mathcal{L}(\mathcal{H})$ is called *completely positive* if its natural extension $\hat{\mathcal{T}} \otimes I$ to the tensor product $\mathcal{L}(\mathcal{H}) \otimes \mathcal{L}(M) = \mathcal{L}(\mathcal{H} \otimes L)$, where L is an arbitrary finite-dimensional complex Hilbert space, namely, $\hat{\mathcal{T}} \otimes I(|\psi\rangle|\phi\rangle) = \hat{\mathcal{T}}|\psi\rangle|\phi\rangle$, is again a positive superoperator on $\mathcal{L}(\mathcal{H}) \otimes \mathcal{L}(L)$.

Complete positivity is the mathematical description of the following physical statement. Let S be a system with the state space $\mathcal{H}$ and let $\hat{\mathcal{T}}$ be a superoperator mapping states into states. We can always add to S another system, say S' with state space M. Assume that systems do not interact. The unit map $I\colon M \to M$ does not change states of S'. Consider now compound system $\mathbf{S} = (S, S')$. Its state space is given by the tensor product $\mathcal{H} \otimes L$. It is natural (from the physical viewpoint) that the superoperator map $\hat{\mathcal{T}} \otimes I$ would map the states of $\mathbf{S}$ into its states, that is, it should preserve positivity as $\hat{\mathcal{T}}$ does. Adding to consideration another system that does not interact with the original one cannot change the physical situation and it would be surprising if a state transformation corresponding to measurement on S would be impossible to extend identically to the state space of the compound system. In terms of observables, violation of complete positivity means that observation on S cannot be treated as observation on the compound system $\mathbf{S}$.

According to the Kraus theorem, for any instrument $\mathcal{I}$, there exists a family $\{\hat{M}_j(x)\}$ of operators, called the *measurement operators* for $\mathcal{I}$, in $\mathcal{H}$ such that

$$\mathcal{I}(x)\rho = \sum_j \hat{M}_j(x)\hat{\rho}\hat{M}_j^\star(x) \tag{1.10}$$

for any state ρ. In this case, we have

$$\sum_{xj} \hat{M}_j^\star(x)\hat{M}_j(x) = I. \tag{1.11}$$

Conversely, any family $\{M_j(x)\}$ of operators in $\mathcal{H}$ satisfying (1.11) defines an instrument $\mathcal{I}$.

Each quantum instrument determines a generalized quantum observable, which is mathematically represented as

$$\hat{\Pi}(x) = \mathcal{I}^\star(x)I. \tag{1.12}$$

Operators $\hat{\Pi}(x), x \in X$, are called *effects*; they are positive semidefinite Hermitian and sum up to the unit operator:

$$\sum_{x \in X} \hat{\Pi}(x) = I.$$

The family of operators $\hat{\Pi} = (\hat{\Pi}(x), x \in X)$ is called a *positive operator valued measure* (POVM); see Section 11.3 for the details. For any $O \subset X$, we set

$$\hat{\Pi}(O) = \sum_{x \in O} \hat{\Pi}(x). \tag{1.13}$$

The map $O \to \hat{\Pi}(O)$ is additive on the collection of all subsets of X, that is, it is really an operator-valued measure. (We remark that we consider a finite of countable set X. Generally it should be endowed with some σ-algebra (Section 1.5) of subsets of X.)

Consider a quantum instrument such that

$$\mathcal{I}_A(x)\hat{\rho} = \hat{E}(x)\hat{\rho}\hat{E}(x), \tag{1.14}$$

where, for any $x \in X, \hat{E}(x)$ is projection. Such an instrument is called the projection instrument. Any Hermitian operator $\hat{A}$ induces the projection instrument based on its spectral projections, $\hat{E}_A = (\hat{E}_A(x) \colon x \in X)$, where $X \subset \mathbb{R}$ is its spectrum. Generally the set of the observable's outcomes X need not be a subset of the real line. A system of mutually orthogonal projections $\hat{E} = (\hat{E}(x) \colon x \in X)$ that is normalized to the unit operator, namely

$$\sum_{x} \hat{E}(x) \hat{=} I, \tag{1.15}$$

is called a projection valued measure (PVM). As earlier, for any $O \subset X$, we set

$$\hat{E}(O) = \sum_{x \in O} \hat{E}(x). \tag{1.16}$$

The map $O \to \hat{E}(O)$ is additive of the collection of all subsets of X, that is, it is PVM.

1.4 Indirect Measurement Model

Construction of quantum instruments is based on the indirect measurement model. This scheme formalizes the following situation. As was pointed out by Bohr (Chapter 3), the measurement's outcomes are created through interaction of a system S with a measurement apparatus M. This apparatus consists of a complex physical device interacting with S and a pointer that shows the result of measurement, say spin up or spin down. An observer can see only outputs of the pointer and they associate these outputs with the values of the observable A for the system S. So, the observer approaches only the pointer, not the system by itself. This is a good place to note that the same observable A can be measured by a variety of apparatuses, so the map $A \to M_A$ is multivalued.

Thus, the indirect measurement scheme involves the following:

- the states of the systems S and the apparatus M_A for measurement of some (physical or mental) observable s;
- the unitary operator $\hat{U}$ representing the interaction dynamics for the system $S + M_A$;
- the meter observable $\hat{M}_A$ giving outputs of the pointer of the apparatus M designed for measurements of observable A.

Formally, an *indirect measurement model*, introduced in [291] as a "(general) measuring process," is a quadruple

$$(\mathcal{K}, \hat{\sigma}, \hat{U}, \hat{M}_A)$$

consisting of a Hilbert space $\mathcal{K}$, a density operator $\hat{\sigma} \in D(\mathcal{K})$, a unitary operator $\hat{U}$ on the tensor product of the state spaces of S and M, $U \colon \mathcal{H} \otimes \mathcal{K} \to \mathcal{H} \otimes \mathcal{K}$, and a self-adjoint operator $\hat{M}_A$ on $\mathcal{K}$. By this measurement model, the Hilbert space $\mathcal{K}$ describes the states of the apparatus M_A, the unitary operator $\hat{U}$ describes the time-evolution of the composite system $S + M_A$, the density operator $\hat{\sigma}$ describes the initial state of the apparatus M_A, and the self-adjoint operator $\hat{M}_A$ describes the meter observable of the apparatus M_A. Then, the output probability distribution $P_\rho(A = x)$ in the system state $\hat{\rho}$ is given by

$$P_\rho(A = x) = \mathrm{Tr}\,[(I \otimes \hat{E}_{M_A}(x))\hat{U}(\hat{\rho} \otimes \hat{\sigma})\hat{U}^\star], \tag{1.17}$$

where $\hat{E}_{M_A}(x)$ is the spectral projection of operator $\hat{M}_A$ for the eigenvalue x.

The change of the state $\hat{\rho}$ of the system S caused by the measurement for the outcome $A = x$ is represented with the aid of the map $\mathcal{I}_A(x)$ in the space of density operators defined as

$$\mathcal{I}_A(x)\hat{\rho} = \mathrm{Tr}_{\mathcal{K}}[(I \otimes \hat{E}_{M_A}(x))\hat{U}(\hat{\rho} \otimes \hat{\sigma})\hat{U}^\star], \tag{1.18}$$

where $\mathrm{Tr}_{\mathcal{K}}$ is the partial trace over $\mathcal{K}$. Then, the map $x \mapsto \mathcal{I}_A(x)$ turns out to be a quantum instrument. Thus, the statistical properties of measurements realized by indirect measurement model $(\mathcal{K}, \hat{\sigma}, \hat{U}, \hat{M}_A)$ are described by a quantum instrument. We remark that conversely any quantum instrument can be represented via the indirect measurement model [291]. Thus, quantum instruments mathematically characterize the statistical properties of all the physically realizable quantum measurements.

Now we make the following remark on terminology. In the textbook quantum theory one speaks about observation performed on systems, where an observable A is mathematically given by Hermitian operator $\hat{A} \in \mathcal{L}(\mathcal{H})$, where $\mathcal{H}$ is the state space of S. This scheme is straightforward and it does not involve the apparatus' state space

Now, we point to a few details which were omitted in the preceding considerations. The measuring interaction between the system S and the apparatus M turns on at time t_0, the time of measurement, and turns off at time $t = t_0 + \Delta t$. We assume that the system S and the apparatus M_A do not interact with each other before t_0 nor after $t = t_0 + \Delta t$ and that the compound system $S + M_A$ is isolated in the time interval (t_0, t).

Now we describe the structure of the apparatus M_A in more detail. The *probe system* P is defined to be the minimal part of apparatus M_A such that the compound system $S + P$ is isolated in the time interval (t_0, t).

Then the preceding scheme is applied to the probe system P instead of the whole apparatus M. The rest of the apparatus M performs the pointer measurement on the probe P. In particular, the unitary evolution operator U describing the state-evolution of the system $S + P$ has the form $\hat{H} = e^{-i\Delta t \hat{H}}$, where

$$\hat{H} = \hat{H}_S + \hat{H}_P + \hat{H}_{SP}$$

is a Hamiltonian of $S + P$ with the terms $\hat{H}_S$ and $\hat{H}_P$ representing the internal dynamics in the subsystems S and P of the compound system and $\hat{H}_{SP}$ describing the interaction between the subsystems.

Introduction of probe systems may be seen as unnecessary complication of the scheme of indirect measurements. However, it is useful if the apparatus M_A is a very complex system that interacts (often in parallel) with many systems $S_j, j = 1, 2, \ldots., m$. Its different probes are involved solely in interaction with the concrete systems, P_j with S_j. And the system $S_j + P_j$ can be considered as an isolated system; in particular, from interactions with other systems S_i and probes P_i.

The indirect measurement scheme is part of the theory of open quantum systems. Instead of a measurement apparatus M_A, we can consider the surrounding environment $\mathcal{E}$ of the system S.

1.5 Classical Probability Space

Classical probability theory was mathematically formalized in a set and measure theoretical framework by Kolmogorov in 1933 [250, 251]. Here we briefly introduce its main notions.

Let Ω be a set of any origin; its points are called elementary events. Consider a collection of subsets $\mathcal{F}$ of Ω forming a σ-algebra, that is, it is closed w.r.t. countable unions and intersections and the operation of complement. (In American literature the term σ-field is used.) If Ω is finite, then $\mathcal{F}$ is the collection of all its subsets. The sets belonging to $\mathcal{F}$ represent events. The set operations correspond to the logical operations as follows:

- union – disjunction,
- intersection – conjunction,
- complement – negation.

So, the σ-algebra of events $\mathcal{F}$ is the set representation of classical Boolean logic.

Let P be a probability measure on $\mathcal{F}$. It assigns to each event $A \in \mathcal{F}$ nonnegative weight $P(A)$; these weights are normalized as $P(\Omega) = 1$. The main property of P is its additivity: if two events A_1, A_2 are disjoint, $A_1 \cap A_2 = \emptyset$, then the probability of disjunction of these events equals the sum of probabilities

$$P(A_1 \cup A_2) = P(A_1) + P(A_2). \tag{1.19}$$

To make probability theory consistent with Lebesgue integration theory, it is supposed that probability is not only additive, but even countably additive. Consider a sequence of pairwise disjoint events, (A_k), that is, $A_i \cap A_j = \emptyset, i \neq j$; then the probability of disjunction of these events equals the sum of probabilities

$$P(\cup_{k=1}^{\infty} A_k) = \sum_{k=1}^{\infty} P(A_k). \tag{1.20}$$

The triple $K = (\Omega, \mathcal{F}, P)$ is called *(Kolmogorov) probability space*.

A random variable is map $A\colon \Omega \to \mathbb{R}$ having a special property – measurability. Measurability is equivalent to the following condition. For any $x \in \mathbb{R}$, the set

$$\Omega_{A<x} = \{\omega \in \Omega\colon A(\omega) < x\} \tag{1.21}$$

belongs to the σ-algebra $\mathcal{F}$. The probability distribution of A, a measure p_A on $\mathbb{R}$, is determined by its values for intervals, $p^A((a,b]), \ldots, p^A((a,b))$, for example,

$$p^A([a,b]) = P(\omega \in \Omega\colon A(\omega) \in [a,b]).$$

The mean value (mathematical expectation, average) of a random variable $A = A(\omega)$ is defined as its integral:

$$E[A] \equiv \langle A \rangle = \int_{\Omega} A(\omega) dP(\omega) = \int_{\mathbb{R}} x dp^A(x). \tag{1.22}$$

For a discrete random variable with the range of values $X = (x_n)$, its probability distribution is determined by the following set of probabilities:

$$p^A(x) = P(\omega \in \Omega : A(\omega) = x), \ x \in X.$$

The mathematical expectation is given by the sum

$$E[A] \equiv \langle A \rangle = \sum_{x \in X} x p_A(x). \tag{1.23}$$

Let A and B be random variables. We recall *the Bayes formula* for conditional probability,

$$P(B = y | A = x) = P(B = y, A = x) / P(A = x), \tag{1.24}$$

for $P(A = x) > 0$. It should be emphasized that the Bayes formula is the *definition* of conditional probability. It is not a theorem; it cannot be derived from other "natural postulates." Here

$$P(B = y, A = x) = P(\omega \in \Omega : A(\omega) = x, B(\omega) = y)$$

is the probability of the event that the variables take the values x and y. In quantum physics such events are not always defined, for example, for position and momentum, or spin projections on different axes. The Bayes formula can't be used. The main difference between classical and quantum probability theories is in conditional probability.

One of the basic classical laws of probability is given by the *formula of total probability* (FTP):

$$P(B = y) = \sum_{x \in X_A} P(B = y | A = x) P(A = x). \tag{1.25}$$

It is one of the key elements of the Bayesian probability inference. FTP is derived on the basis of the Bayes formula for conditional probability and additivity of a probability measure P. As was noted, the Bayes definition of conditional probability is not applicable in quantum probability. Hence, one can expect that FTP can also be violated for quantum observables.

We remark that Kolmogorov considered the condition of countable additivity (1.20) as purely mathematical [250]; it is needed to have the Lebesque integral representation of the mean value of a random variable. He pointed out that this condition is not experimentally testable: one can't realize experimentally an infinite sequence of events. Only the condition of finite additivity (1.19) has the practical meaning. It is also interesting that he did not consider the axioms of his theory as given by God. In particular, in the process of formulation of the axiomatics of probability theory he seriously considered the possibility of proceeding with nonadditive probability measures. But he gave up: such theory would be too complicated. It is interesting that Feynman [120] pointed to the violation of additivity as the main property of quantum probability distinguishing it from classical probability. (He did not know about the Kolmogorov axiomatics of probability theory, so he referred to the Laplace probability theory.)

In my monograph [179] reader can find the references on Kolmogorov's works written in 1920s in which he discussed possible variations of the axiomatics of probability theory. And I was lucky to meet Kolmogorov and discuss the axiomatics and its possible modifications, when I was a graduate student. My supervisor Smolyanov introduced me to Kolmogorov in the process of the discussion on recommendation of our article [342] to Doklady USSR. We developed a generalization of probability theory with complex-valued distributions and proved an analog of the central limit theorem. In this theorem the role of the limiting Gaussian distribution was played by Feynman distribution (see [342] for its mathematical definition). We have some problems with publication of this note due to opposition from the probability community which was rigidly structured within the Kolmogorov theory. So, Smolyanov decided to speak directly with Kolmogorov. And Kolmogorov was open to our theory with "complex-valued probability distributions."

Smolyanov asked once Kolmogorov why he had never tried to work with quantum generalization of probability. Kolmogorov's answer was surprising. He said something as "it was to difficult to proceed deeply at the mathematical level of rigorousness."

PART I

Quantum Foundations

2

Two Faces of Quantum Nonlocality and Two Interpretations of Quantum State

2.1 Intrinsic Quantum and Subquantum Nonlocalities

In the last decades too many futile technical terms in the context of correlation measurements have been created which are neither well defined nor physically interpreted. They evoke different associations in the discussion on entangled systems and these associations are sometimes rather confusing. As was emphasized in a recent paper [232] (see also [224]), the notion "quantum nonlocality" is really misleading. One of the difficulties in struggling with nonlocality is that the present situation in quantum foundations is the real mess. Surprisingly, this mess-problem is commonly ignored. Typically it is claimed that quantum physics is nonlocal without specifying what this means concretely.

Personally, I got the first signal that the nonlocality which is widely discussed in the quantum community is Janus faced from the talk of Alain Aspect at one of the Växjö conferences (see also his papers [29, 31]). He started his talk not from the Bell inequality and its violation (as would be expected from the pioneer in experimental Bell tests), but from the projection postulate in Lüders' form [274] and its nonlocal consequences. He pointed out that Lüders *projection nonlocality* is really counterintuitive and that, to find a proper physical picture, one has to introduce hidden variables. Of course, Aspect follows the EPR (Einstein–Podolsky–Rosen) paper [115], but his reasoning is cleaned from the problem of completeness of QM; the quantum nonlocality disaster is described in all its terrible beauty.

From Aspect's viewpoint (reflecting Bell's viewpoint), quantum theory really cries for hidden variables! By introducing such variables one can reestablish realism. And in the realistic framework there is no need for spooky action at a distance. However, in this way, as is commonly accepted, one confronts the Bell inequality and proceeds toward another sort of nonlocality – *subquantum nonlocality*. The latter is an unfounded claim that the violation of the Bell-type inequalities may not be explained in a causally local way and that any hidden variable description of quantum correlations must invoke instantaneous influences at the distance.[1] The first message of this chapter is that one should sharply distinguish two sorts of nonlocality:

- Intrinsic quantum nonlocality coupled to use of the projection postulate in quantum theory – Lüders nonlocality [115].
- Subquantum nonlocality coupled to hidden variables and the Bell-type inequalities – Bell nonlocality [53, 54].

The aim of the present chapter is to analyze intrinsic quantum nonlocality–Lüders nonlocality. We show that it is fictitious; this is "nonlocality" of probability conditioning; it is similar to "nonlocality" of classical probability conditioning.

We emphasize the role of quantum state interpretations, individual (Copenhagen)[2] versus statistical (ensemble). In the process of the critical analysis of intrinsic quantum nonlocality, we often mention the hidden variables nonlocality, with the emphasis on the sharp distinction between them.

In fact, I started writing article [233], the basic for this chapter, as the reply to the following comment by Johan Summhammer on my previous article [232]:

"I have looked at your 'Get rid of non locality... ' and I agree at the formal level, that you are only dealing with incompatible operators. But the empirical fact remains, that setting a measurement operator on site A has an instantaneous 'influence' on site B, if you later calculate conditional statistics between the data of site A and site B. And this faster-than-light change of statistical correlation is already empirically proven. Sure, no transfer of information. But more than 'no influence'. Which name would you assign to this phenomenon?"

Such often postulated instantaneous influence at a distance is not caused by a real or spooky interaction between the two distant observations. Actually, nothing else happens, other than our knowledge about the entangled quantum state especially about the second photon is increased by measurements on the first photon. The information about the outcome of the first detection process distinctly restricts the probabilities of the possible outcomes of the second measurement process. This reduction of possibilities is an information update process, not a physical one; this is an update of our knowledge.

If the statistical (ensemble) interpretation of QM is adopted, then this knowledge update is a passage from the probabilistic description of all experimental data to the probabilistic description of some particular subsamples of the data. But by adopting

[1] See Chapter 15 for discussion; see also the works of authors belonging to the "probability opposition" and employing the probabilistic arguments against the common interpretation of the violation of the Bell inequalities, see, for example, [71, 81, 107, 108, 112, 174, 179, 181, 203, 229, 257–259, 279, 289].

[2] The term "Copenhagen interpretation" is characterized by variety versions. Plotnitsky even suggested to speak about the interpretation in the spirit of Copenhagen [308, 315]; see Section 2.4 for the version used in this book.

the individual (orthodox Copenhagen) interpretation of QM, one really encounters spooky action at a distance.

We point out that *QBism* (see, e.g., Fuchs et al. [124, 125]) also leads to rejection of the speculations on instantaneous physical influence. In QBism knowledge update is related to subjective probability, update of experience of an individual agent.

The individual interpretation of a quantum state can be merged with the statistical interpretation of probability, Plotnitsky called this combination *the statistical Copenhagen interpretation* and even coupled it to Bohr (at least indirectly); see our joint paper [311].[3]

Finally, we mention *the Växjö interpretation* [193, 203] (Chapter 10). By this interpretation the quantum formalism is not about systems behavior, but about measurement contexts (see Chapter 10 for details). This is the intrinsically contextual interpretation. It reflects the Bohr's complementarity principle with highlighting its contextual component (Chapter 3).

By the Växjö interpretation, QBism, and the statistical Copenhagen interpretation quantum theory is local.

Finally, we remark that often one says that, since within the quantum formalism superluminal signaling and information transfer is impossible, quantum nonlocality in the form of nonlocal influence (spooky action at a distance) is not disastrous; this is some mystical feature of nature which can peacefully coexist with the traditional physical views. Such a position is simply opportunistic. The aim of this book is getting rid of quantum nonlocality from physics in all possible forms.

2.2 Aspect: Action at a Distance from Projection Postulate

We start with the extended citation from the practically unknown preprint of Nobel Prize Laureate Alain Aspect [31] (see also [29]). In this preprint, "projection nonlocality" appears in all its brilliance. Aspect's previously mentioned talk at the Växjö conference on quantum foundations was based on this preprint. Although I do not agree with Aspect's conclusion, I appreciate his theoretical analysis of the Bell experimental situation – the analysis based on the quantum formalism.

"Let us consider the optical variant of the Bohm's version of the E.P.R. Gedankenexperiment. A source S emits a pair of photons with different frequencies ν_1 and ν_2, counterpropagating along Oz. Suppose that the polarization part of the state vector describing the pair is as follows:

[3] I remember that the claim that Bohr explored the statistical interpretation of probability was strongly criticized by Fuchs with whom we discussed the content of this paper in the process of its preparation. QBists consider Bohr as their close ally and, of course, they believe that Bohr assigned probabilities to individual events (may be not subjective ones).

$$|\Psi(\nu_1,\nu_2)\rangle = \frac{1}{\sqrt{2}}\{|x,x\rangle + |y,y\rangle\},$$

where x and y are linear polarizations states.

Let us now consider the probabilities $p_{\pm\pm}(a,b)$ of joint detections of ν_1 and ν_2 in the channels + or − of polarisers I or II, in orientations a and b. Quantum mechanics predicts:

$$p_{++}(a,b) = p_{--}(a,b) = \frac{1}{2}\cos^2(a,b),\ p_{+-}(a,b) = p_{-+}(a,b) = \frac{1}{2}\sin^2(a,b).$$

We are going to show that these quantum mechanical predictions have far-reaching consequences.

As a naive physicist, I like to raise the question of finding a simple image to understand these strong correlations. The most natural way to find an image may seem to be to follow the quantum mechanical calculations leading to $p_{\pm,\pm}(a,b)$. In fact, there are several ways to do this calculation. A very direct one is to project the state vector $|\Psi(\nu_1,\nu_2)\rangle$ onto the eigenvector corresponding to the relevant result. This gives immediately the joint probabilities $p_{\pm,\pm}(a,b)$. However, since this calculation bears on state vectors describing globally the two photons, I do not know how to build a picture in our ordinary space.

In order to overcome this problem, and to identify separately the two measurements happening on both ends of the experiment, we can split the joint measurement in two steps. Suppose, for instance, that the measurement on photon ν_1 takes place first, and gives the result +, with the polarizer I in orientation a. The + result (associated with the polarization state $|a\rangle$) has a probability of 1/2. To proceed with the calculation, we must then use the postulate of reduction of the state vector, which states that after this measurement, the new state vector $|\Psi'(\nu_1,\nu_2)\rangle$ describing the pair is obtained by projection of the initial state vector $|\Psi(\nu_1,\nu_2)\rangle$ (equation 1) onto the eigenspace associated to the result +: this two-dimensional eigenspace has a basis $\{|a,x\rangle, |a,y\rangle\}$. Using the corresponding projector, we find after a little algebra,

$$|\Psi'(\nu_1,\nu_2)\rangle = |a,a\rangle.$$

This means that immediately after the first measurement, photon ν_1 takes the polarization a: this is obvious because it has been measured with a polarizer oriented along a, and the result + has been found. More surprisingly, the distant photon ν_2, which has not yet interacted with any polarizer, has also been projected into the state a with a well-defined polarization, parallel to the one found for photon ν_1. This surprising conclusion, however, leads to the correct final result (3), since a straightforward application of Malus' law shows that a subsequent measurement performed along b on photon ν_2 will lead to

$$p_{++}(a, b) = \frac{1}{2}\cos^2(a, b).(8)$$

The calculation in two steps therefore gives the same result as the direct calculation. But in addition, it suggests a picture for the two steps of measurement:

- i: Photon ν_1, which did not have a well-defined polarization before its measurement, takes the polarization associated to the obtained result, at the moment of its measurement: this is not surprising.
- ii: When the measurement on ν_1 is done, photon ν_2, which did not have a well-defined polarization before this measurement, is projected into a state of polarization parallel to the result of the measurement on ν_1. This is very surprising, because this change in the description of ν_2 happens instantaneously, whatever the distance between ν_1 and ν_2 at the moment of the first measurement.

This picture seems in contradiction with relativity. According to Einstein, what happens in a given region of space-time cannot be influenced by an event happening in a region of space-time that is separated by a spacelike interval. It is therefore not unreasonable to try to find more acceptable pictures for "understanding" the EPR correlations. It is such a picture that we consider now.

For Aspect and Bell, the latter is the hidden variable picture that led them to *subquantum nonlocality.*

Our aim is to show that the above picture does not contradict relativity. Moreover, it is schematically identical to the corresponding picture from classical probability theory: the probability-transformation resulting from conditioning on the concrete observation-output.

2.3 Projection Postulate and State Update

In [274], Lüders formalized in the form of a postulate the operation of the quantum state-transformation resulting from measurement with the concrete output, as projection on the corresponding subspace of the state space. In fact, this postulate was actively used from the first days of quantum theory, for example, in the EPR paper [115] in which authors refer to *"a process known as the reduction of the wave packet."*

Often the projection postulate is misleadingly associated with the name of von Neumann when referring to his book [365]; people even refer to it as the von Neumann projection postulate. We state that von Neumann applied the projection postulate only to observables represented by operators with nondegenerate spectra. This important issue is typically ignored in the foundational discussions (see [202] for the consequences of the general von Neumann approach to state update in comparison with the Lüders special approach). We emphasize that local observables on compound systems typically have degenerate spectrum. For von Neumann, "reduction of the wave packet" need not be represented as projection on a subspace. In fact, in book [365] von Neumann laid the foundations of quantum instrument theory.

We recall that in the original quantum formalism [364, 365] an observable A is represented by a Hermitian operator acting in Hilbert state space $\mathcal{H}$. For simplicity, we consider only operators with purely discrete spectra. The spectral decomposition has the form $\hat{A} = \sum_x x\, \hat{E}_A(x)$, where $\hat{E}_A(x)$ is the orthogonal projector on the subspace $\mathcal{H}_A(x)$ composed of eigenvectors with eigenvalue x.

For a pure initial state $|\psi\rangle$, the postmeasurement state is always again the pure state given by normalized projection:

$$|\psi_{A=x}\rangle = \hat{E}_A(x)|\psi\rangle / \|\hat{E}_A(x)\psi\|.$$

Thus, measurement with output $A = x$ induces the state-transformation: $|\psi\rangle \to |\psi_{A=x}\rangle$. We also recall the Born rule for the probability of A's output:

$$P_\psi(A = x) = \|\hat{E}_A(x)\psi\|^2.$$

At first sight, the projection transformation of the state has nothing to do with nonlocality. However, for compound system $S = (S_1, S_2)$ this state-transformation procedure can give the impression of instantaneous action at a distance.

Let quantum state $|\Psi\rangle$ belong to tensor product $\mathcal{H} = \mathcal{H}_1 \otimes \mathcal{H}_2$ of state spaces $\mathcal{H}_i$ of systems $S_i, i = 1, 2$. Select some observable a on subsystem S_1; it is represented by Hermitian operator $\hat{a}$ where projections $\hat{E}_a(x)$ act in $\mathcal{H}_1$. This observable can be also treated as observable A on compound system S – represented by the operator $\hat{A} = \hat{a} \otimes I$ with projections $\hat{E}_A(x) = \hat{E}_a(x) \otimes I$ in $\mathcal{H}$. By getting output $a = x$, we transform the state of S on the basis of Lüders projection postulate:

$$|\Psi_{a=x}\rangle = \hat{E}_A(x)|\Psi\rangle / \|\hat{E}_A(x)\Psi\|. \tag{2.1}$$

Consider now an observable b on subsystem S_2 with the operator $\hat{b} \colon \mathcal{H}_2 \to \mathcal{H}_2$. It can be treated as an observable B on the compound system $S = (S_1, S_2)$ and represented by the operator $\hat{B} = I \otimes \hat{b}$. Turn to conditional b-measurement under output $a = x$. By the definition of quantum conditional probability we have

$$P_\Psi(b = y|a = x) \equiv P_\Psi(B = y|A = x) = P_{\Psi_{a=x}}(B = y) = \tag{2.2}$$

$$\|\hat{E}_B(y)\Psi_{a=x}\|^2.$$

If state $|\Psi\rangle$ is separable, that is $|\Psi\rangle = |\psi^{(1)}\rangle \otimes |\psi^{(2)}\rangle$, then

$$P_\Psi(B = y|A = x) = \frac{\|\hat{E}_B(y)\hat{E}_A(x)\Psi\|^2}{\|\hat{E}_a(x)\psi^{(1)}\|^2} = \|\hat{E}_b(y)\psi^{(2)}\|^2, \tag{2.3}$$

that is,

$$P_\Psi(b=y|a=x)=P_\Psi(b=y). \tag{2.4}$$

Thus, for such a state, measurement of observable a on S_1 does not change statistics for measurements of observable b on S_2.

However, if a state is entangled, then generally

$$P_\Psi(b=y|a=x)\neq P_\Psi(b=y). \tag{2.5}$$

We remark that any state of S determines the states of its subsystems S_i, $i=1,2$; for the initial state $|\Psi\rangle$,

$$\rho^{(1)}=Tr_{\mathcal{H}_2}|\Psi\rangle\langle\Psi|,\ \rho^{(2)}=Tr_{\mathcal{H}_1}|\Psi\rangle\langle\Psi|, \tag{2.6}$$

and, for the postmeasurement state $|\Psi_{A=x}\rangle$,

$$\rho^{(1)}_{a=x}=Tr_{\mathcal{H}_2}|\Psi_{a=x}\rangle\langle\Psi_{a=x}|,\ \rho^{(2)}_{a=x}=Tr_{\mathcal{H}_1}|\Psi_{a=x}\rangle\langle\Psi_{a=x}|. \tag{2.7}$$

For separable state $|\Psi\rangle$,

$$\rho^{(2)}=\rho^{(2)}_{a=x}, \tag{2.8}$$

that is, a-measurement on subsystem S_1 with the outcome $a=x$, does not change the state of subsystem S_2. For entangled state $|\Psi\rangle$,

$$\rho^{(2)}\neq\rho^{(2)}_{a=x}, \tag{2.9}$$

that is, a-measurement on subsystem S_1, with the outcome $a=x$, changes the state of subsystem S_2. If subsystems are spatially separated, then this process can (but need not) be treated as action at a distance.

The preceding considerations can be represented in the form of probabilities. For separable state $|\Psi\rangle$,

$$P_{\rho^{(2)}}(b=y)=P_{\rho^{(2)}_{a=x}}(b=y), \tag{2.10}$$

and, for an entangled state, generally

$$P_{\rho^{(2)}}(b=y)\neq P_{\rho^{(2)}_{a=x}}(b=y). \tag{2.11}$$

The preceding formulas are just mathematical expressions. To have some physical picture, we should present their interpretation. The main issue (and the foundational problem) is the state's interpretation. Nowadays, some experts in quantum foundations claim that all interpretations are equally useful; so it is meaningless to struggle for a "right interpretation"; others claim that it is even possible to proceed without interpretation at all. I do not think so.

In chapter 13 the notion of entanglement is formalized within the quantum conditional probability framework. The meaning of inequalities (2.11) is discussed in more detail.

2.4 Statistical versus Individual Interpretations

We start this section with an extended citation of Ballentine [43]:

"Quantum theory, and indeed any theory, can be divided into:

- (a) A mathematical formalism consisting of a set of primitive concepts, relations between these concepts (either postulated or obtainable by given rules of deduction), and a dynamical law.
- (b) Correspondence rules which relate the theoretical concepts of (a) to the world of experience.

This division is not absolute – early one must have a formalism in order to make correspondence rules, but unless one has at least some partial idea of correspondence rules, one would not know what one was talking about while constructing the formalism – nevertheless it is convenient for the present task.

The mathematical formalism of quantum theory is well known and can be abstracted from any of several textbooks. The primitive concepts are those of state and of observable . . . .

The correspondence rules must relate the primitive concepts of state and observable to empirical reality. In so doing they will provide a more specific interpretation for the averages and probabilities . . .

The natural requirement placed upon an observable is that we should be able to observe it. More precisely, an observable is a dynamical variable whose value can, in principle, be measured. . . .

The different interpretations of quantum theory are most sharply distinguished by their interpretations of the concept of state. Although there are many shades of interpretation, we wish to distinguish only two:

- **SI** *The statistical interpretation, according to which a pure state (and hence also a general state) provides a description of certain statistical properties of an ensemble of similarly prepared systems, but need not provide a complete description of an individual system.*
- **II** *Interpretations which assert that a pure state provides a complete and exhaustive description of an individual physical system (e.g., an electron).*"

In further considerations we call II *the individual interpretation* of a quantum state. It is commonly known as the *orthodox Copenhagen interpretation.* (The latter term has a variety of flavors – interpretations in the spirit of Copenhagen [315].) In the empirical (operational) approach [106], SI is associated with preparation procedures generating ensembles of similarly prepared systems. We call a preparation procedure (or the corresponding ensemble of systems) *a quantum random system* and denote it by the symbol S (sometimes with indexes), Thus, by SI a density operator ρ represents the state of a quantum random system S. We note that within SI pure states lose their specialty.

So, depending on the interpretation the same symbol S is used to denote a random system (in SI) and an individual system (in II).

One often explores the interpretation of a state given by a density operator ρ as describing a statistical mixture of quantum systems in the concrete pure states. However, as we know, such an interpretation of ρ is ambiguous. In SI all quantum states describe random systems – preparation procedures.

Finally, we remark that researchers using II, the orthodox Copenhagen interpretation, often refer to it as a *statistical quantum theory*. For example, Einstein writes (see, e.g., [116]):

"I now come to what is probably the most interesting subject which absolutely must be discussed in connection with the detailed arguments of my highly esteemed colleagues Born, Pauli, Heitler, Bohr, and Margenau. They are all firmly convinced that the riddle of the double nature of all corpuscles (corpuscular and undulatory character) has in essence found its final solution in the statistical quantum theory."

In the light of this ambiguity in operating with the term "statistical," it might be better to call SI *the ensemble interpretation*. However, the term "statistical interpretation" was so widely advertised by Ballentine that we proceed with it.

Historical remarks

SI is upheld, for example, by Einstein [116], by Popper [319], by Blohintsev [57, 58], and it was actively advertised by Ballentine [43]. We continue to cite Ballentine's paper [43]. Regarding II he pointed out that "This class contains a great variety of members, from Schrödinger's original attempt to identify the electron with a wave packet solution of his equation to the several versions of the Copenhagen Interpretation. Indeed many physicists implicitly make assumption II without apparently being aware that it is an additional assumption with peculiar consequences. It is a major aim of this paper to point out that the hypothesis II is unnecessary for quantum theory, and moreover that it leads to serious difficulties.

Bohr (1935, 1949) argued against Einstein, who rejected II. Heisenberg's position is somewhat unclear. One reference (1930, p. 33) contains a statement fully in accord with the Statistical Interpretation; however, in a later writing specifically in defense of the Copenhagen interpretation (1955, p. 26) he states "an individual atomic system can be represented by a wave function." His interpretation of probability (1958, Chap. 3) was based on the Aristotelian notion of "potentia," which is quite different from the Statistical Interpretation. Messiah (1964) is quite explicit (p. 152, 158) in favoring II over SI. Although both claim orthodoxy, there now seems to be a difference of opinion between what may be called the Copenhagen school represented by Rosenfeld, and the Princeton school represented by Wigner (see Rosenfeld, 1968 and references therein). But since both factions appear to accept hypothesis II, our criticism will apply to both.

We remark that, although both II and SI generate interest for creation of subquantum theories, these interpretations are different in their attitudes toward such theories.

By following II and confronting intrinsic quantum nonlocality (Lüders nonlocality), one considers the quantum formalism as inconsistent and wants a new theory which would not suffer from nonlocality. It was clear from the very beginning that on this pathway one should reject some basic principles of quantum theory. This was done by Bell who suggested a theory with hidden variables which is in the direct contradiction with the Heisenberg uncertainty and Bohr complementarity principles. The Bell project was not successful, since in the common interpretation it led to another sort of nonlocality.

By following SI one has no problem with quantum formalism, in particular, with Lüders nonlocality. The interest for creation of subquantum theories is a consequence of understanding that a quantum state does not provide the complete description of a quantum system and that a finer description is in principle possible. In contrast to a II-user, an SI-user would be very careful in creation of a subquantum theory, not violate the basic principles of QM, the Heisenberg uncertainty and Bohr complementarity principles. Such a theory was a dream of Einstein, but it was never created.

2.5 Interpretations of State and Probability

Each state interpretation should be associated with a probability interpretation. By using II, in any of its versions, and associating a quantum state with a single system it is common to use the individual interpretation of probability giving the possibility of assigning probabilities to individual events. The most natural way to proceed in this direction is to explore the subjective probability theory, as for example, in QBism. Heisenberg treated probability as a measure of potentiality. In principle, one can proceed with comparative probability theory. However, the latter has not yet been employed in QM. A curious reader can find more about probability interpretations in my book [179]. Since QM generates statistical predictions, one working with individual-event probability should justify its connection with statistics, with frequencies. In this book we do not operate with probabilities of individual events. Hence, we are not interested in their statistical justification.

In this book we explore the statistical (frequency) interpretation of probability. We follow Kolmogorov [179, 250, 251]: We may assume that to an event A which may or may not occur under conditions Σ is assigned a real number $P(A)$ which has the following characteristic: one can be practically certain that if the complex of conditions Σ is repeated a large number of times, N, then if $n(A)$ is the number of occurrences of event A, the ratio $\nu_N(A) = n(A)/N$ will differ very slightly from $P(A)$, namely, $P(A) = \lim_{N\to\infty} \nu_N(A)$. In rigorous mathematical terms this process is formalized via the law of large numbers.

The statistical interpretation of probability matches perfectly with the statistical interpretation SI of a quantum state. However, this probability interpretation can also be coupled to II. How it can be done mathematically is a complex problem. However, II endowed with the statistical (frequency) interpretation of probability is used, for example, by von Neumann [364, 365]. He consistently employs II, but at the same time he points out that experimental verification is possible only through von Mises' frequency approach to probability [363]. At that time the Kolmogorov axiomatic approach to probability and coupling between the frequency and probability via the strong law of large numbers were not so widely known. In contrast, Bohr had never expressed precisely his views on probability. Moreover, it seems that on different occasions he interpreted probability differently, from subjective probability (similarly to QBists) to statistical probability. The latter interpretation of Bohr's views is advertized by Plotnitsky and it is discussed in great detail in our joint article [311].

2.6 Individual Interpretation: Intrinsic Quantum Nonlocality

Consider a compound system $S=(S_1,S_2)$. If one employs II, the orthodox Copenhagen interpretation, then according to the Lüders projection postulate a measurement performed on subsystem S_1 can change instantaneously the state of subsystem S_2 at a distant location. This projection nonlocality is intrinsic for QM endowed with II. It has no relation to attempts to go beyond quantum theory, for example, by employing hidden variables. Of course, for one who rejects II, quantum nonlocality is apparent.

For example, consider a compound physical system $S = (S_1, S_2)$ prepared in the entangled state

$$|\Psi\rangle = (|01\rangle + |10\rangle)/\sqrt{2}, \tag{2.12}$$

where the vectors labeled as $|0\rangle, |1\rangle$ are eigenvectors of the operators $\hat{a}\colon \mathcal{H}_1 \to \mathcal{H}_1$ and $\hat{b}\colon \mathcal{H}_2 \to \mathcal{H}_2$, and both $\mathcal{H}_i$ are qubit spaces. These operators represent the observables a and b on the subsystems S_1 and S_2 of the compound system $S = (S_1, S_2)$. The corresponding observables and operators on S are denoted by A, B and $\hat{A} = \hat{a} \otimes I, \hat{B} = I \otimes \hat{b}$.

By measuring observable a on S_1 and getting output $a = 0$, we found the compound system in the state $|\Psi_{a=0}\rangle = |01\rangle$. Thus, according to II the physical state of system S_2 instantaneously becomes $|\phi\rangle = |1\rangle$. Since by II a quantum state has the meaning of the state of the individual physical system, this is nothing else than action at a distance.

II, *the orthodox Copenhagen interpretation, is the source of intrinsic quantum nonlocality that is theoretically expressed by the projection postulate.*

Aspect is sure that a pure quantum state $|\psi\rangle$ is a system's physical state, in the sense that it provides the complete description of a physical system. Therefore, for him, quantum theory is intrinsically nonlocal, that is, without any relation to violation of the Bell inequality. For such II-users, the hidden variables project is an attempt to explain this intrinsic quantum nonlocality by going beyond quantum theory. For Aspect, the violation of the Bell inequalities also leads to nonlocality – subquantum nonlocality. Thus, starting with II one comes to the conclusion that nonlocality is inescapable: QM is nonlocal intrinsically as well as a theory with hidden variables, that is, Lüders as well as Bell nonlocal.

2.7 EPR: Either Individual Interpretation or Locality

We emphasize that *the EPR paper [115] is in fact directed against* **II**. They wrote: "In quantum mechanics it is usually assumed that the wave function does contain a complete description of the physical reality of the system in the state to which it corresponds."

The output of the EPR argument is often referred to as the EPR paradox (and Einstein himself called it a "paradox" [116]). However, this is not a paradox at all. This is merely a discussion on casualties of one concrete interpretation of a quantum state, namely, II. To show this, we follow the line of reasoning of Einstein in article [116] and of Ballentine in article [43].

In order to answer the question posed in the title of the EPR paper [115], *"Can Quantum-Mechanical Description of Physical Reality Be Considered Complete?"*, EPR start with two definitions.

- **D1** A necessary condition for a complete theory is that "every element of physical reality must have a counterpart in the physical theory."
- **D2** A sufficient condition for identifying an element of reality is, "without in any way disturbing a system, we can predict with certainty (i.e., with probability equal to unity) the value of a physical quantity, then there exists an element of physical reality corresponding to this physical quantity."

The EPR argument proceeds by demonstrating with a thought experiment, that two incompatible observables A and B represented by non commuting operators $\hat{A}$ and $\hat{B}$ should, under a special experimental context, both be considered elements of reality. But there is no state vector that can provide the value of both of these observables. EPR conclude that the quantum state vector cannot completely describe an individual system, but only an ensemble of similarly prepared systems.

In 1949 Einstein wrote [116]:

"it becomes evident that the paradox forces us to relinquish one of the following two assertions:

(A) the description by means of the Ψ-function is complete

(B) the real states of spatially separated objects are independent of each other.

On the other hand, it is possible to adhere to (B), if one regards the Ψ-function as the description of a (statistical) ensemble of systems (and therefore relinquishes (A)). However, this view blasts the framework of the orthodox quantum theory."

So, one can assume the validity of either of these two statements, but not both. In this comparison, the majority of physicists would prefer (B). However, typically (A) is not treated as the interpretation of a quantum state, namely, one which can be selected among a variety of interpretations, but as some fundamental feature of QM. As was pointed out by Einstein, if one rejects (A) (which is in fact II) in favor of SI, then he can keep (B) – locality of QM. This position was especially clearly presented in Ballentine's works [42]-[45].

We repeat once again that "quantum nonlocality" is a two-faced Janus. By using SI one escapes the intrinsic quantum nonlocality, the one considered in the EPR paper [115]. However, the appeal to SI opens the door to search of subquantum theories. One of the most popular of them, Bell's theory with hidden variables, is apparently nonlocal and this problem should be analyzed separately. The most surprising for me is that Bell's model and his conclusion on nonlocality is widely explored by II-users. In fact, they need not refer to the Bell inequality; QM endowed with II, the orthodox Copenhagen interpretation, is nonlocal due to the EPR argument (and Aspect honestly recognized this).

Bohr struggled for (A); see his response to the EPR argument [63]. By the laws of logic he should have rejected (B) – locality. However, he did not and he did not go against the laws of logic. In fact, Bohr's version of the orthodox Copenhagen interpretation differs from the version criticized by EPR (see Section 4.1 for my attempt to understand Bohr's position). Operating with a bunch of interpretations unified under the label "the orthodox Copenhagen interpretation" is a source of ambiguity in quantum foundational discussions. This is a good place to remark that the majority of physicists, especially "real physicists" as they call themselves, use the version of the orthodox Copenhagen interpretation which was criticized in the EPR article [115], Einstein criticized this interpretation even in 1949 [116].

2.8 EPR Argument and Heisenberg Uncertainty Principle

We start this section by citing article [115]:

"The usual conclusion from this in quantum mechanics is that when the momentum of a particle is known, its coordinate has no physical reality.... More generally, it is shown in quantum mechanics that, if the operators corresponding to two physical quantities, say $\hat{A}$ and $\hat{B}$, do not commute, that is, if $\hat{A}\hat{B} \neq \hat{B}\hat{A}$, then the precise knowledge of one of them precludes such a knowledge of the other. Furthermore, any attempt to determine the latter experimentally will alter the state of the system in such a way as to destroy the knowledge of the first.... From this follows that either *1) the quantum- mechanical description of reality given by the wave function is not complete or 2) when the operators corresponding to two physical quantities do not commute the two quantifies cannot have simultaneous reality.*"

Statement 2) is a consequence of the preceding statement about quantum observables represented by noncommutative operators – incompatible observables. The existence of such observables is the cornerstone of the EPR argument. We emphasize that *EPR do not question 2) and existence of incompatible observables.* Experimentally existence of incompatible observables is justified by the Heisenberg uncertainty relation in the form of the Robertson inequality.

"They [users of the orthodox Copenhagen interpretation] are apparently all of the opinion that Heisenberg's indeterminacy-relation (the correctness of which is, from my own point of view, rightfully regarded as finally demonstrated) is essentially prejudicial in favour of the character of all thinkable reasonable physical theories in the mentioned sense.... This theory is until now the only one which unites the corpuscular and undulatory dual character of matter in a logically satisfactory fashion; and the (testable) relations, which are contained in it, are, within the natural limits fixed by the indeterminacy-relation, complete. The formal relations which are given in this theory – i.e., its entire mathematical formalism – will probably have to be contained, in the form of logical inferences, in every useful future theory."

So, Einstein said about the Heisenberg uncertainty principle: "the correctness of which is, from my own point of view, rightfully regarded as finally demonstrated," so he had no doubts in its validity. The last sentence in the preceding citation is guidance for possible constructions of subquantum models: they should respect consequences of the mathematical formalism of QM, including the Heisenberg uncertainty principle (Chapter 15). However, Bell's model with hidden variables directly contradicts this principle (see again Chapter 15).

As usual in the Bell's framework, consider the space of hidden variables Λ and the probability measure P on this space. (Bell himself used notation ρ for this measure, i.e., $d\rho(\lambda) \equiv dP(\lambda)$.) Consider Bell's random variables $a_i(\lambda), b_i(\lambda), i = 1, 2$, representing observables of Alice and Bob, respectively. Surprisingly, Bell did not highlight that, besides the probabilities

$$P^{a_i,b_j}(x,y) = P(\lambda \in \Lambda : a_i(\lambda) = x, b_j(\lambda) = y), \tag{2.13}$$

for the pairs of compatible observables a_i, b_j, Bell's model presumes the existence of probabilities

$$P^{a_1,a_2}(x,y) = P(\lambda \in \Lambda : a_1(\lambda) = x, a_2(\lambda) = y), \tag{2.14}$$
$$P^{b_1,b_2}(x,y) = P(\lambda \in \Lambda : b_1(\lambda) = x, b_2(\lambda) = y), \tag{2.15}$$

for the pairs of incompatible observables a_1, a_2 and b_1, b_2 represented by non-commuting operators $\hat{a}_1, \hat{a}_2$ and $\hat{b}_1., \hat{b}_2$. Here we discuss the framework for the derivation of the CHSH-inequality within the hidden variables model. But from the very beginning, that is, without derivation of any Bell-type inequality, this assumption, (2.14), (2.15), contradicts the QM-representation of observables and, hence, the Heisenberg uncertainty principle or more generally the Bohr complementarity principle. In fact, this problem of the existence of JPD for incompatible quantum observables was known from the first days of quantum theory – compare with attempts to define JPD for the position and momentum observables starting with introduction of the Wigner function.

One may proceed toward subquantum models with hidden variables, but *without identification the values of random variables with the outcomes of quantum observables and "hidden correlations" with the experimental correlations for observables.* And de Broglie emphasized [99] this viewpoint. However, people wanted experimental verification, and de Broglie's message was completely ignored. It seems that nowadays de Broglie's viewpoint of the Bell works is mentioned only in my papers.

We conclude this discussion on the role of incompatibility in the EPR argument with the mathematical remark that generally noncommutative operators can have joint eigenvectors. So, in the abstract form the EPR reasoning should be understand as "there exist noncommutative operators that do not have any common eigenvector." In article [115], EPR considered operators of position and momentum. Formally, they satisfy this condition. However, they have continuous spectra. To operate with eigenvectors, one should appeal to the Dirac theory of generalized eigenvectors [109]. In the Bohm-experiment for spin projections, the corresponding operators are bounded and have discrete spectra; spin operators satisfy the aforementioned condition.

Finally, we remark that, for Bohr, the Heisenberg uncertainty principle was unquestionable [63]: "it is a well-known feature of the present formalism of quantum mechanics that it is never possible, in the description of the state of a mechanical system, to attach definite valves to both of two canonically conjugate variables"

2.9 Classical Probability Update

Consider now classical probability theory (CP); see Section 1.5. Consider probability space, a triple $K = (\Omega, \mathcal{F}, P)$.

We consider only random variables with the discrete range of values. For A and its value x, set $\Omega_{A=x} = \{\omega \in \Omega : A(\omega) = x\}$ and define the probability distribution of A as $p^A(x) = P(\Omega_{A=x})$.

In applications of CP to statistical physics set Ω describes an ensemble of systems (physical or virtual). Generally set Ω is infinite. These applications are based on the following mathematical model for states and observables:

- States are represented by probability measures
- Observables are represented by random variables.

Here states are not states of individual systems, but statistical states. A state, probability measure P, represents a preparation context determining statistically the features of an ensemble of systems.

We recall the statistical interpretation of probability is used. However, the coming mathematical formulas can be endowed with the subjective interpretation of probability in the spirit of De Finetti.

It is useful to invent the following terminology: *a random system S is an ensemble of systems Ω*. The states of random systems are mathematically represented by probability measures. The set of all probability measures is the state space.

We highlight that, as in quantum theory, the ensembles do not need to be physical. They can be virtual as in the Gibbs treatment of classical thermodynamics.

Let A, B be random variables (representing classical observables) with probability distributions $p^A(x), p^B(y)$ and conditional probability $P(B = y|A = x)$. We recall that the latter is given by *the Bayes formula:*

$$P(B = y|A = x) = P(B = y, A = x)/P(A = x), \tag{2.16}$$

for $P(A = x) > 0$. This formula can be represented in the form

$$P(B = y|A = x) = P(\Omega_{B=y} \cap \Omega_{A=x})/P(\Omega_{A=x}). \tag{2.17}$$

From the ensemble viewpoint, this formula means that we select subensemble $\Omega_{A=x}$ consisting of systems that generate outputs $\omega \to A(\omega) = x$ and then find proportion (with respect to the weight-measure P) of its elements that generate the output $\omega \to B(\omega) = y$. It is useful to introduce for each value $A = x$ the conditional probability measure $P_{A=x}$:

$$P_{A=x}(E) = P(E \cap \Omega_{A=x})/P(\Omega_{A=x}). \tag{2.18}$$

Then

$$P(B = y|A = x) = P_{A=x}(B = y).$$

The probability measure P plays the role of state. To show this in notation, we set

$$P(A = x) = \Pr(A = x; P), P(B = y) = \Pr(B = y; P), \tag{2.19}$$

$$P(B = y|A = x) = \Pr(B = y; P_{A=x}). \tag{2.20}$$

(The direct analog of the quantum notation would be P_P, but this symbol is not so elegant.)

Consider an ensemble in state P; perform A-observations and select those with the concrete output $A = x$. By using the Bayes formula we obtain the new statistical state given by conditional probability measure $P_{A=x}$. The conditioning transformation in the classical state space – the space of probability measures,

$$P \to P_{A=x}, \tag{2.21}$$

is the direct analog of the conditioning transformation in quantum state space given by the Lüders projection postulate resulting from the A-measurement. This transformation changes the statistical state and, for any observable (random variable) B, its probability distribution is modified:

$$\Pr(B = y; P) \to \Pr(B = y; P_{A=x}). \tag{2.22}$$

Now, consider a compound random system $S = (S_1, S_2)$, with random subsystems $S_i, i = 1, 2$ given by ensembles Ω_i. Then S is given by ensemble $\Omega = \Omega_1 \times \Omega_2$.

If random systems are *independent,* then the state of S is given by $P = P^{(1)} \otimes P^{(2)}$, where $P^i, i = 1, 2$, are states of S_i and the probability measure P is defined as $P(E_1 \times E_2) = P^{(1)}(E_1)P^{(2)}(E_2)$. We call such a probability state *separable.* Separability is just another term for independence of random variables $I_i : \Omega \to \Omega_i, I_i(s) = s_i$. If probability measure P on Ω is not separable, we call this state *entangled* (cf. Chapter 13).

Let a and b be observables on random systems S_1 and S_2 given by random variables, $a : \Omega_1 \to \mathbb{R}$ and $b : \Omega_2 \to \mathbb{R}$. They can be treated as observables on S represented by random variables $A(\omega) = a(\omega_1), B(\omega) = b(\omega_2)$ for $\omega = (\omega_1, \omega_2)$. It is easy to see that for a separable state $P = P^{(1)} \otimes P^{(2)}$,

$$\Pr(B = y; P_{A=x}) = \Pr(B = y; P). \tag{2.23}$$

Compare this with the quantum case, (2.4). So, the state transformation of induced by conditioning on output $A = x$ does not change probability distribution for B-observable. However, if P is entangled, then generally the probability distribution of B is modified:

$$\Pr(B = y; P_{A=x}) \neq \Pr(B = y; P). \tag{2.24}$$

Compare this with the quantum case, (2.5).

We remark that, similarly to the quantum case, in the classical case state P of S determines states of S_i, as the marginal probabilities:

$$P^{(1)}(E_1) = \int_{E_1 \times \Omega_2} dP, \; P^{(2)}(E_2) = \int_{\Omega_1 \times E_2} dP, \tag{2.25}$$

for discrete parameter-spaces, integration is reduced to summation. If a state is separable, $P = P^{(1)} \otimes P^{(2)}$, then its marginals coincide with $P^{(1)}, P^{(2)}$. Hence,

$$P^{(2)}_{A=x} = P^{(2)} \text{ and } P(b = y; P^{(2)}_{A=x}) = P(b = y; P^{(2)}), \tag{2.26}$$

where $P^{(2)}_{A=x}$ is the marginal of the transformed state $P_{A=x}$; compare this with the quantum case, (2.10). For entangled state P, this equality is violated. Generally

$$P^{(2)}_{A=x} \neq P^{(2)} \text{ and } P(b = y; P^{(2)}_{A=x}) \neq P(b = y; P^{(2)}). \tag{2.27}$$

Compare this with the quantum case.

If we operate with theoretical probabilities, then the conditioning transformation is practically instantaneous. In any event, the time of calculation of theoretical probabilities does not depend on the distance between the systems S_1 and S_2. This situation also matches with operating with virtual ensembles in the Gibbs approach. However, the empirical reconstruction of transformed state $P_{A=x}$ time needed to collect the statistics of $A = x$ outputs.

The preceding scheme of conditional observations was based on the statistical interpretation of probability.[4]

2.10 Statistical Interpretation: State's Conditioning

In Section 2.3 we presented the mathematical scheme of the conditioning transformation of a quantum state resulting from observations. Now it will be treated within SI with the emphasis on the analogy with the CP-scheme (within the statistical approach to probability). In Section 2.3 we operated with pure states, since, for II, nonlocality of the Lüders type is straightforwardly associated with conditioning transformations of pure states. For SI, it is natural to proceed with general quantum states represented by density operators. As usual, denote the space of density operators by the symbol $\mathcal{D}$. By SI a quantum state (density operator) provides a description of certain statistical properties of an ensemble of similarily prepared systems.

[4] We briefly describe this scheme from the viewpoint of subjective probability. Here state transition (2.21) really happens practically instantaneously, but in the information space (not in the physical space). And the users of subjective probability do not speak about action at a distance. Everyone understands well that subjective probability theory is about predictions (including "predictions at a distance," see [313, 314]).

For density operators, the Lüders projection postulate loses its special role. In principle, one can consider any (affine) transformation of D as a conditioning transformation, as is done in theory of quantum instruments [98, 298] (Section 1.3).

Let ρ be the state of a quantum random system S and let A, B be quantum observables. Their probability distributions are given by Born's rule:

$$P_\rho(A = x) = \text{Tr } \hat{E}_A(x)\hat{\rho}, \; P_\rho(B = y) = \text{Tr } \hat{E}_B(y)\hat{\rho}. \tag{2.28}$$

Measurement of quantum observable A with output $A = x$ generates the conditioning transformation in space $\mathcal{D}$ of quantum states:

$$\hat{\rho} \to \hat{\rho}_{A=x} = \hat{E}_A(x)\hat{\rho}\hat{E}_A(x)/\text{Tr } (\hat{E}_A(x)\hat{\rho}\hat{E}_A(x)). \tag{2.29}$$

In sequential B-observation the concrete output $B = y$ is obtained with probability given by Born's rule for the transformed state $\hat{\rho}_{A=x}$. This process can be considered as transformation of probability distribution:

$$P_\rho(B = y) \to P_{\rho_{A=x}}(B = y) = \text{Tr } \hat{E}_B(y)\hat{\rho}_{A=x}. \tag{2.30}$$

By SI this mathematical formalism describes the following process:

- ensemble Ω with statistical features described by state ρ is prepared;
- A-measurement is performed;
- systems corresponding to output $A = x$ are selected and new ensemble $\Omega_{A=x}$ is created;
- for systems belonging to $\Omega_{A=x}$, B-observation is performed.

Of course, in the theoretical framework transitions (2.29) and (2.30) can be considered as practically instantaneous. But, empirically creation of ensemble $\Omega_{A=x}$ is not instantaneous at all. So, in the statistical interpretation there is no place for spooky action at a distance.

QM is a special theory of probability transformations. This is the essence of *the Växjö interpretation* (Chapter 10).[5]

Consider theoretical operation with spatially separated quantum observables, namely, let us do mathematical calculation based on the Lüders projection postulate and the Born rule. We can evaluate the conditional marginal distribution for Bob outcomes which are paired with a particular Alice's outcome, $A = x$. This is the distribution $P_{\rho_{A=x}}(B = y)$; see (2.28). Plotnitsky [313, 314] called this evaluation *"spooky prediction at a distance."* By Plotnitsky, "spookiness" of quantum predictions is due to the impossibility of presenting any space-time picture explaining correlations.

[5] We also point to QBism, the subjective probability interpretation of quantum mechanics. The QBists' position is the same as that of classical subjectivists, "faster-than-light change of statistical correlation" is a mental process (well, even this process takes some time) [124, 125]. QBism also treats QM as probability transformation theory.

We also point out that the quantum machinery for the state update based on the projection postulate can be used in classical optics to mimic quantum entanglement up to the violation of Bell inequalities (see Chapter 7).

2.11 Probability: Compatible versus Incompatible Observables

Here is a good place to highlight the difference between classical and quantum probability calculi. First of all, we remark that for compatible observables quantum probability calculus is reduced to classical one. For compatible observables, it is possible to determine the joint probability distribution (JPD) that can be used to represent the state of a random system. Hence, the origin of the difference in the existence of incompatible observables. In probabilistic terms (see [203]), *incompatibility of observables is equivalent to nonexistence of JPD.* The mathematical formalism of quantum mechanics gives a new rule for the probability conditioning resulting from observations. This rule is more general than the classical one. This is a good place to cite Feynman [119]:

"From about the beginning of the twentieth century experimental physics amassed an impressive array of strange phenomena which demonstrated the inadequacy of classical physics. The attempts to discover a theoretical structure for the new phenomena led at first to a confusion in which it appeared that light, and electrons, sometimes behaved like waves and sometimes like particles. This apparent inconsistency was completely resolved in 1926 and 1927 in the theory called quantum mechanics. The new theory asserts that there are experiments for which the exact outcome is fundamentally unpredictable, and that in these cases one has to be satisfied with computing probabilities of various outcomes. But far more fundamental was the discovery that in nature the laws of combining probabilities were not those of the classical probability theory of Laplace."

In particular, the quantum calculus of probabilities for incompatible observables, that is, those without JPD, generates the EPR–Bohm(–Bell) correlations. It is commonly emphasized that these correlations are *nonlocal.* It is important to assign an explicit meaning to this terminology. The plane meaning is "correlations between outcomes of measurements performed for spatially separated systems." In this sense CP-correlations for spatially separated systems are also nonlocal. Of course, this meaning is not assigned to nonlocality of EPR–Bohm(–Bell) correlations. Typically by speaking about nonlocal quantum correlations one emphasizes their generation via spooky action at a distance. However, as was shown in [232], the crucial issue is local incompatibility of observables, for example, incompatibility of spin projections for system S_1 (or S_2). Since these correlations are based on incompatible observables, they are nonclassical from the viewpoint of probability theory. They are not based on common JPD.

2.12 Concluding Remarks on Intrinsic Quantum Nonlocality

The aim of this chapter is disillusion of projection-based nonlocality (Lüders nonlocality). This sort of nonlocality can be considered as intrinsic quantum nonlocality, in contrast to subquantum (hidden variables) nonlocality. It is important to distinguish sharply these two nonlocalities. They are often mixed in the heads of scientists advertising "quantum nonlocality." This two-faced Janus is often seen as having just one face – quantum nonlocality.

This mental nonlocality-mixing is explainable by taking into account coupling between intrinsic and subquantum nonlocalities. This coupling was excellently presented in Aspect's paper [31]. According to him, projection (Lüders) nonlocality makes quantum theory so counterintuitive that any commonsensible scientist would try to find a beyond-quantum explanation. Bell proposed a class of subquantum models, special models with hidden variables [53, 54]. For such models, he derived the inequality and its violation was interpreted by him as the evidence of another sort of nonlocality – subquantum nonlocality

We demonstrated that projection nonlocality is essentially "nonlocality" of probability conditioning, similar to "nonlocality" of CP-conditioning. By claiming that QM is nonlocal one should consistently claim CP and hence classical statistical physics and thermodynamics are nonlocal as well.

In the debate on Lüders nonlocality, we highlighted the role of quantum state interpretation. Our analysis was concentrated on the two basic interpretations, individual (orthodox Copenhagen) versus statistical (ensemble). Assuming that a quantum state provides the complete description of an individual quantum system, one can really confront Lüders nonlocality. On the other hand, SI combined with treatment of quantum mechanics as machinery for probability conditioning implies that projection nonlocality is typical "nonlocality" of probability conditioning.

Following Ballentine [43], we point out that the EPR argument is directed against one special interpretation of a quantum state, namely, II, the orthodox Copenhagen interpretation. One can keep either II or locality. By using II one should recognize nonlocality of QM – at least if one agrees with Einstein's conclusion on the EPR argument. Personally I prefer to reject II that is just an interpretation, to reject it in favor of SI; see Einstein [116]:

"The attempt to conceive the quantum-theoretical description as the complete description of the individual systems leads to unnatural theoretical interpretations, which become immediately unnecessary if one accepts the interpretation that the description refers to ensembles of systems and not to individual systems."

However, not everybody agrees with Einstein's conclusion, either (1) or (2); see section 2.7. In particular, Bohr did not think that by employing the orthodox Copenhagen interpretation one should confront nonlocality. For him, the quantum theory is local and at the same time a quantum state provides the complete description of an individual quantum system. In Chapter 3 this position will be discussed in more detail. We conclude this chapter with the following remark of Einstein [116]:

> I now imagine a quantum theoretician who may even admit that the quantum-theoretical description refers to ensembles of systems and not to individual systems, but who, nevertheless, clings to the idea that the type of description of the statistical quantum theory will, in its essential features, be retained in the future. He may argue as follows: True, I admit that the quantum-theoretical description is an incomplete description of the individual system. I even admit that a complete theoretical description is, in principle, thinkable. But I consider it proven that the search for such a complete description would be aimless. For the lawfulness of nature is thus constituted that the laws can be completely and suitably formulated within the framework of our incomplete description.

As is pointed out by Einstein, this viewpoint is incontestable. However, Einstein expressed a strong hope and wish to go further and to find a finer description than that given by a quantum mechanical state vector (see also [117]).

3

Bohr's Principle of Complementarity and Its Contextual Basis

3.1 Bohr's Contextuality

Contextuality is one of the hottest topics of modern quantum physics, both theoretical and experimental. During the past 20 years, it has been discussed in numerous papers published in top physical journals. Unfortunate for these discussions is that from the very beginning contextuality (in fact, its special form which we call *"join measurement contextuality"* (JMC), see the definition that follows) was coupled to the issue of nonlocality.[1] It was Bell's intention in his analysis of the possible seeds of the violation of the Bell-type inequalities [53]. Surprisingly, Bell had never mentioned [54] general contextuality which we call *"Bohr contextuality."* The latter has no straightforward connection to the Bell inequalities; it is coupled to the notion of incompatibility of observables – the *Bohr principle of complementarity.* One of the explanations for this astonishing situation in quantum foundations is that Bohr presented his ideas in a vague way; moreover, he often changed his vague formulations a few times at different occasions. In this chapter we briefly present Bohr's ideas about contextuality of quantum measurements and their role in his formulation of the complementarity principle. In Chapter 6 we considered the Bell inequalities in the purely quantum mechanical framework, that is, without any relation to hidden variables. This pathway toward these inequalities (i.e., via Bohr's contextuality-complementarity) highlights the role of incompatibility of quantum observables and gives the possibility of operating with the Bell inequalities without mentioning the ambiguous notion of quantum nonlocality (spooky action at a distance).

In quantum measurement theory, selection of observables comeasurable with an observable A is considered as specification of measurement context of A-measurements; the A-value in the B-comeasurement context can differ from the A-value in the C-comeasurement context, for the same state $|\psi\rangle$. This is the essence

[1] In terms of Chapter 2, this is subquantum nonlocality of hidden variables.

of special contextuality playing such an important role in quantum information theory; in [237] we called it *joint measurement contextuality* (JMC).

Definition 3.1 (JMC) If A, B, and C are three quantum observables, such that A is compatible with B and C, a measurement of A might give different results depending upon whether A is comeasured with B or with C.

We note that contextual behavior corresponds to the case of incompatible quantum observables B and C, that is, $[\hat{B}.\hat{C}] \neq 0$. If all observables are pairwise compatible, that is, $[\hat{A}.\hat{B}] = [\hat{A}.\hat{C}] = [\hat{B}.\hat{C}] = 0$, then, for any state $|\psi\rangle$, it is possible to construct the noncontextual model of measurement based on the JPD for the triples of outcomes,

$$P_\psi(A = x, B = y, C = z) = \|\hat{E}_A(x)\hat{E}_B(y)E_C(z)\psi\|^2. \tag{3.1}$$

If B and C are incompatible, such a model is impossible. This is the contextuality scenario. However, JMC contextuality formalized via Definition 3.1 can't be tested experimentally, since it involves counterfactual reasoning. The only possibility is to test contextuality indirectly with the aid of Bell-type inequalities.

In the situation when so many researchers write and speak about quantum contextuality, one should be sure that this notion is well defined and its physical interpretation is clear and well known. In fact, before starting to think about the meaning of contextuality, I was completely sure in this. Strangely enough, I was not able to create a consistent picture. And I was really shocked when, after visiting the institute of Atom Physics in Vienna and having a conversation with Rauch and Hasegawa, I found that they also are disappointed. They asked me about the meaning of contextuality. And they performed brilliant experiments [47, 78] to test contextuality in the framework of neuron interferometry. They had a vague picture of what was tested and the physical meaning of their experimental results.

Then, in Stockholm while on the PhD defense jury of a student who was supervised by Professor Bengtsson (let's call her Alice), I asked Alice about the physical meaning of contextuality. (Her thesis was about it.) Alice answered that she had no idea about the physical interpretation of advanced mathematical results obtained in her thesis. Generally I like discussions. To stimulate a debate, I said that Rauch and Hasegawa had the strange idea that contextuality is just noncommutativity, a sort of the order effect in the sequential measurements (this was the final output of our discussions in Vienna). Unfortunately, in Stockholm the discussion (with comments from a few top experts, including Bengtsson and Lindblad) quickly finished with the conclusion that the question is interesting, but not related to the PhD defense.

Typically, in writing a paper about contextuality in QM one starts by referring to this notion as JMC: dependence of the outcomes of some observable A on its joint measurement with another observable B (see Definition 3.1). However, this definition is countefactual and cannot be used in the experimental framework. Nevertheless, the "universal contextuality writer" is not disappointed by this situation and he immediately jumps to the Bell inequalities [53, 54], which are treated as *noncontextual inequalities* [18]. Moreover, often contextuality is identified with the violation of the Bell inequalities. This identification shadows the problem of the physical meaning of contextuality. One jumps from the problem of understanding to calculation of a numerical quantity, the degree of the violation of some Bell inequality. Such inequalities are numerous. And they can be tested in different experimental situations and generate the permanent flow of highly recognized papers.

I suggested the following critical illustration to this strategy (contextuality = violation of the Bell inequalities) [244, 245]. In the classical probabilistic framework consider the notion of a *random sequence.* Theory of randomness is the result of the intensive research (Mises, Church, Kolmogorov, Solomovov, Chatin, Martin-Löf; see [220]). This theoretical basis led to elaboration of the variety of randomness tests which are used to check whether some sequence of outputs of physical or digital random generator is random. The NIST test (a batch of tests for randomness) is the most widely used. So, here we also use tests, but beyond them there is the well developed theory of randomness. In particular, this leads to understanding that even if a sequence x passed the NIST test, this does not imply that it is random. In principle, there can be found another test such that x would not pass it. The latter would not be a surprise.

In contrast to the preceding illustration by randomness theory, in quantum information theory contextuality is by definition the violation of some noncontextual (Bell) inequality.[2] So, the theoretical notion is identified with a test. This is really bad! Not only from the theoretical viewpoint, but even from the practical one. As was mentioned, by working with randomness people understand well that even passing the NIST test does not guarantee randomness. In QM, passing the Bell test (in the sense of the violation of some Bell-type inequality [53, 54]) by definition is equivalent to contextuality. This is wrong strategy, which led to skews in handling quantum contextuality.

In a series of papers [223, 224, 232, 233] the physical meaning of contextuality was clarified through referring to Bohr's complementarity principle. Often this principle is reduced to wave–particle duality. (In fact, Bohr had never used the latter terminology.) However, Bohr's formulation of the complementarity principle is essentially deeper. Complementarity is not postulated; for Bohr, it is the natural consequence of the irreducible dependence of the observable's outcome on the

[2] We point out to the variety of the Bell-type inequalities; see article [214] on the analysis of equivalence of the statistical tests based on some inequalities.

experimental context. Thus, the outcomes of quantum observables are generated in the complex process of the interaction of a system and a measurement device [65]. This dependence on the complex of experimental conditions is nothing else than a form of contextuality, *Bohr contextuality*. We remark that JMC is its special case. But, in contrast to JMC, the physical interpretation of Bohr contextuality is transparent – dependence of results of measurements on experimental contexts. And it does not involve the use of conterfactuals.[3]

Such contextuality is the seed of complementarity, the existence of incompatible observables. Moreover, contextuality without incompatibility loses its value. *If all observables were compatible, then they might be jointly measured in a single experimental context and multicontextual consideration would be meaningless.*

In 1949, Bohr [65] presented the essence of complementarity in the following widely cited statement:

> This crucial point... implies the impossibility of any sharp separation between the behaviour of atomic objects and the interaction with the measuring instruments which serve to define the conditions under which the phenomena appear. In fact, the individuality of the typical quantum effects finds its proper expression in the circumstance that any attempt of subdividing the phenomena will demand a change in the experimental arrangement introducing new possibilities of interaction between objects and measuring instruments which in principle cannot be controlled. Consequently, evidence obtained under different experimental conditions cannot be comprehended within a single picture, but must be regarded as complementary in the sense that only the totality of the phenomena exhausts the possible information about the objects.

3.2 Dissolving Bohr's Complementarity

Bohr's statement is really a gold mine and we plan to extract a lot of ore from it. The output from this statement can be represented in the form of a few blocks [223, 224, 232, 233].

3.2.1 Contextuality-Incompatibility

We start with the following four blocks presenting interconnection of contextuality and incompatibility:

[3] This is a good place to mention the studies of Grangier, for example, [142–144], as an attempt to suggest a heuristically natural interpretation for contextuality, which is different from JMC and Bell contextualities. Grangier contextuality is in fact also closely coupled to the Bohr complementarity principle, although this was not pointed out. Moreover, Grangier wants to distance from Bohr and especially the complementarity principle. After his talks in Växjö, I pointed out to him the evident coupling of Bohr and Grangier contextualities, but his answer was that "Bohr's principle is so fuzzy and unclear, it is impossible to appeal to it."

- **CONT1** An outcome of any observable is composed of the contributions of a system and a measurement device.[4]
- **CONT2** The whole experimental context has to be taken into account.
- **INCOMP1** Logically there is no reason to expect that all experimental contexts can be combined and all observables can be measured jointly; thus in principle some experimental contexts and hence observables can be incompatible.
- **INCOMP2** Incompatible quantum observables do really exist.

Bohr justified **INCOMP2** on the basis of the Heisenberg uncertainty principle implying that the contexts for measurement of position and momentum observables are incompatible. By using the Schrödinger inequality (Chapter 5) one can also find that, for example, the spin projection observables are also incompatible. It should be pointed out that by referring to the Heisenberg uncertainty principle we refer to the Bohr's quantum postulate (Chapter 5): the existence of the indivisible quantum of action – the Planck constant h. This quantity presents itself in the right-hand side of the Heisenberg uncertainty relation for the position and momentum observables as well as the Schrödinger inequality for the basic quantum observables, as, say, the spin-observable. Thus, in quantum physics the existence of incompatible observables is due to the indivisibility of Planck action.

The statements **CONT1** + **CONT2** and **INCOMP1+INCOMP12** form the contextuality and incompatibility counterparts of Bohr's complementarity principle. So, we have extracted the two basic components of the complementarity principle from the preceding citation of Bohr. They can be treated as two principles:

- *Contextuality Principle.*
- *Incompatibility Principle.*

They are fundamentally interrelated.

3.2.2 Information Complementary-Completeness

In the modern terminology the last part of the aforementioned Bohr's statement is about information. He used the word evidence, but in modern terms this is information:

COMPL is information collected via measurements of incompatible observables is complementary and exhausting.

COMPL is the essence of complementarity. It can be handled as a separate quantum principle: *Information Complementary-Completeness Principle.* To

[4] So, the values of an observable a are not the objective properties of the systems. They are created in the process of the complex interaction between the systems prepared for measurements and the apparatus used for measurement of a.

clarify the notion of complementarity, we also use the following citation of Bohr on the difference between supplementarity of observations in classical physics and complementarity of quantum observations [65]:

> Within the scope of classical physics, all characteristic properties of a given object can in principle be ascertained by a single experimental arrangement, although in practice various arrangements are often convenient for the study of different aspects of the phenomenon. In fact, data obtained in such a way simply supplement each other and can be combined into a consistent picture of the behavior of the object under investigation. In quantum mechanics, however, evidence about atomic objects obtained by different experimental arrangements exhibits a novel kind of complementary relationship. Indeed, it must be recognized that such evidence which appears contradictory when combination into a single picture is attempted, exhausts all conceivable knowledge about the object. Far from restricting our efforts to put questions to nature in the form of experiments, the notion of complementarity simply characterizes the answers we can receive by such inquiry, whenever the interaction between the measuring instruments and the objects forms an integral part of the phenomenon.

3.2.3 Individuality (Phenomenon)

Now we turn to an important component of the citation at the beginning of this section, namely on "the individuality of the typical quantum effects" and the notion of phenomenon. In quantum foundational discussions the individuality component of this principle is completely ignored; maybe with one exception – the works of Plotnitsky [309, 310, 315] from whom I heard about the notion of phenomenon.

For many years I dismissed the importance of this notion. Only recently, by analyzing the difference between classical and quantum entanglement (see Chapter 7), was I really pressed to pay attention to individuality of the quantum phenomena.

By the individuality principle quantum measurements generate discrete events corresponding to interaction of individual quantum systems, say photons or electrons, with measuring devices. Such discrete events are clicks of photodetectors or points on a screen with photo-emulsion in the original diffraction experiments – phenomena in Bohr's terminology. For him, only phenomena can be considered as "elements of reality." This is a good place to present the following citation of Bohr ([65], v. 2, p. 64):

> I advocate the application of the word *phenomenon* exclusively to refer to the observations obtained under specific circumstances including an account of the whole experimental arrangement. In such terminology, the observational problem is free of any special intricacy since, in actual

> experiments, all observations are expressed by unambiguous statements referring, for instance, to the recording of the point at which an electron arrives at a photographic plate. Moreover, speaking in such a way is just suited to emphasize that the appropriate physical interpretation of the symbolic quantum mechanical formalism amounts only to predictions, of determinate or statistical character, pertaining to individual phenomena appearing under conditions defined by classical physical concepts.

The notion of phenomena is crucial in distinguishing entanglements in classical and quantum optics, distinguishing the semiclassical and quantum models for electromagnetic fields (Chapter 7) – the problem of photon existence.

Through the notion of individual phenomenon, Bohr reflected the discrete structure of quantum physics: from discrete energy transitions in atoms to discrete clicks of photodetectors.

3.2.4 Complex Structure of the Complementarity Principle

In summarizing our considerations on the basis of Bohr's statements we represent the complementarity principle as four interrelated principles:

- *Contextuality Principle.*
- *Incompatibility Principle.*
- *Information Complementary-Completeness Principle.*
- *Individuality Principle.*

Unfortunately, the contextual dimension of the complementarity is typically missing in the discussions on quantum foundations. To highlight the role of contextuality, it might be better to call Bohr's complementarity principle *the contextuality-complementarity principle.* And we shall do this sometimes, but not consistently. On the one hand, the terminology "complementarity principle" is firmly established and by saying "contextuality-complementarity principle" we may disappoint the reader. On the other hand, Bohr's complementarity principle is not reduced to contextuality-incompatibility and complementarity-completeness. The individuality principle is also at the heart of quantum physics. Again, unfortunately, the notion of phenomenon is practically unknown for researchers working in quantum physics, even for experts in foundations. This unawareness is the source of rather common surprise over the existence of entangled states and violation of the Bell-type inequalities in classical optics (Chapter 7).

And finally we discuss the following ambiguity in Bohr's reasoning on complementarity. We postponed this issue as far as possible, not to disturb the reasoning's line. Bohr said such things as "exhausts the possible information about the objects" or even "conceivable knowledge about the object." These phrases related to the COMPL component of the complementarity principle seem to be ambiguous. How can one collect exhaustive information about an atomic object via measurements of incompatible observables? Measurement of, say, the position of a quantum system S makes the outcome of momentum measurement totally unpredictable and vice versa.

I suggest to resolve this ambiguity in Bohr's formulation of the complementarity principle by speaking about a quantum state, instead of a quantum object (system).

Consider the statistical interpretation of a quantum state, SI (Chapter 2): a quantum state $|\psi\rangle$ represents statistical features of a large ensemble of identically prepared quantum systems. In the operational approach to QM, $|\psi\rangle$ represents a preparation procedure. By performing measurements of incompatible observables on systems generated by such a procedure the experimenter collects complementary information about the statistical features of these systems; by employing sufficiently many generally incompatible observables, it is possible to collect complete information. In this way the state can be reconstructed from data.

3.3 Complementarity: Bohr versus James, Physics versus Psychology

Many authors (see, e.g., [346, 347] and references herein) underline the similarity of Bohr's complementarity in physics and James' complementarity in psychology. For the latter, see James' book [172]: [I]n certain persons, at least, the total possible consciousness may be split into parts which coexist but mutually ignore each other, and share the objects of knowledge between them. More remarkable still, they are complementary.

Here James wrote about dissociative identity disorder, formerly known as multiple personality disorder; this disorder is characterized by "switching" to alternate identities. Depending on mental context, Alice can express herself as say Karl XII or Napoleon. These behavioral expressions are incompatible: Alice behaves either as Karl XII or Napoleon (with exclusive or). For an external observer, say a psychiatrist, these two expressions play the role of two observables, A and B.

James' and Bohr's complementarities are practically identical and one may imagine that they can be mathematically described by the same formalism – the operator calculus in complex Hilbert space. We remark that this formalism is the tool of the epistemological study and it can be used independently from the ontic basis.

Thus, the mathematical formalism of QM can be explored outside of quantum physics to describe incompatible observables and generally complementarity. And this approach is widely applied to mathematical modeling of cognition, psychology, decision making, economics and finance, social and political sciences (Chapter 18).

It is known as *quantum-like modeling*.[5] Its foundational basis is less solid than that of quantum physics.

One of the differences between the genuine quantum and quantum-like modeling is that the former is supported by the quantization procedure, suggested by Schrödinger: the functions on the phase space representing classical physical quantities are mapped into Hermitian operators representing quantum observables, $A(q,p) \to \hat{A} = A(\hat{q},\hat{p})$. In applications of the quantum formalism to, for example, decision making we cannot appeal to this quantization procedure, since there is no analogue of the classical phase space mechanics for mental phenomena. In quantum physics incompatibility of observables $\hat{A} = A(\hat{q},\hat{p}), \hat{B} = B(\hat{q},\hat{p})$ can be derived from calculation of the Poisson bracket $\{A(q,p), B(q,p)\}$ for the corresponding classical quantities. In quantum-like modeling theoretical incompatibility of, say, two questions cannot be derived so straightforwardly.

Finally, we present some citations from paper [62] in that Bohr used psychological complementarity to motivate physical complementarity:

> For describing our mental activity, we require, on one hand, an objectively given content to be placed in opposition to a perceiving subject, while, on the other hand, as is already implied in such an assertion, no sharp separation between object and subject can be maintained, since the perceiving subject also belongs to our mental content. From these circumstances follows not only the relative meaning of every concept, or rather of every word, the meaning depending upon our arbitrary choice of view point, but also that we must, in general, be prepared to accept the fact that a complete elucidation of one and the same object may require diverse points of view which defy a unique description. Indeed, strictly speaking, the conscious analysis of any concept stands in a relation of exclusion to its immediate application. The necessity of taking recourse to a complementary, or reciprocal, mode of description is perhaps most familiar to us from psychological problems. In opposition to this, the feature which characterizes the so-called exact sciences is, in general, the attempt to attain to uniqueness by avoiding all reference to the perceiving subject. . . .
>
> In considering the resignation with regard to the desires for visualization which give our whole language its character, to which we are compelled by the situation discussed above, it is very instructive that already in simple psychological experiences we come upon fundamental features not only of the relativistic but also of the reciprocal view. The relativity of our perception of motion, with which we become conversant as children when travelling by ship or by train, corresponds to

[5] This terminology was invented by the author and nowadays is widely used. Some authors use the terminology "quantum inspired."

> common-place experiences on the reciprocal character of the perception of touch. . . .
>
> On the whole, the analysis of our sense impressions discloses a remarkable independence of the psychological foundations of the concepts of space and time, on the one hand, and the conceptions of energy and momentum, based upon actions of force, on the other hand. Above all, however, this domain, as already mentioned, is distinguished by reciprocal relationships which depend upon the unity of our consciousness and which exhibit a striking similarity with the physical consequences of the quantum of action. We are thinking here of well-known characteristics of emotion and volition which are quite incapable of being represented by visualizable pictures. In particular, the apparent contrast between the continuous onward flow of associative thinking and the preservation of the unity of the personality exhibits a suggestive analogy with the relation between the wave description of the motions of material particles, governed by the superposition principle, and their indestructible individuality. The unavoidable influence on atomic phenomena caused by observing them here corresponds to the well-known change of the tinge of the psychological experiences which accompanies any direction of the attention to one of their various elements.

The mention of "the preservation of the unity of the personality" echoes the citation of James at the beginning of this section.

4

Information Interpretations of Quantum Mechanics

4.1 Einstein–Bohr Debate

Now we come back to the EPR paradox (Section 2.7). We analyze Bohr's reply to the EPR argument [63]. In this reply Bohr is not constrained by the alternative (Section 2.7):

"[I]t becomes evident that the paradox forces us to relinquish one of the following two assertions:

(A) the description by means of the Ψ-function is complete
(B) the real states of spatially separated objects are independent of each other."

On the other hand, it is possible to adhere to (B), if one regards the Ψ-function as the description of a (statistical) ensemble of systems (and therefore relinquishes (A)). However, this view blasts the framework of the "orthodox quantum theory."

Bohr was able to combine (A) and (B). We stress once again that Bohr never questioned locality of QM. What can be said about Bohr's interpretation of the Ψ-function, the quantum state?

Bohr operated within II – the orthodox Copenhagen interpretation, that is, Ψ provides a complete and exhaustive description of an individual physical system (e.g., an electron). But Bohr's version of the the orthodox Copenhagen interpretation is not the one which is attacked in the EPR article. To distinguish these versions, we shall call Bohr's version "Bohr's Copenhagen interpretation" and reserve the term "the orthodox Copenhagen interpretation" for the one which was attacked in the EPR paper. Of course, such separation is disturbing for those who identify themselves with the supporters of the orthodox Copenhagen interpretation. They are sure that by using the term "the orthodox Copenhagen interpretation," they are in one boat with Bohr (with whom Copenhagen is associated). In fact, the situation is even worse. Initially (in the Como lecture in 1927) Bohr used the interpretation that later was called the orthodox Copenhagen interpretation. However, later Bohr's views evolved under the influence of his development of the idea of complementarity.

By the orthodox Copenhagen interpretation, the values of observables are associated with a system S on which the measurements are performed. However, this is not the case for Bohr. As we have seen, the values of observable A are associated with a system interacting with a measurement apparatus M_A – a measurement context $C_A = (S, M_A)$. (See the indirect measurement scheme, Section 1.4.)

EPR's "elements of reality" are coupled to systems. Bohr's "elements of reality" are coupled to measurement contexts. For any two observables A and B, say position and momentum, their values are coupled to corresponding measurement contexts. From Bohr's viewpoint, one can't associate two different elements of reality with a system without pointing to measurement contexts, so the EPR reasoning collapses at the very beginning. It is unfortunate that the Einstein–Bohr debate did not clarify their positions. Quantum foundations still suffer of this.

The Bohr contextual interpretation was the starting point for creation of the Växjö interpretation, the contextual probabilistic interpretation of QM (Chapter 10). This interpretation emphasizes the role of probability and it treats the quantum probabilistic formalism as one special contextual probabilistic formalism used for nonclassical probability inference.

4.2 Information Interpretations

For Bohr, quantum theory is an epistemic theory; it is not about physical processes in the microworld as it is (see also Chapter 15). It is even not about physics of measurements, because for Bohr measurement devices are classical, but QM does not describe the physical processes in classical apparatuses during measurements. From Bohr's perspective, the quantum methodology and mathematical formalism are designed for the probabilistic predictions of measurements' outcomes. Hence, Bohr definitely supported the *information interpretation* (INI) of QM. Moreover, he was one of its creators together with Heisenberg and Schrödinger. Complementarity principle is an epistemic principle. It is about gaining and structuring knowledge about possible interactions of quantum systems with classical measurement apparatuses. Bohr's view on QM influenced all branches of INI developed by Wheeler [375], Zeilinger and Brukner [74, 75], Fuchs, Mermin, Schack [124]–[131], Plotnitsky [308]–[315], D'Ariano et al. [82, 94, 95], Jaeger [169–171], Khrennikov [193, 203]. This interpretation was loudly highlighted during the Växjö-2001 [184] conference organized by Fuchs and Khrennikov and presented in hundreds of talks during the following Växjö conferences (2001–2023). By the information interpretation INI, the quantum theory is a machine for the prediction of measurements' outcomes, the probabilistic prediction. And the Born rule gives the probabilities. INI has a variety of branches, or sub-interpretations.

4.2.1 Schrödinger and Information Interpretation

The INI viewpoint on quantum theory was especially clearly presented in Schrödinger's article [328]:

> It [ψ-function] is now the means for predicting probability of measurement results. In it is embodied the momentarily-attained sum of theoretically based future expectation, somewhat as laid down in a catalog. It is the relation- and -determinacy-bridge between measurements and measurements... For each measurement one is required to ascribe to the ψ-function (= the prediction-catalog) a characteristic, quite sudden change, which *depends on the measurement result obtained, and so cannot be foreseen*...

In particular, entanglement is treated as "entanglement of predictions," or "entanglement of our knowledge"; see [328]:

> [B]etween these two systems an entanglement can arise, which... can be compactly shown in the two equations: $q = Q$ and $p = -P$. That means: I know, if a measurement of q on the system yields a certain value, that a Q-measurement performed immediately there-after on the second system will give the same value, and vice versa; and I know, if a p-measurement on the first system yields a certain value, that a P-measurement performed immediately thereafter will give the opposite value, and vice verse.

In the EPR paper [115] incompatibility of observables is interpreted as incompatibility of knowledge: "the precise knowledge of one of them precludes such a knowledge of the other." By using INI one does not confront the nonlocality problem (here we speak about intrinsic quantum nonlocality). We note that INI can be combined either with the orthodox Copenhagen interpretation of a quantum state (the individual interpretation II), or with the statistical (ensemble) interpretation SI. This interpretational diversity is completed by the diversity of probability interpretations. The combination II+INI is typically endowed with the subjective interpretation of probability (e.g., D'Ariano, private discussion). As was claimed by Plotnitsky [315], Bohr used II+INI combined with the statistical interpretation of probability. In fact, this is what Plotnitsky called the statistical Copenhagen interpretation – INI is the key component of his interpreting of Bohr.

4.2.2 QBism, Fuchs et al.

INI is the cornerstone of QBism; it is combined with the subjective interpretation of probability. We emphasize that the essence of QBism is not in using this interpretation (as is believed by some experts in foundations), but in the claim that observations are the private experiences of agents who theorize on or experimentally practice with observations (see Chapter 17). QBism is neither a part of II nor of SI. QBism tries to liberate quantum foundations from the complex Hilbert space representation, that is, from states and operators, and operate solely with conditional probabilities. The role of the Hilbert space representation is reduced to a mathematical tool. With this formalism, QBists derived a quantum version of *the formula of total probability* (see Section 11.1 for the classical formula). Quantum predictions can be mathematically described by this formula, without appealing to a quantum state (density operator). Personally, I strongly support this ideology for the foundational justification of QM as a machinery for probability prediction. At the same time it should be remarked that the purely probability machinery is more complicated than the complex Hilbert space machinery. So, for the concrete calculations, it is easier to work in the standard complex Hilbert space framework. For me, the main disturbing feature of QBism is that it associates measurements' outputs with personal experiences of "quantum agents," users of quantum theory – theoretical and experimental.

Surprisingly, the private experience viewpoint of QBists is supported by some leading experimenters in quantum foundations. During the Växjö conferences we had a few debates on QBism's relevance to experiment and a few leading figures in quantum foundational experimenting supported Christopher Fuchs; I can recall, for example, Theodor Hänsch, Anton Zelinger, and Gregor Weihs. Of course, one can see even strongly negative reactions on QBism. During the conference devoted to 50th universary of the Bell inequality (in Vienna) I was sitting near Alain Aspect during the talk of Anton Zeilinger. This talk was extremely QBism biased: the personal experience viewpoint was advertised in absurdum, including the personal experiences of Anton's dog. (I do not know whether he really has a dog?) Alain Aspect was very angry and told me more or less the following: look, for a photon emitted by a source and flying toward a screen with two slits, it is determined in advance that it would never hit the registration screen with photo emulsion at some locations (the white stripes of the interference picture), and this is irrespective of Anton's or his dog's personal experiences. My notes on QBism in this section are so extended not because it is the most important among the information interpretations of QM, but because it is the most provocative. I struggled against QBism by criticizing Fuchs during his talks in Växjö for 20 years (see also [228]). One of the fruits of my critique was acceptance by Christopher Fuchs that a "quantum agent" is not an arbitrary person, but a guy who studied and understood (up to some level) quantum theory.

But after 20 years of critique of QBism. I suddenly found that it may be the best interpretation not for quantum physics, but for quantum-like modeling – applications of the methodology and formalism of quantum theory outside of physics, cognition, psychology, decision making, social and political science, economy and finance [221].

4.2.3 Växjö Interpretation

The Växjö interpretation is also a sub-interpretation of INI; it is very close to SI and, in particular, it is endowed with the statistical interpretation of probability. However, in contrast to the "orthodox statistical interpretation," in the spirit of Einstein and Ballentine, the Växjö interpretation highlights the role of contexts, so systems can be excluded from consideration (Chapter 10). This interpretation is an attempt to merge the views of Bohr and Einstein. Contextual structuring of QM can be considered as a part of empiricism in the spirit of De Muynck [106], but with the emphasis on probability and diminishing the role of linear space representation (LSR).

As well as by QBism, by the Växjö interpretation QM is a machinery for probability prediction. A quantum formula of total probability is derived; this is an additive perturbation of the classical formula. This quantum formula differs from QBism's formula. In the absence of the interference term, the Växjö formula of total probability coincides with the classical one. So, quantum probability inference is smoothly transferred into the classical probability inference. The latter is not the case for QBism. Moreover, the Växjö formula of total probability can be derived without exploring the Hilbert space formalism (in contrast to the QBism formula). The Växjö formula can be derived in the general contextual probabilistic framework (Section 11.2).

The probability prediction viewpoint, demystifies QM that is treated as a formalism for nonclassical probability inference, probability update is non-Bayesian. Here we should comment on the terms "Bayesian" and "Bayesianism." The latter is widely used in quantum information theory (essentially, but not only, due to QBism's influence): QM is interpreted as the machinery for probability update. The former refers to the Bayes' formula for conditional probability and associated with classical probability calculus [251] and classical probability inference.

The Växjö interpretation will be presented in detail in Chapter 10 devoted to the contextual probabilistic measurement model.

4.2.4 Zeilinger and Brukner, D'Ariano

We say just a few words about Zeilinger–Brukner interpretation. It is based on the following postulate:

Principle of Information Quantization. *An elementary system represents the truth value of one proposition.*

A curious reader can follow the articles of Zeilinger and Brukner [74, 75], see also my works [219, 220] for critical analysis.

We also just mention D'Ariano's program of reconstruction of the quantum formalism from the information axioms [82, 94, 95].

4.2.5 Plotnitsky

Now we turn to the complementarity principle and its information content and present the following extended citation of Plotnitsky who, in particular, emphasized the role of free will in the form of the experimenter's free choice of experimental contexts [316]:

> Bohr's complementarity is not only about a mutual exclusivity of certain entities considered in quantum physics, essential and widely used as this aspect of the concept may be, but also about performing quantum experiments and making predictions by human agents, in some of which a mutual exclusivity becomes necessary. That one has a free or, at least a sufficiently free, choice as concerns what kind of experiment to perform is, as noted by Bohr, defines the very idea of experiment in science, including in classical physics. In classical physics or relativity, however, this freedom does not matter fundamentally because our choice only defines which part of the already established reality one decides to consider. At least in principle, all variables necessary for defining the future course of reality, in accord with classical causality, can always be determined at any moment in time, because there is no complementarity or the uncertainty relations. By contrast, quantum physics gives this freedom an essential role, reflected in complementarity. By implementing one's decision concerning what one wants to do, which measurement to perform, ones define the character of physical reality, as each time a single complete reality, and its future course. This determination also allows one to make only certain types of predictions, in general probabilistic, and irrevocably exclude certain other, complementary, types of possible predictions. At the same time, this view still assumes both experimental facts (including the role h, c, the uncertainty relations, and so forth) and the mathematics of QM, which we use to predict (always with h in place in any actual prediction, such as those by Schrödinger's equation), always probabilistically, a future course of events by making one or the other decision entailed by complementarity, as just stated. It is important to keep in mind in this connection that, while one can control the setup of either experiment thus decided on (or any quantum experiment), one cannot control its outcome, which is objective in this sense as well. On this point, I depart from QBism, which sees measurements, rather than only assignments of probabilities as subjective. On probability assignments, my view is that in QM individual events are or at least may be strictly random, and as such cannot be assigned probability, while statistics of repeated experiments are always necessary and are objectively verifiable. Nevertheless, every measurement is still an individual act of the creation of a quantum phenomenon. Wheeler spoke of "elementary act of creation," although my interpretation is not the same as his, connected to the concept of "a participatory universe," which is not part of my or Bohr's view, which influenced Wheeler.

By using **INI**, one may lose the realistic grounds of quantum physics. If QM is merely about information generation and its processing, then subjective treatment of probabilities and personal experience treatment of experimental outputs might be attractive. Physical realism confronts a strong competitor – idealism. To clarify the place of realism within **INI**, Plotnitsky established the reality without realism (RWR) interpretation of QM (see [315] and references herein). This is one of the versions of the interpretations in the spirit of Copenhagen. I now present RWR. This is my view on RWR.

I start by remarking that often Bohr's views are presented as idealism. But this is misunderstanding. He definitely didn't deny the reality of *quantum systems*, say electrons or atoms. However, for him QM doesn't describe the intrinsic properties of quantum systems. Bohr stressed that a measurement's output can't be disassociated from a measurement apparatus and generally the complex of experimental physical conditions, or experimental context. We can point to two common missuses of quantum theory (from the RWR viewpoint). On one hand, one may neglect the role of experimental context and try to assign measurement's output directly to a quantum system. We call this approach "naive realism." From the Bohr–Heisenberg viewpoint, this approach should be rejected as contradicting the Heisenberg uncertainty and Bohr complementarity principles. Another misinterpretation is forgetting about the existence of quantum systems (the reality counterpart of RWR). By Bohr, electron exists! And the output of measurement is assigned to this concrete electron (prepared for measurement), but, of course, the WR-counterpart of the Bohr–Plotnitsky interpretation also has to be taken into account. Therefore quantum theory is about such individual assigning of outputs (quantum phenomena). This is the origin of discreteness of quantum measurements.

5

Quantum Postulate as the Seed of the Complementarity Principle

5.1 Quantum Postulate

The role of the Planck constant in quantum foundations was discussed in articles [235, 236] based on Bohr's works [59]–[61].

The complementarity principle plays a basic role in quantum foundations and it is worthwhile to find the physical ground of this principle. We remark that Bohr highlighted the Planck discovery of the indivisible quantum of action as the starting point for development of the modern QM [60] (see also [59, 61, 65]):

> In the history of science there are few events which, in the brief span of a generation, have had such extraordinary consequences as Planck's discovery of the elementary quantum of action. Not only does this discovery, to an ever increasing degree, form the background for the ordering of our experience concerning atomic phenomena, the knowledge of which has been so amazingly extended in the last thirty years, but, at the same time, it has brought about a complete revision of the foundations underlying our description of natural phenomena. We are dealing here with an unbroken development of points of view and conceptual aids which, beginning with the fundamental works of Planck on black body radiation, has reached a temporary climax, in recent years, in the formulation of a symbolic quantum mechanics.... The more clearly it appears that a uniform formulation of the quantum theory in classical terms is impossible, the more we admire Planck's happy intuition in coining the term "quantum of action" which directly indicates a renunciation of the action principle, the central position of which in the classical description of nature he himself has emphasized on more than one occasion.

And the existence of an indivisible quantum of action (given by Planck constant h) Bohr presented as the *quantum postulate*; he stressed [59] that the essence of quantum theory "may be expressed in the so-called quantum postulate, which attributes to any atomic process an essential discontinuity, or rather individuality, completely foreign to the classical theories and symbolized by Planck's quantum of action."

The existence of *indivisible quantum of action* given by the Planck constant h prevents the total separation of the genuine physical features of a system from the properties of interaction with a measurement apparatus. So, the ontic seed of the Bohr complementarity principle is the Planck constant h. The quantum postulate is an ontic statement about the nature as it is, but the complementarity principle is an epistemic statement. And we treat quantum theory as an epistemic theory. Hence, it is natural to search for an epistemic counterpart of the quantum postulate. To start such a search, we recall Heisenberg's comparison of the existence of the quantum of action ($h \neq 0$) with finiteness of the speed of light ($c < \infty$) and its independence of the inertial frame [59]–[61].

We conclude this section with the following remark. It is meaningless to start a crusade against complementarity with the hidden variables models of the Bell type without trying to understand the origin of this fundamental quantum of action in nature. Neither Einstein nor Bohr nor Bell tried to perform such investigation; for them this is just a feature of nature such as, for example, the constancy of the speed of light. And, for the moment, this position can be considered as the only one possible. Personally I think that quantum physics cries not for an explanation of the long-distance correlations and the violation of the Bell inequalities, but of the existence in Nature of the indivisible action quantum.

Why does Nature restrict divisibility of action up to some fixed value? Where does h come from?

I do not know any serious physical study of this main mystery of quantum physics, the real physical mystery, not spooky action at a distance.

5.2 Quantum Action Invariance Principle

Now we recall that Zeilinger was looking for the fundamental principles of QM [353] which would be heuristically attractive similarly to the principles of special relativity formulated by Einstein:

1. **Relativity principle:** *The laws of physics are invariant (i.e. identical) in all inertial frames of reference.*
2. **Light velocity invariance principle:** *The speed of light in a vacuum is the same for all observers, regardless of the motion of the light source or observer.*

We stress that the second of these principles presumes finiteness of light's velocity. On the basis of Bohr's quantum postulate, we formulate the following principle of QM that can be considered as the analog of Einstein's second principle:

Quantum action invariance principle: *The quantum of action is the same for all observers, regardless of experimental contexts.*

We can say that this principle is the epistemic counterpart of Bohr's quantum postulate. The formulation of the quantum action principle involves observables, but the quantum postulate, the existence in nature of the indivisible quantum of action, is about nature as it is, that is, this is the ontic postulate.

Nowadays, it is practically forgotten that by formulating the uncertainty principle Heisenberg pointed to the analogy with the light velocity constraint in special relativity. This analogy was then emphasized by Bohr [59]. In this paper, Bohr used the term "reciprocal uncertainty" for the Heisenberg uncertainty relation. (This reciprocity is related to position and momentum.)

> Heisenberg has rightly compared the significance of this law of reciprocal uncertainty for estimating the self-consistency of quantum mechanics with the significance of the impossibility of transmitting signals with a velocity greater than that of light for testing the self-consistency of the theory of relativity.... Planck's discovery has brought before us a situation similar to that brought about by the discovery of the finite velocity of light.

For the formulation of the complementarity principle, the concrete value of the Planck constant is not important. It is important only that this quantum of action exists, $h \neq 0$. In the same way, the concrete value of light's velocity is not important for formulation of special relativity, that is, it is only important that it is finite, $c < \infty$. We also stress that constancy of action quantum, its independence of observable (measurement procedure), plays the crucial role in QM, as well as the constancy of the speed of light in special relativity.

We now can formulate two basic principles of quantum theory:

1. **Quantum action invariance principle.**
2. **Complementarity principle.**

We consider QM as an epistemic theory [35], a theory about extraction of knowledge about nature; in the terminology of Hertz and Boltzmann this is an observational theory (Chapter 15). Together these two principles provide the epistemic foundations of quantum theory. The quantum action principle is the direct consequence of the quantum postulate (the ontic principle about nature as it is), and the second quantum principle (complementarity) is coupled to the first quantum principle. But their interrelation is complicated (see [223, 224, 232, 233]). Of course, these principles don't provide representation of QM as a closed formal system. In spite of a rather common opinion, even Einstein's relativity based on the principle of relativity and constancy of light's velocity cannot be treated as such a closed system; see Einstein's own comment on this issue (the citation is taken from book [72]):

> The principle of relativity, or, more exactly, the principle of relativity together with the principle of the constancy of the velocity of light, is not to be conceived as a "complete system," in fact, not as a system at all, but merely as a heuristic principle which, when considered by itself, contains only statements about rigid bodies, clocks, and light signals. It is only by requiring relations between otherwise seemingly unrelated laws that the theory of relativity provides additional statements.

5.3 Concluding Remarks: Complementarity and Quantum Postulate

This chapter is devoted to the analysis of the Bohr complementarity principle. We start by dissolving this complex foundational concept into a few sub-principles, of which the most important are the contextuality and incompatibility principles. The latter is commonly emphasized, often in the form of wave–particle duality. However, the former is essentially less discussable and its recognition and coupling with the modern contextuality theory is due to my works [232, 235, 344]. Another practically ignored component of the complementarity principle is individuality of quantum phenomena, discreteness of measurements outputs detection. And nowadays, during the quantum information revolution, we especially highlight the information complementary-completeness sub-principle.

The ontological core of the complementarity principle is the quantum action postulate. Therefore the rejection of complementarity is impossible without the rejection of the quantum postulate – the existence of the indivisible quantum of action (Planck quantum). The quantum postulate is the basis of the most fundamental exhibitions of quantum theory, as, say, the black body radiation. By questioning the validity of the complementarity principle and, hence, quantum postulate and creating a new subquantum physical model, one should suggest a new description of all physical phenomena which nowadays are described by the conventional quantum theory. From the physical viewpoint, the play-alike Bell with the abstract hidden variables models is totally meaningless.

We accomplished Zeilinger's request [353] for heuristically natural foundational principles of quantum theory. These are (1) the quantum action invariance principle; (2) the complementarity principle. Both these epistemological principles have the same ontic seed, namely, the quantum postulate – the unique ontological base of quantum theory.

Finally, I conclude that if the foundations of quantum theory are presented similarly to the foundations of special relativity, then

the attempts to go beyond the complementarity principle, for example, with hidden variables of the Bell type, can be compared with the attempts to go beyond special relativity, by rejecting Einstein's principle on the constancy of the speed of light.

The main difference between classical and quantum physics is that the latter is constrained by the existence of the Planck constant.

PART II

Bell Inequalities

6

Bell Inequality: Incompatibility versus Nonlocality

6.1 Quantum Physics: Getting Rid of Nonlocality

In this chapter we perform the *genuine quantum mechanical analysis of the CHSH-inequality derivation* considered as an inequality for correlations of quantum observables – *the quantum CHSH-inequality*.[1] Thus, we don't go beyond QM. The quantum mechanical interpretation of the CHSH-inequality experimental violation is the center of our interest. We show that, in fact, the violation degree is straightforwardly coupled to the degree of observables incompatibility, the norms of commutators. In particular, in the scenario with two spatially separated systems, these are tests of local incompatibility, that is, *incompatibility of observables on a single subsystem of a compound system* (Theorem 6.3, Section 6.4.2).

We remark that the CHSH inequality [83] is derived for classical correlations expressed in the hidden variables framework (established by Bell[2] [53, 54]) and by using the calculus of classical probabilities, namely the Kolmogorov probability theory. The quantum CHSH-inequality is derived for quantum correlations by using the operator formalism and the Landau equality [232].

To work with the Landau equality, it is convenient to *normalize the standard CHSH-correlation function with 1/2-multiplier, that is, the right-hand side of the CHSH inequality equals 1, not 2.* So, we operate with the following CHSH combination of correlations:

$$\langle \mathcal{B} \rangle = \frac{1}{2}[\langle A_1 B_1 \rangle + \langle A_1 B_2 \rangle + \langle A_2 B_1 \rangle - \langle A_2 B_2 \rangle]. \tag{6.1}$$

We stress that three different interpretations can be assigned to this quantity:

- Classical (hidden variables) correlation, $\langle \mathcal{B} \rangle_{\mathrm{CL}}$.
- Experimental correlation, $\langle \mathcal{B} \rangle_{\mathrm{EXP}}$.
- Quantum mechanical correlation, $\langle \mathcal{B} \rangle_{\mathrm{QM}}$.

1 This chapter is based on the author's article [232].

2 We remark that the rather common opinion that the Bell inequality framework is the natural experimental representation of the EPR argument [115] is really misleading.

Following Bell, one can compare the classical theoretical correlation $\langle \mathcal{B} \rangle_{\mathrm{CL}}$ with its experimental counterpart $\langle \mathcal{B} \rangle_{\mathrm{EXP}}$. Most of the quantum foundation and information community proceeds in this way. This way leads to operating with the notion of nonlocality and action at a distance.[3] The following natural question arises:

Is there any reason to couple this mysticism with quantum theory? I suggest to start with the quantum theoretical analysis of correlation $\langle \mathcal{B} \rangle_{\mathrm{QM}}$. It is easy to explain under what circumstances it can be bounded by 1 (or exceed 1). Thus, the so-called classical bound has the purely quantum origin. This quantum mechanical explanation is purely local and in quantum theory it is not proper to deal with its nonlocality or associate with it any kind of action at a distance.

By the Bohr complementarity and Heisenberg uncertainty principles (Chapter 3), it is impossible to measure jointly the position and momentum of a quantum system, the spin projections for an electron, or polarization projections for a photon. These are incompatible quantum observables. Therefore, correlation $\langle \mathcal{B} \rangle_{\mathrm{QM}}$ can exceed 1.

If somebody doubts in the validity of QM, it would be natural to check the violation of the complementarity principle for a single quantum system, say an electron or a photon, for example, to check the violation of the Heisenberg uncertainty relation, in the form of the Schrödinger (or Robertson) inequality. We recall that the uncertainty relation stimulated Bohr to derive the complementarity principle.

6.2 CHSH-Inequality as Incompatibility Test

We show that the violation degree of the quantum CHSH inequality can be considered as a measure of interrelation of the incompatibility degrees in two pairs of quantum observables, A_1, A_2 and B_1, B_2 represented by Hermitian operators $\hat{A}_i, \hat{B}_j$.This is the simple consequence of *the Landau identity* [261, 262]. In quantum theory incompatibility is mathematically expressed as noncommutativity. Thus, by testing incompatibility, we test the degree of noncommutativity, or in other words, the "magnitudes" of observables corresponding to commutators,

$$\hat{M}_A = i[\hat{A}_1, \hat{A}_2], \ \hat{M}_B = i[\hat{B}_1, \hat{B}_2]. \tag{6.2}$$

The incompatibility magnitude can be expressed via the maximal value of averages of commutator-operators, that is, by their norms,

$$\sup_{\|\psi\|=1} |\langle \hat{M}_A \rangle_\psi| = \sup_{\|\psi\|=1} |\langle \psi | \hat{M}_A | \psi \rangle| = \|\hat{M}_A\|, \tag{6.3}$$

[3] We stress that this is subquantum nonlocality of the very special model with hidden variables which was suggested by Bell – Bell nonlocality (Chapter 2).

$$\sup_{\|\psi\|=1} |\langle \hat{M}_B \rangle_\psi| = \sup_{\|\psi\|=1} |\langle \psi | \hat{M}_B | \psi \rangle| = \|\hat{M}_B\|. \tag{6.4}$$

By interpreting quantity $\langle M_A \rangle_\psi = \langle \psi | \hat{M}_A | \psi \rangle$ as the theoretical counterpart of experimental average $\langle M_A \rangle_{\psi,\mathrm{EXP}}$ of observable M_A, we can measure experimentally the incompatibility magnitude, that is, norm $\|\hat{M}_A\|$ from measurements of commutator-observable M_A. In the experimental framework, the index ψ labels the preparation procedure for the quantum state ψ. We follow the statistical interpretation (SI) of a quantum state (Einstein and Ballentine, Chapter 2).

A foundational problem is that measurement of such commutator-observables is challenging. Recently some progress is demonstrated on the basis of weak measurements, but generally we are not able to measure commutator-quantities. Therefore, one has to design tricky measurement procedures. In our interpretation measurements leading to calculation of the CHSH-correlation $\langle \mathcal{B} \rangle_{\psi;\mathrm{EXP}}$ is one of such procedures. The CHSH-test estimates the product of incompatibility magnitudes for the A-observables and B-observables, that is, the quantity $\|\hat{M}_A\| \|\hat{M}_B\|$. Now suppose that one is interested to find the degree of incompatibility for observables A_1 and A_2. Then the B-observables are considered as axillary observables. By selecting them in a proper way, so that the B-commutator is a simple operator, we can use the CHSH test to get the experimental value for the incompatibility-magnitude $\|\hat{M}_A\|$.

The norm of the commutator-observable gives the maximal degree of incompatibility w.r.t. all possible states. One can be interested in incompatibility for the fixed state ψ. Then axillary observables B_1 and B_2 should be selected in such a way that $\langle M_B \rangle_\psi$ can be easily calculated.

6.3 Incompatibility as a Necessary Condition of CHSH-Inequality Violation

6.3.1 General case

Consider the Bohm–Bell-type experiment. Four dichotomous observables A_1, A_2, B_1, B_2 taking values ± 1 are involved. It is assumed that observables in each pair $A_i, B_j, i,j = 1,2$, can be measured jointly, that is, A-observables are compatible with B-observables. However, the observables in pairs A_1, A_2 and B_1, B_2 can be incompatible. Thus, probability distributions $p_{A_iB_j}$ are well defined theoretically in QM and they can be verified experimentally; probability distributions $p_{A_1A_2}$ and $p_{B_1B_2}$ may not be well defined in QM.

In the mathematical QM-formalism, compatibility is represented by commutativity of operators, that is, in the Bohm–Bell-type experiments

$$[\hat{A}_i, \hat{B}_j] = 0, i,j = 1,2, \tag{6.5}$$

and generally

$$[\hat{A}_1, \hat{A}_2] \neq 0, \ [\hat{B}_1, \hat{B}_2] \neq 0.$$

The quantum theoretical CHSH-correlation function has the following form:

$$\langle \mathcal{B} \rangle_\psi = \frac{1}{2}[\langle \hat{A}_1 \hat{B}_1 \rangle_\psi + \langle \hat{A}_1 \hat{B}_2 \rangle_\psi + \langle \hat{A}_2 \hat{B}_1 \rangle_\psi - \langle \hat{A}_2 \hat{B}_2 \rangle_\psi]. \tag{6.6}$$

This quantum-theoretical correlation is compared with the experimental CHSH-correlation function.

In the quantum framework, the CHSH-correlation function can be expressed with the aid of the Bell operator:

$$\hat{\mathcal{B}} = \frac{1}{2}[\hat{A}_1(\hat{B}_1 + \hat{B}_2) + \hat{A}_2(\hat{B}_1 - \hat{B}_2)] \tag{6.7}$$

as

$$\langle \mathcal{B} \rangle_\psi = \langle \psi | \hat{\mathcal{B}} | \psi \rangle. \tag{6.8}$$

By straightforward calculation, one can derive at the Landau identity:

$$\hat{\mathcal{B}}^2 = I - (1/4)[\hat{A}_1, \hat{A}_2][\hat{B}_1, \hat{B}_2]. \tag{6.9}$$

Thus, if *at least one of the commutators* equals zero, that is,

$$[\hat{A}_1, \hat{A}_2] = 0, \tag{6.10}$$

or

$$[\hat{B}_1, \hat{B}_2] = 0, \tag{6.11}$$

then the following inequality holds:

$$|\langle \mathcal{B} \rangle|_\psi \leq 1. \tag{6.12}$$

To derive this inequality, we used solely the quantum formalism. The inequality is the consequence of compatibility for at least one pair of observables, A_1, A_2 or B_1, B_2. Thus, although formally inequality (6.12) coincides with the standard CHSH-inequality, it has a totally different meaning. We call inequality (6.12) *the quantum CHSH-inequality.*

Theorem 6.1 *Compatibility of A- or B-observables is sufficient for validity of the quantum CHSH-inequality for all quantum states.*

In other words, *conjunction of incompatibilities of A- and B-observables is the necessary condition for violation of the quantum CHSH-inequality* for some quantum state:

$$|\langle \mathcal{B} \rangle|_\psi > 1 \ \Rightarrow \ [\hat{A}_1, \hat{A}_2] \neq 0 \wedge [\hat{B}_1, \hat{B}_2] \neq 0. \tag{6.13}$$

6.3.2 Tensor Product Case

States of a compound quantum system $S = (S_A, S_B)$ are represented in tensor product $\mathcal{H}_{AB} = \mathcal{H}_A \otimes \mathcal{H}_B$ of the state spaces $\mathcal{H}_A$ and $\mathcal{H}_B$ of subsystems S_A and S_B. Observables are given by operators

$$\hat{A}_i = \hat{a}_i \otimes I, \ \hat{B}_i = I \otimes \hat{b}_i, \tag{6.14}$$

where Hermitian operators $\hat{a}_i$ and $\hat{b}_i$ act in H_A and H_B, respectively. They represent observables a_i, b_i on subsystems S_A, S_B of S. For spatially separated systems, we call them *local observables*.

This tensor representation automatically implies commutativity of operators $\hat{A}_i$ with operators $\hat{B}_j$, that is, Equation (6.5) holds. We remark that the mathematical condition of incompatibility is reduced to condition $[\hat{a}_1, \hat{a}_2] \neq 0$ and $[\hat{b}_1, \hat{b}_2] \neq 0$. For spatially separated systems, it is natural to call incompatibility of the observables on S_A (on S_B) *local incompatibility*. Section 6.3.1 implies that

Corollary *Conjunction of local incompatibilities is the necessary condition for violation of the quantum CHSH inequality.*

We remark that the mathematical formalism of this section is applicable to description of any kind of observables "respecting" the tensor product structure of observables. A physical system S does not need to be composed of two physical subsystems. For example, one can apply this approach to so-called classical entanglement (Chapter 14).

6.4 Incompatibility as a Sufficient Condition of CHSH-Inequality Violation

6.4.1 General Case

Assume that A-observables as well as B-observables are incompatible, that is, corresponding operators do not commute:

$$[\hat{A}_1, \hat{A}_2] \neq 0 \text{ and } [\hat{B}_1, \hat{B}_2] \neq 0, \tag{6.15}$$

that is,

$$\hat{M}_A \neq 0 \text{ and } \hat{M}_B \neq 0, \tag{6.16}$$

where $\hat{M}_A = i[\hat{A}_1, \hat{A}_2]$, $\hat{M}_B = i[\hat{B}_1, \hat{B}_2]$. It is important to note that

$$[\hat{M}_A, \hat{M}_B] = 0.$$

We can write the Landau identity in the form

$$\hat{\mathcal{B}}^2 = I + (1/4)\hat{M}_{AB}, \tag{6.17}$$

where

$$\hat{M}_{AB} = \hat{M}_A \circ \hat{M}_B.$$

If operator $M_{AB} = 0$, then, despite the incompatibility condition (6.15), the quantum CHSH inequality cannot be violated.

So, further we proceed under the nondegeneration assumption:

$$\hat{M}_{AB} \neq 0. \tag{6.18}$$

In our framework, this constraint is not so restrictive. We consider the quantum CHSH inequality as a statistical test of incompatibility. It is natural to estimate the degree of incompatibility in one pair of observables, for example, in the A-pair. In this approach, the B-pair plays the axillary role and we can freely play with its selection. To satisfy constraint (6.18), it is sufficient to select B-operators in such a way that the operator $\hat{M}_B$ is invertible. We especially highlight the case of compound systems (Section 6.3.2). Here incompatibility of the A-observables and the B-observables, see (6.16), automatically implies inequality (6.18).

Under constraint (6.18), there exists a common eigenvector ψ_{AB} of the commutator operators $\hat{M}_A$ and $\hat{M}_B$, such that

$$\hat{M}_A \psi_{AB} = \mu_A \psi_{AB}, \hat{M}_B \psi_{AB} = \mu_B \psi_{AB} \tag{6.19}$$

and both eigenvalues are nonzero.

Suppose that $\mu_A > 0$ and $\mu_B > 0$. Then, this ψ_{AB} is an eigenvector of operator $\hat{\mathcal{B}}^2$ with eigenvalue $(1 + \mu) > 1, \mu = \mu_A \mu_B$. Hence, $\|\hat{\mathcal{B}}^2\| \geq (1 + \mu) > 1$ and

$$1 < (1 + \mu) \leq \|\hat{\mathcal{B}}^2\| = \|\hat{\mathcal{B}}\|^2.$$

Since $\hat{\mathcal{B}}$ is Hermitian, we have

$$\|\hat{\mathcal{B}}\| = \sup_{\|\psi\|=1} |\langle \psi | \hat{\mathcal{B}} | \psi \rangle|.$$

Finally, we get that

$$\sup_{\|\psi\|=1} |\langle \psi | \hat{\mathcal{B}} | \psi \rangle| > \sqrt{1 + \mu} > 1.$$

Thus, there exist pure quantum states such that the quantum CHSH inequality is violated.

Now, suppose that $\mu_A > 0$, but $\mu_B < 0$. To change the sign of μ_B, it is sufficient to interchange the B-observables.

Now consider all CHSH-correlation functions $\langle \mathcal{B}_{\hat{\sigma}} \rangle_\psi$ obtained from $\langle \mathcal{B} \rangle_\psi$ by all possible permutations of indexes for A-observables and for B-observables. Here the symbol $\hat{\sigma}$ denotes such permutation. For example, suppose $\hat{\sigma}$ does not change indexes of the A-observables, but interchanges the indexes of the B-observables; then

$$\langle \mathcal{B}_{\hat{\sigma}} \rangle_\psi = \frac{1}{2}[\langle \hat{A}_1 \hat{B}_1 \rangle_\psi + \langle \hat{A}_1 \hat{B}_2 \rangle_\psi - \langle \hat{A}_2 \hat{B}_1 \rangle_\psi + \langle \hat{A}_2 \hat{B}_2 \rangle_\psi]. \tag{6.20}$$

The inequalities

$$|\langle \mathcal{B}_{\hat{\sigma}} \rangle_\psi| \leq 1 \tag{6.21}$$

are also called quantum CHSH inequalities for observables A_1, A_2, B_1, B_2. Our previous considerations lead to the following result:

Theorem 6.2 *Nondegeneration condition (6.18) is sufficient for the violation of at least one of the quantum CHSH-inequalities for some state ψ, that is,*

$$\max_{\hat{\sigma}} \sup_{\|\psi\|=1} |\langle \mathcal{B}_{\hat{\sigma}} \rangle|_\psi > 1. \tag{6.22}$$

We point out again that generally conjunction of incompatibilities of A- and B-observables is not sufficient for the violation of one of *the quantum* CHSH inequalities for these observables. It may happen that

$$[\hat{A}_1, \hat{A}_2] \circ [\hat{B}_1, \hat{B}_2] = 0, \tag{6.23}$$

in spite of the incompatibility of A- and B-observables. The physical meaning of this equality or its violation is not clear.

6.4.2 Tensor Product Case

Let $\mathcal{H} = \mathcal{H}_A \otimes \mathcal{H}_B$ and $\hat{A}_j = \hat{a}_j \otimes I, \hat{B}_j = I \otimes \hat{b}_j$, where Hermitian operators $\hat{a}_j$ and $\hat{b}_j$ act in $\mathcal{H}_A$ and $\mathcal{H}_B$, respectively.

The joint incompatibility condition in Equation (6.15) is equivalent to incompatibility of observables on subsystems:

$$\hat{m}_A = i[\hat{a}_1, \hat{a}_2] \neq 0 \text{ and } \hat{m}_B = i[\hat{b}_1, \hat{b}_2] \neq 0. \tag{6.24}$$

We have $\hat{M}_{AB} = \hat{M}_A \hat{M}_B = \hat{m}_A \otimes \hat{m}_B$. As mentioned earlier, constraint $\hat{M}_{AB} \neq 0$ is equivalent to Equation (6.24). Section 6.3.1 implies that *conjunction of local incompatibilities* is the sufficient condition for violation of the quantum CHSH inequality. Thus, we obtain:

Theorem 6.3 *Conjunction of local incompatibilities is the necessary condition for the violation of the quantum CHSH inequalities and it is sufficient for the violation of at least one of them (see (6.22)) for some state ψ.*

Now we think more carefully about the eigenvector from consideration of Section 6.4.1. The vector $\psi_{AB} = \psi_A \otimes \psi_B$, where $\psi_A \in H_A, \psi_B \in H_B$, and $\hat{m}_A \psi_A = \mu_A \psi_A, \hat{m}_B \psi_B = \mu_B \psi_B$. We assume that $\mu = \mu_A \mu_B > 0$. Then,

$$\langle \psi_A \otimes \psi_B | \hat{\mathcal{B}}^2 | \psi_A \otimes \psi_B \rangle > 1.$$

Thus, for the squared CHSH-observable $\hat{\mathcal{B}}^2$, the one-boundary can be violated for separable states. Here, entanglement of A and B observables plays no role.

To be more illustrative, let us restrict consideration to finite-dimensional Hilbert spaces. There can be found states Ψ and Φ such that

$$\max_{\|\psi\|=1} |\langle \psi | \hat{\mathcal{B}}^2 | \psi \rangle = |\langle \Psi | \hat{\mathcal{B}}^2 | \Psi \rangle.$$

$$\max_{\|\psi\|=1} |\langle \psi | \hat{\mathcal{B}} | \psi \rangle = |\langle \Phi | \hat{\mathcal{B}} | \Phi \rangle.$$

The tricky thing is that generally $\Psi \neq \Phi$. The equality for norms, $\|\hat{\mathcal{B}}\| = \sqrt{\|\hat{\mathcal{B}}^2\|}$, does not imply equality of the max-optimization states.

Of course, max-states for $\mathcal{B}$ and $\mathcal{B}^2$ are connected: the former can be represented as linear combinations of the latter (the feature of all operators with degenerate spectrum). (As shown in [73], max-states for $\mathcal{B}$ can be represented even as mixtures of max-states for $\mathcal{B}^2$.)

In the experiments to violate the quantum CHSH inequality, tremendous efforts were made to prepare ensembles of entangled states. The main reason for this is that the direct measurement of the observable represented by operator $\hat{\mathcal{B}}^2$ is challenging. In Section 6.7, we present the abstract analog of the Bell experiments treated as experiments to measure the degree of incompatibility. The tensor product structure is excluded and, in particular, an analog of entangled states related to measurement of an observable and its square is considered.

6.5 CHSH-Correlation Function as Measure of Incompatibility

In this section we proceed in the finite-dimensional case and with the observables respecting the tensor product structure on the state space $\mathcal{H} = \mathcal{H}_A \otimes \mathcal{H}_B$. Consider the eigenbases (e_{Ak}) and (e_{Bk}) of operators $\hat{m}_A$ and $\hat{m}_B$ (acting in $\mathcal{H}_A$ and $\mathcal{H}_B$, respectively) and the corresponding eigenvalues μ_{Aj}, μ_{Bj}.

Consideration of finite-dimensional spaces leads to the equalities:

$$\|\hat{m}_A\| = \max_j |\mu_{Aj}| = |\mu_{Ai_a}|, \|\hat{m}_B\| = \max_j |\mu_{Bj}| = |\mu_{Bk_b}|,$$

for some indexes i and k.

Let $\mu_{Ai_a}\mu_{Bk_b} > 0$. Then, $\|\hat{\mathcal{B}}^2\| = (1 + \mu_{Ai_a}\mu_{Bk_b})$. Thus,

$$b = |\langle \mathcal{B} \rangle_\psi| = \|\hat{\mathcal{B}}\| = \sqrt{1 + \frac{1}{4} \|[\hat{a}_1, \hat{a}_2]\| \; \|[\hat{b}_1, \hat{b}_2]\|}, \tag{6.25}$$

where b is the maximal possible value of CHSH correlations. If eigenvalues μ_{Ai_a} and μ_{Bk_b} have different signs, then we interchange the B-observables.

From Equation (6.25), we get that

$$\|[\hat{a}_1, \hat{a}_2]\| \|[\hat{b}_1, \hat{b}_2]\| = 4(b^2 - 1).$$

The norm of commutator can be considered as a measure of incompatibility. Thus, the CHSH-correlation function gives the possibility of checking experimentally the product of degrees of incompatibility for the A and B observables.

One may consider this way of measuring incompatibility as too tricky. However, typically, to measure commutator-observable and then its average is challenging. (By "measuring commutator-observable" we mean measuring an observable represented mathematically by a commutator operator scaled by i.) Therefore, even such a tricky approach to this problem as measurement of the CHSH-correlation function deserves attention.

Now, consider B-observables as axillary. In this way, we are able to determine the degree of incompatibility for the A-observables by using some pair of axillary observables B_1, B_2. We can select the latter in such a way that their commutator is a "good observable," so that it can be easily measured for any state, thus its average and hence the norm can be determined. Then, we can measure incompatibility of observables A_1 and A_2 by using the following formula:

$$\|[\hat{a}_1, \hat{a}_2]\| = 4(b^2 - 1)/\|[\hat{b}_1, \hat{b}_2]\|. \tag{6.26}$$

Why is the use of tensor product states so useful for measuring the degree of incompatibility? By spitting a system into two subsystems it is easy to check compatibility of A and B observables, thus the possibility of defining the CHSH-correlation function which can be measured in experiment.

6.6 Incompatibility Is a Necessary Condition for Violation of Bell Inequalities

Consider a family of quantum observables $D_1, \ldots, D_n$ represented by Hermitian operators $\hat{D}_1, \ldots, \hat{D}_n$. We restrict considerations to observables with discrete values; thus, operators have purely discrete spectra. Denote by $\hat{E}_{D_i}(x)$ the orthogonal projector corresponding to the eigenvalue x of $\hat{D}_i$.

Suppose that the observables are pairwise compatible, namely, (D_i, D_j) can be measured jointly for any quantum state ρ (given by a density operator) and the joint probability distribution (JPD) is well defined:

$$p_{D_iD_j}(x, y; \rho) \equiv P(D_i = x, D_j = y; \rho). \tag{6.27}$$

In QM, compatibility is mathematically represented via commutativity of operators, namely, $[\hat{D}_i, \hat{D}_j] = 0$, and, hence, $[\hat{E}_{D_i}(x), \hat{E}_{D_j}(y)] = 0$. The quantum formalism gives the following formula for JPD (von Neumann [365]):

$$p_{D_iD_j}(x, y; \rho) = \text{Tr } \rho E_{D_i}(x) E_{D_j}(y) = \text{Tr } \rho E_{D_j}(y) E_{D_i}(x). \tag{6.28}$$

Now, we point to the really surprising feature of quantum measurement theory. If observables are pairwise compatible, namely, each pair can be jointly measured with corresponding JPDs $p_{ij}(x, y; \rho)$ given by Equation (6.28), then they are also triple-wise compatible, quadruple-wise compatible, and so on... Any family of observables, $D_{i_1}, \ldots, D_{i_m}$ can be jointly measured, and the joint probability distribution is given by the following formula:

$$p_{D_{i_1} \ldots D_{i_m}}(x_1, \ldots, x_m; \rho) = \text{Tr } \rho E_{D_{i_1}}(x_1) \ldots E_{D_{i_m}}(x_m). \tag{6.29}$$

On the left-hand side of this formula, one can take any permutation of indexes. This implies:

Pairwise compatibility $\Longrightarrow$ *multiple compatibility.*

Consider the most general Bell-type framework. There are K groups of quantum observables:

$$D^k = (D_1^k, \ldots, D_{N_k}^k), k = 1, \ldots, K.$$

Mathematically, they are represented by Hermitian operators:

$$\hat{D}^k = (\hat{D}_1^k, \ldots, \hat{D}_{N_k}^k).$$

Suppose that, for different k, observables are compatible, namely, in the mathematical framework:

$$[\hat{D}_i^n, \hat{D}_j^m] = 0, n \neq m.$$

Thus, JPDs $p_{i_1 \ldots i_K} \equiv p_{D_{i_1}^1 \ldots D_{i_K}^K}$ are well defined and, hence, covariations as well:

$$\langle D_{i_1}^1 \cdots D_{i_K}^K \rangle_\rho = \text{Tr } \rho \hat{D}_{i_1}^1 \cdots \hat{D}_{i_K}^K =$$
$$\sum x_1 \cdots x_K \, p_{i_1 \ldots i_K}(x_1, \ldots, x_K).$$

Consider some Bell-type inequality (e.g., the CHSH inequality [83] or the Mermin inequality [280]),

$$\sum_{i_1 \ldots i_K} t_{i_1 \ldots i_K} \, \langle D_{i_1}^1 \ldots D_{i_K}^K \rangle_\rho + \text{correlations of lower orders} \leq c, \tag{6.30}$$

where $t_{i_1 \ldots i_K}$ are some real constants. This inequality may be violated only if at least one pair of observables, say (D_i^n, D_j^n), is incompatible, namely, in the mathematical formalism

$$[\hat{D}_i^n, \hat{D}_j^n] \neq 0. \tag{6.31}$$

Otherwise, JPD exists and the inequality in Equation (6.30) cannot be violated.

Theorem 6.4 *Incompatibility is a necessary condition for violation of any Bell-type inequality.*

In the standard nonlocality discussions, it is assumed that there are K systems $S^k, k = 1, 2, \ldots, K$, and observables D^k are local observables for S^k. Endowed with this scheme, we analyze the possibility of violating the Bell-type inequality in Equation (6.30). The necessary condition is that Equation (6.31) holds true. This condition is local.

6.7 Measurements of Observable versus Its Square Root

Our previous considerations can be abstractly formalized in the following way. As we have seen, incompatibility of observables is theoretically encoded in the quantum observable with the squared Bell operator $\hat{\mathcal{B}}^2$. However, its direct measurement is challenging. Therefore, experimenters measure the quantum observable with the Bell operator $\hat{\mathcal{B}}$. Its measurement is challenging, although possible. From the outputs of such measurements, one derives information about the degree of incompatibility.

We have considered compound systems and discussed the well-known fact that eigenvectors of the squared Bell operator $\hat{\mathcal{B}}^2$ giving the max-value of its quadratic form can be selected as separable (nonentangled) states. They do not need to eigenvectors of the Bell operator $\hat{\mathcal{B}}$. Its eigenvectors are linear combinations of the aforementioned separable states.

We want to show that the tensor product structure of states and operators is not crucial for the preceding consideration.

Consider any Hermitian operator $\hat{C}$ and its square $\hat{C}^2$. Let u be an eigenvector of the latter, namely, $\hat{C}^2 u = \lambda u, \lambda > 0$, and let u not be an eigenvector of the former. Set

$$v = \hat{C}u/\sqrt{\lambda},$$

that is,

$$\hat{C}u = \sqrt{\lambda}v, \; \hat{C}v = \sqrt{\lambda}u. \tag{6.32}$$

Set

$$\psi_\pm = u \pm v. \tag{6.33}$$

Then, $\hat{C}\psi_\pm = \pm\sqrt{\lambda}\psi_\pm$. Thus, $\psi_\pm$ are eigenvectors of $\hat{C}$. If the quadratic form of $\hat{C}^2$ approaches its max-value on the eigenstate $u/\|u\|$, then the quadratic form of $\hat{C}$ approaches its max-value on the eigenstate $\phi_\pm = \psi_\pm/\|\psi_\pm\|$.

This consideration can be coupled to measurement theory. Consider some quantum observable D represented by Hermitian operator $\hat{D}$. (For simplicity, suppose that $\hat{D} \geq 0$.) Suppose that this observable is simple theoretically, but complex experimentally. The spectrum and eigenvectors of operator $\hat{D}$ can be easily found, but measurement of observable D is really challenging. Consider also the observable C represented by Hermitian operator $\hat{C} \equiv \sqrt{\hat{D}}$. Suppose that this observable is complex theoretically, but rather simple experimentally. The structure of the spectrum and eigenvectors of operator $\hat{C}$ is complicated, but measurement design for C is straightforward.

We are interested in the following problem: *Find experimentally the upper bound for the average $\langle D \rangle_\psi$ of observable D with respect to all possible states.* We stress that we are interested in the experimental verification of a theoretical prediction of QM.

We can easily find state u corresponding to max-eigenvalue λ of operator $\hat{D}$. Then, one of the max-states of operator $\hat{C}$ can be found with the aid of "C-superposition":

$$\phi_+ = (u + \lambda^{-1/2}\hat{C}u)/\|u + \lambda^{-1/2}Cu\|. \tag{6.34}$$

Finally, we prepare an ensemble of systems in quantum state ϕ_+ and perform C-measurement for these systems.

In the Bell-type scenario (for observables respecting the tensor product structure), $\hat{C} = \hat{\mathcal{B}}$ is the Bell operator, $\hat{D} = \hat{\mathcal{B}}^2$. In fact, the degree of incompatibility is encoded in the observable corresponding to operator $\hat{D}$. However, its straightforward measurement would involve measurement of observables corresponding to commutators. The latter is challenging. At the same time, eigenstates of D have the simple tensor product structure (separable states). They can easily be found. Then, eigenstates of the Bell operator can be generated as superpositions of Equation (6.34).

6.8 Interpretations of Bell Inequalities Violation

The point of Bell's theorem is that a local hidden variables theory can't reproduce the results of quantum theory. The implication is that only a nonlocal hidden variable theory, like Bohmian mechanics, can reproduce the correlations found in quantum theory and in the real world. Here, clearly, "nonlocality" refers to hidden variables theories, not to quantum theory. A very common misconception is to incorrectly associate the term with quantum theory (see Chapter 2 for detailed discussion). We hope that the argument presented in this chapter has convincingly demonstrated that this association is wrong.

The deeper message of this chapter is that the Bell inequality can be reinterpreted as *a condition on the quantum compatibility of local observables.* If local commutators vanish, the correlations are bounded just as they are when hidden variables are assumed to be local. The Bell-type inequalities have one interpretation for hidden variables theories (the classical case), and another lesser-known, but very interesting one for quantum theory.

Consequently, the outputs of experiments testing the violation of the Bell-type inequalities can also be interpreted in two different ways. The conventional interpretation is that with such tests one can compare classical and quantum theories. This book advertises the interpretation that such experiments are the tests of local incompatibility of quantum observables. This interpretation does not diminish the foundational value of these breakthrough experiments [28, 138, 159, 338, 369]. Complementarity is the basic feature of quantum observables. Tests on this feature are of great foundational importance.

At the same time, the conventional interpretation of the Bohm–Bell tests, *local realism versus quantum theory*, is opportunistic. To explain this statement, I start by emphasizing that Bell has never used the notion "local realism"; see the collection of his papers in [54]. Integration of realism (R) and locality (L) in the single notion blurs the conventional consequence of the violation of the Bell inequalities. One says that QM is incompatible with $R \wedge L$, that is, the validity of QM implies

$$\overline{R \wedge L} = \bar{R} \vee \bar{L},$$

nonrealism or nonlocality.

Bell was more precise in his conclusion; from the very beginning he claimed that R is the consequence of perfect correlations, that is, for him the validity of QM implies $\bar{L}$, nonlocality. And Bell's position is consistent. In contrast, many researchers interested in quantum foundations do not want to declare directly that the validity of QM implies nonlocality in combination with realism. They operate with the ambiguous notion of local realism and they are fine with its rejection due to the violation of the Bell inequalities.

On the other hand, if one agrees that R contradicts the fundamental quantum principles (and personally I do this), then QM is incompatible with R and the consequence of the Bell project that QM is incompatible with $R \wedge L$ is trivialized.

This is a good place to come back to de Broglie's critique of the Bell argument [100] (see also Lochak [266–268]). Due to de Broglie, Bell could escape the confrontation with the complementarity and uncertainty principles, if he would accept that the values of random variables which are used in derivations of the Bell inequalities in the hidden variable framework need not coincide with the outputs of physical observables measured in experiments. However, this would block the possibility of experimental verification of Bell's argument.

I state once again that the interpretation of the Bohm–Bell tests as the tests of observables' incompatibility presented in this chapter is most consitent one.

We also point out that the Bohm–Bell-type experiments played the crucial role in development of quantum technology: creation of efficient sources of entangled systems and photodetectors as well as transmission of quantum systems to long distances with minimal disturbance. As will be shown in Chapter 7, entanglement can be consistently treated without appealing to nonlocality.

This chapter should not be considered as directed against attempts to go beyond QM, by introducing "hidden variables." However, in such attempts, one should take into account the basic principles of QM – the complementarity and uncertainty principles.

7

Classical Entanglement: Lessons from Optics

7.1 Classical versus Quantum Entanglement and Nonlocality

We start with noting that in this chapter we use the standard definition of entanglement; see for example, [373]: *"A state is said to be entangled if it cannot be written as a convex sum of tensor product states."* (In Chapters 10, 13, and 14, we present generalizations of the notion of entanglement.) The classical electromagnetic field can be successfully used to model some basic features of genuine quantum physical systems (see, e.g., [135, 343, 344] and recent review [11] and the references herein). In particular, classical field modeling is a helpful tool for simulation in quantum information theory. However, the foundational output of quantum-like modeling with classical light and other types of waves is not straightforward. This chapter is directed towards clarification of the foundational meaning of "classical entanglement" and criticizing the widely spread view that "quantum nonlocality" (whatever it means) plays the crucial role in distinguishing classical and genuine quantum entanglement. The "quantum nonlocality" viewpoint was clearly formulated by Spreeuw in a widely cited paper [344] which is often mentioned by experimenters for foundational justification of their activity:

"It is found that the model system (*classical electromagnetic field*) can successfully simulate most features of entanglement, but fails to simulate quantum nonlocality. Investigations of how far the classical simulation can be pushed show that quantum nonlocality is the essential ingredient of a quantum computer, even more so than entanglement. The well known problem of exponential resources required for a classical simulation of a quantum computer, is also linked to the nonlocal nature of entanglement, rather than to the nonfactorizability of the state vector."

And then he continues:

"However, the (*classical-quantum*) analogy fails to produce effects of quantum nonlocality, thus signaling a profound difference between two types of entanglement: (i) 'true,' multiparticle entanglement and (ii) a weaker form of entanglement between different degrees of freedom of a single particle. Although these two types look deceptively similar in many respects, only type (i) can yield nonlocal correlations. Only the type (ii) entanglement has a classical analogy."

However, one can proceed without referring to mysterious quantum nonlocality – by taking into account that quantum theory is about acts of observations. These acts are characterized by *individuality and discreteness.* This crucial point in understanding of quantum theory was missed by authors discussing classical entanglement. They missed that the main deviation of classical light models from quantum theory is not only in the states, but in descriptions of measurement procedures (see also [253]). The classical and semiclassical descriptions of measurements are based on intensities of signals in different channels. The quantum description of measurements is based on counting the discrete events, clicks of detectors, with the aid of Born's rule. Operating with intensities obscures the problem of *coincidence detection.* We recall that quantum theory predicts that the relative probability of coincidence detection given by the coefficient of second-order coherence $g^{(2)}(0)$ is zero (for one-photon states), but in (semi)classical models $g^{(2)}(0) \geq 1$.

Genuine quantum theory differs from classical light models reproducing quantum correlations not by "quantum nonlocality," but by the magnitude of second-order coherence. Classical and semiclassical models were rejected long ago as the result of Grangier et al.'s [140] experiment on coincidence detection.

In fact, the main issue is the difference between classical and quantum superpositions, not entanglements. Our message is that each state-superposition has to be endowed with a proper measurement procedure. Otherwise superposition is just "thing-in-itself." Superposition endowed with the classical measurement procedure crucially differs from superposition endowed with the quantum measurement procedure (counting of discrete events – clicks of photodetectors). This difference in superpositions endowed with classical versus quantum measurement procedures is lifted to the level of entangled states, classical versus quantum: without measurement procedures they are also just "things-in-themselves" and as such they are not interesting for physics.

7.2 Inter–Intra System versus Quantum–Classical Entanglement

In review [11] it is stated:

... the name classical entanglement denotes the occurrence of some mathematical and physical aspects of quantum entanglement in classical beams of light.... the term 'classical' in the name classical entanglement, indicates the non-quantum nature of the excitation of the electromagnetic field.... A typical example thereof is given by a collimated optical beam with nonuniform polarization pattern.

We continue by citing [344]: “It should be noted that the choice of optical waves is not essential for the analogy. Other classical waves such as sound, water waves, or even coupled pendula could be used in principle.”

In short, classical entanglement is associated with the “nonquantum nature of the excitation of the electromagnetic field” or modes of classical waves of any origin. Generally, classical systems of any origin can be considered; see Chapter 14 on entanglement of classical Brownian motions. Following article [234], this chapter is directed against the statement that the difference between classical and genuine quantum entanglements is due to quantum nonlocality.

We remark that comparison classical-quantum entanglements is typically coupled to comparison intra–inter system entanglements. Intra-entanglement is between degrees of freedom of a single system and inter-entanglement is between degrees of freedom of two systems, S_1 and S_2. Classical entanglement modeling is possible only in the intrasystem context [11, 135, 343, 344]. One may conclude that this feature of entanglement plays the crucial role in distinguishing classical and quantum entanglements. This reasoning also leads to the conclusion that only intersystem entanglement is “true quantum entanglement” (since intrasystem entanglement can be generated even with classical fields). In this chapter, we show that classical intrasystem entanglement differs fundamentally not only from quantum intersystem entanglement, but even from quantum intrasystem entanglement. Thus, comparison classical–quantum entanglements has no relation to intra-inter comparison (and, hence, no relation to quantum nonlocality).

This comparative analysis of inter–intra system versus quantum–classical entanglements can be completed by the following remark. The impossibility of classical representation of intersystem entanglement is related only to the very special class of the field models elaborated in the classical entanglement project [11] (compare [208, 212]).

7.3 Photon Exists!

We start by citing the breakthrough paper of Grangier et al. [140]:

However, there has still been no test of the conceptually very simple situation dealing with single-photon states of the light impinging on a beam splitter. In this case, quantum mechanics predicts a perfect anti-correlation for photodetections on both sides of the beam splitter (a single-photon can only be detected once!), while any description involving classical fields would predict some amount of coincidences.

Following [140], we denote by p_1, p_2 the probabilities of detection in two channels after beam splitter and by p_c the coincidence probability. Then, by using the semiclassical model of detection, it is easy to show that

$$p_c \geq p_1 p_2. \tag{7.1}$$

This inequality means clearly that the classical coincidence probability p_c is always greater than the accidental coincidence probability, which is equal to $p_1 p_2$. The violation of inequality (7.1) thus gives an anti-correlation criterion, for characterizing a nonclassical behaviour of light (see [140]).

The crucial theoretical point is that in classical and semiclassical models the basic physical quantity is intensity of a signal. In the Grangier et al. experiment, these are $I(t)$, the intensity of imprinting on the beam splitter, and $I_1(t), I_2(t)$ are intensities of signals in the two output channels. The use of intensities, instead of counting of clicks, obscures the coincidence detection problem. We claim that this is not just a technicality, but a very important foundational issue. And we shall continue discussion in the following sections.

7.4 Quantum Measurements

Consider a quantum observable A represented by Hermitian operator $\hat{A}$ with purely discrete spectrum (a_i). By the spectral postulate of quantum mechanics any measurement of A produces one of the values a_i (as the result of interaction of a quantum system with an apparatus used for A-measurement). Thus quantum measurements are characterized by individuality of outputs. This crucial feature of quantum measurements was emphasized by Bohr who invented the notion of *phenomenon* [65] (see Section 3.2.3 for extended discussion):

> ... in actual experiments, all observations are expressed by unambiguous statements referring, for instance, to the registration of the point at which an electron arrives at a photographic plate.... the appropriate physical interpretation of the symbolic quantum mechanical formalism amounts only to predictions, of determinate or statistical character, pertaining to individual phenomena.... .

Thus, although quantum theory produces statistical predictions, its observables generate individual phenomena. Discreteness of detection events is the fundamental feature of quantum physics justifying existence of quantum systems, carriers of quanta. Although the axiomatic of quantum theory does not contain the special postulate on discrete clicks and the frequency interpretation of probabilities, Bohr's statement highlights that genuine quantum phenomena are individual phenomena. In Chapter 3 we discussed this issue as a part of the Bohr complementarity principle.

One may point to the existence of quantum observables with continuous spectra. The problem of their measurement was analyzed in detail by von Neumann [365].

His analysis implies that measurement of an observable with continuous spectrum has to be reduced to measurements of observables with discrete spectra approximating it. This is the complex foundational issue and we shall not go into a deeper discussion; our considerations are restricted to observables with discrete spectra.

In the classical wave framework, the origin of the analog of the quantum Born's rule, or Born's rule for intensities, is straightforward. If a classical wave has two orthogonal components, that is,

$$\Phi(x) = \Phi_1(x) + \Phi_2(x), \tag{7.2}$$

with intensities I_1 and I_2, then corresponding probabilities can be expressed in the form $p_j = I_j/(I_1 + I_2), j = 1, 2$, and intensities are given by the "classical Born's rule":

$$I_j = \|\Phi_j\|_{L_2} = \int |\Phi_j(x)|^2 dx. \tag{7.3}$$

However, it is a separate question whether the coefficients p_j can really be interpreted as probabilities of discrete events.

In papers on classical entanglement, there are considered expansions of state-vectors with respect to orthonormal bases in complex Hilbert spaces.[1] Such expansions may give the impression that the standard quantum mechanical scheme of measurement can be applied. This is not the case. For classical signals, it is impossible to project the initial state on the state corresponding to one concrete outcome. In the two-slit experiment with classical waves, a wave propagating from the source passes both slits at the same time.

Finally, we remark that in classical field theory the method of complex Hilbert space started to be used even before the appearance of quantum mechanics. We can mention, for example, the *Riemann–Silberstein representation*, $\Psi(x) = E(x) + iH(x)$, for the classical electromagnetic field. In this representation, the Maxwell equation has the form of the Schrödinger equation. So, studies on classical entanglement are consistent with complex Hilbert space analysis of classical signals.

7.5 Classical versus Quantum Superposition

From my viewpoint, the misleading journey toward classical entanglement starts already with identifying classical and quantum superpositions. Physically these superpositions are totally different, in spite of the possibility of representing them by the same mathematical expression.

[1] The tensor product structure is typically emphasized. But this is not the main issue; see Section 7.5 on classical versus quantum interpretation of states' superposition.

Consider a classical electromagnetic field with n orthogonal modes corresponding to frequencies $(\nu_j, j = 1, \ldots, n)$ with complex amplitudes (C_j). This field can be represented in n-dimensional complex Hilbert space with the basis $(e_j \equiv |\nu_1\rangle, j = 1, \ldots, n)$:

$$\Psi = \sum_j C_j |\nu_j\rangle. \tag{7.4}$$

This vector can be normalized:

$$|\psi\rangle = \sum_j c_j |\nu_j\rangle, \tag{7.5}$$

where $c_j = C_j / \sqrt{\sum_j |C_j|^2}$.

What is the main difference between classical field superposition (7.5) and the genuine quantum superposition?

The main difference is in measurement procedures determining probabilities $p_j = |c_j|^2$. For the classical field, it is impossible to detect discrete clicks in n channels without coincidence detections, where the degree of coincidence is determined by coefficient $g^{(2)}(0)$. Thus, to see the difference between classical and quantum light, one need not consider formal entanglement expressions corresponding for different degrees of freedom. It is sufficient to consider one degree of freedom and superposition.

Following papers on classical entanglement, consider two degrees of freedom of the classical electromagnetic field which can be jointly measured. The four-dimensional complex Hilbert space contains states of this field that are nonseparable, and formally they can be treated as entangled. Here "entangled" is understood purely mathematically, as the special form of representation in complex Hilbert space endowed with the tensor product structure. As in the case of superposition, the devil is not in the state, but in measurement. For the classical electromagnetic field, photodetectors cannot produce phenomena, in Bohr's sense. The measurement procedure suffers from coincidence detection.

The preceding critique of attempts to couple studies on classical entanglement with quantum theory can also be applied to attempts to couple classical random field correlations violating Bell-type inequalities with the original Bell argument [53, 54]. Bell applied classical probability theory to derive his inequality. The later was used to compare the classical probabilistic representation of correlations with the quantum theoretical description.

We recall that in classical probability theory observables are represented by random variables, (measurable) functions on sample space. Denote the latter by symbol Λ (although mathematicians typically use the symbol Ω). Then a random variable $\xi : \Lambda \to \mathbb{R}$ and by definition of a function it takes only one value $\xi(\lambda)$ for each $\lambda \in \Lambda$. Thus, generation of clicks in both channels means that it is impossible to describe such measurements by classical random variables.

We remark that originally (following EPR paper [115]) Bell was interested in explanation of perfect correlations. In his original inequality [53, 54], it was assumed that ranges of values of quantum and classical observables should coincide, that is, the range of values is the two-point set $\{\pm 1\}$. A classical random variable is a function $\xi : \Lambda \to \{\pm 1\}$. And if one would accept that for some set of λs, ξ is multivalued, that is, at the same time $\xi(\lambda) = -1$ and $\xi(\lambda) = +1$, then classical probability theory stops working.

7.6 Concluding Remarks on Classical and Quantum Entanglements

The aim of this chapter is to distance the technical impact of "classical entanglement" research (both for theory and experiment) [11] from its misleading interpretation, as supporting "quantum nonlocality" [344]. We list the main conclusions from our analysis of interrelation of classical and quantum entanglements:

- The main issue is the difference between classical and quantum superpositions. It can be explained by Grangier et al.'s experiment [140] on photon existence.
- The distinguishing feature of quantum measurements is discreteness and individuality of outcomes (as expressed in Bohr's notion of phenomenon).
- Derivation of quantum(like) correlations with classical entanglement [11, 232] implies that the Hilbert space mathematical formalism should be distinguished from quantum physics.
- Classical entanglement is not consistent with Bell's hidden variables theory: coincidence detection blocks the use of random variables.

This comparison of classical and quantum entanglements and critique of the "quantum nonlocality" interpretation of their difference is the main output of this chapter.

To support the position expressed in this chapter, we point to the review of Korolkova and Leuchs [253], which is the important step toward resolution of the interpretation problems related to interrelation of classical and quantum entanglements. Its authors no longer refer to quantum nonlocality (compare with the previous review [11]). They recognize that the main issue is not the impossibility of generating intersystem entanglement with classical optics. The main issue is the difference between intra system entanglements, classical versus quantum. And Korolkova and Leuchs, as well as the author of this paper, also emphasize the role of measurement procedures in distinguishing two types of entanglement.

They made the following remark on an intra-entangled state of a genuine quantum system:

> "This state is the quantum entangled state of type $|01>+|10>$ a strict correlation of one photon in one arm and no photon in the other or vice versa."

I would just add that, in fact, the root of the problem lies already in classical versus quantum interpretation of superposition-state $|0>+|1>$.

Finally, we point to a connection with the corresponding counterpart of the Bohr complementarity principle, the individuality principle. By Bohr, photon exists! And the output of a measurement is assigned to this concrete photon; QM is about such individual assigning of outputs (quantum phenomena). This is a origin of discreteness of quantum measurements. That is why measurement of the intensity of á beam of classical light is not a quantum phenomenon. This is a good place to mention RWR (reality without realism) interpretation of QM (Section 4.2). Realization of the classical entanglement project shows the importance of this interpretation emphasizing the role of individual events in quantum physics. Beams of classical electromagnetic radiation cannot be split in such individual events. Such beams are not quantum phenomena.

8

Original Bell Inequality and Experiment

8.1 Original Bell Inequality for Precise Correlations

We emphasize once again that this chapter is directed to experimenters; one may hope that they would be finally interested in testing OB. We do not disturb experimenters by pointing to the related deep foundational problems, as in Chapters 2, 6, and 3. Therefore, in this chapter we present the standard viewpoint on the violation of the Bell inequality and the EPR argument.

Bell proposed [53, 54] the probabilistic test based on the EPR argument [115]. The problem under consideration can be formulated as follows. Einstein, Podolsky, and Rosen proved that quantum mechanics (QM) is incomplete, since its formalism does not represent the EPR elements of reality. Suppose one wants to construct a subquantum theory completing QM. Such a theory should match with statistical predictions of QM and at the same time *it should describe EPR's elements of reality.* Can such a theory be local? (As Einstein dreamed for.)

Bell proposed a test based on an inequality for correlations. This inequality will be called the original Bell (OB) inequality. This inequality was proved under the following crucial assumption about coupling the Bell model with hidden variables and the EPR elements of reality.

For the singlet state (as for the original EPR state), spin projections are EPR's elements of reality. These *elements of reality are equal to measurement outcomes* (elements of reality for S_2 are measurement outcomes for S_1). Hence, *values of variables of a subquantum theory beyond the singlet state can be identified with possible outcomes of measurements.* Therefore, for the singlet state, subquantum and quantum correlations can be identified.

However, this beautiful theoretical scheme supporting nonlocal hidden variable theories did not match the experimental framework of that time, since the degree of (anti-)correlations (for the same setting on both sides) was not so high. This problem was solved by transition from the OB inequality to the CHSH inequality [83] or similar inequalities: the CH74 inequality [84, 85] or the Eberhard inequality [113] (see [214] for comparison of these inequalities). Derivations of such inequalities are not based on the assumption of perfect (anti-)correlations. (For convenience, later we shall compare the OB inequality only with the CHSH inequality, but similar comparison can be done for other "CHSH-like inequalities" such as the CH74 inequality and the Eberhard inequality). Foundational difference between the OB and CHSH-like inequalities is briefly discussed in appendix.

The main issue which we want to highlight in this chapter is that (as was pointed out by Pitowsky [307]) the OB inequality provides a higher degree of violations of classicality than the CHSH inequality. For the CHSH inequality, the fraction of the quantum (Tsirelson) bound

$$Q_{\mathrm{CHSH}} = 2\sqrt{2} \tag{8.1}$$

to the classical bound $C_{\mathrm{CHSH}} = 2$, that is,

$$F_{\mathrm{CHSH}} = \frac{Q_{\mathrm{CHSH}}}{C_{\mathrm{CHSH}}} = \sqrt{2} \tag{8.2}$$

is less than the fraction of the quantum bound for the OB inequality

$$Q_{\mathrm{OB}} = \frac{3}{2} \tag{8.3}$$

to the classical bound $C_{\mathrm{OB}} = 1$, that is,

$$F_{\mathrm{OB}} = \frac{Q_{\mathrm{OB}}}{C_{\mathrm{OB}}} = \frac{3}{2}. \tag{8.4}$$

Thus, by violating the OB inequality it is possible to approach a higher degree of deviation from classicality.

The main problem for performing an experimental test is that the OB inequality is derived under the assumption of perfect (anti-)correlations. Therefore it was impossible to perform experiments to check violation of the OB inequality. However, recent years have been characterized by the amazing development of quantum technologies. Technological improvements led to the loophole-free tests of the Bell-type inequalities (the CHSH, Eberhard, and Clauser-Horne inequalities).[1]

One possible way to test violation of the OB inequality is to follow the quantum optics path initiated by Aspect [30]. Nowadays there exist sources producing with very high probability the pairs of photons in the singlet state. Moreover, the efficiency of photon detectors was improved tremendously. Therefore one can hope to violate the OB inequality, although this is still the real challenge; see Section 8.6.

[1] As was expected by Bell, these experiments did not change the views of those who did not accept the conventional interpretation of experimental outputs.

In any event one can proceed under the *fair sampling assumption,* that is, to solve first the problem of (anti-)correlations.

Another possibility is to test the OB inequality by using the scheme of the Hensen et al. experiment [159]. This experimental scheme does not suffer from inefficiency of detection. However, it seems that the quality of preparation of the singlet state is still insufficient to perform the experimental test to violate the OB inequality; see Section 8.6.

This chapter is based on articles [226, 264, 307, 327]. Its aim is to collect these results in one text and consider their experimental consequences in the light of recent tremendous achievements of modern quantum information technologies.

In Section 8.3 we present probabilistic calculations to estimate the probability of preparation of the singlet state which is sufficient to test violation of the OB-inequality under the assumption of 100% of the detection efficiency. Theorem 8.2 implies that experimenters have to be able to prepare an ensemble in which more than 75% of pairs are in the singlet state, see also Ryff's paper [327]. Thus the existing photon sources of high quality provide the possibility of testing the OB inequality, at least under the assumption of fair sampling. In Section 8.4 we present probabilistic calculations to estimate the minimal efficiency of detection which is sufficient to test violation of the OB-inequality under the assumption of 100% fidelity in preparation of the singlet state. By Theorem 8.3 the efficiency of the joint detection should be higher than 88.9%; see Larsson's paper [264] for the original derivation of this bound. And finally, in Section 8.5 we combined the results of Sections 8.3 and 8.4. By combining 98% level of anti-correlations with 90% level of detection efficiency, one can test violation of the OB inequality.

We remark that generalized (perturbed) Bell's inequalities which are similar to inequalities obtained in Theorems 8.2–8.4 were actively used by one of the coauthors in foundational studies [179]–[183].

Successful experimental testing of violation of the OB inequality would be an important (although very challenging) contribution to clarification of quantum foundations.

8.2 Classical versus Quantum Bounds

We proceed in accordance with Bell [53, 54]. Let p be a probability measure on the space of hidden variables Λ. (Bell used the symbol ρ.) We model measurements on a pair of systems S_1 and S_2 with the aid of random variables $A_s(\lambda)$ and $B_s(\lambda)$, where the parameter s labels settings of measurement devices, $s = a, b, c$.

Consider correlations of these random variables given by the integrals:

$$P(a,b) = \int_\Lambda A_a(\lambda) B_b(\lambda) dp(\lambda). \tag{8.5}$$

It is assumed that these random variables take values ± 1 and that the random variables corresponding to measurements on S_1 and S_2 are anti-correlated:

$$P(a,a) = \int_\Lambda A_a(\lambda)B_a(\lambda)dp(\lambda) = -1. \tag{8.6}$$

Under these assumptions Bell derived [53, 54] the following inequality:

$$|P(a,b) - P(a,c)| - P(b,c) \leq 1, \tag{8.7}$$

see also Section 8.3 for details. We call it the *original Bell inequality* or OB inequality.

This hidden variable model was confronted with spin measurements represented in QM by the spin operators $\hat{\sigma} \cdot s$. In this case s is the unit vector in $\mathbb{R}^3$ representing the axis of spin projection. Thus pairwise correlations for spin operators are compared with correlations for random variables. To distinguish measurements on systems S_1 and S_2, we shall use symbols $\hat{\sigma}_1 \cdot a$ and $\hat{\sigma}_2 \cdot b$.

The OB inequality implies that, for classical correlations, the upper bound C_{OB} for the expression $\Delta = |P(a,b) - P(a,c)| - P(b,c)$ equals one. Now consider the the quantum case. To get perfect anti-correlations, we proceed with the singlet state

$$\Psi = (|+-\rangle - |-+\rangle)/\sqrt{2}. \tag{8.8}$$

For this state, we have

$$P_Q(a,b) = \langle \hat{\sigma}_1 \cdot a \otimes \hat{\sigma}_2 \cdot b \rangle = -\langle a|b\rangle. \tag{8.9}$$

One can find the quantum bound for the expression

$$\Delta_Q(a,b,c) = |P_Q(a,b) - P_Q(a,c)| - P_Q(b,c) = |\langle a|b\rangle - \langle a|c\rangle| + \langle b|c\rangle.$$

Theorem 8.1 $Q_{OB} = \max_{a,b,c} \Delta_Q(a,b,c) = \frac{3}{2}.$

Proof Under the suitable parametrization $\Delta_Q(a,b,c)$ can be represented as

$$\Delta_Q(\phi_1,\phi_2,\theta) = 2|\sin\phi_1 \sin\phi_2 \sin\theta| + 1 - 2\sin^2\phi_1. \tag{8.10}$$

It is easy to find that the maximal value of this function equals 3/2.

Consider, for example three vectors in the same plane, $a=(1,0), b=(1/2, -\sqrt{3}/2), c=(-1/2,-\sqrt{3}/2)$. Then $P(a,b)= -\langle a|b\rangle = -1/2, P(a,c)= -\langle a|c\rangle = 1/2, P(b,c)= -\langle b|c\rangle = -1/2$. Hence, $\Delta_Q(a,b,c) = 3/2$.

Hence, we proved the equality (8.4), $F_{\text{OB}} = 3/2$.

Pitowsky [307] presented the same argument by using a slight modification of the OB inequality (8.7).

8.3 Imperfection of Anti-correlations

Here we proceed in Bell's framework based on classical probability under the assumption that the random variables corresponding to measurements on S_1 and S_2 are anti-correlated. As Bell pointed out, this is possible only if the following equality holds:

$$A_a(\lambda) = -B_a(\lambda), \tag{8.11}$$

except a set of measure zero. Bell derived inequality (8.7) under this assumption of perfect (up to measure zero) anti-correlation. It is easy to modify this equality under assumption of imperfect anti-correlations. Here we follow the original paper [327], but we proceed in measure theoretic framework. Uses of the frequentist approach (as in paper [327]) have been objected to by a few authors; see, for example, [179].

Suppose that, for each a, there exists a subset Λ_a of Λ such that Eq. (8.11) holds for all λ from Λ_a and, for the set $\Lambda_a' = \Lambda \setminus \Lambda_a$, we have:

$$p(\Lambda_a') \leq \epsilon. \tag{8.12}$$

Since random variables are dichotomous, on the set Λ_a',

$$A_a(\lambda) = B_a(\lambda). \tag{8.13}$$

Now on Λ_b we have

$$\begin{aligned} A_a(\lambda)B_c(\lambda) - A_a(\lambda)B_b(\lambda) &= A_a(\lambda)A_b(\lambda)A_b(\lambda)B_c(\lambda) + A_a(\lambda)A_b(\lambda) \\ &= A_a(\lambda)A_b(\lambda)[1 + A_b(\lambda)B_c(\lambda)]. \end{aligned}$$

On Λ_b' we have

$$\begin{aligned} &A_a(\lambda)B_c(\lambda) - A_a(\lambda)B_b(\lambda) \\ &\quad = A_a(\lambda)A_b(\lambda)A_b(\lambda)B_c(\lambda) + A_a(\lambda)A_b(\lambda) - 2A_a(\lambda)A_b(\lambda) \\ &\quad = A_a(\lambda)A_b(\lambda)[1 + A_b(\lambda)B_c(\lambda)] - 2A_a(\lambda)A_b(\lambda). \end{aligned}$$

Thus

$$P(a,c) - P(a,b) = \int_\Lambda A_a(\lambda)A_b(\lambda)[1 + A_b(\lambda)B_c(\lambda)]dp(\lambda) - 2\int_{\Lambda_b'} A_a(\lambda)B_b(\lambda)dp(\lambda).$$

Hence we proved the following theorem:

Theorem 8.2 *Under Assumption (8.12) the following inequality for classical correlations holds:*

$$|P(a,b) - P(a,c)| - P(b,c) \leq 1 + 2\epsilon. \tag{8.14}$$

By introducing the parameter $\gamma = 1 - \epsilon$, we write (8.14) as

$$|P(a,b) - P(a,c)| - P(b,c) \leq 3 - 2\gamma. \tag{8.15}$$

By Theorem 8.1 we have the following inequality: $3 - 2\gamma < 1.5$; namely, $\gamma > 3/4 = 0.75$: Thus to be able to test properly the OB inequality one has to be able to produce an ensemble of pairs of quantum systems in which the percentage of precisely (anti-)correlated pairs will be higher than

$$\gamma = 75\%. \tag{8.16}$$

8.4 Detection Efficiency

For the OB inequality, the issue of the detection efficiency was studied in details by Larsson [264]. Here we present a similar consideration, but in a slightly different form, which is consistent with the preceding presentation of the role of imperfection of correlations. Denote the set of hidden variables for which the pair $A_a(\lambda), B_b(\lambda)$ is detected by the symbol Γ_{ab}. The main parameter of the experimental interest is the probability of joint detection of a pair, $p(\Gamma_{ab})$. For simplicity of considerations, we assume that this probability does not depend on the pair of settings, namely,

$$\eta \equiv p(\Gamma_{ab}). \tag{8.17}$$

Then correlation conditioned on the pairwise detection is given by

$$\tilde{P}(a,b) = \frac{1}{\eta}\int_{\Gamma_{ab}} A_a(\lambda)B_b(\lambda)dp(\lambda). \tag{8.18}$$

Theorem 8.3 *Under the assumptions of 100% perfect anti-correlations and set-independent joint detection efficiency, see (8.17), the following OB inequality for detectable correlations holds:*

$$|\tilde{P}(a,b) - \tilde{P}(b,c)| - \tilde{P}(a,c) \leq \frac{(4-3\eta)}{\eta}. \tag{8.19}$$

Proof From (8.18) we get $\eta\tilde{P}(a,b) = P(a,b) - u_{ab}$, where

$$u_{ab} = \int_{\Lambda\setminus\Gamma_{ab}} A_a(\lambda)B_b(\lambda)dp(\lambda)$$

and, hence, $|u_{ab}| \leq (1-\eta)$. We have

$$\eta\Big(|\tilde{P}(a,b) - \tilde{P}(b,c)| - \tilde{P}(a,c)\Big)$$
$$\leq |P(a,b) - P(b,c)| - P(b,c) + 3(1-\eta) \leq 1 + 3(1-\eta).$$

By dividing both sides of this inequality by η, we obtain (8.19).

To be able to violate inequality (8.19), the experimenter has to have a sufficiently high detection efficiency, such that $\frac{(4-3\eta)}{\eta} < \frac{3}{2}$, namely, $\eta > 8/9 = 0.889$. Thus the *efficiency of the joint detection should be higher than 88.9%*. This result coincides with the corresponding result from [264], p. 57. Thus the detection efficiency should be higher than in the experimental tests for the CHSH inequality [133], [264].

8.5 Imperfection of Anti-correlations and Detection Efficiency

Theorem 8.4 *Under Assumptions (8.17) and (8.12), the following experimentally testable version of the OB inequality holds:*

$$|\tilde{P}(a,b) - \tilde{P}(b,c)| - \tilde{P}(a,c) \le \frac{(4+2\epsilon-3\eta)}{\eta}. \tag{8.20}$$

Proof We have

$$\eta\Big(|\tilde{P}(a,b) - \tilde{P}(b,c)| - \tilde{P}(a,c)\Big) \le |P(a,b) - P(b,c)| - P(b,c) + 3(1-\eta)$$
$$\le 1 + 2\epsilon + 3(1-\eta).$$

To be able to violate inequality (8.20), the experimenter has to have sufficiently high anti-correlations and the detection efficiency, such that $\frac{4+2\epsilon-3\eta}{\eta} < \frac{3}{2}$. It is convenient to introduce a new parameter $\kappa = 1-\epsilon$. Then the generalized OB inequality has the form

$$|\tilde{P}(a,b) - \tilde{P}(b,c)| - \tilde{P}(a,c) \le \frac{(6-2\gamma-3\eta)}{\eta} \tag{8.21}$$

and the condition for possible violation can written as

$$4\gamma + 9\eta > 12. \tag{8.22}$$

For example, let $\gamma = 0.98$. Then $\eta > 0.9$. Thus by approaching the 98% level of anti-correlations and the 90% level of the detection efficiency, the experimenter can test the OB inequality.

8.6 Stimulating Experimental Research on the Original Bell Inequality

Modern quantum technology provides the sources producing photons in the singlet state with very high probability, up to 98% of the generated ensemble of pairs. From this viewpoint it is promising to perform the experimental test for the OB inequality by using entangled photons; compare with experiments [138, 338] to violate the CHSH-like inequalities (the Eberhard and CH inequalities). However, as we have seen, tests for the OB inequality demand higher detection efficiency than tests for the CHSH inequality. We recall (see also [214] for discussion) that the detection efficiency is not reduced to the efficiency of photodetectors. Although nowadays there are available photodetectors. having an efficiency close to 100%, this does not solve the problem of the detection efficiency. A weak element of the experimental setup based on quantum optics is polarization beam splitter, where one can lose 8–13% of photons. This loss can play a crucial role in attempts to lift the detection efficiency from 83% [133, 264] in the tests for the CHSH inequality to approximately 90% in the planned experimental test for the OB inequality.

It may be reasonable to proceed under the assumption of *fair sampling*. And such a project seems to be realizable.

If one wants to proceed without the fair sampling assumption, then it is very promising to test violation of the OB inequality by using entangled electron spins, namely the scheme of Hensen et al.'s [159] experiment which was performed for the CHSH inequality. As reported in [159], the parameter γ in inequality (8.21) can be selected as $\gamma \approx 0.92$. It exceeds the bound $\gamma = 0.75$; see Eq. (8.16). So, it seems that such an experiment can be performed already today.

9

Maximal Violation of the Original Bell Inequality: Two-Qubit and -Qutrit States

9.1 Singlet and Nonsinglet States

As in Chapter 8, we start with pointing to a practically unknown paper of Pitowsky [307] where he claims that by violating the original Bell inequality and its generalizations it would be possible to approach a higher degree of nonclassicality than for the CHSH-like inequalities.

This claim is built upon the fact that, for the CHSH inequality, the fraction $F_{\text{CHSH}}^{(\hat{\rho}_d)}$ of the quantum (Tsirelson) upper bound [349, 350] $2\sqrt{2}$ to the classical one is equal to $F_{\text{CHSH}}^{(\hat{\rho}_d)} = \sqrt{2}$ for a bipartite state $\hat{\rho}_d$ of an arbitrary dimension $d \geq 2$, whereas, for the original Bell inequality, the fraction $F_{\text{OB}}^{(\hat{\rho}_{singlet})}$ of the quantum upper bound for the two-qubit singlet ($d = 2$) to the classical bound (equal to one; see Section 9.2) is given by

$$F_{\text{OB}}^{(\hat{\rho}_{singlet})} = \frac{3}{2} > \sqrt{2} = F_{\text{CHSH}}^{(\hat{\rho}_d)}, \quad \forall d \geq 2. \tag{9.1}$$

The rigorous mathematical proof of the least upper bound $\frac{3}{2}$ on the violation of the original Bell inequality by the two-qubit singlet was presented in Chapter 8, where the considerations were restricted only to the two-qubit singlet case.

However, for the violation $F_{\text{OB}}^{(\hat{\rho}_d)}$ of the original Bell inequality by a two-qudit state $\hat{\rho}_d$ exhibiting perfect correlations/anti-correlations, the CHSH inequality implies for all $d \geq 2$ the upper bound $(2\sqrt{2} - 1)$ (see Section 9.3) and the latter upper bound is more than the least upper bound $\frac{3}{2}$ proved [226, 307] for the two-qubit singlet.

We stress that quantum nonlocality is not equivalent [373] to quantum entanglement and that larger violations of Bell inequalities can be reached [175] by states with less entanglement. Therefore, the proof [226] that, for the two-qubit singlet state (which is maximally entangled), the least upper bound on violation of the original Bell inequality is equal to $\frac{3}{2}$ does not automatically mean that $\frac{3}{2}$ is the least upper bound on violation of the original Bell inequality for all two-qubit states. Moreover, the proof of the least upper bound $\frac{3}{2}$ on violation of the original Bell inequality by the singlet state has no consequence for quantifying violation of this inequality by a two-qudit state of an arbitrary dimension $d \geq 2$.

In this chapter we follow article [227] and rigorously prove that under Alice and Bob spin measurements, the least upper bound $\frac{3}{2}$ on the violation of the original Bell inequality holds for all two-qutrit and all two-qubit states exhibiting perfect correlations/anti-correlations.

In article [227], this result was essentially strengthened. For an even qudit dimension $d \geq 2$, there was specified (Proposition 9.1) a class of two-qudit states exhibiting perfect correlations/anti-correlations and it was proven (Theorem 9.3) that, for each two-qudit state from this class, the maximal violation of the OB inequality over all traceless qudit observables with eigenvalues ± 1 is also bounded from above by the value 3/2. However, the technique used in [271] is very advanced and we do not include the results of the article in this book. The results obtained in [271] are important steps towards proving the 3/2 upper bound conjecture on OB inequality violation for all bipartite quantum states with perfect correlations/anti-correlations.

In Section 9.2, we present the condition [270] on perfect correlations or anti-correlations for joint probabilities and prove, under this condition, the validity of the original Bell inequality in the local hidden variable (LHV) frame. This general condition is true for any number of outcomes at each site and reduces to Bell's perfect correlation/anti-correlation condition [53] on the correlation function only in the case of Alice and Bob outcomes ± 1.

In Section 9.3, we analyse violation of the original Bell inequality by a two-qudit quantum state and show that, for all dimensions of a two-qudit state exhibiting perfect correlations/anti-correlations and any three qudit observables, the maximal violation of the original Bell inequality cannot exceed the value $(2\sqrt{2} - 1)$.

In Section 9.4, we introduce (Proposition 9.2) the general class of symmetric two-qubit density operators which guarantee perfect correlation or anti-correlation of Alice and Bob outcomes whenever some (the same) spin observable is measured at both sites. We prove (Theorem 9.3) that, for all states from this class, the maximal violation of the original Bell inequality is upper bounded by $\frac{3}{2}$ and specify the two-qubit states for which this quantum upper bound is attained.

In Section 9.5, we consider Alice and Bob spin measurements on two-qubit states. This case is more complicated. Here, we are also able to prove the upper bound $\frac{3}{2}$ for all spin measurements on an arbitrary two-qubit state, but we have not yet been able to find two-qubit states for which this upper bound is attained. In the future, we plan to study this problem as well as to consider spaces of higher dimensions.

9.2 Original Bell Inequality in the General Case

Both Bell's proofs [53, 54] of the original Bell inequality in a local hidden variable (LHV) frame are essentially built up on two assumptions: a dichotomic character of Alice's and Bob's measurements plus the perfect correlation or anti-correlation of their outcomes for a definite pair of their local settings. Specifically, the latter assumption is abbreviated in quantum information as the condition on perfect correlations or anti-correlations.

In this section, we present the proof [270] of the original Bell inequality in the LHV frame for any numbers of Alice and Bob outcomes in $[-1, 1]$ and under the condition which is more general than the one introduced by Bell.

Consider an arbitrary bipartite correlation scenario with two measurement settings a_i, b_k, $i, k = 1, 2$, and any numbers of discrete outcomes $\lambda_a, \lambda_b \in [-1, 1]$ at Alice and Bob sites, respectively. This bipartite scenario is described by four joint measurements (a_i, b_k), $i, k = 1, 2$, with joint probability distributions $P_{(a_i,b_k)}$ of outcomes in $[-1, 1]^2$. Notation $P_{(a_i,b_k)}(\lambda_a, \lambda_b)$ means the joint probability of the event that, under a measurement (a_i, b_k), Alice observes an outcome λ_a while Bob—an outcome λ_b. For the general framework on the probabilistic description of an arbitrary N-partite correlation scenario with any numbers of measurement settings and any spectral type of outcomes at each site, discrete or continuous, see [269].

For a joint measurement (a_i, b_k), we denote by

$$\langle \lambda_{a_i} \rangle = \sum_{\lambda_a, \lambda_b \in [-1,1]} \lambda_a P_{(a_i,b_k)}(\lambda_a, \lambda_b), \quad \langle \lambda_{b_k} \rangle = \sum_{\lambda_a, \lambda_b \in [-1,1]} \lambda_b P_{(a_i,b_k)}(\lambda_a, \lambda_b) \tag{9.2}$$

the averages of outcomes, observed by Alice and Bob, and by

$$\langle \lambda_{a_i} \lambda_{b_k} \rangle = \sum_{\lambda_a, \lambda_b \in [-1,1]} \lambda_a \lambda_b P_{(a_i,b_k)}(\lambda_a, \lambda_b) \tag{9.3}$$

the average of the product $\lambda_a \lambda_b$ of their outcomes.

Let, under a joint measurement (a_i, b_k), Alice and Bob outcomes satisfy the conditions that either the event

$$\{\lambda_a = \lambda_b\} := \{(\lambda_a, \lambda_b) \in [-1, 1]^2 \mid \lambda_a = \lambda_b\} \tag{9.4}$$

or the event

$$\{\lambda_a = -\lambda_b \neq 0\} := \{(\lambda_a, \lambda_b) \in [-1, 1]^2 \mid \lambda_a = -\lambda_b \neq 0\} \tag{9.5}$$

is observed with certainty, that is [270],

$$P_{(a_i,b_k)}(\{\lambda_a = \lambda_b\}) = \sum_{\lambda_a = \lambda_b} P_{(a_i,b_k)}(\lambda_a, \lambda_b) = 1 \tag{9.6}$$

or

$$P_{(a_i,b_k)}(\{\lambda_a = -\lambda_b \neq 0\}) = \sum_{\lambda_a = -\lambda_b \neq 0} P_{(a_i,b_k)}(\lambda_a, \lambda_b) = 1, \tag{9.7}$$

respectively.

To demonstrate that, under conditions (9.6) or (9.7) on probabilities, outcomes of Alice and Bob are perfectly correlated or anti-correlated, consider, for example, the plus sign case (9.6). From (9.6) it follows that, for arbitrary $\lambda_a \neq \lambda_b$, the joint probability

$$P_{(a_i,b_k)}(\lambda_a, \lambda_b)|_{\lambda_a \neq \lambda_b} = 0. \tag{9.8}$$

Hence, under a joint measurement (a_i, b_k), the marginal probabilities at Alice and Bob sites are given by

$$P_{a_i}(\lambda_a) = \sum_{\lambda_b} P_{(a_i,b_k)}(\lambda_a, \lambda_b) = P_{(a_i,b_k)}(\lambda_a, \lambda_b)|_{\lambda_b = \lambda_a}, \quad \forall \lambda_a, \tag{9.9}$$

$$P_{b_k}(\lambda_b) = \sum_{\lambda_a} P_{(a_i,b_k)}(\lambda_a, \lambda_b) = P_{(a_i,b_k)}(\lambda_a, \lambda_b)|_{\lambda_a = \lambda_b}, \quad \forall \lambda_b.$$

Therefore, under this joint measurement, at Alice and Bob sites the marginal probability distributions of observed outcomes $\lambda \in [-1, 1]$ coincide $P_{a_i}(\lambda) = P_{b_k}(\lambda)$ and, given, for example, that Alice observes an outcome $\lambda_a = \lambda_0$, Bob observes the outcome $\lambda_b = \lambda_0$ with certainty, that is, the conditional probability $P_{b_k}(\lambda_b = \lambda_0 \mid \lambda_a = \lambda_0) = 1, \forall \lambda_0$. Also, under condition (9.6), the Pearson correlation coefficient γ_{cor}, considered in statistics, is given by

$$\gamma_{cor} = \frac{\sum_{\lambda_a,\lambda_b} (\lambda_a - \langle \lambda_a \rangle)(\lambda_b - \langle \lambda_b \rangle) P_{(a_i,b_k)}(\lambda_a, \lambda_b)}{\sqrt{\sum_{\lambda_a} (\lambda_a - \langle \lambda_a \rangle)^2 P_{a_i}(\lambda_a)} \sqrt{\sum_{\lambda_b} (\lambda_b - \langle \lambda_b \rangle)^2 P_{b_k}(\lambda_b)}} = 1. \tag{9.10}$$

Therefore, under the plus sign condition (9.6), Alice and Bob outcomes are perfectly correlated also in the meaning generally accepted in statistics.

The minus sign case (9.7) is considered quite similarly and results in the relation $P_{a_i}(\lambda) = P_{b_k}(-\lambda), \forall \lambda \in [-1, 1]$, for marginal distributions of Alice and Bob, the relation $P_{b_k}(\lambda_b = -\lambda_0 \mid \lambda_a = \lambda_0) = 1, \forall \lambda_0$, for the conditional probability and the Pearson correlation coefficient $\gamma_{cor} = -1$. All this means the perfect anti-correlation of Alice and Bob outcomes.

For a joint measurement with outcomes ± 1, the general conditions (9.6), (9.7) are equivalently represented by the condition on the product expectation

$$\langle \lambda_a \lambda_b \rangle = \pm 1. \tag{9.11}$$

respectively, introduced originally in Bell [54]. However, for any number of outcomes in $[-1, 1]$ at both sites, Alice and Bob outcomes may be correlated or anti-correlated in the sense of (9.6) or (9.7), respectively, but their product expectation $\langle \lambda_a \lambda_b \rangle \neq \pm 1$.

Thus, under a bipartite scenario with any number of different outcomes in $[-1, 1]$, relations (9.6) and (9.7), introduced in [270], constitute the general condition on perfect correlation or anti-correlation of outcomes observed by Alice and Bob. This general perfect correlations/anti-correlations condition reduces to the Bell one (9.11) only in a dichotomic case with $\lambda_a, \lambda_b = \pm 1$.

Let a 2×2-setting correlation scenario with joint measurements $(a_i, b_k,)$, $i, k = 1, 2$ and outcomes $\lambda_{a_i}, \lambda_{b_k} \in [-1, 1]$ admit a local hidden variable (LHV) model for joint probabilities; for details, see section 4 in [269], that is, all joint distributions $P_{(a_i,b_k)}$, $i, k = 1, 2$, admit the representation

$$P_{(a_i,b_k)}(\lambda_a, \lambda_b) = \int_\Omega P_{a_i}(\lambda_a|\omega) P_{b_k}(\lambda_b|\omega)\ \nu(\mathrm{d}\omega), \quad \forall \lambda_{a_i}, \lambda_{b_k}, \tag{9.12}$$

via a single probability distribution ν of some variables $\omega \in \Omega$ and conditional probability distributions $P_{a_i}(\cdot\ |\omega)$, $P_{b_k}(\cdot\ |\omega)$ of outcomes at Alice's and Bob's sites. The latter conditional probabilities are usually referred to as "local" in the sense that each of them depends only on a measurement setting at the corresponding site.

Then all scenario product expectations $\langle \lambda_{a_i} \lambda_{b_k} \rangle$, $i, k = 1, 2$, admit the LHV representation

$$\langle \lambda_{a_i} \lambda_{b_k} \rangle = \int_\Omega f_{a_i}(\omega) f_{b_k}(\omega)\ \nu(\mathrm{d}\omega) \tag{9.13}$$

with

$$f_{a_i}(\omega) := \sum_{\lambda_a \in [-1,1]} \lambda_a P_{a_i}(\lambda_a|\omega) \in [-1, 1], \tag{9.14}$$
$$f_{b_k}(\omega) := \sum_{\lambda_b \in [-1,1]} \lambda_b P_{b_k}(\lambda_b|\omega) \in [-1, 1].$$

If an LHV model (9.12) for joint probabilities is deterministic [123, 269], then the values of functions f_{a_i}, f_{b_k}, $i, k = 1, 2$, constitute outcomes under Alice and Bob corresponding measurements with settings a_i and b_k, respectively. However, in a stochastic LHV model [123, 269], functions f_{a_i}, f_{b_k} may take any values in $[-1, 1]$ even in a dichotomous case.

On the other side, if, for a scenario admitting an LHV model (9.12) and having outcomes λ_{a_i}, $\lambda_{b_k} = \pm 1$, the Bell perfect correlation/anti-correlation restriction $\langle \lambda_{a_{i_0}} \lambda_{b_{k_0}} \rangle = \pm 1$ is fulfilled under some joint measurement (a_{i_0}, b_{k_0}), then, in this

LHV model, the corresponding functions $f_{a_{i_0}}, f_{b_{k_0}}$ take only two values ± 1 and, moreover, $f_{a_{i_0}}(\omega) = \pm f_{b_{k_0}}(\omega)$, ν-almost everywhere (a.e.) on Ω.

We have the following statement [270]:

Proposition 9.1 *Let, under a 2×2-setting correlation scenario with joint measurements $(a_i, b_k,)$, $i, k = 1, 2$ and any number of outcomes $\lambda_{a_i}, \lambda_{b_k}$ in $[-1, 1]$, Alice's and Bob's outcomes under the joint measurement (a_2, b_1) be perfectly correlated or anti-correlated:*

$$P_{(a_2,b_1)}(\{\lambda_a = \lambda_b\}) = 1 \tag{9.15}$$

or

$$P_{(a_2,b_1)}(\{\lambda_a = -\lambda_b \neq 0\}) = 1 \tag{9.16}$$

If this scenario admits an LHV model (9.12), then its product expectations satisfy the original Bell inequality:

$$\left| \langle \lambda_{a_1} \lambda_{b_1} \rangle - \langle \lambda_{a_1} \lambda_{b_2} \rangle \right| \pm \langle \lambda_{a_2} \lambda_{b_2} \rangle \leq 1, \tag{9.17}$$

in its perfect correlation (plus sign) or perfect anti-correlation (minus sign) forms, respectively.

Proof Let, for a joint measurement (a_2, b_1), the perfect anti-correlation (9.16) be fulfilled and this scenario admit an LHV model (9.12). This and (9.14) imply

$$\begin{aligned}
0 &\leq \int_\Omega \left| f_{a_2}(\omega) + f_{b_1}(\omega) \right| \nu(d\omega) \\
&= \int_\Omega \left| \sum_{\lambda_a, \lambda_b} (\lambda_a + \lambda_b)\, P_{a_2}(\lambda_a|\omega) P_{b_1}(\lambda_b|\omega) \right| \nu(d\omega) \\
&\leq \int_\Omega \sum_{\lambda_a, \lambda_b} |\lambda_a + \lambda_b|\ P_{a_2}(\lambda_a|\omega) P_{b_1}(\lambda_b|\omega) \nu(d\omega) \\
&\leq 2 \sum_{\lambda_a \neq -\lambda_b} P_{(a_2,b_1)}(\lambda_a, \lambda_b) = 0.
\end{aligned}$$

Thus, under condition (9.16) on scenario joint probabilities, the LHV functions $f_{a_2}(\omega) = -f_{b_1}(\omega)$, ν-*a.e.* on Ω. Quite similarly, for the case of perfect correlations (9.15) we derive $f_{a_2}(\omega) = f_{b_1}(\omega)$, ν-*a.e.* on Ω. These relations and the number inequality

$$|x - y| \leq 1 - xy, \quad \forall\, x, y \in [-1, 1],$$

give

$$\begin{aligned}
&\left|\langle\lambda_{a_1}\lambda_{b_1}\rangle - \langle\lambda_{a_1}\lambda_{b_2}\rangle\right| \pm \langle\lambda_{a_2}\lambda_{b_2}\rangle \\
&= \left|\int_\Omega f_{a_1}(\omega)f_{b_1}(\omega) - f_{a_1}(\omega)f_{b_2}(\omega)\ \nu(\mathrm{d}\omega)\right| \pm \int_\Omega f_{a_2}(\omega)f_{b_2}(\omega)\nu(\mathrm{d}\omega) \\
&\leq \int_\Omega \left|(f_{b_1}(\omega) - f_{b_2}(\omega))\right|\ \nu(\mathrm{d}\omega) \pm \int_\Omega f_{a_2}(\omega)f_{b_2}(\omega)\ \nu(\mathrm{d}\omega) \leq 1.
\end{aligned}$$

This proves the proposition.

We stress that, for the validity of the original Bell inequality (9.17) in the LHV frame, it suffices that condition (9.15) or condition (9.16) on perfect correlations or anti-correlations be fulfilled only under a joint measurement (a_2, b_1).

9.3 Quantum Violations: CHSH versus Original Bell Inequalities

Consider Alice and Bob projective measurements of quantum qudit observable $\hat{X}_{a_1}, \hat{X}_{a_2} = \hat{X}_{b_1}, \hat{X}_{b_2}$ in an arbitrary two-qudit state $\hat{\rho}$ on $\mathbb{C}^d \otimes \mathbb{C}^d$.

In this case, Alice and Bob outcomes coincide with eigenvalues λ_a, λ_b of these observables and restriction $\lambda_a, \lambda_b \in [-1, 1]$ implies the restriction on operators norms $\left\|\hat{X}_{a_i}\right\|, \left\|\hat{X}_{b_k}\right\| \leq 1$. The joint probability $P_{(a_i,b_k)}(\lambda_a, \lambda_b)$ that, under a joint measurement (a_i, b_k), Alice observes an outcome λ_a, while Bob and outcome λ_b are given by

$$\mathrm{tr}[\hat{\rho}\{\hat{E}_{X_{a_i}}(\lambda_a) \otimes \hat{E}_{X_{b_k}}(\lambda_b)\}] \tag{9.18}$$

where $\hat{E}_{X_{a_i}}(\lambda_a)$, $\hat{E}_{X_{b_k}}(\lambda_b)$, $i, k = 1, 2$, are the spectral projections of observables $\hat{X}_{a_i}$ and $\hat{X}_{b_k}$, corresponding to eigenvalues λ_a and λ_b, respectively. The averages in (9.2) and (9.3) take the form

$$\langle\lambda_{a_i}\rangle = \mathrm{tr}[\hat{\rho}\hat{X}_{a_i}], \quad \langle\lambda_{b_k}\rangle = \mathrm{tr}[\hat{\rho}\hat{X}_{b_k}], \quad \langle\lambda_{a_i}\lambda_{b_k}\rangle = \mathrm{tr}[\hat{\rho}\{\hat{X}_{a_i} \otimes \hat{X}_{b_k}\}],\ i, k = 1, 2. \tag{9.19}$$

The general conditions (9.15) and (9.16) on perfect correlations or anti-correlations of Alice and Bob outcomes under a joint measurement (a_2, b_1) reduce to

$$\sum_{\lambda_a = \lambda_b} \mathrm{tr}[\hat{\rho}\{\hat{E}_{X_{b_1}}(\lambda_a) \otimes \hat{E}_{X_{b_1}}(\lambda_b)\}] = 1, \tag{9.20}$$

$$\sum_{\lambda_a = -\lambda_b \neq 0} \mathrm{tr}[\hat{\rho}\{\hat{E}_{X_{b_1}}(\lambda_a) \otimes \hat{E}_{X_{b_1}}(\lambda_b)\}] = 1, \tag{9.21}$$

respectively, and for observables with eigenvalues ± 1, these conditions are equivalent to

$$\mathrm{tr}[\hat{\rho}\{\hat{X}_{b_1} \otimes \hat{X}_{b_1}\}] = \pm 1. \tag{9.22}$$

Thus, under the considered quantum scenario, the left-hand-side $W_{\hat{\rho}_d}^{(\pm)}$ of the original Bell inequality (9.17) takes the form

$$W_{\hat{\rho}}^{(\pm)}(\hat{X}_a, \hat{X}_{b_1}, \hat{X}_{b_2}) = \left| \mathrm{tr}[\hat{\rho}\{\hat{X}_a \otimes \hat{X}_{b_1}\}] - \mathrm{tr}[\hat{\rho}\{\hat{X}_a \otimes \hat{X}_{b_2}\}] \right| \pm \mathrm{tr}[\hat{\rho}\{\hat{X}_{b_1} \otimes \hat{X}_{b_2}\}], \tag{9.23}$$

where, for short, we changed the index notation $a_1 \to a$, and the general condition on perfect correlations/anti-correlations of Alice and Bob outcomes under a joint measurement (b_1, b_1) is given by (9.20)/(9.21).

It is, however, well known that the two-qubit singlet state $\hat{\rho}_{singlet}$ satisfies the perfect anti-correlation (minus sign) condition (in the form (9.22)) whenever the same qubit observable $\hat{X}_b$ with eigenvalues ± 1 is measured at both sites but, depending on a choice of qubit observables $\hat{X}_a, \hat{X}_{b_1}, \hat{X}_{b_2}$, this state may, however, violate [53, 54] the perfect anti-correlation form of the original Bell inequality (9.17).

As has been proven in [226, 307], for the singlet $\hat{\rho}_{singlet}$, the maximal value of the left hand side (9.23) of the original Bell inequality (9.17) over qubit observables with eigenvalues ± 1 is equal to $F_{\mathrm{OB}} = \frac{3}{2}$.

This value is beyond the maximal value $F_{\mathrm{CHSH}} = \sqrt{2}$ for the Clauser–Horne–Shimony–Holt (CHSH) inequality [83] $\left|\mathcal{B}_{lhv}^{CHSH}\right| \leq 2$.

On the other side, the Tsirelson bound $2\sqrt{2}$ on the quantum violation of the CHSH inequality [83] holds for a bipartite quantum state of an arbitrary dimension. For different choices of signs, this implies

$$\begin{aligned}
&\mathrm{tr}[\hat{\rho}\{\hat{X}_a \otimes \hat{X}_{b_1}\}] - \mathrm{tr}[\hat{\rho}\{\hat{X}_a \otimes \hat{X}_{b_2}\} + \mathrm{tr}[\hat{\rho}\{\hat{X}_{b_1} \otimes \hat{X}_{b_1}\} + \mathrm{tr}[\hat{\rho}\{\hat{X}_{b_1} \otimes \hat{X}_{b_2}\}] \leq 2\sqrt{2}\\
&\mathrm{tr}[\hat{\rho}\{\hat{X}_a \otimes \hat{X}_{b_1}\}] - \mathrm{tr}[\hat{\rho}\{\hat{X}_a \otimes \hat{X}_{b_2}\} - \mathrm{tr}[\hat{\rho}\{\hat{X}_{b_1} \otimes \hat{X}_{b_1}\} - \mathrm{tr}[\hat{\rho}\{\hat{X}_{b_1} \otimes \hat{X}_{b_2}\}] \leq 2\sqrt{2}\\
&-\mathrm{tr}[\hat{\rho}\{\hat{X}_a \otimes \hat{X}_{b_1}\}] + \mathrm{tr}[\hat{\rho}\{\hat{X}_a \otimes \hat{X}_{b_2}\} + \mathrm{tr}[\hat{\rho}\{\hat{X}_{b_1} \otimes \hat{X}_{b_1}\} + \mathrm{tr}[\hat{\rho}\{\hat{X}_{b_1} \otimes \hat{X}_{b_2}\}] \leq 2\sqrt{2}\\
&-\mathrm{tr}[\hat{\rho}\{\hat{X}_a \otimes \hat{X}_{b_1}\}] + \mathrm{tr}[\hat{\rho}\{\hat{X}_a \otimes \hat{X}_{b_2}\} - \mathrm{tr}[\hat{\rho}\{\hat{X}_{b_1} \otimes \hat{X}_{b_1}\} - \mathrm{tr}[\hat{\rho}\{\hat{X}_{b_1} \otimes \hat{X}_{b_2}\}] \leq 2\sqrt{2}.
\end{aligned} \tag{9.24}$$

Combining the first line with the third one, for a two-qudit state exhibiting perfect correlations (condition (9.20)), we get the following upper bound,

$$\begin{aligned}
&W_{\hat{\rho}}^{(+)}(\hat{X}_a, \hat{X}_{b_1}, \hat{X}_{b_2})|_{perfect} \\
&= \left| \mathrm{tr}[\hat{\rho}\{\hat{X}_a \otimes \hat{X}_{b_1}\}] - \mathrm{tr}[\hat{\rho}\{\hat{X}_a \otimes \hat{X}_{b_2}\} \right| + \mathrm{tr}[\hat{\rho}\{\hat{X}_{b_1} \otimes \hat{X}_{b_2}\}] \\
&\leq 2\sqrt{2} - \left| \mathrm{tr}[\hat{\rho}\{\hat{X}_{b_1} \otimes \hat{X}_{b_1}\}] \right|,
\end{aligned} \tag{9.25}$$

on the left-hand side of the original Bell inequality. Similarly, combining the second line with the fourth one under condition (9.21) on perfect anti-correlations, we derive

$$W_{\hat{\rho}}^{(-)}(\hat{X}_a, \hat{X}_{b_1}, \hat{X}_{b_2})|_{perfect} \tag{9.26}$$
$$\left| \operatorname{tr}[\hat{\rho}\{\hat{X}_a \otimes \hat{X}_{b_1}\}] - \operatorname{tr}[\hat{\rho}\{\hat{X}_a \otimes \hat{X}_{b_2}\} \right| - \operatorname{tr}[\hat{\rho}\{\hat{X}_{b_1} \otimes \hat{X}_{b_2}\}]$$
$$\leq 2\sqrt{2} - \left| \operatorname{tr}[\hat{\rho}\{\hat{X}_{b_1} \otimes \hat{X}_{b_1}\}] \right|.$$

Thus, for an arbitrary two-qudit state exhibiting perfect correlations or anti-correlations whenever the same quantum observable $\hat{X}_{b_1}$ is measured at both sites, we have

$$\begin{aligned} &W_{\hat{\rho}}^{(\pm)}(\hat{X}_a, \hat{X}_{b_1}, \hat{X}_{b_2})|_{perfect} \\ &= \left| \operatorname{tr}[\hat{\rho}\{\hat{X}_a \otimes \hat{X}_{b_1}\}] - \operatorname{tr}[\hat{\rho}\{\hat{X}_a \otimes \hat{X}_{b_2}\}] \right| \pm \operatorname{tr}[\hat{\rho}\{\hat{X}_{b_1} \otimes \hat{X}_{b_2}\}] \\ &\leq 2\sqrt{2} - \left| \operatorname{tr}[\hat{\rho}\{\hat{X}_{b_1} \otimes \hat{X}_{b_1}\}] \right|. \end{aligned} \tag{9.27}$$

If observable $\hat{X}_{b_1}$ has only eigenvalues ± 1, then conditions (9.20) and (9.21) reduce to the Bell condition (9.22) and the upper bound (9.27) takes the form

$$W_{\hat{\rho}}^{(\pm)}(\hat{X}_a, \hat{X}_{b_1}, \hat{X}_{b_2})|_{perfect} \leq 2\sqrt{2} - 1 \tag{9.28}$$

and holds for a two-qudit state $\hat{\rho}$ of an arbitrary dimension $d \geq 2$. For $d = 2$, this upper bound is more than the maximal value $\frac{3}{2}$ proved [226, 307] for the two-qubit singlet.

Therefore, in the following section, we proceed to analyze the maximal value which the left hand of $W_{\hat{\rho}}^{(\pm)}(\hat{X}_a, \hat{X}_{b_1}, \hat{X}_{b_2})|_{perfect}$ over all qubit observables $\hat{X}_a, \hat{X}_{b_1}, \hat{X}_{b_2}$ with eigenvalues ± 1 and all two-qubit states $\hat{\rho}$, satisfying the perfect correlation/anti-correlation condition (9.22).

9.4 Two-Qubit Case

Consider the violation of the original Bell inequality (9.17) by a two-qubit state exhibiting perfect correlations/anti-correlations whenever the same qubit quantum observable with eigenvalues ± 1 is projectively measured at both sites.

We further consider only symmetric two-qubit states $\hat{\rho}$ (identical quantum particles), that is, states on $\mathbb{C}^2 \otimes \mathbb{C}^2$ which do not change under the permutation of the Hilbert spaces $\mathbb{C}^2$ in the tensor product $\mathbb{C}^2 \otimes \mathbb{C}^2$, and, for simplicity, change index notations $b_1 \to r, b_2 \to c$ in (9.23).

For $d = 2$, a generic qubit observable $\hat{X}$ on $\mathbb{C}^2$ admits the representation

$$\hat{X} = \alpha \mathbb{I}_{\mathbb{C}^2} + r \cdot \hat{\sigma}, \tag{9.29}$$

$$r \cdot \hat{\sigma} = r_1\hat{\sigma}_1 + r_2\hat{\sigma}_2 + r_3\hat{\sigma}_3 \tag{9.30}$$

where $\alpha = \frac{1}{2}\mathrm{tr}[X]$, $r = (r_1, r_2, r_3)$ is a vector in $\mathbb{R}^3$ with components

$$r_1 = \frac{1}{2}\mathrm{tr}[X\hat{\sigma}_1], \quad r_2 = \frac{1}{2}\mathrm{tr}[X\hat{\sigma}_2], \quad r_3 = \frac{1}{2}\mathrm{tr}[X\hat{\sigma}_3], \tag{9.31}$$

and

$$\hat{\sigma}_1 = |e_1\rangle\langle e_2| + |e_2\rangle\langle e_1|, \quad \hat{\sigma}_2 = i(|e_2\rangle\langle e_1| - |e_1\rangle\langle e_2|), \quad \hat{\sigma}_3 = |e_1\rangle\langle e_1| - |e_2\rangle\langle e_2| \tag{9.32}$$

are self-adjoint operators on $\mathbb{C}^2$ with eigenvalues ± 1, represented in the standard orthonormal basis $\{e_1, e_2\}$ in $\mathbb{C}^2$ by the Pauli matrices

$$\hat{\sigma}_1 = \begin{pmatrix} 0 & 1 \\ 1 & 0 \end{pmatrix}, \ \hat{\sigma}_2 = \begin{pmatrix} 0 & -i \\ i & 0 \end{pmatrix}, \ \hat{\sigma}_3 = \begin{pmatrix} 1 & 0 \\ 0 & -1 \end{pmatrix}. \tag{9.33}$$

Every qubit observable with eigenvalues ± 1 is represented in (9.29) by some unit vector $\|r\| = 1$ and constitutes projection $\hat{\sigma}_r := r \cdot \hat{\sigma}$ of the qubit spin along a unit vector (direction) r in $\mathbb{R}^3$.

Therefore, for Alice and Bob measurements of qubit observables with eigenvalues ± 1, the left-hand side (9.23) of the original Bell inequality takes the form

$$W_{\hat{\rho}}^{(\pm)}(\hat{\sigma}_a, \hat{\sigma}_r, \hat{\sigma}_c) = \left| \mathrm{tr}[\hat{\rho}\{\hat{\sigma}_a \otimes \hat{\sigma}_r\}] - \mathrm{tr}[\hat{\rho}\{\hat{\sigma}_a \otimes \hat{\sigma}_c\}] \right| \pm \mathrm{tr}[\hat{\rho}\{\hat{\sigma}_r \otimes \hat{\sigma}_c\}], \tag{9.34}$$

where a, r, c are unit vectors in $\mathbb{R}^3$ and the relation

$$\mathrm{tr}[\hat{\rho}\{\hat{\sigma}_r \otimes \hat{\sigma}_r\}] = \pm 1 \tag{9.35}$$

constitutes the perfect correlation/anti-correlation of Alice and Bob outcomes whenever the same spin observable $\hat{\sigma}_r$ – the projection of qubit spin along the same direction r in $\mathbb{R}^3$ – is measured at both sites.

Substituting representation (9.30) into (9.34) and (9.35), we rewrite these relations via scalar products of vectors in $\mathbb{R}^3$:

$$W_{\hat{\rho}}^{(\pm)}(\hat{\sigma}_a, \hat{\sigma}_r, \hat{\sigma}_c) = \left| (a, \hat{T}^{(\rho)}r) - (a, \hat{T}^{(\rho)}c) \right| \pm (r, \hat{T}^{(\rho)}c), \tag{9.36}$$

$$(r, \hat{T}^{(\rho)}r) = \pm 1, \tag{9.37}$$

where $(a, \hat{T}^{(\rho)}r) := \sum_{i,j} T_{ij}^{(\rho)} a_i r_j$ and $\hat{T}^{(\hat{\rho})}$ is the linear operator on $\mathbb{R}^3$, defined in the canonical basis in $\mathbb{R}^3$ by the matrix with real elements

$$T_{ij}^{(\rho)} := \mathrm{tr}[\hat{\rho}\{\hat{\sigma}_i \otimes \hat{\sigma}_j\}, \ \ i,j = 1,2,3. \tag{9.38}$$

This correlation matrix is symmetric (since $\hat{\rho}$ is symmetric), has eigenvalues λ_m, $m = 1, 2, 3$, where all $|\lambda_m| \leq 1$, and is similar in its form to the matrix considered in [168].

Let us first analyze when an arbitrary symmetric two-qubit state $\hat{\rho}$ may satisfy condition (9.37). By decomposing a unit vector $r = \sum_m \beta_m \mathrm{v}_m$, $\sum_m \beta_m^2 = 1$, in the orthonormal basis $\{\mathrm{v}_j, j = 1, 2, 3\}$ of eigenvectors of $T^{(\rho)}$, we rewrite condition (9.37) in the form

$$\sum_m \beta_m^2(\lambda_m \mp 1) = 0. \tag{9.39}$$

Since all eigenvalues $|\lambda_m| \leq 1$, relation (9.39) implies the following statement.

Proposition 9.2 *A symmetric two-qubit state $\hat{\rho}$ exhibits perfect correlation/anti-correlations*

$$\mathrm{tr}[\hat{\rho}\{\hat{\sigma}_r \otimes \hat{\sigma}_r\}] = \pm 1 \tag{9.40}$$

if and only if its correlation matrix $\hat{T}^{(\rho)}$ has at least one eigenvalue equal to ± 1, respectively. In this case:

(1) if only one of eigenvalues of $\hat{T}^{(\rho)}$ is equal to ± 1, say $\lambda_{m_0} = \pm 1$, then $\hat{\rho}$ satisfies the perfect correlation/anti-correlation condition (9.40), respectively, only for the unit vector $r = \mathrm{v}_{m_0}$;

(2) if $\hat{T}^{(\rho)}$ has two eigenvalues equal to ± 1, say $\lambda_{m_1}, \lambda_{m_2} = \pm 1$, then $\hat{\rho}$ satisfies the perfect correlation/anti-correlation condition (9.40), respectively for every unit vector $r = \beta_{m_1}\mathrm{v}_{m_1} + \beta_{m_2}\mathrm{v}_{m_2}$, $\beta_{m_1}^2 + \beta_{m_2}^2 = 1$ in the plane determined by the eigenvectors $\{\mathrm{v}_{m_1}, \mathrm{v}_{m_2}\}$ of $\hat{T}^{(\rho)}$;

(3) if all three eigenvalues of $\hat{T}^{(\rho)}$ are equal to ± 1, then $\hat{\rho}$ satisfies the perfect correlation/anti-correlation condition (9.40), respectively, for any unit vector r in $\mathbb{R}^3$.

For the two-qubit Bell states

$$\phi_{(\pm)} = \frac{1}{\sqrt{2}}(e_1 \otimes e_1 \pm e_2 \otimes e_2), \quad \psi_{(\pm)} = \frac{1}{\sqrt{2}}(e_1 \otimes e_2 \pm e_2 \otimes e_1), \tag{9.41}$$

we have

$$\begin{aligned} \hat{T}^{(\phi_+)} &= \begin{pmatrix} 1 & 0 & 0 \\ 0 & -1 & 0 \\ 0 & 0 & 1 \end{pmatrix}, \quad \hat{T}^{(\phi_-)} = \begin{pmatrix} -1 & 0 & 0 \\ 0 & 1 & 0 \\ 0 & 0 & 1 \end{pmatrix} \\ \hat{T}^{(\psi_+)} &= \begin{pmatrix} 1 & 0 & 0 \\ 0 & 1 & 0 \\ 0 & 0 & -1 \end{pmatrix}, \quad \hat{T}^{(\psi_-)} = \begin{pmatrix} -1 & 0 & 0 \\ 0 & -1 & 0 \\ 0 & 0 & -1 \end{pmatrix} \end{aligned} \tag{9.42}$$

and this implies the following.

Corollary *(1) The Bell state ϕ_+ exhibits perfect anti-correlations under spin measurements at both sites along the coordinate axis Y and perfect correlations* *under spin measurements at both sites along the same arbitrary direction in the coordinate plane XZ;*

(2) The Bell state ϕ_- exhibits perfect anti-correlations under spin measurements at both sites along the coordinate axis X and perfect correlations—under spin measurements at both sites along the same arbitrary direction in the coordinate plane YZ;

(3) The Bell state ψ_+ exhibits perfect anti-correlations under measurements at both sites of spin projections along the coordinate axis Z and perfect correlations—under spin measurements at both sites along the same arbitrary direction in the coordinate plane XY;

(4) The Bell state (singlet) ψ_- exhibits perfect anti-correlations under spin measurements at both sites along the same arbitrary direction in $\mathbb{R}^3$.

Let us now analyze the maximal value of the left-hand side (9.36) of the original Bell inequality for a two-qubit state $\hat{\rho}$ exhibiting perfect correlations/*anti-correlations* (9.37).

Under condition $\|a\| = 1$, the maximum of $W_{\hat{\rho}}^{(\pm)}(\hat{\sigma}_a, \hat{\sigma}_r, \hat{\sigma}_c)$ over a is reached on the unit vector

$$a = \pm \frac{\hat{T}^{(\rho)}(r-c)}{\left\| \hat{T}^{(\rho)}(r-c) \right\|} \tag{9.43}$$

and is given by

$$\left\| \hat{T}^{(\rho)}(r-c) \right\| \pm (r, \hat{T}^{(\rho)}c). \tag{9.44}$$

Expanding vectors $r = \sum_m \beta_m \mathrm{v}_m$, $\sum \beta_m^2 = 1$, $c = \sum_m \gamma_m \mathrm{v}_m$, $\sum_m \gamma_m^2 = 1$, in terms of the orthonormal eigenvectors $\{\mathrm{v}_m\}$ of $\hat{T}^{(\rho)}$, we rewrite (9.44) in the form

$$\sqrt{\sum_{m=1,2,3} \lambda_m^2 (\beta_m - \gamma_m)^2} \pm \sum_{m=1,2,3} \lambda_m \beta_m \gamma_m, \tag{9.45}$$

where, due to perfect correlations/*anti-correlations* condition (9.37), the coefficients β_m are specified in Proposition 9.2.

Consider the maximum of expression (9.45) over coefficients γ_m. By Proposition 9.2, expression (9.45) reduces to

$$\begin{aligned} &\sqrt{\textstyle\sum_{\lambda_m^2=1} (\beta_m - \gamma_m)^2 + \sum_{\lambda_m^2 \neq 1} \lambda_m^2 {\gamma_m}^2} + \textstyle\sum_{\lambda_m^2=1} \beta_m \gamma_m \\ = \; &\sqrt{2(1 - \textstyle\sum_{\lambda_m^2=1} \beta_m \gamma_m) - \sum_{\lambda_m^2 \neq 1} (1 - \lambda_m^2) \gamma_m^2} + \textstyle\sum_{\lambda_m^2=1} \beta_m \gamma_m \end{aligned} \tag{9.46}$$

since $\sum_{\lambda_m^2=1} \beta_m^2 = 1$. From (9.46) it follows that, for all choices of a direction r—coefficients β_m in (9.46) specified in Proposition 9.2, we have

$$\sup_{a,c} W_{\hat{\rho}}^{(\pm)}(\hat{\sigma}_a, \hat{\sigma}_r, \hat{\sigma}_c)|_{perfect} \leq \max_{z\in[-1,1]} \left(\sqrt{2(1-z)} + z\right) = \frac{3}{2} \tag{9.47}$$

where the upper bound $\frac{3}{2}$ is, for example, reached on every Bell state where all eigenvalues of the correlation matrices $\lambda_m \in \{-1, 1\}, m = 1, 2, 3$.

Also, if a two-qubit state, exhibiting perfect correlations/anti-correlations (see Proposition 9.2), has the correlation matrix with at least two eigenvalues, say $\lambda_{m_1}, \lambda_{m_2}$, with $\left|\lambda_{m_1}\right|, \left|\lambda_{m_2}\right| = 1$, then the upper bound $\frac{3}{2}$ is reached on the unit vector c which is in the plane of eigenvectors $\mathrm{v}_{m_1}, \mathrm{v}_{m_2}$ corresponding to these eigenvalues (vector r is in this plane; see Proposition 9.2) and satisfies condition $c \cdot r = \sum_{\lambda_m^2=1} \beta_m \gamma_m = \frac{1}{2}$, that is, at angle $\pi/3$ to vector r.

Thus, we have proved the following new result.

Theorem 9.3 *Let $\hat{\rho}$ be a symmetric two-qubit states on $\mathbb{C}^2 \otimes \mathbb{C}^2$ exhibiting perfect correlations/anti-correlations whenever the same qubit observable $\hat{\sigma}_r$ is measured at both sites. Then the maximal value of the left-hand side $W_{\hat{\rho}}^{(\pm)}(\hat{\sigma}_a, \hat{\sigma}_r, \hat{\sigma}_c)$ of the original Bell inequality is given by*

$$\max_{\hat{\rho},a,r,c} W_{\hat{\rho}}^{(\pm)}(\hat{\sigma}_a, \hat{\sigma}_r, \hat{\sigma}_c)|_{perfect} = \frac{3}{2} \tag{9.48}$$

and is reached on symmetric two-qubit states discussed in the lines following Equation (9.47).

We stress that this maximal value is less than the upper bound (9.28) following from the CHSH inequality.

9.5 Two-Qutrit Case

Consider now the violation of the original Bell inequality under Alice and Bob spin measurements on a symmetric two-qubit state $\hat{\rho}$ on $\mathbb{C}^3 \otimes \mathbb{C}^3$, exhibiting perfect correlations or anti-correlations.

For Alice and Bob spin measurements in a two-qutrit state $\hat{\rho}$, the left-hand side (9.23) of the original Bell inequality and the condition on perfect correlations and anti-correlations take the forms

$$W_{\hat{\rho}}^{(\pm)}(\hat{S}_a, \hat{S}_r, \hat{S}_c) \tag{9.49}$$

$$= \left| \operatorname{tr}[\hat{\rho}\{\hat{S}_a \otimes \hat{S}_r\}] - \operatorname{tr}[\hat{\rho}\{\hat{S}_a \otimes \hat{S}_c\}] \right| \pm \operatorname{tr}[\hat{\rho}\{\hat{S}_r \otimes \hat{S}_c\}],$$

$$\operatorname{tr}[\hat{\rho}\{\hat{S}_r \otimes \hat{S}_r\}] = \pm 1, \tag{9.50}$$

where a, r, c are unit vectors in $\mathbb{R}^3$ and

$$\hat{S}_r = r \cdot \hat{S} = r_1\hat{S}_1 + r_2\hat{S}_2 + r_3\hat{S}_3, \quad \hat{S} = (\hat{S}_1, \hat{S}_2, \hat{S}_3), \tag{9.51}$$

is the qutrit observable with eigenvalues $\{1,0,-1\}$, describing projection of qutrit spin along a unit vector r in $\mathbb{R}^3$.

Note that if a two-qutrit state $\hat{\rho}$ exhibits perfect correlations/anti-correlations (9.49) under measurements in this state at both sites of spin projection along a direction r, the probability of event that either Alice or Bob observe at their site the outcome $\lambda = 0$ is equal to zero.

In the standard orthonormal basis $\{e_1, e_2, e_3\}$ in $\mathbb{C}^3$ these operators have the following matrix representations:

$$\hat{S}_1 = \frac{1}{\sqrt{2}}\begin{pmatrix} 0 & 1 & 0 \\ 1 & 0 & 1 \\ 0 & 1 & 0 \end{pmatrix}, \quad \hat{S}_2 = \frac{1}{\sqrt{2}}\begin{pmatrix} 0 & -i & 0 \\ i & 0 & -i \\ 0 & i & 0 \end{pmatrix}, \quad \hat{S}_3 = \begin{pmatrix} 1 & 0 & 0 \\ 0 & 0 & 0 \\ 0 & 0 & -1 \end{pmatrix} \tag{9.52}$$

and

$$\hat{S}_r = \begin{pmatrix} r_3 & \frac{r_1 - ir_2}{\sqrt{2}} & 0 \\ \frac{r_1 + ir_2}{\sqrt{2}} & 0 & \frac{r_1 - ir_2}{\sqrt{2}} \\ 0 & \frac{r_1 + ir_2}{\sqrt{2}} & -r_3 \end{pmatrix} \tag{9.53}$$

In view of (9.51), quite similarly to our techniques in Section 9.4 we introduce for a symmetric two-qutrit state $\hat{\rho}$ the correlation matrix $Z^{(\rho)}$ with real elements

$$Z_{ij}^{(\rho)} = \operatorname{tr}[\hat{\rho}\{\hat{S}_i \otimes \hat{S}_j\}], \tag{9.54}$$

which is symmetric, diagonalized, and has eigenvalues $|\lambda_m| \leq 1$, and this allows us to rewrite (9.50) and (9.49) in the following form:

$$\begin{aligned} W_{\hat{\rho}}^{(\pm)}(\hat{S}_a, \hat{S}_r, \hat{S}_c) &= \left|(a, Z^{(\rho)}r) - (a, Z^{(\rho)}c)\right| \pm (r, Z^{(\rho)}c), \\ (r, Z^{(\rho)}r) &= \pm 1. \end{aligned} \tag{9.55}$$

These expressions are quite the same by their form to expressions (9.36), (9.37) for a two-qubit state. By using the same techniques as in a qubit case, we derive

$$\sup_{a,c} W_{\hat{\rho}}^{(\pm)}(\hat{S}_a, \hat{S}_r, \hat{S}_c)|_{perfect} \leq \frac{3}{2}. \tag{9.56}$$

We, however, do not know whether under the considered measurements this supremum is reached.

Theorem 9.4 *Let $\hat{\rho}$ be a symmetric two-qutrit state on $\mathbb{C}^3 \otimes \mathbb{C}^3$ exhibiting perfect correlations/anti-correlations whenever spin projection $\hat{S}_r$ along a direction r is measured at both sites. Then, under Alice and Bob spin measurements on these* *two-qutrit states, the maximal value of the left-hand-side* $W_{\hat{\rho}}^{(\pm)}(\hat{S}_a, \hat{S}_r, \hat{S}_c)$ *of the original Bell inequality (9.17) is upper bounded as*

$$\sup_{\hat{\rho},a,r,c} W_{\hat{\rho}}^{(\pm)}(\hat{S}_a, \hat{S}_r, \hat{S}_c)|_{perfectBell} \leq \frac{3}{2}. \tag{9.57}$$

This two-qutrit upper bound is less than the upper bound (9.28) following from the CHSH inequality.

PART III

Contextuality: Mathematical Modeling and Interpretation

10

Contextual Approach to Probability and Measurement

10.1 Historical Remarks on Generalized Probability and Contextuality

The creation of the Växjö interpretation of QM [189, 190, 193, 203] was stimulated by the Växjö series of conferences on foundations of quantum mechanics, information, and probability (see, for example, [8, 10, 184, 210, 211] as a sample of proceedings). The second conference (2001) was the epochal event in establishing the information interpretation of QM in hot debates that involved the leading experts in quantum foundations (Caves, Folse, Greenberger, Jaeger, Jozsa, Lahti, Mermin, Pitowsky, Plotnitsky, Schack, Vaidman, and many others). This conference strongly influenced my views on the physical meaning of the quantum formalism and methodology. My previous probabilistic analysis of the quantum foundations got the information component and in this way connection with the rapidly growing quantum information community.

Since 1990s I tried to understand the interrelation of quantum and classical probability theories. This is a very complex foundational problem involving interpretational, mathematical, and philosophic questions. Research in this area is characterized by a diversity of views, opinions, and mathematical formalisms.[1] Slowly I came to understand that quantum probability is a machinery for probability update, analogous to classical Bayesian inference [189, 190, 193, 203], [152, 153]. In contrast to the latter, quantum probability inference is not based on the Bayes formula for conditional probability. So, quantum probability theory is a theory of probability inference with a special class of probability update transformations (given by projections or quantum instruments). It is natural to create a general probabilistic framework that covers both the classical and quantum ones. Such generalization would provide a global panorama similar to that from the top of a mountain where one can enjoy a view of the whole city and through it connect districts which otherwise look as totally separated. In this way it is easier to

[1] See, e.g., [97, 119, 120, 146, 147, 179, 252, 276, 277, 296, 298, 348, 364], [1]–[9], [43]–[45].

find similarities and differences in districts' plans and architecture of buildings. As just one of the possible machines for probability update, the quantum probabilistic formalism loses its mystery.

This framework is the *contextual measurement model* (CMM) based on contextual probability space; it was developed in a series of my works [189]–[203]. It is the basis of the *Växjö interpretation* of QM – one of the contextual probabilistic interpretations. Since probability update is at the same time information update, the Växjö interpretation is part of the information interpretation (**II**) of QM (see Section 4.2 for a brief review on **II**).

10.1.1 Contextual Probability

From the mathematical viewpoint, the essence of the problem is in generalizations of conditional probability and probability update. On the way to such rethinking of the interrelation between the classical and quantum probability theories, I was strongly influenced by Ballentine [42]–[45] and Accardi [2, 3] who treated all quantum probabilities as conditional probabilities. Later I learned that this was also Koopman's viewpoint [252]. It is interesting that Kolmogorov (who in 1933 formalized classical probability in the measure-theoretic framework [250, 251]) advertised this conditional viewpoint even for classical probability. This viewpoint was especially clearly described in his early works preceding monograph [250]. Unfortunately, these works (in Russian and published in proceedings) are practically unknown; see [179] for references and details. I got to know about these "pre-axiomatic" studies of Kolmogorov from Shiryaev and Bulinski, former students of Kolmogorov. But even monograph [251] contains the statement which in modern words can be formulated as a statement about *contextuality of probability*. Kolmogorov's message is that it is meaningless to speak about probability without determining a complex of experimental conditions, measurement context. This Kolmogorov position matches well with Bohr's statement on contextuality of a measurement's outcomes that is the cornerstone of his complementarity principle [65] (which is better to call contextuality-complementarity principle [223, 224]).

Unfortunately, Kolmogorov's original message on contextuality of probability was practically ignored in further development of classical probability theory. A mathematical work on probability typically starts with fixing one probability measure, without mentioning that it corresponds to some measurement context. The contextuality component of Bohr's statements on complementarity was neither emphasized in quantum foundational research; typically the Bohr complementarity principle is reduced to the wave–particle duality. Following Bohr [65], Kolmogorov [250, 251], Koopman [252], and Ballentine [42]–[45], I introduced the contextual probability space and CMM based on it [203]. The main idea behind this approach is operating solely with contexts and observables and to exclude physical systems from consideration.

A *measurement context* consists of a premeasurement context C, an observable A, and postmeasurement context $C_{A=x}$ corresponding to the outcome $A = x$, namely a triple $(C, A, C_{A=x})$. Transformation $C \to C_{A=x}$ can be described as a map $T_A(x)\colon \mathcal{C} \to \mathcal{C}$, where $\mathcal{C}$ is the set of premeasurement contexts. Alike theory of quantum instruments (Section 1.3), we call the pair $I_A = (A, T_A)$ a contextual instrument. The latter is the basic mathematical component of measurement theory. It is meaningless to formulate it solely in terms of observables. The same observable A can be a component of a variety of instruments describing different measurement procedures for A. So, an observable is a theoretical quantity expressing some features of premeasurement contexts.

This is a good place to mention Feynman's contextual analysis of the two-slit experiment in the book [119, 120]. He presented the purely probabilistic picture of this fundamental experiment of QM and expressed the interference phenomenon as interference of probabilities. Mathematically he described this situation as the violation of additivity of probability, the classical formula is disturbed by an additional term, the interference term. In classical probability the combination of additivity and the Bayes formula for conditional probability leads to the *formula of total probability* (FTP) playing the important role in probability inference. Following this line, Feynman's conclusion can be rewritten as the violation of classical FTP. The difference between classical and quantum probability models can be moved from the violation of additivity of probability to the violation of the Bayes formula for conditional probability – quantum probability is additive, but conditioning is non-Bayesian. The quantum FTP is a perturbation of classical FTP with an additional term, the interference term. The main distinguishing feature of Feynman's presentation of the two-slit experiment and generally quantum interference is its contextual structure. He operates with three contexts C_1, C_2, C_{12}; the first slit is open and the second is closed, vice versa, and both slits are open. Quantum probabilistic specialty is expressed not via the linear space representation of states and observables, but in the purely contextual probabilistic way.

We make a remark on the notion of contextuality. In the modern quantum information literature, the notion of contextuality is reduced to contextuality of the joint measurement of a few quantum observables. This sort of contextuality was considered by Bell in [53, 54] in his analysis of the violation of the Bell inequalities (although he did not use the term "contextuality"). Feynman's contexuality [119, 120] is more general and in fact coincides with Bohr's contextuality [65]. In my works, including those on the Växjö interpretation, I followed Bohr and Feynman: context as a complex of all conditions involved in an experiment.

10.1.2 Linear Space versus Contextual Frameworks for Probability

Our foundational pathway is opposite to the pathway leading to the generalized probability theories[2] which are directed to creation of general linear space representation (LSR) of probability and measurement process. Linear space representation also provides a panoramic view which covers both classical and quantum probabilities and observables. This is the linear panorama illuminating the place of the quantum probability and measurement formalism among other linear models.

We remark that there are several different approaches to generalized probability theory and corresponding measurement theory, but all of them are either equivalent or only slightly different. One of them is the Davies–Lewis [97] operational probability theory grounded on LRS with the base norm spaces, a class of partially ordered linear spaces. The corresponding measurement theory is formulated within instrument theory [97, 98]; in particular, observables are mathematically represented as positive operator-valued measures (POVMs). They are widely used in quantum information theory [291] - [299], [82, 94, 95] (see also [218, 246], [302, 303] and Chapter 18 for applications to cognition and decision making). Another approach is to start with an abstract definition of state space, a convex subset of a linear space. It goes back to Gudder's work [146] who constructed the operational representation of quantum states and observables starting with preconvex structure. Under natural conditions, this approach leads to a convex state space and LRS for the latter. As was shown by Ozawa [291], these two formulations (Davies–Lewis and Gudder) are actually equivalent. In Section 11.4 we explore the Davies–Lewis approach for operational LRS of the measurement model with states given by classical probability measures. Then we express this model in the form of CMM.

The closest to CMM is the model in which one starts with all possible probabilities that can be generated in an experiment, conditioned on preparation and measurement procedures (see Mackey's book [277]). Then one proceeds to LSR. This construction can be employed to construct LSR for CMM (Section 11.5). However, this is done just to show connection with previously developed theories.[3]

Why is it useful to proceed in a contextual probabilistic framework as far as possible without appealing to LSR?

I start with some remarks on an uncritical use of LSR:

- LSR shadows the essence of the quantum probability formalism as the machinery for probability inference.

[2] See [55, 97, 98, 146, 147, 165, 276, 277, 296, 298, 348].

[3] Another approach to generation of the complex Hilbert space representation of CMM is developed in a series of the author's works; see, for example, [198, 199, 203]. It is based on the contextual version of FTP, FTP with interference term.

- LSR for classical probability via the use of the linear space of measures with the positive cone of nonnegative measures and convex state space of probability measures seems to be inadequate to Kolmogorov's theory [250, 251] based on conditioning (contextualization) with Bayes' formula (Section 11.1).
- LSR generates (via creation of convex linear hull and its closure) plenty of unphysical states and observables (see Ballentine [43]); operating with them led for example to von Neumann's no-go theorem [365].[4]
- The picture that quantum probability theory is just one of LSR of probability diminishes the exclusiveness of linearity of QM. One loses the physical ground for the latter; LSR becomes just a part of the mathematical apparatus of QM (see Section 12.3 for my views on the physical ground for QM-linearity).
- Linking of entanglement to the LSR tensor product structure shadows its contextual probabilistic nature and supports the ambiguous statements on quantum nonlocality.
- Recently the mathematical formalism of quantum theory, especially probability, started to be widely applied outside of physics, for example, in cognition, psychology, social and political sciences, economics and finance, so-called quantum-like modeling (Chapter 18). In such models, the set of possible states (premeasurement contexts) is not so rich as in physics. In quantum-like modeling even the possibility of preparing statistical mixtures is not evident, that is, proceeding toward convex structures might be misleading.

CMM development is important for quantum foundations. CMM demystifies quantum theory by reducing its probabilistic counterpart to the tool for probability update and inference (compare this with QBism [124] – [131]; see Section 12.3); CMM diminishes the role of pure states, in the complete agreement with the statistical (ensemble) interpretation of QM; in CMM, quantum interference is just an additive perturbation of classical FTP due to interplay between a few measurement contexts; the violation of the Bell inequality has the same origin; contextual entanglement is naturally coupled to classical dependence of random variables. The latter demystifies entanglement. This is very important for resolution of the century-long debate on quantum nonlocality.

On the other hand, LSR is very convenient from the mathematical viewpoint (simply linear algebra), and operating within the LSR framework is useful in concrete mathematical calculations. However, the calculations should be completed by the critical analysis on connection of the mathematical LSR-constructions with physics. In quantum-like modeling a similar problem arises – the problem of

[4] Generally some outputs of quantum information theory obtained in the abstract LSR framework might be its artifacts, without coupling to physical reality. A critical analysis of connection of LSR mathematics and physics is needed.

matching between the output of the Hilbert space formalism and, say, psychological effects in decision making (Chapter 18).

Our CMM can be considered as the most general probabilistic framework for measurement, in particular, the notion of contextual probability space is based on the first three axioms of Mackey's theory [277]. Then Mackey moves toward quantum logic by constraining the model with additional axioms. This path makes theory mathematically elegant, but at the same time more complex and the basic probabilistic components are blurred by additional mathematical constructions.

10.1.3 Terminology: Context versus State

We make the following remark about the terminology "context versus state." Since QM operates with the notion of "state," generalized probability theory also employs this terminology. However, even in QM the term "state" is ambiguous. It matches the orthodox Copenhagen interpretation by which a state is treated as the state of an individual quantum system, say the state of an electron – one concrete electron. Many experts consider this interpretation of the quantum state as leading to paradoxes and mismatching with the statistical nature of quantum phenomena. This is a complicated foundational issue, since the leading supporters of the orthodox Copenhagen interpretation also consider QM to be a statistical theory in that the state of an individual system encodes the statistics of the coming experimental runs.

For example, Einstein, Koopman, Margenau, Blohintzev, Ballentine, and nowadays, for example, Ballian, Nieuwenhuizen, and Khrennikov use the so-called statistical (or ensemble) interpretation of QM. By this interpretation a quantum state represents statistical properties of an ensemble of identically prepared systems. So, whose state? The state of an ensemble? In the operational approach, "state" corresponds to a preparation procedure. It seems that the term "state" borrowed from the orthodox Copenhagen interpretation does not match the statistical and operational interpretations of QM. In the generalized probability theory the term "state" is typically associated with a preparation procedure or a class of equivalent preparation procedures. However, this meaning of the term state is not highlighted and the output of the generalized probability theory is often projected onto the orthodox Copenhagen interpretation, that is, this theory is interpreted as a theory about the structure of the state space of individual quantum systems. Therefore, in the Växjö interpretation we prefer to use the notion of context as a complex of experimental conditions; premeasurement context can be associated with a class of equivalent preparation procedures (as is done in the consistent presentation of the generalized probability theory), and measurement context is the combination of the preparation, measurement, and state update generated by measurement feedback with the fixed outcome.

In contrast to the generalized probability theory employing LSR, we do not assume that the set of premeasurement contexts $\mathcal{C}$ contains contexts generated by statistical mixtures (see Axiom 4 in Mackey's book [277]), that is, for $C_1, C_2 \in \mathcal{C}$ and $p_1, p_2 \geq 0, p_1 + p_2 = 1$, the set $\mathcal{C}$ need not contain a context which can be identified with $p_1 C_1 + p_2 C_2$. Proceeding without the mixture axiom illuminates the difference between state and context; consider for example, the "basic contextual probability representation" of the classical Kolmogorov probability space (Section 11.1). Here contexts are not probability distributions, but elements of the (σ-)algebra. Generally a context provides a finer description of the measurement setup than a probability distribution.

10.2 Contextual Measurement Model (CMM)

Contextual Probability Space

Definition 10.1 A contextual probability space is a triple $\Sigma = (\mathcal{C}, \mathcal{O}, \mathcal{P})$, where $\mathcal{C}$ and $\mathcal{O}$ are sets of premeasurement contexts and observables and $\mathcal{P}$ is the space of the corresponding probability distributions.

In physics premeasurement contexts can be associated with preparation procedures.[5] Each observable A has its range of values X_A; for simplicity, consider discrete observables, that is, having finite ranges of values. The following considerations are straightforwardly extended to observables with arbitrary ranges of values. For an observable A and premeasurement context C, denote the probability of an outcome $x \in X_A$ as $P_C^A(x) \equiv P_C(A = x)$. By definition of a probability distribution:

$$P_C^A(x) \geq 0, \sum_{x \in X_A} P_C^A(x) = 1. \tag{10.1}$$

The range set X_A can be endowed with the algebra of all its subsets $\mathcal{F}_A$. We set $P_C^A(G) = \sum_{x \in G} P_C^A(x), G \in \mathcal{F}_A$. This is a probability measure on $\mathcal{F}_A$. In the definition of Σ, the symbol $\mathcal{P}$ denotes the collection of such probability measures:

$$\mathcal{P} = \{P_C^A : C \in \mathcal{C}, A \in \mathcal{O}\}$$

(see Axiom 1 in Mackey's book [277]). Elements of $\mathcal{P}$ are called contextual probabilities. These are the analogs of the conditional probabilities in the classical (Kolmogorov) probability model. But we reserve the term "conditional probability" for a special class of contextual probabilities generated by the context's updates.

[5] In applications outside of physics, so called qunatum-like modeling (Chapter 18), not all premeasurement contexts can be straightforwardly represented in the form of a preparation procedure; here we operate with mental, social, and financial premeasurement contexts. Within the statistical (ensemble) interpretation of QM, contexts are represented as ensembles of similarly prepared systems.

It is natural to assume (see Axiom 2 in Mackey's book [277]) that two observables having the same probability distribution for all contexts should coincide, that is,

$$P_C^{A_1} = P_C^{A_2} \text{ for any } C \in \mathcal{C} \Rightarrow A_1 = A_2. \tag{10.2}$$

We also assume (see Axiom 2 in Mackey's book [277]) that two contexts having the same probability distribution for all observables should coincide, that is,

$$P_{C_1}^{A} = P_{C_2}^{A} \text{ for any } A \in \mathcal{O} \Rightarrow C_1 = C_2. \tag{10.3}$$

The average of an observable $A \in \mathcal{O}$ (with $X_A \subset \mathbb{R}$) w.r.t. a premeasurement context $C \in \mathcal{C}$ is defined as

$$\langle A \rangle_C \equiv E[A|C] = \sum_{x \in X_A} x P_C(A = x). \tag{10.4}$$

Context Update and Conditional Probability

Measurement of an observable A with the concrete outcome x in a pre-measurement context C updates this premeasurement context:

$$C \to C_{A=x}. \tag{10.5}$$

In terms of preparation procedures, we can consider a measurement as a subsequent preparation procedure; context $C_{A=x}$ is measurement of A and filtering w.r.t. the fixed outcome x.[6] It is natural to consider this map only for contexts belonging to the set

$$\mathcal{C}_A(x) = \{C \colon P_C(A = x) > 0\}. \tag{10.6}$$

If $P_C(A = x) = 0$, then the postmeasurement context is not well defined. Thus, each observable A and its outcome x determine a map

$$T_A(x) : \mathcal{C} \to \mathcal{C},\ C \to C_{A=x} = T_A(x)C, \tag{10.7}$$

with the domain of definition $\mathcal{C}_A(x)$.

The delicate point of measurement theory is that generally an observable does not determine the context update map unequally. An observable A can be measured via different measurement procedures and each procedure generates its own context update map. A pair I_A = (observable, context update map) = $(A; T_A)$ is called a *contextual instrument* (cf. Section 11.3); a pair (C, I_A), or a triple (C, A, T_A), is called a *measurement context.* We stress once again that a variety of instruments can be associated with the same observable A: $I_A = (A, T_A), I'_A = (A, T'_A), I''_A =$

[6] In terms of ensembles, A-measurement is performed for systems of initially prepared ensemble and then systems generating the outcome $A = x$ form new ensemble.

$(A, T''_A), \ldots$. We stress that all these update maps have the same domain of definition determined by the observer A; see (10.6).

Typically one fixes some class of context update maps. In the von Neumann measurement theory (Section 11.2), this is the class of normalized projections. In quantum instrument theory (Section 11.3) these are quantum channels or more generally, in the theory of Davies–Levis instruments, these are positive trace preserving maps.

We emphasize that von Neumann's measurement theory is very special: here by fixing an observable, a Hermitian operator $\hat{A}$, we automatically fix the update map – via the operator's spectral family. This special situation leads to the illusion that an observable determines the update map. We repeat that generally this is not the case.

Definition 10.2 Let $\Sigma = (\mathcal{C}, \mathcal{O}, \mathcal{P})$ be a contextual probability space. A contextual measurement model (CMM) is a pair $M = (\Sigma, \mathcal{I})$, where Σ is a contextual probability space and $\mathcal{I}$ is a collection of contextual instruments.

CMM is a set of measurement contexts, namely triples (C, A, T_A). CMM generates the notion of the conditional probability:

Definition 10.3 Consider a measurement context (C, A, T_A). Let it generate the output $A = x$ and the corresponding context update, $C \to C_{A=x} = T_A(x)C$. Consider measurement of another observable B under condition $A = x$, that is, w.r.t. context $C_{A=x}$. The conditional probability is given by the formula:

$$P_{C,I_A}(B = y|A = x) \equiv P_{C_{A=x}}(B = y) = P_{T_A(x)C}(B = y), \; C \in \mathcal{C}_A(x). \tag{10.8}$$

We note that this definition involves context update only for the A-observable; different contextual instruments $I_A, I'_A, \ldots$, induce their own probabilistic conditioning, $P_{C,I_A}(B = y|A = x), P_{C,I'_A}(B = y|A = x), \ldots$. For simplicity, we shall typically omit the index I_A of dependence on the concrete instrument.

10.2.1 Contextual Formula of Total Probability

Now we point out that generally contextual probability differs from the classical Kolmogorov probability. One of the basic classical laws of probability is the law of total probability formulated in the form of FTP, *formula of total probability* (Section 1.5):

$$P(B = y) = \sum_{x \in X_A} P(B = y|A = x)P(A = x). \tag{10.9}$$

In classical probability theory, contextual probability is identified with the conditional one (Section 11.1), and the contextual-conditional analog of FTP has the form

$$P_C(B=y) = \sum_{x \in X_A} P_C(B=y|A=x)P_C(A=x). \tag{10.10}$$

See (11.4). However, in a general contextual probability space this formula can be violated:

$$\begin{aligned} P_C(B=y) &\neq \sum_{x \in X_A} P_C(B=y|A=x)P_C(A=x) \\ &= \sum_{x \in X_A} P_{C_{A=x}}(B=y)P_C(A=x). \end{aligned} \tag{10.11}$$

The difference between the LHS and RHS determines the degree of context-disturbance due to its update; it can serve as a measure of nonclassicality of a contextual model,

$$\delta_C(B=y|A) = P_C(B=y) - \sum_{x \in X_A} P_C(B=y|A=x)P_C(A=x). \tag{10.12}$$

We call this quantity *the interference term* (see [179, 189, 198, 199, 203]). This consideration can be formalized in the contextual formula of total probability (FTP with an interference term):

$$P_C(B=y) = \sum_{x \in X_A} P_C(B=y|A=x)P_C(A=x) + \delta_C(B=y|A). \tag{10.13}$$

The equality of the interference term to zero is the necessary condition of the classical probabilistic representation of a contextual probability model, but it is not the sufficient condition [203].

Let us jump for the moment to Section 11.2 in that the von Neumann quantum measurement model is treated as CMM. Consider dichotomous observables $A = x_1, x_2$ and $B = y_1, y_2$. In this case the interference term has the form

$$\begin{aligned} &\delta_C(B=y|A) \\ &\quad = 2\cos\theta\sqrt{P_C(B=y|A=x_1)P_C(A=x_1)P_C(B=y|A=x_2)P_C(A=x_2)}, \end{aligned} \tag{10.14}$$

where context C is identified with quantum state ψ (for simplicity consider contexts corresponding to pure states) and the angle $\theta = \theta(B = y|A;\psi)$. For dichotomous observables, even in general CMM it is useful to write the interference term as

$$\begin{aligned} &\delta_C(B=y|A) \\ &\quad = 2\lambda\sqrt{P_C(B=y|A=x_1)P_C(A=x_1)P_C(B=y|A=x_2)P_C(A=x_2)}, \end{aligned} \tag{10.15}$$

where $\lambda = \lambda(B = y|A; C)$. If $|\lambda| \leq 1$, then this is the trigonometric interference and the interference term has the form (10.14). If $|\lambda| \geq 1$, then this is the hyperbolic interference and the interference term can be represented as

$$\delta_C(B = y|A) = \pm 2 \cosh\theta \sqrt{P_C(B = y|A = x_1)P_C(A = x_1)P_C(B = y|A = x_2)P_C(A = x_2)}. \tag{10.16}$$

In the quantum framework such interference can be generated by quantum instruments [198]. In general CMM we can employ the hyperbolic version of QM [203].

10.2.2 Conditional JPD: Order and Response Replicability Effects

For observables $A_1, A_2 \in \mathcal{O}$, the conditional joint probability distribution (JPD) is defined by

$$P_C(A_1 = x_1, A_2 = x_2) = P_C(A_1 = x_1)P_C(A_2 = x_2|A_1 = x_1). \tag{10.17}$$

We remark that this is really a probability distribution, that is, $\sum_{x_1,x_2} P_C(A_1 = x_1, A_2 = x_2) = 1$. We can also define JPD for inverse order of measurements, $P_C(A_2 = x_2, A_1 = x_1) = P_C(A_2 = x_2)P_C(A_1 = x_1|A_2 = x_2)$. Observables A_1, A_2 show the *order effect* (OE) w.r.t. context C, if

$$P_C(A_1 = x_1, A_2 = x_2) \neq P_C(A_2 = x_2, A_1 = x_1), \tag{10.18}$$

for at least one pair of outcomes (x_1, x_2); otherwise, there is no OE in context C. We remark OE was actively investigated in decision making and psychology, both theoretically and experimentally; in particular, within quantum-like modeling – the applications of quantum methodology and formalism to decision making and psychology [367, 368, 303, 304].

In the absence of OE we have:

$$P_C(A_1 = x_1, A_2 = x_2) = P_C(A_2 = x_2, A_1 = x_1),\ x_j \in X_{A_j}, \tag{10.19}$$

that is,

$$P_C(A_1 = x_1)P_C(A_2 = x_2|A_1 = x_1) = P_C(A_2 = x_2)P_C(A_1 = x_1|A_2 = x_2). \tag{10.20}$$

In this case we call the observables *conditionally compatible* for context $C \in \mathcal{C}$ and their JPD is defined by (10.19). We remark that the marginals of JPD coincide with the probability distributions $P_C^{A_i}$.

In the von Neumann CMM M_{QVN} (Section 11.2) with observables and state updates of the projection type, conditional compatibility for all possible pre-measurement contexts (given by density operators) is equivalent to commonly considered compatibility of observables and their representation by commutative Hermitian operators.

By considering conditional JPD, we do not assume that the observables A_1 and A_2 are jointly measurable. We consider sequential measurements, say first A_1 then A_2 or vice versa. We remark that, in fact, precisely this experimental setup is realized in the Bell experiments. Here the instances of time for measurement's outputs on subsystems coincide with zero probability; always the click of the photodetector for the subsytem S_1 is before the click of the photodetector for the subsystem S_2 or vice versa and the time window serves for clicks pairing.

Conditional compatibility implies the Bayes formula for the conditional probability:

$$P_C(A_2 = x_2|A_1 = x_1) = \frac{P_C(A_1 = x_1, A_2 = x_2)}{P_C(A_1 = x_1)}, \tag{10.21}$$

$$P_C(A_1 = x_1|A_2 = x_2) = \frac{P_C(A_1 = x_1, A_2 = x_2)}{P_C(A_2 = x_2)}. \tag{10.22}$$

The equality (10.20) implies the Bayes theorem for the probability inference. Let the outcomes of the observable A_2 label hypotheses, $H_1, \ldots, H_m$. Then (10.20) is written as

$$P_C(A_2 = H_j|A_1 = x_1) = \frac{P_C(A_2 = H_j)P_C(A_1 = x_1|A_2 = H_j)}{P_C(A_1 = x_1)}. \tag{10.23}$$

The Bayes formula for conditional probability (10.21),(10.22) implies the validity of the classical FTP, that is, the Bayes theorem can be written in the standard form:

$$P_C(A_2 = H_j|A_1 = x_1) = \frac{P_C(A_2 = H_j)P_C(A_1 = x_1|A_2 = H_j)}{\sum_{H_i} P_C(A_1 = x_1|H_i)P_C(A_2 = H_i)}. \tag{10.24}$$

Observable A shows *replicability* for context C, if

$$P_C(A = x, A = x) = P_C(A = x), \tag{10.25}$$

or

$$P_{C_{A=x}}(A = x) = 1. \tag{10.26}$$

Observable A shows replicability if (10.25) holds for any $C \in \mathcal{C}_A(x), x \in X_A$.

In quantum-like modeling, the following effect plays the important role. Observables A_1 and A_2 show *the response replicability effect* (RRE) w.r.t. context C, if

$$P_C(A_1 = x_1, A_2 = x_2, A_1 = x_1) = P_C(A_1 = x_1, A_2 = x_2), \tag{10.27}$$

$$P_C(A_2 = x_2, A_1 = x_1, A_2 = x_2) = P_C(A_2 = x_2, A_1 = x_1) \tag{10.28}$$

for all pairs of outcomes (x_1, x_2). This is a kind of memory effect. The challenging problem for quantum-like modeling was the combination of OE and RRE [213]. It was solved in articles [302, 303] within quantum instrument theory.

10.2.3 Contextual Correlations and Bell-Type Inequalities

Consider a pair of conditionally compatible observables $A, B \in \mathcal{O}$ (with $X_A, X_B \subset \mathbb{R}$). Their correlation w.r.t. a context $C \in \mathcal{C}$ is defined as

$$\langle AB \rangle_C \equiv E[AB|C] = \sum_{x,y} xyP_C(A = x, B = y). \tag{10.29}$$

The most popular Bell-type inequality is the CHSH inequality. We consider this inequality within CMM. There are given four observables $A_i, B_j, i, j = 1, 2$, valued in $[-1, 1]$; observables in the pairs (A_i, B_j) are conditionally compatible for some context C with JPDs $P_C^{A_i,B_j}$. The CHSH inequality has the form:

$$|\langle A_1B_1 \rangle_C + \langle A_2B_1 \rangle_C + \langle A_1B_2 \rangle_C - \langle A_2B_2 \rangle_C| \le 2, \tag{10.30}$$

that is,

$$|\sum_{x,y} xy(P_C^{A_1,B_1}(x,y) + P_C^{A_2,B_1}(x,y) + P_C^{A_1,B_2}(x,y) - P_C^{A_2,B_2}(x,y))| \le 2. \tag{10.31}$$

If there exists a probability measure P_C such that JPDs $P_C^{A_i,B_j}$ can be obtained as its marginals, for example,

$$P_C^{A_1,B_1}(x,y) = \sum_{x_2,y_2} P_C(x, x_2, y, y_2),$$

then the CHSH inequality holds true. However, if such P_C does not exist, then this inequality can be violated and the maximum of its left-hand side w.r.t. contexts $C \in \mathcal{C}$ and observables $A_1, A_2, B_1, B_2 \in \mathcal{O}$ valued in $[-1, 1]$ can approach the value four, $\max_{\text{CHSH}} = 4$. This maximum depends on CMM. For von Neumann CMM M_{QVN}, this is

$$\max_{\text{CHSH}} = 2\sqrt{2}.$$

10.2.4 Entanglement of Contextual Instruments

Entanglement is typically considered as one of the distinguished features of LSR; from my viewpoint the association of entanglement with LSR and the tensor product structures shadows its physical nature; its mathematical description is identified with physics. As was shown in articles [51, 52], the notion of entanglement can be formalized in the purely probabilistic framework and dissociated from the tensor product and generally from LSR.

By starting with such a probabilistic approach to the notion of entanglement, the authors of [51, 52] proceed towards its complex Hilbert space realization. Now we present the purely probabilistic picture of entanglement. The main value of the contextual probabilistic realization of entanglement is in clarification of its foundational meaning. At the same time, the use of LSR can essentially simplify the concrete calculations. However, one should be careful with connection of the mathematical structures of LSR with physics (or in qunatum-like modeling with, for example, psychology and decision making).

Consider CMM $M = (\Sigma, \mathcal{I})$, where $\Sigma = (\mathcal{C}, \mathcal{O}, \mathcal{P})$ is a contextual probability space and $\mathcal{I}$ is a collection of contextual instruments of this model, namely, pairs (observable, state update map). Consider two contextual instruments $I_A = (A, T_A)$ and $I_B = (B, T_B)$.

Definition 10.4 In premeasurement context $C \in \mathcal{C}$, the outcome $B = \beta$ depends on the outcomes of A if for at least two values of $A, \alpha = \alpha_i, \alpha_j$, the corresponding conditional probabilities don't coincide,

$$P_C(B = \beta | A = \alpha_i) \neq P_C(B = \beta | A = \alpha_j). \tag{10.32}$$

Thus, the probability of getting the outcome $B = \beta$ if the preceding A-measurement had the outcome $A = \alpha_i$ differs from the probability of getting the same outcome $B = \beta$ if the preceding A-measurement had the outcome $A = \alpha_j$. We remark that the update map T_B is not involved in this definition, that is, one can consider just an observable B without referring to the corresponding instrument I_B. We consider two instruments by symmetry reason.

We note that the outcome $B = \beta$ does not depend on the outcomes of the observable A if and only if

$$P_C(B = \beta | A = \alpha_i) = P_C(B = \beta | A = \alpha_j), \text{ for all pairs } \alpha_i, \alpha_j, \tag{10.33}$$

that is, the conditional probability for this outcome is constant w.r.t. the outcomes of A. Denote it $P_C(B = \beta | A)$. The following natural question arises: Does the probability $P_C(B = \beta | A)$ coincide with unconditional probability $P_C(B = \beta)$?

Definition 10.5 Two instruments I_A and I_B are called AB-entangled in $C \in \mathcal{C}$ (or C is AB-entangled), if all outcomes of the B-observable depend on outcomes of the A-observable, that is, for all β, condition (10.32) holds for some α_i, α_j.

Concerning the notation "AB-entangled," it would be better to write "$I_A I_B$-entangled" to emphasize that this is entanglement of instruments and not simply the observables, but to make the notation compact, we proceed with "AB-entangled." The order of observables is important. Generally AB-entanglement does not imply BA-entanglement. This is the purely probabilistic definition; it does not involve LSR, and it can be applied to any statistical physical theory. This definition formalizes dependence of observables.

We introduce the following quantitative measure of entanglement:

Definition 10.6 For contextual instruments I_A and I_B and premeasurement context C, AB-concurrence of conditional probabilities is defined as

$$\lambda_{AB}(\psi) = \sum_{\beta} \sum_{\alpha \neq \alpha'} |P_C(B = \beta | A = \alpha) - P_C(B = \beta | A = \alpha')|. \tag{10.34}$$

(See Chapter 13 on comparison with the standard concurrence measure of entanglement.) The crucial issue is that AB-concurrence depends on a pair of instruments.

Proposition 10.7 *For dichotomous observables $A, B = \pm$, dependencies of the values $B = -$ and $B = +$ on the outcomes of A are equivalent. Thus, each dependence is equivalent to AB-entanglement.*

Proof In the state context C, the value $B = -$ depends on the outcomes of A if

$$P_C(B = -|A = +) \neq P_C(B = -|A = -). \tag{10.35}$$

This automatically implies that even the value $B = +$ depends on the outcomes of A,

$$P_C(B = +|A = +) = 1 - P_C(B = -|A = +) \neq$$
$$1 - P_C(B = -|A = -) = P_C(B = +|A = -),$$

that is, instruments I_A and I_B are entangled in context C.

As was already pointed out, in articles [51, 52] contextual probabilistic entanglement can be realized in the complex Hilbert space and in this way connected with the ordinary notion of entanglement (Chapter 7). In the LSR representation, the main distinguishing feature of AB-entanglement (Definition 10.5) is that it is associated with a pair of instruments, I_A, I_B. The standard definition of entanglement is coupled with the tensor product structure and not with two concrete instruments (observables).

For simplicity consider CMM M_{QVN} (Section 11.2) with von Neuman observables [365]; in this CMM an observable $\hat{A}$, a Hermitian operator, automatically determines the state update map, via its spectral family. So, there is no need to operate with instruments; one can operate solely with observables. In this CMM (which is typically used in entanglement studies), contextual probabilistic entanglement is associated with the pairs of observables, i.e., two observables are entangled or disentangled in some state (context) $C = \hat{\rho}$. The main mathematical features of AB-entanglement and ordinary, tensor product based, entanglement are similar, but some essential differences can be found [51, 52] (Chapter 13).

The probabilistic viewpoint on the "EPR paradox" [115] is presented in Schrödinger's papers [328, 329] that initiated the modern theory of entanglement. However, this theory ignores the important message of Schrödinger: entanglement characterizes probability update for the outcomes of observable B conditioned on the outcomes of observable A. In the framework of [328, 329] it is meaningless to speak about entanglement without specifying the observables. The state update – the Hilbert space representation of the probability update – encodes the procedure of conditional prediction. For Schrödinger, the quantum formalism is a mathematical machinery for probability prediction (as in the Växjö interpretation or QBism) and a quantum state is a part of such machinery. We can say that Schrödinger interpreted quantum probabilities as conditional (contextual) probabilities and entanglement as contextual probabilistic entanglement. But this is my private interpretation of Schrödinger's views and many experts in quantum foundations may disagree with me.

By following Schrödinger [328, 329], in article [51] we considered a special sort of the contextual probabilistic entanglement that matches perfectly the Schrödinger analysis of the EPR argument.

Definition 10.8 For $C \in \mathcal{C}$, instruments I_A and I_B are perfectly conditionally correlated for values $(A = \alpha, B = \beta)$ if the conditional probability of getting the outcome $B = \beta$ if the preceding A-measurement had the outcome $A = \alpha$ equals to 1,

$$P_C(B = \beta | A = \alpha) = 1. \tag{10.36}$$

More generally, consider observables with values (α_i) and (β_i) and some set Γ of pairs (α_i, β_j).

Definition 10.9 (EPR-entanglement) Let $C \in \mathcal{C}$. If instruments I_A and I_B are perfectly conditionally correlated for all pairs belonging to Γ, then they are called EPR-entangled w.r.t. set G in context C.

We are interested in sets Γ such that each of the α and β values appears in the pairs once and only once. We call such EPR-entanglement complete.

For example, for two dichotomous observables with $\alpha, \beta = \pm$, we consider, for example, the set of the pairs $(A=+, B=-), (A=-, B=+)$, in short, $A= -B$ EPR-entanglement, or the pairs $(A=+, B=+), (A=-, B=-)$, $A=B$ EPR-entanglement. We analyze such EPR-entanglements.

Let us start with $A=-B$ EPR-entanglement, that is, $P_C(B=-|A=+)=1$ and $P_C(B=+|A=-)=1$. Thus, $P_C(B=+|A=+)=0$ and $P_C(B=-|A=-)=0$, and $P_C(B=-|A=+)=1 \neq P_C(B=-|A=-)=0$ and $P_C(B=+|A=-)=1 \neq P_C(B=+|A=+)=0$. In this case EPR-entangled instruments are automatically entangled in the sense of Definition 10.5.

Now turn to $A=B$ EPR-entanglement, that is, $P_C(B=+|A=+)=1$ and $P_C(B=-|A=-)=1$. Thus, $P_C(B=-|A=+)=0$ and $P_C(B=+|A=-)=0$, and $P_C(B=+|A=+) = 1 \neq P_C(B=+|A=-)=0$ and $P_C(B=-|A=-)=1 \neq P_C(B=-|A=+)=0$. And again EPR-entangled instruments are automatically entangled in the sense of Definition 10.5.

Thus, EPR-entanglement is just the very special case of AB-entanglement. For dichotomous observables, AB-concurrence of conditional probabilities has the form

$C\wr\backslash_{AB}(\psi)=|P(B=+|A=-)-P(B=+|A=+)|+|P(B=-|A=-)-P(B=-|A=+)|$, hence it can written as

$$\lambda_{AB}(\psi) = 2|P(B=+|A=-) - P(B=+|A=+)|. \tag{10.37}$$

From this formula, we immediately obtain the following characterization of maximally AB-entangled states:

Proposition 10.10 *AB-concurrence of conditional probabilities approaches its maximal value, $\lambda_{AB}(\psi) = 2$, if and only if the instruments are EPR-entangled in premeasurement context C.*

Functions of Observables

Suppose that all observables are valued in multidimensional real space and we remove (for the moment) the restriction that observables' ranges of values are finite. We consider probability measures on the σ-algebra $\mathcal{B}$ of the Borel sets; it is generated by all semi-open intervals, $(\alpha_1, \beta_1] \times \ldots \times (\alpha_n, \beta_n]$. So, P_C^A is a probability measure on $\mathcal{B}$. A function $f\colon \mathbb{R}^n \to \mathbb{R}^m$ is called $\mathcal{B}$-measurable if for any Borel subset G of $\mathbb{R}^m$ its preimage $f^{-1}(G)$ is a Borel subset of $\mathbb{R}^n$. Only such functions are considered.

Following Mackey [277] (Axiom 3), we assume that, for each $A \in \mathcal{O}$ with the range of values $\mathbb{R}^n$ and a Borel function $f\colon \mathbb{R}^n \to \mathbb{R}^m$, there exists $B = B_f \in \mathcal{O}$ such that, for any $C \in \mathcal{C}$ and Borel set $G \subset \mathbb{R}^m$,

$$P_C^B(G) = P_C^A(f^{-1}(G)). \tag{10.38}$$

Such observable is uniquely defined, due to condition (10.2) (Mackey's Axiom 2) and it can be denoted as $B = f(A)$.

Two observables A_1 and A_2 are called functionally compatible (jointly measurable) if there exists an observable A and functions f_i such that $A_i = f_i(A)$. For the von Neumann CMM M_{QVN} (Section 11.2) functional compatibility is equivalent to compatibility and hence conditional compatibility. Generally, in CMM interrelation between these two notions is complex and we shall not proceed to detailed comparison.

10.2.5 Distinguishing Features of Contextual Measurement Models

We list the probabilistic constraints which can be used to distinguish different CMMs (theoretically and experimentally):

- violation of FTP
- OE
- RRE
- OE+RRE
- violation of Bell inequalities

10.3 Interpretations of Contextual Probability

Probability is characterized by the diversity of interpretations [179]. We now discuss the interpretations of contextual probability. We start with the remark that mathematically a contextual probability space $\Sigma = (\mathcal{C}, \mathcal{O}, \mathcal{P})$ cannot be described as the single Kolmogorov probability space: Σ is a bunch of such spaces. However, fixed $C \in \mathcal{C}$ and $A \in \mathcal{O}$ can be realized within some probability space $K = (\Omega, \mathcal{F}, P)$ with realization of observable A by a random variable $a \colon \Omega \to X_A$; its probability distribution coincides with P^A_C, that is,

$$P^A_C(\alpha) = P(\omega \in \Omega \colon a(\omega) = \alpha), \ \alpha \in X_A.$$

We note that a contextual probability space Σ can be represented as

$$\Sigma = \cup_{C,A} K_{C,A}, \tag{10.39}$$

where $K_{C,A}$ is a Kolmogorov probability space for describing measurement of the observable A in the premeasurement context C.

Therefore, one can assign to the contextual probability any interpretation used for probability defined in the measure-theoretic framework. The main interpretation employed in classical and quantum physics is the frequency interpretation. In the Kolmogorov theory [250, 251], this interpretation is mathematically rooted to the strong law of large numbers. Another basic interpretation of probability in physics is the statistical (ensemble) interpretation. By this interpretation, $\Omega = \Omega_C$ represents an ensemble of systems prepared for measurement and the probability measure $P = P_A$ depends on the observable A. Finally, we mention the subjective interpretation. It is widely employed in decision making and psychology, but was not used in physics until QBism was invented.

In contextual probability, we need not represent a premeasurement context C by an ensemble of systems. Instead, we can consider a sequence of measurements of an observable A in the same premeasurement context C,

$$\mathbf{x} \equiv \mathbf{x}_{C,A} = (x_1, \ldots, x_N, \ldots, \ldots), \tag{10.40}$$

where $x_j (\in X_A = \{\alpha_1, \ldots, \alpha_m\})$ are the measurement's outcomes. Such a sequence determines the frequencies of realizations of the concrete values,

$$\nu_N(\alpha_j) = n_N(\alpha_j)/N, \tag{10.41}$$

where $\nu_N(\alpha_j)$ is the number of measurement outcomes with the fixed value α_j. The probability of obtaining the value α_j in a sequence of measurements $\mathbf{x}$ is defined as the limit

$$P_C^A(\alpha_j) \equiv \lim_{N\to\infty} \nu_N(\alpha_j). \tag{10.42}$$

This is the straightforward frequency introduction of probability. The deep mathematical theory of frequency probability was developed by von Mises [363] (see also [179] for an introduction). A sequence $\mathbf{x}$ generated by observations is called a *collective*. Von Mises' theory is a theory of collectives. Instead of operations on sets, as done in the Kolmogorov measure-theoretic theory [251], von Mises [363] constructed a probability theory based on operations with collectives. We remark that $P_C(A = \alpha_j) \equiv P_C^A(\alpha_j)$ can be considered as the probability generated by the collective $\mathbf{x} \equiv \mathbf{x}_{C,A}$, that is, $P_C^A(\alpha_j) = P_{\mathbf{x}_{C,A}}(\alpha_j)$. It is important to note that not all collectives are compatible – combinable in von Mises' terminology. Two collectives, $\mathbf{x}$ and $\mathbf{y}$, are combinable if their combination,

$$\mathbf{z} = (z_1, \ldots, z_N, \ldots),\ z_j = (x_j, y_j), \tag{10.43}$$

is also a collective and the probability distributions $P_{\mathbf{x}}$ and $P_{\mathbf{y}}$ are marginal for the probability distribution $P_{\mathbf{z}}$, that is,

$$P_{\mathbf{x}}(\alpha_j) = \sum_{\beta_i} P_{\mathbf{z}}(\alpha_j, \beta_i),\ P_{\mathbf{z}}(\beta_i) = \sum_{\alpha_j} P_{\mathbf{z}}(\alpha_j, \beta_i). \tag{10.44}$$

The frequency probability theory contains the notion of conditional probability that is similar to CMM's conditioning, the postmeasurement context $C_{A=a}$ corresponds to the postmeasurement collective $\mathbf{x}_{C,A=a}$ (see [179, 203] for the details).

In contrast to the Kolmogorov measure-theoretic probability theory, in the von Mises frequency probability theory classical FTP is violated [179, 203]; the probabilities can interfere and generate the aditive perturbation of FTP in the form of the interference term. The presence of uncombinable collectives leads to the violation of the Bell-type inequalities [179, 203].

The notion of collective was the seed for future growing theory of random sequences. Besides the existence of the limits for frequencies, (10.42), a collective is characterized by stability of these limits w.r.t. place selections within a sequence $\mathbf{x}$, that is, the limit-probability is the same for all subsequences of $\mathbf{x}$ for a special class of place selections. However, von Mises' definition of place selection was criticized for nonrigorousness. Its critical analysis was very fruitful and led to the modern theory of randomness; see, for example, monograph [220]. In particular, the monograph presents a "light version" of the von Mises theory. In physics (at least in quantum physics) one does not analyze in the random structure of the sequences of measurement outcomes. In principle, in QM one can proceed with "light-collectives" determined solely by the existence of limits (10.42). The calculus of such "light-collectives" can be explored for the description of the probabilistic structure of QM [220]. The first version of the contextual probability was presented in such a "light frequency" framework.

Finally, we note that von Neumann pointed [364, 365] to the von Mises frequency probability [363] as the probabilistic foundation of QM. This is the complex foundational issue, because von Neumann used the individual interpretation (**II**) of a quantum state.

11

Contextual Structuring of Classical and Quantum Physics

11.1 Contextual Measurement Model for Kolmogorov Theory

Let $K = (\Omega, \mathcal{F}, P)$ be a Kolmogorov probability space. Here Ω is a set of any origin, $\mathcal{F}$ is a collection of its subsets forming σ-algebra, that is, $\mathcal{F}$ is closed w.r.t. countable unions and intersections, and the operation of complement. (If Ω is finite, then $\mathcal{F}$ is a collection of all its subsets.) And P is a probability measure on $\mathcal{F}$.

Set $\mathcal{C} = \{C \in \mathcal{F} : P(C) \neq 0\}$. This is the set of contexts. For each context C, the Bayes formula defines the conditional probability measure

$$P_C(G) = P(G \cap C)/P(C), G \in \mathcal{F}.$$

We highlight that the statistical mixtures of contexts are not determined, that is, for subsets C_1, C_2 of Ω and weights $p_1, p_2 \geq 0, p_1 + p_2 = 1$, there is no subset C of Ω which can be identified with the weighted sum $p_1 C_1 + p_2 C_2$.

As an illustrative example, consider some agricultural region Ω and as contexts consider its sub-fields (some areas). Generally there is no field of the form $p_1 C_1 + p_2 C_2$. In applications to decision making and cognition, one can meet situations such that $p_1 C_1 + p_2 C_2$ is determined only for a few pairs of weights p_1, p_2. This situation is related to poorness of the set of possible experimental contexts.

The set of observables $\mathcal{O}$ is the set of (discrete) random variables, $a\colon \Omega \to X_a$, where X_a is a finite set. (Discrete random variables are considered for simplicity.) Denote this set by the symbol $\mathcal{R}_d$. For $x \in X_a$, we set $\Omega_{a=x} = \{\omega \in \Omega\colon a(\omega) = x\}$. Contextual probability coincides with conditional probability given by the Bayes formula:

$$P^a_C(x) \equiv P_C(a = x) = P(\Omega_{a=x} \cap C)/P(C). \tag{11.1}$$

Thus, the set of probability distributions $\mathcal{P} = \{P^a_C\colon C \in \mathcal{C}, a \in \mathcal{O}\}$.

For any set $D \in \mathcal{F}$ and random variable $a \in \mathcal{R}_d, x \in X_a$, we define set

$$D_{a=x} = \{\omega \in D : a(\omega) = x\} = D \cap \Omega_{a=x}$$

and the map

$$T_a(x)\colon \mathcal{F} \to \mathcal{F},\ D \to D_{a=x} = T_a(x)D. \tag{11.2}$$

For any context $C \in \mathcal{C}$ and random variable $a \in \mathcal{R}_d, x \in X_a$, we define the family of contexts

$$\mathcal{C}_a(x) = \{C \in \mathcal{C}\colon P_C(a = x) > 0\} = \{C \in \mathcal{C}\colon P(C_{a=x}) > 0\}.$$

Each random variable a and its outcome x determine a map

$$T_a(x)\colon \mathcal{C} \to \mathcal{C},\ C \to C_{a=x} = T_a(x)C, \tag{11.3}$$

with the domain of definition $\mathcal{C}_a(x)$.

Thus, classical CMM M_{cl} consists of measurement contexts composed of pre-measurement contexts – elements of $\mathcal{F}$ with nonzero probabilities, observables – (discrete) random variables, and context update maps, $\mathcal{T} = \{T_a(x)\}$. Here each observable a, a random variable, determines uniquely the context update maps $T_a(x)$, and, hence, the contextual instrument.

The conditional probability is given by the Bayes formula:

$$\begin{aligned} P_C(b = y|a = x) &= P(\Omega_{b=y} \cap \Omega_{a=x} \cap C)/P(\Omega_{a=x} \cap C) \\ &= P(\Omega_{b=y} \cap C_{a=x})/P(C_{a=x}) = P_{C_{a=x}}(b = y). \end{aligned}$$

Since, for each $C \in \mathcal{C}$, P_C is a probability measure, for any pair of random variables a, b, we have the following version of FTP, formula of total probability (Section 10.2.1),

$$\begin{aligned} P_C(b = y) &= \sum_{x \in X_a} P_C(b = y|a = x)P_C(a = x) \\ &= \sum_{x \in X_a} P_{C_{a=x}}(b = y)P_C(a = x). \end{aligned} \tag{11.4}$$

In this measurement model all observables are compatible, conditional JPD coincides with JPD (again by Bayes' formula); no pair of observables show OE, since

$$\begin{aligned} P_C(a_1 = x_1, a_2 = x_2) &= P_C(a_2 = x_2, a_1 = x_1) \\ &= P_C(\omega \in \Omega : a_1(\omega) = x_1, a_2(\omega) = x_2). \end{aligned}$$

All observables show repeatability, since $P_C(a = x, a = x) = P_C(a = x)$, and RRE, since

$$\begin{aligned} P_C(a_1 = x_1, a_2 = x_2, a_1 = x_1) &= P_C(\omega \in \Omega : a_1(\omega) = x_1, a_2(\omega) = x_2, a_1(\omega) = x_1) \\ &= P_C(a_1 = x_1, a_2 = x_2). \end{aligned}$$

The Bell inequalities are not violated, since their derivation is based on the existence of JPD.

We can summarize properties of classical CMM by referring to the aforementioned list of possibilities:

- violation of FTP – no
- OE – no
- violation of replicability – no
- RRE – yes
- OE+RRE – yes
- violation of Bell inequalities – no

One of the problems of the preceding contextual representation of the classical probability is that the uniqueness conditions (10.2), (10.3), and Mackey's axiom 2, can be violated, that is, generally

$$P_C^{a_1} = P_C^{a_2} \text{ for any } C \in \mathcal{C} \not\Rightarrow a_1 = a_2, \tag{11.5}$$

$$P_{C_1}^{a} = P_{C_2}^{a} \text{ for any } a \in \mathcal{O} \not\Rightarrow C_1 = C_2. \tag{11.6}$$

This problem can be easily resolved in the standard way; see the following.

Let us consider a Kolmogorov probability space $K = (\Omega, \mathcal{F}, P)$ with a complete probability measure, that is, any subset of a set $D \in \mathcal{F}, P(D) = 0$, also belongs to $\mathcal{F}$. We recall that the symmetric difference of two sets D_1 and D_2 is defined as

$$D_1 \Delta D_2 = (D_1 \setminus D_2) \cup (D_2 \setminus D_1) = (D_1 \cup D_2) \setminus (D_1 \cap D_2).$$

For $D_1, D_2 \in \mathcal{F}$, we set $D_1 \sim D_2$ if $P(D_1 \Delta D_2) = 0$. This is an equivalence relation on $\mathcal{F}$; it splits $\mathcal{F}$ into disjoint equivalence classes. Denote the set of equivalence classes by the symbol $\tilde{\mathcal{F}}$; denote the equivalence class of zero probability sets by the symbol $\tilde{Z}$. The set of premeasurement contexts is $\tilde{\mathcal{C}} = \tilde{\mathcal{F}} \setminus \tilde{Z}$, that is, all equivalent classes of sets from $\mathcal{F}$ of nonzero measure.

We also modify the class of observables. Two random variables are equivalent, $a_1 \sim a_2$, if $P(\omega \in \Omega \colon a_1(\omega) \neq a_2(\omega)) = 0$. This is the equivalence relation on the space of random variables; in our consideration these are discrete random variables $\mathcal{R}_d$. So, $\mathcal{R}_d$ is split into disjoint classes of equivalent random variables, denote the set of these classes by the symbol $\tilde{\mathcal{R}}_d$, and set $\tilde{\mathcal{O}} = \tilde{\mathcal{R}}_d$.

For a premeasurement context $\tilde{C} \in \tilde{\mathcal{C}}$ and observable $\tilde{a} \in \tilde{\mathcal{R}}_d$ we define the probability distribution $P_{\tilde{C}}^{\tilde{a}}(x) = P_C(a = x)$ for some representatives $C \in \tilde{\mathcal{C}}$ and $a \in \tilde{\mathcal{R}}_d$ (the correctness of this definition is proved in what follows), and set $\tilde{\mathcal{P}} = \{P_{\tilde{C}}^{\tilde{a}}\}$. The modified classical contextual probability space is the triple $\tilde{\Sigma} = (\tilde{\mathcal{C}}, \tilde{\mathcal{R}}_d, \tilde{\mathcal{P}})$.

The map $T_a(x)$, see (11.3), generates a map of $\tilde{\mathcal{F}}$ into itself

$$\tilde{T}_a(x) \colon \tilde{\mathcal{F}} \to \tilde{\mathcal{F}}. \tag{11.7}$$

Set $\tilde{C}_a(x) = \{\tilde{C} \in \tilde{C}: \tilde{T}_a(x)\tilde{C} \in \tilde{C}\}$. Then $\tilde{T}_a(x): \tilde{C}_a(x) \to \tilde{C}$. A measurement context is a triple $(\tilde{C}, \tilde{a}, \tilde{T}_a)$. Modified classical CMM $\tilde{M}_{\rm cl}$ is given by the set of such measurement contexts.

Now we demonstrate that in $\tilde{M}_{\rm cl}$ the uniqueness conditions (10.2), (10.3), and Mackey's axiom 2, hold true. We assume that the ranges of values of random variables are subsets of a set X; for simplicity let $X = \mathbb{R}$. First we note that if $D_1, D_2 \in \tilde{D}$, then $P(D_1) = P(D_2)$. We have $P(D_1) = P((D_1 \cap D_2) \cup (D_1 \setminus D_2)) = P(D_1 \cap D_2) = P(D_2)$. Let now $C_1, C_2 \in \tilde{C}$, then, as we have seen, $P(C_1) = P(C_2)$; also, for any $G \in \mathcal{F}, P(G \cap C_1) = P(G \cap C_1 \cap C_2) = P(G \cap C_2)$. Hence, $P_{C_1}(G) = P_{C_2}(G)$. So, conditional probability measure P_C does not depend on the choice of a representative $C \in \tilde{C}$ and it can be denoted as $P_{\tilde{C}}$.

We show that implication (10.2) holds. Let for random variables a_1, a_2, $P_C^{a_1} = P_C^{a_2}$ for any context C. Let, for some $x, P(\Omega_{a_1=x}) > 0$. Take $C = \Omega_{a_1=x}$; then

$$1 = P_{\Omega_{a_1=x}}(\Omega_{a_1=x}) = P(\Omega_{a_2=x} \cap \Omega_{a_1=x})/P(\Omega_{a_1=x}),$$

that is,

$$P(\Omega_{a_2=x} \cap \Omega_{a_1=x}) = P(\Omega_{a_1=x}) = P((\Omega_{a_2=x} \cap \Omega_{a_1=x}) \cup (\Omega_{a_1=x} \setminus \Omega_{a_2=x}).$$

Hence, $P(\Omega_{a_1=x} \setminus \Omega_{a_2=x}) = 0$ and symmetrically $P(\Omega_{a_2=x} \setminus \Omega_{a_1=x}) = 0$. The sets $\Omega_{a_i=x}, i = 1, 2$, belong to the same equivalence class. This implies that the random variables also belong to the same equivalence class.

Now we turn to implication (10.3). Let, for any random variable a, $P^a_{C_1} = P^a_{C_2}$. Select a as the characteristic function of the set C_1. Then $P^a_{C_1}(1) = 1 = P(C_1 \cap C_2)/P(C_2)$, that is, $P(C_1 \cap C_2) = P(C_2) = P((C_1 \cap C_2) \cup (C_2 \setminus C_1)$, that is, $P(C_2 \setminus C_1) = 0$ and symmetrically $P(C_1 \setminus C_2) = 0$, that is, contexts belong to the same equivalence class.

11.2 Contextual Measurement Model for von Neumann Theory

We restrict consideration to finite-dimensional Hilbert state spaces. The space of premeasurement contexts is mathematically represented as the space of density operators $\mathcal{D}$, that is, $\mathcal{C} = \mathcal{D}$, observables are Hermitian operators (von Neumann observables [365]). Denote the space of Hermitian operators by the symbol $\mathcal{L}_H$, that is, $\mathcal{O} = \mathcal{L}_H$. This real linear space is endowed with the scalar product $\langle \hat{A} | \hat{B} \rangle = \mathrm{Tr}\, \hat{A}\hat{B}$.

Operator $\hat{A} \in \mathcal{L}_H$ has the spectral decomposition: $\hat{A} = \sum_{x \in X_A} x\, \hat{E}_A(x)$, where $\hat{E}_A(x)$ is the orthogonal projection on the subspace $\mathcal{H}_A(x)$ composed of eigenvectors with eigenvalue x, and X_A is the operator's spectral set. Then

$$P^A_\rho(x) \equiv P_\rho(A = x) = \mathrm{Tr}\, \hat{E}_A(x)\hat{\rho} = \mathrm{Tr}\, \hat{E}_A(x)\hat{\rho}\hat{E}_A(x), \qquad (11.8)$$

$$T_A(x)\colon \mathcal{D} \to \mathcal{D}, \hat{\rho}_{A=x} \equiv T_A(x)\hat{\rho} = \frac{\hat{E}_A(x)\hat{\rho}\hat{E}_A(x)}{\mathrm{Tr}\,\hat{E}_A(x)\hat{\rho}\hat{E}_A(x)}, \tag{11.9}$$

with the domain of definition $\mathcal{C}_A(x) = \{\hat{\rho} \in D\colon P_\rho(A = x) > 0\}$.

Thus, $\mathcal{P} = \{P^A_\rho\colon \rho \in \mathcal{D}, A \in \mathcal{L}_H\}$ and $\mathcal{T} = \{T_A(x)\colon A \in \mathcal{L}_H, x \in X_A\}$. We remark that observable A uniquely determines the family of maps $T_A(x), x \in X_A$ by equality (11.9). So, measurement contexts can be represented by pairs $(\hat{\rho}, \hat{A}), \hat{\rho} \in \mathcal{D}, \hat{A} \in \mathcal{L}_H$. Denote this CMM by the symbol M_{QVN}.

In this CMM one need not separately define context update maps; they are automatically encoded in observables. On the one hand, this simplifies theory. On the other hand, this is the misleading path in measurement theory; compare with quantum instrument theory.

In CMM M_{QVN}, the conditional probability is given by the following formula:

$$P_\rho(B = y|A = x) = \mathrm{Tr}\,\hat{E}_B(y)T_A(x)\rho = \frac{\mathrm{Tr}\,\hat{E}_B(y)\hat{E}_A(x)\hat{\rho}\hat{E}_A(x)}{\mathrm{Tr}\,\hat{E}_A(x)\hat{\rho}\hat{E}_A(x)}. \tag{11.10}$$

This can be rewritten as

$$P_\rho(B = y|A = x) = \frac{\mathrm{Tr}\,\hat{E}_B(y)\hat{E}_A(x)\hat{\rho}\hat{E}_A(x)\hat{E}_B(y)}{\mathrm{Tr}\,\hat{E}_A(x)\hat{\rho}\hat{E}_A(x)}. \tag{11.11}$$

In this LSR-based CMM, it is convenient to introduce the maps:

$$\mathcal{I}_A(x)\hat{\rho} = \hat{E}_A(x)\hat{\rho}\hat{E}_A(x). \tag{11.12}$$

Then the preceding formulas can be rewritten as

$$P_\rho(A = x) = \mathrm{Tr}\,\mathcal{I}_A(x)\hat{\rho}, \tag{11.13}$$

$$T_A(x)\hat{\rho} = \frac{1}{P_\rho(A = x)}\mathcal{I}_A(x)\hat{\rho}, \tag{11.14}$$

or

$$\mathcal{I}_A(x)\hat{\rho} = P_\rho(A = x)T_A(x)\hat{\rho}; \tag{11.15}$$

and the conditional probability is written in the form

$$P_\rho(B = y|A = x) = \frac{\mathrm{Tr}\,\mathcal{I}_B(y)\mathcal{I}_A(x)\hat{\rho}}{\mathrm{Tr}\,\mathcal{I}_A(x)\hat{\rho}}, \tag{11.16}$$

and conditional JPD as

$$P_\rho(A = x, B = y) = \mathrm{Tr}\,\mathcal{I}_B(y)\mathcal{I}_A(x)\hat{\rho}. \tag{11.17}$$

These formulas lead to quantum instrument theory (Section 11.3): $(A, \mathcal{I}_A(x))$ is a special quantum instrument and $(A, T_A(x))$ is the corresponding contextual instrument.

We note that in CMM $M_{\rm QVN}$ probabilities determine contexts (states) and observables (operators), that is, (10.2) and (10.3) hold (Mackey's axiom 2).

Let $P_\rho^{A_1}(x) = P_\rho^{A_1}(x)$ for all $\hat{\rho} \in \mathcal{D}$ and real x. Then $\mathrm{Tr}\,(E_{A_1}(x) - E_{A_2}(x))\hat{\rho} = \langle E_{A_1}(x) - E_{A_2}(x)|\hat{\rho}\rangle = 0$. Hence, $E_{A_1}(x) = E_{A_2}(x)$ for any x; so $\hat{A}_1 = \hat{A}_2$. Thus, an observable can be identified with the set of probability distributions $\mathcal{P}_A = \{P_\rho^A\colon \rho \in \mathcal{D}\}$.

Now let $P_{\rho_1}^A(x) = P_{\rho_2}^A(x)$ for all $\hat{A} \in \mathcal{L}_H$ and real x. Then $\mathrm{Tr}\,E_A(x)(\hat{\rho}_1 - \hat{\rho}_2) = \langle E_A(x)|\hat{\rho}_1 - \hat{\rho}_2\rangle = 0$. Hence, $\hat{\rho}_1 = \hat{\rho}_2$. Thus, a quantum state can be identified with the set of probability distributions $\mathcal{P}_\rho = \{P_\rho^A\colon A \in \mathcal{L}_H\}$.

In the von Neumann measurement theory, two observables are *compatible* if they are represented by commuting operators $\hat{A}, \hat{B}\colon\ [\hat{A}, \hat{B}] = 0$. Compatibility is interpreted as guaranteeing the possibility of joint measurement of these observables; their JPD is given by the formula:

$$P_\rho^{A,B} = \mathrm{Tr}\,\hat{E}_A(x)\hat{E}_B(y)\rho = \mathrm{Tr}\,\hat{E}_B(y)\hat{E}_A(x)\rho. \tag{11.18}$$

In fact, this is the separate axiom – a complement to the Born rule [365]. For compatible observables, JPD and conditional JPD coincide. The conditional JPD is given by

$$\begin{aligned} &P_\rho(A = x, B = y) = P_\rho(A = x)P_\rho(B = y|A = x) = \\ &P_\rho(A = x)P_{\rho_{A=x}}(B = y) == \mathrm{Tr}\,\hat{E}_B(y)\hat{E}_A(x)\rho\hat{E}_A(x) = \\ &\mathrm{Tr}\,\hat{E}_A(x)\hat{E}_B(y)\hat{E}_A(x)\rho = \mathrm{Tr}\,\hat{E}_B(y)\hat{E}_A(x)\rho. \end{aligned} \tag{11.19}$$

In particular, for compatible observables there is no OE for any state ρ. If operators $\hat{A}_1, \hat{A}_2$, do not commute, then there exists a state $\hat{\rho}$ showing OE for these observables, that is, $P_\rho(A = x, B = y) \neq P_\rho(B = y, A = x)$.

Each observable shows replicability, for example,

$$P_\rho(A = x, A = x) = \mathrm{Tr}\,\hat{E}_A(x)\hat{E}_A(x)\hat{\rho}\hat{E}_A(x) = \mathrm{Tr}\,\hat{E}_A(x)\hat{\rho} = P_\rho(A = x).$$

If observables are compatible, then for any $\hat{\rho}$ they show RRE, for example,

$$\begin{aligned} &P_\rho(A_1 = x_1, A_2 = x_2, A_1 = x_1) = \mathrm{Tr}\,\hat{E}_{A_1}(x_1)\hat{E}_{A_2}(x_2)\hat{E}_{A_1}(x_1)\hat{\rho}\hat{E}_{A_1}(x)\hat{E}_{A_2}(x_2) = \\ &\mathrm{Tr}\,\hat{E}_{A_2}(x_2)\hat{E}_{A_1}(x_1)\hat{\rho}\hat{E}_{A_1}(x) = P_\rho(A_1 = x_1, A_2 = x_2). \end{aligned}$$

We highlight that it is impossible to combine OE and RRE within CMM $M_{\rm QVN}$ [213].

FTP (the formula of total probability) can be violated; classical FTP is additively perturbed by the interference term; consider a pure state $|\psi\rangle$, then

$$P_\psi(B=\beta)=\sum_{\alpha,\alpha'}\langle\psi|\hat{E}_A(\alpha)\hat{E}_B(\beta)\hat{E}_A(\alpha')|\psi\rangle, \tag{11.20}$$

that is,

$$P_\psi(B=\beta)=\sum_{\alpha}P_\psi(B=\beta|A=\alpha)P_\psi(A=\alpha)+\delta_\psi(B=y|A), \tag{11.21}$$

where

$$\delta_\psi(B=y|A)=\sum_{\alpha\neq\alpha'}\langle\psi|\hat{E}_A(\alpha)\hat{E}_B(\beta)\hat{E}_A(\alpha')|\psi\rangle. \tag{11.22}$$

In RHS, the first summand corresponds to classical FTP and the second one is the interference term; it quantifies the degree of nonclassicality for this CMM. See Section 10.2.1 for FTP in the general contextual probabilistic framework.

Consider dichotomous observables, $A = x_1, x_2$ and $B = y_1, y_2$, of the von Neumann type. In this case the interference term has the form

$$\delta_\psi(B=y|A)=$$
$$2\cos\theta\sqrt{P_\psi(B=y|A=x_1)P_\psi(A=x_1)P_\psi(B=y|A=x_2)P_\psi(A=x_2)}, \tag{11.23}$$

where the angle $\theta=\theta(B=y|A;\psi)$.

We can summarize properties of von Neumann CMM with the list of possibilities just presented:

- violation of FTP – yes
- OE – yes
- violation of replicability – yes
- RRE – no
- OE+RRE – no
- violation of Bell inequalities – yes

11.3 Contextual Measurement Model for Quantum Instruments

In this section we present CMM M_{QI} in that a measurement process is mathematically described by quantum instrument theory (Section 1.3). This CMM extends CMM M_{QVN} in that a measurement process is mathematically described by a Hermitian operator (von Neumann observable). The space of linear Hermitian operators $\mathcal{L}_H$ is a real Hilbert space. We consider linear operators acting in it to be *superoperators*. A superoperator T is called positive if it maps the set of positive semidefinite operators onto itself: for $g \geq 0,\ T(g) \geq 0$.

Consider an observable A with a finite range of values X. Its measurements can be performed with various apparatuses; for each apparatus, the corresponding measurement procedure is mathematically described in the following way.

Any map $x \to \mathcal{I}(x)$, where for each $x \in X$, the map $\mathcal{I}(x)$ is a positive superoperator and

$$\mathcal{I}(X) \equiv \sum_{x \in X} \mathcal{I}(x) \colon \mathcal{D} \to \mathcal{D} \tag{11.24}$$

is called a *quantum instrument.* It determines some observable; denote it by the symbol A; see mathematical formula (11.32).

The probability of the output $A = x$ is given by the generalized Born rule in the form

$$P_\rho(A = x) = \text{Tr } [\mathcal{I}(x)\rho]. \tag{11.25}$$

We note that measurement with the output $A = x$ generates the state update by transformation

$$\rho \to \rho_{A=x} = T_A(x)\rho \equiv \frac{\mathcal{I}(x)\rho}{Tr\mathcal{I}(x)\rho}, \tag{11.26}$$

with the domain of definition $C_A(x) = \{\hat{\rho} \in D \colon P_\rho(A = x) > 0\}$. Let

$$\mathcal{I}(x)\rho = \hat{E}(x)\rho\hat{E}(x), \tag{11.27}$$

where $(\hat{E}(x))$ are projections giving the orthogonal decomposition of I. Such an instrument is called a projection instrument.

The most natural generalization of projection instruments is an atomic instrument. Let $(\hat{V}(x))$ be a family of linear operators constrained by the normalization condition:

$$\sum_x \hat{V}(x)\hat{V}^\star(x) = I. \tag{11.28}$$

An atomic quantum instrument is a superoperator of the form:

$$\rho \to \mathcal{I}(x)\rho = \hat{V}(x)\rho\hat{V}^\star(x). \tag{11.29}$$

Applications of the quantum instrument theory to quantum information are typically restricted by the use of atomic instruments.

The space of Hermitian operators $\mathcal{L}_H$ is the real Hilbert space, that is, for each linear operator acting in $\mathcal{L}_H$ (superoperator) its adjoint is well defined. The generalized Born rule can written as

$$P_\rho(A = x) = \langle \mathcal{I}(x)\rho | I \rangle = \langle \rho | \mathcal{I}^\star(x) I \rangle = \langle \mathcal{I}^\star(x) I | \rho \rangle, \tag{11.30}$$

where I is the unit operator and $\mathcal{I}^\star(x)$ is the superoperator that is adjoint to $\mathcal{I}(x)$ in Hilbert space $\mathcal{L}_H$. Hence, the generalized Born rule has the form:

$$P_\rho(A = x) = \text{Tr } \hat{\Pi}(x)\rho, \tag{11.31}$$

where

$$\hat{\Pi}(x) = \mathcal{I}^\star(x)I. \tag{11.32}$$

Operators $\hat{\Pi}(x), x \in X$, are called *effects*; they are positive semidefinite Hermitian and sum up to the unit operator:

$$\sum_{x \in X} \hat{\Pi}(x) = I.$$

The family of operators $\hat{\Pi} = (\hat{\Pi}(x), x \in X)$ is called a *positive operator valued measure* (POVM): for a subset Δ of X, we set

$$\hat{\Pi}(\Delta) = \sum_{x \in \Delta} \hat{\Pi}(x) \geq 0;$$

this is an additive operator-valued measure, that is, for $\Delta_1, \Delta_2 \subset X, \Delta_1 \cap \Delta_2 = \emptyset$,

$$\hat{\Pi}(\Delta_1 \cup \Delta_2) = \hat{\Pi}(\Delta_1) + \hat{\Pi}(\Delta_2).$$

Instruments of the projection type, see (11.27), determine the special class of POVMs, *projection valued measures* (PVMs).

Two POVMs $\hat{\Pi}_A = (\hat{\Pi}(x), x \in X)$ and $\hat{\Pi}_B = (\hat{\Pi}_B(y), y \in Y)$ are called *compatible* if there exists a POVM $C = (\hat{C}(x,y), (x,y) \in X \times Y)$ such that

$$\hat{\Pi}_A(x) = \sum_{y \in Y} \hat{C}(x,y), \ \hat{\Pi}_B(x) = \sum_{x \in X} \hat{C}(x,y). \tag{11.33}$$

Compatibility is interpreted as guaranteeing the possibility of the joint measurement of these observables; their JPD is given by generalization of the Born rule for compatible von Neumann observables:

$$P_\rho^{A,B}(x,y) = \operatorname{Tr} \hat{C}(x,y)\hat{\rho}. \tag{11.34}$$

In contextual probability space, contexts are mathematically represented by density operators (quantum states), $\mathcal{C} = \mathcal{D}$, observables by POVMs (also known as generalized quantum observables), the probability distributions are determined by the generalized Born rule (11.25). CMM M_{QI} is endowed by quantum instrument maps updating quantum states (contexts) due to the measurement feedback (11.26).

Consider POVM $\hat{\Pi} = (\hat{\Pi}(x))$ and all quantum instruments generating it via (11.32). Then the corresponding state (context) update maps are defined by equality (11.26). The same POVM, generalized observable, can be coupled to a variety of such maps. Therefore the commonly used approach highlighting POVMs as generalized observables is ambiguous. POVMs are just biproducts generated by quantum instruments.

The quantum instruments just considered were invented in article [97] and these are Davies–Levis instruments, so $M_{\mathrm{QI}} = M_{\mathrm{QI;DL}}$. In quantum information theory, one uses the special class of quantum instruments given by *completely positive maps* $\mathcal{I}(x)$; denote the corresponding CMM $M_{\mathrm{QI;O}}$, where I use the index "O" to mention Masanao Ozawa who contributed so much to the theory of such quantum instruments [293]-[295]. It is commonly assumed that the instruments belonging to $M_{\mathrm{QI;DL}} \setminus M_{\mathrm{QI;O}}$ are nonphysical. I debated this question with Masanao Ozawa and he firmly stays on this position. As was proved by him, only completely positive instruments can be realized via the indirect measurement scheme (Section 1.4). This scheme is adequate to quantum measurement processes and any deviation from this scheme is nonphysical. Nevertheless, it might be that in quantum-like modeling, even instruments which are not completely positive can find applications. Such applications would lead to modification of the indirect measurement scheme, maybe via consideration of nonunitary interactions.

We remark that instrument maps $\mathcal{I}(x)$ are linear in the Hilbert space $\mathcal{L}_H$. In terms of context (state) update, these maps can be written as

$$\mathcal{I}(x) = P_\rho(A = x)T_A(x). \tag{11.35}$$

Hence, in quantum instrument CMM scaling of the update map $T_A(x)$ by probability $P_\rho(A = x)$ is a linear map. Generally, LSR-CMM with the context (state) space $\mathcal{C} = \mathcal{D}$ can be endowed with context (state) update maps such that scaling (11.35) can be nonlinear. CMM $M_{\mathrm{QI;NL}}$ with nonlinear context update maps might be useful for quantum-like modeling. It is interesting to find concrete applications of $M_{\mathrm{QI;NL}}$. Of course, such applications would lead to modification of the indirect measurement scheme (Section 1.4).

We can summarize properties of quantum instrument CMM, both models $M_{\mathrm{QI;DL}}$ and $M_{\mathrm{QI;O}}$:

- violation of FTP – yes
- OE – yes
- violation of replicability – yes
- RRE – no
- OE+RRE – generally no, but can be realized by special instruments
- violation of Bell inequalities – yes

It is interesting to find a property distinguishing $M_{\mathrm{QI;DL}}$ and $M_{\mathrm{QI;O}}$ via an experimental test, namely, some experimentally testable property such that only completely positive instruments have it.

One of the important features of the von Neumann model is coincidence of JPD and conditional JPD for compatible observables. In contrast, the instrument model shows that generally the situation is not simple at all. Consider two instruments $\mathcal{I}_A(x)$ and $\mathcal{I}_B(y)$ such that their observables are POVMs of PVM type, that is, $\hat{E}_A = (\hat{E}_A(x))$ and $\hat{E}_B = (\hat{E}_B(y))$. They are jointly measurable and the JPD is given by formula (11.18). The conditional JPD is given by

$$P_\rho(A = x, B = y) = P_\rho(A = x)P_\rho(B = y|A = x) = \\ P_\rho(A = x)P_{\rho_{A=x}}(B = y) = \operatorname{Tr} \mathcal{I}_B(x)\mathcal{I}_A(x)\rho. \tag{11.36}$$

The right-hand sides of (11.18) and (11.36) coincide only if the instrument superoperators are of the projection type, namely, $\mathcal{I}(x)\rho = \hat{E}(x)\rho\hat{E}(x)$.

Moreover, in this case two projection-type observables, PVMs, can have a variety of conditional probability distributions corresponding to the different instruments generating them by the rule $\hat{E}(x) = \mathcal{I}^\star(x)I$.

11.4 Ordered Space Measurement Model with Measure States

In this section we connect the generalized probability theory (the Davies–Lewis approach [97]) for probability measures with CMM. Here we use the ordered linear space approach. This is the concrete application of the universal scheme based on the abstract framework of ordered linear spaces. Consider the space $\mathcal{M}$ of all real-valued measures on some set Ω with a σ-algebra of subsets $\mathcal{F}$, that is, $\mathcal{M} \equiv \mathcal{M}(\Omega, \mathcal{F})$. Real linear space $\mathcal{M}$ has the natural order structure and the positive cone $\mathcal{M}^+$ consisting of nonnegative measures. Consider the elements of this cone given by probability measures, that is, $\mu \geq 0$ and $\mu(\Omega) = 1$; denote this set by the symbol $\mathcal{S}$; this is the set of states; $\mathcal{S}$ is a convex subset of $\mathcal{M}$. The latter is endowed with the variation norm, $||\mu|| = \mathrm{var}(\mu)$ and it is a Banach space. Consider its dual space $\mathcal{M}'$, the space of continuous linear functionals $f\colon \mathcal{M} \to \mathbb{R}$. Denote by $\mathcal{A}$ the subset of $\mathcal{M}'$, consisting of functionals mapping $\mathcal{S}$ into [0, 1]. Elements of $\mathcal{A}$ are called effects; these are basic observables. They can be described solely in the terms of the state space $\mathcal{S}$ as affine functionals valued in [0, 1], that is, $\mathcal{A} \equiv \mathcal{A}(\mathcal{S})$.

Consider the functional $u \in \mathcal{M}'$ defined as $\mu \to \langle u|\mu\rangle = \mu(\Omega)$. Its characteristic property is that $\langle u|\mu\rangle = 1$ for any state $\mu \in \mathcal{S}$.

Let $X = \{x_1, .., x_m\}$ be a finite set and let $A = (A(x_i), i = 1, \ldots, m)$ where $A(x_i) \in \mathcal{A}(\mathcal{S})$ and $A(X) \equiv \sum_{x \in X} A(x) = u$. Such vectors of functionals are analogs of POVMs; we call them $\mathcal{M}$-POVMs. These are observables of the contextual probability space Σ_{measure} with contexts $\mathcal{C} = \mathcal{S}$ and the set of probability distributions $\mathcal{P}$ defined as

$$P_\mu^A(x) \equiv P_C(A = x) = \langle A(x)|\mu\rangle.$$

As we learn from the quantum instrument theory, the basic elements of measurement procedures are not observables, but instruments. Let $\mathcal{L}(\mathcal{M})$ denote the space of continuous linear operators, $J\colon \mathcal{M} \to \mathcal{M}$. The $\mathcal{M}$-instrument with the range of values X is a map $\mathcal{I}\colon X \to \mathcal{L}(\mathcal{M})$ such that $\mathcal{I}(x)\mathcal{M}^+ \subset \mathcal{M}^+$ and

$$\mathcal{I}(X) \equiv \sum_x \mathcal{I}(x)\colon \mathcal{S} \to \mathcal{S}.$$

Each instrument determines the state update map

$$\mu \to T_A(x)\mu \equiv \frac{1}{\mathcal{I}(x)\mu(\Omega)}\mathcal{I}(x)\mu = \frac{1}{\langle u|\mathcal{I}(x)\mu\rangle}\mathcal{I}(x)\mu,$$

and the probability distribution

$$P_\mu(A = x) = \langle u|\mathcal{I}(x)\mu\rangle.$$

The domain of definition of the state update map $T_A(x)$ is given by the set of probability measures $\mathcal{C}_A(x) = \{\mu \in S\colon P_\mu(A = x) > 0\}$.

Let $J\colon \mathcal{M} \to \mathcal{M}$ be a continuous linear operator. Then its adjoint operator $J^\star$ is well defined, $J^\star\colon \mathcal{M}' \to \mathcal{M}'$, and

$$\langle J^\star f|\mu\rangle = \langle f|J\mu\rangle.$$

Set

$$\hat{\Pi}(x) = \mathcal{I}^\star(x)u.$$

Then, for each x, $\hat{\Pi}(x)$ is an effect, that is, $\hat{\Pi}(x) \in \mathcal{A}(\mathcal{S})$. So, each $\mathcal{M}$-instrument determines a $\mathcal{M}$-POVM.

The $\mathcal{M}$-CMM consists of context (states) given by probability measures and POVM-observables with state updates given by instruments.

11.5 Linear Space Representation for Contextual Probability Space

The state space is given by the set S, the set of possible measurement outcomes of an observable quantity is denoted by X. Let a system be in a state $s \in S$. A probability $p(x, s)$ is assigned to any possible outcome $x \in X$. Thus, we have a function

$$p\colon X \times S \to [0, 1].$$

To each outcome $x \in X$ and state $s \in S$, this function is a probability of the outcome x for the system that is in the state s. The generalized probability model is a triple $(S, p(\cdot, \cdot), X)$. Denote by $\Phi_{[0,1]}$ the space of function from X to $[0, 1]$. By considering state s as a variable, we obtain the map

$$S \to \Phi_{[0,1]},\ s \to s(\cdot) = p(\cdot, s). \tag{11.37}$$

It is natural to assume that each state $s \in S$ determines the probability distribution uniquely, that is,

$$p(x, s_1) = p(x, s_2) \text{ for any } x \in X \Rightarrow s_1 = s_2. \tag{11.38}$$

Under this assumption the map (11.37) is injection. Thus, each state s can be mapped to a function belonging to space $\Phi_{[0,1]}$; it will be denoted by the same symbol, s. Consider now the vector space Φ of all real-valued functions on X. So, S is identified with a subset of this functional space. Consider its closed convex hull $\bar{S}$. The vectors from it are all possible probabilistic mixtures (convex combinations) of states in S. Each $x \in X$ defines a linear functional on Φ, $\phi \to f_x(\phi) = \phi(x)$. If $\phi = s \in \bar{S}$, then $f_x(s) = s(x) \in [0, 1]$, that is, $f_x : \bar{S} \to [0, 1]$. This is an affine functional on the convex set $\bar{S}$. It describes a measurement outcome and $f_x(s) = p(x, s)$ is the probability for this outcome in state s.

Denote by $A(\bar{S})$ to the space of all affine functionals

$$f : \bar{S} \to [0, 1].$$

In particular, for any $x \in X$, $f_x \in A(\bar{S})$. Any functional $f \in A(\bar{S})$ describes an outcome of some observable, and thus $f(s)$ is the probability for that outcome in state s.

In QM, $\bar{S}$ is the set of density operators and elements of $A(\bar{S})$ are effects – components of POVMs, $f(\rho) = \operatorname{Tr} \hat{E}_f \hat{\rho}$, where $\hat{E}_f$ is the effect corresponding to the affine functional f.

The elements of $A(\bar{S})$ are called *effects*. It is typically assumed that there exists an element u of $A(\bar{S})$ such that $u(s) = 1$ for any $s \in \bar{S}$. It is an analog of quantum observable given by the unit operator I. Consider the pointwise order structure on $A(\bar{S}), f \leq g$ if and only if $f(s) \leq g(s)$ for any state s. Thus, any observable $f \in A(\bar{S})$ is majorated by $u, 0 \leq f \leq u$. A discrete measurement is represented by a set of effects (f_i) such that $\sum_i f_i = u$.

We now connect LSR for the contextual probabilistic model. We assume that all observables have the same range of values X. The straightforward intention is to set $S = \mathcal{C} \times \mathcal{O}$. Let, as earlier, $\Phi_{[0,1]}$ denote the space of functions from X to $[0, 1]$. We map S into $\Phi_{[0,1]}$, $s = (C, A) \to P^A_C$. However, generally this map is not injection: $P^{A_1}_{C_1}(x) = P^{A_2}_{C_2}(x)$ for all $x \in X$ does not imply that $C_1 = C_2$ and $A_1 = A_2$. So, such straightforward construction seems to be nonproper for our aim.

We modify it by setting $\mathbf{X} = \mathcal{O} \times X$, the elements of $\mathbf{X}$ are pairs $\mathbf{x}$ =(observable, outcome)= (A, x). We now use the symbols $\Phi_{[0,1]}$ and Φ for functions from $\mathbf{X} \to [0, 1]$ and to the real line respectively. Each context C can be represented as a vector belonging to $\Phi_{[0,1]}$, $C(\mathbf{x}) = C(A, x) = P^A_C(x)$. Due to (10.3), embedding of the set of contexts $\mathcal{C}$ into $\Phi_{[0,1]}$ is injection. Again denote by $\bar{\mathcal{C}}$ the convex hull of $\mathcal{C}$. Each point $\mathbf{x}$ =(observable, outcome)= (A, x) determines the affine functional $C \to f_{A,x}(C) = P^A_C(x) \in [0, 1]$. Now fix $A \in \mathcal{O}$ and consider the family of functionals $F = (F_A(x) = f_{A,x} : x \in X)$. This is a representation of observable A. So, any contextual probability model can be realized like a COM – an observational COM.

12

Växjö Interpretation

12.1 Bayesian Classical Probability Inference

Bayesian inference is the machinery for obtaining the posterior probability from two antecedents: a prior probability and a likelihood function derived from a statistical model for the observed data. The posterior probability is calculated from Bayes' theorem:

$$P(H \mid E) = \frac{P(E \mid H)P(H)}{P(E)}, \tag{12.1}$$

where

- H is any hypothesis whose probability is updated by observational data. Typically, there are a few competing hypotheses, and the aim of Bayesian inference is to determine which of them is the most probable.
- $P(H)$, the prior probability, is the estimate of the probability of H before the data E, the post-observation evidence, is observed.
- evidence E corresponds to newly collected data.
- $P(H \mid E)$, the posterior probability, is the probability of the hypothesis H given new observed data E.
- $P(E \mid H)$ is the probability of observing the event E given the hypothesis .H It is called the likelihood.[1]
- $P(E)$ is sometimes called model evidence, it is the probability of evidence E which is calculated on the basis of FTP.

The task of Bayesian inference is to find $P(H \mid E)$, the probability of a hypothesis H given the observed data E.

[1] As a function of E with the hypothesis H fixed, $P(E \mid H)$ represents the compatibility of the evidence and the hypothesis. It is a function of the evidence E, while the posterior probability $P(H \mid E)$ is a function of the hypothesis, H.

Bayes' inference formula is typically represented with expressing the model evidence through FTP; denote negation of a hypothesis H by $\bar{H}$, then

$$P(H \mid E) = \frac{P(E \mid H)P(H)}{P(E \mid H)P(H) + P(E \mid \bar{H})P(\bar{H})}. \tag{12.2}$$

12.2 Contextual Probability Inference

Consider Bayesian inference within CMM. Hypothesis choice is represented as an observable $H \in \mathcal{O}, X_H = \{H_1, \dots, H_n\}$. Evidence collection is also represented as an observable $E \in \mathcal{O}$ with range of values $X_E = \{E_1, \dots, E_m\}$, where in the simplest case E_j is a real number x_j, the measurement's outcome. Since CMM determines the notion of conditional probability, we can rewrite the left and right sides of formula (12.1) in a contextual framework by coupling all probabilities to some context $C \in \mathcal{C}$, an inference context:

$$P_C(H = H_i|E = E_j) = P_{C_{E=E_j}}(H = H_i) = P_{T_E(j)C}(H = H_i),$$
$$P_C(E = E_j|H = H_i) = P_{C_{H=H_i}}(E = E_j) = P_{T_H(i)C}(E = E_j),$$

and $P_C(H = H_i), P_C(E = E_j)$. However, the Bayes theorem of the classical probability theory does not hold true, since generally:

$$P_{C_{E=E_j}}(H = H_i) \neq P_{C_{H=H_i}}(E = E_j)P_C(H = H_i)/P_C(E = E_j). \tag{12.3}$$

In the quantum projection measurement model, this formula is true for compatible observables, namely, if $[\hat{E}, \hat{H}] = 0$.

Consider now the quantum instrument model two instruments $\mathcal{I}_H$ and $\mathcal{I}_E$ with observables given by projection-valued measures (PVMs), that is, $\hat{F}(i) = \mathcal{I}_H(i)^\star I, \hat{G}(j) = \mathcal{I}_E(j)^\star I$ are projections giving the orthogonal decompositions of the state space. Suppose that these observables are compatible in the sense that $[\hat{F}(i), \hat{G}(j)] = 0$. Generally

$$P_{C_{E=E_j}}(H = H_i)P_C(E = E_j) = \mathrm{Tr}\,\mathcal{I}_H(i)\mathcal{I}_E(j)\rho \neq \tag{12.4}$$
$$P_{C_{H=H_i}}(E = E_j)P_C(H = H_i) = \mathrm{Tr}\,\mathcal{I}_E(j)\mathcal{I}_H(i)\rho.$$

So, for quantum instruments with observables of projection type that are compatible and have JPD, classical probability inference machinery is not applicable. This happens, since commutativity of PVMs is not equivalent to commutativity of instruments. On the other hand, if the superoperators commute, $[\mathcal{I}_H(i), \mathcal{I}_E(j)] = 0$, the classical Bayesian inference machinery can be employed, even if the corresponding observables are incompatible or they are not PVMs.

Hence, in CMM, instead of the classical Bayesian inference based on the Bayes formula for conditional probability, we should employ non-Bayesian inference based on estimation of the probability $P_C(H = H_i|E = E_j) = P_{C_{E=E_j}}(H = H_i)$ on the basis of the E-measurement and its comparison with the prior probability $P_C(H = H_i)$. In a given context C, one first guesses on the value of the probability $P_C(H = H_i)$ for the hypothesis H_i. Then the E-observable is measured and an outcome $E = E_j$ is obtained. This measurement updates the initial context, $C \to C_{E=E_j} = T_E(j)C$. Finally, the probability for the hypothesis H_i is updated as $P_{C_{E=E_j}}(H = H_i)$. The quantity

$$Q(H_i|E_j) = \frac{P_{C_{E=E_j}}(H = H_i)}{P_C(H = H_i)} \tag{12.5}$$

reflects the dynamics of probability update. If $Q(H_i|E_j) > 1$ the observation of the outcome $E = E_j$ increases the degree of the belief in the validity of the hypothesis H_j. To compare different hypotheses, one calculates odds, for example, in the simplest case of a hypothesis and its negation, namely, $X_H = \{H, \bar{H}\}$:

$$O(H, \bar{H}|E_j) = \frac{P_{C_{E=E_j}}(H)}{P_{C_{E=E_j}}(\bar{H})}. \tag{12.6}$$

12.3 Växjö Interpretation and Contextual Probability Inference

Now we use the preceding constructions to form the Växjö interpretation of QM [189, 203]. As already pointed out, this is one of the sub-interpretations of the information interpretation (**INI**). It is based on the Bohr complementary principle with the emphasis on its contexual counterpart, the principle of contextuality (Section 3.2.1). It is also one of the versions of the statistical interpretation (SI), that is, a quantum state provides a description of certain statistical properties of an ensemble of similarly prepared systems. In CMM M_{QVN}, a quantum state represents a pre-measurement context $C = C_\rho$.

- The probabilistic structure of QM is given by a particular contextual probability space.
- Quantum measurement model is a part of CMM.
- Quantum formalism is a machinery for probability inference.

This inference is nonclassical, but there is nothing exotic in it: within the Växjö interpretation this is just one of the contextual inferences. In our contextual framework, the quantum and classical probabilistic structures are similar, although they differ in some specialties, as say the questions about the violations of FTP or the Bell inequalities. The long-term debate of classical versus quantum is not central for quantum foundations, although it plays an important role in their development. It might be that the main foundational question is the following one:

Why is the quantum model fruitful in the description of statistics of measurements in the micro-world?

On Physical Origin of Linear Space Representation

The basic physical principle of QM is the Bohr complementarity principle, so the existence of incompatible observables is the main distinguishing property of QM. However, plenty of contextual probability spaces can be found with this property. How can one determine the proper probability model for micro-phenomena among them? The answer to this question cannot be found through mathematical analysis of the quantum formalism. It can be found only from further physical studies. The complex Hilbert space formalism matches perfectly the wave modeling, as is known from the classical field theory. One may guess that such structure can be found beyond the quantum theory. The wave viewpoint on QM was guiding for Schrödinger in the derivation of his equation and for de Broglie in his association of a wave with any quantum particle. Schrödinger's attempt to treat QM as a wave mechanics was not successful. "Schrödinger's wave" associated with a compound system, for example, a pair of electrons cannot be described as a wave on the physical space.[2] De Broglie's double solution theory and its relative, Bohmian mechanics, are nonlocal. Nevertheless, Einstein was sure that a classical field model reproducing QM could be constructed (see the book by Einstein and Infeld [117]). In a series of works (see, for example, monograph [212] and Chapter 16) I and the coauthors developed *prequantum classical statistical field theory* (PCSFT). This is a random field theory reproducing the basic statistical predictions of QM, in particular, the representation of quantum states by density operators and observables by Hermitian operators. Density operators correspond to normalized correlation operators and Hermitian operators to quadratic forms of prequantum random fields. PCSFT is a local theory, but it presumes the presence of a background field carrying correlations. In particular, entanglement in a compound quantum system is generated with the help of this background field; for example, the electromagnetic zero point field can serve as such a background field. However, for the moment picturing QM as a special image of some classical field model is merely a speculation.

Växjö Interpretation and Quantum Foundations

Now we come back to the general contextual probability and measurement theory. As stressed a few times, this framework demystifies QM, at least its probabilistic part. As one of the possible mathematical formalisms for probability inference, it is not more mystical than classical probability theory. Of course, some properties of

[2] Nowadays, in the light of development of string theory, employing of multidimensional spaces in physics is not so unusual, but in 1920 and 1930s waves on multidimensional spaces, $n > 3$, were treated as unphysical.

the classical and quantum models are different, but they are based on the same ideology of contextual probability and its update due to measurement. CMM does not suffer of nonlocality: for observations employing spatially separated complexes of physical conditions (e.g., measurement apparatuses), the context update machinery works consistently without appealing to spooky action at a distance. The violation of the Bell inequalities as well as FTP is the consequence of the nonclassical state update that is not based on the Bayes formula for conditional probability.

Contextual theory is useful for testing the possibility that quantum measurement theory does not cover completely statistical phenomena in the micro-world, that maybe more complex probabilistic models should be explored. A good example of comparing the quantum probability with other probabilistic models is testing the validity of Sorkin's inequalities. In the universal contextual framework they can be violated, but they are valid in the quantum formalism. And a series of advanced experiments [340, 341] confirmed their validity.

Another application of the contextual approach is to quantum-like modeling, the applications of the quantum methodology and formalism outside of physics, for example, in cognition, psychology, decision making, economics and finance, and social and political sciences (Chapter 18). In quantum-like modeling coupling to the wave phenomena and, hence, to the complex Hilbert space formalism, is not as convincing as in quantum physics.[3] More general contextual probabilistic models can match better the mental and social phenomena.

Quantum CMMs, especially M_{QVN}, are attractive due to their simplicity – working with linear algebra in complex Hilbert spaces (typically finite-dimensional). Therefore, development of abstract CMMs does not diminish the role of the conventional complex Hilbert space formalism. Moreover, as is shown, it is possible to construct LSR of an arbitrary contextual model and hence to operate in a linear space. However, one should remember that the latter is just a special mathematical representation of CMM. Linearity should be taken with caution. I guess that in quantum physics linearity is objective and it represents the physical field structure beyond QM.

Växjö Interpretation and QBism: Similarities and Differences

Finally, we point out to QBism. Similarly to the Vẍjö interpretation, QBism treats QM, at least its probabilistic part, as the special machinery for probability update. Mathematically, QBism is aimed to exclude quantum states from consideration and operate solely with probabilities. (This is my understanding of QBism.) In contrast to the Växjö interpretation, QBism interprets probabilities subjectively, and what is even more distinctive, it interprets measurements' outcomes as private

[3] However, in cognition and modeling of brain function linking with electromagnetic fields in the brain and surrounding environment, up to linking with the zero point field, is discussed in [205].

experiences of observers. Although contextual probability theory is neutral w.r.t. the probability interpretation, by the Växjö interpretation of QM probabilities are interpreted statistically; they are connected with the frequencies of realization of measurements' outcomes through the law of large numbers,

$$P_C^A(x) = \lim_{n \to \infty} \frac{n_C^A(x)}{\sum_y n_C^A(y)},$$

where $n_C^A(y)$ is the number of observations with the outcome $A = y$ for the (pre-observation) context C. One of the crucial mathematical differences between the Växjö interpretation and QBism is in quantum modifications of FTP. In CMM, generally FTP is an additive deformation of FTP of classical probability theory. If the interference term equals zero, the contextual FTP coincides with the classical one. QBism's analog of FTP differs from classical FTP crucially; this is a totally different formula. Classical FTP cannot be obtained as the limiting case of QBism's FTP.

PART IV

Contextual Entanglement in Quantum and Classical Physics

13
Probabilistic Entanglement of Quantum Observables

13.1 Introduction to Observables' Entanglement

In Section 10.2, we defined the notion of contextual probabilistic entanglement. This approach to mathematical formalization of this notion has no direct connection to the linear space representation (LRS) of probability and, in particular, to the tensor product structure. Probabilistic entanglement is introduced for any contextual measurement model (CMM) and it formalizes dependence of observables (generally instruments) within CMM. As might be expected, it is natural to describe connection between probabilistic and LSR entanglements. This is done in the present chapter (based on articles [51, 52]). The aim is not only to connect two mathematical frameworks for entanglement, but also to clarify and demystify the notion of entanglement by highlighting the observables' role and exploring the quantum conditional probability calculus. As highlighted in Section 10.2, probabilistic entanglement is entanglement of observables. It expresses the degree of dependence between two quantum observables A and B for a quantum state (premeasurement context).[1] We repeat that in our approach it is meaningless to speak about entanglement without selecting the fixed observables A and B. In this aspect our work is close to the papers [351, 352]; we cite paper [352]: "Here we propose that a partitioning of a given Hilbert space is induced by the experimentally accessible observables.... In this sense entanglement is always relative to a particular set of experimental capabilities."

The distinguishing feature of our approach is exploring the calculus of quantum conditional probabilities for formalization of the notion of entanglement and not properties of the tensor product. Observables' entanglement differs from the standard tensor product nonfactorization approach, which is widely employed in modern quantum physics, especially quantum information theory; see, for example, [373] for the definition:

[1] We restrict considerations to CMM M_{QVN} with von Neumann observables given by Hermitian operators and the state update described by the projection postulate (Section 11.2).

"A state is said to be entangled if it cannot be written as a convex sum of tensor product states."

In this chapter we proceed with pure states; the scheme of extension of this approach to mixed states is straightforward. We start with an overview of the main distinguishing characteristics of observables' entanglement.

Dependence of Observables

Dependence of observables is formalized on the basis of the calculus of quantum conditional probabilities (the special case of CMM). For dichotomous observables A and B, a state $|\psi\rangle$ is AB-entangled if and only if

$$P_{\psi}(B = +|A = +) \neq P_{\psi}(B = +|A = -). \tag{13.1}$$

(Following the LSR traditional terminology, we say that "a state $|\psi\rangle$ is AB-entangled," although in our framework it would be more natural to say "observables A and B are entangled in state (context) $|\psi\rangle$"; so we shall use the later terminology as well.) There is nothing mystical in this notion of AB-entanglement. The difference between conditional probabilities quantifies the degree of dependence of B-outcomes on A-outcomes. This measure of AB-entanglement can be expressed in Hilbert space terms and it is similar to the concurrence-measure for the usual entanglement.

AB-entanglement is not coupled to the tensor product structure. It can be considered, for example, for a state space $\mathcal{H}$ having nonfactorizable dimension, say $\dim \mathcal{H} = 5$. Generally AB entanglement is introduced for noncommuting operators (incompatible observables). But, in the main part of this chapter we work with commuting operators (with dichotomous spectrum). As can be expected, for noncommuting operators, AB and BA-entanglements do not coincide. The main part of this chapter is devoted to the Hilbert space characterization of AB-entangled quantum states. The standard entangled states are defined as nonfactorizable w.r.t. the tensor product structure on $\mathcal{H}$. We demonstrate that, for compatible dichotomous observables, AB-entangled states can be characterized as amplitude nonfactorizable states. Here amplitudes are defined as norms of the state projections on subspaces $\mathcal{H}_{AB}(\alpha\beta)$ consisting of common eigenvectors of the operators (with eigenvalues α, β). Such expansion w.r.t. to the joint eigenvectors was actively explored in works of Ozawa, for example, [297], devoted to a similar problem.

Conditional probability treatment of entanglement can be considered an indirect contribution to critique of nonlocal interpretation of quantum mechanics (QM) and connection of the violation of the Bell inequalities with spooky action at a distance.

Correlations of Observables

The description of entangled states as amplitude nonfactorizable states leads to the following result. For compatible dichotomous observables, a quantum state $|\psi\rangle$ is

AB-entanglement if and only if correlation of these observables in this state is not factorisable, that is,

$$\langle AB \rangle_\psi \neq \langle A \rangle_\psi \langle B \rangle_\psi, \tag{13.2}$$

or observables covariance is nonzero:

$$\mathrm{cov}\,(A, B; \psi) \equiv \langle (A - \langle A \rangle_\psi)(B - \langle B \rangle_\psi) \rangle_\psi = \langle AB \rangle_\psi - \langle A \rangle_\psi \langle B \rangle_\psi \neq 0. \tag{13.3}$$

Let $\mathcal{H} = \mathcal{H}_1 \otimes \mathcal{H}_2$. In this case, it is worthwhile to compare AB-entanglement and usual entanglement w.r.t. to the correlation property. If a quantum state $|\psi\rangle$ is factorisable, that is, $|\psi\rangle = |\phi_1\rangle \otimes |\phi_2\rangle$, then, of course, correlation is factorisable for any pair of operators,

$$\langle AB \rangle_\psi = \langle A \rangle_\psi \langle B \rangle_\psi, \text{ i.e. } \mathrm{cov}\,(A, B; \psi) = 0. \tag{13.4}$$

However, the inverse statement is not correct. For some entangled states, the equality (13.4) holds true. The latter is impossible for AB-entangled states.

Entanglement versus Dependence

We recall that in classical probability theory (Section 1.5) independence of random variables A and B implies correlation factorization

$$\langle AB \rangle = E[AB] = E[A]E[B] = \langle A \rangle \langle B \rangle, \text{ i.e. } \mathrm{cov}\,(A, B; K) = 0 \tag{13.5}$$

where E denotes the mathematical expectation for a classical (Kolmogorov) probability space K (Section 1.5). For classical dichotomous random variables, independence is equivalent to uncorrelation; hence, for such random variables dependence is equivalent to correlation, that is, nonfactorization of product's average. We obtain the same characterization of compatible dichotomous AB-entangled quantum observables. For such observables, the notion of observables entanglement is simply the Hilbert space representation of the notion of dependence for classical random variables, cov $(A, B;) \neq 0$. Generally observables entanglement extends the notion of dependence of random variables, within the Hilbert space formalism.

In Section 13.6, we study $A_u B_u$-entanglement for families of dichotomous pairwise compatible observables, that is, represented by commuting operators, $[\hat{A}_u, \hat{B}_u] = 0$, $\mathcal{U} = (\hat{u})$ is a family of unitary transformations. Here generally $[\hat{A}_u, \hat{A}_v] \neq 0, [\hat{B}_u, \hat{B}_v] \neq 0$. And we found an analog of the singlet state $|\psi_{\mathcal{U}}\rangle$, namely, a state that is $A_u B_u$-entangled for any transformation $\hat{u} \in \mathcal{U}$.

This construction highlights the essence of quantum entanglement, as formalized within quantum probability calculus in complex Hilbert space. For each concrete pair of compatible dichotomous observables, $A_u B_u$-entanglement is just the Hilbert space representation of dependence of classical random variables or equivalently (since this is the dichotomous case) correlation of random variables. So for each pair, there exists a classical probability space (Section 1.5) K_u such that cov $(A_u, B_u; K_u) \neq 0$, that is, correlation of these two observables can be described within classical probability theory. However, generally it is impossible to construct a probability space K which would describe dependence for all pairs of observables $\hat{A}_u, \hat{B}_u$. In contrast to this classical impossibility, quantum probability can describe all these correlations as correlations in the state $|\psi_{\mathcal{U}}\rangle$.

This example illustrates the general advantage of quantum entanglement in the description of observables' correlations. Thus, the nonclassical essence of quantum observables entanglement is the possibility of modeling correlations for a wider class of observables than in classical probability theory. Hence, our approach to entanglement highlights not the strength of correlations, but the possibility of realizing a wide class of correlated observables for the same quantum state.

13.2 Observables' Entanglement

All probabilities under consideration depend on quantum state $|\psi\rangle$, that is, $P \equiv P_\psi$. We recall once again that (in CMM M_{QVM}, section 11.2) an observable A (with a discrete range of values) is represented by a Hermitian operator $\hat{A} = \sum_\alpha \alpha \hat{E}_A(\alpha)$, where $\hat{E}_A(\alpha)$ is projection onto the space $\mathcal{H}_A(\alpha)$ of eigenvectors for the eigenvalue α. For a pure state $|\psi\rangle$, the probability of getting the outcome $A = \alpha$ is given by Born's rule: $P_\psi(A = \alpha) = \|\hat{E}_A(\alpha)\psi\|^2$. A measurement with the outcome $A = x$ generates back-action onto the system's state: $|\psi\rangle \to |\psi\rangle_{A=\alpha} = \hat{E}_A(\alpha)|\psi\rangle / \|\hat{E}_A(\alpha)\psi\|$. Measurements of another observable, say B, conditioned on the outcome $A = \alpha$ lead to probability: $P_\psi(B = \beta | A = \alpha) = \|\hat{E}_B(\beta)\psi_{A=\alpha}\|^2 = \frac{\|\hat{E}_B(\beta)\hat{E}_A(\alpha)\psi\|^2}{\|\hat{E}_A(\alpha)\psi\|^2}$.

Consider two (generally incompatible) observables A and B represented by Hermitian operators $\hat{A}$ and $\hat{B}$ with eigenvalues $X_a = \{\alpha_1, \ldots, \alpha_m\}$ and $X_b = \{\beta_1, \ldots, \beta_k\}$. We recall the notion of contextual probabilistic entanglement accommodated to the complex Hilbert space.

Definition 13.1 In the state $|\psi\rangle$, the outcome $B = \beta$ depends on the outcomes of A if, for at least two values of $A, \alpha = \alpha_i, \alpha_j$, the corresponding conditional probabilities don't coincide,

$$P_\psi(B = \beta | A = \alpha_i) \neq P_\psi(B = \beta | A = \alpha_j). \tag{13.6}$$

Definition 13.2 For observables A and B, a state $|\psi\rangle$ is called AB-entangled (or observables A and B are entangled in this state), if all outcomes of B depend on outcomes of A, that is, for all β, condition (13.6) holds for some α_i, α_j.

For incompatible observables, AB-entanglement does not imply BA-entanglement. A state $|\psi\rangle$ is called $A \leftrightarrow B$-entangled if it is both AB- and BA-entangled. For compatible observables, AB- and BA-entanglements in some state are equivalent (see Section 13.7).

We remark that condition (13.6) characterizing AB-entanglement should be completed by the conditional probability existence constraints:

$$P_\psi(A = \alpha) \neq 0, \ \alpha \in X_a. \tag{13.7}$$

In the Hilbert space formalism, the basic inequality (13.6) determining outcome dependence can be written as

$$\frac{\|\hat{E}_B(\beta)\hat{E}_A(\alpha_i)\psi\|^2}{\|\hat{E}_A(\alpha_i)\psi\|^2} \neq \frac{\|\hat{E}_B(\beta)\hat{E}_A(\alpha_j)\psi\|^2}{\|\hat{E}_A(\alpha_j)\psi\|^2}. \tag{13.8}$$

It implies that

$$||\hat{E}_B(\beta)\hat{E}_A(\alpha_i)\psi|| \ ||\hat{E}_A(\alpha_j)\psi|| \neq ||\hat{E}_B(\beta)\hat{E}_A(\alpha_j)\psi|| \ ||\hat{E}_A(\alpha_i)\psi||. \tag{13.9}$$

Proposition 13.3 *Inequality (13.9) is a necessary and sufficient condition for dependence of the outcome $B = \beta$ on the outcomes $A = \alpha_i, \alpha_j$.*

Proof 1. Inequality (13.6) implies inequality (13.9).
2. Now let inequality (13.9) hold for some β and α_i, α_j.
Generally we have

$$P_\psi(A = \alpha_i) = ||\hat{E}_A(\alpha_i)\psi||^2 = ||\sum_\beta \hat{E}_B(\beta)\hat{E}_A(\alpha_i)\psi||^2 = \sum_\beta ||\hat{E}_B(\beta)\hat{E}_A(\alpha_i)\psi||^2.$$

Hence, if $P_\psi(A = \alpha_i) = 0$, then, for any β, $||\hat{E}_B(\beta)\hat{E}_A(\alpha_i)\psi||^2 = 0$. Hence, inequality (13.9) is violated. Thus, inequality (13.9) implies that $P(A = \alpha_i) \neq 0$ and $P(A = \alpha_j) \neq 0$; conditional probabilities are well defined. Hence, inequality (13.9) can be written in the form (13.6). So, the outcome $B = \beta$ depends on the outcomes $A = \alpha_i, \alpha_j$.

Proposition 13.4 *Let observable A be dichotomous. Then inequalities*

$$||\hat{E}_B(\beta)\hat{E}_A(+)\psi|| \ ||\hat{E}_A(-)\psi|| \neq ||\hat{E}_B(\beta)\hat{E}_A(-)\psi|| \ ||\hat{E}_A(+)\psi||, \quad \beta \in X_b, \tag{13.10}$$

are the necessary and sufficient conditions for AB-entanglement.

Proposition 13.5 *AB-entanglement is invariant under unitary transformations.*

Proof Let $\hat{U}$ be a unitary operator and let $\hat{A}_U = \hat{U}\hat{A}\hat{U}^\star, \hat{B}_U = \hat{U}\hat{B}\hat{U}^\star, |\psi_U\rangle = \hat{U}|\psi\rangle$. Then $||\hat{E}_{B_U}(\beta)\hat{E}_{A_U}(\alpha)\psi_U\rangle = ||\hat{E}_B(\beta)\hat{E}_A(\alpha)\psi||$ and $||\hat{E}_{A_U}(\alpha)\psi_U|| = ||\hat{E}_A(\alpha_i)\psi||$.

Proposition 10.7 (Section 10.2) implies that, for dichotomous observables A and B, AB-entanglement is characterized by the single constraint:

$$||\hat{E}_B(+)\hat{E}_A(+)\psi|| \; ||\hat{E}_A(-)|| \neq ||\hat{E}_B(+)\hat{E}_A(-)\psi|| \; ||\hat{E}_A(+)||. \tag{13.11}$$

A state $|\psi\rangle$ is AB-disentangled, if and only if

$$||\hat{E}_B(+)\hat{E}_A(+)\psi|| \; ||\hat{E}_A(-)|| = ||\hat{E}_B(+)\hat{E}_A(-)\psi|| \; ||\hat{E}_A(+)||. \tag{13.12}$$

Quantifying Observables' Entanglement

In Section 10.2 we introduced a quantitative measure of contextual probabilistic entanglement, it can be used in the quantum framework

Definition 13.6 For quantum observables A and B, AB-concurrence in the state $|\psi\rangle$ is defined as

$$\lambda_{AB}(\psi) = \sum_{\beta} \sum_{\alpha \neq \alpha'} |P_\psi(B=\beta|A=\alpha) - P_\psi(B=\beta|A=\alpha')|. \tag{13.13}$$

See Section 13.5 on comparison with the standard concurrence measure of entanglement. The crucial difference from the standard, tensor product entanglement, is that AB-concurrence depends on a pair of observables. The maximally AB-entangled states are considered in Section 10.2, these are the EPR states invented in [51].

In operator terms, AB-concurrence in state $|\psi\rangle$ is given by

$$\lambda_{AB}(\psi) = \sum_{\beta} \sum_{\alpha \neq \alpha'} \left| \frac{||\hat{E}_B(\beta)\hat{E}_A(\alpha)\psi||^2}{||\hat{E}_A(\alpha)\psi||^2} - \frac{||\hat{E}_B(\beta)\hat{E}_A(\alpha')\psi||^2}{||\hat{E}_A(\alpha')\psi||^2} \right|. \tag{13.14}$$

We also define re-normalized AB-concurrence as

$$\zeta_{AB}(\psi) = \sum_{\beta} \sum_{\alpha \neq \alpha'} \left| \, ||\hat{E}_B(\beta)\hat{E}_A(\alpha)\psi|| \; ||\hat{E}_A(\alpha')\psi|| - ||\hat{E}_B(\beta)\hat{E}_A(\alpha')\psi|| \; ||\hat{E}_A(\alpha)\psi|| \, \right|. \tag{13.15}$$

13.3 Joint Eigenvectors

Our aim is to find interrelation between AB-entanglement and the standard one. The latter is considered in the case of compatible observables. From now we assume (except Section 13.7) that observables are dichotomous and compatible, that is, represented by commuting Hermitian operators $\hat{A}$ and $\hat{B}$, $[\hat{A}, \hat{B}] = 0$, with eigenvalues $\alpha, \beta = \pm 1$.

The state space can be represented as the direct sum of subspaces for the joint eigenvectors, $\hat{A}|\phi_{\alpha\beta}\rangle = \alpha|\phi_{\alpha\beta}\rangle, \hat{B}|\phi_{\alpha\beta}\rangle = \beta|\phi_{\alpha\beta}\rangle, \alpha, \beta = \pm$,

$$\mathcal{H} = \mathcal{H}_{ab}(++) \oplus \mathcal{H}_{ab}(+-) \oplus \mathcal{H}_{ab}(-+) \oplus \mathcal{H}_{ab}(--). \tag{13.16}$$

Any (normalized) vector can be represented as superposition

$$|\psi\rangle = |\psi_{++}\rangle + |\psi_{+-}\rangle + |\psi_{-+}\rangle + |\psi_{--}\rangle, \tag{13.17}$$

where $|\psi_{\alpha\beta}\rangle \in \mathcal{H}_{ab}(\alpha\beta)$ and

$$1 = \|\psi\|^2 = \|\psi_{++}\|^2 + \|\psi_{+-}\|^2 + \|\psi_{-+}\|^2 + \|\psi_{--}\|^2. \tag{13.18}$$

To be able to determine the A-conditional probabilities, we have to assume that

$$|\psi_{++}\rangle + |\psi_{+-}\rangle \neq 0, \; |\psi_{-+}\rangle + |\psi_{--}\rangle \neq 0. \tag{13.19}$$

This is the constraint for operating with AB-entanglement.

We calculate the quantities for formula (13.10):

$$\langle\psi|\hat{E}_A(+)|\psi\rangle = \|\psi_{++}\|^2 + \|\psi_{+-}\|^2, \; \langle\psi|\hat{E}_A(-)|\psi\rangle = \|\psi_{-+}\|^2 + \|\psi_{--}\|^2.$$
$$\langle\psi|\hat{E}_B(-)\hat{E}_A(+)|\psi\rangle = \||\psi_{+-}\rangle\|^2, \; \langle\psi|\hat{E}_B(-)\hat{E}_A(-)|\psi\rangle = \||\psi_{--}\rangle\|^2.$$

Inequality (13.9) has the following form:

$$\|\psi_{+-}\|^2\Big(\|\psi_{-+}\|^2 + \|\psi_{--}\|^2\Big) \neq \|\psi_{--}\|^2\Big(\|\psi_{++}\|^2 + \|\psi_{+-}\|^2\Big).$$

And it is reduced to the inequality

$$\|\psi_{+-}\|^2\|\psi_{-+}\|^2 \neq \|\psi_{--}\|^2\|\psi_{++}\|^2. \tag{13.20}$$

By Proposition 13.3, this is the necessary and sufficient condition for AB-entanglement. We note that the constraint (13.19) is automatically encoded in (13.19).

13.4 Tensor Product Space

Let us consider the state space with a tensor product structure $\mathcal{H} = \mathcal{H}_1 \otimes \mathcal{H}_2$ and represent the operators $\hat{A}$ and $\hat{B}$ as tensor products, that is, $\hat{A} = \hat{a} \otimes I$ and $\hat{B} = I \otimes \hat{b}$, where $\hat{a}\colon \mathcal{H}_1 \to \mathcal{H}_1$ and $\hat{b}\colon \mathcal{H}_2 \to \mathcal{H}_2$. We remark that $\mathcal{H}_1 = \mathcal{H}_1(+) \oplus \mathcal{H}_1(-)$ and $\mathcal{H}_2 = \mathcal{H}_2(+) \oplus \mathcal{H}_2(-)$. Each $|\phi_A\rangle \in \mathcal{H}_1$ and $|\phi_B\rangle \in \mathcal{H}_2$ can be represented as superpositions: $|\phi_A\rangle = |\phi_{a,-}\rangle + |\phi_{c.+}\rangle$ and $|\phi_B\rangle = |\phi_{b,-}\rangle + |\phi_{b.+}\rangle$. Hence, for a factorisable state $|\psi\rangle = |\phi_A\rangle|\phi_B\rangle$, superposition (13.17) has the following form:

$$|\psi\rangle = |\phi_{a,+}\rangle|\phi_{b,+}\rangle + |\phi_{a,+}\rangle|\phi_{b,-}\rangle + |\phi_{a,-}\rangle|\phi_{b,+}\rangle + |\phi_{a,-}\rangle|\phi_{b,-}\rangle. \tag{13.21}$$

Hence,

$$\|\psi_{+-}\|^2\|\psi_{-+}\|^2 = \|\psi_{--}\|^2\|\psi_{++}\|^2. \tag{13.22}$$

So, we proved the following simple, but important theorem:

Theorem 13.7 *Let $\mathcal{H} = \mathcal{H}_1 \otimes \mathcal{H}_2$ and $\hat{A} = \hat{a} \otimes I, \hat{B} = I \otimes \hat{b}$, where operators a, b have dichotomous spectra. Then any AB-entangled state $|\psi\rangle$ is also entangled w.r.t. the tensor product structure.*

Hence, *AB*-entangled states form the subclass $\mathcal{E}_{AB}$ of the class of all entangled states $\mathcal{E}$ for the tensor product decomposition $\mathcal{H} = \mathcal{H}_1 \otimes \mathcal{H}_2$. We shall see that $\mathcal{E}_{AB}$ is a proper subclass of $\mathcal{E}$.

13.5 Amplitude Factorization and Correlation

The standard entanglement is characterized via the condition of state nonfactorization w.r.t. the tensor product structure. We want to find an analogous condition for *AB*-entanglement. We start with studying the case of four-dimensional state space.

Four-Dimensional State Space

Consider now two commuting operators $\hat{A}$ and $\hat{B}$ acting in four-dimensional space $\mathcal{H}$ having dichotomous spectra $\alpha, \beta = \pm$ and common eigenvectors for all possible combinations of α, β, $\hat{A}|\alpha\beta\rangle = \alpha|\alpha\beta\rangle, \hat{B}|\alpha\beta\rangle = \beta|\alpha\beta\rangle$, that is, all joint eigensubspaces are one-dimensional. In this case $\mathcal{H}$ and the operators can be represented in the tensor form, $\mathcal{H} = \mathcal{H}_1 \otimes \mathcal{H}_2$ (two-qubit state space) and $\hat{A} = \hat{a} \otimes I, \hat{B} = I \otimes \hat{b}$; we can use the standard notion of entanglement and compare it with *AB*-entanglement.

Any state can be represented as superposition

$$|\psi\rangle = c_{++}| + +\rangle + c_{+-}| + -\rangle + c_{-+}| - +\rangle + c_{--}| - -\rangle, \tag{13.23}$$

where $|c_{++}|^2 + |c_{+-}|^2 + |c_{-+}|^2 + |c_{--}|^2 = 1$.

This state is factorisable if and only if the coefficients satisfy the following equality:

$$c_{+-}c_{-+} = c_{--}c_{++}. \tag{13.24}$$

One of the measures of standard entanglement is concurrence and, for $|\psi\rangle$, concurrence is given by

$$\zeta(\psi) = |c_{+-}c_{-+} - c_{--}c_{++}|. \tag{13.25}$$

So, a state is entangled if and only if $\zeta(\psi) \neq 0$. Renormalized *AB*-concurrence (13.15) is written as

$$\zeta_{AB}(\psi) = |\,|c_{+-}c_{-+}| - |c_{--}c_{++}|\,|. \tag{13.26}$$

So, a state is AB entangled iff $\zeta_{AB}(\psi) \neq 0$.

Definition 13.8 A state $|\psi\rangle$ is called AB amplitude-factorisable w.r.t. the pair of compatible observables A and B with dichotomous spectrum if the absolute values of the coordinates in the joint eigenvectors basis are factorizable, that is, for (13.23),

$$|c_{\alpha\beta}| = \lambda_\alpha \, \mu_\beta, \text{ where } \alpha, \beta = \pm, \tag{13.27}$$

where $\lambda_\alpha, \ \mu_\beta \geq 0$ *and*

$$\lambda_+^2 + \lambda_-^2 = 1, \ \mu_+^2 + \mu_-^2 = 1. \tag{13.28}$$

The state factorization implies the amplitude factorization, but not vice versa: condition (13.27) does not take into account the presence of phases, but the latter can generate the standard entanglement even for amplitude-factorizable states.

Let

$$|\psi\rangle = (\lambda_+|+\rangle + e^{i\theta_1}\lambda_-|-\rangle) \otimes (\mu_+|+\rangle + e^{i\theta_2}\mu_-|-\rangle) =$$
$$\lambda_+\mu_+| + +\rangle + e^{i\theta_2}\lambda_+\mu_-| + -\rangle + e^{i\theta_1}\lambda_-\mu_+| - +\rangle + e^{i(\theta_1+\theta_2)}\lambda_-\mu_-| - -\rangle.$$

Now consider the state

$$|\psi\rangle = | + +\rangle + | + -\rangle - | - +\rangle + | - -\rangle,$$

which is entangled in the ordinary sense, although it is amplitude factorizable.

The violation of relation (13.20) has the form of the following equality:

$$|c_{+-}|^2|c_{-+}|^2 = |c_{--}|^2|c_{++}|^2. \tag{13.29}$$

This is a good place to recall that the existence of conditional probabilities induces the following constraints:

$$|c_{++}|^2 + |c_{+-}|^2 \neq 0, \ |c_{-+}|^2 + |c_{--}|^2 \neq 0. \tag{13.30}$$

Theorem 13.9 *Let* $\dim \mathcal{H} = 4$. *For a pair of compatible dichotomous observables A and B, a state* $|\psi\rangle$ *is AB-entangled if and only if it is amplitude nonfactorizable.*

Proof As we have seen, AB-disentangled is characterized by the equality (13.29). We prove that this equality is equivalent to the amplitude factorization. It is clear that amplitude factorization implies (13.29). Now let this equality hold true. We set (under the condition that the denominators are not equal to zero)

$$\lambda_+ = \frac{|c_{++}|}{\sqrt{|c_{++}|^2 + |c_{-+}|^2}}, \lambda_- = \frac{|c_{-+}|}{\sqrt{|c_{++}|^2 + |c_{-+}|^2}}. \tag{13.31}$$

$$\mu_+ = \frac{|c_{++}|}{\sqrt{|c_{++}|^2 + |c_{+-}|^2}}, \mu_- = \frac{|c_{+-}|}{\sqrt{|c_{++}|^2 + |c_{+-}|^2}}. \tag{13.32}$$

It is easy to check that these quantities factorize the modules of the coordinates $c_{\alpha,\beta}$. For example,

$$\lambda_+\mu_- = |c_{++}||c_{+-}|/\sqrt{(|c_{++}|^2 + |c_{-+}|^2)(|c_{++}|^2 + |c_{+-}|^2)}.$$

So, the square of the denominator is given by

$$|c_{++}|^4 + |c_{++}|^2|c_{+-}|^2 + |c_{++}|^2|c_{-+}|^2 + |c_{+-}|^2|c_{-+}|^2.$$

On the other hand, the AB-disentanglement equation gives us

$$|c_{++}|^2(1 - |c_{++}|^2 - |c_{+-}|^2 - |c_{-+}|^2) = |c_{+-}|^2|c_{-+}|^2.$$

Hence, the denominator equals $|c_{++}|$ and $\lambda_+\mu_- = |c_{+-}|$.

Now consider the case that one of the denominators in (13.31), (13.32) equals zero, say $c_{++} = 0, c_{-+} = 0$. In this case, amplitude factorization can be shown directly. We just set $\lambda_- = |c_{--}|, \lambda_+ = |c_{+-}|, \mu_- = 1, \mu_+ = 0$.

Since some states belonging to $\mathcal{E}$ can be amplitude factorizable, embedding $\mathcal{E}_{AB} \to \mathcal{E}$ is not surjection.

Generally for a state $|\psi\rangle$, we set $\langle A\rangle_\psi = \langle\psi|\hat{A}|\psi\rangle, \langle B\rangle_\psi = \langle\psi|\hat{B}|\psi\rangle, \langle AB\rangle_\psi = \langle\psi|\hat{A}\hat{B}|\psi\rangle$. Covariance of observables A and B represented by Hermitian operators $\hat{A}$ and $\hat{B}$ is defined as

$$\mathrm{cov}(A, B) = \langle(\hat{A} - \langle\hat{A}\rangle_\psi)(\hat{B} - \langle\hat{B}\rangle_\psi)\rangle_\psi = \langle AB\rangle_\psi - \langle A\rangle_\psi\langle B\rangle_\psi. \tag{13.33}$$

Observables are called AB-uncorrelated in state $|\psi\rangle$ if $\mathrm{cov}(A, B) = 0$ or the correlation factorization condition holds:

$$\langle AB\rangle_\psi = \langle A\rangle_\psi\langle B\rangle_\psi. \tag{13.34}$$

Otherwise, they are called correlated, that is, if $\mathrm{cov}(A, B) \neq 0$.

Theorem 13.10 *Compatible dichotomous observables A and B are AB-entangled in a state $|\psi\rangle$ if and only if they are correlated:* $\mathrm{cov}(A, B) \neq 0$, *that is,*

$$\langle AB\rangle_\psi \neq \langle A\rangle_\psi\langle B\rangle_\psi. \tag{13.35}$$

Proof Set $a = |c_{++}|^2, b = |c_{--}|^2, c = |c_{+-}|^2, d = |c_{-+}|^2$. Set $x = a + b, y = c + d$. We remark that $x + y = 1$. The latter implies that $x^2 - y^2 = x - y$. Hence,

$$\langle AB\rangle_\psi = a + b - c - d = x - y = x^2 - y^2 = a^2 + b^2 - c^2 - d^2 + 2ab - 2cd.$$

Then

$$\begin{aligned}\langle A\rangle_\psi \langle B\rangle_\psi &= ((a+c)-(d+b))((a+d)-(b+c))\\ &= a^2+b^2-c^2-d^2-2ab+2cd\\ &= x^2-y^2+4ab-4cd.\end{aligned}$$

Hence, the correlation $\langle AB\rangle_\psi$ is factorizable if and only if $ab = cd$, that is, the amplitude factorization condition (13.29) is satisfied.

This proposition supports our invention of entanglement as dependence of observables. Disentanglement corresponds to the classical probabilistic situation of independence of observables given by random variables A and B. For such random variables, independence implies $\langle AB\rangle_P = \langle A\rangle_P \langle B\rangle_P$, where P is the probability measure (it plays the role of a quantum state). For dichotomous random variables, independence is equivalent uncorrelation (see Section 13.7).

It is easy to show that in the quantum theory, factorization of correlations does not imply the standard disentanglement, because the amplitude factorization condition can hold for an entangled state. In contrast, by Theorem 3 AB-disentanglement is equivalent to correlation factorization for these observables.

State Space of Arbitrary Dimension

Now we turn to Section 13.3. Let in superposition (13.17) all vectors $|\psi_{\alpha\beta}\rangle \neq 0$. In particular, this presumes that all subspaces $\mathcal{H}_{AB}(\alpha\beta)$ are nontrivial.

We rewrite this superposition as (13.23), where

$$c_{\alpha\beta} = ||\psi_{\alpha\beta}|| \tag{13.36}$$

and $|\alpha\beta\rangle = |\psi_{\alpha\beta}\rangle/||\psi_{\alpha\beta}||$. Hence, we again can restrict consideration to the four-dimensional case. Instead of the expansion of a state in the basis of joint eigenvectors of the operators $\hat{A}$ and $\hat{B}$, we consider its expansion in its projections on joint eigensubspaces $\mathcal{H}_{AB}(\alpha\beta), \alpha\beta = \pm$. We remark that here we do not consider any tensor product structure.

The violation of AB-entanglement condition (13.20) has the form of equality:

$$\|\psi_{+-}\|^2\|\psi_{-+}\|^2 = \|\psi_{--}\|^2\|\psi_{++}\|^2. \tag{13.37}$$

Definition 13.11 A state $|\psi\rangle$ is called AB amplitude-factorizable if

$$||\psi_{\alpha\beta}|| = \lambda_\alpha \mu_\beta,\ \alpha,\beta = \pm, \tag{13.38}$$

where $\lambda_\alpha, \mu_\beta \geq 0$ *and*

$$\lambda_+^2 + \lambda_-^2 = 1,\ \mu_+^2 + \mu_-^2 = 1. \tag{13.39}$$

Theorem 13.12 *For compatible dichotomous observables A and B, a state $|\psi\rangle$ is AB-entangled if and only if it is amplitude nonfactorizable.*

Proof AB-disentangled is characterized by the equality (13.37). This equality is equivalent to the amplitude factorization. The proof is in line with the proof for the four-dimensional case. It is evident that amplitude factorization implies equality (13.37). Now let the latter hold true. We set

$$\lambda_+ = \frac{||\psi_{++}||}{\sqrt{||\psi_{++}||^2 + ||\psi_{-+}||^2}}, \lambda_- = \frac{||\psi_{-+}||}{\sqrt{||\psi_{++}||^2 + ||\psi_{-+}||^2}}, \tag{13.40}$$

$$\mu_+ = \frac{||\psi_{++}||}{\sqrt{||\psi_{++}||^2 + ||\psi_{+-}||^2}}, \mu_- = \frac{||\psi_{-+}||}{\sqrt{||\psi_{++}||^2 + ||\psi_{+-}||^2}}. \tag{13.41}$$

It is easy to check that these quantities factorize the modules of the coordinates $c_{\alpha,\beta}$. The cases of zero denominators are analyzed similarly to the four-dimensional case.

As in the four-dimensional case, we obtain characterization of AB-entanglement as the observables' correlation:

Theorem 13.12 is valid in the general case.

13.6 Singletlike State

Consider a family of pairs of operators $(\hat{A}_u, \hat{B}_u)$, where u is some parameter. Can one find a quantum state $|\psi\rangle$ that is $A_u B_u$-entangled for all these pairs?

We assume that, for any u, $[\hat{A}_u, \hat{B}_u] = 0$, but it may be that, for some pairs u, v, $[\hat{A}_u, \hat{A}_v] \neq 0$ or (and) $[\hat{B}_u, \hat{B}_v] \neq 0$.

Let us consider the tensor product case $\mathcal{H} = H \otimes H$, where dim $H = 2$, and two operators in $H, \hat{a}, \hat{b}$, with the eigenbases (f_+, f_-) and (g_+, g_-), where $af_\pm = \pm f_\pm, bg_\pm = \pm g_\pm$. Set $\hat{A} = \hat{a} \otimes I, \hat{B} = I \otimes \hat{b}$.

Take now any unitary transformation $\hat{u}$ in H and consider operators of the form

$$\hat{a}_u = \hat{u}\hat{a}\hat{u}^\star, \hat{b}_u = \hat{u}\hat{b}\hat{u}^\star, \tag{13.42}$$

which are diagonal w.r.t. the bases $(f_{u,+}, f_{u,-})$ and $(g_{u,+}, g_{u,-})$, obtained with the unitary transformation $\hat{u}$ from the bases (f_+, f_-) and (g_+, g_-).

Then consider state $|\psi\rangle$ having the form

$$|\psi\rangle = c_{+-}|f_+ g_-\rangle + c_{-+}|f_- g_+\rangle, \ c_{-+}, c_{-+} \neq 0. \tag{13.43}$$

This state is AB-entangled. And it preserves its form under the unitary transformation $\hat{U} = \hat{u} \otimes \hat{u}$ if and only if $c_{+-} = -c_{-+}$, that is,

$$|\psi\rangle = (|f_+ g_-\rangle - |f_- g_+\rangle)/\sqrt{2} = (|f_{u,+} g_{u,-}\rangle - |f_{u,-} g_{u,+}\rangle)/\sqrt{2}. \tag{13.44}$$

Hence, this state is observables entangled both for the pair $\hat{A}, \hat{B}$ and the pair $\hat{A}_u = \hat{a}_u \otimes I, \hat{B}_u = I \otimes \hat{b}_u$ for any unitary transformation $\hat{u}: H \to H$.

In particular, let us consider the standard singlet state:

$$|\psi_{\text{singlet}}\rangle = (|f_+ f_-\rangle - |f_- f_+\rangle)/\sqrt{2}. \tag{13.45}$$

This state is AA'-entangled for the operators $\hat{A} = \hat{a} \otimes I$ and $\hat{A}' = I \otimes \hat{a}$. And it is also $A_u A'_u$-entangled for the operators $\hat{A}_u = \hat{a}_u \otimes I$ and $\hat{A}'_u = I \otimes \hat{a}_u$.

The singlet state is entangled (in the standard nonfactorization sense) independently from the selection of bases, that is, the operators $\hat{A}$ and $\hat{B}$. However, this is not the case of the operator-based entanglement. It is easy to find $\hat{u}$ such that $|\psi_{\text{singlet}}\rangle$ is AA'_u-disentangled, that is, amplitude factorizable (compare the discussion in Section 13.5). Let

$$|f_-\rangle = u_{-+}|f_{u,+}\rangle + u_{--}|f_{u,-}\rangle,\ |f_+\rangle = u_{++}|f_{u,+}\rangle + u_{+-}|f_{u,-}\rangle,$$

where $u_{-+}\bar{u}_{++} + u_{--}\bar{u}_{+-} = 0,\ |u_{-+}|^2 + |u_{--}|^2 = 1,\ |u_{++}|^2 + |u_{+-}|^2 = 1$. Then

$$|\psi_{\text{singlet}}\rangle = (u_{-+}|f_+ f_{u,+}\rangle + u_{--}|f_+ f_{u,-}\rangle + u_{++}|f_- f_{u,+}\rangle - u_{+-}|f_- f_{u,-}\rangle)/\sqrt{2}.$$

Now select $u_{-+} = u_{+-} = u_{--} = -u_{++} = 1/\sqrt{2}$. Then

$$|\psi_{\text{singlet}}\rangle = (|f_+ f_{u,+}\rangle + |f_+ f_{u,-}\rangle + |f_- f_{u,+}\rangle - |f_- f_{u,-}\rangle)/2.$$

This state is amplitude factorizable (but not tensor product factorizable) and hence $A\hat{A}'_u$-disentangled.

Starting with the conditional probabilistic definition of AB-entanglement, we analyzed its representation in the complex Hilbert space formalism and its interrelation (in the tensor product case) with the standard entanglement.

For compatible dichotomous observables, AB-entanglement is equivalent to amplitude nonfactorization, where amplitudes are the state's projections on joint eigensubspaces of the operators. This characterization leads to the equivalence of AB-entanglement to correlation of observables. With such construction, we can proceed in an arbitrary Hilbert space, that is, without referring to the tensor product structure.

13.7 Entanglement of Observables versus Dependence of Random Variables

We recall that in classical probability (Section 1.5) theory discrete random variables A and B are independent if and only if their joint probability distribution (JPD) is factorizable, that is, for all α, β,

$$P(A = \alpha, B = \beta) = P(A = \alpha)P(B = \beta). \tag{13.46}$$

This definition is equivalent to reduction of conditional probabilities to "absolute probabilities":

$$P(B = \beta | A = \alpha) = P(B = \beta),\ \text{if}\ P(A = \alpha) \neq 0, \tag{13.47}$$

$$P(A = \alpha | B = \beta) = P(A = \alpha), \text{ if } P(B = \beta) \neq 0. \tag{13.48}$$

Thus, for nonzero probabilities $P(A = \alpha)$ and $P(B = \beta)$, independence – probability factorization (13.46) is equivalent to conditional independence, (13.47) and (13.48).

In quantum theory, AB-disentanglement means that there exists an outcome $B = \beta$ such that

$$P_\psi(B = \beta | A = \alpha) = P_\psi(B = \beta | A = \alpha') \tag{13.49}$$

for all α, α' (in the case of $P_\psi(A = \alpha) \neq 0, \alpha \in X_a$).

In classical probability, for random variables A and B, equality (13.49) is equivalent to equality (13.47) – for the concrete outcome $B = \beta$ (for nonzero probabilities for A-outcomes). We prove this to illustrate the role of the *formula of total probability* (FTP):

$$P(B = \beta) = \sum_\alpha P(B = \beta | A = \alpha) P(A = \alpha). \tag{13.50}$$

Denote the quantity determined by (13.49) as $P(B = \beta | A)$. We want to show that it coincides with $P(B = \beta)$. We put $P(B = \beta | A)$ in (13.50):

$$P(B = \beta) = P(B = \beta | A) \sum_\alpha P(A = \alpha) = P(B = \beta | A). \tag{13.51}$$

Thus, we obtain (13.47) and hence (13.48) as well as (13.46) – for the concrete value β.

By trying to perform the preceding classical probabilistic calculations, in the quantum framework we obtain

$$P\psi(B = \beta) - P_\psi(B = \beta | A)) = \delta_\psi(B = y | A), \tag{13.52}$$

where RHS is equal to the interference term in the quantum FTP: Section 11.2, formulas (11.21), (11.22).

Thus, generally the notion of AB-disentanglement differs from the classical notion of independence, formulated as (13.47). What is the basic definition of classical independence via probability factorization? See (13.46).

In quantum probability, conditional JPD is defined as

$$P_\psi(A = \alpha, B = \beta) \equiv P_\psi(A = \alpha) P_\psi(B = \beta | A = \alpha). \tag{13.53}$$

In operator terms:

$$P_\psi(A = \alpha, B = \beta) = \langle \psi | \hat{E}_A(\alpha) \hat{E}_B(\beta) \hat{E}_A(\alpha) | \psi \rangle = || \hat{E}_B(\beta) \hat{E}_A(\alpha) \psi ||^2. \tag{13.54}$$

For noncommuting observables, generally

$$P_\psi(A = \alpha, B = \beta) = || \hat{E}_B(\beta) \hat{E}_A(\alpha) \psi ||^2 \neq \\ || \hat{E}_A(\alpha) \hat{E}_B(\beta) \psi ||^2 = P_\psi(B = \beta, A = \alpha). \tag{13.55}$$

Definition 13.13 Observables $\hat{A}$ and $\hat{B}$ are called *AB*-independent in state $|\psi\rangle$, if their conditional JPD is factorizable, for example, for all pairs α, β,

$$P_\psi(A = \alpha, B = \beta) = P_\psi(A = \alpha)P_\psi(B = \beta). \tag{13.56}$$

Otherwise, observables are called *AB*-dependent in $|\psi\rangle$. Since JPD is the conditional JPD, *AB*- and *BA*-(in)dependence are not equivalent. (For commuting observables, *AB*- and *BA*-(in)dependence are equivalent).

It is easy to check that two sorts of *AB*-independence, that is, via (13.47), (13.48) and (13.56) are equivalent even for quantum probability (for nonzero probabilities of outcomes). They can be used as another approach to the notion of *AB*-conditional probability (dis)entanglement.

The most interesting for application is entanglement for compatible observables, $[\hat{A}, \hat{B}] = 0$. Here quantum FTP (11.21), Section 11.2, is reduced to classical FTP (13.51), since the interference term (11.22) equals zero.

Proposition 13.14 *For compatible observables A and B and state $|\psi\rangle$, the notions of AB-entanglement and AB-dependence are equivalent.*

Proof (1) Let quantum observables be dependent. Then there exist outcomes β, α such that

$$P_\psi(A = \alpha, B = \beta) \neq P_\psi(A = \alpha)P_\psi(B = \beta). \tag{13.57}$$

In particular, this implies that $P_\psi(A = \alpha) \neq 0, P_\psi(B = \beta) \neq 0$, since for commuting operators $P_\psi(A = \alpha, B = \beta) = ||\hat{E}_B(\beta)\hat{E}_A(\alpha)\psi||^2 \leq ||\hat{E}_A(\alpha)\psi||^2 = P_\psi(A = \alpha)$ and in the same way $P_\psi(A = \alpha, B = \beta) \leq P_\psi(B = \beta)$. Thus, we have

$$P_\psi(B = \beta|A = \alpha) \neq P_\psi(B = \beta). \tag{13.58}$$

Now suppose that $|\psi\rangle$ is not *AB*-entangled. Then there exists an outcome β such that the equality $P_\psi(B = \beta|A = \alpha) = P_\psi(B = \beta|A = \alpha')$ holds for all pairs α, α'. Denote the quantity determined by this constraint by $P_\psi(B = \beta|A)$. Now we apply FTP, and obtain that $P_\psi(B = \beta) = P_\psi(B = \beta|A)$. But this contradicts (13.58).

(2) Let quantum observables be *AB*-entangled. Then, for each β, $P_\psi(B = \beta|A = \alpha) \neq P_\psi(B = \beta|A = \alpha')$ for some α, α'. Suppose that quantum observables are independent, that is, $P_\psi(A = \gamma, B = \beta) = P_\psi(A = \gamma)P_\psi(B = \beta)$ for all γ, β. This implies that $P_\psi(B = \beta|A = \gamma) = P_\psi(B = \beta)$, and hence $P_\psi(B = \beta|A = \alpha) = P_\psi(B = \beta|A = \alpha')$ for α, α' considered earlier.

We remark that compatible observables $\hat{A}$ and $\hat{B}$ can represented by random variables ξ_A and ξ_B on classical probability space. In this case dependence of observables corresponds to dependence of random variables. By Proposition 5, *AB*-entanglement also corresponds to dependence of random variables. We formulate this coupling as

Proposition 13.15 *For compatible observables, AB-entanglement can be treated as the complex Hilbert space representation of dependence of classical random variables.*

Now we compare the notion of uncorrelated classical random variables and quantum observables in the dichotomous case, that is, $A, B = pm1$. We recall the following result.

Proposition 13.16 *Dichotomous random variables are independent if and only if they are uncorrelated.*

This is classical probabilistic counterpart part of Theorem 13.12.

13.8 Concluding Discussion on Probabilistic Entanglement

In this article we extended the conditional probability approach to entanglement initiated in [51] for probability one predictions. This approach can be considered as the probabilistic formalization of the original Schrödinger's viewpoint on entanglement [328, 329]. We recall that he considered entanglement as a part of complex Hilbert space machinery for prediction of outcomes of observables.

It (ψ-function) is now the means for predicting probability of measurement results. In it is embodied the momentarily-attained sum of theoretically based future expectation, somewhat as laid down in a catalog. It is the relation-and-determinacy-bridge between measurements and measurements... [328, 329].

And he treated the EPR argument [115] in this way. Hence, for Schrödinger, entanglement is related to a pair of observables *A* and *B*, that is, a state is entangled w.r.t. a selected pair; this is *AB*-entanglement. This original Schrödinger's treatment of entanglement as observables entanglement was deformed into state-entanglement and highlighting the role of the tensor product structure (with just a few exceptions; see [351, 352]).

In this article entanglement is defined as dependence of observables *A* and *B*. This viewpoint on entanglement was reflected in Schrödinger's
"Beschränkung" by using for it the meaning "restriction." Of course, it would be better if Schrödinger would from the very beginning use the word "dependence" and associate dependence of classical random variables with *AB*-entanglement, as is done in this chapter.

14

Brownian Motion: Classical and Semiclassical Entanglement

14.1 Quantum versus Classical Statistical Mechanics

In this chapter we follow the article of Allahverdyan, Khrennikov, and Nieuwenhuizen [16].

There is a long tradition of seeking connections between quantum mechanics and classical statistical physics. A list of known examples includes

- *i)* analogies between the notion of complementarity in QM and statistical thermodynamics, in particular, between quantum mechanical uncertainty relations and the energy-temperature uncertainty relation in statistical thermodynamics [64, 260];
- *ii)* mathematical relations between Schrödinger and Fokker–Planck equations [323], which makes quantum intuition very useful, for example, for polymers [145];
- *iii)* attempts to derive Schrödinger equations from the classical kinetic picture [284] or from classical stochastic electrodynamics [103].

More generally, both QM and statistical physics are essentially probabilistic theories and already at this level one expects to find certain analogies between their concepts. The progress in this direction has been for a long time plagued by statements on incompatibility between QM and classical probability theory, which lies in the basis of classical statistical physics. As an example, see Feynman [120] for statements that quantum interference phenomena contradict to classical probability theory. It was, however, noted that such opinions are not warranted [41], and recently it was shown explicitly that several basic relations of QM can be derived from the classical probability theory provided the contexts of physical conditions (measurements) are properly taken into account [191, 195]. Both quantum and classical probability theories can be embedded in contextual probability theory, and the corresponding measurement theories are covered by CMM (contextual measurement model); see Chapter 10.

Once the basic probabilistic ground of QM and classical statistical physics is recognized to be the same, one wonders whether some unusual aspects of QM, such as entanglement, can find analogies in classical statistical physics. We see at least two reasons for seeking such analogies. First, it is going to refresh our understanding of classical statistical physics, and may imply in the future that advantages offered by QM in certain tasks of information processing and transfer are not unique to quantum mechanics, and can be looked for in classical areas of physics as well [191, 195]. Second, it is useful for understanding QM that certain concepts believed to be purely quantum (that is, incomprehensible in classical terms) can find natural classical analogies. In fact, the works and ideas mentioned in the preceding points *i)–iii)* were partially directed toward this goal. More recent results along these lines are classical analogies to quantum entanglement found in optics [134, 343] (Chapter 7) and classical information theory [86] (secret classical correlations), and a classical probabilistic model for certain aspects of (multi-time) quantum measurements [249].

The purpose of this chapter (based on paper [15], see also [16]) is to show that quantum entanglement can have a natural analogy in the physics of Brownian particles (Brownian entanglement). The reason for the existence of this analogy can be qualitatively explained as follows. It is known that the dynamics of a Brownian particle can be observed at two levels [323]. Within the first, more fundamental level, the Brownian particle coupled to a thermal bath at temperature T is described via definite coordinate x and momentum p and moves under influence of external potential, friction force, and an external random force. The latter two forces are generated by the bath. The second, overdamped regime applies when the characteristic relaxation time of the coordinate τ_x is much larger than that of the momentum τ_p,

$$\tau_x \gg \tau_p$$

(overdamped regime). On times much larger than τ_p one is interested in the change of the coordinate and defines the *coarse-grained* velocity as $v = \Delta x/\Delta t$ for $\tau_x \gg \Delta t \gg \tau_p$. This definition of v is the only operationally meaningful one for the (effective) velocity within the overdamped regime. It appears that the coarse-grained velocity, though pertaining to single particles, is defined in the context of the whole systems of coupled Brownian particles. Together with uncertainty relations between the coordinate and the change of the coarse-grained velocity—the role of Planck's constant is being played by the temperature of the bath—this contextuality feature will be shown to cause a phenomenon similar to quantum entanglement: The common probability distribution of the two coordinates and the corresponding coarse-grained velocities—for two Brownian particles which interacted in the past, but need not interact in the present—cannot be prepared via mixing of any factorized distributions referring to the two particles separately. This Brownian entanglement is a consequence of a coarse-grained description and disappears within the first (more fundamental) level of description, simply because entanglement is absent in classical mechanics.

In this chapter we adhere to the statistical (ensemble) interpretation of QM (see Chapter 2), where a quantum 'state' is described by a density matrix $\hat{\rho}$, and any state, including a pure state $|\psi\rangle\langle\psi|$, refers to an ensemble $\mathcal{E}(\hat{\rho})$ of identically prepared systems; see, for example, [43, 46, 58, 102, 105, 167, 285][1]. To put it succinctly: QM makes[2]

- statistical statements on
- the results of measurements done
- on ensembles of identically prepared systems.

As was stressed repeatedly [43, 46, 58, 102, 105, 167, 285], in particular by experimentalists [176], the experimentally relevant statements of QM do not require more than the minimal statistical interpretation.[3] Moreover, this interpretation deals more successfully with the measurement problem, as instanced by a recent exactly solvable model [13], and allows one to reconcile QM with classical probability theory [191, 195]. The fact that discussions on quantum entanglement (and on Bell inequalities and related matters) do not employ the statistical interpretation is a mere prejudice; see Refs. [13, 43] for examples of such discussions.

The chapter is organized as follows. In Section 14.2 we recall the phenomenon of (continuous-variable) quantum entanglement, focusing especially on the relations between the entanglement and the uncertainty relations. Section 14.3 discusses coarse-grained velocities and uncertainty relations for the classical Brownian motion. The next two sections define and study the phenomenon of Brownian entanglement. Sections 14.5 and 14.6 offer detailed comparison between the features of quantum entanglement and those of its Brownian counterpart. In Section 14.7 we discuss possibilities of experimental realization of Brownian entanglement. Our conclusions are presented in the last section. Some technical questions are worked out in Section 14.8.

[1] The minimal statistical interpretation should, of course, be distinguished from various hidden-variable theories and assumptions, in particular, from the preassigned initial value assumptions, where quantum measurements are viewed as merely revealing preexisting values of all observables. Unfortunately, some proponents of the statistical interpretation were unclear at this point, a fact that for a while discredited this interpretation. For a well-balanced discussion on this and related points, see [102, 105].

[2] This way of putting the message of statistical interpretation is adopted from [256].

[3] Still, as correctly pointed out in [173], the choice of interpretation can and does influence one's estimates of importance for various scientific problems.

14.2 Entanglement: Quantum, Classical, and Semiclassical Systems

Quantum Systems

Consider a quantum system S consisting of two subsystems S_1 and S_2. A state $\hat{\rho}$ of S is called entangled, see for example [306], (or nonseparable) with respect to S_1 and S_2, if it *cannot* be represented as

$$\hat{\rho} = \sum_{k=1}^{n} p_k\, \hat{\rho}_k^{(1)} \otimes \hat{\rho}_k^{(2)}, \qquad \sum_{k=1}^{n} p_k = 1, \qquad p_k \geq 0, \tag{14.1}$$

where $\hat{\rho}_i^{(1)}$ and $\hat{\rho}_i^{(2)}$ are arbitrary density matrices living in the Hilbert spaces of S_1 and S_2, respectively, n is an integer, and where $\{p_k\}_{k=1}^{n}$ is a probability distribution.

According to definition (14.1), a separable quantum state can always be prepared by means of mixing[4] noncorrelated states $\hat{\rho}_k^{(1)} \otimes \hat{\rho}_k^{(2)}$ of the two subsystems. For a pure state $\hat{\rho} = |\psi\rangle\langle\psi|$ we return to the more known definition of entanglement: $|\psi\rangle$ cannot be represented as $|\psi\rangle = |\psi\rangle_1 \otimes |\psi\rangle_2$. For this particular case the absence of entanglement implies the absence of any correlation. In contrast, for the more general situation given by (14.1), a nonentangled state can still possess certain (classical) correlations, since the totally uncorrelated situations will be given as $\hat{\rho} = \hat{\rho}^{(1)} \otimes \hat{\rho}^{(2)}$.

Two important features of quantum entanglement should be noted. First, it can be tested only through measuring some correlations between the subsystems S_1 and S_2. Observables pertaining to S_1 or to S_2 alone are obviously insensitive to entanglement. Second, if the state (14.1) is prepared by two different observers 1 and 2, then this preparation is seen to involve correlated actions of them, and thus the corresponding observers have to communicate classically.

Classical systems

In classics the representation (14.1) is apparently always possible. Indeed, let us have a probability distribution $P(x_1, x_2)$ of two classical systems S_1 and S_2 represented by random variables x_1 and x_2. Assume for simplicity that x_1 can take values $a_1, .., a_n$, while x_2 can take values $b_1, .., b_n$. Then $P(x_1, x_2)$ can be written as

$$P(x_1, x_2) = \sum_{\alpha=a_1,\ldots,a_n} \; \sum_{\beta=b_1,\ldots,b_n} P(\alpha, \beta)\, \delta_{\alpha x_1}\, \delta_{x_2 \beta}, \tag{14.2}$$

where $\delta_{\alpha\beta} = 1$ if $\alpha = \beta$, and $\delta_{\alpha\beta} = 0$ otherwise. *Provided* that $\delta_{\alpha x_1}$ and $\delta_{x_2\beta}$ are two legitimate probability distributions belonging to S_1 and S_2, that is, provided

[4] Mixing ensembles $\mathcal{E}(\hat{\rho}_1)$ and $\mathcal{E}(\hat{\rho}_2)$ with probabilities p_1 and p_2, respectively, means that one throws a dice with probabilities of outcomes equal to p_1 and p_2, and depending on the outcome one picks up a system from $\mathcal{E}(\hat{\rho}_1)$ or $\mathcal{E}(\hat{\rho}_2)$, keeping no information on where the system came from. Alternatively, one can join together Np_1 systems from $\mathcal{E}(\hat{\rho}_1)$ and Np_2 systems from $\mathcal{E}(\hat{\rho}_2)$ ($N \gg 1$), so that no information is kept on where a single system came from.

there are no mechanisms prohibiting the realization of $\delta_{\alpha x_1}$ and $\delta_{x_2\beta}$ as physically acceptable distributions usable in the actual preparation, the representation (14.1) is realized for the classical situation: there is no entanglement.

In fact, the quantum situation goes to the classical one right in (14.1) if we assume that the involved density matrices are always diagonal. So it is the presence of nondiagonal elements of the given density matrix that makes the situations different.

Semiclassical Systems

We shall now localize the cause of quantum entanglement for systems which behave (semi)classically in several other respects. This will help us to understand the way of searching for entanglement in nonquantum situations.

Let S_1 and S_2 be two quantum (noninteracting) harmonic oscillators, and the overall density matrix (in the coordinate representation) $\rho(x_1, x_1'; x_2, x_2')$ be a Gaussian function of its variables. The corresponding coordinate and momenta operators are denoted by $\hat{x}_1$, $\hat{p}_1$ and $\hat{x}_2$, $\hat{p}_2$, respectively. As is well known, Gaussian states are conveniently dealt with by Wigner functions $W(x,p)$, which are functions of x and p.[5] The Wigner function is equivalent to the density matrix, that is, it allows one to calculate all averages, but in contrast to the density matrix it has several properties expected for the common probability distribution of the coordinate and momentum. In particular, the analysis via Wigner functions will provide us in what follows with a richer intuition on the relations between entanglement and uncertainty relations; thus, between entanglement and noncommutativity.

Since the overall Hamiltonian of S_1 and S_2 is assumed to be harmonic, the initially Gaussian state remains Gaussian for all times and the corresponding Wigner function will be positive and thus will *partially* admit a classical interpretation in terms of common probability distribution of the coordinate and momentum [163]. A related fact is that for two harmonic oscillators the Ehrenfest equations of motion can be recast into a classical form [163]. So, many aspects of this system can be accounted for in classical terms.

In order to see why, in spite of these classical features, the system of two oscillators can be entangled, note that for the Wigner functions one can rewrite the condition (14.1) as

$$W(x_1,p_1;x_2,p_2) = \int \mathrm{d}\lambda\, \mathcal{P}(\lambda)\, W_1(x_1,p_1|\lambda)\, W_2(x_2,p_2|\lambda), \tag{14.3}$$

$$\mathcal{P}(\lambda) \geq 0, \qquad \int \mathrm{d}\lambda\, \mathcal{P}(\lambda) = 1,$$

[5] For good reviews on the properties of Wigner functions, see [163].

where W_1 and W_2 are separate Wigner functions for S_1 and S_2 respectively, and $\mathcal{P}(\lambda)$ is some probability distribution. If S_1 and S_2 were classical oscillators, then instead of Wigner functions we would have distribution functions, and one can always write down the analog of (14.2) for the common distribution function $P(x_1,p_1;x_2,p_2)$ and $\lambda = (\alpha_1\,\beta_1,\alpha_2,\beta_2)$:

$$P(x_1,p_1;x_2,p_2) = \tag{14.4}$$
$$\int \mathrm{d}\alpha_1\,\mathrm{d}\beta_1\,\mathrm{d}\alpha_2\,\mathrm{d}\beta_2\,P(\alpha_1,\beta_1;\alpha_2,\beta_2)\,\delta(\alpha_1-x_1)\,\delta(\beta_1-p_1)\,\delta(\alpha_2-x_2)\,\delta(\beta_2-p_2).$$

In the classical situation, this means that there is no entanglement. The same formula (14.4) can formally be written down also for the positive Wigner functions $W(x_1,p_1;x_2,p_2)$. However, in the quantum situation $\delta(\alpha_1-x_1)\,\delta(\beta_1-p_1)$ and $\delta(\alpha_2-x_2)\,\delta(\beta_2-p_2)$ are *not* legitimate Wigner functions, since they prescribe definite values to both coordinate and momentum and thus do not respect the uncertainty relations.

We conclude that the uncertainty relations are necessary—but not sufficient—for the existence of entanglement in semiclassical systems.

A Simple Sufficient Condition for Quantum Entanglement

The definition of the entanglement as given by Eq. (14.1) is not practical (except for a pure density matrix $\hat{\rho}$, when no entanglement means no correlations). It is therefore useful to have certain sufficient conditions for the presence of entanglement which will be easy to handle in applications and which will have a transparent physical meaning.

We choose the units in such a way that the coordinate and the momentum have the same dimension as $\sqrt{\hbar}$. For a harmonic oscillator with mass m and frequency ω it will suffice to make the following canonical transformation: $\hat{x} \to \sqrt{m\omega}\,\hat{x}$, $\hat{p} \to \hat{p}/\sqrt{m\omega}$.

For a two-particle system S with coordinate and momenta operators $\hat{x}_1$, $\hat{x}_2$ and $\hat{p}_1$, $\hat{p}_2$, respectively, one can propose the following sufficient condition for the entanglement [166].

Let us first note that the standard uncertainty relation

$$\langle \Delta\hat{x}^2\rangle\,\langle \Delta\hat{p}^2\rangle \geq \frac{\hbar^2}{4}, \tag{14.5}$$

$$\Delta\hat{x} \equiv \hat{x} - \langle\hat{x}\rangle, \qquad \Delta\hat{p} \equiv \hat{p} - \langle\hat{p}\rangle, \tag{14.6}$$

implies

$$\langle \Delta\hat{x}^2\rangle + \langle \Delta\hat{p}^2\rangle \geq \langle \Delta\hat{x}^2\rangle + \frac{\hbar^2}{4\,\langle \Delta\hat{x}^2\rangle}, \tag{14.7}$$

and then one gets via minimizing the RHS of (14.7) over $\langle \Delta \hat{x}^2 \rangle$ (which produces $\langle \Delta \hat{x}^2 \rangle \to \hbar/2$, to be put in the RHS of (14.7)):

$$\langle \Delta \hat{x}^2 \rangle + \langle \Delta \hat{p}^2 \rangle \geq \hbar. \tag{14.8}$$

Now assume that the two-particle system is described by a factorized density matrix:

$$\hat{\rho} = \hat{\rho}^{(1)} \otimes \hat{\rho}^{(2)}. \tag{14.9}$$

Then, due to (14.8), one can write

$$\begin{aligned} &\langle (\Delta \hat{x}_1 - \Delta \hat{x}_2)^2 \rangle + \langle (\Delta \hat{p}_1 + \Delta \hat{p}_2)^2 \rangle = \\ &\langle \Delta \hat{x}_1^2 \rangle + \langle \Delta \hat{p}_1^2 \rangle + \langle \Delta \hat{x}_2^2 \rangle + \langle \Delta \hat{p}_2^2 \rangle \geq 2\hbar, \end{aligned} \tag{14.10}$$

just because for noncorrelated systems the corresponding variances add up. If the overall system is in a separable state, then (14.10) is even strengthened, since the variance of *any* observable $\langle \Delta \hat{A}^2 \rangle$ increases under mixing, that is, under a transformation:

$$\{p_k, \hat{\rho}_k\} \to \sum_k p_k \hat{\rho}_k, \qquad p_k \geq 0, \sum_k p_k = 1, \tag{14.11}$$

where $\hat{\rho}_k$ are normalized ($\mathrm{tr}\, \hat{\rho}_k = 1$) density matrices. Indeed,

$$\begin{aligned} \langle \Delta \hat{A}^2 \rangle &= \sum_k p_k \mathrm{tr}\,[\, \hat{\rho}_k (\hat{A} - \langle \hat{A} \rangle)^2 \,] = \\ &\sum_k p_k \left\{ \mathrm{tr}\,[\, \hat{\rho}_k \hat{A}^2 \,] - \mathrm{tr}\,[\, \hat{\rho}_k \hat{A} \,]^2 \right\} + \sum_k p_k \left\{ \mathrm{tr}\,[\, \hat{\rho}_k \hat{A} \,] - \langle \hat{A} \rangle \right\}^2 \\ &= \sum_k p_k \langle \Delta \hat{A}^2 \rangle_k + \sum_k p_k \left\{ \mathrm{tr}\,[\, \hat{\rho}_k \hat{A} \,] - \langle \hat{A} \rangle \right\}^2 \geq \sum_k p_k \langle \Delta \hat{A}^2 \rangle_k. \end{aligned} \tag{14.12}$$

Employing (14.12) for $\hat{A} = \Delta \hat{x}_1 - \Delta \hat{x}_2$ and for $\hat{A} = \Delta \hat{p}_1 - \Delta \hat{p}_2$, we get that though (14.10) was obtained for the factorized state (14.9), it remains valid for an arbitrary nonentangled state $\sum_k p_k \hat{\rho}_k^{(1)} \otimes \hat{\rho}_k^{(2)}$. Thus the violation

$$\langle (\Delta \hat{x}_1 - \Delta \hat{x}_2)^2 \rangle + \langle (\Delta \hat{p}_1 + \Delta \hat{p}_2)^2 \rangle \leq 2\hbar \tag{14.13}$$

of (14.10) is a sufficient condition for entanglement.

Eq. (14.13) has a transparent physical meaning:

- entanglement is present, if fluctuations do not sum up additively, that is, if the changes $\Delta \hat{x}_1$ and $\Delta \hat{x}_2$ of the coordinates tend to correlate with each other, while those of the momenta tend to anti-correlate.

Alternative sufficient conditions for entanglement can be built based on $\langle(\Delta\hat{x}_1 + \Delta\hat{x}_2)^2\rangle + \langle(\Delta\hat{p}_1 - \Delta\hat{p}_2)^2\rangle$ or $\langle(\Delta\hat{x}_1 + \Delta\hat{x}_2)^2\rangle + \langle(\Delta\hat{p}_1 + \Delta\hat{p}_2)^2\rangle$, and so on. Various conditions obtained in this way are obviously not equivalent to each other. An obvious way to strengthen it is to demand that

$$\langle(\Delta\hat{x}_1 + \epsilon\,\Delta\hat{x}_2)^2\rangle + \langle(\Delta\hat{p}_1 + \zeta\,\Delta\hat{p}_2)^2\rangle \le 2\hbar, \tag{14.14}$$

at least for one of four independent choices $\epsilon = \pm 1$, $\zeta = \pm 1$.

For some special states (e.g. Gaussian states) there exist in literature necessary and sufficient conditions for entanglement [374]. However, they are technically involved and do not have such a straightforward physical meaning as (14.13). For our purposes conditions (14.13) and (14.14) are sufficient.

Operational issues

Let us finally recall how the condition (14.13) is checked operationally. To this end, we rewrite this inequality as

$$\begin{aligned} &\langle\Delta\hat{x}_1^2\rangle + \langle\Delta\hat{x}_2^2\rangle + \langle\Delta\hat{p}_1^2\rangle + \langle\Delta\hat{p}_2^2\rangle \\ &-2\langle\Delta\hat{x}_1\,\Delta\hat{x}_2\rangle + 2\langle\Delta\hat{p}_1\,\Delta\hat{p}_2\rangle \le 2\hbar. \end{aligned} \tag{14.15}$$

One now needs two ensembles of systems S, each one consisting of identically prepared correlated subsystems S_1 and S_2. Assuming that S_1 and S_2 are in possession of observers 1 and 2, respectively, observer 1 measures on the first (second) ensemble $\hat{x}_1$ ($\hat{p}_1$), while observer 2 measures on the first (second) ensemble $\hat{x}_2$ ($\hat{p}_2$). Note that the preceding two ensembles are necessary, since $\hat{x}_1$ and $\hat{p}_1$ (respectively, $\hat{x}_2$ and $\hat{p}_2$) do not commute: $[\hat{x}_k, \hat{p}_l] = i\hbar\delta_{kl}$, with δ_{kl} being Kronecker's delta, and thus cannot be measured simultaneously. Now the quantities in (14.15) can be estimated by each observer separately, while for the quantities in (14.16) the observers need to put the results of their measurements together (or to communicate in any other classical way) and to count the coinciding events.

14.3 Coarse-Grained Velocities and Uncertainty Relations for Brownian Particles

Consider N identical Brownian particles with coordinates $\mathbf{x} = (x_1, \ldots, x_N)$ and mass m interacting with N independent thermal baths at temperatures T_i and subjected to a potential $U(x_1, \ldots, x_N)$. The overdamped limit is defined by the following two conditions [323] (for a more detailed discussion, see Section 14.8):

- The characteristic relaxation time[6] of the (real) momenta $m\dot{x}_i$ is much smaller than the one of the coordinates.

[6] The characteristic relaxation time τ_θ of a random time-dependent variable θ (e.g. coordinate or momentum) is an *ensemble notion* and is defined as the time necessary for the conditional distribution of θ to become

- One is interested in times which are much larger than the relaxation time of the momenta, but which can be much smaller than or comparable to the relaxation time of the coordinates.

Under these conditions the dynamics of the system are described by the following Langevin equations [323]:

$$\dot{x}_i = f_i(\mathbf{x}) + \eta_i(t), f_i(\mathbf{x}) = -\partial_{x_i} U(\mathbf{x}), \langle \eta_i(t)\, \eta_j(t') \rangle = 2\, T_i\, \delta_{ij}\, \delta(t-t'), \quad (14.16)$$

where for our convenience both the mass m and the damping constant (coupling constant of the particles to the bath) are in the main text taken equal to one (they are recorded in Section 14.8).[7]

The conditional probability $P(\mathbf{x}, t|\mathbf{x}', t')$ is known to satisfy the following Fokker–Planck equation [323]:

$$\partial_t P(\mathbf{x}, t|\mathbf{x}', t') = -\sum_i \partial_{x_i} [f_i(\mathbf{x})\, P(\mathbf{x}, t|\mathbf{x}', t')] + \sum_i T_i\, \partial^2_{x_i x_i} P(\mathbf{x}, t|\mathbf{x}', t'), t \geq t'. \quad (14.17)$$

Eq. (14.17) is associated with the following initial condition:

$$P(\mathbf{x}, t|\mathbf{x}', t) = \delta(\mathbf{x} - \mathbf{x}') \equiv \prod_{i=1}^{N} \delta(x_i - x_i'), \quad (14.18)$$

as follows from the very definition of the conditional probability.

Consider an ensemble $\Sigma(\mathbf{x}, t)$ of all realizations of the whole N-particle system which at time t have a coordinate vector $\mathbf{x}$. Such an ensemble can be selected out of all possible realizations by measuring $\mathbf{x} = (x_1, \ldots, x_N)$. For this ensemble the average coarse-grained velocity for the Brownian particle with index j might naively be defined as:

$$v_j(\mathbf{x}, t) = \lim_{\varepsilon \to 0} \int \mathrm{d}\mathbf{y} \frac{y_j - x_j}{\varepsilon} P(\mathbf{y}, t + \varepsilon|\mathbf{x}, t). \quad (14.19)$$

memoryless: $P(\theta, t' + \tau_\theta|\theta', t') = P(\theta, t' + \tau_\theta)$. For the system of Brownian particles sufficiently strongly coupled to their thermal baths, the arbitrarinesses of the preceding definition (e.g., the precise choice of θ' and t') do not change the physical content of the definition [323].

[7] Equation (14.16) can be obtained from the complete Langevin equations of motion: [with the same definition of $\eta_i(t)$ as in (14.16)] $m\ddot{x}_i + \dot{x}_i = f_i(\mathbf{x}) + \eta_i(t)$, by disregarding the first term $m\ddot{x}_i$ corresponding to acceleration. One might make this heuristically by requiring that the friction force $\dot{x}_i$ dominates, and that the second-derivative $\ddot{x}_i$ is small for long times. More rigorous derivation of this classical problem is presented in [323], and is recalled in Section 14.8 for a simple model. In their turn the preceding complete Langevin equations can be rigorously derived from the Newton equations of motion for the Brownian particles and the baths. Here are the basic conditions for this derivation: (1) The thermodynamical (macroscopic) limit for the baths. (2) A reasonable model for the particle-bath interaction. (3) The initially equilibrium state of the baths, described by the corresponding Gibbs distributions at temperature T_i (this is how the random noises $\eta_i(t)$ come into existence). The details of the derivation can be looked up, for example, in [132, 357].

However, the absence of regular trajectories enforces us to define different velocities for different directions of time [284]:

$$v_{+,j}(\mathbf{x},t) = \lim_{\varepsilon\to+0} \int \mathrm{d}y_j \, \frac{y_j - x_j}{\varepsilon} \, P(y_j, t+\varepsilon|\mathbf{x},t), \tag{14.20}$$

$$v_{-,j}(\mathbf{x},t) = \lim_{\varepsilon\to+0} \int \mathrm{d}y_j \, \frac{x_j - y_j}{\varepsilon} \, P(y_j, t-\varepsilon|\mathbf{x},t). \tag{14.21}$$

The physical meaning of these expressions can be explained as follows.

1. As seen from the definitions, $v_{+,j}(\mathbf{x},t)$ is the average velocity to move anywhere starting from $(\mathbf{x},t)$, whereas $v_{-,j}(\mathbf{x},t)$ is the average velocity to come from anywhere and to arrive at $\mathbf{x}$ at the moment t.

   Since these velocities are defined already in the overdamped limit, ε is assumed to be much larger than the characteristic relaxation time of the (real) momentum, which is small in the overdamped limit. Therefore, we call (14.20, 14.21) coarse-grained velocities. It is known that for the overdamped Brownian motion almost all trajectories are not smooth. This is connected to the chaotic influences of the bath(s), which randomize the real momenta on much smaller times, and this is also the reason for $v_{+,j}(\mathbf{x},t) \neq v_{-,j}(\mathbf{x},t)$. The difference $v_{+,j}(\mathbf{x},t) - v_{-,j}(\mathbf{x},t)$ thus characterizes the degree of the above nonsmoothness.

   Recall that if one were to take ε much smaller than the characteristic relaxation time of the momentum—which will amount to applying definitions (14.20) and (14.21) to a smoother trajectory—then $v_{+,j}(\mathbf{x},t)$ and $v_{-,j}(\mathbf{x},t)$ will be equal to each other and equal to the average momentum. These points are discussed in detail in Section 14.8.
2. Here is an operational procedure for measuring $v_{+,j}(\mathbf{x},t)$ and $v_{-,j}(\mathbf{x},t)$. Consider the overall ensemble of the Brownian particles. The single members of this ensemble are N coupled Brownian particles. For each such single member one measures
   i) the coordinate of the Brownian particle with index j at the moment $t-\varepsilon$;
   ii) the coordinates of *all* Brownian particles at the moment t;
   iii) the coordinate of the same Brownian particle with index j at the moment $t+\varepsilon$.

   Repeating these points many times on various members of the preceding overall ensemble, ignoring all the results from *i)* and *iii)*, and selecting events from the second step, we reconstruct the ensemble $\Sigma(\mathbf{x},t)$. Employing the results from *i)* and *iii)* and conditioning repeatedly upon a single member from $\Sigma(\mathbf{x},t)$, we estimate the conditional probabilities $P(y_j, t\pm\varepsilon|\mathbf{x},t)$ and then finally calculate $v_{+,j}(\mathbf{x},t)$ and $v_{-,j}(\mathbf{x},t)$ via Eqs. (14.20, 14.21).
3. The preceding ensemble $\Sigma(\mathbf{x},t)$ is different from an ensemble $\Sigma_j(x_j,t)$ obtained by measuring x_j (at time t) *irrespective* of other coordinates. Indeed, $\Sigma(\mathbf{x},t)$

is defined globally, and if there are different observers in possession of each Brownian particle, they have to communicate to each other in order to be able to construct $\Sigma(\mathbf{x},t)$. In contrast, $\Sigma_j(x_j,t)$ is defined exclusively with respect to the Brownian particle with index j. This point will be discussed in more detail later on in Section 14.5.

The calculation of $v_{+,i}(x,t)$ and $v_{-,i}(x,t)$ is straightforward upon using the following three things: relation

$$P(\mathbf{x},t+\varepsilon|\mathbf{y},t) = \delta(\mathbf{x}-\mathbf{y}) + \varepsilon \sum_i \left[-f_i(\mathbf{y})\partial_{x_i}\delta(\mathbf{x}-\mathbf{y}) + T_i\,\partial^2_{x_ix_i}\delta(\mathbf{x}-\mathbf{y})\right], \quad (14.22)$$

which follows from (14.17, 14.18), Bayes formula, and partial integration assuming natural boundary conditions at infinity: $P(x,t) \to 0$ if $x_j \to \pm\infty$. Starting from the definitions (14.20, 14.21), we obtain

$$\begin{aligned}
v_{+,j}(\mathbf{x},t) &= \lim\nolimits_{\varepsilon\to+0} \int \mathrm{d}\mathbf{y}\, \frac{y_j - x_j}{\varepsilon}\, P(\mathbf{y},t+\varepsilon|\mathbf{x},t) \\
&= \int \mathrm{d}\mathbf{y}\,(y_j - x_j) \sum_i \left[-f_i(\mathbf{x})\partial_{y_i}\delta(\mathbf{x}-\mathbf{y}) + T_i\partial^2_{y_iy_i}\delta(\mathbf{x}-\mathbf{y})\right] \\
&= f_j(\mathbf{x}), \qquad (14.23) \\
v_{-,j}(\mathbf{x},t) &= \lim\nolimits_{\varepsilon\to+0} \int \mathrm{d}\mathbf{y}\, \frac{x_j - y_j}{\varepsilon}\, P(\mathbf{y},t-\varepsilon|\mathbf{x},t) \\
&= \lim\nolimits_{\varepsilon\to+0} \int \mathrm{d}\mathbf{y}\, \frac{x_j - y_j}{\varepsilon}\, P(\mathbf{x},t|\mathbf{y},t-\varepsilon)\, \frac{P(\mathbf{y},t-\varepsilon)}{P(\mathbf{x},t)} \\
&= \int \mathrm{d}\mathbf{y}\,(x_j - y_j)\, \frac{P(\mathbf{y},t)}{P(\mathbf{x},t)} \sum_i \left[-f(\mathbf{y})\,\partial_{x_i}\delta(\mathbf{x}-\mathbf{y}) + T_i\partial_{x_ix_i}\delta(\mathbf{x}-\mathbf{y})\right] \\
&= f_j(\mathbf{x}) - 2T_j\,\partial_{x_j}\ln P(\mathbf{x},t). \qquad (14.24)
\end{aligned}$$

The difference between the coarse-grained velocities $v_{+,j}(\mathbf{x},t)$ and $v_{-,j}(\mathbf{x},t)$ is

$$u_j(\mathbf{x},t) = \frac{v_{-,j}(\mathbf{x},t) - v_{+,j}(\mathbf{x},t)}{2} = -T_j\partial_{x_j}\ln P(\mathbf{x},t). \quad (14.25)$$

Recall that this quantity is nonzero due to the action of the thermal bath (the factor T_j in (14.25)), and due to the fact that at the coarse-grained level of description almost all trajectories of the Brownian particles are not smooth. Sometimes $u_j(\mathbf{x},t)$ is referred to as osmotic velocity of the Brownian particle with index j [284]. We shall use this word as a useful shorthand for the more precise term "change of the coarse-grained velocity."

Note that once the interactions between the Brownian particles are absent: $U(\mathbf{x}) = \sum_k U(x_k)$—and consequently correlations are absent as well, $P(\mathbf{x},t) = \prod_k P(x_k,t)$, if they were absent initially—the coarse-grained velocities $v_{\pm,j}$ depend only on the corresponding coordinate: $v_{\pm,j}(\mathbf{x},t) = v_{\pm,j}(x_j,t)$.

The regular (or mechanical) counterpart of the coarse-grained velocity for the Brownian particle with unit mass and index j can be naturally associated with the Newtonian force $f_j(\mathbf{x})$ as defined by (14.16). When the bath(s) are absent, this is the only contribution to the coarse-grained velocity. The fact that velocity appears to be proportional to force reminds us that we are in the overdamped regime of description, where, in particular, friction dominates over acceleration.

Uncertainty Relations and Their Interpretation

Uncertainty relations exist not only in QM but also in physics of overdamped Brownian motion [118]. As the very subject of statistical physics, they arise out of ignorance reasons, or more precisely due to separation of timescales: though the real classical particles involved in the classical Brownian motion certainly do have sharply defined coordinates and momenta at the microscopical level, at the coarse-grained (overdamped) level of description a Brownian particle does not have a well-defined trajectory, and cannot possess sharply defined coordinate and coarse-grained velocity, as seen in what follows.

We saw previously that the coarse-grained velocities $v_{+,j}(\mathbf{x},t)$ and $v_{-,j}(\mathbf{x},t)$ are defined with respect to the ensemble $\Sigma(\mathbf{x},t)$ of all Brownian particles which at the moment t pass via coordinate vector $\mathbf{x}$. On the other hand, $\Sigma(\mathbf{x},t)$ is by itself a subensemble embedded with probability (weight) $P(\mathbf{x},t)$ into the ensemble of all realizations of the random coordinate vector $\mathbf{x}$ at the time t. Thus, both $v_{+,j}(\mathbf{x},t)$ and $v_{-,j}(\mathbf{x},t)$, as well as the osmotic velocity $u_j(\mathbf{x},t)$ are random quantities as functions of the random configuration of N Brownian particles $\mathbf{x} = (x_1,\cdots,x_N)$. In other words, this randomness enters via the context $(\mathbf{x},t)$ which was chosen to define the ensemble $\Sigma(\mathbf{x},t)$.

More specifically, let us focus on the common distribution function $P(\mathbf{x},\mathbf{u};t)$ of

$$\mathbf{x} = (x_1,\ldots,x_N) \quad \text{and} \quad \mathbf{u} = (u_1,\ldots,u_N), \tag{14.26}$$

$$P(\mathbf{x},\mathbf{u};t) = P(\mathbf{x},t)\prod_{j=1}^{N}\delta(\,u_j - u_j(\mathbf{x},t)\,), \tag{14.27}$$

where $u_j(\mathbf{x},t)$ is defined by (14.25).

One has, after averaging over $P(\mathbf{x},\mathbf{u};t)$,

$$\begin{aligned}\langle u_j\rangle &= \int \mathrm{d}u_j\mathrm{d}\mathbf{x}u_j\,P(\mathbf{x},t)\delta(\,u_j - u_j(\mathbf{x},t)) \\ &= -T_j\int \mathrm{d}\mathbf{x}\partial_{x_j}P(\mathbf{x},t) = 0,\end{aligned} \tag{14.28}$$

$$\begin{aligned}\langle\,(x_k - \langle x_k\rangle\,)\,(\,u_j - \langle u_j\rangle\,)\rangle &= \int \mathrm{d}\mathbf{x}\,u_j(\mathbf{x},t)\,P(\mathbf{x},t)\,(x_k - \langle x_k\rangle) \\ &= -T_j\int \mathrm{d}\mathbf{x}\,(x_k - \langle x_k\rangle)\,\partial_{x_j}P(\mathbf{x},t) = T_j\delta_{kj}.\end{aligned} \tag{14.29}$$

The fact of $\langle u_j \rangle = 0$ is natural, since u_j is the (coarse-grained) velocity difference generated due to the interaction to the bath. It is also seen from (14.29) that u_j of the corresponding Brownian particle correlates only with its own coordinate. This is related to the independence of the thermal baths which act on different Brownian particles.

Applying to (14.29) the standard Cauchy–Schwartz inequality, one deduces:

$$\langle (x_k - \langle x_k \rangle)^2 \rangle \langle (u_j - \langle u_j \rangle)^2 \rangle \geq \\ |\langle (x_k - \langle x_k \rangle)(u_j - \langle u_j \rangle) \rangle|^2 = T_j^2 \delta_{kj}, \tag{14.30}$$

which implies an uncertainty relation between the uncertainty of the coordinate and that of the osmotic velocity [118].

Obviously, the regular counterpart $f_j(x, t)$ of the coarse-grained velocity does not (and should not) have any uncertainty relation with the coordinate.

Discussion on the Brownian Uncertainty Relation

Let us discuss in more detail the difference between the physical meaning of the Brownian uncertainty relation (14.30) as compared to the standard quantum mechanical uncertainty relation (14.5). The latter refers to two identical quantum ensembles $\mathcal{E}(\hat{\rho}(t))$ such that on the first (second) ensemble one measures $\hat{x}$ ($\hat{p}$), and then combines into the single relation (14.5) the obtained dispersions.

This is a good place to note that the frequent interpretation of (14.5) as referring to disturbance of the momentum when measuring the coordinate and vice versa is in general incorrect [105, 296], since, in particular, the quantities $\langle \Delta\hat{x}^2 \rangle$ and $\langle \Delta\hat{p}^2 \rangle$ clearly refer to different measurements done on different ensembles, rather than to a simultaneous measurement of the coordinate and momentum done on a single ensemble. More refined arguments and explicit counterexamples for this point can be found in [105, 296].

In contrast to this, the quantities entering into Eq. (14.30) refer to consecutive measurements done on the same ensemble. The involved coarse-grained velocity is operationally well-defined and characterizes the change of the coordinate. This is in contrast to the situation with the quantum mechanical observable of the momentum which (in the Heisenberg representation) in general does *not fully* characterize the change of the coordinate for any initial state; see footnote 14 for more details.

The quantity $P(\mathbf{x}, \mathbf{u}; t)$ is similar to the conditional joint probability distribution; see (10.17) (Section 10.2), but these quantities are methodologically different.

14.4 Brownian Entanglement

Consider now two Brownian particles. An analog of entanglement can be introduced in the following way. We call the state of two particles nonentangled (separable) if the common distribution function (14.27) of the coordinates and the osmotic velocities can be prepared by mixing noncorrelated distributions[8]

$$P(x_1, u_1, x_2, u_2) = \int d\lambda\, \mathcal{P}(\lambda) P_1(x_1, u_1|\lambda) P_2(x_2, u_2|\lambda), \tag{14.31}$$

$$\mathcal{P}(\lambda) \geq 0, \int d\lambda\, \mathcal{P}(\lambda) = 1.$$

We shall naturally require that the separate distributions $P_1(x_1, u_1|\lambda)$ and $P_2(x_2, u_2|\lambda)$ ensure the basic properties (14.28, 14.29) of the osmotic velocity u_j. For the rest they can be completely arbitrary. The requirements (14.28, 14.29) make the basic difference as compared to the (naive) classical discussion in Section 14.2, where any distribution was admissible. More detailed discussion on the physical meaning of (14.31), and, in particular, on its similarities and differences with the definition (14.1) of quantum entanglement, is postponed till Section 14.6.

For simplicity we choose units such that x and u have the same dimension as $\sqrt{T}$ (e.g., we measure time in units of square root of the damping constant; see Section 14.8). We take also $T_1 = T_2 = T$ again for simplicity. In the same way as we derived (14.5, 14.13), one can write the Brownian uncertainty relation (14.30) as

$$\langle\, \Delta u_1^2 \,\rangle + \langle\, \Delta x_1^2 \,\rangle \geq 2T, \qquad \langle\, \Delta u_2^2 \,\rangle + \langle\, \Delta x_2^2 \,\rangle \geq 2T, \tag{14.32}$$

and proceed to derive

$$\langle (\, \Delta u_1 + \Delta u_2 \,)^2 \,\rangle + \langle (\, \Delta x_1 - \Delta x_2 \,)^2 \,\rangle < 4T, \tag{14.33}$$

as a sufficient condition for the entanglement. More general relations (14.14) can also be obviously transferred to the Brownian situation:

$$\langle (\, \Delta u_1 + \zeta \Delta u_2 \,)^2 \,\rangle + \langle (\, \Delta x_1 + \epsilon \Delta x_2 \,)^2 \,\rangle < 4T, \tag{14.34}$$

where ζ and ϵ can independently assume values ± 1.

Gaussian State of Two Interacting Brownian Particles

It will be useful to work out a simple model for two coupled Brownian particles, where the preceding general concepts can be visualized and studied in detail.

Consider now two harmonically interacting Brownian particles with an overall potential energy

$$U(x_1, x_2) = \frac{ax_1^2}{2} + \frac{ax_2^2}{2} + gx_1x_2, \tag{14.35}$$

[8] We consider only the common distribution function of the coordinates and the osmotic velocities, and not, for example, the common distribution of the coordinates and forward velocities $v_{+,j}(\mathbf{x}, t)$, simply because, due to the Brownian uncertainty relation (14.29), only the former can lead to a nontrivial definition of entanglement. Recall in this context that the quantum uncertainty relations are necessary for the existence of the quantum entanglement.

where $a > 0$ and g characterize the interaction between the particles.[9] The particles interact with an independent bath at the same temperature T.

The Fokker–Planck equation (14.17) for the dynamics of these Brownian particles can be solved directly and the outcome is known to be given by the following two-dimensional Gaussian (provided the initial distribution was Gaussian):

$$P(x_1,x_2;t) = \frac{\sqrt{\det C}}{2\pi} \exp\left[-\frac{1}{2}\sum_{i,j=1}^{2} C_{i,j}x_ix_j\right], \tag{14.36}$$

$$C^{-1} = \begin{pmatrix} \langle x_1^2(t)\rangle & \langle x_1(t)x_2(t)\rangle \\ \langle x_1(t)x_2(t)\rangle & \langle x_2^2(t)\rangle \end{pmatrix} \equiv \begin{pmatrix} \sigma_{11}(t) & \sigma_{12}(t) \\ \sigma_{12}(t) & \sigma_{22}(t) \end{pmatrix},$$

where for simplicity we assumed that

$$\langle x_1(t)\rangle = \langle x_2(t)\rangle = 0, \tag{14.37}$$

for all times. It is also useful to note the marginal distributions of $P(x_1,x_2;t)$:

$$P_j(x_j;t) = \exp\left[-\frac{x_j^2}{2\sigma_{jj}(t)}\right], \qquad j = 1,2. \tag{14.38}$$

The easiest way to obtain $\sigma_{11}(t)$, $\sigma_{22}(t)$, and $\sigma_{12}(t)$ is to look directly at the Langevin equations (14.16) with the potential (14.35):

$$\dot{x}_1(t) + ax_1(t) + gx_2(t) = \eta_1(t), \dot{x}_2(t) + ax_2(t) + gx_1(t) = \eta_2(t), \tag{14.39}$$

which can be more conveniently rewritten in terms of the relative coordinate $r_- = (x_1 - x_2)/2$ and the center of mass coordinate $r_+ = (x_1 + x_2)/2$:

$$\dot{r}_\pm(t) = -(a \pm g)r_\pm(t) + \frac{1}{2}(\eta_1(t) \pm \eta_2(t)), \tag{14.40}$$

and then solved directly:

$$r_\pm(t) = e^{-(a\pm g)t}r_\pm(0) + \frac{1}{2}\int_0^t \mathrm{d}s\, e^{-(a\pm g)s}(\eta_1(t-s) \pm \eta_2(t-s)). \tag{14.41}$$

Eqs. (14.25, 14.36) produce for the coarse-grained velocity difference (osmotic velocity) of the first and the second particle, respectively,

$$u_1(\mathbf{x},t) = \frac{T\,(\sigma_{22}x_1 - \sigma_{12}x_2)}{\sigma_{11}\sigma_{22} - \sigma_{12}^2}, \qquad u_2(\mathbf{x},t) = \frac{T\,(\sigma_{11}x_2 - \sigma_{12}x_1)}{\sigma_{11}\sigma_{22} - \sigma_{12}^2}. \tag{14.42}$$

The sufficient condition (14.34) reduces to

$$\langle\, (\,\Delta u_1 + \zeta\,\Delta u_2\,)^2\,\rangle + \langle\, (\,\Delta x_1 + \epsilon\,\Delta x_2\,)^2\,\rangle = \tag{14.43}$$

[9] Note that the positivity of $U(x_1,x_2)$ is implied by $g^2 < a^2$.

$$\frac{T^2(\sigma_{22}+\sigma_{11}-2\zeta\sigma_{12})}{\sigma_{11}\sigma_{22}-\sigma_{12}^2}+\sigma_{22}+\sigma_{11}+2\epsilon\sigma_{12}<4T.$$

A simpler expression is gotten for $\sigma_{11}=\sigma_{22}$:[10]

$$\frac{T^2}{\sigma_{11}+\zeta\sigma_{12}}+\sigma_{11}+\epsilon\sigma_{12}<2T. \tag{14.44}$$

This condition is naturally not satisfied for $\sigma_{12}=0$, as the minimal value of $\sigma_{11}+T^2/\sigma_{11}$ over σ_{11} is just equal to $2T$. For the correlated situation $\sigma_{12}>0$ and we apply (14.44) for $\zeta=1, \epsilon=-1$, while for the anti-correlated situation $\sigma_{12}<0$ we use in the same way (14.44) for $\zeta=-1, \epsilon=1$. With these we get from (14.44) the following condition

$$(\sigma_{11}-T)^2<\sigma_{12}^2+2T|\sigma_{12}|. \tag{14.45}$$

It is seen that this sufficient condition for the Brownian entanglement is symmetric with respect to correlation and anti-correlation. and it can be satisfied if σ_{11} is sufficiently close to T and $|\sigma_{12}|$ is finite. Eq. (14.45) can also be satisfied when the particles are either correlated or anticorrelated sufficiently strongly, for example, when $|\sigma_{12}|$ is sufficiently large and sufficiently close to its upper bound σ_{11}. Note that (14.45) is not satisfied for $T\to 0$. This is natural, since there are no Brownian uncertainty relations in this limit.

Thus the Brownian entanglement can exist for sufficiently strongly fluctuating and/or sufficiently strongly interacting Brownian particles.

Let us check (14.44) more specifically with the stationary (equilibrium) state of the Brownian particles which is established for long times provided the potential energy $U(x_1,x_2)$ is (strictly) positive, and where

$$\sigma_{11}=\sigma_{22}=\frac{Ta}{a^2-g^2}, \sigma_{12}=-\frac{Tg}{a^2-g^2}. \tag{14.46}$$

These relations can be obtained either by directly solving (14.41) and then taking the limit $t\to\infty$, or directly via the stationary distribution for the particles which is known to be Gibbsian: $P(x_1,x_2)\propto \exp[-U(x_1,x_2)/T]$.

Eq. (14.45) reduces to[11]

$$|g|>-1+\sqrt{1+(a-1)^2}. \tag{14.47}$$

Thus if $|g|$ is large enough, there can be entanglement in the equilibrium state.

[10] This equality will be satisfied for two identical Brownian particles if their initial conditions are the same.

[11] Conditions (14.47) are compatible with the stability requirement $a>|g|$, which comes from demanding positivity of $U(x_1,x_2)$. Indeed, once $|g|$ satisfies (14.47), $a>|g|$ becomes $a(2+a)>(a-1)^2$ and is satisfied provided a is not very small.

Entanglement for Two Noninteracting Brownian Particles

In the quantum case two subsystems can be entangled even if they interacted in the past, and they do not interact at the moment when the entanglement is tested. This is also the case with the Brownian particles, as we show now.

Return to the preceding system of two harmonically interacting Brownian particles and assume that they did interact for $t < 0$ but are not interacting for positive times.[12] One has from (14.39) (or from (14.41)):

$$\sigma_{jj}(t) = e^{-2at}\sigma_{jj}(0) + \frac{T}{a}(1 - e^{-2at}),\ \sigma_{12}(t) = e^{-2at}\sigma_{12}(0), j = 1, 2. \quad (14.48)$$

We assume that the condition (14.44) (with $\sigma_{11} = \sigma_{22}$ at all times) was satisfied at the initial time, so that the Brownian entanglement was present. For positive times and $a > 0$, $\sigma_{12}(t)$ gradually disappears. This means that for long times the sufficient condition (14.44) for the Brownian entanglement will not be valid. Clearly, no Brownian entanglement can persist for very long times, since the particles become in this limit completely noncorrelated, that is, the common probability distribution for $t \to \infty$ factorizes $P(x_1, x_2, t) = P(x_1, t)P(x_2, t)$ (compare with our discussion after (14.25)).

It is, however, to be stressed that for certain not very long times we shall surely have the condition (14.44) satisfied, at least once it was satisfied initially. Moreover, even if this condition was not satisfied at $t = 0$, it can curiously get valid for some (not very long) times. Here is an example. For free Brownian particles with $a = 0$, the correlation $\sigma_{12}(t)$ does not change in time at all, but instead the dispersion of the coordinates increases linearly with time: $\sigma_{jj}(t) = \sigma_{jj}(0) + 2Tt$. Condition (14.45) for the presence of entanglement now reads:

$$|\sigma_{11}(0) - T + 2Tt| < \sqrt{\sigma_{12}^2(0) + 2T|\sigma_{12}(0)|}. \quad (14.49)$$

Even if this condition was not valid at $t = 0$, $|\sigma_{11}(0) - T| > \sqrt{\sigma_{12}^2(0) + 2T|\sigma_{12}(0)|}$, and additionally $\sigma_{11}(0) < T$, Eq. (14.49) can still be satisfied for

$$t_- < t < t_+, \quad (14.50)$$

where

$$t_\pm = \frac{T - \sigma_{11}(0) \pm \sqrt{\sigma_{12}^2 + 2T|\sigma_{12}|}}{2T}. \quad (14.51)$$

Our general conclusion is that the Brownian entanglement is possible if the two Brownian particles do interact or had interacted strongly enough. The absence of

[12] To avoid possible misunderstanding, recall that the thermal baths acting on two particles were taken to be completely independent of each other, so that there is no influence via them.

interactions at present need not always destroy the Brownian entanglement, for some finite times it may even facilitate the sufficient conditions for its existence, as we saw earlier.

14.5 Local Osmotic Velocities

In Section 14.3 we have seen that the definition of the coarse-grained velocities (14.20, 14.21, 14.25) is given via the ensemble $\Sigma(x_1, x_2, t)$ in the common context of the two Brownian particles.

Let us now turn to the local velocities determined with respect to the ensemble $\Sigma_1(x_1, t)$, which is obtained by measuring at time t only the coordinate x_1 of the first particle.

There are two completely equivalent ways for determining the coarse-grained velocities over this ensemble. The first way amounts to repeating definitions (14.20, 14.21) for the ensemble $\Sigma_1(x_1, t)$:

$$v_{\pm,1}(x_1, t) = \pm \lim_{\varepsilon\to+0} \int \mathrm{d}y_1 \frac{y_1 - x_1}{\varepsilon} P(y_1, t \pm \varepsilon | x_1, t), \tag{14.52}$$

where all the involved probability distributions contain no references on the second Brownian particle. The corresponding osmotic velocity reads in complete analogy to (14.25):

$$\mu_1(x_1, t) = v_{-,1}(x_1, t) - v_{+,1}(x_1, t). \tag{14.53}$$

The second way is to note that if observer 1 has no information at all from observer 2, then he effectively sums the ensemble $\Sigma(x_1, x_2, t)$ over all possible results of the second coordinate x_2 at time t with his result x_1 being fixed. Each value of x_2 during this summation is then met with probability $P(x_2, t|x_1, t)$, and this results in

$$v_{\pm,1}(x_1, t) = \int \mathrm{d}x_2\, v_{\pm,1}(x_1, x_2, t)\, P(x_2, t|x_1, t). \tag{14.54}$$

The equivalence between definitions (14.52) and (14.54) can be established via (14.23, 14.24) and the Bayes formula.[13]

[13] Note that for obtaining the local coarse-grained velocities $v_{\pm,1}(x_1, t)$ we had to average $v_{\pm,1}(x_1, x_2, t)$ over the conditional distribution $P(x_2, t|x_1, t)$, and *not* over the unconstrained probability of the second coordinate $P(x_2, t)$. This in-equivalence can be illustrated with the help of the following fact of probability theory concerning three random variables A, B, C: $\sum_C P(C)\, P(A|BC) \neq \sum_C P(C|B)\, P(A|BC) = P(A|B)$. Were we to adopt the second possibility, for example, $\tilde{u}(x_1, t) = \int \mathrm{d}x_2\, u(x_1, x_2, t)\, P(x_2, t)$, this would lead us to an explicitly common-context (nonlocal) quantity, for example, for the preceding example of harmonic oscillators we would get $\tilde{u}(x_1, t) = \frac{T\sigma_{22}(t)x_1}{\sigma_{11}(t)\sigma_{22}(t) - \sigma_{12}^2(t)}$, which explicitly depends on the dispersion $\sigma_{22}(t)$ of the second particle. If the two particles do not interact for positive times, but did interact in the past (for $t < 0$), $\tilde{u}(x_1, t)$ can still be influenced, for example, by external fields which apply on the second particle for positive times (apparent or false nonlocality).

Completely similar definitions can be given for the second particle, employing the ensemble $\Sigma(x_2)$ obtained by measuring the second particle's coordinate only. For the corresponding osmotic component, one now has from (14.53, 14.54, 14.25):

$$\mu_j(x_j,t) = -T_j\partial_{x_j}P(x_j,t), \qquad j=1,2. \tag{14.55}$$

For the Brownian particle in the harmonic potential, as described, for example, by (14.38), one has

$$\mu_j(x_j,t) = T_j\frac{x_j}{\sigma_{jj}(t)}. \tag{14.56}$$

If now the definition (14.31) of nonentanglement is applied to the common distribution function

$$P(x_1,\mu_1,x_2,\mu_2) = P_1(x_1)\,\delta(\,\mu_1-\mu_1(x_1)\,)P_2(x_2)\,\delta(\,\mu_2-\mu_2(x_2)\,), \tag{14.57}$$

then it is seen to be satisfied trivially: No entanglement occurs with locally defined osmotic velocities

14.6 Discussion

Let us compare in more detail the physical meaning of the quantum mechanical entanglement versus its Brownian analog.

- In analogy to continuous-variable quantum entanglement, Brownian entanglement is defined as a type of correlation between the coordinates and the changes of the coarse-grained velocities (osmotic velocities) of two Brownian particles, which is impossible to reproduce by mixing noncorrelated [that is, referring to each particle separately] distributions.
- The Brownian uncertainty relation (14.29) between the coordinate and the change of the coarse-grained velocity is necessary for the very existence of the Brownian entanglement. This is again similar to the quantum situation, where the analogous role of a necessary condition is being played by the quantum uncertainty relations.
- For checking quantum entanglement via sufficient condition (14.13) one needs to make measurements of noncommuting coordinate $\hat{x}_k$ and momentum $\hat{p}_k$ for each quantum particle ($k=1,2$). To this end one has to have two different ensembles of the particle S_1 and S_2 (compare with our discussion in Section 14.2). In contrast, the Brownian entanglement involves only consecutive coordinate measurements done on a single ensemble of the Brownian pairs.

- The momentum operator $\hat{p}$ in QM |though being equal in Heisenberg representation to the time-derivative of the coordinate operator $\hat{p} = m\frac{\mathrm{d}}{\mathrm{d}t}\hat{x}$| does not in general fully characterize the intuitive notion of "change of the coordinate for an infinitesimal time."[14]
  In contrast, the Brownian entanglement is about the coordinates and the coarse-grained velocities which do characterize the change of the coordinates on the coarse-grained scale of time.
- Both quantum entanglement and its Brownian analog can exist for subsystems which interacted in the past, but do not interact at the present.
- In QM the sufficient condition (14.13) for entanglement involves correlations between the coordinate operators $\hat{x}_1$ and $\hat{x}_2$, and the momentum operators $\hat{p}_1, \hat{p}_2$ of the subsystems S_1 and S_2. Quantum entanglement can only be found after the results got via measurements on the corresponding subensembles for the subsystems S_1 and S_2 are put together. The situation with the Brownian entanglement is similar, since it also requires correlation experiments.
- Here is finally the main conceptual difference between the quantum entanglement and its Brownian analog. In QM the preceding operators of coordinate and momentum pertain to the corresponding subsystems S_1 and S_2 *independently* of the full system context. By this we mean that all the statistics of, for example, $\hat{p}_1$ can be collected via local measurements on the corresponding quantum subensemble, whether or not this subensemble forms a part of any larger ensemble.

  In contrast, the very definition of the coarse-grained velocities (14.20, 14.21, 14.25) involves a global (that is, depending on the two subsystems) ensemble $\Sigma(\mathbf{x}, t)$. As seen in Section 14.5, the purely local definition of coarse-grained velocities can also be given, but there will not be any entanglement on that level, for the same reason as there is no entanglement in other classical systems (see Section 14.2).

  This conclusion on the main difference is close to the analogous conclusion of Ref. [343], which discusses similarities between quantum entanglement and certain correlations in classical optics.

14 This is a general point. A difference $\hat{A}(t) - \hat{A}(0) = \int_0^t \mathrm{d}t \frac{\mathrm{d}}{\mathrm{d}t}\hat{A}$ of Heisenberg operators does not fully characterize the change in time of the observable $\hat{A}$, because there are ensembles described by (time-independent in the Heisenberg representation) states $|\psi\rangle\langle\psi|$ for which $(\hat{A}(t) - \hat{A}(0))|\psi\rangle = 0$. This seems to imply that the value of $\hat{A}$ did not change at all, but this is not correct, since the preceding eigenvalue relation may be still compatible—due to noncommutativity $[\hat{A}(t), \hat{A}(0)] \neq 0$—with different statistics of $\hat{A}(t)$ and $\hat{A}(0)$, for example, $\langle\psi|\hat{A}^3(t)|\psi\rangle \neq \langle\psi|\hat{A}^3(0)|\psi\rangle$. For more elaborated discussion and concrete examples see [14]. For unbound operators, such as $\hat{x}$ and $\hat{p}$, the preceding reasonings may need to be technically modified, since eigenstates of unbound operators are not normalizable. In this context it may suffice to require $(\hat{A}(t) - \hat{A}(0))|\psi\rangle \approx 0$.

14.7 Possibilities of Experimental Realization

One hopes that an experimental verification of the Brownian entanglement is going to be easier than that of its quantum analog: since Brown's discovery in 1828, various examples of Brownian motion are routinely observed in many systems. On the other hand, the basic conditions needed for observation of the Brownian entanglement amount to two coupled Brownian particles and a resolution of the Brownian motion sufficient for observing the osmotic velocities. Recall that on the coarse-grained time scale the osmotic velocities can be visualized as average kicks got by the Brownian particle due to its interaction with bath particles.

Here we discuss two experimentally studied examples of Brownian motion. The pecularity of these examples is that the Brownian motion can be detected with human eyes (without microscopes). These are thus examples of macroscopic Brownian motion. Recall that the typical examples of Brownian motion involve much smaller scales: pollen molecules in water observed originally by Brown had a size $\approx 10^{-3}$ cm, and even smaller lengths are typical for many other realizations of Brownian motion.

The first experiment [320] considers a two-dimensional circular container with elastic walls. Inside the container there are motorized balls with a mass of 120 g and a diameter of 8 cm. Each ball is driven by a battery-powered motor inside and moves chaotically due to elastic collisions with the walls of the container and with other motorized balls. The set of balls (with concentration 50 balls/m^2) models the bath particles. The Brownian particles are modelled with a set of ping-pong balls (with a mass 2 g and diameter 4 cm) which undergo random collisions with the motorized balls. The primary purpose of [320] was to study polymer statistics, so that the ping-pong balls were joined into a long chain by means of springs. The resulting coupled Brownian motion produced a probability distribution for the end-to-end distance of the ping-pong chain in close agreement with the existing theories of two-dimensional (self-avoiding) random walk. In this experiment the Langevin equation as such was not tested directly, but there were several indirect reasons supporting its validity [320]. This system would fit for observing the Brownian entanglement, because almost all needed ingredients of this phenomenon are present.

A curious pedagogical analog of the preceding experiment is the motion of a child toy called Bumble Ball [355]: a plastic sphere about 11 cm in diameter with a symmetrical pattern of rubber knobs extending about 3 cm from its surface. An internal motor rotates an off-axis mass such that when activated and placed on a hard surface, a Bumble Ball simulates a random walk [355]. Joining two such balls with a spring and placing them into a container creates a toy model which might provide at least the most rough indications of the Brownian entanglement.

The second experiment [283] studied a single sphere (a ping-pong ball of a mass 2.5 g and diameter 4 cm) rolling stochastically in an upflow of gas under flow speed 280 cm/c. The sphere rolls stochastically due to the turbulence it generates in the gas stream (the corresponding Reynolds number is ≈40). The study focused on the full time-dependent dynamics of the sphere, as captured by video imaging. It appears that the dynamics is that of the underdamped Brownian particle in a harmonic potential. This correspondence was thoroughly checked from various perspectives. When looking for the Brownian entanglement in such a situation, it would be necessary to go to the overdamped motion regime (e.g., by decreasing the mass of the sphere and by increasing the flow speed of the stream) and to add there a second sphere.

14.8 Coarse-Grained Velocity: Technicalities

This section has two closely related purposes. First, we shall apply the definition (14.20, 14.21) of the coarse-grained velocities in a non-overdamped situation, that is, we shall take ε in those definitions much smaller than the characteristic relaxation time of the (real) momenta. We shall convince ourselves that the expected answer is obtained, relating the coarse-grained velocities to the average (real) momentum. In addition, we shall see that the coarse-grained velocity difference (osmotic velocity) (14.25) disappears in this situation, that is, the definitions (14.20, 14.21) become equivalent. Second, based on an exactly solvable situation, we shall follow in detail to the behavior of the coarse-grained velocity as a function of ε. For simplicity we operate only with one Brownian particle.

14.8.1 Reduction of Coarse-Grained Velocity to Average Momentum

The evolution of the common probability distribution of the coordinate x and the momentum p of the Brownian particle is described by the following Fokker–Planck–Kramers–Klein equation [323]:

$$\partial_t \mathcal{P} = -\frac{p}{m}\partial_x \mathcal{P} + \partial_p(\frac{\gamma}{m}p + V'(x) + \gamma T \partial_p)\mathcal{P}, \tag{14.58}$$

where m, γ, and $V(x)$ are mass, damping constant, and potential, respectively, and where

$$\mathcal{P}(x,p,t|x',p',t'), t \geq t', \mathcal{P}(x,p,t|x',p',t) = \delta(x-x')\,\delta(p-p') \tag{14.59}$$

is the conditional probability of moving from (x',p') at time t' to (x,p) at time t. This equation corresponds to the Langevin equation

$$m\ddot{x} = -ax - \gamma\dot{x} + \eta(t), \langle\eta(t)\rangle = 0, \langle\eta(t)\eta(t')\rangle = 2\gamma T\delta(t-t'), \tag{14.60}$$

where $-\gamma\dot{x}$ is the friction force, and where $\eta(t)$ is the random Gaussian white noise.

Applying the definition (14.20) of the coarse-grained velocity and using (14.58,14.59), one has

$$
\begin{aligned}
v_+(x,t) &= \lim_{\varepsilon\to+0} \int \mathrm{d}y \, \frac{y-x}{\varepsilon} \, \mathcal{P}(y,t+\varepsilon|x,t) \\
&= \lim_{\varepsilon\to+0} \int \mathrm{d}y \, \mathrm{d}p \, \mathrm{d}p' \, \frac{y-x}{\varepsilon} \, \mathcal{P}(y,p,t+\varepsilon|x,p',t) \, \mathcal{P}(p',t|x,t) \\
&= -\int \mathrm{d}y \, \mathrm{d}p \, (y-x) \, \mathcal{P}(p,t|x,t) \, \frac{p}{m} \, \partial_y \delta(x-y) \\
&= \int \mathrm{d}p \, \mathcal{P}(p,t|x,t) \, \frac{p}{m},
\end{aligned}
\tag{14.61}
$$

where $\mathcal{P}(p,t|x,t)$ is the conditional probability for having the momentum equal to p at time t, provided the coordinate was equal to x at the same time.

Likewise,

$$
\begin{aligned}
v_-(x,t) &= \lim_{\varepsilon\to+0} \int \mathrm{d}y \, \frac{x-y}{\varepsilon} \, P(y,t-\varepsilon|x,t) \\
&= \lim_{\varepsilon\to+0} \int \mathrm{d}y \, \mathrm{d}p \, \mathrm{d}p' \, \frac{x-y}{\varepsilon} \, \mathcal{P}(x,p,t|y,p',t-\varepsilon) \, \frac{\mathcal{P}(y,p',t-\varepsilon)}{\mathcal{P}(x,t)} \\
&= \int \mathrm{d}y \, \mathrm{d}p \, (y-x) \, \frac{\mathcal{P}(y,p,t)}{\mathcal{P}(x,t)} \, \frac{p}{m} \, \partial_x \delta(x-y) \\
&= \int \mathrm{d}p \, \mathcal{P}(p,t|x,t) \, \frac{p}{m} = v_+(x,t).
\end{aligned}
\tag{14.62}
$$

The fact that $v_-(x,t) = v_+(x,t)$ can be easily generalized to any number of interacting Brownian particles.

Now if there are no correlations between the momentum and coordinate $\mathcal{P}(p,t|x,t) = \mathcal{P}(p,t)$, then $v_-(x,t) = v_+(x,t)$ reduce to the usual average momentum. This is the case, in particular, for the deterministic situation, where $\mathcal{P}(x,p,t) = \delta(x - x(t))\,\delta(p - p(t))$.

Thus, provided ε has been taken much smaller than any relevant time-scale related to the coordinate and/or the momentum, the coarse-grained velocity reduces to the average momentum as it should. In this situation the osmotic velocity is zero.

14.8.2 Harmonic Potential

Now let us show on the exactly solvable situation with a harmonic potential

$$
V(x) = \frac{ax^2}{2}, \tag{14.63}
$$

that if the characteristic relaxation times of the momentum

$$\tau_p = \frac{m}{\gamma}, \tag{14.64}$$

and the coordinate

$$\tau_x = \frac{\gamma}{a}, \tag{14.65}$$

are well separated:

$$\tau_p \ll \tau_x, \tag{14.66}$$

and if additionally

$$\tau_x \gg \varepsilon \gg \tau_p, \tag{14.67}$$

one gets back the values for osmotic velocity which were obtained in the main text by means of the overdamped Fokker–Planck equation (14.17).

The case with $V(x) = ax^2/2$ can be solved either directly from (14.58), or using the equivalent Langevin equation (14.60). The solution of the latter—obtained, for example, via Laplace transformation—reads:

$$x(t) = x(0)g(t) + \frac{1}{m}p(0)f(t) + \frac{1}{m}\int_0^t \mathrm{d}t' f(t-t')\eta(t'), \tag{14.68}$$

where

$$f(t) = \frac{e^{-\omega_2 t} - e^{-\omega_1 t}}{\omega_1 - \omega_2}, \qquad g(t) = \frac{\omega_1 e^{-\omega_2 t} - \omega_2 e^{-\omega_1 t}}{\omega_1 - \omega_2}, \tag{14.69}$$

$$\omega_{1,2} = \frac{\gamma}{2m}\left(1 \pm \sqrt{1 - \frac{4am}{\gamma^2}}\right). \tag{14.70}$$

For simplicity reasons we shall put $x(0) = p(0) = 0$. This can be done, since these quantities are assumed to be independent of the noise by the very definition of the considered stochastic process.

As the process (14.68) is Gaussian, the two-time probability distribution of the coordinate can be written down as follows:

$$\mathcal{P}(y, s; x, t) = \frac{\sqrt{d}}{2\pi}\exp -\frac{1}{2}\left[A_{11}y^2 + A_{22}x^2 + 2A_{12}xy\right], \tag{14.71}$$

where

$$d \equiv \sigma(t,t)\sigma(s,s) - \sigma^2(s,t), \tag{14.72}$$

$$\begin{pmatrix} A_{11} & A_{12} \\ A_{12} & A_{22} \end{pmatrix}^{-1} = \begin{pmatrix} \sigma(s,s) & \sigma(s,t) \\ \sigma(s,t) & \sigma(t,t) \end{pmatrix} = \frac{1}{d}\begin{pmatrix} \sigma(t,t) & -\sigma(s,t) \\ -\sigma(s,t) & \sigma(s,s) \end{pmatrix}, \tag{14.73}$$

and where $\sigma(s,t)$ is the correlation function of the coordinate,

$$\sigma(s,t) = \langle x(s)x(t)\rangle = \frac{2\gamma T}{m^2}\int_0^s\int_0^t \mathrm{d}t_1\mathrm{d}t_2 f(s-t_2)f(t-t_2)\delta(t_1-t_2) = \frac{2\gamma T}{m^2}\int_0^{\min(s,t)} \mathrm{d}t' f(t')f(t'+|s-t|). \tag{14.74}$$

Using

$$\int \mathrm{d}y\, y\, \mathcal{P}(y,s|x,t) = \int \mathrm{d}y\, y\, \frac{\mathcal{P}(y,s;x,t)}{\mathcal{P}(x,t)} = -x\frac{A_{12}}{A_{11}} = x\frac{\sigma(s,t)}{\sigma(t,t)}, \tag{14.75}$$

one gets, keeping ε finite and fixed,

$$v_+(x,t) = \int \mathrm{d}y \frac{y-x}{\varepsilon}\mathcal{P}(y,t+\varepsilon|x,t) = \frac{x}{\varepsilon}\left(\frac{\sigma(t+\varepsilon,t)}{\sigma(t,t)}-1\right), \tag{14.76}$$

$$v_-(x,t) = \int \mathrm{d}y \frac{x-y}{\varepsilon}P(y,t-\varepsilon|x,t) = \frac{x}{\varepsilon}\left(1-\frac{\sigma(t-\varepsilon,t)}{\sigma(t,t)}\right), \tag{14.77}$$

for any $\varepsilon \geq 0$ smaller than t: $t-\varepsilon > 0$.

If now ε is at the smallest time-scale,

$$v_-(x,t) - v_+(x,t) = \frac{2\gamma T}{m^2}\frac{x\varepsilon}{\sigma(t,t)}\left[\int_0^t \mathrm{d}t' \dot{f}^2(t') - f(t)\dot{f}(t)\right] + \mathcal{O}(\varepsilon^2) \tag{14.78}$$

goes to zero with $\varepsilon \to 0$, as was predicted earlier.

As seen from (14.70), the overdamped limit is realized with the dimensionless parameter $4am/\gamma^2$ being small:

$$\frac{4am}{\gamma^2} \ll 1, \tag{14.79}$$

and then the charactersitc relaxation times of the momentum and the coordinate [obtained from (14.68, 14.69, 14.70)] are well-separated:

$$\tau_p = \frac{m}{\gamma} \approx \frac{1}{\omega_1} \ll \tau_x = \frac{\gamma}{a} \approx \frac{1}{\omega_2}. \tag{14.80}$$

Now if

$$\omega_1 t \gg 1, \qquad \omega_1 s \gg 1, \qquad \omega_1|s-t| \gg 1, \tag{14.81}$$

one gets from (14.74)

$$\sigma(s,t) = \frac{T}{a}\left(e^{-\omega_2|s-t|} - e^{-\omega_2(t+s)}\right), \tag{14.82}$$

which coincides with the corresponding correlator obtained from the overdamped Fokker–Planck equation directly. Taking now $s = t \pm \varepsilon$ in (14.76, 14.77) and employing (14.82) with the limit $\varepsilon \to 0$—which now should be understood in the context of conditions (14.80, 14.81)—one can get the expressions for $v_{\pm}(x,t)$ which can alternatively be derived from (14.23, 14.24), if we put there $N = 1$ (only one Brownian particle), $f(x) = -ax/\gamma$ (the linear force divided over the friction constant), and $P(x,t) \propto \exp[-\frac{x^2}{2\sigma(t,t)}]$ (Gaussian distribution function).

14.8.3 Random-Context Osmotic Velocities and False Nonlocality

Let us now again consider the situation, where the two Brownian particles interacted for $t > 0$, but do not interact for positive times. In view of this fact, one can think there is no influence of one particle on the other, and thus the local osmotic velocity, for example for the first particle, can be defined simply by averaging the global osmotic velocity over the coordinate of the second Brownian particle, that is, over all possible values of x_2 at the time $t > 0$:

$$\tilde{u}_1(x_1,t) = \int \mathrm{d}x_2\, P_2(x_2,t)\, u_1(x_1,x_2,t). \tag{14.83}$$

The similar construction can be obviously designed by interchanging 1 and 2, and will result in

$$\tilde{u}_2(x_2,t) = \int \mathrm{d}x_1\, P_1(x_1,t)\, u_2(x_1,x_2,t). \tag{14.84}$$

It is seen from (14.83, 14.84) that

$$\tilde{u}_1(x_1,t) = \frac{T\sigma_{22}x_1}{\sigma_{11}\sigma_{22} - \sigma_{12}^2}, \qquad \tilde{u}_2(x_2,t) = \frac{T\sigma_{11}x_2}{\sigma_{11}\sigma_{22} - \sigma_{12}^2}. \tag{14.85}$$

However, $\tilde{u}_1(x_1)$ is not equal to the truly local velocity and $\mu_1(x_1)$, as the former one was obtained in the random common context, while the latter one is defined without any common context, that is, without any connection or communication between the observers 1 and 2.

This inequivalence can be illustrated with the help of the following fact of probability theory concerning three events A, B, C:

$$\sum_C P(C)\, P(A|BC) \neq \sum_C P(C|B)\, P(A|BC) = P(A|B) \tag{14.86}$$

random context C is still different as compared to having no context at all.

It remains to stress that no nonlocality arises when operating with the truly local coarse-grained velocities.

To see how one can get incorrect conclusions by insisting on separability of noninteracting objects, let us apply independent external fields on each Brownian particle. The total potential energy is now

$$U(x_1, x_2) = \frac{a_1(t)x_1^2}{2} + \frac{a_2(t)x_2^2}{2} + gx_1x_2, \tag{14.87}$$

and the corresponding variances read from the solution of the Langevin equation (14.16):

$$\sigma_{jj}(t) = e^{-2\bar{a}_j(t)}\sigma_{jj}(0) + 2Te^{-\bar{a}_j(t)} \int_0^t \mathrm{d}s\, e^{\bar{a}_j(s)}, \bar{a}_j(t)$$

$$= \int_0^t \mathrm{d}s\, a_j(s), j = 1, 2, \tag{14.88}$$

$$\sigma_{12}(t) = e^{-\bar{a}_1(t)-\bar{a}_2(t)}\sigma_{12}(0). \tag{14.89}$$

It is seen from (14.83, 14.84) that

$$\tilde{u}_1(x_1, t) = \frac{T\sigma_{22}x_1}{\sigma_{11}\sigma_{22} - \sigma_{12}^2}, \qquad \tilde{u}_2(x_2, t) = \frac{T\sigma_{11}x_2}{\sigma_{11}\sigma_{22} - \sigma_{12}^2}. \tag{14.90}$$

Now, for example, the average square of $\tilde{u}_1(x_1, t)$ for the first particle (not interacting with the second one) does depend on the field $a_2(t)$ exerted on the second particle for $t > 0$:

$$\langle \tilde{u}_1^2 \rangle = \frac{T^2}{\sigma_{11}} \left[1 - \frac{\sigma_{12}^2}{\sigma_{11}\sigma_{22}} \right]^{-2}, \tag{14.91}$$

provided the parameter

$$\frac{\sigma_{12}^2(t)}{\sigma_{11}(t)\sigma_{22}(t)} = \frac{\sigma_{12}^2(0)}{\sigma_{11}(0)\sigma_{22}(0)} \left[1 + \frac{2T}{\sigma_{11}(0)} \int_0^t \mathrm{d}s\, e^{a_1(s)+a_1(t)} \right]^{-1} \times \left[1 + \frac{2T}{\sigma_{22}(0)} \int_0^t \mathrm{d}s\, e^{a_2(s)+a_2(t)} \right]^{-1} \tag{14.92}$$

is not very close to zero. This is a "false nonlocality," since $\tilde{u}_1$, though referring to the first particle, is not truly a local quantity, since random context differs from having no context at all. Put differently, the absence of direct interactions cannot be a reason for neglecting correlations between the two particles.

14.9 Brownian Entanglement from Uncertainty Relations and Contextuality

We have uncovered the phenomenon of Brownian entanglement: a correlation effect between the coordinates and the coarse-grained velocities of two classical Brownian particles, which resembles the quantum entanglement. In contrast to the latter, which is presently given a fundamental status, the Brownian entanglement (as the very subject of statistical physics) arises out of coarse-graining (incomplete description) reasons. In that respect it is similar to other basic relations of the statistical physics, such as the second law [14]. In the present situation the coarse-graining comes due to the time-scale separation: the evolution of the momenta of the Brownian particles is very fast and cannot be resolved on the time-scales available to the experiment. The idea of time-scale separations is one of the most pertinent ones in nonequilibrium statistical physics. In a qualitative form, it appears already in good textbooks on this subject [260, 275], and has been since then formalized in various contexts and on various levels of generality [12, 87, 263, 287, 339].

Once there is time-scale separation between the (stochastic) motion of the coordinate and the momentum of the Brownian particles, the operational definition of velocity via the rate of the coordinate leads to the coarse-grained velocity which is not equal to the real momentum. The change of the coarse-grained velocity is controlled by the Brownian uncertainty relation. Moreover, the coarse-grained velocities appear to be contextual random quantities, that is, they depend on the concrete setting of the coordinate measurement used to define them. These two aspects (uncertainty relations and contextuality) suffice to define Brownian entanglement, similarly to quantum entanglement, as an impossibility to prepare the common distribution of the coordinates and the coarse-grained velocities of two Brownian particles via mixing locally independent (noncorrelated) distributions referring to the two particles separately. Again in analogy to quantum entanglement, Brownian entanglement can be witnessed via the uncertainty relations.

Time-scales separation and the corresponding coarse-graining of the velocities is a good example of two-level structure of scientific theories which was emphasized by Hertz and Boltzmann (see Chapter 15). The phase space (coordinate-momentum, x, p) representation of the Brownian motion is a causal theoretic model (CTM, chapter 15) and the coarse-grained representation is an observational model (OM, chapter 15). Generally coupling between CTM and corresponding OM is not so straightforward as in the Bell-like theories with hidden variables. Coarse-grained velocities are not simply functions on hidden variables, as Bell dreamed for quantum particles.

Following paper [15] in this chapter, the Brownian entanglement was demonstrated on the simplest, exactly solvable, one-dimensional models of two interacting Brownian particles. It should, however, be kept in mind that interacting Brownian particles (random walks) are basic for several fields of modern statistical physics, such as colloids [326] or polymers [145]. One of the most transparent examples from the polymer science is a DNA macromolecule which consists of two interacting random walks (strands) [145]. As we argued, the experimental observation of the Brownian entanglement might be easier than those of the quantum entanglement, since there are realizations of macroscopic Brownian motion that are visible with the human eye.

PART V

Hertz, Boltzmann, Schrödinger, and de Broglie on Hidden Parameters

15

Bild Conception for Scientific Theory

15.1 Bild Conception

The Bild conception of scientific theory was developed by Hertz [160, 161] starting with Helmholtz's analysis [359] of interrelation between physical reality and scientific theory. This line of thinking was continued by Boltzmann [66, 67] and in the 1950s by Schrödinger [330]–[336]. The articles of D'Agostino [88]–[93] contain philosophically deep reviews on their works.

Typically the German word "Bild" is translated to English as "picture." But, for the Bild conception in science, the meaning "model conception" is more appropriate; see also Patton [305]:

"Bild is translated variously in the literature as "picture," "image," or "model." A Bild should not be understood as a visual or mental image or picture, however. I tend to use the German word, to underscore the fact that Hertz and Wittgenstein use the same terminology."

Patton also pointed out that the Bild conception is a precursor of model theory that influenced Einstein, Hilbert, and Schrödinger. Her paper [305] also reviews Wittgenstein's works [379]-[380], which are closely related to Hertz's works. But we would not consider Wittgenstein's contribution to development of the Bild conception.

Helmholtz pointed out that a scientific theory does not describe reality as it is. A scientific theory structures our sensations and perceptions within a priori forms of intuition (cf. with Kant). Such structuring leads to models of reality reflecting some features of the environment of observers. Therefore the dream for creation of a "true theory" matching perfectly with natural phenomena is in contradiction with Helmholtz's philosophy of science.

Hertz, Boltzmann, and Schrödinger were the followers of Helmholtz and for them the Bild conception was not about the pictures of reality, but about creation of models of reality. The term "picture" presumes that this is a picture of something existing in reality. But due to the Bild conception, the human mind constructs a model of natural phenomena. In any event, Helmholtz's philosophic position supports the use of the English word "model" in the presentation of the Bild conception.

Observational data should be taken with caution. Helmholtz highlighted causality of the natural phenomena and for him the main task of a scientific theory is to reflect this causality. Thus, from his viewpoint the main aim of scientific studies is construction of the causal theoretical models (CTMs) of natural phenomena. Theoretical causality is an image of natural causality. In terms of cognition, causality of human reasoning reflects causality of natural processes and it was developed during biological evolution, from the primitive forms of life to humans.

We remark that causality is a complex notion with the deep philosophic and physical counterparts. The current viewpoint in physics on causality was formed under the strong influence of development of special relativity. Helmholtz's works were written earlier and he, as well as Hertz and Boltzmann, had a more general viewpoint on causality (see Section 15.2 for a discussion).

Hertz followed Helmholtz's approach to scientific theory, but he claimed that generally CTM can't be designed solely on the basis of observational data and it typically contains hidden quantities. (So, in physics hidden variables were employed long before the quantum revolution.) Experimental data is described by an observational model (OM) which is often acausal. The CTM-OM interrelation can be tricky. This framework Hertz presented [160, 161] as a Bild conception (model-concept). He highlighted the role of mathematics and treatment of a scientific model as a mathematical model (see also Plotnitsky [317]). In particular, Hertz presented Maxwell's theory as the system of the Maxwell equations.

Later the Bild conception was resurrected in the foundational studies of Schrödinger [330]–[336] (see especially [330]) who tried to create CTM for quantum mechanics (QM) and QM was treated as OM. He advertised the two-level structuring of the description of microphenomena. We follow him and suggest a special CTM for QM, the so-called *prequantum classical statistical field theory* (PCSFT) [207, 212, 225]. QM treated as OM can be considered as a PCSFT image, but not as straightforward as in Bell's model with hidden variables [53, 54].

We analyze Bell's model with hidden variables within the Bild-framework and criticize identification of subquantum (hidden) quantities with quantum observables and hidden probability distributions with quantum probability distributions. The evident barrier to such identification is the Heisenberg uncertainty principle and the Bohr complementarity principle (Chapter 3). The same viewpoint was presented long ago by de Broglie [100] (see also Lochak [266–268]) who justified the legitimacy of his double solution theory [40, 99], in fact, within the Bild conception (although it seems that he was not aware about it). He pointed to the inconsistency of the no-go interpretation of the von Neumann [365] and Bell [53, 54] theorems. De Broglie's double solution model is an CTM for QM. (Its structuring within the Bild conception deserves separate study as well as the Bild conception presentation of Bohmian mechanics.)[1]

We also use the Bild-approach for the two level CTM-OM modeling of Brownian motion: the overdamped regime corresponds to OM [16] (chapter 14). Coarse gained velocities are observable quantities. This example represents clearly the physical origin of the two-level structuring of the mathematical description of Brownian motion. This is the time-scale separation technique. The evolution of the momenta of the Brownian particles is very fast and cannot be resolved on the time-scales available to the experiment. We notice that the OM model for the Brownian motion shows some distinguished properties of QM, see, for example, article [16] (chapter 14) for the corresponding uncertainty relations and Brownian entanglement theory.

The idea of time-scale separations is one of the most pertinent ones in non-equilibrium statistical physics. In a qualitative form it appears already in good textbooks on this subject [260, 275], and since then it has been formalized in various contexts and on various levels of generality [12, 87, 263, 287, 339].

In classical mechanics CTM = OM; on the one hand, this is very convenient, on the other hand, this exceptional coincidence blurred the general CTM-OM structuring of scientific theories.

We also briefly discuss ontic-epistemic structuring of scientific theories, Primas-Atmanspacher [33, 34, 321], and its relation to the Bild concept.

This chapter is a continuation of my works [207, 225]. I hope that in this chapter the Bild conception and its implementation for quantum and classical mechanics are presented more clearly. Presentation of the two-level CTM-OM description for Brownian motion is a good complement to such description for quantum phenomena. The CTM-OM viewpoint on the Bell inequality project clarifies the difference in the positions of Schrödinger [330]–[336] and Bell [53, 54] on the possibility of constructing a subquantum model with hidden variables.

[1] This is a good place to note that one should not identify de Broglie's and Bohm's theories, not consider the latter as just an extension and improvement of the former. De Broglie did not consider his theory as nonlocal, rather he noted that its nonlocality is only apparent; this is nonlocality of mathematical equations and not physical processes [99]. Deep foundational studies on de Broglie's double solution model are presented in works of Bacciagaluppi, for example, [38, 39]; see also appendixes written by Bacciagaluppi and Valentini in book [40].

15.2 Two-Level Structuring of Scientific Theories

Von Helmholtz: Scientific Theory Is Not a Faithful Image of Nature

We start by citing the article of D' Agostino [93]:

"Hermann von Helmholtz (1821–1894) was one of the first scientists to criticize the objective conception of physical theory by denying that theoretical concepts describe real physical objects. He realized that Immanuel Kant's a priori forms of intuition should be taken into account in analyzing problems that were emerging at the end of the nineteenth century in the new formulations of physics.

The objective conception of physical theory also was criticized by such physicists as Heinrich Hertz (1857–1894) and Ludwig Boltzmann (1844–1906), who adopted the Kantian term Bild[2] to designate the new conception of physical theory, which they took to mean not a faithful image of nature but an intellectual construct whose relationship to empirical phenomena was to be analyzed."

The works of von Helmholtz, Hertz, and Boltzmann [66, 67, 160, 161, 359] played the crucial role in development of a novel scientific methodology. Since the time of Galileo and Newton, scientific theories varied essentially in their content, but nobody questioned their "ontological significance," their adequacy to represent physical reality.

In 1878, von Helmholtz posed the following philosophical questions [359]:

"What is true in our intuition and thought? In what sense do our representations correspond to actuality?"

Von Helmholtz' answers to these questions were based on his physiological, especially visual, research that led him to the following conclusion [359]:

"Inasmuch as the quality of our sensation gives us a report of what is peculiar to the external influence by which it is excited, it may count as a symbol of it, but not as an image. For from an image one requires some kind of alikeness with the object of which it is an image . . . ."

We point out that if the fathers of QM had taken this statement into account, then the surprise by "unusual features" of the quantum mechanical description of micro-phenomena would not have been so strong. We also note that Bohr's views on QM match with this conclusion of Helmholtz. Surprisingly, it seems that Bohr had never referred to his works.

Helmholtz's viewpoint on interrelation of sensations and generally observations and real objects led to the well-known statement on the parallelism of the laws of nature and science [359]:

"Every law of nature asserts that upon preconditions alike in a certain respect, there always follow consequences which are alike in a certain other respect. Since like things are indicated in our world of sensations by like signs, an equally regular

[2] Here the German word "Bild" ("picture") is used in the sense of a model.

sequence will also correspond in the domain of our sensations to the sequence of like effects by the law of nature [that like effects follow from] ... like causes."

Causality

We point out that Helmholtz's statement "that upon preconditions alike in a certain respect, there always follow consequences which are alike in a certain other respect" is about physical causality. So, for Helmholtz, nature is causal, that is, laws in nature really exist and laws presented in scientific theories are mental representations of laws of nature. The laws expressed by our sensation and through them by our perception are "parallel" to natural laws, but only parallel, not identical, since our mind operates not with precise images of real objects, but only with symbols assigned to them.

In modern physics the notion of causality was strongly influenced by creation of relativity theory. An effect can occur only from a cause that is in its past light cone and a cause can lead only to an effect inside its future light cone. This viewpoint on causality was not present in the time of Helmholtz and Hertz.

In philosophy the notion of causality is based on the categories of the cause and effect formalizing the genetic connection between events (or states) such that one event (or state) induces another event (or state) – an effect (e.g., [68]). A few causes can generate the same effect.

Many philosophers claim that causality is a primary notion preceding the notions of space and time; see, for example, Robb [324] and Whitehead [376].

In the present chapter we consider the special form of causality which is mathematically described via the functional representation,

$$y = f(x), \tag{15.1}$$

where x and y are interpreted as the cause and effect variables (see [370]). The cause-effect interpretation is crucial; in mathematical physics, a non-arbitrary equation (15.1) has the causal meaning.

In coming considerations we select the x-variable as the system's state. The latter is interpreted as a context, a complex of physical conditions (see [193]). A physical quantity A is mathematically represented by a function f_A. A state plays the role of preconditions in the preceding citation of Helmholtz. Then these preconditions, or context, contribute to generation of the A-value via (15.1), this value-generation is an event. A state precedes this event; function f_A is considered as an effect-generator. For example, x is a sensation and $y = f_p(x)$ the corresponding perception.

Consider the classical phase space model. Here states are given by points of the phase space, $x = (q,p)$. Consider the Hamilton function $H(q,p) = p^2/2m + V(q)$. Then $x = (q,p)$ is the cause and $y = H(q,p)$ is the effect – the event of assigning energy to a system in this state. In this framework it is meaningless to speak about the energy-quantity without specifying the system's state, the context for energy determination. The Hamilton function describes generation of the energy-value events from states.

Hertz: Need of Hidden Quantities

Hertz questioned Helmholtz's parallelism of laws. Hertz believed that Helmholtz's parallelism of laws not only was indeterminate but in general even impossible if theory were limited to describing observable quantities [161]:

"If we try to understand the motions of bodies around us, and to refer them to simple and clear rules, paying attention only to what can be directly observed, our attempt will in general fail. We soon become aware that the totality of things visible and tangible do not form an universe conformable to law, in which the same results always follow from the same conditions. We become convinced that the manifold of the actual universe must be greater than the manifold of the universe which is directly revealed to us by our senses."

By Hertz a causal theory cannot be based solely on observable quantities [161] : "— do not form an universe conformable to law, in which the same results always follow from the same conditions." Only by introducing hidden quantities can Helmholtz's parallelism of laws become a general principle in physical theory. But such hidden quantities (concepts that correspond to no perceptions) bring too much freedom in the choice of theoretical concepts. To limit this freedom of choice, Hertz introduced special requirements for the validation of a physical theory. Besides causality, the most important was theory's simplicity [161]:

"It is true we cannot a priori demand from nature simplicity, nor can we judge what in her opinion is simple. But with regard to images [Bilder] of our own creation we can lay down requirements. We are justified in deciding that if our images are well adapted to the things, the actual relations of the things must be represented by simple relations between the images."

So, Helmholtz and Herz questioned the ontological status of scientific theories as describing reality as it is. Scientific theories are only "Bilder," models of reality. Outputs of sensations and observations are just symbols encoding external phenomena. Hence, one should not sanctify observational quantities and their role in scientific theories. Moreover, an observational theory, that is, a theory operating with solely observables, cannot be causal. Causality demands introduction of hidden (unobservable) quantities. Of course, a theory with hidden quantities should be coupled to observational data. However, this coupling need not be straightforward.

According to Helmholtz, a scientific theory should be causal. Hertz claimed [161] that generally the causality constraint requires invention of hidden quantities; a causal description cannot be done solely in terms of observational quantities. This approach unties scientists' hands; by introducing hidden quantities they can generate a variety of theoretical causal models coupled to the same observational quantities. How can one select a true causal model? Hertz suggested to use the model's simplicity as a criterion for such selection. We note that even a true model does not describe reality as it is, it provides just a mathematical symbolic representation involving a variety of elements having no direct relation with observational quantities.

Is It Possible to Construct the True Model Describing Reality As It Is?

It is natural to search for such a (causal) theoretical model that would describe what nature really is, a true model (an ontic model). It is not clear whether Hertz might hope to design such a model for the electromagnetic phenomenon.[3] Schrödinger who later contributed to development of the Bild concept of scientific theories, especially in relation to the quantum foundations, claimed [331] that no true model can be formulated on the basis of our large-scale experience, because

"we find nature behaving so entirely differently from what we observe in visible and palpable bodies of our surroundings.... A completely satisfactory model of this type is not only practically inaccessible, but not even thinkable. Or, to be precise, we can, of course, think it, but however we think it, it is wrong; not perhaps quite as meaningless as a "triangular circle," but much more so than a "winged lion."

Creation of a causal theoretical model coupled to some observed natural phenomena is a complex and long process. Moreover, there is always a chance that such a model would be never found – due to the intellectual incapacity of humankind. Therefore it is natural to design models matching observations, but not satisfying the causality constraint. We call such models observational models.

Observational versus Causal Theoretical Model

Thus, we distinguish two classes of models, *observational models* (OMs) and *causal theoretical models* (CTMs). We remark that both kinds of scientific models are mental constructions, providing symbolic mathematical descriptions of natural phenomena. One may say that any model is theoretical, so OM is also theoretical. And they would be right. So, the main difference between OM and CTM is in causality. If OM is causal by itself, then there is no need to go beyond it with some CTM.

Interrelation between CTM and OM, $\mathbf{M}_C$ and $\mathbf{M}_O$, depends on the present stage of development of science. If $\mathbf{M}_C$ rightly reflects the real physical processes, then development of measurement technology can lead to novel observational

[3] He tried to model it with systems of mechanical oscillators, i.e., to go beyond the electromagnetic field representation [160]. But he did not succeed with this project. His project was not meaningless. It has some degree of similarity with the representation of the quantum electromagnetic field as a system of quantum oscillators - photons.

possibilities and some hidden quantities of $\mathbf{M}_C$ can become measurable. Hence, $\mathbf{M}_C$ becomes OM, $\mathbf{M}_C \rightarrow \mathbf{M}'_O$. In principle, $\mathbf{M}'_O$ need not cover all observations described by the previous OM $\mathbf{M}_O$. New theoretical efforts might be needed to merge $\mathbf{M}_O$ and $\mathbf{M}'_O$. This abstract discussion will be illustrated by the concrete example from classical statistical physics – the two-level modeling of the Brownian motion (Section 15.8.1).

Schrödinger: Bild Conception Viewpoint on Quantum Mechanics

The ideas of Helmholtz and Hertz were further developed (and modified) in the works of Boltzmann [66, 67]. Then, 60 years later, Schrödinger [330]–[336] contributed to development of the Bild viewpoint on quantum theories. He was confronted with the special case of the aforementioned problem.

The observational model for microphenomena was developed (in particular, due to his own efforts): this is QM. But QM suffered from acausality. The impossibility of solving the measurement problem (which was highlighted by von Neumann [365]) generates a gap in the quantum description of microphenomena. Schrödinger came back to this problem in the 1950s [330]–[336]; this comeback was stimulated by development of quantum field theory and the method of second quantization.

He saw in quantum field theory a possibility of justifying his attempts of the purely wave (continuous) approach to modeling of the microphenomena. In complete agreement with the Bild concept, he considered QM as an observational model. As well as von Neumann, Schrödinger highlighted its acausality. But it was not treated as a property of nature as it is, that is, quantum acausality (of measurements and spontaneous quantum events) is not ontic. We notice that, for von Neumann, it is ontic; he wrote about irreducible quantum randomness [365]. Quantum acausality is just a property of special OM – QM. Schrödinger claimed that quantum acausality is related to ignoring of the Bild concept and assigning the ontological status to quantum particles; see his article "What is an elementary particle?" [330]. We remark that Bohr did not question the ontological status of quantum systems, atoms, electrons and maybe even photons [65, 309, 310]. Schrödinger considered indistinguishability of quantum particles as a sign that they do not have the ontological status. Hence, instead of OM (= QM), one can hope to develop CTM for microphenomena by liberating it from particles and operating solely with waves.

Since waves propagate in space, for Schrödinger causality (in fact, the wave causality) is coupled to continuity in the space, so the waves should be continuous (see Plotnitsky [317] on analysis of continuity versus discontinuity in physics). We remark that he considered continuity of waves on multidimensional space $\mathbb{R}^{3n}$. In the 1920s the fact that the multiparticle Schrödinger equation describes the waves not on the physical space represented as $\mathbb{R}^3$, but on the mathematical space $\mathbb{R}^{3n}$, was disturbing for him. This was the main reason for Schrödinger to accept the probabilistic interpretation of the wave function. At that time he did not use the Bild concept for scientific theories (maybe he was not aware of the works of Helmholtz, Herz, and Boltzmann?). By the Bild concept, the wave representation of QM is just a symbolic mathematical representation of the microphenomena. The use of the multidimensional space $\mathbb{R}^{3n}$ has the same descriptive status as the use of $\mathbb{R}^3$.

Schrödinger dreamed for creation of CTM for microphenomena, and his concrete intention was toward a wave-type model. He also highlighted the principle of continuity for quantum waves, but he suspected that it would be valid only at the microlevel. He pointed to quantum field theory as a good candidate to proceed in this direction. Since he coupled causality and continuity, it became possible to relax the causality-continuity constraint and restrict this constraint to the level of infinitesimals. In a theoretical model completing QM (an observational model) for which Schrödinger dreamed, causality need not be global.

Schrödinger's continuous wave completion project for QM has some degree of similarity with Einstein's project on designing a classical field model of microphenomena which he announced with Infeld in a popular form in book [117].[4] However, in contrast to Schrödinger, Einstein did not appeal to the Bild concept on the two-level modeling of natural phenomena, observational and causal theoretical (OM and CTM), and a possible gap between these two models. The presence of such a gap, in particular, implies that CTM need not describe the observational data straightforwardly.

Einstein's project on reconsideration of quantum foundations starting with the EPR paper [115] was not directed to the two-level structuring of the mathematical description of microphenomena. He dreamed for creation of CTM which would match perfectly with quantum observations. Einstein argued that the current theory of QM was incomplete. According to this interpretation, quantum theory is incomplete as some variables are present in the theory which have not been known. Such variables are known as hidden variables. A special theory with hidden variables corresponding to Einstein's dream was suggested by Bell [53, 54]. Its main distinguishing property is identification of the outcomes of quantum observables with functions of hidden variables and, hence, identification of quantum and subquantum averages and correlations. This identification contradicts the Bild conception and the two-level structuring of physical theories. But the latter quantities in a subquantum model with hidden variables are not straightforwardly coupled to quantum observables. In particular, such a viewpoint was advertised by de Broglie [100] (Section 15.7).

4 Einstein's intention was that a complete theory beyond QM should be nonlinear field theory. Later Infeld contributed a lot into this project. In contrast to Einstein, Schrödinger dreamed for a linear model.

Schrödinger understood [331] that CTM of microphenomena of the wave type is not “the observed or observable facts; and still less do we claim that we thus describe what nature (matter, radiation, etc.) really is. In fact we use this picture (the so-called wave picture) in full knowledge that it is neither.”

This statement expresses the extreme view on the Bild concept; Schrödinger [331] also pointed out that “observed facts... appear to be repugnant to the classical ideal of a continuous description in space and time.” Such highlighting of decoupling of theory and observations was too provocative and played the negative role. The idea of using the Bild concept in quantum foundations was rejected by the majority of experts in quantum foundations.

Primas and Atmanspacher: Ontic-Epistemic Modelling

However, the Bild concept did not disappear completely and its trace can be found in the philosophy of the ontic-epistemic structuring of physical theories that was developed by Primas and Atmanspacher [33–35, 321]). They tried to find an answer [34] to the old question:

“Can nature be observed and described as it is in itself independent of those who observe and describe – that is to say, nature as it is ‘when nobody looks?’”

As well as Helmholtz, Hertz, Boltzmann, and Schrödinger, they pointed out that observations give to observers only some knowledge about systems; this knowledge is incomplete. This knowledge is mathematically structured within an epistemic (=observational) model . For them, QM is such a model, that is, w.r.t. QM the views of Schrödinger and Primas-Atmanspacher coincide. Then, in the same way as Schrödinger, they want to have a complete model of microphenomena. The crucial difference from the Bild concept is that Primas and Atmanspacher were seeking for an ontic model, a model of reality as it is, the true model in terms of Schrödinger. Generally Primas and Atmanspacher also supported the idea of the two-level structure of scientific theories: epistemic (observational) and ontic. As well as Schrödinger, they pointed out that the connection between epistemic and ontic models is not straightforward. Causality is the basic property of the ontic model. So, if one would ignore the term ontic,[5] then formally (and mathematically) Primas-Atmanspacher structuring of the scientific description of nature is similar to the Bild concept. (In contrast to Schrödinger, they did not emphasize the continuous wave structure of an ontic model beyond QM.)

However, by pointing to formal mathematical similarity of the ontic-epistemic and Bild approaches, one should remember that they differ crucially from the foundational perspective. We recall [33] that

[5] Its use would be very disturbing for Helmholtz, Hertz, Boltzmann, and Schrödinger.

"Ontological questions refer to the structure and behavior of a system as such, whereas epistemological questions refer to the knowledge of information gathering and using systems, such as human beings."

From the Bild perspective, it is totally meaningless even to "refer to the structure and behavior of a system as such ... " The essence of the ontic-epistemic approach is expressed in the following quote from Atmanspacher [34] (for more details, the reader is referred to Primas [321]):

"Ontic states describe all properties of a physical system exhaustively. ("Exhaustive" in this context means that an ontic state is "precisely the way it is," without any reference to epistemic knowledge or ignorance.) Ontic states are the referents of individual descriptions; the properties of the system are treated as intrinsic properties.Their temporal evolution (dynamics) is reversible and follows universal, deterministic laws. As a rule, ontic states in this sense are empirically inaccessible. Epistemic states describe our (usually nonexhaustive) knowledge of the properties of a physical system, that is, based on a finite partition of the relevant phase space. The referents of statistical descriptions are epistemic states, the properties of the system are treated as contextual properties. Their temporal evolution (dynamics) typically follows phenomenological, irreversible laws. Epistemic states are, at least in principle, empirically accessible."

From the Bild perspective, the statement: "Ontic states are the referents of individual descriptions; the properties of the system are treated as intrinsic properties," is meaningless, since systems do not have intrinsic properties; a theoretical causal model beyond the quantum observational (epistemic) model still describes not the properties of the systems, but our mental pictures.

In short, for Primas and Atmanspacher, CTM describes reality as it is; for them, by searching an CTM one searches for a true model of reality. From the Bild viewpoint, no CTM describes reality as it is; therefore Schrödinger was against searching true CTM beyond QM.

Nietzsche

And we conclude this section by the quote from Nietzsche[6] (written in 1873, but published later); his statement is very similar to Helmholtz's statements, but it is more passionate or even poetic! It seems that Nietzsche was influenced by Helmholtz, especially on nerve stimulus. Nietzsche wrote about language, but the point is more general [286]:

"The various languages placed side by side show that with words it is never a question of truth, never a question of adequate expression; otherwise, there would not be so many languages. The 'thing in itself' (which is precisely what the pure

[6] I would like to thank Arkady Plotnitsky for mentioning this quote in our discussion on the works of Helmholtz, Hertz, Boltzmann, and Schrödinger in the context of the present chapter.

truth, apart from any of its consequences, would be) is likewise something quite incomprehensible to the creator of language and something not in the least worth striving for. This creator only designates the relations of things to men, and for expressing these relations he lays hold of the boldest metaphors. To begin with, a nerve stimulus is transferred into an image: first metaphor. The image, in turn, is imitated in a sound: second metaphor. And each time there is a complete overleaping of one sphere, right into the middle of an entirely new and different one. One can imagine a man who is totally deaf and has never had a sensation of sound and music. Perhaps such a person will gaze with astonishment at Chladni's sound figures; perhaps he will discover their causes in the vibrations of the string and will now swear that he must know what men mean by "sound." It is this way with all of us concerning language; we believe that we know something about the things themselves when we speak of trees, colors, snow, and flowers; and yet we possess nothing but metaphors for things-metaphors which correspond in no way to the original entities. In the same way that the sound appears as a sand figure, so the mysterious of the thing in itself first appears as a nerve stimulus, then as an image, and finally as a sound. Thus the genesis of language does not proceed logically in any case, and all the material within and with which the man of truth, the scientist, and the philosopher later work and build, if not derived from never-never land, is at least not derived from the essence of things."

15.3 Coupling of Theoretical and Observational Models

Models explored in natural science are mainly mathematical. Therefore coupling between CTM and OM corresponding to the same natural phenomena is a mapping of one mathematical structure to another.

Let **M** be some mathematical model of natural phenomena, either CTM or OM; **M** is based on two spaces, the space of states S and the space of quantities V. For OM, V is the space of observables, instead of states one can consider measurement contexts (see Chapter 10).

Consider OM model $\mathbf{M}_O = (S_O, V_O)$ and its causal theoretical completion $\mathbf{M}_C = (S_C, V_C)$. It is natural to have a mathematical rule establishing correspondence between them. We recall that CTMs are causal and OMs are often acausal; if it happens that OM is causal, then there is no need for a finer description given by some CTM. Thus, the task is to establish correspondence between causal and acausal models. It is clear that such correspondence cannot be straightforward. We cannot map directly states from S_C to states from S_O. Causality can be transformed into acausality through consideration of probability distributions. So, consider some space of probability distributions P_C on the state space S_C and construct a map from P_C to S_O, the state space of OM. This approach immediately implies that the states of OM are interpreted statistically. We also should establish correspondence between quantities of $\mathbf{M}_C$ and $\mathbf{M}_O$. Thus, we need to define two physically natural maps:

$$J_S\colon P_C \to S_O,\ J_V\colon V_C \to V_O. \tag{15.2}$$

Since J_S is not defined for states of CTM, but only for probability distributions, physically natural means coupling between the probability structures of $\mathbf{M}_C$ and $\mathbf{M}_O$; the minimal coupling is the equality of averages

$$\langle J_V(f)\rangle_{J_S(P)} = \langle f\rangle_P \tag{15.3}$$

and correlations between quantities

$$\langle J_V(f)J_V(g)\rangle_{J_S(P)} = \langle fg\rangle_P. \tag{15.4}$$

Generally the correlation need not be defined, so (15.4) should hold for quantities $f, g \in V_C$ and observables $A_f = J_V(f)$ and $A_g = J_V(g)$ for which the correlations in the states P and $J_S(P)$ are defined.

Mathematically causality can be realized as *functional representation* of quantities [370] (15.1). The mathematical formula (15.1) is endowed with the causal interpretation: a state determines the value of a physical variable, $x \to y = f_A(x)$: first state x and then the variable's value y. It is meaningless to speak about the values of quantities before the determination system's state x. Without such interpretation, mathematical formula (15.1) encodes just a deterministic connection between the state x and the variable's value y. Since CTM is a theoretical model, this discussion is about the theoretical causality of connection between the states and the values of quantities.

Therefore, we assume that V_C can be represented as a space of functions $f\colon S_C \to \mathbb{R}$. Such model is causal, the state ϕ uniquely determines the values of all quantities belonging to V_C: $\phi \to f(\phi)$. The state space S_C can be endowed with a σ-algebra of subsets $\mathcal{F}$. Elements of P_C are probability measures on $\mathcal{F}$. The minimal mathematical restriction on elements of V_C is that they are measurable functions, $f\colon S_C \to \mathbb{R}$. In such a framework,

$$\langle f\rangle_P = \int_{S_C} f(\lambda)P(d\lambda),\ \langle fg\rangle_P = \int_{S_C} f(\lambda)g(\lambda)P(d\lambda), \tag{15.5}$$

if the integrals exist, for example, if CTM-quantities are square integrable:

$$\int_{S_C} |f(\lambda)|^2 P(d\lambda) < \infty.$$

Since in $\mathbf{M}_O$ quantities have the experimental statistical verification, we can establish some degree of experimental verification for $\mathbf{M}_C$ through mapping of $\mathbf{M}_C$

to $\mathbf{M}_O$. But such verification is only indirect; one should not expect direct coupling between quantities of $\mathbf{M}_C$ and experiment (as Einstein, Bell, and all their followers wanted to get). Generally these maps are neither one-to-one nor onto.

- A cluster of probability distributions on S_C can be mapped into the same state from S_O.
- $J_S(P_C)$ need not coincide with S_O.
- A cluster of elements of V_C can be mapped into a single variable (observable) from V_O.
- $J_V(V_C)$ need not coincide with V_O.

Moreover, the model-correspondence maps J_S, J_V need not be defined on whole spaces P_C and V_C. They have their domains of definition, $\mathcal{D}_{J_S} \subset P_C$ and $\mathcal{D}_{J_V} \subset V_C$. (In principle, one can reduce P_C to $P'_C = \mathcal{D}_{J_S}$ and V_C to $V'_C = \mathcal{D}_{J_V}$ and operate with maps J_S, J_V which are defined everywhere on these reduced spaces of CTM's states and quantities.)

We remark that the same $\mathbf{M}_O$ can be coupled to a variety of CTMs. We also remark that the same observational data can be mathematically described by a variety of OMs.

We also remark that similarly to the deformation quantization (here we discuss just the mathematical similarity), CTM may depend on some small parameter κ (in the deformation quantization this is action; roughly speaking, the Planck constant h). Thus, $\mathbf{M}_C = \mathbf{M}_C(\kappa)$. In such a more general framework, the correspondence maps also depend on κ, that is, $J_S = J_S(\kappa), J_V = J_V(\kappa)$. The probabilistic coupling constraints (15.3), (15.4) can be weakened:

$$\langle J_V(\kappa; f)\rangle_{J_S(\kappa;P)} = \langle f\rangle_P + o(\kappa), \kappa \to 0, \tag{15.6}$$

$$\langle J_V(\kappa; f) J_V(\kappa; g)\rangle_{J_S(\kappa;P)} = \langle fg\rangle_P + o(\kappa), \kappa \to 0. \tag{15.7}$$

The problem of identification of the parameter κ with some physical scale is complex (see, e.g., [200, 201] for an attempt of such identification within PCSFT).

15.4 Prequantum Classical Statistical Field Theory as a Causal Theoretical Model

We illustrate the general scheme of CTM-OM correspondence by two theories of microphenomena, QM as $\mathbf{M}_O$ and PCSFT as $\mathbf{M}_C$. Re-denote these model with the symbols $\mathbf{M}_{\mathrm{QM}}$ and $\mathbf{M}_{\mathrm{PCSFT}}$. We briefly recall the basic elements of PCSFT (see [207, 212, 225]).

In $\mathbf{M}_{QM}$ states are given by density operators acting in complex Hilbert space $\mathcal{H}$ (with scalar product $\langle\cdot|\cdot\rangle$) and observables are represented by Hermitian operators in $\mathcal{H}$. Denote the space of density operators by S_{QM} and the space of Hermitian operators by V_{QM}.

In $\mathbf{M}_{PCSFT}$ states are vectors of $\mathcal{H}$, that is, $S_{PCSFT} = \mathcal{H}$. Subquantum quantities are quadratic forms

$$\phi \to f(\phi) = \langle \phi | \hat{A} | \phi \rangle,$$

where $\hat{A} \equiv \hat{A}_f$ is a Hermitian operator. The space of quadratic forms is denoted by the symbol V_{PCSFT}. Consider probability measures on the σ-algebra of Borel subsets of $\mathcal{H}$ (i.e., generated by balls in this space) having zero first momentum (mean value), that is,

$$\int_{\mathcal{H}} \langle \phi | a \rangle dp(\phi) = 0 \tag{15.8}$$

for any vector $a \in H$, and finite second momentum, that is,

$$\mathcal{E}_p \equiv \int_{\mathcal{H}} \|\phi\|^2 dp(\phi) < \infty. \tag{15.9}$$

Denote the space of such probability measures by the symbol P_{PCSFT}.

We can start not with probability measures, but with $\mathcal{H}$-valued random vectors with zero mean value and finite second moment: $\phi = \phi(\omega)$, such that $E[\phi] = 0$ and $E[\|\phi\|^2] < \infty$.[7] The space of such random vectors is denoted by the symbol R_{PCSFT}. In the finite-dimensional case, these are complex vector-valued random variables; if $\mathcal{H}$ is infinite-dimensional, then the elements of R_{PCSFT} are random fields.

An example of random fields is given by selection $\mathcal{H} = L_2(\mathbb{R}^n; \mathbb{C})$ of square integrable complex valued functions. Each $\mathbf{M}_C$ state ϕ is an L_2-function, $\phi : \mathbb{R}^n \mapsto \mathbb{C}$. Random fields belonging to R_{PCSFT} are functions of two variables, $\phi = \phi(x; \omega)$: chance parameter ω and space coordinates x.

We remark that, for the state space $\mathcal{H} = L_2(\mathbb{R}^n; \mathbb{C})$, the quantity $\mathcal{E}_p$ can be represented as

$$\mathcal{E}_p = \int_{\mathcal{H}} \mathcal{E}(\phi) dp(\phi),$$

where

$$\mathcal{E}(\phi) = \|\phi\|^2 = \int_{\mathbb{R}^n} |\phi(x)|^2 dx$$

is the energy of the field. The quantity $\mathcal{E}_p$ can be interpreted as the average of the field energy with respect to the probability distribution p on the space of fields. We

[7] Random vectors are defined on some Kolmogorov probability space $(\Omega, \mathcal{F}, P)$; these are functions $\phi : \Omega \to \mathcal{H}$ which are measurable w.r.t. to the Borel σ-algebra of $\mathcal{H}$, that is, for any Borel subset B of $\mathcal{H}$, $\phi^{-1}(B) \in \mathcal{F}$. A map is measurable if and only if, for any $c > 0$, the set $\Omega_{\phi,c} = \{\omega \in \Omega : ||\phi(\omega)|| < c\} \in \mathcal{F}$.

can also use the random field representation. Let $\phi = \phi(x;\omega)$ be a random field. Then its energy is the random variable

$$\mathcal{E}_\phi(\omega) = \int_{\mathbb{R}^n} |\phi(x;\omega)|^2 dx$$

and $\mathcal{E}_p$ is its average.

For any $p \in P_{\mathrm{PCSFT}}$, its (complex) covariance operator $\hat{B}_p$ is defined by its bilinear (Hermitian) form:

$$\langle a|\hat{B}_p|b\rangle = \int_{\mathcal{H}} \langle a|\phi\rangle\langle\phi|b\rangle\, dp(\phi),\ a,b \in \mathcal{H}, \tag{15.10}$$

or, for a random field ϕ, we have

$$\langle a|\hat{B}_\phi|b\rangle = E[\langle a|\phi\rangle\langle\phi|b\rangle].$$

We note that

$$\mathcal{E}_p = \int_{\mathcal{H}} \|\phi\|^2 dp(\phi) = \mathrm{Tr}\,\hat{B}_p \tag{15.11}$$

or in terms of a random field:

$$\mathcal{E}_p = E[||\phi||^2] = E[\int_{\mathbb{R}^n} |\phi(x;\omega)|^2 dx] = \mathrm{Tr}\,\hat{B}_p. \tag{15.12}$$

Thus, the average energy of a random field $\phi = \phi(\omega, x)$ can be expressed via its covariance operator.

Generally a probability measure ($\mathcal{H}$-valued random variable) is not determined by its covariance operator (even under the constraint given by zero average).

A complex covariance operator has the same properties as a density operator, besides normalization by the trace one; a covariance operator $\hat{B}_p$ is

- Hermitian,
- positively semidefinite,
- trace class.

A physically natural coupling of the models $\mathbf{M}_{\mathrm{QM}}$ and $\mathbf{M}_{\mathrm{PCSFT}}$ is based on the following formula coupling mathematically the averages for these models. For a probability measure $p \in P_{\mathrm{PCSFT}}$ and a variable $f \in V_{\mathrm{PCSFT}}$, we have

$$\langle f\rangle_p = \int_{\mathcal{H}} f(\phi)dp(\phi) = \mathrm{Tr}\,\hat{A}_f\hat{B}_p, \tag{15.13}$$

where $f(\phi) = \langle\phi|\hat{A}_f|\phi\rangle$. This formula is obtained through expansion of the quadratic form $\langle\phi|\hat{A}_f|\phi\rangle$ w.r.t. the basis of eigenvectors of the Hermitian operator $\hat{A}_f$.

Let us consider the following maps $J_S\colon P_{\mathrm{PCSFT}} \to S_{\mathrm{QM}}$ and $J_V\colon V_{\mathrm{PCSFT}} \to V_{\mathrm{QM}}$,

$$J_S(p) = \hat{\rho}_p = \hat{B}_p/\mathrm{Tr}\,B_p,\ J_V(f) = \hat{A}_f. \tag{15.14}$$

This correspondence connects the averages given by the causal theoretical and observational models:

$$\frac{1}{\mathcal{E}_p}\langle f\rangle_p = \mathrm{Tr}\,\hat{\rho}_p\hat{A}_f, \tag{15.15}$$

that is, the QM and PCSFT averages are coupled with the scaling factor which is equal to the inverse of the average energy of the random field (for $\mathcal{H} = L_2$).

Thus, density operators representing quantum states correspond to covariance operators of random fields normalized by the average energy of a random field and the Hermitian operators representing quantum observables correspond to quadratic forms of fields.

Let us rewrite (15.15) in the following form:

$$\langle\frac{f}{\mathcal{E}_p}\rangle_p = \mathrm{Tr}\,\hat{\rho}_p\hat{A}_f.$$

If random fields have low energy, that is, $\mathcal{E}_p << 1$, the quantity

$$g_p(\phi) \equiv \frac{f(\phi)}{\mathcal{E}_p}$$

can be interpreted as an amplification of the PCSFT physical variable f. Hence, by connecting QM with PCSFT, QM can be interpreted as an observational theory describing averages of amplified subquantum quantities – quadratic forms of random fields. The subquantum random fields are unobservable and they can be experimentally verified only indirectly, via coupling with the observational model – QM.

In contrast to QM, PCSFT is causal: selection of a vector ('field') $\phi \in \mathcal{H}$ determines the values of all PCSFT-quantities, quadratic forms of classical fields: $\phi \to \langle\phi|\hat{A}|\phi\rangle$.

For physical quantities, the correspondence map J_V is one-to-one, but the map J_S is not one-to-one. But it is a surjection, that is, it is onto the map.

Interpretations of Quantum Mechanics and the Bild Conception

As was discussed in Chapters 2, 3, and 10, QM is characterized by the diversity of interpretations. Some experts consider this situation as a sign of the crises in quantum foundations. Others struggle to convince the community that one fixed interpretation (say orthodox Copenhagen, or many worlds, or QBism) is the only true interpretation of QM. Pragmatic researchers typically proceed without even thinking about interpretations. Some experts claim that all interpretations have equal rights and it is merely the researcher's test that determines the interpretation in the use. In any event, the problem of possible interpretations of QM is very complicated and up to now we have tried to escape highlighting the interpretational side of applications of the Bild conception to quantum phenomena. Now we briefly discuss this issue. As was pointed out in Chapter 2. There are many shades of interpretation, but following Ballentine we wish to distinguish only two:

- **SI** *The statistical interpretation, according to which a pure state (and hence also a general state) provides a description of certain statistical properties of an ensemble of similarly prepared systems, but need not provide a complete description of an individual system.*
- **II** *Interpretations which assert that a pure state provides a complete and exhaustive description of an individual physical system (e.g., an electron).*

We call II *the individual interpretation* of a quantum state. This interpretation is especially clear as presented in von Neumann's book [365]. It is commonly known as the *orthodox Copenhagen interpretation.* The latter term has a variety of flavors; as was pointed out by Plotnitsky [308], it is more natural to speak about interpretations in the spirit of Copenhagen. Unfortunately, Bohr expressed his views not so clearly. One might even guess that he used SI (see [311]). Moreover, the following terminological ambiguity contributed into the interrelation II versus SI debate. The majority of II-users would also say that they use the statistical interpretation and accept that QM produces only statistical prediction. They combine (in a way somewhat unclear to me) the individual interpretation of a quantum state and the probabilistic description of experiments outcomes. For example, von Neumann suggested to describe quantum probabilities by using the von Mises frequency probability theory. But it is unclear how the quantum state, as the state of an individual electron, can generate a random sequence (a collective in von Mises terminology).

We remark that SI was actively advertised, for example, by Blokhintsev [57, 58], Ballentine [42]–[44], and De Muynck [106].

In fact, in Section 15.4 we employed **SI** without even mentioning the interpretation issue. However, from the section's content it is clear that QM is endowed with SI. Within SI, statistical models similar to PCSFT are natural, as they combine the subquantum causality with the probabilistic structure of QM.

By employing II and especially following von Neumann [365], one might extract from QM a causal component. This is the model which handles solely quantum states and their Schrödinger dynamics; so mathematically the quantum CTM is given by the space of quantum states S_{QM}, the space of density operators. The corresponding OM is then given by the space of Hermitian operators V_{QM}. However, such a split of QM into counterparts and treating them as two separate models seems to be inconsistent even with von Neumann's viewpoint, who considered QM as the integral and consistent model of quantum phenomena. Nor would

Schödinger consider such a split of QM as an application of the Bild conception. And Einstein in his search of a true subquantum model was looking for something different from the aforementioned possibility of combining the Bild conception with II.

15.5 On Usefulness of Causal Theoretical Models

The preceding presentation of the possible two-level description of the microphenomena, QM versus PCSFT, can be used as the initial point for the discussion on usefulness of CTMs. To be provocative, we start by noting that for Bohr and other fellows of the orthodox Copenhagen interpretation of QM attempts to construct CTM for QM are meaningless [63, 65, 309, 310]. At the same time, Bohr never claimed that such CTM can't be constructed [309, 310]; he was not interested in no-go theorems. In his writings I did not find a word about the von Neumann no-go theorem. I am sure that he would ignore the Bell no-go theorem [54] and be surprised by interest in it among members of the modern quantum foundational community. Bohr highlighted the observational status of QM, but for him any kind of CTM is metaphysical. For physicists, it is meaningless to spend time by trying to design prequantum CTM. This position is very common among physicists.[8] For Bohr, it is impossible to complete QM in the causal way by operating with quantities which have a direct connection to observations. And he is completely right: the complementarity principle and Heisenberg uncertainty relation block searching of a finer OM for QM (Chapters 3, 6). It seems that in contrast to von Neumann, Bohr was not disturbed by acausality; observational acausality is a consequence of the contextuality and complementarity of quantum observations. We also repeat that Bohr did not deny the possibility of constructing CTMs beyond QM, but for him the introduction of hidden variables was a metaphysical and totally meaningless exercise.

Bohr's position can be questioned and on the questioners' side are Helmholtz, Herz, Boltzmann, and Schrödinger. As we have seen, Schrödinger agreed with Bohr that QM is a good OM for microphenomena. He did not think that the acausal structure of QM prevents construction of corresponding CTM. For him causality is closely coupled to continuity and hence to his original wave approach to microphenomena.

As for Helmholtz, Herz, Boltzmann, and Schrödinger, I think that consideration of acausality as a property of nature (at least at the micro-level) destroys completely the methodology of science. If Helmholtz did make a mistake by saying [359] that

[8] At the beginning of this century, academician Kamil Akhmetovich Valiev (one of the originators of the quantum computer project in Russia) told me: "Andrei, why do you consume your days in such totally meaningless activity? You are talented and can contribute so much to real physics – quantum computing!" He pointed to my attempts to find a prequantum CTM.

"every law of nature asserts that upon preconditions alike in a certain respect, there always follow consequences which are alike in a certain other respect," then physics becomes a science about gambling (as, e.g., QBists claim). It is difficult (at least for me) to accept this position. Thus, the main impact of creation of CTMs is a reestablishing of causality that might be violated in OM.

Now we turn to QM. Reestablishing of causality of microphenomena (even without the direct coupling to observations) would demystify quantum theory. We do not claim that PCSFT is true CTM for QM; as Schrödinger claimed [331], it is meaningless and even dangerous for science development to search for such a model. But PCSFT can be used as a causal Bild of quantum processes. One of the main advantages of this Bild is that it is local. PCSFT reproduces not only QM averages, but even its correlations [212]; hence, the Bell inequalities can be violated for PCSFT-quantities (they are hidden from the observational viewpoint). PCSFT demystifies quantum entanglement by connecting it with correlations of subquantum (classical) random fields (compare [207, 212, 225]). PCSFT can be considered as a step toward merging of QM with general relativity, but within some CTM.

Can one learn from CTM something that might be lifted to the observational level? In our concrete case, can some theoretical elements of PCSFT be realized experimentally (maybe in the future)? The basic element of PCSFT is a random field $\phi = \phi(x; \omega)$. Measurement of such a subquantum field would be the real success of PCFT. However, it seems that one cannot expect this. As was pointed out by Bohr (in the 1930s), even the classical electromagnetic field cannot be measured in a fixed point.

Another component of PCSFT which can be connected with real physics is the need in the background field. This component was not discussed in the preceding brief presentation, so see [206, 212] for details. Such random background field $\phi_{\text{bground}}(x, \omega)$ is the necessary element of the mathematical model $\mathbf{M}_{\text{PCSFT}}$ for generation of entangled states in S_{QM}. In this way PCSFT is related to stochastic electrodynamics and supports it. Unfortunately, the background (zero point field) is not a component of conventional QM; stochastic electrodynamics is commonly considered an unconventional model of microphenomena. From the Bild viewpoint this model should be treated as one of possible CTM for QM; this viewpoint would clarify the interrelation between these two models. But in this chapter we do not plan to go deeper into this issue. We note that the background field carries long-distance correlations which contribute to violation of the Bell-type inequalities. However, these are classical field correlations having nothing to do with the spooky action at a distance.

The model closest to experimental verification is the PCSFT representation of Born's rule as an approximate rule for calculation of probabilities; the standard Born rule is perturbed by additional terms which in principle can be verified (see

[212] and especially article [206] suggesting the concrete experimental test). For a totally different reason this prediction was tested by the research group of Weihs [340, 341] – testing Sorkin's inequality in the triple-slit experiment. Surprisingly, transition from two slits to three slits is not trivial and Weih's group confronted difficulties related to nonlinearity of detection processes. For the moment, no deviations from the Born rule were observed.

For me, the main message of PCSFT as one of possible CTMs for QM is that quantum nonlocality is an artifact of OM (= QM). The presence of such artifacts in OMs is natural from the Bild-viewpoint. This is one of the reasons to construct CYMs.

15.6 Bell's Project from the Bild Viewpoint

Unfortunately, Bell (as well as other icons of modern quantum foundational studies) did not read the works on the Bild concept for scientific theories. By introducing hidden variables he suggested some CTM for OM = QM. However, he considered too naive coupling of his CTM, $\mathbf{M}_{\text{Bell}}$, with OM – $\mathbf{M}_{\text{QM}}$. The sub-quantum quantities, functions of hidden variables, $A = A(\lambda)$, were identified with quantum observables. In particular the ranges of values of quantities from $\mathbf{M}_{\text{Bell}}$ coincide with the ranges of values of quantum observables (and this is not the case for example in PCSFT). As was pointed out in Chapter 2, $\mathbf{M}_{\text{Bell}}$ confronts the complementarity principle. The latter point will be clarified in what follows.

We recall the mathematical structure of $\mathbf{M}_{\text{Bell}}$ by connecting it with the framework of Section 15.3. Bell considered [54] an arbitrary set of hidden variables Λ; this is the set of states of his CTM, that is, $S_{Bell} = \Lambda$. To put this model in the mathematical framework of probability theory, Λ should be endowed with some σ-algebra of its subsets, say $\mathcal{F}$. Denote by P_{Bell} the space of all probability measures on $(\Lambda, \mathcal{F})$. The space of subquantum quantities consists of all measurable functions $A\colon \Lambda \to \mathbb{R}, A = A(\lambda)$, that is, random variables in terms of the Kolmogorov probability theory. We stress that the correspondence map $J_S\colon P_{Bell} \to S_{QM}$ is not specified, it is just assumed that such a map exists. For Bell's reasoning [54], this map need not be onto S_{QM} (so it need not be a surjection).

To make his model with hidden variables straightforwardly experimentally verifiable, Bell identified the values of CTM quantities $A = A(\lambda)$ with the values of QM quantities, the outcomes of quantum observables. First we discuss the mathematical side of this assumption and then its foundational side.

Mathematically, identification of quantities of $\mathbf{M}_{\text{Bell}}$ with QM observables means that the range of values of $A \in V_{Bell}$ coincides with the spectrum of the corresponding Hermitian operator $\hat{A}$. This is the important mathematical constraint on the map $J_V\colon V_{Bell} \to V_{QM}$ (we recall that V_{QM} is the set of density operators). Purely mathematical relaxation of this assumption destroys the Bell inequality argument, for exampleas in PCSFT.

However, Bell *should* proceed with this assumption on the coincidence of ranges of values of the subquantum quantities and quantum observables, since he dreamed for straightforward experimental verification of his model with hidden variables [53, 54]. He was not accustomed to the Bild concept of a scientific theory. In particular, Herz's (and Schrödinger's) statement on hidden quantities which could not be observed directly was totally foreign for Bell. For him, as well as for Bohr, a theory in which quantities can't be directly verified is a part of metaphysics, not physics [63, 65, 309, 310].

However, by identifying the outcomes of subquantum quantities with the outcomes of quantum observables Bell confronts the complementarity principle (Chapter 2). This can be clearly seen in the CHSH-framework [83]. There are two pairs of observables: (A_1, A_2), in Alice's lab, and (B_1, B_2), in Bob's lab represented by Hermitian operators $(\hat{A}_1, \hat{A}_2)$ and $(\hat{B}_1, \hat{B}_2)$. Observables corresponding to cross measurements for Alice-Bob are compatible, that is, they can be jointly measurable, but local observables of Alice as well as of Bob are incompatible, that is, they cannot be jointly measurable. In the operator terms,

$$[\hat{A}_i, \hat{B}_j] = 0, \ [\hat{A}_1, \hat{A}_2] \neq 0, \ [\hat{B}_1, \hat{B}_2] \neq 0. \tag{15.16}$$

This is the quantum mechanical description of the CHSH experimental context. We note that if local observables are compatible for at least one lab (in the operator terms, at least one of commutators $[\hat{A}_1, \hat{A}_2]$, $[\hat{B}_1, \hat{B}_2]$ equals zero), then the CHSH inequality can't be violated; see Chapter 6. For the reader's convenience, we briefly repeat this result.

For quantum observables, the correlation used in the CHSH framework can be represented via the Bell-operator,

$$\hat{\mathcal{B}} = \frac{1}{2}[\hat{A}_1(\hat{B}_1 + \hat{B}_2) + \hat{A}_2(\hat{B}_1 - \hat{B}_2)], \tag{15.17}$$

as

$$\langle \mathcal{B} \rangle_\psi = \langle \psi | \hat{\mathcal{B}} | \psi \rangle. \tag{15.18}$$

By the Landau identity, we have

$$\hat{\mathcal{B}}^2 = I - (1/4)[\hat{A}_1, \hat{A}_2][\hat{B}_1, \hat{B}_2]. \tag{15.19}$$

Hence, if *at least one of the commutators* is zero,

$$[\hat{A}_1, \hat{A}_2] = 0, \tag{15.20}$$

or

$$[\hat{B}_1, \hat{B}_2] = 0, \tag{15.21}$$

then the following inequality holds:

$$|\langle \mathcal{B} \rangle|_{\psi} \leq 1. \tag{15.22}$$

We remark that such purely quantum mechanical treatment of the Bell-type inequalities does not involve the issue of nonlocality at all. The crucial point of the preceding consideration is the existence of incompatible observables. The latter is a consequence of the Heisenberg uncertainty principle or more generally the Bohr complementarity principle.

Bell considered quantities of his CTM $\mathbf{M}_{\text{Bell}}$ as representing physical observables, hence all observables can be represented as functions $A_i = A_i(\lambda), B_j = A_i(\lambda)$ and their values are identified with outcomes of observations. Besides the pairs $(A_i(\lambda), B_j(\lambda))$ of compatible observables, one can consider the pairs $(A_i(\lambda), A_j(\lambda))$ and $(B_i(\lambda), B_j(\lambda))$. By treating the latter two pairs as representing the outcomes of physical observables, it is natural to assume the possibility of their joint measurability (maybe not nowadays, but in the future, when the measurement technologies would be improved). So, the complementarity principle loses its fundamental value. By keeping Bell's model $\mathbf{M}_{\text{Bell}}$ as representing physical reality, one confronts treating complementarity as the fundamental property of (observational) microphenomena.

At the level of correlations,

$$\langle A_i B_j \rangle_{\rho} = \text{Tr}\, \hat{\rho} \hat{A}_i \hat{B}_j = \lim_{N \to \infty} \frac{1}{N} \sum_{k=1}^{N} A_{ik} B_{jk}, \tag{15.23}$$

where $(A_{ik}), (B_{jk})$ are observables' outcomes. At the same time, for the probability distribution P_{ρ} such that $\hat{\rho} = J_S(P_{\rho})$, we have

$$\langle A_i B_j \rangle_{P_{\rho}} = \int_{\Lambda} A_i(\lambda) B_j(\lambda) P_{\rho}(d\lambda) = \lim_{N \to \infty} \frac{1}{N} \sum_{k=1}^{N} A'_{ik} B'_{jk}, \tag{15.24}$$

where A'_{ik} and B'_{jk} are outcomes of subquantum quantities $A_i = A_i(\lambda)$ and $B_j = B_j(\lambda)$. However, since these outcomes can be identified with the outcomes of the quantum observables, $A'_{ik} = A_{ik}, B'_{jk} = B_{jk}$, we can write

$$\langle A_i B_j \rangle_{P_{\rho}} = \lim_{N \to \infty} \frac{1}{N} \sum_{k=1}^{N} A_{ik} B_{jk}. \tag{15.25}$$

But the same reasoning is applicable to the subquantum quantities $A_1 = A_1(\lambda)$, $A_2 = A_2(\lambda)$ and $B_1 = B_1(\lambda), B_2 = B_2(\lambda)$ representing the incompatible quantum observables:

$$\langle A_1 A_2 \rangle_{P_{\rho}} = \int_{\Lambda} A_1(\lambda) A_2(\lambda) P_{\rho}(d\lambda) == \lim_{N \to \infty} \frac{1}{N} \sum_{k=1}^{N} A'_{1k} A'_{2k}, \tag{15.26}$$

$$\langle B_1 B_2 \rangle_{P_\rho} = \int_\Lambda B_1(\lambda) B_2(\lambda) P_\rho(d\lambda) = \lim_{N\to\infty} \frac{1}{N} \sum_{k=1}^{N} B'_{1k} B'_{2k}. \tag{15.27}$$

Again by identifying the values of subquantum and quantum observables, we obtain

$$\langle A_1 A_2 \rangle_{P_\rho} = \lim_{N\to\infty} \frac{1}{N} \sum_{k=1}^{N} A_{1k} A_{2k}, \tag{15.28}$$

$$\langle B_1 B_2 \rangle_{P_\rho} = \lim_{N\to\infty} \frac{1}{N} \sum_{k=1}^{N} B_{1k} B_{2k}. \tag{15.29}$$

Representations (15.28) and (15.29) of subquantum correlations (within CTM $\mathbf{M}_{\text{Bell}}$) via outcomes of observables supports the assertion that the subquantum correlations $\langle A_1 A_2 \rangle_{P_\rho}$, $\langle B_1 B_2 \rangle_{P_\rho}$ should be measurable (at least in principle and in the future). It is not clear how Bell would treat this objection to his argument. I guess that he would agree that his model with hidden variables, $\mathbf{M}_{\text{Bell}}$, collides with the complementarity principle. However, he might choose to move between Scylla and Charybdis:

- **S:** identifying the values of subquantum quantities with the values of quantum observables,
- **Ch:** the complementarity principle,

and claim that **S** and **Ch** can peacefully coexist. He might say that (15.25) is legal and hence the experimental verification of $\mathbf{M}_{\text{Bell}}$ is possible, but (15.28) and (15.29) are illegal and treatment of these correlations as experimentally verifiable is forbidden. For me the latter position is inconsistent (although logically possible). This inconsistency was the basis of de Broglie's critique of Bell's argument [100] (Section 15.7).

This is a good place to recall that the physical seed of the complementarity principle is in Bohr's quantum postulate on the existence of indivisible quantum of action given by the Planck constant h: incompatible observables exist due to the existence in nature of the minimal action [59]–[61]. Thus, Bell's conflict with the complementarity principle is in fact conflict with the quantum postulate – the existence of h. Hence, this is the conflict with the very foundation of quantum physics, for example, the quantum model of the black body radiation and processes of spontaneous and stimulated emissions.

We conclude this section by noticing that PCSFT reproduces quantum correlations (Chapter 16) and, hence, the PCSFT correlations violate the Bell inequalities. But this is not surprising, since the ranges of the values of the subquantum quantities (represented as quadratic forms of the fields) and quantum observables (represented as Hermitian operators) do not coincide. And this is a good place to recall that the proofs of the Bell-type theorems are based on the assumption that the functions of hidden variables are valued in the segment $[-1, +1]$.

15.7 De Broglie's Critique of the No-Go Theorems of von Neumann and Bell

Nowadays it is practically forgotten that de Broglie considered all no-go theorems by which hidden variables for QM do not exist as totally misleading [100]. His double solution model [99] can be considered as a model with hidden variables of the field type. The pilot wave is a hidden variable.[9] As we shall see, de Broglie followed the Bild conception without knowing about it.

To justify his double solution model and its peaceful coexistence with QM, de Broglie criticized the most famous no-go theorems, the von Neumann and Bell theorems [54, 365]. He did not criticize the mathematical derivation of these theorems, but their interpretation and straightforward identification of hidden and observational quantities. His interpretation of these theorems was presented in detail by Lochak [266–268]. Paper [267] is available for free reading via Google books; we cite it:

Von Neumann proved a theorem which claims that there are no pure states without statistical dispersion. This result is indeed intuitively obvious because the absence of dispersion in a pure state means that it would be possible to measure simultaneously the physical quantities attached to a system described by this state. But in fact we know that this is impossible for non-commuting quantities. In this sense, the theorem is nothing but a consequence of Heisenberg's uncertainties.

From this one can conclude that a pure state of QM can't be considered as representing an ensemble of systems following the laws of classical probability theory (see [267] for detailed discussion) and quantum observables as classical random variables. From this, von Neumann made the conclusion that generally it is impossible to create any model with hidden variables behind QM. As pointed out in [267],

De Broglie's answer consists essentially in asserting that if any hidden parameters do exist, they cannot obey quantum mechanics because if you try to imagine hidden parameters it is of course in order to restore the classical scheme of probabilities. Now if you need a classical scheme of probabilities for objective(but hidden) values of physical quantities which are introduced in quantity of hidden

9 In this aspect the double solution model and PCSFT are similar. Schrödinger's attempt to find a subquantum model of the wave type as well as Einstein's search for a subquantum nonlinear field model were also steps in the same direction, although, as was noted, methodologically their positions are different. Schrödinger tried to realize the Bild approach for QM, treated as OM, but Einstein dreamed for a model carrying jointly the features of CTM and OM.

parameters, these probabilities cannot be probabilities observed on the result of a measurement: simply because the observed probabilities do obey the quantum scheme and not the classical one!

Hence, not only hidden parameters $\lambda \in \Lambda$, but even subquantum quantities $A = A(\lambda)$ are hidden and the probability distributions of these quantities P_A should not be identified with quantum probability distributions.

For de Broglie, it was evident that classical and quantum probability calculi differ crucially; by attempting to apply the former to quantum observables one immediately confronts the Heisenberg uncertainty principle (and Bohr's complementarity principle [65, 309, 310]). This is precisely my viewpoint, which was presented in article [233] (see Section 15.6).

Hence, de Broglie's viewpoint on the interrelation of subquantum and quantum models matches perfectly the CTM-OM approach. In fact, it matches the ontic-epistemic framework, since de Broglie considered hidden variables and physical quantities, $A = A(\lambda)$, as objective entities. But, as was already noted, schematically CTM-OM and the ontic-epistemic frameworks are similar.

De Broglie's statement that quantities of a subquantum theory are hidden and their probability distributions should not be identified with the probability distributions of quantum observables matches the PCSFT-QM coupling considered in Section 15.4. PCFT quantities are quadratic forms of fields playing the role of hidden variables. Such quantities have a continuous range of values, but say the quantum spin observables have discrete spectra. Of course, they cannot have the same probability distribution, even without consideration of correlations. The correspondence between the classical probability calculus of PCSFT and the quantum probability calculus is fuzzy, classical covariance operators are mapped to density operators; see (15.14).

De Broglie and Lochak used the same argument for critical analysis of the Bell theorem: one should sharply distinguish subquantum and quantum quantities and not identify their outcomes and probability distributions. Not only are hidden parameters hidden, but also quantities dependent on them and their probability distributions. So, Bell's model is a very special CTM for QM. Yes, it should be rejected, as follows from the quantum formalism and experiments. But its rejection does not prevent the search for other CTMs for QM with more complicated connection between subquantum and quantum quantities and their probability distributions. From this viewpoint, the foundational value of the Bell theorem is overestimated.

I again repeat that it is a pity that the fathers of QM, including de Broglie, were not aware of the works of Helmholtz, Hertz, and Boltzmann. The Bild conception would provide the rigid philosophic basis for establishing proper interrelation between subquantum models with hidden variables and QM.

15.8 Classical Mechanics

For classical mechanics, CTM and OM coincide. On one hand, this was fortunate for the development of physics, since it simplified so much its philosophic basis and highlighted the role of observation. On the other hand, identification of one special mathematical model, Newtonian mechanics, with reality supported similar ontic treatment of all physical models. The ontic viewpoint of a scientific theory dominated for a few hundred years, up to the works of Helmholtz, Hertz, Boltzmann, and Schrödinger. However, these works did not revolutionize the philosophy of science. For example, acausality of QM is still considered as a property of nature, that is to say irreducible quantum randomness is ontic.

We note that it seems that Hertz didn't consider classical mechanics as OM; see again [161]: "If we try to understand the motions of bodies around us, and to refer them to simple and clear rules, paying attention only to what can be directly observed, our attempt will in general fail."

This statement definitely refers to classical mechanics. Similarly to Hertz, Atmanspacher [34] also considered the two-level description even for classical physics and suggested the corresponding mathematical examples.

The need for a separate OM model for classical mechanics became clear in the process of creating the mathematical description of the Brownian motion, which will be considered in the next section, and structuring CTM-OM in accordance with the Bild conception.

15.8.1 Brownian Motion: Two Levels of Description from the Time-Scale Separation

Here we follow Chapter 14:

The dynamics of a Brownian particle can be observed at two levels [323]. Within the first, more fundamental level, the Brownian particle coupled to a thermal bath at temperature T is described via definite coordinate x and momentum p and moves under influence of external potential, friction force, and an external random force. The latter two forces are generated by the bath. The second, overdamped regime applies when the characteristic relaxation time of the coordinate τ_x is much larger than that of the momentum τ_p,

$$\tau_x \gg \tau_p$$

(overdamped regime). On times much larger than τ_p one is interested in the change of the coordinate and defines the *coarse-grained* velocity as $v = \Delta x/\Delta t$ for $\tau_x \gg \Delta t \gg \tau_p$. This definition of v is the only operationally meaningful one for the (effective) velocity within the overdamped regime. It appears that the coarse-grained velocity, though pertaining to single particles, is defined in the context of the whole systems of coupled Brownian particles.

The evolution of the momenta of the Brownian particles is very fast and cannot be resolved on the time-scales available to the experiment. To obtain experimentally accessible quantities, one employs the technique of the time-scale separation and measurement of the coarse-grained velocity and osmotic velocity. These quantities can be measured. They are assigned not to an individual Brownian particle, but to an ensemble of particles coupled to bath, so these are statistical quantities.

In terms of the present article, Brownian motion is described by CTM $\mathbf{M}_{CTMB}$ with the coarse-grained (x, p) and OM $\mathbf{M}_{\mathrm{OMB}}$ with the coarse-grained velocities v_+, v_- or the osmotic velocity $u = v_+ - v_-$, The latter description is based on observational quantities (x, u). As was shown in Chapter 14, $\mathbf{M}_{\mathrm{OMB}}$ shows some properties of QM, for example, there are analogs of the Heisenberg uncertainty relations and entanglement; in particular, for a pair of Brownian particles the joint probability distribution $P(t, x_1, u_1, x_2, u_2)$ does not exist. Of course, the OMs $\mathbf{M}_{\mathrm{OMB}}$ and $\mathbf{M}_{\mathrm{QM}}$ differ essentially. For example, for a single particle the probability distribution $P(t, x, u)$ is well-defined: incompatibility appears only in compound systems.

Nowadays the preceding two-level structuring of scientific theory of the Brownian motion is shaken by the novel experimental possibilities for the measurement of momentum p of a Brownian particle. A variety of experiments has been performed in recent years (see, for example, [265]). In spite of some diversity in experimental outputs, it is clear that experimental science is on the way to establishing robust procedures for measurement of the Brownian momentum p. Through experimental research, CTM $\mathbf{M}_{\mathrm{CTMB}}$ is getting OM-status. However, new theoretical efforts are needed to merge $\mathbf{M}_{\mathrm{CTMB}}$ and $\mathbf{M}_{\mathrm{OMB}}$ treated as OMs. The osmotic velocity u (an element of $\mathbf{M}_{\mathrm{OMB}}$) is not straightforwardly derived within $\mathbf{M}_{\mathrm{CTMB}}$. At least for me, connection between the velocity and coarse-grained velocity is not clear. How is the latter derived from the former?

This special example supports the search for CTMs for QM (see the discussion at the end of Section 15.5). Some hidden quantities of such models can serve as the candidates for future experimental verification. One of the problems of such a project is that, since the creation of QM, physicists (mathematicians and philosophers) have created too many subquantum models operating with a variety of hidden quantities, as say the quantum potential in Bohmian mechanics or the random field in PCSFT. What are the most probable candidates for future experimental verification? The Bell hidden variable model [53, 54] is one of CTMs for QM; it can be directly tested experimentally. It was tested and rejected.

15.9 Bild Conception and Science Foundations

My aim is to remind physicists and especially experts in foundations (not only of quantum physics, but also classical mechanics and field theory, statistical mechanics, and thermodynamics) about the works of Helmholtz, Hertz, and Boltzmann [66, 67, 160, 161, 359] on the meaning of a scientific theory which led to the Bild conception – the mathematical model concept of a scientific theory. By appealing to the two-level description of natural phenomena, CTM-OM description, it is possible to resolve many foundational problems, including acausality of QM. Moreover, the Bild conception demystifies quantum foundations. The quantum foundational problems such as the possibility of introducing hidden variables were discussed long ago. The latter problem was analyzed by Hertz, who tried to reduce the classical electromagnetic field to an ensemble of mechanical oscillators [160]. From the viewpoint of the Bild conception, Bell's attempt to invent hidden variables for QM is very naive; if such variables existed, their coupling with quantum observables might not be as straightforward as in the Bell model. Within the Bild conception, it becomes clear why Schrödinger did not consider acausality of quantum observations as the barrier on the way toward a causal description of quantum phenomena [330]–[333]. It seems that similarly to Bell [54], von Neumann also was not aware of development of the philosophy of science by the German school of physicists in the nineteenth century. He treated the quantum measurement problem too straightforwardly, and acausality and irreducible quantum randomness appeared as consequences of such treatment [365]. He did not appeal to the two-level CTM-OM description of microphenomena.

In a series of works [200, 201, 206, 212], Khrennikov et al. developed PCSFT, CTM with classical random fields, reproducing QM interpreted as OM for microphenomena. However, PSCFT-QM coupling is not as simple as in the Bell framework.

The two-level description of physical phenomena is in fact widely used in statistical physics and it is based on the timescales separation technique and consideration of coarse quantities. All such descriptions are well accommodated within the Bild conception. The Brownian motion in the overdamped regime is described by OM, which is not directly coupled to CTM based on the classical mechanical description.

Finally, we remark that the Primas-Atmanspacher [33, 321] ontic-epistemic approach to physical theories (see also, e.g., [33–35, 321], which is formally similar to the Bild conception). But in accordance with the Bild concept, no model describes reality as it is.

16

Prequantum Classical Statistical Field Theory (PCSFT)

16.1 PCSFT from Bild-Conception

By the statistical interpretation quantum mechanics (QM) is a statistical theory. Its mathematical formalism can't describe an individual quantum system, for example, an electron or a photon. Quantum mechanics predicts only probabilities of measurements' outcomes for large ensembles of quantum systems. We point out that classical statistical mechanics (CSM) does the same. Why are QM and CSM based on different mathematical probabilistic models?

In CSM, averages of physical variables are expressed via integrals with respect to probability measures, but in QM via operator traces. In CSM we have

$$\langle f \rangle_\mu = \int_X f(\phi) d\mu(\phi), \tag{16.1}$$

where X is the state space. In probabilistic terms: there is given a random vector $\phi = \phi(\omega)$ valued in the state space X. And

$$\langle f \rangle_\mu = Ef(\phi(\omega)).$$

In QM, the average is expressed via the operator trace-formula:

$$\langle \hat{A} \rangle_\rho = \mathrm{Tr}\, \rho \hat{A}. \tag{16.2}$$

This formal mathematical difference induces the widespread prejudice that the classical and quantum worlds differ fundamentally.

In Chapter 15 it was shown that the seed of this foundational prejudice is in ignoring the message of Herz and Boltzmann on the two-level description of physical phenomena. The Bild-conception implies that an observational model (OM) should be completed by a causal theoretical model (CTM). The latter can contain hidden variables which are not present in OM. The complete and consistent understanding of physical phenomena is possible only on the basis of CTM-OM duality.[1]

As was noted in Chapter 15, the Bild-conception based on two-level structuring of scientific theories is closely related to ontic-epistemic structuring (Primas–Atmanspacher), especially if one would not interpret the ontic model as the model of reality as it is, say "ontic" is just as containing unobservational quantities.

Within the Bild-conception (or ontic-epistemic conception) one can easily proceed towards creation of CTMs for QM. One of them is *prequantum classical statistical field theory* (PCFT) [212] which is a kind of CSM beyond QM. Within PCFT, quantum averages can be mathematically represented as classical averages and, moreover, even correlations between entangled systems can be expressed as classical correlations. But the correspondence between PCSFT and QM variables is not as straightforward as in the Bell model with hidden variables.

This is the good place to recall *Einstein's dream.* Albert Einstein did not accept that acausality of QM is the intrinsic feature of nature; he did not believe in irreducible randomness. He dreamed of a better, so to say "prequantum," model [117]:

- **Dream 1.** Reduction of quantum randomness to classical.
- **Dream 2.** Renaissance of causal description.
- **Dream 3.** Instead of particles, classical fields will provide a complete description of reality, a reality of fields.

If I rightly understand his ideas from reading book [117],[2] in his dreams the new fieldlike CTM should replace QM. He didn't follow the Bild-conception and he didn't dream about creation of the two-level modeling of micro-phenomena.

Einstein's dreams came true in PCSFT, a version of CSM in which fields play the role of particles. One might say that

A quantum particle is the symbolic representation of a "prequantum" classical field fluctuating on the space-time scale which is essentially finer than the space-time scale of present measurements.

[1] The example of Brownian motion modeling (Chapter 14) shows that generally the CTM-OM structure depends on the level of development of experimental technology. For approximately one hundred years, experimenters were not able to measure the velocity of a Brownian particle. Moreover, the common opinion was that it is impossible. So, the particle's velocity played the role of a hidden variable. Experimenters were satisfied by measurement of coarse-grained velocity and the osmotic velocity. However, recently some experimental groups reported about successful measurement of the velocity of a Brownian particle (Chapter 14). So, the velocity lost its status as a hidden variable and became an observational quantity.

[2] I also got to know a lot about Einstein's dreams from one of his relatives whom I met at a conference and who owned the collection of Einstein's writings during the last years of his life. Was this collection finally published? I do not know.

But one should remember about the Bild-conception and take this heuristic interpretation of subquantum random fields with caution. For the moment, such fields are only mathematical entities giving the possibility of restoring causality of quantum processes. This causality, however, is only theoretical, since these fields are unapproachable for the present measurement technology and, moreover, maybe they would never be approached experimentally.

As is well known, the most striking difference between CSM and QM is in the mathematical descriptions of compound systems, in the notions of classical correlation versus entanglement (compare, however, with Chapter 13). The latter is typically quantified in terms of violation of the Bell inequality.

In CSM a compound system $S = (S_1, S_2)$ is mathematically described by the *Cartesian product* of the state spaces of its parts S_1 and S_2. In QM a compound system $S = (S_1, S_2)$ is described by the *tensor product.* Typically this difference in the mathematical models is considered as crucial for foundations. And entanglement in its conventional tensor product structuring is interpreted as the basic exhibition of quantumness (see chapters 13 and 14). We study this controversy between the Cartesian and tensor product representations from the viewpoint of the PCFT-QM interrelation in Section 16.4. The Cartesian product–based PCFT description is mathematically and interpretationally tricky, so it is postponed to Section 16.4. To show the possibility of PCSFT description of compound quantum systems, we start with the straightforward description based on random fields valued in the tensor product (Section 16.3). The Cartesian product model (Section 16.4) matches better classical physics. However, if PCSFT is treated as just a mathematical causal model beyond QM, then both Cartesian and tensor products models for compound systems are equally interesting.

16.2 PCSFT: Formalism and Interpretation

For the reader's convenience, we briefly remind them of the PCSFT-QM coupling scheme (Section 15.4). See monograph [212]).

In the quantum model $\mathbf{M}_{QM}$, states are given by density operators and observables are represented by Hermitian operators. Denote the space of density operators by S_{QM} and the space of Hermitian operators by V_{QM}.

In the subquantum model $\mathbf{M}_{\text{PCSFT}}$ states are vectors of Hilbert space; physical variables are quadratic forms on this space; the corresponding spaces are denoted S_{PCSFT} and V_{PCSFT}. Since QM is a statistical theory, one cannot expect direct coupling between the state spaces S_{PCSFT} and S_{QM}. Such coupling can be established between the space of probability measures P_{PCSFT} and the space of density operators S_{QM}. Here P_{PCSFT} is the space of probability measures (on the σ-algebra of

Borel subsets of Hilbert space S_{PCSFT}) having zero mean value and finite second momentum, namely,

$$\mathcal{E}_p \equiv \int_{\mathcal{H}} \|\phi\|^2 dp(\phi) < \infty. \tag{16.3}$$

The PCSFT-QM maps, $J_S\colon P_{\mathrm{PCSFT}} \to S_{\mathrm{QM}}$ and $J_V\colon V_{\mathrm{PCSFT}} \to V_{\mathrm{QM}}$, are given by

$$J_S(p) = \hat{\rho}_p = \hat{B}_p/\mathrm{Tr}\,\hat{B}_p,\; J_V(f) = \hat{A}_f, \tag{16.4}$$

where B_p is the (complex) covariance operator of measure p and $\hat{A}_f$ is the Hermitian operator determined by quadratic form f. We remark that $\mathrm{Tr}\,\hat{B}_p = \mathcal{E}_p$.

This correspondence connects the averages given by the causal theoretical and observational models:

$$\frac{1}{\mathrm{Tr}\,\hat{B}_p}\langle f\rangle_p = \frac{1}{\mathrm{Tr}\,\hat{B}_p}\int_{\mathcal{H}} f(\phi)dp(\phi) = \mathrm{Tr}\,\hat{\rho}_p\hat{A}_f = \langle \hat{A}_f\rangle_{\rho_p}. \tag{16.5}$$

Subquantum Random Fields

As in Chapter 15, instead of probability measures, we can consider $\mathcal{H}$-valued random vectors with zero mean value and finite second moment: $\phi = \phi(\omega)$, such that

1. $E[\phi] = 0$,
2. $\mathcal{E}_\phi = E[\|\phi\|^2] < \infty$.

Denote the space of such random vectors by the symbol R_{PCSFT}. They will be called random fields. This terminology is motivated by the case $\mathcal{H} = L_2(\mathbb{R}^m, \mathbb{C})$. Here each vector is a field, $x \to \phi(x)$; a random vector is a function $\phi = \phi(x;\omega)$, where $x \in \mathbb{R}^m$ and ω is a random parameter.

The first condition determining space R_{PCSFT} means that on average the fluctuations of a subquantum random field compensate each another. If the spatial and temporal scales of a potential field measurement roughly compare with the scales of the field fluctuations, then a potential observer would see nothing, just empty space-time. In fact, the zero average condition is mathematically written as

$$E[\langle\phi|u\rangle] = 0, \tag{16.6}$$

for any vector u; that is, a field's fluctuations have zero average in any direction. The statistical and physical meaning of the second condition determining R_{PCSFT} will be discussed later.

For a random field belonging to space R_{PCSFT}, the complex covariance operator is defined by its bilinear form

$$\langle u|\hat{B}_\phi|v\rangle = E[\langle u|\phi\rangle\langle\phi|v\rangle],\; u, v \in \mathcal{H}. \tag{16.7}$$

We remark that

$$\mathcal{E}_\phi = \mathrm{Tr}\,\hat{B}_\phi. \tag{16.8}$$

In terms of the random field, the map $J_S\colon R_{\mathrm{PCSFT}} \to S_{\mathrm{QM}}$ is given by

$$J_S(\phi) = \hat{\rho}_\phi = \hat{B}_\phi/\mathrm{Tr}\,\hat{B}_\phi. \tag{16.9}$$

The quantity $\mathcal{E}_\phi$ is the dispersion of a random field and $\sigma_\phi = \sqrt{\mathcal{E}_\phi}$ is its standard deviation from the zero field (the field representing "totally empty space"). For the space $\mathcal{H} = L_2(\mathbb{R}^3, \mathbb{C})$, the field can be assigned with the physical dimension of energy$^{1/2}$. Hence, in this case the normalized field

$$r(x;\omega) = \phi(x;\omega)/\sqrt{\mathcal{E}_\phi} \tag{16.10}$$

is the dimensionless quantity and its covariance operator $\hat{B}_r = \hat{B}_\phi/\mathrm{Tr}\,\hat{B}_\phi$ has the unit trace and it can be identified with the density operator $\hat{\rho}_\phi$.

We remark that the covariance operator $\hat{B}$ of a subquantum field corresponding to a pure quantum state ψ has the form $\hat{B} = |\Psi\rangle\langle\Psi|$, where Ψ is a vector belonging to $\mathcal{H}$; here $\mathrm{Tr}\,\hat{B} = ||\Psi||^2$ and $J_S(\hat{B}) = \frac{|\Psi\rangle\langle\Psi|}{||\Psi||^2}$. We remark that the corresponding random field is concentrated on the one-dimensional subspace of vector Ψ. So, $\phi(\omega) = \xi(\omega)\Psi$, where $\xi = \xi(\omega)$ is a one-dimensional random variable. Hence, subquantum random fields corresponding to pure quantum states are special: they are concentrated on one-dimensional subspaces.

Circularly Symmetric Gaussian Random Vectors

Generally the covariance operator does not determine uniquely the probability distribution of a random vector with zero average. This implies that the map $J_S\colon P_{\mathrm{PCSFT}} \to V_{\mathrm{QM}}$ is degenerate. One can try to diminish this degeneration by consideration of a special class of probability distributions. The most natural candidate is the class of Gaussian distributions. In the real case the covariance operator determines the Gaussian distribution with zero average uniquely. But this is not correct in the complex case: generally the complex covariance operator corresponds to a family of zero average Gaussian distributions. To find the "right class" of Gaussian distributions, we turn to the class of random fields corresponding to a pure quantum state. In QM a it is determined up to the phase factor, that is, normalized vectors ψ and $\psi_\theta = e^{i\theta}\psi$ represent the same pure state. We would like to find a subquantum analog of this equivalence relation.

In classical probability theory a complex random vector $\phi = \phi(\omega)$ is called *circularly symmetric* if for every phase $\theta \in [-\pi, \pi]$ the distribution of the random vector $\phi_\theta = e^{i\theta}\phi(\omega)$ equals the distribution of ϕ. We remark that, for any circularly symmetric random vector ϕ, $E[\phi] = 0$.

Consider now Gaussian circularly symmetric random vector. Its probability distribution is uniquely determined by its complex covariance operator (16.7).

Now turn to PCSFT and consider the one-dimensional random field $\phi_\Psi = \phi_\Psi(\omega)$ concentrated in one-dimensional space based on a vector $\Psi \in \mathcal{H}$. Let ϕ be circularly symmetric and consider the class of random fields ϕ_{Ψ_θ}, where $\Psi_\theta = e^{i\theta}\Psi$. In QM this class is represented by the state with representative $\psi = \Psi/||\Psi||$.

Denote now by $P_{\mathrm{PCSFT;gcs}}$ the space of Gaussian circularly symmetric probability measures. The map

$$J_S \colon P_{\mathrm{PCSFT;gcs}} \to S_{\mathrm{QM}} \tag{16.11}$$

is one-to-one.

Schrödinger Equation: QM versus PCSFT

Now we turn to the mathematical description of the subquantum random field dynamics. First, we consider the Schrödinger equation in the standard QM-formalism:

$$ih\frac{\partial \psi}{\partial t}(t) = \hat{H}\psi(t), \tag{16.12}$$

$$\psi(t_0) = \psi_0, \tag{16.13}$$

where $\hat{H}$ is Hamiltonian. Originally, for $\mathcal{H} = L_2(\mathbb{R}^3; \mathbb{C})$, Schrödinger interpreted the wave function $\psi = \psi(x)$ as a classical field (e.g., the electron field; the distribution of electron charge in space). But later he accepted the conventional probabilistic interpretation suggested by Born.

We recall that a time-dependent random variable $\phi(t;\omega)$ is called a *stochastic process*; we consider processes valued in complex Hilbert space $\mathcal{H}$. In real physical modeling, $\mathcal{H} = L_2$.

Dynamics of the prequantum random field ($\mathcal{H}$-valued random variable) is described by the simplest stochastic process which is given by *deterministic dynamics with random initial conditions.*

In PCSFT the Schrödinger equation, but with the random initial condition, describes dynamics of the subquantum random field, namely the subquantum stochastic process can be obtained from the mathematical equation which is used in QM for dynamics of the wave function:

$$ih\frac{\partial \phi}{\partial t}(t;\omega) = \hat{H}\phi(t;\omega), \tag{16.14}$$

$$\phi(t_0;\omega) = \phi_0(\omega), \tag{16.15}$$

where the initial random field $\phi_0(x,\omega)$ is determined (but not uniquely) by the quantum pure state ψ_0 as follows. Consider the corresponding density operator

$\hat{\rho}_{\psi_0} = |\psi_0\rangle\langle\psi_0|$. Consider now a random field $\phi_0(\omega)$ with the covariance operator $\hat{B}_{\phi_0}$ such that $J_S(\hat{B}_{\phi_0}) = \hat{B}_{\phi_0}/\mathrm{Tr}\,\hat{B}_{\phi_0} = \hat{\rho}_{\psi_0}$. So, the quantum state dynamics $t \to \psi(t)$ is the formal mathematical image of a variety of subquantum dynamics corresponding to a variety of initial random fields which are encoded in QM by the same quantum state.

We remark that up to the normalization factor the covariance operator of the subquantum random field $\phi(t;\omega)$ coincides with the density operator $\hat{\rho}_{\psi(t)} = |\psi(t)\rangle\langle\psi(t)|$. Hence, the PCSFT and QM dynamics match each other, but only on the level of covariance versus density operators.

Set $\hat{U}(t) = e^{-i\hat{H}/h}$; this is the unitary group determining the random field dynamics as

$$\phi(t;\omega) = \hat{U}(t)\phi_0(\omega). \tag{16.16}$$

First, we remark that if the mean value of the initial random field equals zero, $E[\phi_0] = 0$, then $E[\phi(t)] = 0$, for any t. Finiteness of the second momentum is also preserved,

$$E[||\hat{U}(t)\phi||^2] = E[||\phi_0||^2] < \infty. \tag{16.17}$$

Thus, $\phi\colon [0, +\infty) \to R_{\mathrm{PCSFT}}$.

By (16.17) we have

$$\mathrm{Tr}\,\hat{B}_{\phi(t)} = E[||\hat{U}(t)\phi||^2] = E[||\phi_0||^2] = \mathrm{Tr}\,\hat{B}_{\phi_0}, \tag{16.18}$$

that is, the normalization factor in the map J_S does not depend on t.

The previous considerations based on matching of the Schrödinger dynamics of pure quantum states and subquantum random fields can be generalized to matching of the von Neumann dynamics of a density operator and subquantum random field; let $u, v \in \mathcal{H}$, then

$$\langle u|\hat{B}_{\phi(t)}|v\rangle = E[\langle u|\phi(t;\omega)\rangle\langle\phi(t;\omega)|v\rangle = \langle u|\hat{U}(t)\hat{B}_{\phi_0}\hat{U}^\star(t)|v\rangle, \tag{16.19}$$

that is,

$$\hat{B}_{\phi(t)} = \hat{U}(t)\hat{B}_{\phi_0}\hat{U}^\star(t). \tag{16.20}$$

By taking into account (16.18) we see that the corresponding density operator $\hat{\rho}(t) = \hat{B}_{\phi(t)}/\mathrm{Tr}\,\hat{B}_{\phi_0}$ satisfies the von Neumann equation.

As an illustration, consider the case $\mathcal{H} = L_2(\mathbb{R}^3, \mathbb{C})$. Its vectors are fields $\phi = \phi(x)$ and random fields are complex functions of $x \in \mathbb{R}^3$ and a random parameter, that is, $\phi(x;\omega)$. Consider Hamiltonian

$$\hat{H} = \hat{p}^2/2m + V(x), \tag{16.21}$$

where $V(x)$ is a potential. Then the subquantum stochastic process is a solution of the linear partial differential equation with random initial field:

$$ih\frac{\partial\phi}{\partial t}(t,x;\omega) = -\frac{h^2}{2m}\frac{\partial^2}{\partial x^2}\phi(t,x;\omega) + V(x)\phi(t,x;\omega), \tag{16.22}$$

$$\phi(t_0,x;\omega) = \phi_0(x;\omega). \tag{16.23}$$

16.3 Compound Systems: Tensor Product–Valued Random Fields

The simplest way to handle compound systems within PCSFT is to use the preceding scheme without any modification. Consider a compound system $S = (S_1, S_2)$. Its quantum description is based on the Hilbert space $\mathcal{H} = \mathcal{H}_1 \otimes \mathcal{H}_2$, the tensor product of the subsystems' Hilbert spaces. So, S_{QM} and V_{QM} are the spaces of density operators and Hermitian operators in the tensor product. Then, S_{PCSFT} equals $\mathcal{H}$ and V_{PCSFT} are quadratic forms on this space; the space P_{PCSFT} consists of probability measures defined on the tensor product and having zero mean value and finite second momentum. Here, for quadratic form $f_A(\Phi) = \langle\Phi|\hat{A}|\Phi\rangle$,

$$\langle f_A\rangle_p = \int_{\mathcal{H}_1\otimes\mathcal{H}_2} f_A(\Phi)dp(\Phi).$$

This presentation can be given in terms of tensor product–valued random variables, or random fields.

CHSH Inequality and Its Violation for Classical Random Fields

Turn now to the CHSH inequality. In the QM formalism, there are considered Hermitian operators $(\hat{A}_1, \hat{A}_2)$ acting in $\mathcal{H}_1$ and $(\hat{B}_1, \hat{B}_2)$ acting in $\mathcal{H}_2$ such that

$$[\hat{A}_i, \hat{B}_j] = 0, \ [\hat{A}_1, \hat{A}_2] \neq 0, \ [\hat{B}_1, \hat{B}_2] \neq 0.$$

In quantum terms, the CHSH correlation function has the form

$$C = C(\psi) = \langle\psi|\hat{A}_1 \otimes \hat{B}_1|\psi\rangle + \langle\psi|\hat{A}_1 \otimes \hat{B}_2|\psi\rangle \tag{16.24}$$
$$+\langle\psi|\hat{A}_2 \otimes \hat{B}_1|\psi\rangle - \langle\psi|\hat{A}_2 \otimes \hat{B}_2|\psi\rangle,$$

where $\psi \in \mathcal{H}$ is a pure quantum state. The main point of the CHSH no-go statement for hidden variables is that for some states,

$$|C(\psi)| > 2, \tag{16.25}$$

but, for the models with hidden variables à la Bell, the CHSH inequality holds:

$$|C| \leq 2. \tag{16.26}$$

Now we write the CHSH correlation function in PCSFT representation. Let ψ be a pure quantum state and let Ψ be a vector such that $\psi = \Psi/||\Psi||$. Denote by

p_Ψ one of probability measures with zero mean value and the complex covariance operator $\hat{B}_\Psi = |\Psi\rangle\langle\Psi|$. For generality, we consider four such measures which are used in the measure-theoretic representation of four correlations combining the CHSH correlation function, namely, $p_{11;\Psi}, p_{12;\Psi}, p_{21;\Psi}, p_{22;\Psi}$. Then we have

$$\begin{aligned} C &= C(p_{ij;\Psi}) \qquad (16.27)\\ &= \frac{1}{||\Psi||^2}\Big[\int_{\mathcal{H}} \langle\Phi|\hat{A}_1 \otimes B_1|\Phi\rangle dp_{11;\Psi}(\Phi) + \int_{\mathcal{H}} \langle\Phi|\hat{A}_1 \otimes B_2|\Phi\rangle dp_{12;\Psi}(\Phi) \\ &+ \int_{\mathcal{H}} \langle\Phi|\hat{A}_2 \otimes B_1|\Phi\rangle dp_{21;\Psi}(\Phi) - \int_{\mathcal{H}} \langle\Phi|\hat{A}_2 \otimes B_2|\Phi\rangle dp_{22;\Psi}(\Phi)\Big], \end{aligned}$$

where $\mathcal{H} = \mathcal{H}_1 \otimes \mathcal{H}_2$.

There is no reason to expect that correlation function $C(p_{ij;\psi})$ should satisfy the CHSH inequality. The Bell-like proofs are blocked by two factors:

- *Range of values*: quadratic forms normed by the dispersion $\sigma^2 = ||\Psi||^2$ are not valued in the segment $[-1,+1]$;
- *Contextuality*: for a quantum state ψ, subquantum probability measure is not uniquely determined by the covariance operator $\hat{B}_\psi$.

The *Contextuality* barrier for derivation of the CHSH and other Bell-type inequalities is closely related to the hypothesis of ensemble nonreproducibility that was discussed by de Muynck, De Baere, and Martens (see, e.g., article [104] and the corresponding references herein) and by Khrennikov (see, e.g., [181]). The main distinguishing feature of the Bell–Bohm experiment is that the correlation function is calculated from statistical data collected in four different experimental runs. Ensembles of systems prepared for these runs might have different statistical features. The only statistical constraint checked in this experiment is reproduction of correlations. This ensemble variability blocks both the frequency and measure theoretical derivations of the Bell inequalities. In the frequency derivation (à la Stapp [345]) one identifies the values for outcomes of an observable measured in different runs.[3] The measure theoretic derivation was suggested by Bell, and in the CHSH framework it is based on the assumption that one can use the same probability measure to express all pairwise correlations. The PCSFT (and generally Bild-conception) viewpoint on the hidden variable theories makes clear that this Bell assumption is not so natural.

In fact, PCSFT contextuality is even more complex, since measures used for correlations can correspond to different vectors Ψ_{ij} such that $\psi = \Psi_{ij}/||\Psi_{ij}||$. We

[3] See article [179] for critical analysis of such derivation. We also remark it was criticized in the series of papers of Hess and Philipp; see, for example, [162].

remark that $\sigma = ||\Psi||$ is the standard deviation of the corresponding subquantum random field. So, in the preceding considerations we assumed that in four experiments the standard deviations are the same.

If the space of probability measures is reduced to circularly symmetric Gaussian measures, then the map J_S becomes one-to-one up to the scalar factor, and the *Contextuality* (ensemble nonreproducibility) counterargument can be omitted, under the assumption that the standard deviation $\sigma = ||\Psi||$ is fixed. In this case the same measure p_Ψ can be considered for the measure-theoretic description of all experiments determining counterparts of the CHSH correlation function:

$$
\begin{aligned}
C &= C(p_\Psi) \qquad (16.28)\\
&= \frac{1}{\operatorname{Tr}\hat{B}_\psi} \int_{\mathcal{H}} [\langle\Phi|\hat{A}_1 \otimes B_1|\Phi\rangle + \langle\Phi|\hat{A}_1 \otimes B_2|\Phi\rangle \\
&\quad + \langle\Phi|\hat{A}_2 \otimes B_1|\Phi\rangle - \langle\Phi|\hat{A}_2 \otimes B_2|\Phi\rangle] dp_\Psi(\Phi),
\end{aligned}
$$

where $\mathcal{H} = \mathcal{H}_1 \otimes \mathcal{H}_2$.

However, we do not know whether subquantum random fields are circularly symmetric Gaussian, although monograph [212] presents the argument based on the central limit theorem in favor of Gaussianity. Nor do we know if subquantum random fields generated in different experimental runs are characterized by the same standard deviation.

The preceding presentation can be easily modified to density operators and their PCSFT preimages.

16.4 Compound Systems: Cartesian Product–Valued Random Fields

Consider a compound system $S = (S_1, S_2)$, where systems $S_i, i = 1, 2$, have the state spaces $\mathcal{H}_i$ which are complex Hilbert spaces. Within CSM the state space of S is

$$\mathbf{H} = \mathcal{H}_1 \times \mathcal{H}_2.$$

By PCSFT an ensemble of compound systems is described by a probability distribution on this Cartesian product or by a vector random field:

$$\phi(\omega) = (\phi_1(\omega), \phi_2(\omega)) \in \mathbf{H}.$$

In our approach each subsystem is described by its own random field: S_i by $\phi_i(\omega), i = 1, 2$. These fields can be *correlated* in the sense of classical probability theory. The initial correlation propagates matching the laws of QM (Section 16.4.3). This is classical dynamics of two stochastic processes which were correlated at the beginning.

16.4.1 Coupling of Subquantum and Quantum Correlations

Now we turn to the QM-model. Take an arbitrary vector Ψ belonging to the tensor product

$$\mathcal{H} = \mathcal{H}_1 \otimes \mathcal{H}_2.$$

To show PCSFT-QM connection, Ψ should be interpreted not as a vector belonging to a complex Hilbert space, but as a linear operator. This linear operator is the off-diagonal block of the covariance operator of the corresponding prequantum random field $\phi(\omega)$ valued in Cartesian (not tensor) product $\mathbf{H} = \mathcal{H}_1 \times \mathcal{H}_2$. Thus, the states of a single quantum system S and a compound quantum system $S = (S_1, S_2)$ have different PCSFT-preimages: the covariance operator versus the off-diagonal block of the covariance operator. The PCSFT-QM correspondence map (15.14), chapter 15.4, should be modified).

First we restrict the mathematical construction to the case of the L_2-space, that is, $\mathcal{H}_i = L_2(\mathbb{R}^{n_i}; \mathbb{C}), i = 1, 2$, and $\mathcal{H} == L_2(\mathbb{R}^{n_1+n_2}; \mathbb{C})$. Any function $\Psi(x, y) \in L_2(\mathbb{R}^{n_1+n_2}; \mathbb{C})$ determines the integral operator

$$\hat{\Psi}\phi(x) = \int \Psi(x, y)\phi(y)dy, \tag{16.29}$$

acting from $L_2(\mathbb{R}^{n_2}; \mathbb{C}) \to L_2(\mathbb{R}^{n_1}; \mathbb{C})$. This operator is of the Hilbert–Schmidt type; moreover, each operator of the Hilbert–Schmidt type in the L_2-space can be represented as such an integral operator.

Now turn to the abstract Hilbert scheme. Take an arbitrary vector $\Psi \in \mathcal{H}$ and define the operator $\hat{\Psi}$ with the aid of its bilinear form:

$$\langle \phi_2 | \hat{\Psi} | \phi_1 \rangle = \langle \Psi | \phi_1 \otimes \bar{\phi_2} \rangle, \phi_k \in \mathcal{H}_k, k = 1, 2,$$

and, for $\phi = q + ip$, its complex conjugate is given by $\bar{\phi} = q - ip$. We remark that a complex Hilbert space, say W, can be represented as the direct sum of two real Hilbert spaces.

For simplicity of mathematical presentation, we consider the special class of observables given by *real operators*. An operator $\hat{A}$ is called real, if the following equality holds:

$$\langle u | \hat{A} | v \rangle = \overline{\langle \bar{u} | \hat{A} | \bar{v} \rangle}. \tag{16.30}$$

Hermitian operator $\hat{A}$ is real if its matrix elements are real in any real basis of the complex Hilbert space. Let $\{e_j\}$ be an orthonormal basis which is real, that is, $\bar{e}_j = e_j$. Then, for matrix elements, equality (16.30) implies

$$a_{jk} = \langle e_j | \hat{A} | e_k \rangle = \overline{\langle e_j | \hat{A} | e_k \rangle} = \overline{a_{jk}}.$$

Any Hamiltonian (acting in L_2-space) of the form

$$\hat{H} = \hat{p}^2/2m + V(q),$$

where $\hat{p}$ is the momentum operator and V is a real-valued potential, is real; the momentum operator $\hat{p}$ is not real. In general, one can proceed without this rather restrictive condition that observables are represented by real operators by deriving slight modifications of coming equalities [212]. But for the sake of mathematical simplicity, we assume that in this section the operators $\hat{A}_j, j = 1, 2$, are *real.*

Theorem 16.1 *[212] Let $\hat{A}_j \colon \mathcal{H}_j \to \mathcal{H}_j, j = 1, 2$, be Hermitian real operators and let Ψ be an arbitrary vector belonging to the tensor product $\mathcal{H} = \mathcal{H}_1 \otimes \mathcal{H}_2$. Then the following equalities hold:*

$$\langle \Psi | \hat{A}_1 \otimes I | \Psi \rangle = \mathrm{Tr}\, \hat{\Psi} \hat{\Psi}^\star \hat{A}_1, \tag{16.31}$$

$$\langle \Psi | I \otimes \hat{A}_2 | \Psi \rangle = \mathrm{Tr}\, \hat{\Psi}^\star \hat{\Psi} \hat{A}_2, \tag{16.32}$$

and

$$\langle \Psi | \hat{A}_1 \otimes \hat{A}_2 | \Psi \rangle = \mathrm{Tr}\, \hat{\Psi} \hat{A}_2 \hat{\Psi}^\star \hat{A}_1. \tag{16.33}$$

This mathematical theorem [212] can be proven by expanding the vector $\Psi \in \mathcal{H}$ w.r.t. the orthonormal basis consisting of the tensor products of the eigenvectors of the operators $\hat{A}_j, j = 1, 2$.

We remark that these operator equalities are not directly coupled to PCSFT: just operators are used, instead of vectors from the tensor product. This operator representation will be used for mathematical reformulation of the quantum formalism for compound systems. For a quantum state $\psi \in \mathcal{H}$, that is, $||\psi|| = 1$, the averages and correlations can be represented as

$$\langle \hat{A}_1 \rangle_\psi \equiv \langle \psi | \hat{A}_1 \otimes I | \psi \rangle = \mathrm{Tr}\, \hat{\psi} \hat{\psi}^\star \hat{A}_1, \tag{16.34}$$

$$\langle \hat{A}_2 \rangle_\psi \equiv \langle \psi | I \otimes \hat{A}_2 | \psi \rangle = \mathrm{Tr}\, \hat{\psi}^\star \hat{\psi} \hat{A}_2, \tag{16.35}$$

$$\langle \hat{A}_1 \otimes \hat{A}_2 \rangle_\psi \equiv \langle \psi | \hat{A}_1 \otimes \hat{A}_2 | \psi \rangle = \mathrm{Tr}\, \hat{\psi} \hat{A}_2 \hat{\psi}^\star \hat{A}_1. \tag{16.36}$$

In PCSFT a compound system $S = (S_1, S_2)$ is mathematically represented by a random field $\phi(\omega) = (\phi_1(\omega), \phi_2(\omega))$ valued in the Cartesian product $\mathbf{H}$. Field's covariance operator $\hat{D}$ has the block structure

$$\hat{D} = \begin{pmatrix} \hat{D}_{11} & \hat{D}_{12} \\ \hat{D}_{21} & \hat{D}_{22} \end{pmatrix}, \tag{16.37}$$

where $\hat{D}_{ii} \colon \mathcal{H}_i \to \mathcal{H}_i, \hat{D}_{ij} \colon \mathcal{H}_j \to \mathcal{H}_i$. The covariance operator is Hermitian. Hence $\hat{D}_{ii}^\star = \hat{D}_{ii}$, and $\hat{D}_{12}^\star = \hat{D}_{21}$. The diagonal blocks $\hat{D}_{ii}$ describe internal correlations in subsystems. The block $\hat{D}_{12}$ describes cross-correlations between subsystems. Here, by definition,

$$\langle u_j|\hat{D}_{ij}|v_i\rangle = E\langle u_j|\phi_j(\omega)\rangle\langle\phi_i(\omega)|v_i\rangle, u_i \in \mathcal{H}_i, v_j \in \mathcal{H}_j.$$

Theorem 16.2 *[212] Let $\phi(\omega) = (\phi_1(\omega), \phi_2(\omega))$ be a random vector with the components belonging to R_{PCSFT} and let $f_{A_i}(\phi_i)$ be the quadratic forms with Hermitian real operators, $\hat{A}_i\colon \mathcal{H}_i \to \mathcal{H}_i, i = 1, 2$. Then the following equalities hold:*

$$E[f_{A_i}(\phi_i(\omega))] = Tr\hat{D}_{ii}\hat{A}_i, i = 1, 2, \tag{16.38}$$

$$\langle f_{A_1} f_{A_2}\rangle \equiv Ef_{A_1}(\phi_1(\omega))f_{A_2}(\phi_2(\omega)) = (Tr\hat{D}_{11}\hat{A}_1)(Tr\hat{D}_{22}\hat{A}_2) + Tr\hat{D}_{12}\hat{A}_2\hat{D}_{21}\hat{A}_1. \tag{16.39}$$

This mathematical theorem [212] can be proven by expanding the random fields $\phi_i(\omega), i = 1, 2$, w.r.t. the orthonormal bases consisting of the eigenvectors of the operators $\hat{A}_i$.

By combining equalities (16.38) and (16.39), we obtain the following basic equality:

$$\langle f_{A_1}(\phi_1(\omega))f_{A_2}(\phi_2(\omega))\rangle = E[f_{A_1}(\phi_1(\omega))]E[f_{A_2}(\phi_3(\omega))] + \mathrm{Tr}\hat{D}_{12}\hat{A}_2\hat{D}_{21}\hat{A}_1. \tag{16.40}$$

Select now a random field such that $\hat{D}_{12} = ||\Psi||\hat{\Psi}$, where Ψ is an arbitrary vector belonging to $\mathcal{H}$. The role of the scaling factor $||\Psi||$ will be discussed later. Then combining (16.40) with the operator wave-function representation of quantum correlations, we obtain the following equality:

$$\frac{1}{||\Psi||^4}E[(f_{A_1} - Ef_{A_1})(f_{A_2} - Ef_{A_2})] = \langle\psi|\hat{A}_1 \otimes \hat{A}_2|\psi\rangle \equiv \langle\hat{A}_1 \otimes \hat{A}_2\rangle_\psi, \tag{16.41}$$

where $\psi = \Psi/||\Psi||$. The LHS and RHS quantities belong to the domains of PCSFT and QM, respectively.

Hence, for covariance of two classical random variables f_{A_1}, f_{A_2}, we have

$$\frac{1}{||\Psi||^4}\mathrm{cov}\,(f_{A_1}, f_{A_2}) = \langle\hat{A}_1 \otimes \hat{A}_2\rangle_\psi. \tag{16.42}$$

We remark that this equality establishing the PCSFT-QM correspondence suffers from asymmetry: the covariance in LHS and correlation in RHS. In Section 16.4.2 we show that it can be rewritten symmetrically with covariances in both sides. However, for the moment we proceed with (16.42).

We want to construct a random field such that the PCSFT averages and correlations match those given by QM. We have done this for correlations; see (16.42). Now we want to have a similar correspondence for averages of f_{A_i}. They are determined by the diagonal blocks of the covariance operator. Let us select $\hat{D}_{11} = \hat{\Psi}\hat{\Psi}^\star$ and $\hat{D}_{22} = \hat{\Psi}^\star\hat{\Psi}$. Then

$$\frac{1}{||\Psi||^2}E[f_{A_1}(\phi_1(\omega))] = \langle \hat{A}_1\rangle_\psi = \langle\psi|\hat{A}_1 \otimes I|\psi\rangle = \mathrm{Tr}(\hat{\psi}\hat{\psi}^\star)\hat{A}_1; \tag{16.43}$$

$$\frac{1}{||\Psi||^2}E[f_{A_2}(\phi_2(\omega))] = \langle \hat{A}_2\rangle_\psi = \langle\psi|I \otimes \hat{A}_2|\psi\rangle = \mathrm{Tr}(\hat{\psi}^\star\hat{\psi})\hat{A}_2, \tag{16.44}$$

where I denotes the unit operator.

By associating $||\Psi||$ with the mean amplitude of a subquantum field we can interpret the factors $\frac{1}{||\Psi||^4}$ and $\frac{1}{||\Psi||^4}$ in equalities (16.42) and (16.43), (16.44) as amplification factors for subquantum quantities,

$$f_A \to \frac{f_A}{||\Psi||^2}.$$

So, QM can be interpreted as the mathematical formalism describing amplifications of subquantum quantities.

Thus, it would be natural to use the following operator as the covariance operator of a subquantum random field represented in QM (the observational model) by the state $\psi = \Psi/||\Psi||$,

$$\hat{D}_\Psi = \begin{pmatrix} \hat{\Psi}\hat{\Psi}^\star & ||\Psi||\hat{\Psi} \\ ||\Psi||\hat{\Psi}^\star & \hat{\Psi}^\star\Psi \end{pmatrix}. \tag{16.45}$$

The scalar normalization factor $||\Psi||$ in the off-diagonal term is needed to assign the same "field dimension" to all blocks of the operator $\hat{D}_\Psi$. The presence of this scaling factor leads to the following natural interrelation between normalizations of the vector Ψ and the corresponding covariance operator $\hat{D}_\Psi$,

$$\frac{1}{||\Psi||^2}\hat{D}_\Psi = \hat{D}_{\Psi/||\Psi||}, \tag{16.46}$$

or in terms of projection operator $\hat{\rho}_\Psi = |\Psi\rangle\langle\Psi|$ and the corresponding labeling of the covariance operator, that is, $\hat{D}_\Psi \equiv \hat{D}_{\hat{\rho}_\Psi}$, we have

$$\frac{1}{\mathrm{Tr}\,\hat{\rho}_\Psi}\hat{D}_{\hat{\rho}_\Psi} = \hat{D}_{\hat{\rho}_\Psi/\mathrm{Tr}\,\hat{\rho}_\Psi}. \tag{16.47}$$

However, generally *the operator $\hat{D}_\Psi$ is not positively defined!* It does not determine any probability distribution on the space of classical fields. To obtain a positively defined operator, we modify $\hat{D}_\Psi$ to include the noise component. We show that the operator

$$\hat{D}_{\Psi;\epsilon} = \begin{pmatrix} \hat{\Psi}\hat{\Psi}^\star + \epsilon I & ||\Psi||\hat{\Psi} \\ ||\Psi||\hat{\Psi}^\star & \hat{\Psi}^\star\hat{\Psi} + \epsilon I \end{pmatrix} \tag{16.48}$$

is positively semidefinite if $\epsilon \geq 0$ is large enough, with the low bound depending on vector Ψ, or in QM terms, on the state $\psi = \Psi/||\Psi||$.

Heuristically the additional term in the diagonal blocks of the covariance operator might be interpreted as a subquantum noise with dispersion ϵ. One may (but need not) interpret this noise as associated with the preparation of the quantum state ψ. However, this is a subquantum noise having no impact on quantum measurements. For Gaussian random fields, this is the white noise. So, subquantum random fields are noisy.

Theorem 16.3 *For any vector $\Psi \in H_1 \otimes H_2$, there exists the low bound $\epsilon_\Psi \geq 0$, such that for any $\epsilon \geq \epsilon_\Psi$, the operator $\hat{D}_{\Psi;\epsilon}$ is positively semidefinite. A rough estimate for this low bound is given by the inequality*

$$\epsilon_\Psi \leq ||\Psi||^2. \tag{16.49}$$

Proof The space $\mathbf{H} = H_1 \times H_2$ is endowed with the scalar product

$$\langle u|v\rangle = \langle u_1|v_1\rangle + \langle u_2|v_2\rangle,$$

for vectors $u = (u_1, u_2), v = (v_1, v_2)$. For any vector $\phi = (\phi_1, \phi_2) \in \mathbf{H}$, we have

$$\langle \hat{D}_{\Psi;\epsilon}\phi|\phi\rangle = \epsilon(||\phi_1||^2 + ||\phi_2||^2) + (||\hat{\Psi}^\star\phi_1||^2 + ||\hat{\Psi}\phi_2||^2) + ||\Psi||\langle\hat{\Psi}\phi_2|\phi_1\rangle + ||\Psi||\langle\hat{\Psi}^\star\phi_1|\phi_2\rangle.$$

Consider the operator

$$\hat{\Gamma}_\psi = \begin{pmatrix} 0 & \hat{\Psi} \\ \hat{\Psi}^\star & 0 \end{pmatrix} \tag{16.50}$$

mapping $\mathbf{H} \to \mathbf{H}$, namely,

$$\hat{\Gamma}_\psi \begin{pmatrix} \phi_1 \\ \phi_2 \end{pmatrix} = \begin{pmatrix} \hat{\Psi}\phi_2 \\ \hat{\Psi}^\star\phi_1 \end{pmatrix}. \tag{16.51}$$

We have

$$||\hat{\Gamma}_\psi\phi||^2 = ||\hat{\Psi}^\star\phi_1||^2 + ||\hat{\Psi}\phi_2||^2 \quad .$$

and

$$\langle\hat{\Gamma}_\psi\phi|\phi\rangle = \langle\hat{\Psi}\phi_2|\phi_1\rangle + \langle\hat{\Psi}^\star\phi_1|\phi_2\rangle.$$

Thus,

$$\langle \hat{D}_{\Psi;\epsilon}\phi|\phi\rangle = \epsilon||\phi||^2 + ||\hat{\Gamma}_\psi\phi||^2 + ||\Psi||\langle\hat{\Gamma}_\psi\phi|\phi\rangle. \tag{16.52}$$

Hence, the inequality $\langle \hat{D}_{\Psi;\epsilon}\phi|\phi\rangle \geq 0$ can be written as

$$\epsilon \geq \sup_{\phi\neq 0} \frac{1}{||\phi||^2}[-||\hat{\Gamma}_\psi\phi||^2 - ||\Psi||\langle\hat{\Gamma}_\psi\phi|\phi\rangle]. \tag{16.53}$$

But the quantity in RHS is bounded, since the operator $\hat{\Gamma}_\psi$ is bounded, that is,

$$\epsilon_\psi = \sup_{\phi\neq 0} \frac{1}{||\phi||^2}[-||\hat{\Gamma}_\psi\phi||^2 - ||\Psi||\langle\hat{\Gamma}_\psi\phi|\phi\rangle] < +\infty. \tag{16.54}$$

So, if

$$\epsilon \geq \epsilon_\psi, \tag{16.55}$$

then the covariance operator $\hat{D}_{\Psi;\epsilon}$ is positively semidefinite.

To get a simple sufficient condition (in fact, rather rough one), we estimate ϵ_ψ from above. We remark that

$$|\langle\hat{\Gamma}_\psi\phi|\phi\rangle| \leq ||\hat{\Gamma}_\psi\phi||||\phi||.$$

Hence,

$$\epsilon_\psi \leq ||\Psi|| \sup_{\phi\neq 0} \frac{1}{||\phi||^2} ||\hat{\Gamma}_\psi\phi||\ ||\phi|| = ||\Psi||\ ||\hat{\Gamma}_\psi||. \tag{16.56}$$

We now estimate the norm of the operator $\hat{\Gamma}_\psi$. We have

$$||\hat{\Gamma}_\psi||^2 = \sup_{\phi\neq 0}[||\hat{\Psi}^\star\phi_1||^2 + ||\hat{\Psi}\phi_2||^2]/||\phi||^2$$
$$= \sup_{\phi\neq 0}\left[\frac{||\hat{\Psi}^\star\phi_1||^2}{||\phi_1||^2}\frac{||\phi_1||^2}{||\phi_1||^2+||\phi_2||^2} + \frac{||\hat{\Psi}\phi_2||^2}{||\phi_1||^2}\frac{||\phi_2||^2}{||\phi_1||^2+||\phi_2||^2}\right] \leq ||\hat{\Psi}||^2.$$

We recall that, for any Hilbert–Schmidt operator $\hat{K}$ (in particular, any operator in finite-dimensional space), the following inequality between the operator norm and Hilbert–Schmidt norm holds: $||K|| \leq ||K||_{\mathrm{HS}} = \sqrt{\mathrm{Tr}\,\hat{K}^\star\hat{K}}$. We also recall that $||\hat{\Psi}||_{\mathrm{HS}} = ||\Psi||$, where $||\Psi|| = ||\Psi||_{\mathcal{H}}$. Finally, we obtain the following inequality:

$$\epsilon_\psi \leq ||\Psi||^2, \tag{16.57}$$

which implies the sufficient condition for $D_{\Psi;\epsilon} \geq 0$,

$$\epsilon \geq ||\Psi\ |^2. \tag{16.58}$$

In particular, the operator

$$\hat{D}_\Psi = \begin{pmatrix} (\hat{\Psi}\hat{\Psi}^\star + ||\Psi||^2 I) & ||\Psi||\hat{\Psi} \\ ||\Psi||\hat{\Psi}^\star & (\hat{\Psi}^\star\hat{\Psi} + ||\Psi||^2 I) \end{pmatrix} \tag{16.59}$$

is positively semidefinite and hence corresponds to a subquantum random field.

In the previous works on PCSFT (see, e.g., [212]) we did not use the scaling factor $||\Psi||$ for the off-diagonal blocks, that is, instead of (16.45), we started with the operator

$$\hat{D}_\Psi = \begin{pmatrix} \hat{\Psi}\hat{\Psi}^\star & \hat{\Psi} \\ \hat{\Psi}^\star & \hat{\Psi}^\star\hat{\Psi} \end{pmatrix}. \tag{16.60}$$

In this operator the diagonal and off-diagonal blocks have different "Ψ-dimension" and scaling-normalization laws (16.46) or (16.47) do not hold. We think that the present model with the scaling factor for off-diagonal blocks is proper for subquantum-quantum coupling. In particular, starting with operator (16.60) on obtains the state-independent condition of positive semi-definiteness for operator

$$\hat{D}_{\Psi;\epsilon} = \begin{pmatrix} \hat{\Psi}\hat{\Psi}^\star + \epsilon I & \hat{\Psi} \\ \hat{\Psi}^\star & \hat{\Psi}^\star\hat{\Psi} + \epsilon I \end{pmatrix}, \tag{16.61}$$

it is positively semidefinite if $\epsilon \geq 1/4$. This approach [212] led to consideration of state-independent noise background field, a kind of zero-point field. Although the author is not totally against consideration of such a field, it is more natural to proceed without it. So, in the previous works on PCSFT, we used the background field interpretation of the subquantum noise.

Proposition 16.4 *Let vector $\Psi \in H_1 \otimes H_2$ be factorizable, that is, $\Psi = \psi_1 \otimes \psi_2$, where $\psi_i \in \mathcal{H}_i$ (in the Dirac notation $|\Psi\rangle = |\psi_1\psi_2\rangle$). Then the operator*

$$\hat{D}_{\Psi;0} = \begin{pmatrix} \hat{\Psi}\hat{\Psi}^\star & ||\Psi||\hat{\Psi} \\ ||\Psi||\hat{\Psi}^\star & \hat{\Psi}^\star\hat{\Psi} \end{pmatrix} \tag{16.62}$$

is positively semidefinite.

Proof Here we have

$$\begin{aligned} \hat{\Psi}^\star\phi_1 &= \langle\psi_1|\phi_1\rangle\psi_2,\ \hat{\Psi}^\star\phi_2 = \langle\psi_2|\phi_2\rangle\psi_1, \\ \langle\hat{\Psi}^\star\phi_1|\phi_2\rangle &= \overline{\langle\psi_1|\phi_1\rangle}\langle\psi_2|\phi_2\rangle,\ \langle\hat{\Psi}\phi_2|\phi_1\rangle = \overline{\langle\psi_2|\phi_2\rangle}\langle\psi_1|\phi_1\rangle, \\ \langle\hat{\Psi}\hat{\Psi}^\star\phi_1|\phi_1\rangle &= |\langle\psi_1|\phi_1\rangle|^2||\psi_2||^2,\ \langle\hat{\Psi}^\star\hat{\Psi}\phi_2|\phi_2\rangle = |\langle\psi_2|\phi_2\rangle|^2||\psi_1||^2. \end{aligned}$$

Hence,

$$\begin{aligned} \langle\hat{D}_\Psi\phi|\phi\rangle &= |\langle\psi_1|\phi_1\rangle|^2||\psi_2||^2 + 2||\Psi||\ |\langle\psi_2|\phi_2\rangle\langle\psi_1|\phi_1\rangle|\cos\theta + |\langle\psi_2|\phi_1\rangle|^2||\psi_1||^2 \\ &\geq |\langle\psi_1|\phi_1\rangle|^2||\psi_2||^2 - 2|\langle\psi_2|\phi_2\rangle||\langle\psi_1|\phi_1\rangle|||\psi_1||||\psi_2|| + |\langle\psi_2|\phi_1\rangle|^2||\psi_1||^2 \\ &= (|\langle\psi_1|\phi_1\rangle|||\psi_2|| - |\langle\psi_2|\phi_2\rangle|||\psi_1||)^2 \geq 0 \end{aligned}$$

(here $\theta = \theta(\psi_1, \psi_2, \phi_1, \phi_2)$ is some phase).

Operator $\hat{D}_{\Psi;\epsilon}, \epsilon \geq \epsilon_\Psi$, determines a cluster of measures belonging to space $P_{\rm PCSFT}$ or a cluster of random fields belonging to space $R_{\rm PCSFT}$. This operator determines uniquely the circularly symmetric Gaussian measure or the circularly symmetric Gaussian random field.

In the L_2-case the covariance operator is an integral operator with the kernel

$$\hat{D}_{\Psi;\epsilon}(x,y) = \begin{pmatrix} \hat{D}_{11}(x,y) & \hat{D}_{12}(x,y) \\ \hat{D}_{21}(x,y) & \hat{D}_{22}(x,y) \end{pmatrix}, \tag{16.63}$$

where

$$\hat{D}_{12}(x,y) = \Psi(x,y), \tag{16.64}$$

$$\hat{D}_{11}(x,y) = \int dz \Psi(x,z)\bar{\Psi}(y,z)dz + \epsilon\delta(x-y), \tag{16.65}$$

$$\hat{D}_{22}(x,y) = \int dz \bar{\Psi}(x,z)\Psi(y,z) + \epsilon\delta(x-y). \tag{16.66}$$

Renormalization

Suppose now that $\phi(\omega)$ is a random vector with the covariance operator $\hat{D}_{\Psi;\epsilon}$ given by (16.59), that is, $\epsilon = ||\Psi||^2$. This choice is natural from the "Ψ-dimension" viewpoint – for dimensional homogeneity of all blocks. Then

$$E[f_{A_1}] = \mathrm{Tr}\,(\hat{\Psi}\hat{\Psi}^\star + ||\Psi||^2 I)\hat{A}_1 = \langle\Psi|\hat{A}_1 \otimes I|\Psi\rangle + ||\Psi||^2 \mathrm{Tr}\hat{A}_1, \tag{16.67}$$

$$E[f_{A_2}] = \mathrm{Tr}\,(\hat{\Psi}^\star\hat{\Psi} + ||\Psi||^2 I))\hat{A}_2 = \langle\Psi|I \otimes \hat{A}_2|\Psi\rangle + ||\Psi||^2 \mathrm{Tr}\hat{A}_2. \tag{16.68}$$

Thus, for $i = 1, 2$,

$$\langle\hat{A}_i\rangle_\psi = \frac{1}{||\Psi||^2}(E[f_{A_i}] - ||\Psi||^2 \mathrm{Tr}\,\hat{A}_i), \tag{16.69}$$

or

$$\langle\hat{A}_i\rangle_\psi = \frac{E[f_{A_i}]}{||\Psi||^2} - \mathrm{Tr}\,\hat{A}_i. \tag{16.70}$$

These relations for averages and relation (16.42) for the correlation provide coupling between PCSFT and QM. Quantum statistical quantities can be obtained from corresponding quantities for a classical random field. One may say that *"irreducible quantum randomness" is reduced to randomness of classical prequantum fields..*

However, this reduction is nontrivial and it is based on additive renormalization proportional to $\mathrm{Tr}\hat{A}$, the trace of an observable, and amplification of subquantum quantities by the factor $\frac{1}{||\Psi||^2}$. It is natural to consider $||\Psi||^2$ the amplitude of a subquantum field and assume that

$$||\Psi|| << 1. \tag{16.71}$$

From the physical viewpoint, the additive renormalization is not straightforward. For real physical systems, the state space is infinite-dimensional and the trace of an observable is generally infinite. Thus, coupling between the causal theoretical model (CTM) given by PCSFT and the observational model given by QM (see Chapter 15) is based on infinite renormalization; this situation reminds the quantum field theory. We remark that, for correlations between observables on subsystems ("local observables"), such renormalization is not needed; subsystems' renormalizations somehow compensate each other.

This formal mathematical renormalization procedure can be heuristically represented as follows. We remark that $\mathrm{Tr}\hat{A} = \mathrm{Tr}I\hat{A} = E[f_A(\eta)]$, where η is a random field with zero average and the covariance operator I. Assume that this field is Gaussian, then this is white noise. We can interpret the additive renormalization as subtraction of the average w.r.t. white noise from the amplified subquantum average. We can say that the A-measurement apparatus filters this white noise contribution.

This coupling between PCSFT and QM highlights the role of the operator trace. For quantum observables with zero trace, we have:

$$\langle \hat{A}_i \rangle_\psi = \frac{E[f_{A_i}]}{||\Psi||^2}. \tag{16.72}$$

In this case, quantum average is amplification of the classical one with the factor $\frac{1}{||\Psi||^2}$. However, the physical meaning of the condition $\mathrm{Tr}\,\hat{A} = 0$ is not clear.

PCSFT-QM Correspondence

Coming back to establishing correspondence between PCSFT and QM (Chapter 15), we describe the PCSFT-QM correspondence map for compound systems with the Cartesian product representation of the subquantum state space:

1. random fields valued in space $\mathcal{H}_1 \times \mathcal{H}_2$ and having covariance operator (16.48) are mapped into the pure state $\psi \in \mathcal{H}_1 \otimes \mathcal{H}_2, \psi = \Psi/||\Psi||$,
2. a pair of quadratic forms $(f_{A_1}(\phi_1), f_{A_2}(\phi_2))$, where $\hat{A}_i$ are Hermitian operators in $\mathcal{H}_i, i = 1, 2$, is mapped into a quantum observable with Hermitian operator $\hat{A}_1 \otimes \hat{A}_2$.
3. subquantum and quantum averages are coupled by equalities (16.42), (16.70).

By restricting the class of subquantum fields to circularly symmetric Gaussian random fields, we make the first map one-to-one.

The previous consideration was initialized with the vector from the tensor product $\mathcal{H} = \mathcal{H}_1 \otimes \mathcal{H}_2$ only to make direct coupling to the QM formalism. However, we can start with the special class of subquantum random fields having the covariance operators of the form

$$\hat{D} = \begin{pmatrix} \hat{D}_{11} & \hat{D}_{12} \\ \hat{D}_{21} & \hat{D}_{22} \end{pmatrix}, \tag{16.73}$$

where

$$\hat{D}_{11} = \hat{D}_{12}\hat{D}_{12}^{\star} + \epsilon I,\ \hat{D}_{22} = \hat{D}_{12}^{\star}\hat{D}_{12} + \epsilon I.$$

We take the off-diagonal block $\hat{D}_{12} \in \mathcal{L}(\mathcal{H}_2, \mathcal{H}_1)$ (representing correlations between the field's components). Then this operator is realized as a vector belonging to the tensor product $\mathcal{H}$ and in this way we come to the standard QM representation of the state of a compound quantum system as a normalized tensor product vector ψ corresponding to the operator

$$\hat{\psi} = \hat{D}_{12}/\sqrt{\mathrm{Tr}\,\hat{D}_{12}\hat{D}_{12}^{\star}}.$$

The normalization has the clear statistical meaning of normalization of a random field by its standard deviation.

CHSH Inequality and Its Violation for Classical Random Fields

We remark that if $\mathrm{Tr}\,\hat{A}_i = 0$, as say for Pauli matrices, then equality (16.70) is reduced to equality (16.72).

Now consider once again the CHSH framework with operators such that, for $i = 1, 2$,

$$\mathrm{Tr}\,\hat{A}_i = 0,\ \mathrm{Tr}\,\hat{B}_i = 0$$

and a quantum state $\psi \in \mathcal{H}$ such that

$$\langle \hat{A}_i \rangle_{\psi} = \langle \psi | \hat{A}_i \otimes I | \psi \rangle = 0,\ \langle \hat{B}_i \rangle_{\psi} = \langle \psi | I \otimes \hat{B}_i | \psi \rangle = 0.$$

For example, $\hat{A}_i, \hat{B}_i$ are Pauli matrices and ψ is the singlet state. Then

$$E[f_{A_i}] = 0\ , E[f_{B_i}] = 0,$$

and the equality (16.42) is reduced to

$$\frac{1}{||\Psi||^4} E[f_{A_i} f_{B_j})] = \langle \hat{A}_i \otimes \hat{B}_j \rangle_{\psi}. \tag{16.74}$$

Let Ψ be a vector and $\psi = \Psi/||\Psi||$. Denote by p_{Ψ} one of the probability measures with zero mean value and the complex covariance operator $\hat{D}_{\Psi;\epsilon}$ given by (16.48). For generality, we consider four such measures which are used in the measure-theoretic representation of four correlations combining the CHSH correlation function, that is, $p_{11;\Psi}, p_{12;\Psi}, p_{21;\Psi}, p_{22;\Psi}$. In the PCSFT formalism with the Cartesian product representation of compound systems, the CHSH correlation function has the form:

$$C = C(p_{ij;\Psi}) \tag{16.75}$$
$$= \frac{1}{||\Psi||^4}\Big[\int_{\mathcal{H}_1\times\mathcal{H}_2}\langle\phi|\hat{A}_1\otimes\hat{B}_1|\phi\rangle dp_{11;\Psi}(\phi) + \int_{\mathcal{H}_1\times\mathcal{H}_2}\langle\phi|\hat{A}_1\otimes\hat{B}_2|\phi\rangle dp_{12;\Psi}(\phi)$$
$$+\int_{\mathcal{H}_1\times\mathcal{H}_2}\langle\phi|\hat{A}_2\otimes\hat{B}_2|\phi\rangle dp_{21;\Psi}(\phi) - \int_{\mathcal{H}_1\times\mathcal{H}_2}\langle\phi|\hat{A}_2\otimes\hat{B}_2|\phi\rangle dp_{22;\Psi}(\phi).$$

So, the CHSH correlation function has the classical probabilistic representation matching the quantum one, for the state $\psi = \Psi/||\Psi||$. Moreover, in contrast to the representation of Section 16.3, (16.27), this formula matches with the CSM description of a compound system by the Cartesian product of subsystems' state spaces. The same barriers, as in Section 16.3, block the derivation of the CHSH inequality, $|C| \leq 2$.

By relaxing the condition of contextuality via consideration of only circularly symmetric Gaussian measures, the CHSH correlation can be written as

$$C = C(p_\Psi) = \frac{1}{||\Psi||^4}\int_{\mathcal{H}_1\times\mathcal{H}_2}[\langle\phi|\hat{A}_1\otimes\hat{B}_1|\phi\rangle + \langle\phi|\hat{A}_1\otimes\hat{B}_2|\phi\rangle \tag{16.76}$$
$$+\langle\phi|\hat{A}_2\otimes\hat{B}_2|\phi\rangle - \langle\phi|\hat{A}_2\otimes\hat{B}_2|\phi\rangle]dp_\Psi(\phi).$$

This Gaussian integral can violate the CHSH inequality.

16.4.2 Correspondence between Subquantum and Quantum Covariances

Equality (16.42) describes correspondence between subquantum and quantum correlations. However, one might be not completely satisfied by this correspondence, since this equality is asymmetric: its LHS contains covariance of subquantum quantities f_{A_1} and f_{A_2} and its RHS contains correlation of quantum observables $\hat{A}_1$ and $\hat{A}_2$. In this section we rewrite the PCSFT-QM correspondence in the symmetric form. To simplify notation, we assume that the vector (operator) determining the subquantum field $\phi = \phi(\omega)$ is normalized by one; denote it by ψ. Hence, equality (16.42) has the following form:

$$\text{cov}\,(f_{A_1}, f_{A_2}) = \langle\hat{A}_1\otimes\hat{A}_2\rangle_\psi. \tag{16.77}$$

We recall that the covariances of two quantum observables are given by

$$\text{cov}\,(\widehat{A}_1,\widehat{A}_2) = \langle A_1\otimes A_2\rangle - \langle A_1\rangle\langle A_2\rangle = \langle(\widehat{A}_1 - \langle\widehat{A}_1\rangle I)\otimes(\widehat{A}_2 - \langle\widehat{A}_2\rangle I)\rangle, \tag{16.78}$$

where $\langle\ldots\rangle \equiv \langle\ldots\rangle_\psi$. We set

$$\widehat{A}_{0i} = \widehat{A}_i - \langle A\rangle I, i = 1, 2. \tag{16.79}$$

Then the average of observable $\widehat{A}_{0i}$ w.r.t. the state ψ equals zero, $\langle\widehat{A}_{0i}\rangle = 0$ and, hence,

$$\text{cov}\,(\widehat{A}_{01},\widehat{A}_{02}) = \langle\widehat{A}_{01}\otimes\widehat{A}_{02}\rangle = \text{cov}\,(\widehat{A}_1,\widehat{A}_2). \tag{16.80}$$

Now we invent a new rule for correspondence between classical and quantum variables. Instead of the correspondence rule,

$$\widehat{A}\to f_A(\phi) = \langle\widehat{A}\phi|\phi\rangle, \tag{16.81}$$

we introduce a new mapping. And it is more convenient to map not PCSFT quantities to QM observables, but the other way around: QM observables to PCSFT quantities,

$$\widehat{A}_0\to f_{A_0}(\phi) = \langle\widehat{A}_0\phi|\phi\rangle,\ \widehat{A}_0 = \widehat{A} - \langle A\rangle I. \tag{16.82}$$

This new map is state-dependent. This dependence can be interpreted as a form of contextuality: ψ determines the measurement context and subquantum quantities depend on this context.

By using (16.42) for $\widehat{A}_{0i}$ instead of $\widehat{A}_i$, we obtain

$$\text{cov}\,(f_{A_{01}},f_{A_{02}}) = \langle\widehat{A}_{01}\otimes\widehat{A}_{02}\rangle. \tag{16.83}$$

Thus

$$\text{cov}\,(f_{A_{01}},f_{A_{02}}) = \text{cov}\,(\widehat{A}_{01},\widehat{A}_{02}). \tag{16.84}$$

In contrast to equality (16.77), both sides of equality (16.84) are expressed as covariances.

16.4.3 Cartesian Representation of Random Field Dynamics for Compound System

We consider the finite-dimensional case (with remarks about the infinite-dimensional case). As before, we set $\mathbf{H} = \mathcal{H}_1\times\mathcal{H}_2$; this is the *Cartesian product* of Hilbert state spaces of subsystems of a compound system $S = (S_1, S_2)$.

In QM, Schrödinger's equation describes the dynamics of the wave function (pure quantum state in the abstract framework). We remark that, although the QM descriptions of the state dynamics of noncompound and compound systems are formally the same, but the latter case is characterized by the special axiom, the state space of a compound system is the tensor product of subsystems' state spaces. We made this remark to prepare the reader that in PCSFT the mathematical representations of dynamics of noncompound and compound systems are not identical. In this section we speak about the CSM-like model for a compound system, based on the state space $\mathbf{H}$. For the PCSFT model of a compound system based on the tensor product state space (Section 16.3), the subquantum dynamics of noncompound and compound systems are described by the same formalism.

In the CSM-like model for a compound system, the standard Schrödinger equation describes the dynamics of the off-diagonal block of the covariance operator of a subquantum random field $\phi = (\phi_1, \phi_2)$ beyond a compound quantum system $S = (S_1, S_2)$. The time variable t is the "quantum time variable," that is, related to the scale of measurements (compare with time scale separation for Brownian motion, Chapters 14, 15).

In the QM model, consider Schrödinger's equation for a compound quantum system $S = (S_1, S_2)$.

$$i\frac{d\Psi}{dt}(t) = \hat{H}\Psi(t), \;\; \Psi(0) = \Psi_0, \tag{16.85}$$

where $\hat{H}$ is Hamiltonian of S.

Now, to make the transition to PCSFT, we don't assume that Ψ_0 is normalized by one, so it is an arbitrary vector belonging to $\mathcal{H} = \mathcal{H}_1 \otimes \mathcal{H}_2$, and $t \to \Psi(t)$ is the map $[0, +\infty) \to \mathcal{H}$. In the operator terms, $t \to \hat{\Psi}(t)$ is the map $[0, +\infty) \to \mathcal{L}(\mathcal{H}_2, \mathcal{H}_1)$. (In the infinite-dimensional case, instead of the space of linear operators $\mathcal{L}(\mathcal{H}_2, \mathcal{H}_1)$, we consider the space of the Hilbert–Schmidt operators.)

Hence, at the instant t, the covariance operator of the prequantum random field $\phi(t, \omega) = (\phi_1(t, \omega), \phi_2(t, \omega))$ has the form (for $\epsilon = ||\Psi||^2$):

$$\hat{D}_{\Psi(t);\epsilon} \equiv \hat{D}(t) = \begin{pmatrix} \hat{\Psi}(t)\hat{\Psi}(t)^\star + ||\Psi_0||^2 I & ||\Psi_0||\hat{\Psi}(t) \\ ||\Psi_0||\hat{\Psi}(t)^\star & \hat{\Psi}(t)^\star\hat{\Psi}(t) + ||\Psi_0||^2 I \end{pmatrix}. \tag{16.86}$$

We remark that $||\Psi(t)|| = ||\Psi_0||$.

The following fundamental question, having both mathematical and physical counterparts, immediately arises:

Can one construct a stochastic process valued in the Cartesian product **H** *such that at each* $t \in [0, \infty)$ *its covariance operator coincides with* $\hat{D}(t)$*?*

Bernoulli Process

The formal mathematical answer to this question is yes! It is easy to construct such a stochastic process. Take space $\Omega = \prod_{t\in[0,\infty)} \mathbf{H}$ as the space of random parameters; points of this space $\omega = (\omega_t)$ can be considered as functions $\omega\colon [0, \infty) \to \mathbf{H}$, trajectories. Consider the family of Gaussian measures $\mu_t \equiv \mu_{\hat{D}(t)}$ on **H** having zero mean value and covariance operators $\hat{D}(t)$, $t \in [0, \infty)$. Consider the direct product of these measures $P = \prod_{t\in[0,\infty)} \mu_t$ on the space Ω.

Proposition 16.5 *Let $\phi(t, \omega)$ be a stochastic process having the probability distribution P on Ω. Then, for any pair of vectors $y_1, y_2 \in \boldsymbol{H}$ and any instant of time $t \geq 0$,*

$$E\langle y_1|\phi(t,\omega)\rangle\langle\phi(t,\omega)|y_2\rangle = \langle y_1|\hat{D}(t)|y_2\rangle, \tag{16.87}$$

where $\hat{D}(t)$ is given by (16.87).

Proof First we remark that the measure P is σ-additive by generalization of the Kolmogorov theorem [250]. The product space Ω can be represented as the space of functions $u\colon [0,\infty) \to \mathbf{H}$. Any fixed instant of time t and any vector $y \in \mathbf{H}$ determine a linear functional $u \to \langle u(t), y\rangle$ on Ω. This is a cylindrical function on Ω. By the definition of a product measure we have

$$\begin{aligned} E\langle y_1|\phi(t,\omega)\rangle\langle\phi(t,\omega)|y_2\rangle &= \int_\Omega \langle y_1|u(t)\rangle\langle u(t)|y_2\rangle dP(u) \\ &= \int_{\mathbf{H}} \langle y_1|\phi\rangle\langle\phi|y_2\rangle d\mu_t(\phi) = \langle y_1|\hat{D}(t)|y_2\rangle, \end{aligned}$$

by the definition of the covariance operator.

For any compound system prepared initially in a pure state, there exists a classical stochastic process, inducing the standard Schrödinger evolution of correlations. This equation describes dynamics of the off-diagonal block of the covariance matrix of this prequantum stochastic process.

The PCSFT is a tricky theory: a *prequantum stochastic process is not uniquely determined* by dynamics of the covariance operator $\hat{D}(t)$! To determine uniquely a Gaussian process (up to natural equivalence), one has to define not only the covariance for each instant of time, that is,

$$E\langle y_1|\phi(t,\omega)\rangle\langle\phi(t,\omega)|y_2\rangle,$$

but also the so-called *covariance kernel* $\hat{D}(t_1, t_2)$

$$E\langle y_1|\phi(t_1,\omega)\rangle\langle\phi(t_2,\omega)|y_2\rangle = \langle y_1|\hat{D}(t_1,t_2)|y_2\rangle.$$

Subquantum Dynamics for Noninteracting Systems

We restrict our consideration to dynamics in the absence of interactions between S_1 and S_2 after the preparation procedure. Thus we are interested in propagation of initially correlated random fields. Although this is a rather special dynamics, it plays an important role in quantum foundations. In particular, it describes the evolution of entanglement in the EPR–Bohm-type experiments.

We start with the standard quantum description of two noninteracting systems and consider Hamiltonian

$$\hat{H} = \hat{H}_1 \otimes I + I \otimes \hat{H}_2, \tag{16.88}$$

where $\hat{H}_j$ is Hamiltonian of S_j. Here we use the QM terminology. Now we represent quantum dynamics of a compound system by using the operator representation of the wave function. We state again that this is still the standard QM, but in a new mathematical representation.

Lemma 16.6 *Let Hamiltonian have the form (16.88). Then*

$$\widehat{\Psi(t)} = e^{-i\hat{H}_1 t}\widehat{\Psi_0}e^{-i\hat{H}_2 t}. \tag{16.89}$$

$$\widehat{\Psi(t)}^\star = e^{i\hat{H}_2 t}\widehat{\Psi_0}^\star e^{-i\hat{H}_1 t}. \tag{16.90}$$

Proof In this case

$$\Psi(t) = e^{-it(\hat{H}_1 \otimes I + I \otimes \hat{H}_2)}\Psi_0.$$

We expand the initial state Ψ_0: $\Psi_0 = \sum_{ij}^k \Psi_{ij} e_i \otimes f_j$, $\Psi_{ij} \in \mathbf{C}$, where $\{e_i\}$ and $\{f_i\}$ are two orthonormal bases, in H_1 and H_2, respectively. Then

$$\Psi(t) = \sum_{ij} \Psi_{ij} e^{-i\hat{H}_1 t} e_j \otimes e^{-i\hat{H}_2 t} f_j.$$

Thus, for $v \in H_2$, we get

$$\widehat{\Psi(t)}v = \sum_{ij}^k \Psi_{ij}\langle v, e^{i\hat{H}_2 t}\bar{f}_j\rangle e^{-i\hat{H}_1 t} e_j = e^{-i\hat{H}_1 t}[\sum_{ij}^k \Psi_{ij}\langle e^{-i\hat{H}_2 t} v, \bar{f}_j\rangle e_j].$$

Lemma 16.7 *Let the condition of Lemma 16.6 hold. Then*

$$\widehat{\Psi(t)}\widehat{\Psi(t)}^\star = e^{-i\hat{H}_1 t}\widehat{\Psi_0}\widehat{\Psi_0}^\star e^{i\hat{H}_1 t}, \tag{16.91}$$

$$\widehat{\Psi(t)}^\star\widehat{\Psi(t)} = e^{i\hat{H}_2 t}\widehat{\Psi_0}^\star\widehat{\Psi_0}e^{-i\hat{H}_2 t}. \tag{16.92}$$

Proof We prove, for example, (16.91). By (16.89) we have

$$\widehat{\Psi(t)}\widehat{\Psi(t)}^\star = e^{-i\hat{H}_2 t}\widehat{\Psi_0}^\star e^{i\hat{H}_1 t} e^{-i\hat{H}_2 t}\widehat{\Psi_0}^\star e^{i\hat{H}_1 t} = e^{-i\hat{H}_1 t}\widehat{\Psi_0}\widehat{\Psi_0}^\star e^{i\hat{H}_1 t}.$$

Now we leave the domain of standard QM and come back to the PCSFT-model; we have

Lemma 16.8 *Let the condition of Lemma 16.6 hold. Then the operator $D(t)$ given by (16.86) can be represented in the form*

$$\hat{D}(t) = \begin{pmatrix} e^{-i\hat{H}_1 t}\widehat{\Psi_0}\widehat{\Psi_0}^\star e^{i\hat{H}_1 t} + ||\Psi_0||^2 I & ||\Psi_0|| e^{-i\hat{H}_1 t}\widehat{\Psi_0}e^{-i\hat{H}_2 t} \\ ||\Psi_0|| e^{i\hat{H}_2 t}\widehat{\Psi_0}^\star e^{i\hat{H}_1 t} & e^{i\hat{H}_2 t}\widehat{\Psi_0}^\star\widehat{\Psi_0}e^{-i\hat{H}_2 t} + ||\Psi_0||^2 I \end{pmatrix} \tag{16.93}$$

Proof We put (16.91), (16.91), and (16.89) into the matrix (16.86).

By using representation (16.93) it is easy to prove

Proposition 16.9 *Let the condition of Lemma 16.6 hold. Then the operator $D(t)$ given by (16.86) is the covariance operator of the vector process with coordinates*

$$\phi_1(t,\omega) = e^{-i\hat{H}_1 t}\xi_{01}(\omega), \phi_2(t,\omega) = e^{i\hat{H}_2 t}\xi_{02}(\omega), \tag{16.94}$$

where the initial random vector, $\xi_0(\omega) = (\xi_{01}(\omega), \xi_{02}(\omega)) \sim N(0, D(0))$, is Gaussian with zero mean value and the covariance operator $D(0)$.

Proof We will find not only the covariance operator for a fixed instant of time, but even the covariance kernel. We have, for any pair of vectors $u, w \in \mathcal{H}_1$,

$$\begin{aligned} E\langle u|\phi_1(t_1,\omega)\rangle\langle\phi_1(t_2,\omega)|w\rangle &= E\langle e^{i\hat{H}_1 t_1} u|\xi_{01}(\omega)\rangle\langle\xi_{01}(\omega))|e^{i\hat{H}_1 t_2} w\rangle \\ &= \langle e^{-i\hat{H}_1 t_2}(\widehat{\Psi_0}^\star \widehat{\Psi_0} + ||\Psi_0||^2 I_1) e^{i\hat{H}_1 t_1} u|w\rangle. \end{aligned}$$

The same calculations can be done for the second diagonal block. Thus the diagonal blocks of the covariance operator of the stochastic process given by (16.94) coincide with the diagonal blocks of the operator $D(t)$. We now consider the off-diagonal blocks. Let now $u \in \mathcal{H}_1, v \in \mathcal{H}_2$. We have

$$\begin{aligned} E\langle u|\phi_1(t_1,\omega)\rangle\langle\phi_2(t_2,\omega)|v\rangle &= E\langle e^{i\hat{H}_1 t_1} u, \xi_{01}(\omega)\rangle\langle\xi_{02}(\omega))|e^{-i\hat{H}_2 t_2} v\rangle \\ &= \langle\widehat{\Psi_0}^\star e^{i\hat{H}_1 t_1} u|e^{-i\hat{H}_2 t_2} v\rangle = \langle e^{i\hat{H}_2 t_2}\widehat{\Psi_0}^\star e^{i\hat{H}_1 t_1} u|v\rangle. \end{aligned}$$

The same calculations can be done for the second off-diagonal block. Thus the covariance kernel has the form

$$\begin{aligned} &\hat{D}(t_1,t_2) \\ &= \begin{pmatrix} e^{-i\hat{H}_1 t_2}\widehat{\Psi_0}\widehat{\Psi_0}^\star e^{i\hat{H}_1 t_1} + ||\Psi_0||^2 e^{i\hat{H}_1(t_1-t_2)} I & ||\Psi_0||e^{-i\hat{H}_1 t_2}\widehat{\Psi_0} e^{i\widehat{-H_2} t_1} \\ ||\Psi_0||e^{i\hat{H}_2 t_2}\widehat{\Psi_0}^\star e^{i\hat{H}_1 t_1} & e^{i\hat{H}_2 t_2}\widehat{\Psi_0}^\star\widehat{\Psi_0} e^{-i\hat{H}_2 t_1} + ||\Psi_0||^2 e^{-i\hat{H}_2(t_1-t_2)} I \end{pmatrix} \end{aligned} \tag{16.95}$$

And hence, for $t_1 = t_2 = t, \hat{D}(t,t) = \hat{D}(t)$.

Dynamics (16.94) is purely deterministic, stochasticity is generated by initial conditions.

Propositions 16.5, 16.9 describe two different stochastic processes. The covariance kernel (16.95) differs from the covariance kernel of the process which has been constructed by considering the product of Gaussian distributions μ_t, Section 16.4.3. The latter has the covariance kernel

$$D(t_1, t_2) = D(t_1)\delta(t_2 - t_2).$$

For this process, its realizations at different instants of time are independent. The process defined by (16.94) contains nontrivial dependence between its realizations at different times. The following interesting problem arises:

For an arbitrary Hamiltonian, to construct a stochastic process such that in the case of the absence of interactions this construction gives the process (16.94).

PART VI

Further Developments

17

QBism versus Ozawa's Intersubjectivity Theorem?

17.1 Critical Analysis of QBism

I continue my critical analysis of QBism. Why am I so interested in QBism? I like its interpretation of the quantum formalism as a machinery for probability update based on the measurement outcomes (see, e.g., [127]–[131] for QBism basics). For QBists, quantum theory represents nonclassical probability update. In fact, this update is not based on the Bayes formula for conditional probability. So, there is some terminological ambiguity, since quantum probability inference is not the Bayesian probability inference. In QBism "Bayesianism" is understood widely as treating probability theory as the formalism of probability inference.

However, I am not sure that the individual agent perspective on a measurement's outcome can be accepted as the interpretation of QM. On the other hand, the QBism interpretation can be employed in applications of the quantum formalism outside of physics, in so-called quantum-like modeling (Chapter 18), especially in decision making.

This chapter is based on my paper [248]. This paper as well as two previous articles [228, 230] directly critiques the individual agent perspective on a measurement's outcomes [219]. My previous appraisal convinced QBists to specify the level of the agent's individuality. In contrast to the general subjective probability theory, the class of agents should be restricted, at least to agents who were educated in basics of quantum theory. So, Ivan who lives in a Siberian village, a busy hunter, can't be treated as a QBism agent.

Now I intend to offend QBism by using Ozawa's Intersubjectivity Theorem (OIT) [301]. QBism's statement that "the outcome of a measurement of an observable is personal" is in direct contradiction to OIT. This theorem is practically unknown and this chapter advertises it. OIT states that two observers, agents within the QBism terminology, performing joint measurements of the same observable A on a system S in the state ψ should register the same outcome $A = x$ with probability one. Hence, the outcome is intersubjective [301], and it's unnatural to consider outcomes of quantum observations as the agent's personal experiences.

Ozawa's Intersubjectivity Theorem is proven within quantum measurement theory (quantum instruments and indirect measurement model). This is the rigorous mathematical statement. But, as many theorems having the quantum foundational impact, its interpretation is not straightforward. The analysis of the OIT-impact on QBism is coupled to the foundations of quantum measurement theory and especially the notion of quantum observable. Therefore, we discuss foundations of quantum measurement theory by highlighting the difference between the accurate, von Neuman, and inaccurate, noisy, quantum observables, mathematically represented by projection valued measures (PVMs) and positive operator valued measures (POVMs), respectively. Ozawa's Intersubjectivity Theorem is about the agents who are able to perform the joint accurate measurements. For such agents, a measurement's outcome loses its personalization, in favor of intersubjectivity.

The conclusion of our analysis is that QBism should update its ideology by taking OIT into consideration. But, how? See Section 17.5. Thus, I am in line with the criticism of QBism presented in article [301]. However, I depart from its conclusion that OIT contradicts the Copenhagen interpretation; in contrast, OIT peacefully coexists with this interpretation. It is relevant to recall here that QBism fundamentally differs from the Copenhagen interpretation [129].

Right away we start with the mathematical formulation of OIT and its proof. We set out to make the presentation very shortly (see [301] for details). The indirect measurement model (see Section 1.4) is the heart of OIT. We go ahead with the recollection of the notion of quantum observable, namely, Hermitian operator or PVM, and generalized quantum observable (POVM) and the indirect measurements scheme for their generation.

17.2 Generalized Quantum Observables

In QM axiomatics, von Neumann [365] introduced quantum observables as Hermitian operators acting in complex Hilbert space $\mathcal{H}$, the state space of a system.[1] The spectral decomposition is the essential part in this framework. We restrict our consideration to observables represented by the operators with totally discrete spectra $X \subset \mathbb{R}$. Here $\hat{A} = \sum_x x\hat{E}_A(x)$, where $\hat{E}_A(x)$ is a projection on the eigensubspace corresponding to the eigenvalue x; these projectors form the resolution of unity,

[1] Why did von Neumann select the Hermitian operators for mathematical representation of observables in quantum theory? Moreover, he considered only such observables as the genuine quantum observables. I guess that he followed Schrödinger's quantization rule for the position and momentum observables which are realized by Hermitian operators in L_2-space. This rule implies that each classical observable given by the real-valued function $A = A(q, p)$ on the phase space is represented as a Hermitian operator in L_2-space.

$I = \sum_x \hat{E}_A(x)$. The Born rule determines the probabilities of the outcomes of measurements for a system S in the state ψ, $P(A = x|\psi) = \langle\psi|E_A(x)|\psi\rangle$.

Later, generalized quantum observables were invented. Such observables are represented by POVMs. We restrict our consideration to POVMs with a discrete domain of definition X. POVM is a map $x \to \hat{\Pi}(x)$ such that for each $x \in X$, $\hat{\Pi}(x)$ is a semipositive contractive Hermitian operator, namely $0 \leq \hat{\Pi}(x) \leq I$) (called an *effect*), and effects form the resolution of unity:

$$\sum_x \hat{\Pi}(x) = I. \tag{17.1}$$

This map defines an operator valued measure on algebra of all subsets of set X. For $O \subset X$,

$$\hat{\Pi}(O) = \sum_{x \in O} \hat{\Pi}(x).$$

The condition (17.1) is the operator-measure counterpart of the condition normalization by 1 for usual probability measures.

POVM $\hat{\Pi}$ represents statistics of measurements for observable A with the following generalization of Born's rule:

$$P_\psi(\Pi = x) = \langle\hat{\Pi}(x)\psi|\psi\rangle. \tag{17.2}$$

Any quantum observable $\hat{A}$ can also be represented as POVM of the special type – PVM $\hat{E}_A = (\hat{E}_A(x))$.

Quantum observables given by PVMs were interpreted by von Neumann [365] as describing *accurate measurements*. And generalized observables given by POVMs which are not PVMs are interpreted as representing *inaccurate measurements*. In von Neumann's [365], the notion of measurement's precision was not completely formalized. Only recently the consistent formalization of this notion was presented in [299].

We shall keep firmly the expression "quantum observable" for observable axiomatically introduced by von Neumann [365] and represented by PVMs and the expression "generalized quantum observable" for POVMs.

We recall that the indirect measurement scheme (Section 1.4) involves the following components: (a) the state spaces $\mathcal{H}$ and $\mathcal{K}$ of the systems S and the apparatus M_A for measurement of some observable A; (b) the evolution operator $\hat{U} = \hat{U}(t)$ representing the interaction-dynamics for the system $S + M_A$; (c) the meter observable $\hat{M}_A$ giving outputs of the pointer of the apparatus M_A.

Here the quantum observables $\hat{A}$ and $\hat{M}_A$ can be represented as PVMs, $\hat{E}_A = (\hat{E}_A(x)), \hat{E}_{M_A} = (\hat{E}_{M_A}(x))$, where $\hat{E}_A(x), \hat{E}_{M_A}(x)$ are projections in Hilbert spaces $\mathcal{H}$ and $\mathcal{K}$ respectively. It is assumed that the compound system's evolution is driven by the Schrödinger equation, so the evolution operator is unitary. Formally, an *indirect measurement model* for an observable A, introduced in [291] as a "measuring process," is a quadruple $(\mathcal{K}, |\xi\rangle, \hat{U}, \hat{M})$ where $|\xi\rangle \in \mathcal{K}$ represents the apparatus state (here we restrict consideration to pure states; see Section 1.4 for mixed states).

We explore the Heisenberg picture. To describe the meter's evolution, we represent it in the state space of the compound system, that is, as $I \otimes \hat{M}_A$. The meter observable evolves as

$$\hat{M}(t) = \hat{\hat{U}}^{\star}(t)(I \otimes \hat{M}_A)\hat{U}(t). \tag{17.3}$$

By the Born rule

$$P_{\psi\xi}(M(t) = x) = \langle \hat{E}_{M_A(t)}(x)\psi\xi \,|\, \psi\xi \rangle. \tag{17.4}$$

This is the probability distribution for the outputs of measurements done by the apparatus and given by the meter. In principle, one can ignore the representation of the measurement process as the system-apparatus interaction and operate solely with the system's states. In this picture one proceeds with generalized observables given by POVMs. The meter observable generates the POVM $\hat{\Pi} = (\hat{\Pi}(x))$

$$\hat{\Pi}(x) = \langle \hat{E}_{M_A(T)}(x)\xi \,|\, \xi \rangle, \tag{17.5}$$

where T is the time needed to complete the experiment. The probability distribution of the generalized observable given by a POVM is determined by (17.2).

Generally the probability distribution generated by a measurement process does not coincide with the probability distribution of the quantum observable A for which this process was constructed, that is, generally

$$P_{\psi}(\Pi = x) = \langle \hat{\Pi}(x)\psi \,|\, \psi \rangle \neq P_{\psi}(A = x) = \langle \hat{E}_A(x)\psi \,|\, \psi \rangle. \tag{17.6}$$

We remark that, as was proven by Ozawa [291], any generalized observable (POVM) can be generated via the indirect measurement scheme. Typically one operates solely with generalized observables by ignoring the indirect measurement scheme. This simplifies considerations, but it can lead to misunderstanding of the foundations of the quantum measurement theory.

17.3 Probability Reproducibility Condition

Definition 17.1 A measurement process $(\mathcal{K}, |\xi\rangle, \hat{U}, \hat{M})$ reproduces the probability distribution for quantum observable $\hat{A}$ (accurate von Neumann observable) if

$$P_{\psi}(A = x) = P_{\psi\xi}(M_A(T) = x). \tag{17.7}$$

In this case

$$\langle \hat{E}_{M_A(T)}(x)\psi\xi \,|\, \psi\xi \rangle = \langle \hat{E}_A(x)\psi \,|\, \psi \rangle \tag{17.8}$$

or

$$\langle\hat{\Pi}(x)\psi|\psi\rangle = \langle\hat{E}_A(x)\psi|\psi\rangle, \tag{17.9}$$

and hence,

$$\hat{\Pi}(x) = \hat{E}(x).$$

Proposition 17.2 *Probability reproducibility condition for a measurement process is equivalent to the representation of the corresponding generalized observable by the PVM $\hat{E}_A$ of measured quantum observable $\hat{A}$.*

17.4 Intersubjectivity of Outcomes of Quantum Observables

Following [301], consider two remote observers O_1 and O_2 who perform joint measurements on a system S; in mathematical terms this means that the meter quantum observables of the corresponding measurement processes commute,

$$[\hat{M}_{1A}(t), \hat{M}_{2A}(t)] = 0.$$

Here each apparatus has its own state space, namely, $\mathcal{K} = \mathcal{K}_1 \otimes \mathcal{K}_2$. We call such measurements local. In this situation the joint probability distribution is well defined:

$$P_{\psi\xi_1\xi_2}(M_{1A}(t) = x, M_{2A}(t) = y) = \langle\psi\xi_1\xi_2|\hat{E}_{M_{1A}(t)}(x)\hat{E}_{M_{2A}(t)}(y)|\psi\xi_1\xi_2\rangle. \tag{17.10}$$

Suppose that both observers perform the accurate measurements of the quantum observable $\hat{A}$ given by PVM $\hat{E}_A = (\hat{E}_A(x))$. Then the corresponding POVMs $\hat{\Pi}_{jA}, j = 1, 2$ coincide with $\hat{E}_A$:

$$\hat{\Pi}_{1A}(x) = \hat{\Pi}_{2A}(x) = \hat{E}_A(x). \tag{17.11}$$

This equality implies

Theorem 17.3 *(OIT [301]) Two observers performing the joint local and probability reproducible measurements of the same quantum observable A on the system S should get the same outcome with probability 1:*

$$P_{\psi\xi_1\xi_2}(M_{1A}(T) = x, M_{2A}(T) = y) = \delta(x - y)\|\hat{E}_A(x)\psi\|^2. \tag{17.12}$$

17.5 Intersubjectivity Challenges QBism

We start with the following citation of Fuchs and Schack [129]:

"The fundamental primitive of QBism is the concept of experience. According to QBism, quantum mechanics is a theory that any agent can use to evaluate her expectations for the content of her personal experience."

See also [126]: "In QBism, a measurement is an action an agent takes to elicit an experience. The measurement outcome is the experience so elicited. The measurement outcome is thus personal to the agent who takes the measurement action."

However, OIT implies that, for accurate local observables, a measurement's outcome is intersubjective, which is the strong objection to QBism. There is nothing concerning personal experiences and QBists should respond to this objection.

My suggestion (see also [219]) is to proceed not with individual agents and their personal experiences, but with the *universal agent* and consider measurement outcomes as the experiences of this agent (up to [239]). I remark that consideration of universal agents is common in the general theory of decision making. However, for QBists, such a solution seems to be unacceptable, since it would destroy the consistency of QBism's private agency perspective. QBism rejects even the possibility of two agents sharing the experience [126]: "... quantum theory provides a calculus for gambling on each agent's own experiences — it doesn't give anything else than that. It certainly doesn't give one agent the ability to conceptually pierce the other agent's personal experience."

The OIT-objection to QBism is foundationally interesting and generates the discussion on the notion of quantum observable. Due to the efforts of Helström, Holevo, and Ozawa [158, 164, 291, 292], [291], generalized quantum observables which are mathematically represented by POVMs became one of the basic tools of quantum information theory. Nowadays the special role of accurate observables represented by PVMs is not emphasized. In particular, the notion of observables in QBism is identified with generalized quantum observables given by POVM. However, the clash between QBism and OIT stimulates highlighting of the accurate PVM as the genuine quantum observables, and treating the generalized quantum observables which are not accurate POVM as imprecise and noisy ones. Of course, this is a well-known fact, but the clash between OIT and QBism is a good opportunity to emphasize this difference.

What does this difference between accurate PVM and noisy POVM observables mean for QBism?

I have the following picture of the situation. OIT holds only for the accurate PVM-observables; for generalized quantum observables, it can be violated and generally it is impossible to assign the same value for measurements' outcomes for observers O_1 and O_2. Thus, QBism ideology of the personal experiences of observers (agents) can still be kept for such generalized observables. But, where does individuality come from? *The personal experiences come from noise!* So, different observers performing inaccurate measurements are coupled to different noisy environments. This is just my personal view on the consequences of OIT for QBism.

In conclusion, QBism might respond to the OIT-challenge by considering the universal agent who is able to perform accurate measurements; another possibility is to proceed without referring to the universal agent, but then individuality of experience is due to noise generated in the process of measurement. Since this noise is generated by physical processes in the measurement apparatus, this is merely the "personal experience of the apparatus." Of course, it can be treated as the personal experience of the observer performing this measurement, but this treatment loses the flavor of subjectivity.

17.6 Intersubjectivity and Copenhagen Interpretation

We start the discussion with the following important citation [315][2]:

Bohr's interpretation, in any of its versions, will be distinguished in this study from "the Copenhagen interpretation," because there is no single such interpretation, as even Bohr has changed his a few times. It is more suitable to speak, as Heisenberg did [157], of "the Copenhagen spirit of quantum theory" or, as a handier shorthand, "the spirit of Copenhagen," referring to certain common features of a group of interpretations, which may be different in their other features.

It seems that one of such common features is the statement that measurements' outcomes cannot be treated as the objective properties of a system S. They are the results of the complex process of interaction of a system and an apparatus; this is a good place to cite again Bohr [65]:

This crucial point... implies the impossibility of any sharp separation between the behaviour of atomic objects and the interaction with the measuring instruments which serve to define the conditions under which the phenomena appear. In fact, the individuality of the typical quantum effects finds its proper expression in the circumstance that any attempt of subdividing the phenomena will demand a change in the experimental arrangement introducing new possibilities of interaction between objects and measuring instruments which in principle cannot be controlled. Consequently, evidence obtained under different experimental conditions cannot be comprehended within a single picture, but must be regarded as complementary in the sense that only the totality of the phenomena exhausts the possible information about the objects.

The indirect measurement scheme matches perfectly with the Copenhagen interpretation. Therefore it is surprising that OIT contradicts it. The clash between OIT and the the Copenhagen interpretation was highlighted in the conclusion section of OIT-article [301]:

[2] Arkady Plotnitsky presented this viewpoint at the second Växjö conference in 2002 (see [308]).

Schrödinger [329] argued that a measurement does not ascertain the pre-existing value of the observable and is only required to be repeatable. Since the inception of quantum mechanics, this view has long been supported as one of the fundamental tenets of quantum mechanics. In contrast, we have shown that any probability reproducible measurement indeed ascertains the value that the observable has, whether the repeatability is satisfied or not.

I disagree with the author of [301]. The seed of this misunderstanding is in ignoring the two-level structure of physical theories, ontic and epistemic [33, 34]. The former is about reality as it is and the latter is about knowledge about reality. Bohr and Schrödinger wrote about the ontic reality, about the impossibility of assigning to quantum systems preexisting values, and here "preexisting" is a synonym for "objective," "ontic." But OIT is not about such values; it is about epistemic reality, reality of knowledge about the possible outcome of measurement.

Hence, in my opinion *OIT can peacefully coexist with the Copenhagen interpretation.*

But, as was stressed, OIT is a challenge for QBism which operates at the epistemic level of scientific description of quantum phenomena. This is a good place to recall that QBism should be sharply separated from the Copenhagen interpretation; see again Fuchs and Schack [129]:

According to QBism, quantum mechanics can be applied to any physical system. QBism treats all physical systems in the same way, including atoms, beam splitters, Stern–Gerlach magnets, preparation devices, measurement apparatuses, all the way to living beings and other agents. In this, QBism differs crucially from various versions of the Copenhagen interpretation.

17.7 QBism versus Bayesian Framework for Quantum Probability

QBism is often presented as a Bayesian, subjective probability interpretation of quantum mechanics. However, the Bayesian viewpoint on quantum probabilities is not coupled solely to QBism. I think that quantum Bayesianism preliminaries were already presented by Schrödinger in his "Cat Paradox" paper [329]:

It (ψ-function) is now the means for predicting probability of measurement results. In it is embodied the momentarily-attained sum of theoretically based future expectation, somewhat as laid down in a catalog. It is the relation- and -determinacy-bridge between measurements and measurements...

For each measurement one is required to ascribe to the ψ-function (= the prediction-catalog) a characteristic, quite sudden change, which depends on the measurement result obtained, and so cannot be foreseen...

Similar interpretation of the wave function is explored in the works of M. D'Ariano (see, e.g., [96]) and my papers (see, e.g., [203]) within the Växjö interpretation of quantum mechanics [190, 193]. D'Ariano uses subjective probabilities and I use statistical ones.[3]

As was pointed out a few times, the essence of QBism is in treatment of measurements' outcomes as the personal experiences of observers and OIT is the objection for this key point of QBism (and not for the Bayesian approach to quantum probabilities).

[3] By the Växjö interpretation the quantum formalism is treated as a special calculus of probability update based on the quantum analog of the formula of total probability – the classical formula perturbed by the interference term.

18

Quantum-like Modeling in Biology, Cognition, and Decision Making

18.1 Quantum-like Modeling

The year 2022 was a celebrated one for quantum information studies. Aspect, Clauser, and Zeilinger got the Nobel Prize for experimental and theoretical studies on quantum foundations supporting quantum computing, cryptography, and teleportation. This is a good time to highlight the not so commonly known output of the quantum information revolution that is often called the second quantum revolution, namely, the project on applications of the quantum foundations and formalism outside of physics – *quantum-like modeling* [178, 318] . One can also speak about a quantum-like revolution [318].

This chapter is on *quantum-like modeling* and its applications, with the emphasis on the role of theory of open quantum systems (see review [247] and monograph [246]). Such modeling is built on the methodology and the mathematical apparatus of quantum theory and it is directed toward applications to biology, cognition, psychology, decision making, economics and finances, social and political sciences, and artificial intelligence. It is of essential importance to signify that this approach can be explored for macroscopic systems and the system's size is not significant. The quantum-like framework is applicable on all scales, that is to say from proteins and genes to animals, humans, and ecological and social systems. The crucial role is played by the character of information processing by a system and matching with the laws of quantum information theory. Systems are treated as information processors. Metaphorically one may say that a system's "hardware," its physical and biological structures, are not so significant, but the system's "software" plays the central role. We can speak about *quantum bioinformatics* [22]–[25] which should not be confused with *quantum biophysics* [20]. The latter studies the genuine quantum physical processes in biosystems, for example, in cells.

This chapter can't reflect all publications on quantum-like modeling. A Google search generates 790,000 results on "quantum-like modeling" in 0.47 s. (this is really surprising for me). Partially this review represents the content of monograph [246] by reflecting related references, especially on applications theory of open quantum systems to cognition and decision making. The author was the pioneer in employing quantum information and open systems outside of physics [178].

It is important to point out the immense influence of mathematics in physics, emphasized by many scientists and in particular by E. Wigner [378]. However, mathematical tools commonly used in theoretical and mathematical biology, cognition, and psychology are not so efficient as in theoretical and mathematical physics. Gelfand pointed out the "ineffectiveness of mathematics in biology" [21]. From my point of view Gelfand's statement has to be reformulated and one would speak about the ineffectiveness of mathematics that is *commonly used* in biology, cognition and psychology. I presume someone might have an intention to model micro-systems behavior, say electrons, atoms, photons, within classical analysis of functions defined on phase space, $A = A(q,p)$. In this case one would confront difficulties and soon would notice either the impossibility of such a description of quantum phenomena or at least its ineffectiveness.

Personally I stress the effectiveness of the quantum description and do not highlight various no-go statements concerning the impossibility of the classical description.

Physicists explored a new branch of mathematics, theory of operators in complex Hilbert space in order to describe the quantum phenomena in an effective way. And in quantum physics the noncommutative operator calculus works very well. Similarly, one should search for novel mathematics which is proper for biological and mental phenomena. The quantum-like approach advertises the same mathematics that was employed in quantum physics, noncommutative operator calculus in complex Hilbert space. Why is it so attractive for discussed applications? Personally I was mainly driven by specialties of quantum probability (QP) calculus which matches mental phenomena very well. This point will be discussed later in some detail. But one can look at an even deeper level. It is useful to extract the basic problems in mathematical modeling of mental phenomena highlighted by the experts in the field.

The central objective of this chapter is to illuminate some directions in development of quantum-like modeling. First we introduce the motivation for operating with the quantum formalism and especially probability outside of physics. Then we compare classical and quantum probability (CP and QP) theories and set forth the principles of quantum-like modeling of decision making. In particular, consideration of a special quantum-like model, "decision making via decoherence" [23], leads to coupling with theory of open quantum systems. The latter is discussed in more detail, highlighting its applications to behavior of complex biosystems (especially the problem of stability), cognition, brain function, theories of consciousness, and emotional coloring of perceptions.

Theory of open quantum systems accommodates the formalism of quantum instruments. This formalism realizes the most general quantum state updates. We stress that the combinations of basic psychological effects such as the question order effect (QOE) and the response replicability effect (RRE) lead us to inquire into the possibility of proceeding using the standard quantum measurement theory (with representation of observables by Hermitian operators and the quantum state update via the projection postulate) [213]. We demonstrate that this psychological effects combination can be modeled with quantum instruments [302, 303].

For newcomers to the field of quantum-like modeling searching for its motivation, I can recommend two handbooks [27, 155] (on quantum models in social science and mathematical psychology), especially article [222] in the first one and the preface of the second one. In article [222] I argue that functioning of biosystems should include cognition and in particular unconscious-conscious interaction should be modeled within open quantum systems theory (see also [218]).

Edgar Allan Poe's Reasoning in Favor of Quantum-like Modeling

The preface to a handbook on mathematical psychology [27] starts with a brilliant citation from a story written by Edgar Allan Poe (1845) entitled "The Purloined Letter." In this story a protagonist, Mr. C. Auguste Dupin, discussed the limits of mathematics' applicability:

> *Mathematical axioms are not axioms of general truth. What is true of relation — of form and quantity — is often grossly false in regard to morals, for example. In this latter science it is very usually untrue that the aggregated parts are equal to the whole. [. . .] two motives, each of a given value, have not, necessarily, a value when united, equal to the sum of their values apart.*

One can be surprised by Poe's doubts in the applicability of mathematics (of the nineteenth) century to moral phenomena (compare with attempts of, say, Freud to proceed with "classical mathematics"). He also expressed doubts in the validity of the value-additivity law. This is a very deep statement, and in quantum mathematics it is formulated as "eigenvalues of the sum of operators $C = A + B$ are not equal to the sums of the eigenvalues of the summands," that is, generally

$$c_i \neq a_i + b_i.$$

In fact, the violation of the value-additivity law is the key point of von Neumann's no-go theorem, the first statement on impossibility of classical reduction of quantum theory. Then, the authors of [27] also pointed out the noncommutativity effect in conjunctions,

$$A\&B \neq B\&A.$$

This order effect is also naturally formalized in the quantum framework. In fact, these two effects, the value-nonadditivity and the order ones, are closely connected. In probabilistic terms they are jointly expressed in the violation of the formula of total probability, interference of probabilities.

The essential part of quantum-like modeling is devoted to the order effect. Its QP-realization in decision making has been done in articles [367, 368].

Mathematical Models of Mental Phenomena: Why Quantum?

We remark that in [27] the discussion is not coupled to quantum-like modeling: the authors searched for novel mathematical tools for psychology, but their considerations really cry for the appeal to the quantum formalism. The main message [27] was that a variety of mathematical methods could be explored to solve the problems mentioned by Edgar Allan Poe which I completely agree with. The quantum formalism should not be treated as pretending to be the unique mathematical tool for modeling of mental phenomena. A while ago, in response to the developments of using quantum formalism outside of quantum mechanics, the eminent quantum physicist Anton Zeilinger (the Nobel Prize Laureate 2022) told me, "Why should it be precisely the quantum mechanics formalism? Maybe its generalization would be more adequate for mathemtical modeling of mental phenomena ..."

And he is right, for the moment; despite its tremendous success, quantum-like modeling is still at the testing stage. Maybe one day a new, more advanced mathematical formalism will be suggested for modeling in cognition, psychology, and decision making.

Simplicity, Elegance, and Generality

However, from my viewpoint, quantum formalism is the most successful due to its simplicity. The reader may be surprised: "Simplicity? But the quantum theory is the mysterious and very complicated!" One would immediately recall the famous statement which is commonly attributed to Richard Feynman: "I think I can safely say that nobody understands quantum mechanics."

But here "understanding" is related to the interpretation problem of quantum mechanics; its formalism is very simple; it is linear algebra. And in quantum information theory, which is the most useful for applications, including quantum engineering, linear state spaces are finite dimensional. So, this is the matrix calculus in $\mathcal{H} = \mathbb{C}^n$. Linear evolution is very rapid and this is the advantage of the quantum-like representation for mental states and the corresponding linear processing of them.

In engineering, linear models often appear as approximations of essentially more complex nonlinear ones. Hence, the quantum-like representation of biological and mental phenomena might be just an approximation for more complex nonlinear processes in the living systems.

We also emphasize the generality of mathematical modeling based on the quantum formalism: the same formalism and methodology cover a variety of biological and mental processes.

Classical versus Quantum Probability

Now we motivate employing quantum probability (QP), instead of classical probability (CP), in the mathematical modeling in cognitive psychology and decision making. We refer to the Kahneman (the Nobel Prize Laureate in economics) and Tversky (the most cited psychologist) works pointing out that CP used as the basis of decision theory leads to inconsistencies and paradoxes (such as Allai and Ellsberg paradoxes). Such motivation is not grounded on foundational principles. Here we proceed in parallel with quantum physics, created to resolve inconsistencies between classical electrodynamics and experimental data for the black body radiation.

In cognitive psychology and decision making, the experiments on irrational behavior and probability fallacies generated a lot of statistical data which does not match, at least straightforwardly, with main CP-laws [204].

To support the use of QP, we highlight the quantum-like paradigm [178] by which context-sensitive systems, including humans, process information in the form of superpositions, that is, without ambiguity resolution. Such processing can be effectively described by the quantum formalism. Once again, we appeal to the effectiveness of the QP description, not to the impossibility of exploring CP. Generally proving various no-go statements is counterproductive, both in physics and decision making. If one is able to describe physical or mental phenomena with CP versus QP and this description is effective, then one can proceed with CP versus QP.

Quantum Formalism for Decision Making

We recount the fundamentals of the quantum-like modeling of decision making. The basic scheme explores the standard quantum measurement theory. The basic components are represented as follows:

- questions, problems, and tasks as quantum observables – Hermitian operators;
- belief or mental states of decision makers by quantum states – normalized vectors in a complex Hilbert space or generally density operators;
- the quantum state update via the projection postulate.

From the very beginning, we highlight that in applications of the quantum measurement theory to cognition and generally biology the crucial role is played by finding the proper formalization of the state update resulting from decision making. The projection state update is the simplest one and it can't cover all cognitive phenomena. More general state updates are explored in quantum information theory; they are formalized within quantum instruments theory.

We also mention a special quantum-like model, *"decision making via decoherence,"* coupled with open quantum systems theory.

The problem of the belief-state interpretation is discussed by paying attention to the diversity of possible interpretations [220]. The latter is one of the disturbing problems of quantum foundations. We recount the basic quantum state interpretations, namely individual and statistical. We also mention QBism as maybe the most useful framework for quantum-like decision making.

This is a good place to point out the applications of QBism to decision making in geology within the project on determination of the perspective for intelligent petroleum reservoir characterization, monitoring, and management [272, 273]

18.2 Quantum and Classical Logic of Reasoning

We start this section with the remark that Boole designed classical (Boolean) logic for "investigation of the laws of thought" [70].

Although I put a lot of effort into justification of quantum-like modeling through QP analysis and especially its contextual nature, slowly I started to understand that the seed of cognition's quantumness (not only of humans, but also other biosystems) is in the logic structure of information processing (see, e.g., articles on quantum-like modeling of the problem of common knowledge and violation of the Aumann theorem [36, 37] on the impossibility of agreeing to disagree [215, 216]).

Quantum logic corresponds to the linear representation of information. The basic law distinguishing classical (Boolean) and quantum logic is the distributivity law; it is violated in quantum logic (see article [304] for the details). We now briefly recall the basics of quantum logic [55] (compare with classical logic [69, 70]).

The logical operations are defined on subspaces of complex Hilbert space $\mathcal{H}$ or equivalently on the set of orthogonal projectors $\mathcal{P}(\mathcal{H})$. Subspaces (projections) are represents of propositions (events). Let P be a projection and let L_P be its image, $L_P = P\mathcal{H}$. For a subspace L, P_L is the projection on L. Denote the projection onto the orthogonal complement to the subspace L_P by the symbol $\overline{P}$, that is, $\mathcal{H} = L_P \oplus L_{\overline{P}}$.

Negation of proposition P is represented by $\overline{P}$. The operations of conjunction $\wedge$ and disjunction $\vee$ are defined as follows.

Let P and Q be orthogonal projections representing some propositions. The conjunction-proposition (event) $P \wedge Q$ is defined as the projector on intersection of subspaces L_P and L_Q, that is, $L_{P \wedge Q} = L_P \cap L_Q$. We remark that this operation is well defined even for noncommuting projectors, namely, incompatible quantum observables. Moreover, it is commutative:

$$P \wedge Q = Q \wedge P. \tag{18.1}$$

The same can be said about the operation of disjunction. Here subspace $L_{P \vee Q}$ is defined as the subspace generated by the union of subspaces L_P and L_Q, that is, $P \vee Q$ is projector on this subspace. This operation is also well defined for noncommuting projectors and, moreover, it is commutative:

$$P \vee Q = Q \vee P. \tag{18.2}$$

Thus, quantum logic is commutative logic. This fact is never highlighted. Thus, in quantum reasoning noncommutativity is not present at the level of the basic operations of quantum logic, conjunction and disjunction. In light of this fact, the following natural question arises:

What is the logical meaning of noncommutativity of quantum operators?

We recall that noncommutativity is commonly considered as the basic mathematical feature of the quantum theory. Hence, it should play the crucial role also at the level of quantum logic. The answer is rather unexpected and it is given by Theorem 18.1.

As one knows, classical Boolean logic is distributive [69, 70], that is, for any three propositions (events) X, Y, Z, for example, represented by subsets of some set,

$$X \wedge (Y \vee Z) = (X \wedge Y) \vee (X \wedge Z). \tag{18.3}$$

Theorem 18.1 *[304] Let P, Q, R be projections. They are pairwise commutative if and only the distributivity law (18.3) holds for $X, Y, Z = P, Q, R, \bar{P}, \bar{Q}, \bar{R}$.*

Thus, noncommutativity encodes nondistributivity of quantum logic! Hence, the existence of incompatible quantum observables, namely represented by noncommuting operators, is equivalent to nontransitivity of logical relations between quantum propositions. In particular, Bohr's complementarity principle [65] reflects this logical structure.

This statement is especially important for quantum-like modeling of cognition. The existence of incompatible propositions or questions is a consequence of nondistributivity of logic used by a quantum reasoner.

Are humans classical or quantum reasoners?

It depends on the context of information processing. In some situations humans use the distributive Boolean logic, but in other situations they violate distributivity law and quantum logic can be employed to describe mathematically the latter form of reasoning.

At first sight, it seems to be impossible to characterize distributive versus nondistributive (classical *versus* nonclassical) reasoning in an experimentally testable way. However, such characterization was obtained in [304] and it is based on testing of the response replicability effect (RRE).

The notion of *response replicability* plays an important role in quantum physics. This property of observations is also the common feature of human behavior. Suppose that Alice is asked some question A and she replies, for example, "yes." If immediately after answering she is asked this question again, then she replies "yes" with probability one. This is *$A - A$ response replicability.* This is the feature of quantum observables of the projection type.

Commonly human decision making has another property: *$A{-}B{-}A$ response replicability.* Suppose that after answering the A-question with say the "yes"-answer, Alice is asked another question B. She replies to it with some answer. And then she is asked A again. In the social opinion pools and other natural decision making experiments, Alice definitely repeats her original answer to A, "yes." This is $A - B - A$ response replicability. Combination of $A - A$ with $A - B - A$ and $B - A - B$ response replicability is called *the response replicability effect* (RRE).

Theorem 18.2 *[304] The projection observables P and Q show RRE in a state ψ if and only if the distributive law holds for this state, that is,*

$$[X \wedge (Y \vee Z)]\psi = [(X \wedge Y) \vee (X \wedge Z)]\psi, \qquad (18.4)$$

for $X, Y, Z = P, Q, \bar{P}, \bar{Q}$.

As was shown in article [304], RRE can be checked experimentally. For the moment, only one experimental test was done [77] and its design and methodology was questioned by a few researchers; see comments on the PLOS One web-page for this article. Maybe article [304] would attract the attention of experimenters to RRE as the test for nonclassicality of hymen logic.

18.3 Quantum versus Quantum-like Cognition

We emphasize that quantum-like modeling of cognition should be sharply distinguished from the quantum brain studies (e.g. of Penrose) attempting to reduce information processing by cognitive systems including "generation of consciousness" to quantum physical effects in the brain. However, we do not criticize the quantum brain project, although its difficulties are well known: for example, the brain is too hot and big, the scales of neurons' operations are too rough compared with the quantum physical scales.

In quantum-like modeling, it is simply not important whether the genuine quantum physical processes in the brain's cells contribute to cognition or not. Generally quantum-like modeling is performed on the meta-level of cognition; it does not concern the biophysical processes in neurons. In this framework a biosystem, in particular, the brain is treated as a *black box* in which information processing cannot be described by the *classical probability* (CP) and, hence, the classical information theory. Nonclassical probability and information theories are on demand. In particular, in decision making exploring of CP leads to various paradoxes which are typically coupled to irrational behavior of humans. My suggestion [178] was to employ quantum probability (QP) and quantum information theories, instead of the classical ones. Why should specifically quantum theory be involved? This is the complex problem.

Classical, Quantum or More General Probability Theories?

There exist plenty of other models different from CP and QP. The use of QP in, for example, decision making, was not derived from some basic principles for cognition and psychology. Commonly QP is used pragmatically – to resolve paradoxes and to have a general probabilistic framework applicable to decision making, in all areas of humanities and economics, as well as in biology. There is no a priori reason to hope that QP would cover all problems which arise in decision making. One might find paradoxes even in QP-based decision theory. Maybe other probabilistic models different from both CP and QP should be employed.

Surprisingly, physicists have the same problem. In contrast to relativity theory, QM was not derived from natural physical principles (see Zeilinger [353] for the discussion on this problem). There is no reason to expect that all experiments in the micro-world would match QP-constraints.

In physics, one typically debates CP versus QP, classical versus quantum physics. However, one can even test whether physics of microsystems can violate the QP-laws, that is, whether electrons and photons can behave exotically even from the QP-viewpoint. The corresponding test is given by the *Sorkin equality* for the three-slit experiment. It is really surprising that two- and three- slit experiments have such different probabilistic structures. The three-slit experiment was done by the Weihs' group (Austria). They did not find deviations from QP, and the Sorkin inequality was not violated [340, 341]. Similar experiments can be done for decision making by humans [322] by using the theoretical formalism of article [49] – presentation of the Sorkin equality in terms of quantum probability theory.

18.4 Biosystems as Open Quantum-like Systems

Any living biosystem is an open system, and to analyze its behavior it is reasonable to take advantage of the open quantum systems theory, where the biosystems are acknowledged as information processors and the open quantum systems theory is treated as a part of the quantum information theory. The latter is the most general information theory comprising the classical information theory as a particular case. Thus, the open quantum systems part of quantum-like modeling concerns information processing in complex biosystems interacting with their environments. From the information viewpoint, even a cell or a protein is a very complex system.

The challenging problem of mathematical formalization of unconsciousness–consciousness interrelation can also be handled within the open quantum systems theory [218, 238]. Consciousness plays the role of an apparatus performing measurements over unconsciousness. This formalism matches well to the higher-order theory of consciousness. It is applied to model mathematically the emotional coloring of conscious experiences. Such coloring is framed as contextualization. So, the theory of emotions is coupled to such a hot topic in quantum foundations as contextuality. Finally, we discuss the Bell-type experiments for emotional coloring [238] (see also [80]–[50] for such experiments in cognition and decision making).

What Is Life?

Treating biosystems as quantum-like information processors can expound order stability in them, that is, present the quantum-like formalization of Schrödinger's speculations in his notorious book *What Is Life?* [334] – see article [240]. Schrödinger stressed that order stability is one of the characteristic features of biosystems. Entropy can be used as a quantitative measure of order. Then he noted that in physical systems, entropy has a tendency to increase (the Second Law of Thermodynamics for isolated classical systems and dissipation in open classical and quantum systems). In contrast, biosystems beat this tendency. Schrödinger asked: "How?" Quantum information and open systems theory may give the answer to this fundamental question of modern science.

In [240] the process of biosystem's adaptation to the surrounding environment is described by the *Gorini–Kossakowski–Sudarshan–Lindblad equation*, where the von Neumann and linear quantum entropies are employed as measures of the disorder degree. This equation describes Markovean evolution, so we work with quantum Markov dynamics. Markoveanity possesses the strong constraint on the class of the mental state dynamics. The description of information processing in biosystems with quantum non-Markovean dynamics is more promising, but at the same time it is more complicated.

We highlight the role of the special class of quantum dynamics – generating the camel-like shape for quantum entropies. The camel's hump represents:

- the entropy increase in the process of the initial adaptation to the environment;
- the entropy decrease at the post-adaptation stage of the dynamics.

Our analysis [240] is based on numerical simulation and to describe such class of quantum dynamics analytically is must-have.

Order-Stability in Biosystems

Once again theory of open quantum systems is used (see paper [241]) in attempting to bring more clarification on the question "What is life?" We consider a complex biosystem S composed of many subsystems, say proteins, or cells, or neural networks in the brain, that is, $S = (S_i)$. We study the following problem:

If the composed system S can preserve the "global order" in the situation of increase of local disorder and if S can preserve its entropy while some of the subsystems S_i increase their entropies.

It is shown that within quantum information theory the answer is positive [241]. Entanglement of the subsystems' states plays the crucial role. In the absence of entanglement, the increase of local disorder generates the increase of disorder in the compound system S (as in the classical regime).

Brain: From Electrochemical to Quantum Information States

As an application of the open quantum systems theory to cognition, we suggest a quantum-like model of brain function (see [218, 246]). In this model the general approach of quantum-like modeling – to start directly with the quantum information representation of biosystems' states – is broken. We start with consideration of electrochemical states of neurons encoded in action potentials. Such states generate the brain's mental states which are processed with the open quantum systems dynamics.

The model does not refer to the genuine quantum physical processes in the brain. Hence, it does not suffer from the well-known problems of matching of the quantum and neural scales, temporal, spatial, temperature. In this model, uncertainty generated by the action potential of a neuron is represented as quantum-like superposition of the basic mental states corresponding to some neural code, for example, quiescent/firing neural code.

Mathematically a neuron's state space is described as a complex Hilbert space of quantum information states. The state of a neural network is presented in the tensor product of single-neuron state spaces.

The brain's mental functions perform self-measurements by extracting from the quantum information states concrete answers to questions. This extraction is modeled in the framework of theory of open quantum systems.

In this way, it is possible to proceed without appealing to state's collapse. The dynamics of the state of mental function F is described by the quantum dynamical equation. Its stationary states represent classical statistical mixtures of possible outputs of F (decisions). Thus, through interaction with an electrochemical environment, F (considered as an open system) resolves uncertainty that was originally encoded in superposition representing action potentials of neurons.

Decision Making via Decoherence

The above scheme of resolution of uncertainty through interaction with environment is known as the quantum dynamical decision making or *decision via decoherence* [23]; see also recent article [325] on the experimental study on eye tracking in the process of decision making and its modeling with Gorini–Kossakowski–Sudarshan–Lindblad equation. Here the experimentally observed stabilization in eye tracking matches perfectly with stabilization of the solution of this equation. One of the specialties of this model is employing a three-dimensional space of mental states. The equation is phenomenological, that is, it is not derived from neurophysiology beyond eye moving in the process of decision making. We hope that this study may stimulate further cooperation for finding the physical and physiological signatures of the mental state stabilization in the process of decision making. It is also interesting to perform new experiments with the design similar to the experiment in [325] and check whether the same phenomenological equation as in [325] can be in use.

The most promising realization of this scheme is the "differentiation" model [26] by which the mental state experiences step-by-step state transitions under the influence of surrounding electrochemical environmental factors. The differentiation leads to stabilization of the biosystem's state. This model is applicable on all biological and social scales, from cells to ecological and social systems. Decoherence is a deep foundational notion. Heuristically it can be interpreted as the loss of quantumness, transition from QP to CP, washing out of interference of probabilities (see, e.g., Zurek [356]). It is quantified with linear entropy, the measure of a state's purity. In quantum information theory, typically decoherence is considered as a negative factor disturbing information processing. In quantum dynamical decision making, decoherence plays the constructive role as decisions' generator.

Emotional Coloring of Conscious Experiences

The open quantum systems theory is also used for mathematical formalization of the consciousness–unconsciousness interaction, the information exchange between them. Consciousness plays the role of a measurement device, it performs observations over the states of unconsciousness. These observations can be interpreted as the brain's self-observations. So, human's thoughts and decisions are generated in the complex process of interaction between unconscious and conscious states. From the viewpoint of quantum foundations, we use Bohr's interpretation of the outcomes of quantum measurements as generated in the complex process of interaction between a system and a measurement apparatus [65]. In particular, these outcomes are not objective properties of a system which could be associated with it before measurement. In the same way the mind is not objective. Bohr's ideology structured within open quantum systems theory matches the higher-order theory of consciousness.

The unconscious–conscious framework can be explored for quantum-like modeling of interconnected dynamics of perceptions and emotions [238]. More generally, this framework describes the emotional coloring process for a variety of conscious experiences, including decision making. There are considered two classes of observables: perceptions and emotions. These observables are represented by Hermitian operators acting in the corresponding unconscious state spaces (or, more generally, by projection valued measures – PVMs). The total unconscious state space is their tensor product. Emotional coloring is formalized within the quantum contextuality formalism: emotional observables determine contexts. Such contextualization reduces degeneration of spectra for observables representing conscious experiences, for instance perceptions or decision making. Quantum measurement theory serves as the basis for mathematical modeling of generation of conscious experiences as observations over unconscious states. To model self-measurements performed by the brain, brain function is treated as the cooperative work of two subsystems, unconsciousness and consciousness. They correspond to a system and an observer. Article [238] is concluded with an experimental test of contextual emotional coloring of conscious experiences (cf. [80]–[50]), namely, on the violation of the CHSH inequality.

18.5 Question Order and Response Replicability Effects and QQ-Equality

One of the specialties of my research in recent years is exploring quantum instruments in applications to psychology and decision making – starting with article [218] devoted to modeling unconscious–conscious interaction. Quantum instruments are the basic tools of the modern theory of quantum measurements and open quantum systems. They describe the probability distributions of measurements' outcomes and the quantum state transformations generated by measurements' feedback. Thus, a quantum instrument describes both probability and its update – via quantum state update.

Such updates are not reduced to ones based on the projections, that is, given by the Lüders projection postulate. More general state space transformations are also needed, both in quantum physics, especially quantum information theory, and in quantum-like modeling, for example, in decision making and psychology.

Instruments cause representation of quantum observables by a *positive operator valued measures* (POVM)s, generalized quantum observables. Typically POVMs are considered as the basic entities of the modern theory of quantum measurements, especially in quantum information theory. However, POVMs are just byproducts of quantum instruments. POVM does not determine uniquely a state transformation coupled to a measurement's feedback on the system's state.

In physics, quantum instruments and, in particular, POVMs associated with them were introduced at the advanced stage of quantum theory's development. However, modeling of cognition and decision making should be based on quantum instruments even for the basic psychological effects; for example, the combination of the question order and response replicability effects [302, 303]. The von Neumann measurement theory has a restricted domain of applications [48, 213].

The quantum instrument formalism is derived from open quantum systems theory through the indirect measurement scheme employing the unitary operator realization of the interaction between a system and a measurement apparatus [298]. The question order effect (QOE) is an effect of the dependence of the sequential joint probability distribution of answers on the questions' order:

$$p_{AB} \neq p_{BA}.$$

We remark that for classical probability:

$$\begin{aligned} p_{AB}(x,y) = P(\omega \in \Omega\colon A(\omega) = x, B(\omega) = y) = \\ P(\omega \in \Omega\colon B(\omega), A(\omega) = x) = p_{BA}(y,x). \end{aligned}$$

So, no QOE. But the experimental statistical data collected in social opinion pools demonstrated QOE. A simple and natural example, the Clinton–Gore opinion pool [282]. In this opinion-polling experiment, people were asked one question at a time, for example,

- A = "Is Bill Clinton honest and trustworthy?"
- B = "Is Al Gore honest and trustworthy?"

Two sequential probability distributions were calculated on the basis of the experimental statistical data, p_{AB} and p_{BA} (first question A and then question B and vice versa). Wang and Busemeyer [367] modelled QOE with projective instruments – PVMs, that is, with state update given by the Lüders projection postulate (see also [368]). However, this strategy was challenged in paper [213]. This challenge is related to RRE.

As was demonstrated in [48, 213], QOE+RRE combination cannot be modeled by von Neumann observables (with projection state update). The authors of article [302] have overcome this difficulty within quantum instrument theory by using the mathematical construction based on the indirect measurement scheme [298]. The unitary operators describing the interaction of a system and an observable (question) are directly written via their actions. In this model, the systems are humans and the observables are questions asked to them.

Article [302] contains rather long calculations in Hilbert space. They are not complicated, but might be boring for inexperienced readers. One may be satisfied by the statement that, for QOE+RRE, it is possible to construct quantum instruments (by using the indirect measurement scheme). For more experienced readers, the calculations can serve as the basis example of construction of instruments for various combinations of psychological or social effects.

We remark that, in contrast to QOE that attracted a lot of attention in experimental psychology and decision making and was supported by plenty of statistical data [282], RRE is not supported by experimental studies. We can mention just one experiment [77]. We hope that the result and discussion in paper [304] would stimulate psychologists to perform experiments to check RRE.

Wang and Busemeyer derived the famous QQ-equality (QQE), the amazing (and unexpected) property of cognitive and social data [368]. This is a nonparametric inequality to which the probabilities of an experiment must satisfy in order for a quantum model to exist for them, as follows:

$$p(AyBn) + p(AnBy) - p(ByAn) - p(BnAy) = 0,$$

where A and B correspond to questions with two possible outcomes: 'Yes' and 'No'. The joint probabilities are the probabilities of receiving given answers to questions A and B in the same order as they appear, for example $P(AyBn)$ means the probability of obtaining a negative answer to question B before obtaining an affirmative answer to question A. The QQ-equality can be easily derived within the standard von Neumann measurement theory, with Hermitian operators and the projection state update. However, von Neumann theory is not powerful enough to describe the combination of QQE with other psychological effects, for example QOE + RRE.

Ozawa and Khrennikov [303] proved that the combination QOE+RRE+QQE can be modeled within the theory of quantum instruments.

19

Noncommutative Probability in Classical Systems from Experimental Contextuality

19.1 Classical Systems

It is widely believed that classical systems should be described by commutative, or Kolmogorovian, probability space. In this chapter the following question is considered:

Is it possible to observe correlation of noncommutative observables in a purely classical situation?

The positive answer would mean that the classical system will be described by a noncommutative probability space.

In Chapter 10 a contextual probabilistic approach to the statistical theory of measurements is presented. This general approach is neutral to the size of a system, macro or micro. The contextual measurement model (CMM) operates with measurement contexts. By taking into account dependence of probabilities on complexes of experimental physical conditions, measurement contexts, one can derive quantum-like interference for probabilities of alternatives – formula of total probability (FTP) with an interference term. The contextual derivation is not directly related to the special quantum (e.g. superposition) features of the states of physical systems or generally on the possibility of constructing linear space representation (LSR) for probability. The same can be said about violation of the Bell inequalities. The contextual approach to interference and the Bell inequalities stimulated the search for noncommutative structures in classical statistical physical models, for example disordered systems; see article [194] serving as the base for this chapter. Following this article, we discuss two examples of *classical statistical mechanical systems,* where we obtain the correlation functions in *noncommutative probability space.*

Our approach, in principle, may be compared with experiments in the following standard way. The traditional way to distinguish between classical and quantum systems is the Bell inequality, satisfied by classical correlation functions. Thus if we would find violation of the Bell inequality, we will prove that the system is described by noncommutative probability space. However, our present theoretical considerations are still far from such experimental study.

19.2 Time Averaging

In the present section we discuss the following problem. Consider the dynamics of a quantum system, described by some Hamiltonian $\hat{H}_0$ and the algebra of observables $\mathcal{A}$. Let this (noncommutative) algebra of observables $\mathcal{A}$ contain some (commutative) classical subalgebra $\mathcal{A}_{\text{cl}}$. This classical subalgebra is not conserved by time evolution, but for $X, Y \in \mathcal{A}_{\text{cl}}$ the time evolutions

$$\hat{X}(t) = e^{it\hat{H}_0}\hat{X}e^{-it\hat{H}_0}, \quad \hat{Y}(t) = e^{it\hat{H}_0}\hat{Y}e^{-it\hat{H}_0}$$

will commute by definition.

Let us assume that the time evolution, defined by Hamiltonian $\hat{H}_0$ is very fast, and in experiment we observe some time-averaged observables. These time-averaged observables, in general, already will not commute, since the classical subalgebra is not conserved by time evolution. This means that, in principle, we might expect that these time-averaged operators for classical physical variables $\hat{X}(t)$, $\hat{Y}(t)$ from the classical subalgebra $\mathcal{A}_{\text{cl}}$ will have nonclassical correlations. A natural example of this kind of behavior is observed in *the quantum stochastic limit* approach [7]. In this approach we consider the quantum system with the Hamiltonian in the form

$$\hat{H} = \hat{H}_0 + \lambda\hat{H}_I,$$

where $\hat{H}_0$ is called the free Hamiltonian, $\hat{H}_I$ is called the interaction Hamiltonian, and $\lambda \in \mathbf{R}$ is the coupling constant.

We investigate the dynamics of the system in the new slow time scale of the stochastic limit, taking the van Hove time rescaling [358]

$$t \mapsto t/\lambda^2$$

and considering the limit $\lambda \to 0$. In this limit [7] the free evolutions of the suitable collective operators

$$\hat{A}(t,k) = e^{it\hat{H}_0}\hat{A}(k)e^{-it\hat{H}_0}$$

will become quantum white noises:

$$\lim_{\lambda\to 0}\frac{1}{\lambda}\hat{A}\left(\frac{t}{\lambda^2},k\right) = \hat{b}(t,k).$$

The convergence is understood in the sense of correlators. The $\lambda \to 0$ limit describes the time averaging over infinitesimal intervals of time and allows us to investigate the dynamics on large time scales, where the effects of interaction with the small coupling constant λ are important.

For the details of the procedure, see [7].

The collective operators describe joint excitations of different degrees of freedom in systems with interaction, and may have the form of polynomials over creations and annihilations of the field, or may look like combinations of the field and particles operators and so on. For example, for nonrelativistic quantum electrodynamics without the dipole approximation the collective operator is

$$\hat{A}_j(k) = e^{ik\hat{q}}\hat{a}_j(k) \tag{19.1}$$

where $\hat{a}_j(k)$ is the annihilation of the electromagnetic (Bose) field with wave vector k and polarization j, $\hat{q} = (\hat{q}_1, \hat{q}_2, \hat{q}_3)$ is the position operator of quantum particle (say electron), $\hat{q}k = \sum_i \hat{q}_i k_i$.

The nontrivial fact is that, after the $\lambda \to 0$ limit, depending on the form of the collective operator, the statistics of the noise $b(t, k)$ depend on the form of the collective operator and may be nontrivial.

Consider the following examples.

(1) We may have the following possibility:

$$[\hat{b}_i(t,k), \hat{b}_j^\dagger(t',k')] = 2\pi\delta_{ij}\delta(t-t')\delta(k-k')\delta(\omega(k)-\omega_0), \tag{19.2}$$

which corresponds to the quantum electrodynamics in the dipole approximation, describing the interaction of the electromagnetic field with a two-level atom with the level spacing (energy difference of the levels) equal to ω_0. Here $\omega(k)$ is the dispersion of quantum field.

In this case the quantum noise will have the Bose statistics, and different annihilations of the noise will commute:

$$[\hat{b}_i(t,k), \hat{b}_j(t',k')] = 0.$$

(2) The other possibility is the relation

$$\hat{b}_i(t,k)\hat{b}_j^\dagger(t',k') = 2\pi\delta_{ij}\delta(t-t')\delta(k-k')\delta(\omega(k)+\varepsilon(p)-\varepsilon(p+k)) \tag{19.3}$$

which corresponds to the quantum electrodynamics without the dipole approximation (19.1). Here $\omega(k)$ and $\varepsilon(p)$ are dispersion functions of the field and of the particle correspondingly.

In this case the quantum noise will have the quantum Boltzmann statistics [7], [4], [5], and different annihilations of the noise will not commute:

$$\hat{b}_i(t,k)\hat{b}_j(t',k') \neq \hat{b}_j(t',k')\hat{b}_i(t,k).$$

The commutation relations of the types (19.2), (19.3) are universal in the stochastic limit approach (a lot of systems will have similar relations in the stochastic limit $\lambda \to 0$).

Take two operators $\hat{b}_i(t,k)$ for the same time and fixed polarization and consider the combinations

$$\hat{X}(k) = \hat{b}_i(t,k) + \hat{b}_i^\dagger(t,k), \qquad \hat{X}(k') = \hat{b}_i(t,k') + \hat{b}_i^\dagger(t,k'), \tag{19.4}$$

which correspond to the coordinate operator.

Then for the case of quantum Boltzmann relations (19.3) we have

$$\hat{X}(k)\hat{X}(k') \neq \hat{X}(k')\hat{X}(k).$$

Of course, we need some regularization of the product of generalized functions.

Operators $\hat{X}(k)$ before the stochastic limit belonged to the classical subalgebra. More precisely, corresponding combinations $\hat{x}(k)$ of interacting operators (19.1) would belong to the classical subalgebra:

$$\hat{x}(k) = \hat{A}_i(k) + \hat{A}_i^\dagger(k)$$
$$[\hat{x}(k), \hat{x}(k')] = 0.$$

We proved that for the case when, after the stochastic limit, the statistics of the field become quantum Boltzmannian, operators (19.4) will not commute even if we take them for equal time. This is not a mystery, since in the stochastic limit we work with the time-averaged observables. But in real experiments we may observe the result of time averaging. The discussed example shows that for quantum electrodynamics beyond the dipole approximation we may observe quantum correlations for time-averaged observables in the classical subalgebra.

The situation considered in the present section is similar in some sense to the results of [187], [188]. In these papers the quantum-like interference fringes were observed for the discretization of the classical dynamical system. The discretization is an analog, in some sense, of the time-averaging procedure. Probably the results of [187], [188] might be possible to embed into the frameworks of the approach considered in the present section.

We would like to mention that long ago De Muynck, see for example [105], discussed the analogy between thermodynamics and quantum mechanics. They considered in the EPR–Bohm framework quantum expectations as a kind of thermodynamic averages. This should induce violation of Bell's inequality.

19.3 Disordered Systems

In the present section we discuss the possibility of using noncommutative probability to describe (classical) disordered systems, following [254], [255]. In these papers the new procedure, called the noncommutative replica procedure, which is an analog of the replica procedure of Edwards and Anderson [114], was proposed to describe the statistical mechanics of quenched disordered systems (for example, spin glasses).

We will not discuss here the standard replica approach; see for an introduction to spin glasses and the replica method [114], [281], [56].

Consider the disordered system with Hamiltonian $H[\sigma, J]$ which depends on the random parameter J, which in the most interesting cases (for spin glasses, for instance) is the large random $N \times N$ matrix with independent Gaussian matrix elements J_{ij}, considered in the thermodynamic $N \to \infty$ limit.

To describe the system with quenched disorder in [254], [255] it was proposed to consider the state described by the noncommutative replica statistic sum

$$Z^{(p)} = \int \sum_{\{\sigma\}} \exp\left(-\beta H[\sigma, \Delta J]\right) \prod_{a=0}^{p-1} \exp\left(-\frac{1}{2} \sum_{i \leq j}^{N} J_{ij}^{(a)2}\right) \prod_{i \leq j}^{N} dJ_{ij}^{(a)}, \qquad (19.5)$$

where Δ is the following coproduct operation:

$$\Delta : J_{ij} \mapsto \frac{1}{\sqrt{p}} \sum_{a=0}^{p-1} J_{ij}^{(a)}, \qquad (19.6)$$

which maps the matrix element J_{ij} into the linear combination of independent replicas $J_{ij}^{(a)}$, enumerated by the replica index a. This operation was called the quenching in [255].

In the large N limit, by the Wigner theorem, see [6, 19, 366, 377], the system of p random matrices with independent variables will give rise to the quantum Boltzmann algebra with p degrees of freedom with the generators $\hat{A}_a$, $\hat{A}_a^\dagger$, where $a = 0, \ldots, p-1$ and the relations

$$\hat{A}_a \hat{A}_b^\dagger = \delta_{ab}.$$

These operators are the limits of the large random matrices

$$\lim_{N \to \infty} \frac{1}{N} J_{ij}^{(a)} = \hat{Q}_a = \hat{A}_a + \hat{A}_a^\dagger,$$

where the convergence is understood in the sense of correlators (as in the central limit theorem).

Then in the thermodynamic limit $N \to \infty$ the noncommutative replica procedure (19.6) will take the form of the following map of the quantum Boltzmann algebra with one degree of freedom into quantum Boltzmann algebra with p degrees of freedom:

$$\Delta : \hat{Q} \mapsto \frac{1}{\sqrt{p}} \sum_{a=0}^{p-1} \hat{Q}_a.$$

Note that different $\hat{Q}_a$ do not commute. We see that we again have obtained noncommutative probability in a purely classical system.

Actually the picture is more complicated, compared to that discussed above. The correlations of the system in the noncommutative replica approach will be given by

$$\lim_{N\to\infty} \langle(\Delta J)^k\rangle = \left\langle \left(\frac{1}{\sqrt{p}} \sum_{a=0}^{p-1} \hat{Q}_a \right)^k \right\rangle,$$

where the state $\langle\cdot\rangle$ is generated by the noncommutative replica statistic sum (19.5). In principle, it is not clear how to extract noncommutativity from this set of correlators, since different degrees $\left(\Delta\hat{Q}\right)^k$ commute.

To distinguish noncommutative and commutative systems we have to consider the set of correlation functions which will be large enough. In the present case this set should contain correlations of different linear combinations of $\hat{Q}_a$, more general than $\Delta\hat{Q}$.

This problem may be discussed in the following way. The quenching (19.6) in principle may be related to a particular way of preparation of the disordered system under consideration. If we were to use different physical preparation of the disordered system, this might result in the different quenching procedure. The example of quenching different from (19.6) was discussed in [255]. This example has the following form:

$$\Delta' : J \mapsto \frac{1}{\sqrt{p}} \sum_{a=0}^{p-1} c_a J_a, \tag{19.7}$$

where c_a are real-valued coefficients, which should satisfy the condition

$$\sum_{a=0}^{p-1} c_a^2 = p.$$

Varying coefficients c_a, we will obtain different quenchings.

Then, using different physical preparations of the system, we measure the correlation functions which will correspond to different quenchings. After we (at least in principle) use the Bell inequality to prove, do we really have a noncommutative probability space which describes the behavior of the disordered system under investigation.

Actually the most natural example of this setup is the experiments with spin glasses, where the glass transition with a different external magnetic field was investigated, and nontrivial behavior of magnetization on preparation was observed [281], [56].

We propose to analyze these experiments, taking into account the correlations between the systems with different preparations (i.e. frozen in the presence of different external magnetic fields, which in our approach should correspond to different quenchings), and to check the validity of the Bell inequalities.

This would help to check the validity of the noncommutative replica approach itself, since there is no direct way to introduce noncommutativity in the standard replica approach.

The experimental situation discussed here could also be described by using a contextual probabilistic approach; see [185], [186]. In the contextual framework probabilities (which are interpreted as conventional ensemble probabilities) depend on physical contexts — complexes of experimental physical conditions. Mathematically this means that we could not use one fixed Kolmogorov probability space and we should work with a system of probability spaces depending on physical contexts; see (10.39), section 10.2. In our case *various contexts are defined by choosing various external magnetic fields.*

We recall once again that by using the contextual probabilistic approach it is possible to obtain a "quantum rule" for interference of probabilities of alternatives; see [185], [186]. Such an interference of probabilities is just another way to describe noncommutativity of probabilities (induced in the conventional formalism by using Hilbert space calculus). Therefore we could expect that, by taking into account contextuality of statistics for disordered systems frozen in the presence of different external magnetic fields, it would be possible to find experimental confirmations of the presence of a noncommutative structure for classical disordered systems (in particular, spin glasses).

References

[1] Accardi, L. Accardi, L. Topics in quantum probability. *Physics Reports*, **77**(3): (1981), 169–192.

[2] Accardi, L. (1984). The probabilistic roots of the quantum mechanical paradoxes. In *The Wave-Particle Dualism: A Tribute to Louis de Broglie on his 90th Birthday* (pp. 297–330). Springer Netherlands.

[3] Accardi, L. *Urne e camaleonti*. Il Saggiatore (1997).

[4] Accardi, L., Lu, Y. G., Volovich, I. V. Interacting Fock spaces and Hilbert module extensions of the Heisenberg commutation relations, Publications of IIAS, 1997, pp. 21–44.

[5] Accardi, L., Kozyrev, S. V., Volovich, I. V. Dynamical q-deformation in quantum theory and the stochastic limit, *Journal of Physics A*, **32**: (1999), 3485–3495.

[6] Accardi, L., Lu, Y. G., Volovich, I. V. Noncommutative (quantum) probability, master fields and stochastic bosonisation, arXiv:hep-th/9412241.

[7] Accardi, L., Lu, Y. G., Volovich, I. V. *Quantum Theory and Its Stochastic Limit*, Texts and Monographs in Physics, Springer (2002).

[8] Accardi, L., Adenier, G., Fuchs, C., Jaeger, G., Khrennikov, A., Larsson, J.-A., and Stenholm, S. (eds.). Foundations of Probability and Physics 5, **1101**. AIP Conference Proceedings, Melville, NY (2009).

[9] Accardi, L. New challenges for classical and quantum probability. *Entropy*, **24**(10): (2022), 1502.

[10] Adenier, G., Fuchs, C. A., and Khrennikov, A. Yu. (eds.). Foundations of Probability and Physics 4 **889**. AIP Conference Proceedings, Melville, NY (2007).

[11] A. Aiello et al., Quantum-like nonseparable structures in optical beams. *New Journal of Physics* 17: (2015), 043024.

[12] Allahverdyan, A. E., and Nieuwenhuizen, T. M. Steady adiabatic state: Its thermodynamics, entropy production, energy dissipation, and violation of Onsager relations. *Physical Review E*, **62**(1): (2000), 845.

[13] Allahverdyan, A. E., Balian, R., and Nieuwenhuizen, T. M. Curie-Weiss model of the quantum measurement process *Europhysics Letters*, **61**: (2003), 452.

[14] Allahverdyan, A. E., and Nieuwenhuizen, T. M. The second law and fluctuations of work: The case against quantum fluctuation theorems. arXiv:cond-mat/0408697.

[15] Allahverdyan, A. E., Khrennikov, A., and Nieuwenhuizen, T. M. (2005). Brownian entanglement. arXiv:quant-ph/0412132.

[16] Allahverdyan, A. E., Khrennikov, A., and Nieuwenhuizen, T. M. Brownian entanglement. *Physical Review A*, **72**(3): (2005), 032102.

[17] Allahverdyan, A. E., Balian, R., and Nieuwenhuizen, T. M. Understanding quantum measurement from the solution of dynamical models. *Physics Reports*, 525(1): (2013), 1–166.

[18] Araujo, M., Quintino, M. T., Budroni, C., Cunha, M. T., and Cabello, A. All noncontextuality inequalities for then-cycle scenario. *Physical Review A* **88**: (2013), 022118.

[19] Aref'eva, I. Ya., and Volovich, I. V. The master field for QCD and q-deformed quantum field theory, *Nuclear Physics B* **462**: (1996), 600–612.

[20] Arndt, M., Juffmann, T., and Vedral, V. Quantum physics meets biology. *HFSP Journal* **3**: (2009), 386–400.

[21] Arnol'd, V. I. On teaching mathematics. *Russian Mathematical Surveys*, **53**(1): (1998), 229–248.

[22] Asano, M., Ohya, M., Tanaka, Y., Khrennikov, A., and Basieva, I. On application of Gorini–Kossakowski–Sudarshan–Lindblad equation in cognitive psychology. *Open Systems & Information Dynamics* **18**(01): (2011), 55–69.

[23] Asano, M., Ohya, M., Tanaka, Y., Basieva, I., and Khrennikov, A. Quantum-like model of brain's functioning: Decision making from decoherence. *Journal of Theoretical Biology* **281**(1): (2011), 56–64.

[24] Asano, M., Hashimoto, T., Khrennikov, A., Ohya, M., and Tanaka, Y. Violation of contextual generalization of the Leggett–Garg inequality for recognition of ambiguous figures. *Physica Scripta*, **T163**: (2014), 014006.

[25] Asano, M., Khrennikov, A., Ohya, M., Tanaka, Y., and Yamato, I. *Quantum Adaptivity in Biology: From Genetics to Cognition*. Springer (2015).

[26] Asano, M., Basieva, I., Khrennikov, A., and Yamato, I. A model of differentiation in quantum bioinformatics. *Progress in Biophysics and Molecular Biology Part A* **130**: (2017), 88–98.

[27] Ashby, G., Colonius, H., and Dzhafarov, E. N. (eds.). Preface: Mathematical psychology in a quest for conceptual clarity. *New Handbook of Mathematical Psychology* **3**, Perceptual and Cognitive Processes (Cambridge Handbooks in Psychology). Cambridge University Press (2023).

[28] Aspect, A., Dalibard, J., and Roger, G. *Physical Review Letters* **49**: (1982), 180.

[29] Aspect, A. Experimental tests of Bell's inequalities in atomic physics, in *Atomic Physics* 8, Proceedings of the Eighth International Conference on Atomic Physics, edited by I. Lindgren, A. Rosen and S. Svanberg (pp. 103–128) (1982).

[30] Aspect, A. (1983). Three experimental tests of Bell inequalities by the measurement of polarization correlations between photons (Orsay). https://inis.iaea.org/search/search.aspx?orig_q=RN:16018802.

[31] Aspect, A. Bell's Theorem: The naive view of an experimentalist. arXiv:quant-ph/0402001.

[32] Aspect, A., and Grangier, P. Closing the door on Einstein and Bohr's quantum debate. Video, special session Big event – Final Bell test, A. Khrennikov (chairman), Quantum and Beyond conference, Växjö, 2016. https://play.lnu.se/media/t/0_61nm165n.

[33] Atmanspacher, H. (1994). Is the ontic/epistemic-distinction sufficient to describe quantum systems exhaustively. In *Symposium on the Foundations of Modern Physics*, edited by K. V. Laurikainen. C. Montonen, and K. Sunnarborg (pp. 15–32). Editions Frontières.

[34] Atmanspacher, H., and Primas, H. (2003). Epistemic and ontic quantum realities. In *Time, Quantum and Information*, edited by L. Castell and O. Ischebeck (pp. 301–321). Springer.

[35] Atmanspacher, H. Determinism is ontic, determinability is epistemic, in *Between Chance and Choice: Interdisciplinary Perspectives on Determinism*, edited by H. Atmanspacher and R. C. Bishop (pp. 49–74). Imprint Academic (2002).

[36] Aumann, R. J. Agreeing to disagree In Arló-Costa, H., Hendricks, V., van Benthem, J. (eds.) *Readings in Formal Epistemology*. Springer Graduate Texts in Philosophy, vol. 1. (pp. 859–862). Springer International Publishing (2016).

[37] Aumann, R. J. Backward induction and common knowledge of rationality. *Games and Economic Behavior* **8**(1): 1995, 6–19.

[38] Bacciagaluppi, G. Delocalized properties in the modal interpretation of a continuous model of decoherence. *Foundations of Physics*, **30**: (2000), 1431–1444.

[39] Bacciagaluppi, G. (2005). A conceptual introduction to Nelson's mechanics. In *Endophysics, Time, Quantum and the Subjective* (with CD-ROM) (pp. 367–388).

[40] Bacciagaluppi, G., and Valentini, A. *Quantum Theory at the Crossroads: Reconsidering the 1927 Solvay Conference*. Cambridge University Press (2009).

[41] Ballentine, L. E. (1986). Probability theory in quantum mechanics. *American Journal of Physics*, **54**(10): (1986), 883–889.

[42] L. Ballentine, Probability in quantum mechanics. *Annals of New York Academy of Science, Techniques and Ideas in Quantum Measurement Theory*, **480**(1): (1986), 382–392.

[43] Ballentine, L. E. The statistical interpretation of quantum mechanics, *Reviews of Modern Physics*, **42**: (1989), 358–381.

[44] Ballentine, L. E. *Quantum Mechanics: A Modern Development*. WSP, (2014).

[45] Ballentine, L. E. Interpretations of probability and quantum theory. In Khrennikov, A. Yu. (ed.). *Foundations of Probability and Physics*, Quantum Probability and White Noise Analysis **13**: (2001), 71–84. WSP.

[46] Balian, R. On the principles of quantum mechanics and the reduction of the wave packet. *American Journal of Physics*, **57**(11): (1989), 1019–1027.

[47] H. Bartosik, J. Klepp, C. Schmitzer, S. Sponar, A. Cabello, H. Rauch, and Y. Hasegawa, Experimental test of quantum contextuality in neutron interferometry. *Physical Review Letters* **103**: (2009), 040403.

[48] Basieva, I., and Khrennikov, A. On the possibility to combine the order effect with sequential reproducibility for quantum measurements. *Foundations of Physics* **45**(10): (2015), 1379–1393.

[49] Basieva, I., and Khrennikov, A. Testing boundaries of applicability of quantum probabilistic formalism to modeling of cognition: Metaphors of two and three slit experiments. In: *International Symposium on Quantum Interaction*, pp. 49–56. Springer (2016).

[50] Basieva, I., Cervantes, V. H., Dzhafarov, E. N., and Khrennikov, A. True contextuality beats direct influences in human decision making. *Journal of Experimental Psychology: General*, **148**(11): (2019), 1925.

[51] Basieva, I., and Khrennikov, A. Conditional probability framework for entanglement and its decoupling from tensor product structure. *Journal of Physics A: Mathematical and Theoretical*, **55**(39): (2022), 395302.

[52] Basieva, I., and Khrennikov, A. Entanglement of observables: Quantum conditional probability approach. *Foundations of Physics*, **53**: (2023), 84.

[53] Bell, J. S. On the problem of hidden variables in quantum theory. *Reviews of Modern Physics* **38**: (1966), 450.

[54] Bell, J. S. *Speakable and Unspeakable in Quantum Mechanics*, 2nd ed.; Cambridge University Press (2004).

[55] Beltrametti, E. G., and Cassinelli, C. The logic of quantum mechanics. *SIAM* **25**: (1983), 429–431.

[56] Binder, K., and Young, A. P. Spin glasses: experimental facts, theoretical concepts, and open questions, *Reviews of Modern Physics* **58**(4): (1992), 801–976.

[57] Blokhintsev, D. I. *Quantum Mechanics*, Gordon and Breach Publishing, New York, 1964.

[58] Blokhintsev, D. I. *Philosophy of Quantum Mechanics*, D. Reidel Publishing (1968).

[59] Bohr, N. The quantum postulate and the recent development of atomic theory. *Supplement to Nature*, April 14, 580–590 (1928).

[60] Bohr, N. Wirkungsquantum und Naturbeschreibung. Naturwissenschaft, 17, (1929).

[61] Bohr, N. The quantum of action and the description of nature. In: *Foundations of Quantum Physics I (1926–1932)*. Niels Bohr collected works. J. Kalckar (ed.), v. 6, Elsevier B.V., 1985, pp. 201–217.

[62] Bohr, N. Light and life. *Nature* **133** (1933), 421–423 (Pt. 1), 457–459 (Pt. 2).

[63] Bohr, N. Can quantum-mechanical description of physical reality be considered complete? *Physical Review*, **1935**, 48, 696–702.

[64] Bohr, N. in *Collected Works*, ed. J. Kalckar (North-Holland, Amsterdam, 1985), Vol. 6, pp. 316–330, 376–377.

[65] Bohr, N. *The Philosophical Writings of Niels Bohr*. Ox Bow Press (1987).

[66] Boltzmann, L. (1905). Uber die Frage nach der objektiven Existenz der Vorgnge in der unbelebten Natur. In Barth, J. A. (Ed.) *Populre Schriften*: Leipzig.

[67] Boltzmann, L. (1974). On the development of the methods of theoretical physics in recent times. In McGuinness, B. (Ed.) *Theoretical Physics and Philosophical Problems*. Vienna circle collection, Vol. 5. Springer.

[68] Bunge, M. *Causality and Modern Science*. Routledge (2017).

[69] Boole, G. On the theory of probabilities. *Philosophical Transactions of the Royal Society* **152** (1862), 225–242.

[70] Boole, G. *An Investigation of the Laws of Thought*. Dover (1958)

[71] Boughn, S. Making sense of Bell's theorem and quantum nonlocality. *Foundations of Physics*, **47** (2017), 640–657.

[72] Brown, K. *Reflections on Relativity*. Lulu.com Publ. (2020)

[73] Braunstein, S. L., Mann, A., and Revzen, M. Maximal violation of Bell inequalities for mixed states. *Physical Review Letters* **68** (1992), 3259–3261.

[74] Brukner, C., and Zeilinger, A. Operationally invariant information in quantum mechanics. *Physical Review Letters* **83** (1999), 3354–3357.

[75] Brukner, C., and Zeilinger, Quantum physics as a science of information. In: Elitzur, A., Dolev, S., Kolenda, N. (eds.) *Quo Vadis Quantum Mechanics?* Springer (2005).

[76] Bruza, P. D., Kitto, K., Ramm, B. J., and Sitbon, L. A probabilistic framework for analysing the compositionality of conceptual combinations. *Journal of Mathematical Psychology*, **67** (2015), 26–38.

[77] Busemeyer, J. R. and Wang, Zh. Is there a problem with quantum models of psychological measurements? *PLOS ONE* **2**(11): (2017), e0187733.

[78] Cabello, A., Filipp, S., Rauch, H., and Hasegawa, Y. Proposed experiment for testing quantum contextuality with neutrons. *Physical Review Letters*, **100** (2008), 130404.

[79] Cervantes, V. H., and Dzhafarov, E. N. Snow queen is evil and beautiful: Experimental evidence for probabilistic contextuality in human choices. *Decision* **5**(3): (2018), 193.

[80] Conte, E., Khrennikov, A., Todarello, O., Federici. A., Mendolicchio, L., and Zbilut, J. P. A preliminary experimental verification on the possibility of Bell inequality violation in mental states. *NeuroQuantology* **6** (2008), 214–221.

[81] Cetto, A. M., Valdes-Hernandez, A., and de la Pena, L. On the spin projection operator and the probabilistic meaning of the bipartite correlation function. *Found. Phys.* **50** (2020), 27–39.

[82] Chiribella, G., D'Ariano, G. M., and Perinotti, P. Realization schemes for quantum instruments in finite dimensions. *Journal of Mathematical Physics*, **50**(4): (2009), 042101.

[83] Clauser, J. F., Horne, M. A., Shimony, A., and Holt, R. A. Proposed experiment to test local hidden-variable theories. *Physical Review Letters* **23** (1969), 880.

[84] Clauser, J. F., and Horne, M. A. Experimental consequences of objective local theories. *Physical Review Research* D **10** (1974), 526–535.

[85] Clauser, J. F., and Shimony, A. Bell's Theorem. Experimental Tests and Implications. *Reports on Progress in Physics*, **41** (1978), 1881–1927.

[86] Collins, D., and Popescu, S. Classical analog of entanglement. *Physical Review A*, **65**(3): (2002), 032321.

[87] Cugliandolo, L. F., Kurchan, J., and Peliti, L. Energy flow, partial equilibration, and effective temperatures in systems with slow dynamics. *Physical Review E*, **55**(4): (1997), 3898.

[88] D'Agostino, S. Boltzmann and Hertz on the Bild conception of physical theory, *History of Science* **28** (1990), 380–398.

[89] D'Agostino, S. Continuity and completeness in physical theory: Schrödinger's return to the wave interpretation of quantum mechanics in the 1950s, in Michel Bitbol and Olivier Darrigol (eds.), *Erwin Schrödinger: Philosophy and the Birth of Quantum Mechanics* (Gifsur-Yvette: Editions Frontieres, 1992)

[90] D'Agostino, S. Schrödinger on continuity and completeness in physical theory, *Physis* **29** (1992), 539–561.

[91] D'Agostino, S. On Boltzmann's mechanics and his Bild conception of physical theory, in S. D'Agostino, *A History of the Ideas of Theoretical Physics: Essays on the Nineteenth and Twentieth Century Physics* (Dordrecht: Kluwer, 2000), pp. 201-211.

[92] D'Agostino, S. Hertz's researches and their place in nineteenth century theoretical physics, *Centaurus* **36** (1993), 46–82.

[93] D'Agostino, S. Bild conception of physical theory. *Physics in Perspective* **6** (2004), 372–389.

[94] D'Ariano, G., M., Chiribella, G., and Perinotti, P. *Quantum Theory from First Principles: An Informational Approach*. Cambridge University Press (2017).

[95] D'Ariano, G. M., Perinotti, P., and Tosini, A. Incompatibility of observables, channels and instruments in information theories. *Journal of Physics A: Mathematical and Theoretical*, **55**(39): (2022), 394006.

[96] D'Ariano, G. M. No purification ontology, no quantum paradoxes. *Foundations of Physics* **50**(2020): 1921–1933.

[97] Davies, E. B., and Lewis, J. T. An operational approach to quantum probability. *Communications in Theoretical Physics* **17** (1970), 239–260.

[98] Davies, E. B. *Quantum Theory of Open Systems* (Academic Press, London, 1976).

[99] De Broglie, L. *The Current Interpretation of Wave Mechanics: A Critical Study*. Elsevier (1964).

[100] De Broglie, L. Sur la réfutation du théoreme de Bell. *Comptes Rendus de l'Academie des Sciences B*, **278** (1974), 721–722.

[101] DeBrota, J. B., Fuchs, C. A., Pienaar, J. L., and Stacey, B. C. Born's rule as a quantum extension of Bayesian coherence. *Physical Review A*, **104**(2): (2021), 022207.

[102] D'Espagnat, B. *Veiled Reality*. Addison-Wesley (1995).

[103] de la Peña, L., and Cetto, A. M. *The Quantum Dice: An Introduction to Stochastic Electrodynamics*. Kluwer. (1996).
[104] De Muynck, M., De Baere, W., and Martens, H. Interpretations of quantum mechanics, joint measurement of incompatible observables, and counterfactual definiteness. *Foundations of Physics* **24**(12): (1994), 1589–1664.
[105] De Muynck, W. M. Interpretations of quantum mechanics, and interpretations of violation of Bell's inequality, in PQ-QP: Quantum Probability and White Noise Analysis, 13: In: *Foundations of Probability and Physics*, A. Khrennikov (ed.), World Scientific (2001).
[106] De Muynck, W. M. *Foundations of Quantum Mechanics: An Empiricist Approach*. Springer (2006).
[107] De la Pena, L., Cetto, A. M., and Brody, T. A. On hidden variable theories and Bell's inequality. *Lettere al Nuovo Cimento* **5** (1972), 177.
[108] De Raedt, H., Hess, K., and Michielsen, K. Extended Boole-Bell inequalities applicable to quantum theory. *Journal of Computational and Theoretical Nanoscience* **8** (2011), 10119.
[109] Dirac, P. *The Principles of Quantum Mechanics*, 4th ed. Clarendon Press (2012).
[110] Dzhafarov, E. N., and Kujala, J. V. Selectivity in probabilistic causality: Where psychology runs into quantum physics. *Journal of Mathematical Psychology* **56** (2012), 54–63.
[111] Dzhafarov, E. N., Kujala, J. V., Cervantes, V. H., Zhang, R., and Jones, M. On contextuality in behavioral data. *Philosophical Transactions of the Royal Society A* **374** (2016), 20150234.
[112] Dzhafarov, E. Contextuality and Random Variables. In: Plotnitsky, A., Haven, E. (eds.) *The Quantum-Like Revolution*. Springer (2023).
[113] Eberhard, P. H. Background level and counter efficiencies required for a loophole-free Einstein–Podolsky–Rosen experiment. *Physical Review A*, **47** (1993), 477–750.
[114] Edwards, S. F., and Anderson, P. W. (1975). Theory of spin glasses. *Journal of Physics F: Metal Physics*, **5**(5): (1975), 965.
[115] Einstein, A., Podolsky, B., and Rosen, N. Can quantum-mechanical description of physical reality be considered complete? *Physical Review*, **1935**, *47*, 777–780.
[116] Einstein, A. Reply to criticism, In: P. A. Schilpp, *Albert Einstein: Philosopher-Scientist. (The Library of Living Philosophers)*. Cambridge University Press, 1949, pp. 665–688
[117] Einstein, A., and Infeld, L. *Evolution of Physics: The Growth of Ideas from Early Concepts to Relativity and Quanta*. Simon and Schuster (1961).
[118] Fenyes, I. Eine wahrscheinlichkeitstheoretische begründung und interpretation der quantenmechanik. *Zeitschrift für Physik*, **132** (1952), 81–106.
[119] Feynman, R. P. (1951). *The concept of probability in quantum mechanics*, Proceedings of the Second Berkeley Symposium on Mathematical Statistics and Probability (University of California Press), 533–541.
[120] Feynman, R. and Hibbs, A. *Quantum Mechanics and Path Integrals*. McGraw-Hill (1965).
[121] Filipp, S., and Svozil, K. Tracing the bounds on Bell-type inequalities. *AIP Conference Proceedings* **2005**, 750, 87–94.
[122] Filipp, S., and Svozil, K. Generalizing Tsirelson's bound on Bell inequalities using a min-max principle. *Physical Review Letters*, **2004**, 93, 130407.
[123] Fine, A. Hidden variables, joint probability, and the Bell inequalities. *Physical Review Letters* **48** (1982), 291–295.

[124] Fuchs, C. A. Quantum mechanics as quantum information (and only a little more). In: Khrennikov, A. (ed.), *Quantum Theory: Reconsideration of Foundations*, Ser. Math. Model. **2**, pp. 463–543. Växjö Univ. Press, Växjö (2002)

[125] Fuchs, C. A. The anti-Växjö interpretation of quantum mechanics. In: Khrennikov, A. (ed.), *Quantum Theory: Reconsideration of Foundations*, Ser. Math. Model. **2**, pp. 99–116. Växjö Univ. Press, Växjö (2002)

[126] Fuchs, C. A. QBism, the perimeter of quantum Bayesianism. 2010, https://arxiv.org/pdf/1003.5209.pdf.

[127] Fuchs, C. A., and Schack, R. A quantum-Bayesian route to quantum-state space. *Foundations of Physics* **41** (2011), 345–356.

[128] Fuchs, C. A., and Schack, R. QBism and the Greeks: why a quantum state does not represent an element of physical reality. *Physica Scripta* **90** (2014), 015104.

[129] Fuchs, C. A., Mermin, N. D., and Schack, R. An introduction to QBism with an application to the locality of quantum mechanics. *American Journal of Physics* **82** (2014), 749.

[130] Fuchs, C. A. (2017). Notwithstanding Bohr, the reasons for QBism. *Mind and Matter*, **15**(2): (2017), 245–300.

[131] Fuchs, C. A. (2023). *QBism, Where Next?* arXiv preprint arXiv:2303.01446.

[132] Gardiner, C. W. *Quantum Noise*. Springer (1991).

[133] Garg, A., and Mermin, N. D. Detector inefficiencies in the Einstein-Podolsky-Rosen experiment, *Physical Review D*, **25**(12): (1987), 3831–3835.

[134] Gatti, A., Brambilla, E., Bache, M., and Lugiato, L. A. (2004). Ghost imaging with thermal light: Comparing entanglement and classical correlation. *Physical Review Letters*, **93**(9): (2004), 093602.

[135] Ghose, P., and Mukherjee, A. Entanglement in classical optics. *Reviews in Theoretical Science*, **2014**, *2*, 1–14.

[136] Gisin, N. Non-realism: Deep thought or a soft option? *Foundations of Physics*, **42** (2012), 80–85.

[137] Giustina, M., Mech, A., Ramelow, S., Wittmann, B., Koer, J., Beyer, J., Lita, A., Calkins, B., Gerrits, T., Woo Nam, S., Ursin, R., and Zeilinger, A. Bell violation with entangled photons, free of the fair-sampling assumption. *Nature* **2013**, 497, 227–230.

[138] Giustina, M., Versteegh, M. A. M., Wengerowsky, S., Handsteiner, J., Hochrainer, A., Phelan, K., Steinlechner, F., Kofler, J., Larsson, J.-A., Abellan, C., Amaya, W., Pruneri, V., Mitchell, M. W., Beyer, J., Gerrits, T., Lita, A. E., Shalm, L. K., Woo Nam, S., Scheidl, T., Ursin, R., Wittmann, B., and Zeilinger, A., A significant-loophole-free test of Bell's theorem with entangled photons, *Physical Review Letters* **115** (2015), 250401.

[139] Giustina, M., and Versteegh, M. Significant-loophole-free test of local realism with entangled photons. Video, special session *Big event – Final Bell test*, A. Khrennikov (chairman), Conf. *Quantum and Beyond*, Växjö, 2016. https://play.lnu.se/media/t/0_iasiwe06.

[140] Grangier, P., Roger, G., and Aspect, A. Experimental evidence for a photon anticorrelation effect on a beam splitter: A new light on single-photon interferences. *Europhysics Letters*, **1**(4): (1986), 173–179.

[141] Grangier, P. Violation of Bell's Inequalities in a quantum realistic framework. Video, special session *Big event – Final Bell test*, A. Khrennikov (chairman), Conf. *Quantum and Beyond*, Växjö, 2016. https://play.lnu.se/media/t/0_3wrv1e5c

[142] Grangier, P. Contextual objectivity and the quantum formalism. Int. J. Quantum Inf., **3**(01): (2005), 17–22.

[143] Grangier, P. Contextual inferences, nonlocality, and the incompleteness of quantum mechanics. *Entropy*, **23**(12): (2021), 1660.

[144] Grangier, P. (2023). Revisiting quantum mysteries. In: *The Quantum-Like Revolution: A Festschrift for Andrei Khrennikov* (pp. 181–189). Springer International Publishing.

[145] Grosberg, A. Yu., and Khokhlov, A. R. *Statistical Physics of Macromolecules*, American Institute of Physics (1994).

[146] Gudder, S. P. Convex structures and operational quantum mechanics. *Communications in Mathematical Physics* **29** (1973), 249–264.

[147] Gudder, S. P. *Stochastic Methods in Quantum Mechanics*. Courier Corporation (2014).

[148] Hanson, R. From the first loophole-free Bell test to a quantum Internet. Video, special session *Big event – Final Bell test*, A. Khrennikov (chairman), Conf. *Quantum and Beyond*, Växjö, 2016. https://play.lnu.se/media/t/0_q2r30syq

[149] Haven, E. Pilot-wave theory and financial option pricing. *International Journal of Theoretical Physics*, **44**(11): (2005), 1957–1962.

[150] Haven, E. and Khrennikov, A. Quantum mechanics and violation of the sure-thing principle: the use of probability interference and other concepts. *Journal of Mathematical Psychology*, **53**: (2009), 378–388.

[151] Haven, E. and Khrennikov, A. *Quantum Social Science*. Cambridge University Press (2013).

[152] Haven, E. and Khrennikov, A. Quantum probability and the mathematical modelling of decision-making. *Philosophical Transactions of the Royal Society A* **374**(2058): (2016), 20150105.

[153] Haven, E., and Khrennikov, A. Statistical and subjective interpretations of probability in quantum-like models of cognition and decision making. *Journal of Mathematical Psychology* **74**: (2016), 82–91.

[154] Haven, E., Khrennikov, A., and Robinson, T. R. *Quantum Methods in Social Science: A First Course*. WSP (2017).

[155] Haven, E., and Khrennikov, A. *The Palgrave Handbook of Quantum Models in Social Science*. Macmillan (2017).

[156] Haven, E. Quantum mechanical pragmatic rules and social science. *Activitas Nervosa Superior* **61**(1): (2019), 83–85.

[157] Heisenberg, W., *The Physical Principles of Quantum Theory*, Chicago University Press (1930).

[158] Helstrom, C. W. *Quantum Detection and Estimation Theory*. Academic (1976).

[159] Hensen, B., Bernien, H., Dréau, A. E., Reiserer, A., Kalb, N., Blok, M. S., Ruitenberg, J., Vermeulen, R. F. L., Schouten, R. N., Abellán, C., Amaya, W., Pruneri, V., Mitchell, M. W., Markham, M., Twitchen, D. J., Elkouss, D., Wehner, S., Taminiau, T. H., and Hanson, R. Experimental loophole-free violation of a Bell inequality using entangled electron spins separated by 1.3 km. *Nature* **526**: (2015), 682–686.

[160] Hertz, H. Untersuchungen ueber die Ausbreitung der elektrischen Kraft (J. A. Barth, 1892); translated by D. E. Jones as *Electric Waves being Researches on the Propagation of Electric Action with Finite Velocity through Space* (Macmillan, 1893; reprinted: Dover, 1962), p. 21.

[161] Hertz, H. *Die Prinzipien der Mechanik in neuen Zusammenhange dargestellt* (J. A. Barth, 1894), pp. xxi–xxii; translated by D. E. Jones and J. T. Walley as *The Principles of Mechanics Presented in a New Form* (Dover, 1956).

[162] Hess, K., and Philipp, W. A possible loophole in the theorem of Bell. *PNAS* **98**(25): (2001), 14224–14227.

[163] Hillery, M. O. S. M., O'Connell, R. F., Scully, M. O., and Wigner, E. P. Distribution functions in physics: Fundamentals. *Physics Reports* **106**(3): (1984), 121–167.
[164] Holevo, A. S. *Probabilistic and Statistical Aspects of Quantum Theory*. North-Holland, Amsterdam (1982).
[165] Holik, F., Massri, C., Plastino, A., and Saenz, M. Generalized probabilities in statistical theories. *Quantum Reports* **3**(3): (2021), 389–416.
[166] Hofmann, H. F., and Takeuchi, S. Violation of local uncertainty relations as a signature of entanglement. *Physical Review A*, **68**(3): (2003), 032103.
[167] Home, D., and Whitaker, M. A. Ensemble interpretations of quantum mechanics. A modern perspective. *Physics Reports*, **210**(4): (1992), 223–317.
[168] Horodecki, R., Horodecki, P., and Horodecki, M. Violating Bell inequality by mixed spin-1/2 states: Necessary and sufficient condition. *Physical Review A*, **200**: (1995), 340–344.
[169] Jaeger, G. *Quantum Information: An Overview*. Springer (2007).
[170] Jaeger, G. *Entanglement, Information, and the Interpretation of Quantum Mechanics (The Frontiers Collection)*. Springer (2009).
[171] Jaeger, G. Quantum contextuality and indeterminacy. Entropy **22**(8): (2020), 867.
[172] James, W. *The Principles of Psychology*. Henry Holt and Co. (1890); Reprinted: Harvard University Press (1983)
[173] Jaynes, E. T. Information theory and statistical mechanics. *Physical Review* **106**(4): (1957), 620.
[174] Jung K. Violation of Bell's inequality: Must the Einstein locality really be abandoned? *Journal of Physics: Conference Series* **880**: (2017), 012065, 1–8.
[175] Junge, M., and Palazuelos, C. Large Violation of Bell inequalities with low entanglement. *Communications in Mathematical Physics* **306**: (2011), 695–746.
[176] Itano, W. M., et al. Quantum projection noise: Population fluctuations in two-level systems. *Physical Review A*, **47.5:** (1993), 3554.
[177] Khrennikov, A. On the physical interpretation of negative probabilities in Prugovecki's empirical theory of measurement. *Canadian Journal of Physics*, **75**(5): (1997), 291–298.
[178] Khrennikov, A. Classical and quantum mechanics on information spaces with applications to cognitive, psychological, social and anomalous phenomena. Found. Physics **29**: (1999), 1065–1098.
[179] Khrennikov, A. Yu. *Interpretations of Probability*. VSP International Science Publishers (1999); 2nd ed., De Gruyter (2009).
[180] Khrennikov, A. Statistical measure of ensemble nonreproducibility and correction to Bell's inequality. *Il Nuovo Cimento,* **B 115**: (1999), 179–184.
[181] Khrennikov, A. Non-Kolmogorov probability models and modified Bell's inequality, *Journal of Mathematical Physics*, **41** (2000), 1768–1777.
[182] Khrennikov, A. A perturbation of CHSH inequality induced by fluctuations of ensemble distributions. *Journal of Mathematical Physics*, **41**: (2000), 5934–5944.
[183] Khrennikov, A. Frequency analysis of the EPR-Bell argumentation. *Foundations of Physics,* **32**: (2002), 1159–1174.
[184] Khrennikov, A. (ed.) *Quantum Theory: Reconsideration of Foundations*, Series: Mathematical Modelling in Physics, Engineering and Cognitive Science, vol. 2. **2**. Växjö University Press (2002).
[185] Khrennikov, A. Ensemble fluctuations and the origin of quantum probabilistic rule, *Journal of Mathematical Physics* **43**: (2002), 789–802.
[186] Khrennikov, A. Quantum statistics via perturbation effects of preparation procedure. *Il Nuovo Cimento*, **B 117**: (2002), 267–281.

[187] Khrennikov, A., and Volovich, J. I. Interference effect for probability distributions of determinsitic particles. In: *Quantum Theory: Reconsideration of Foundations*. Växjö University Press, 2002, pp. 455–462; http://xxx.lanl.gov/abs/quant-ph/0111159

[188] Khrennikov, A., and Volovich, J. I. Discrete time leads to quantum-like interference of deterministic particles, In: *Quantum Theory: Reconsideration of Foundations*. Växjö University Press, 2002, pp. 441–454; http://xxx.lanl.gov/abs/quant-ph/0203009.

[189] Khrennikov, A. Origin of quantum probabilities. In: Khrennikov, A.(ed.) Foundations of Probability and Physics, pp. 180–200. Växjö-2000, Sweden; WSP (2001).

[190] Khrennikov, A. Växjö interpretation of quantum mechanics. In: Khrennikov, A. (ed.) Quantum Theory: Reconsideration of Foundations, Ser. Math. Modelling **2**, pp. 163–170. Växjö University Press (2002); arXiv:quant-ph/0202107

[191] Khrennikov, A. Contextual viewpoint to quantum stochastics. *Journal of Mathematical Physics* **44**(6): (2003), 2471–2478.

[192] Khrennikov, A. Representation of the Kolmogorov model having all distinguishing features of quantum probabilistic model. *Physics Letters A* **316**(5): (2003), 279–296.

[193] Khrennikov, A. Växjö interpretation-2003: Realism of contexts. In *Quantum Theory: Reconsideration of Foundations*; Khrennikov A., Ed.; Växjö University Press, 2004, pp. 323–338.

[194] Khrennikov, A., and Kozyrev, S. V. Noncommutative probability in classical disordered systems. *Physica A*, **326**(3–4): (2003), 456–463.

[195] Khrennikov, A. Contextual approach to quantum mechanics and the theory of the fundamental prespace. *Journal of Mathematical Physics*, **45**(3): (2004), 902–921.

[196] Khrennikov, A. On quantum-like probabilistic structure of mental information, *OSID*, **11**(3): (2004), 267–275.

[197] Khrennikov, A. *Information Dynamics in Cognitive, Psychological, Social, and Anomalous Phenomena. Ser.: Fundamental Theories of Physics*. Kluwer (2004).

[198] Khrennikov, A. Y., and Loubenets, E. R. On relations between probabilities under quantum and classical measurements. *Foundations of Physics*, **34**: (2004), 689–704.

[199] Khrennikov, A. Y. A formula of total probability with the interference term and the Hilbert space representation of the contextual Kolmogorovian model. *Theory of Probability and Its Applications*, **51**(3): (2007), 427–441.

[200] Khrennikov, A. Quantum mechanics as the quadratic Taylor approximation of classical mechanics: The finite-dimensional case. *Theoretical and Mathematical Physics*, **152**: (2007) 1111–1121.

[201] Khrennikov, A. To quantum averages through asymptotic expansion of classical averages on infinite-dimensional space. *Journal of Mathematical Physics*, **48**(1): (2007), 013512.

[202] Khrennikov, A. The role of von Neumann and Lüders postulates in the Einstein, Podolsky, and Rosen considerations: Comparing measurements with degenerate and nondegenerate spectra. *Journal of Mathematical Physics*, **49**(5): (2008), 052102.

[203] Khrennikov, A. *Contextual Approach to Quantum Formalism*. Springer (2009).

[204] Khrennikov, A. *Ubiquitous Quantum Structure: from Psychology to Finances*. Springer (2010).

[205] Khrennikov, A. Quantum-like model of processing of information in the brain based on classical electromagnetic field. *Biosystems* **105**: (2011), 250–262.

[206] Khrennikov, A. Towards violation of Born's rule: Description of a simple experiment. In: *Advances in Quantum Theory*, G, Jaeger, A. Khrennikov, M. Schlosshauer, G. Weihs (eds.), *AIP Conference Proceedings*, **1327**: (2011), 387–393.
[207] Khrennikov, A., Hertz's viewpoint on quantum theory. *Activitas Nervosa Superior*, **61**: (2019), 24–30; arXiv:1807.06409 [quant-ph].
[208] Khrennikov, A. Prequantum classical statistical field theory: Schrödinger dynamics of entangled systems as a classical stochastic process, *Foundations of Physics*, **41**: (2011), 317–329.
[209] Khrennikov, A. Bell argument: Locality or realism? Time to make the choice. *AIP Conference Proceedings*, **1424**: (2012), pp. 160–175; https://arxiv.org/pdf/1108.0001v2.pdf
[210] Khrennikov, A., Atmanspacher, H., Migdall, A., and Polyakov, S. (eds.): *Quantum Theory: Reconsideration of Foundations 6*, **1508**. American Institute of Physics, Conference Proceedings, Melville, NY (2012)
[211] Khrennikov, A., and Gregor Weihs, G. Preface of the special issue quantum foundations: Theory and experiment. *Foundations of Physics*, **42**(6): (2012), 721–724.
[212] Khrennikov, A. *Beyond Quantum*. Pan Stanford. (2014).
[213] Khrennikov, A., Basieva, I., Dzhafarov, E. N., and Busemeyer, J. R. Quantum models for psychological measurements: An unsolved problem. *PLOS ONE* **9**, e110909 (2014).
[214] Khrennikov, A., Ramelow, S., Ursin, R., Wittmann, B., Kofler, J., and Basieva, I. On the equivalence of the Clauser-Horne and Eberhard inequality based tests. *Physica Scripta*, **T163**: (2014), 014019.
[215] Khrennikov, A., and Basieva, I. Possibility to agree on disagree from quantum information and decision making. *Journal of Mathematical Psychology*, **62**: (2014), 1–15.
[216] Khrennikov, A. Quantum version of Aumann's approach to common knowledge: Sufficient conditions of impossibility to agree on disagree. *Journal of Mathematical Economics*, **60**: (2015), 89–104.
[217] Khrennikov, A. CHSH inequality: quantum probabilities as classical conditional probabilities. *Foundations of Physics*, **45**: (2015), 711–725.
[218] Khrennikov A. Quantum-like model of unconscious-conscious dynamics. *Frontiers in Psychology*, **6**: (2015), 997–1010.
[219] Khrennikov, A. Reflections on Zeilinger–Brukner information interpretation of quantum mechanics. *Foundations of Physics*, **46**: (2016), 836–844.
[220] Khrennikov, A. *Probability and Randomness: Quantum versus Classical*. WSP (2016).
[221] Khrennikov, A. Quantum Bayesianism as the basis of general theory of decision-making, *Philosophical Transactions of the Royal Society*, **374**(2068): (2016), 20150245.
[222] Khrennikov, A. Why quantum? In: Haven, E. and Khrennikov, A. (eds). *The Palgrave Handbook of Quantum Models in Social Science: Applications and Grand Challenges* (pp. 321–334). Palgrave Macmillan Ldt (2017).
[223] Khrennikov, A., Bohr against Bell: complementarity versus nonlocality. *Open Phys.* **15**: (2017), 734–773.
[224] Khrennikov, A. After Bell. *Fortschritte der Physik (Progress in Physics)*, **65**(6–8): (2017), 1600014.
[225] Khrennikov, A. Quantum epistemology from subquantum ontology: Quantum mechanics from theory of classical random fields. *Annals of Physics*, **377**: (2017), 147–163.

[226] Khrennikov, A., and Basieva, I. Towards experiments to test violation of the original Bell inequality. *Entropy*, **20**: (2018), 280.

[227] Khrennikov, A. Y., and Loubenets, E. R. Evaluating the maximal violation of the original Bell inequality by two-qudit states exhibiting perfect correlations/anticorrelations. *Entropy*, **20**(11): (2018), 829.

[228] Khrennikov, A. (2018). External observer reactions on QBism, its possible modifications, and novel applications. In: *Quantum Foundations*, STEAM-H: Science, Technology, Engineering, Agriculture, Mathematics & Health; Khrennikov A. and Toni B. Eds.; Springer, pp. 93–118.

[229] Khrennikov, A., and Alodjants, A. Classical (local and contextual) probability model for Bohm–Bell type experiments: No-signaling as independence of random variables, *Entropy*, **21**: (2018), 157.

[230] Khrennikov, A. Towards better understanding QBism, *Foundations of Science*, **23**(1): (2018), 181–195.

[231] Khrennikov, A. Hertz's viewpoint on quantum theory. *Activitas Nervosa Superior*, **61**: (2019), 24. https://doi.org/10.1007/s41470-019-00052

[232] Khrennikov, A. Get rid of nonlocality from quantum physics. *Entropy*, **21**(8): (2019), 806.

[233] Khrennikov, A. Two faced Janus of quantum nonlocality. *Entropy*, **22**: (2020), 303.

[234] Khrennikov, A. Quantum versus classical entanglement: eliminating the issue of quantum nonlocality. *Foundations of Physics*, **50**(12): (2020), 1762–1780.

[235] Khrennikov, A. Quantum postulate vs. quantum nonlocality: On the role of the Planck constant in Bell's argument. *Foundations of Physics*, **51** (2021), 16.

[236] Khrennikov, A. Is the Devil in h? *Entropy*, **23**(5): (2021), 632.

[237] Khrennikov, A. Can there be given any meaning to contextuality without incompatibility? *International Journal of Theoretical Physics*, **60**: (2021), 106–114.

[238] Khrennikov, A. Quantum-like model for unconscious-conscious interaction and emotional coloring of perceptions and other conscious experiences. *Biosystems*, **208**: (2021), 104471.

[239] Khrennikov, A. (2023). God as decision maker and Quantum Bayesianism. In *Open Quantum Systems in Biology, Cognitive and Social Sciences* (pp. 333–337). Springer International Publishing.

[240] Khrennikov, A., and Basieva, I. "What is life?": Open quantum systems approach. *Open Systems & Information Dynamics* **29**(04), (2022), 2250016. Preprint, DOI:10.21203/rs.3.rs-143923/v1 (2021)

[241] Khrennikov, A., and Watanabe, N. Order-stability in complex biological, social, and AI-systems from quantum information theory. *Entropy* **23**: 2021, 355.

[242] Khrennikov, A. Order stability via Fröhlich condensation in bio, eco, and social systems: The quantum-like approach. *Biosystems* **212** (2022), 104593.

[243] Khrennikov, A. Social Fröhlich condensation: preserving societal order through sufficiently intensive information pumping. *Kybernetes* **51**: (2022), 138–155.

[244] Khrennikov, A. Against identification of contextuality with violation of the Bell inequalities: lessons from theory of randomness. *Journal of Russian Laser Research*, **43**(1): (2022), 48–59.

[245] Khrennikov, A. Contextuality, complementarity, signaling, and Bell tests. *Entropy*, **24**(10): (2022), 1380.

[246] Khrennikov, A. *Open Quantum Systems in Biology, Cognitive and Social Sciences*. Springer Nature (2023).

[247] Khrennikov, A. Open systems, quantum probability, and logic for quantum-like modeling in biology, cognition, and decision-making. *Entropy*, **25**(6): 2023, 886.

[248] Khrennikov, A. Ozawa's Intersubjectivity Theorem as objection to QBism individual agent perspective. https://arxiv.org/abs/2301.04014

[249] Kirkpatrick, K. A. "Quantal" behavior in classical probability. *Foundations of Physics Letters*, **16**: (2003), 199–224.

[250] Kolmogoroff, A. N. *Grundbegriffe der Wahrscheinlichkeitsrechnung*. Springer (1933).

[251] Kolmogorov, A. N. *Foundations of the Theory of Probability*. Chelsea Publishing Company (1956).

[252] Koopman, B. O. Quantum theory and the foundations of probability. In *Applied Probability*. L. A. MacColl, Ed.: 97–102. McGraw-Hill (1952).

[253] Korolkova, N., and Leuchs, G. Quantum correlations in separable multi-mode states and in classically entangled light. *Reports on Progress in Physics*, **82** (2019), 056001.

[254] Kozyrev, S. V. The noncommutative replica approach, http://xxx.lanl.gov/abs/cond-mat/0110238.

[255] Kozyrev, S. V. The noncommutative replica procedure, http://xxx.lanl.gov/abs/cond-mat/0210196.

[256] Krüger, T. Ensemble teleportation. quant-ph/0404091.

[257] Kupczynski, M. Bertrand's paradox and Bell's inequalities. *Physics Letters A*, **121**: (1987), 205–207.

[258] Kupczynski, M. Entanglement and quantum nonlocality demystified. In: Eds. A. Khrennikov, H. Atmanspacher, A. Migdall and S. Polyakov. *AIP Conf. Proc.* Melville, New York, **1508**: (2012), 253–264.

[259] Kupczynski, M. Bell Inequalities, Experimental Protocols and Contextuality. *Foundations of Physics* **45**: (2015), 73.

[260] Landau, L. D., and Lifshitz, E. M. *Statistical Physics, I*. Pergamon Press Oxford (1978).

[261] Landau, L. J. Experimental tests of general quantum theories. *Letters in Mathematical Physics* **14**: (1987), 33–40.

[262] Landau, L. J. On the violation of Bell's inequality in quantum theory. *Physics Letters A*, **120**: (1987), 54–56.

[263] Landauer, R., and J. W. Entropy changes in the steady state far from equilibrium. *Physical Review A*, **6**(6): (1972), 2204.

[264] Larsson, J.-A. Bell's inequality and detector inefficiency. *Physical Review A*, **57**: (1998), 3304–3305.

[265] Li, T., Kheifets, S., Medellin, D., and Raizen, M. G. Measurement of the instantaneous velocity of a Brownian particle. *Science*, **328**(5986): (2010), 1673–167.

[266] Lochak, G. Has Bell's inequality a general meaning for hidden-variable theories? *Foundations of Physics*, **6**: (1976), 173–184.

[267] Lochak, G. De Broglie's initial conception of De Broglie waves. In: *The Wave-Particle Dualism: A Tribute to Louis de Broglie on his 90th Birthday* S. Diner, D. Fargue, G. Lochak, F. Selleri (eds.), Springer, pp. 1–25 (1984).

[268] Lochak, G. (1977). Hidden parameters, hidden probabilities. In: Lopes, J. L., Paty, M. (eds.) *Quantum Mechanics, A Half Century Later*. Episteme, vol 5. Springer, pp. 245–259.

[269] Loubenets, E. R. On the probabilistic description of a multipartite correlation scenario with arbitrary numbers of settings and outcomes per site. *Journal of Physics A: Mathematical and Theoretical*, **41**: (2008), 445303.

[270] Loubenets, E. R. Local hidden variable modelling, classicality, quantum separability and the original Bell inequality. *Journal of Physics A: Mathematical and Theoretical*, **44**: (2011), 035305.

[271] Loubenets, E. R., and Khrennikov, A. Y. Quantum analog of the original Bell inequality for two-qudit states with perfect correlations/anticorrelations. *Journal of Physics A: Mathematical and Theoretical*, **52**: (2019), 435304.

[272] Lozada Aguilar, M. Á., Khrennikov, A., and Oleschko, K. From axiomatics of quantum probability to modelling geological uncertainty and management of intelligent hydrocarbon reservoirs with the theory of open quantum systems. *Philosophical Transactions of the Royal Society A: Mathematical, Physical and Engineering Sciences*, **376**(2118): (2018). 20170225.

[273] Lozada Aguilar, M. Á., Khrennikov, A., Oleschko, K., and de Jesús Correa, M. Quantum Bayesian perspective for intelligence reservoir characterization, monitoring and management. *Philosophical Transactions of the Royal Society A: Mathematical, Physical and Engineering Sciences*, **375**(2106): (2017), 20160398.

[274] Lüders, G. Über die Zustandsänderung durch den Messprozess. *Annalen der Physik* (Leipzig) **8**: (1951), 322–328.

[275] Ma, S.-K. *Statistical Mechanics*. World Scientific (1985).

[276] Mackey, G. W. Quantum mechanics and Hilbert space. *American Mathematical Monthly*, **64**: (1957), 45–57.

[277] Mackey, G. W. *Mathematical Foundations of Quantum Mechanics*. Benjamin (1963).

[278] Mackie, J. L. *The Cement of the Universe: a Study of Causation*. Oxford University (2002).

[279] Man'ko, M. A., and Man'ko, V. I. (2023). Probability distributions as alternatives of the density operator and wave function for describing quantum states. In: Plotnitsky, A., Haven, E. (eds.) *The Quantum-Like Revolution*. Springer.

[280] Mermin, N. D. Extreme quantum entanglement in a superposition of macroscopically distinct states. *Physical Review Letters*, **65**: (1990), 1838–1840.

[281] Mezard, M., Parisi, G., and Virasoro, M. *Spin-Glass Theory and Beyond*. World Scientific (1987).

[282] Moore, D. W. Measuring new types of question-order effects. *Public Opinion Quarterly*, **60**: (2012), 80–91.

[283] Mielnik, B. The screen problem. *Foundations of Physics*, **24**: (1994), 1113–1129.

[284] Nelson, E. *Quantum Fluctuations*. Princeton University Press (1985).

[285] Newton, R. G. Probability interpretation of quantum mechanics. *American Journal of Physics*, **48(12)**: (1980), 1029–1034.

[286] Nietzsche, F. On truth and lie in an extra-moral sense. In: F. Nietzsche, *The Portable Nietzsche*. Translated by W. Kaufmann, Penguin Books (1977).

[287] Nieuwenhuizen, T. M. Thermodynamic picture of the glassy state gained from exactly solvable models. *Physical Review E*, **61**(1): (2000), 267.

[288] Nieuwenhuizen, T. M. Thermodynamic picture of the glassy state gained from exactly solvable models. *Physical Review E*, **61**: (2000), 267.

[289] Nieuwenhuizen, T. M. Is the contextuality loophole fatal for the derivation of Bell inequalities? *Foundations of Physics*, **41**: (2011), 580–591.

[290] Nieuwenhuizen, T. M. (2023). Contra multos verbos: on scandals of quantum mechanics. In: Plotnitsky, A., Haven, E. (eds.) *The Quantum-Like Revolution*. Springer. DOI: https://doi.org/10.1007/978-3-031-12986-5.

[291] Ozawa, M. Quantum measuring processes for continuous observables. *Journal of Mathematical Physics* **25**: (1984), 79–87.

[292] Ozawa, M. Realization of measurement and the standard quantum limit. In: Tombesi, P. and R. Pike,R. (eds.) *Squeezed and Nonclassical Light*, NATO Science Series B **190**, pp. 263–286. Plenum (1989).

[293] Ozawa, M. Mathematical characterizations of measurement statistics, *Quantum Communication and Measurement*, Plenum, 109–117 (1995).

[294] Ozawa, M. On information gain by quantum measurements of continuous observables, *Journal of Mathematical Physics* 27, 759–763 (1986).

[295] Ozawa, M. An operational approach to quantum state reduction. *Annals of Physics* (N.Y.) **259**: (1997), 121–137.

[296] Ozawa, M. Position measuring interactions and the Heisenberg uncertainty principle. *Physics Letters A*, **299**(1): (2002), 1-7.

[297] Ozawa, M. Quantum reality and measurement: A quantum logical approach. *Foundations of Physics* **41** (2009), 592–607.

[298] Ozawa, M. Probabilistic interpretation of quantum theory. *New Generation Computing* **34**: (2016) 125–152.

[299] Ozawa, M. Soundness and completeness of quantum root-mean-square errors. *Quantum Information* **5**: (2019), article number: 1.

[300] Ozawa, M. Quantum measurement theory for systems with finite dimensional state spaces. In: Plotnitsky, A., Haven, E. (eds) *The Quantum-Like Revolution*. Springer (2023).

[301] Ozawa, M. (2019). Intersubjectivity of outcomes of quantum measurements. https://arxiv.org/abs/1911.10893

[302] Ozawa, M., and Khrennikov, A. Application of theory of quantum instruments to psychology: Combination of question order effect with response replicability effect. *Entropy*, **22**(1): (2020), 1–9436.

[303] Ozawa, M., and Khrennikov, A. Modeling combination of question order effect, response replicability effect, and QQ-equality with quantum instruments. J. Math. Psych. **100**: (2021), 102491.

[304] Ozawa, M., and Khrennikov, A. Nondistributivity of human logic and violation of response replicability effect in cognitive psychology. *Journal of Mathematical Psychology* **112**: (2023), 102739.

[305] Patton, L. Signs, toy models, and the a priori: from Helmholtz to Wittgenstein. *Studies in History and Philosophy of Science* **40**: (2009), 281–289.

[306] Peres, A. *Quantum Theory: Concepts and Methods*. Kluwer Academic Publishers (1993).

[307] Pitowsky, I. New Bell inequalities for the singlet state: Going beyond the Grothendieck bound. *Journal of Mathematical Physics* **49**: (2008), 012101-11.

[308] Plotnitsky, A. *The Spirit and the Letter of Copenhagen: A Response to Andrei Khrennikov*. arXiv:quant-ph/0206026v2 7 Jun 2002

[309] Plotnitsky, A. *Epistemology and Probability: Bohr, Heisenberg, Schrödinger and the Nature of Quantum-Theoretical Thinking*. Springer (2009).

[310] Plotnitsky, A. *Niels Bohr and Complementarity: An Introduction*. Springer (2012).

[311] Plotnitsky, A., and Khrennikov, A. Reality without realism: on the ontological and epistemological architecture of quantum mechanics. *Foundations of Physics* **45** (2015), 1269-1300.

[312] Plotnitsky, A. *The Principles of Quantum Theory, from Planck's Quanta to the Higgs Boson: The Nature of Quantum Reality and the Spirit of Copenhagen*. Springer (2016).

[313] Plotnitsky, A. Spooky predictions at a distance: Reality, complementarity and contextuality in quantum theory. *Philosophical Transactions of the Royal Society A* **377**: 2019, 20190089.

[314] Plotnitsky, A. "Without in any way disturbing the system": Illuminating the issue of quantum nonlocality. arXiv:1912.03842 [quant-ph].

[315] Plotnitsky, A. *Reality without Realism. Matter, Thought, and Technology in Quantum Physics*. Springer (2021).

[316] Plotnitsky, A. "The agency of observation not to be neglected": Complementarity, causality and the arrow of events in quantum and quantum-like theories. *Philosophical Transactions of the Royal Society A*, **381**2256): (2023), 20220295.

[317] Plotnitsky, A. Continuity and discontinuity in, and between, mathematics and physics. In: *Essays in Geometry dedicated to Nobert A'Campo*, A. Papadopoulos (ed.), EMS Press, pp. 883–948 (2023).

[318] Plotnitsky, A., and Haven, E. *The Quantum-Like Revolution. A Festschrift for Andrei Khrennikov* with a foreword by 2022 Nobel Laureate Anton Zeilinger. Springer Nature (2023).

[319] Popper, K. R. Quantum mechanics without "the observer." In Quantum theory and reality (pp. 7-44). Springer (1967).

[320] Prentis, J. J. Experiments in statistical mechanics. *American Journal of Physics* **68**(12): (2000), 1073–1083.

[321] Primas, H. (1990). Mathematical and philosophical questions in the theory of open and macroscopic quantum systems. In *Sixty-Two Years of Uncertainty*, ed. by A. I. Miller (Plenum, New York), pp. 233–257.

[322] Rafai, I., Duchene, S., Guerci, E., Basieva, I., and Khrennikov, A.: The triple-store experiment: a first simultaneous test of classical and quantum probabilities in choice over menus. *Theory and Decision* **92**(2): (2022), 387–406.

[323] H. Risken, *The Fokker–Planck Equation*, Springer (1984).

[324] Robb, A. A. *Optical Geometry of Motion. A New View of the Theory of Relativity*. Kessinger Publishing, 1911. Reprinted: W. Heffer and Sons Ltd., 2021.

[325] Rosner, A., Basieva, I., Barque-Duran, A., Glčkner, A., von Helversen, B., Khrennikov, A., and Pothos, E. M. (2022). Ambivalence in decision making: An eye tracking study. *Cognitive Psychology*, **134**: (2022), 101464.

[326] Russel, W. B., Saville, D. A., and Schowalter, W. R. *Colloidal Dispersions*. Cambridge University Press (1989).

[327] Ryff, L. C. Bell and Greenberger, Horne, and Zeilinger theorems revisited. Am. J. Phys. **65**(12): (1997), 1197–1199.

[328] Schrödinger, E. Die gegenwärtige Situation in der Quantenmechanik, *Naturwissenschaften* **23**: (1935), 807–812, 823–828, 844–849.

[329] Schrödinger, E. The present situation in quantum mechanics: A translation of Schrödinger's "Cat Paradox" paper (by: J. D. Trimmer). *Proceedings of the American Philosophical Society* **124**: (1980) 323–338.

[330] Schrödinger, E. What is an elementary particle? Endeavour **9**: (1950), 109–116. reprinted in: E. Schödinger, Gesammelte Abhandlungen. Band 4. Allgemein wissenschaftliche und populäe Aufsätze. Verlag der Österreichischen Akademie der Wissenschaften and Braunschweig/Friedr.Vieweg & Sohn (1984), pp. 456–463.

[331] Schrödinger, E. *Science and Humanism: Physics in Our Time* Cambridge University Press (1951).

[332] Schrödinger, E. The meaning of wave mechanics: La signification de la mecanique ondularoire, in Andre George, ed., *Louis de Broglie: Physicien et Penseur* (Paris: Editions Alain Michel, 1953), pp. 16–32; reprinted in Erwin Schrödinger, Gesammelte Abhandlungen. Band 3. (ref. 35), pp. 694–710.

[333] Schrödinger, E. *Nature and the Greeks*. Cambridge University Press (1956).

[334] Schrödinger, E. *What Is Life? and Other Scientific Lectures*. Doubleday (1956).

[335] Schrödinger, E. *Science Theory and Man*. Dover (1957).

[336] Schrödinger, E. *Mind and Matter*. Cambridge University Press (1958).

[337] Schrödinger, E. *Statistical Thermodynamics*. Dover Publications (1989).

[338] Shalm, L. K., Meyer-Scott, E., Christensen, B. G., Bierhorst, P., Wayne, M. A., Stevens, M. J., Gerrits, T., Glancy, S., Hamel, D. R., Allman, M. S., Coakley, K. J., Dyer, S. D., Hodge, C., Lita, A. E., Verma, V. B., Lambrocco, C., Tortorici, E., Migdall, A. L., Zhang, Y., Kumor, D. R., Farr, W. H., Marsili, F., Shaw, M. D., Stern, J. A., Abellán, C., Amaya, W., Pruneri, W., Jennewein, T., Mitchell, M. W., Kwiat, P. G., Bienfang, J. C., Mirin, R. P., Knill, E., and Woo Nam, S. A strong loophole-free test of local realism, Phys. Rev. Lett. **115**: 2015, 250402. [arXiv:quant-ph1511.03189].

[339] Schulman, L. S., and Gaveau, B. (2001). Coarse grains: The emergence of space and order. *Foundations of Physics*, **31**: (2001), 713–731.

[340] Sinha, U., Couteau, C., Medendorp, Z., Sillner, I., Laflamme, R., Sorkin, R., and Weihs, G. Testing Born's rule in quantum mechanics with a triple slit experiment. In: Accardi, L., Adenier, G., Fuchs, C., Jaeger, G., Khrennikov, A., Larsson, J.-A., and Stenholm, S. (eds)., Foundations of Probability and Physics **5**(1101): (2009), pp. 200–207. American Institute of Physics, Ser. Conf. Proc., Melville, NY.

[341] Sinha, U., Couteau, C., Jenewein, T., Laflamme, R., and Weihs, G. Ruling out multi-order interference in quantum mechanics. *Science* **329**: (2010), 418–421.

[342] Smolyanov, O. G., and Khrennikov, A. Y. E. The central limit theorem for generalized measures on infinite-dimensional spaces. *Doklady Akademii Nauk* **281**(2): (1985), 279–283.

[343] Spreeuw, R. J. C. A classical analogy of entanglement, *Foundations of Physics* **28**: (1998), 361–374.

[344] Spreeuw, R. J. C. Classical wave-optics analogy of quantum information processing. *Physical Review A*, 63, 062302 (2001).

[345] Stapp, H. P. The Copenhagen interpretation. *American Journal of Physics* **40**: (1972), 1098.

[346] Stephenson, W. William James, Niels Bohr, and complementarity: I - *Psychological Record* **36**: (1986), 519–529.

[347] Stephenson, W. William James, Niels Bohr, and complementarity: II – Pragmatics of a thought. *Psychological Record* **36**: (1986), 529–543.

[348] Svozil, K. *Quantum Logic*. Springer Science and Business Media (1998).

[349] Tsirelson, B. S. Quantum generalizations of Bell's inequality. *Letters in Mathematical Physics* **4**: (1980), 93–100.

[350] Tsirelson, B. S. Quantum analogues of the Bell inequalities: The case of two spatially separated domains. *Journal of Soviet Mathematics* **36**: (1987), 557–570.

[351] Zanardi, P. Virtual quantum subsystems. *Physical Review Letters*, **87**(7): (2001), 077901.

[352] Zanardi, P., Lidar, D. A., and Lloyd, S. Quantum tensor product structures are observable induced. *Physical Review Letters*, **92**(6): (2004), 060402.

[353] Zeilinger, A. A foundational principle for quantum mechanics. *Foundations of Physics* **29**(4): (1999), 631–643.

[354] Zeilinger, A. The future of Bell experiments. Video, special session *Big event – Final Bell test*, A. Khrennikov (chairman), Conf. *Quantum and Beyond*, Växjö, 2016. https://play.lnu.se/media/t/0_hgz7034c

[355] Zeilik, M. Modeling energy outflow in stars. *Physics Teacher* **37**: (1999), 236.

[356] Zurek, W. H. Decoherence and the Transition from Quantum to Classical – Revisited. In: Duplantier, B., Raimond, J. M., Rivasseau, V. (eds.) *Quantum Decoherence*. Progress in Mathematical Physics **48** (2006).

[357] Ullersma, P. An exactly solvable model for Brownian motion: I. Derivation of the Langevin equation. *Physica*, **32**(1): (1966), 27–55.

[358] Van Hove, L. Quantum-mechanical perturbations giving rise to a statistical transport equation. *Physica*, **21**: (1954), 517–540.

[359] von Helmholtz, H. Die Thatsachen in der Wahrnehmung, in Vorträge und Reden, Fünfte Auflage, Zweiter Band (Friedrich Vieweg und Sohn, 1903), pp. 215–247, 387–406; translated as The Facts in Perception, in R. S. Cohen and Y. Elkana (eds.), *Hermann von Helmholtz Epistemological Writings: The Paul Hertz/Moritz Schlick Centenary Edition of 1921 with Notes and Commentary by the Editors* (D. Reidel, 1977), pp. 115–185; on p. 117.

[360] von Helmholtz, H. Über das Ziel und die Fortschritte der Naturwissenschaft. In idem, *Das Denken in der Naturwissenschaft* (pp. 31–61). Wissenschaftliche Buchgesellschaft (1869/1968).

[361] von Helmholtz, H. (1977a). On the origin and significance of the axioms of geometry. In idem, *Epistemological Writings: The Paul Hertz/Moritz Schlick Centenary Edition of 1921, with Notes and Commentary by the Editors* (R. Cohen, & Y. Elkana, Eds.; M. Lowe, Trans.) (pp. 1–38). *Boston Studies in the Philosophy of Science*, **37**. D. Reidel. (1870/1977a).

[362] von Helmholtz, H. (1977b). The facts in perception. In idem, *Epistemological Writings: The Paul Hertz/Moritz Schlick Centenary Edition of 1921, with Notes and Commentary by the Editors* (R. Cohen & Y. Elkana, Eds.; M. Lowe, Trans.) (pp. 115–185). *Boston Studies in the Philosophy of Science*, **37**. D. Reidel. (Talk first given in 1878)

[363] von Mises, R. (1957). *Probability, Statistics and Truth.* Macmillan.

[364] Von Neumann, J. *Mathematische Grundlagen der Quantenmechanik.* Springer (1932).

[365] Von Neumann, J. *Mathematical Foundations of Quantum Mechanics*; Princeton University Press (1955).

[366] Voiculescu, D., Dykema, K. J., and Nica, A. *Free Random Variables.* CRM Monograph Series, **1**, American Mathematical Society.

[367] Wang, Z., and Busemeyer, J. R. A quantum question order model supported by empirical tests of an a priori and precise prediction. *Topics in Cognitive Science* **5** (2013), 689–710.

[368] Wang, Z., Solloway, T., Shiffrin, R. M., and Busemeyer, J. R. Context effects produced by question orders reveal quantum nature of human judgments. PNAS **111**: (2014), 9431–9436.

[369] Weihs, G., Jennewein, T., Simon, C., Weinfurther, H., and Zeilinger, A. Violation of Bell's Inequality under Strict Einstein Locality Conditions. *Physical Review Letters* **81**: (1998) 5039–5043.

[370] Wagner, G. Function and causality. *Revue internationale de philosophie* **1**(259): (2012), 35–53.

[371] Weihs, G. Violation of Bell's inequality under strict Einstein locality conditions. Video, special session *Big event – Final Bell test*, A. Khrennikov (chairman), Conf. *Quantum and Beyond*, Växjö, 2016. https://play.lnu.se/media/t/0_6jkhpefo

[372] Weinfurter, H. Event ready loophole free Bell test using heralded atom-atom entanglement. Video, special session *Big event – Final Bell test*, A. Khrennikov (chairman), Conf. *Quantum and Beyond*, Växjö, 2016. https://play.lnu.se/media/t/0_xprww7jj

[373] Werner, R. F. Quantum states with Einstein–Podolsky–Rosen correlations admitting a hidden-variable model. *Physical Review A* **40** (1989), 4277–4281.

[374] Werner, R. F., and Wolf, M. M. Bound entangled Gaussian states. *Physical Review Letters* **86**(16): (2001), 3658.
[375] Wheeler, J. A. Information physics, quantum: The search for links. *Proceedings of the Third International Symposium Foundations of Quantum Mechanics* Tokyo, 1989, pp. 364–368.
[376] Whitehead, A. N. (1929). Process and Reality. An Essay in Cosmology. Gifford Lectures Delivered in the University of Edinburgh During the Session 1927–1928. Cambridge University Press (1929).
[377] Wigner, E. P. Characteristic vectors of bordered matrices with infinite dimensions, *Annals of Mathematics* 62 (1955) 548–564.
[378] Wigner, E. P. The unreasonable effectiveness of mathematics in the natural sciences. *Communications on Pure and Applied Mathematics* **13**: (1960), 1–14.
[379] Wittgenstein, L. (1922). *Tractatus logico-philosophicus* (C. K. Ogden, Trans.). London: Routledge & Kegan Paul. (First published 1921 as *Logisch-Philosophische Abhandlung*. Annalen der Naturphilosophie, **14**(3/4)).
[380] Wittgenstein, L. (1961). *Notebooks 1914–1916*. Basil Blackwell.

Index

Printed in the USA
CPSIA information can be obtained
at www.ICGtesting.com
CBHW081521101224
18738CB00005B/97